Handbook of Water Harvesting and Conservation: Basic Concepts and Fundamentals

Handbook of Water Harvesting and Conservation: Basic Concepts and Fundamentals

Edited By

Saeid Eslamian
Isfahan University of Technology
IR

Faezeh Eslamian
McGill University
CA

The New York Academy of Sciences

WILEY

This edition first published 2021

Registered Offices
John Wiley & Sons, Inc., 111 River Street, Hoboken, NJ 07030, USA
John Wiley & Sons Ltd, The Atrium, Southern Gate, Chichester, West Sussex, PO19 8SQ, UK

Editorial Office
The Atrium, Southern Gate, Chichester, West Sussex, PO19 8SQ, UK

For details of our global editorial offices, customer services, and more information about Wiley products visit us at www.wiley.com.

Wiley also publishes its books in a variety of electronic formats and by print-on-demand. Some content that appears in standard print versions of this book may not be available in other formats.

Library of Congress Cataloging-in-Publication Data

Names: Eslamian, Saeid, editor. | Eslamian, Faezeh A., editor. | Wiley-Blackwell (Firm), publisher.
Title: Handbook of water harvesting and conservation / [edited by] Saeid Eslamian, Isfahan University of Technology, IR.
Description: First edition. | Hoboken, NJ : Wiley-Blackwell, 2021. | Includes index.
Identifiers: LCCN 2020025967 (print) | LCCN 2020025968 (ebook) | ISBN 9781119478959 (cloth) | ISBN 9781119478942 (adobe pdf) | ISBN 9781119478928 (epub)
Subjects: LCSH: Water harvesting. | Water conservation. | Water–Storage.
Classification: LCC TD418 .H36 2021 (print) | LCC TD418 (ebook) | DDC 628.1/42–dc23
LC record available at https://lccn.loc.gov/2020025967
LC ebook record available at https://lccn.loc.gov/2020025968

Cover Design: Wiley
Cover Images: Rainy day in the mountain © Peter Zelei Images/Getty Images, Close-Up Of Raindrops On Green Leaf, © Maciej Rawluk/EyeEm/Getty Images

Set in 9.5/12.5pt STIXTwoText by SPi Global, Chennai, India
Printed and bound by CPI Group (UK) Ltd, Croydon, CR0 4YY

10 9 8 7 6 5 4 3 2 1

Contents

List of Contributors

Hossein Abbasizadeh
Department of Water Engineering
Science and Research Branch
Islamic Azad University
Tehrans
Iran

Fayez Abdulla
Jordan University of Science and Technology
Irbid, 22110
Jordan

Cealeen Abdulla
Jordan University of Science and Technology
Irbid, 22110
Jordan

Maria Do Céu Almeida
Department of Hydraulics and Environment
National Laboratory of Civil Engineering
1700-066 Lisboa
Portugal

Vaibhav Bahadur
Walker Department of Mechanical Engineering
The University of Texas at Austin
Austin, TX
USA

Bruna Peres Battemarco
Programa de Engenharia Civil, COPPE
Universidade Federal do Rio de Janeiro
Rio de Janeiro
Brazil

Anna Blanta
Department of Civil Engineering
University of Thessaly
Volos
Greece

Elena Bresci
Department of Agriculture, Food, Environment
and Forestry (DAGRI)
Università degli Studi di Firenze
Firenze
Italy

Nelson Carriço
Barreiro School of Technology
Polytechnic Institute of Setubal
2839-001 Lavradio
Portugal

Giulio Castelli
Department of Agriculture, Food, Environment
and Forestry (DAGRI)
Università degli Studi di Firenze
Firenze
Italy

Tanya Clements
Department of Microbiology, Faculty of Science
Stellenbosch University
Stellenbosch
South Africa

Irvan Dahlan
School of Chemical Engineering
Universiti Sains Malaysia
Nibong Tebal, Pulau Pinang
Malaysia

Nicolas R. Dalezios
Department of Civil Engineering
University of Thessaly
Volos
Greece

Jew Das
Discipline of Civil Engineering
Indian Institute of Technology
Indore
India

Adebayo Eludoyin
Department of Geography
Obafemi Awolowo University
Ife-Ife
Nigeria

Oyenike Eludoyin
Department of Geography and Planning Sciences
Adekunle Ajasin University
Akungba-Akoko, Ondo State
Nigeria

Duygu Erten
Department of Civil Engineering
İstanbul Medipol University
34810 İstanbul
Turkey

and

Vice Director of Climate Change Research Center
34810 İstanbul
Turkey

Saeid Eslamian
Department of Water Engineering, College of Agriculture
Isfahan University of Technology
Isfahan
Iran

Ioannis Faraslis
Department of Civil Engineering
University of Thessaly
Volos
Greece

Hossein Farraji
School of Physical and Chemical Department
University of Canterbury
Canterbury
New Zealand

and

Institute of Scientific Research Aminazmay Shargh (ISRAS)
Mashhad
Iran

Neda Torabi Farsani
Department of Museum and Tourism
Art University of Isfahan
Isfahan
Iran

Siavash Fasihi
Ahvarzan Consulting Engineers
Tehrah
Iran

Hadis Salehi Gahrizsangi
School of Food
Plant and Animal Industries
Melbourne Polytechnic
Melbourne
Australia

Maryam Ghashghaie
Department of Water Resources Engineering
Faculty of Agriculture
Bu-Ali Sina University
Hamedan
Iran

Sayed Salman Ghaziaskar
Department of Geography
Islamic Azad University
Najafabad
Iran

Dillip K. Ghose
Department of Civil Engineering
NIT Silchar
788010, Assam
India

Manish Kumar Goyal
Discipline of Civil Engineering
Indian Institute of Technology
Indore
India

Assed Naked Haddad
Programa de Engenharia Ambiental, Escola Politécnica
Universidade Federal do Rio de Janeiro
Rio de Janeiro
Brazil

and

Programa de Pós-Graduação em Engenharia Civil
Universidade Federal Fluminense
Niterói
Brazil

Boutaghane Hamouda
Department of Hydraulics
Universite Badji Mokhtar
Annaba
Algeria

Muna Yacoub Hindiyeh
Civil and Environmental Engineering Department
School of Natural Resources Engineering and Management
German Jordanian University
Jordan

Monzur A. Imteaz
Swinburne University of Technology
Melbourne, VIC
Australia

Homa Moazzen Jamshidi
Department of Economics and Arts Entrepreneurship
Art University of Isfahan
Isfahan
Iran

A. Jebamalar
Velammal Engineering College
Chennai, Tamil Nadu
India

M.V.V. Prasad Kantipudi
Department of ECE
Sreyas Institute of Engineering and Technology
Hyderabad, Telangana
India

Aritra Kar
Walker Department of Mechanical Engineering
The University of Texas at Austin
Austin, TX
USA

S. Sreenath Kashyap
Department of ECE
Kommuri Pratap Reddy Institute of Technology
Hyderabad, Telangana
India

Sehaam Khan
Faculty of Health Sciences
University of Johannesburg
Doornfontein
South Africa

Wesaal Khan
Department of Microbiology, Faculty of Science
Stellenbosch University
Stellenbosch
South Africa

Billy Kniffen
Biological and Agricultural Engineering Department
Texas A&M University Water Texas AgriLife
Extension Service Resource
College Station, Texas
USA

Rina Kumari
School of Environment and Sustainable Development
Central University of Gujarat
Gandhinagar
India

Zhong Liu
Goddard Earth Sciences Data and Information
Services Center
NASA Goddard Space Flight Center
Greenbelt, MD
USA

and

Center for Spatial Information Science and Systems
George Mason University
Fairfax, VA
USA

Mohadaseh Madadi
Department of Architecture
Islamic Azad University
Shahinshahr Branch
Isfahan
Iran

Jose Francisco Maestre-Valero
Escuela Técnica Superior de Ingeniería Agronómica
Universidad Politécnica de Cartagena
30203 Cartagena
Spain

Mousa Maleki
Department of Civil Engineering
Islamic Azad University
Najafabad Branch, Isfahan
Iran

Bernardo Martin-Gorriz
Escuela Técnica Superior de Ingeniería Agronómica
Universidad Politécnica de Cartagena
30203 Cartagena
Spain

Victoriano Martínez-Alvarez
Escuela Técnica Superior de Ingeniería Agronómica
Universidad Politécnica de Cartagena
30203 Cartagena
Spain

Tanimola Martins
Medical School
University of Exeter
EX1 2LU, UK

Mohammed Matouq
Chemical Engineering Department
Al-Balqa Applied University
Amman
Jordan

David Meyer
Goddard Earth Sciences Data and Information Services Center
NASA Goddard Space Flight Center
Greenbelt, MD
USA

Marcelo Gomes Miguez
Programa de Engenharia Civil, COPPE
Universidade Federal do Rio de Janeiro
Rio de Janeiro
Brazil

and

Programa de Engenharia Urbana, Escola Politécnica
Universidade Federal do Rio de Janeiro
Rio de Janeiro
Brazil

and

Programa de Engenharia Ambiental, Escola Politécnica
Universidade Federal do Rio de Janeiro
Rio de Janeiro
Brazil

Muhammad Moniruzzaman
Swinburne University of Technology
Melbourne, VIC
Australia

Mohammad Mortazavi
Department of Conservation and Restoration
Art University of Isfahan
Isfahan
Iran

Sara Nazif
School of Civil Engineering, College of Engineering
University of Tehran
Iran

John Ndiritu
University of the Witwatersrand
South Africa

Thando Ndlovu
Department of Microbiology
Faculty of Science
Stellenbosch University
Stellenbosch
South Africa

Kaveh Ostad-Ali-Askari
Department of Civil Engineering
Islamic Azad University
Najafabad, Branch, Isfahan
Iran

Dana Ostrenga
Goddard Earth Sciences Data and Information Services Center
NASA Goddard Space Flight Center
Greenbelt, MD
USA

and

Adnet Systems Inc.
Bethesda, MD
USA

Mayowa Oyinloye
Department of Geography
Obafemi Awolow University
Ife-Ife
Nigeria

Onur Ozkan
Walker Department of Mechanical Engineering
The University of Texas at Austin
Austin, TX
USA

Saeideh Parvizi
Department of Water Engineering, College of Agriculture
Isfahan University of Technology
Isfahan
Iran

Dhruvesh Patel
Department of Civil Engineering
Pandit Deendayal Petroleum University
Gandhinagar, Gujarat, 382007
India

Cristina Prieto
Environmental Hydraulics Institute, IH Cantabria
University of Cantabria
Santander
Spain

G. Ravikumar
College of Engineering Guindy
Anna University
Chennai, Tamil Nadu
India

Seyed Ghasem Razavi
School of Civil Engineering
College of Engineering
University of Tehran
Iran

Brandon Reyneke
Department of Microbiology
Faculty of Science
Stellenbosch University
Stellenbosch
South Africa

Negin Sadeghi
Department of Architecture
Isfahan (Khorasgan) Branch
Islamic Azad University
Isfahan, 81595-158
Iran

Dipak R. Samal
Faculty of Technology
CEPT University
Ahmedabad, Gujarat, 380009
India

Sandeep Samantaray
Department of Civil Engineering
NIT Silchar, 788010, Assam
India

João Santos
School of Management and Technology
Polytechnic Institute of Beja
7800-295 Beja
Portugal

Andrey Savtchenko
Goddard Earth Sciences Data and Information
Services Center
NASA Goddard Space Flight Center
Greenbelt, MD
USA

and

Adnet Systems Inc.
Bethesda, MD
USA

Sisuru Sendanayake
Department of Civil Engineering
South Asian Institute of Technology and Medicine
Sri Lanka

Pouria Soleimani
School of Civil Engineering
College of Engineering
University of Tehran
Iran

R. Sudharsanan
Velammal Engineering College
Chennai, Tamil Nadu, 600066
India

Hamed Tavakolifar
School of Civil Engineering
College of Engineering
University of Tehran
Tehran
Iran

William Teng
Goddard Earth Sciences Data and Information Services Center
NASA Goddard Space Flight Center
Greenbelt, MD
USA

and

Adnet Systems Inc.
Bethesda, MD
USA

Fernanda Rocha Thomaz
Programa de Engenharia Civil, COPPE
Universidade Federal do Rio de Janeiro
Rio de Janeiro
Brazil

N.V. Umamahesh
Department of Civil Engineering
National Institute of Technology
Warangal
India

Elaine Garrido Vazquez
Programa de Engenharia Urbana, Escola Politécnica
Universidade Federal do Rio de Janeiro
Rio de Janeiro
Brazil

Aline Pires Veról
Programa de Pós-Graduação em Arquitetura
Faculdade de Arquitetura e Urbanismo
Universidade Federal do Rio de Janeiro
Rio de Janeiro
Brazil

and

Programa de Pós-Graduação em Engenharia Civil
Universidade Federal Fluminense, Niterói
Brazil

Bruce Vollmer
Goddard Earth Sciences Data and Information Services Center
NASA Goddard Space Flight Center
Greenbelt, MD
USA

Monique Waso
Department of Microbiology
Faculty of Science
Stellenbosch University
Stellenbosch
South Africa

Jennifer Wei
Goddard Earth Sciences Data and Information Services Center
NASA Goddard Space Flight Center
Greenbelt, MD
USA

Enakshi Wikramanayake
Walker Department of Mechanical Engineering
The University of Texas at Austin
Austin, TX
USA

Abdullah Yilmaz
La Trobe University
Bendigo, VIC
Australia

About the Editors

Saeid Eslamian is a full professor of environmental hydrology and water resource engineering in the Department of Water Engineering at Isfahan University of Technology, where he has been since 1995. His research focuses mainly on statistical and environmental hydrology in a changing climate. In recent years, he has worked on modeling natural hazards, including floods, severe storms, wind, drought, pollution, water reuses, sustainable development and resiliency, etc. Formerly, he was a visiting professor at Princeton University, New Jersey, and the University of ETH Zurich, Switzerland. On the research side, he started a research partnership in 2014 with McGill University in Montreal, Canada. He has contributed to more than 600 publications in journals, books, and technical reports. He is the founder and chief editor of the *International Journal of Hydrology Science and Technology* (IJHST). Eslamian is now associate editor of four important publications: *Journal of Hydrology* (Elsevier), *Eco-Hydrology and Hydrobiology* (Elsevier), *Journal of Water Reuse and Desalination* (IWA), and *Journal of the Saudi Society of Agricultural Sciences* (Elsevier). Professor Eslamian is the author of approximately 35 books and 180 chapters.

Dr. Eslamian's professional experience includes membership on editorial boards, and he is a reviewer of approximately 100 Web of Science (ISI) journals, including the *ASCE Journal of Hydrologic Engineering*, *ASCE Journal of Water Resources Planning and Management*, *ASCE Journal of Irrigation and Drainage Engineering*, *Advances in Water Resources*, *Groundwater*, *Hydrological Processes*, *Hydrological Sciences Journal*, *Global Planetary Changes*, *Water Resources Management*, *Water Science and Technology*, *Eco-Hydrology*, *Journal of American Water Resources Association*, *American Water Works Association Journal*, etc. UNESCO has also nominated him for a special issue of the *Eco-Hydrology and Hydrobiology Journal* in 2015.

Professor Eslamian was selected as an outstanding reviewer for the *Journal of Hydrologic Engineering* in 2009 and received the EWRI/ASCE Visiting International Fellowship in Rhode Island (2010). He was also awarded outstanding prizes from the Iranian Hydraulics Association in 2005 and Iranian Petroleum and Oil Industry in 2011. Professor Eslamian has been chosen as a distinguished researcher of Isfahan University of Technology (IUT) and Isfahan Province in 2012 and 2014, respectively. In 2016, he was a candidate for national distinguished researcher in Iran.

He has also been the referee of many international organizations and universities. Some examples include the U.S. Civilian Research and Development Foundation (USCRDF), the Swiss Network for International Studies, the Majesty Research Trust Fund of Sultan Qaboos University of Oman, the Royal Jordanian Geography Center College, and the Research Department of Swinburne University of Technology of Australia. He is also a member of the following associations: American Society of Civil Engineers (ASCE), International Association of Hydrologic Science (IAHS), World Conservation Union (IUCN), GC Network for Drylands Research and Development (NDRD), International Association for Urban Climate (IAUC), International Society for Agricultural Meteorology (ISAM), Association of Water and Environment Modeling (AWEM), International Hydrological Association (STAHS), and UK Drought National Center (UKDNC).

Professor Eslamian finished Hakimsanaei High School in Isfahan in 1979. After the Islamic Revolution, he was admitted to IUT for a BS in water engineering and graduated in 1986. After graduation, he was offered a scholarship for a master's degree program at Tarbiat Modares University, Tehran. He finished his studies in hydrology and water resources engineering in 1989. In 1991, he was awarded a scholarship for a PhD in civil engineering at the University of New South Wales, Australia. His supervisor was Professor David H. Pilgrim, who encouraged him to work on "Regional Flood Frequency Analysis Using a New Region of Influence Approach." He earned a PhD in 1995 and returned to his home country and IUT. In 2001, he was promoted to associate professor and in 2014 to full professor. For the past 25 years, he has been nominated for different positions at IUT, including university president consultant, faculty deputy of education, and head of department. Eslamian is now director for center of excellence in Risk Management and Natural Hazards (RiMaNaH).

Professor Eslamian has made three scientific visits to the United States, Switzerland, and Canada in 2006, 2008, and 2015, respectively. In the first, he was offered the position of visiting professor by Princeton University and worked jointly with Professor Eric F. Wood at the School of Engineering and Applied Sciences for one year. The outcome was a contribution in hydrological and agricultural drought interaction knowledge by developing multivariate L-moments between soil moisture and low flows for northeastern U.S. streams.

Recently, Professor Eslamian has published the editorship of nine handbooks published by Taylor & Francis (CRC Press): the three-volume *Handbook of Engineering Hydrology* in 2014, *Urban Water Reuse Handbook* in 2016, *Underground Aqueducts Handbook* (2017), the three-volume *Handbook of Drought and Water Scarcity* (2017), and *Constructed Wetlands: Hydraulic Design* (2020). *An Evaluation of Groundwater Storage Potentials in a Semiarid Climate* by Nova Science Publishers is also his joint book publication in 2019.

Faezeh Eslamian is a PhD Holder of Bioresource Engineering from McGill University. Her research focuses on the development of a novel lime-based product to mitigate phosphorus loss from agricultural fields. Faezeh completed her bachelor and master's degrees in Civil and Environmental Engineering from Isfahan University of Technology, Iran, where she evaluated natural and low-cost absorbents for the removal of pollutants such as textile dyes and heavy metals. Furthermore, she has conducted research on the worldwide water quality standards and wastewater reuse guidelines. Faezeh is an experienced multidisciplinary researcher with interest in soil and water quality, environmental remediation, water reuse, and drought management.

Part A

Concepts and Standards for a Secure Water Harvesting

1

Concept and Technology of Rainwater Harvesting

Fayez Abdulla[1], Cealeen Abdulla[1], and Saeid Eslamian[2]

[1]Jordan University of Science and Technology, Irbid 22110, Jordan
[2]Department of Water Engineering, College of Agriculture, Isfahan University of Technology, Isfahan, Iran

1.1 Introduction

"Rainwater harvesting" (RWH) is defined in this chapter as the collection of rainfall, usually collected and stored in either artificial reservoirs known as cisterns or natural surfaces used in micro-catchment rainwater harvesting.

Rainwater harvesting has been used in many locations in the world to provide water that is suitable for various domestic and irrigation uses. People continue to collect rainwater despite the availability of water distribution systems due to the shortage of water. In early civilization, people in deserts and semi-arid regions have relied on collecting rainwater from land surfaces and storing it in cisterns (Gorokhovich et al. 2011). A number of distinctive historical examples that incorporate effective water harvesting systems survive in different climatic zones, and can be found in areas which receive between 100 and 1000 mm of precipitation annually (Prinz 1995). The capture and utilization of rainwater today is based on an ancient tradition and uses techniques similar to those used around 5000 years ago. Building dams to tap stream water, channeled through canals and stored in reservoirs, was practiced by ancient civilizations about 5000 years ago to provide drinking water to the old city of Jawa (Abdelkhaleq and Ahmed 2007) and Umm el-Jimal city in the Early Roman period (de Vries 1997). In the Nabataean civilization that emerged in the arid region of southern Jordan more than 2500 years ago, people built dams to provide their capital city Petra and other settlements with water for drinking and irrigation (Abdelkhaleq and Ahmed 2007; Oleson 1995). The earliest known evidence of the use of the technology in Africa comes from northern Egypt, where tanks ranging from 200 to 2000 m^3 have been used for at least 2000 years – many are still operational today. Indians, over centuries, developed a range of techniques to harvest every possible form of water – from rainwater to groundwater, stream to river water, and floodwater. In the seventeenth century the small island of Malta built an aqueduct to collect rainwater for its growing population. In Meghalaya, a 200-year-old system of tapping stream and spring water for irrigating plants by using bamboo still exists (Agrawal and Narain 1997).

During the time of the Roman Empire, rainwater collection became something of an art and science, with many new cities incorporating state of the art for the time. The Romans were masters at these new developments and great progress was made right up until the sixth century AD. One of the most impressive rainwater harvesting construction can be found in Istanbul in the Sunken Palace which was used to collect rainwater from streets above.

Although rainwater harvesting has existed since the days of the early Romans, its popularity declined once central treatment facilities were able to supply treated drinking water. Today, rainwater harvesting is gaining popularity again for a variety of environmental and economic reasons. Obviously, technology and techniques have changed considerably from the days of the early Romans, but the theory remains the same.

Arid and semi-arid countries suffer deficits in water resources. These countries are the most water-stressed countries. The total renewable freshwater resources of these countries are less than the demand required by different sectors. The current water availability in these countries is less than the poverty level of 1000 m^3 per capita per year. The scarcity of water resources in these countries seems to be dictated by climatic conditions, such as aridity and abundance of high solar radiation, and by population pressure which grew at a location distant from water resources (Salameh and Bannayan 1993). Urban development and increasing demand are putting stress on existing water resources. This highlights the need to implement measures to ensure that the rain falling on rooftops is tapped as fully

Handbook of Water Harvesting and Conservation: Basic Concepts and Fundamentals, First Edition. Edited by Saeid Eslamian and Faezeh Eslamian.

as possible through rooftop rainwater harvesting, either by recharging it into the groundwater aquifers or storing it for direct use. Attention is now focusing on alternative water sources, such as rainwater harvesting systems as supplementary water sources with multi-purpose functions. The increasing pressure on the available resources represent a challenge for scientists, engineers, and policy makers because the entire development of the country in a variety of fields depends on the availability of this vital resource (Abdulla and Al-Shareef 2009).

For many years, RWH was highly accepted and adopted in many places around the world (Okhravi et al. 2014). The practice of RWH systems is highly related to cultural values that were gained and transferred from one generation to another. The preliminary cost–benefit analysis suggests that implementing a rainwater collection system is not economically feasible at low water price. But this should not prevent implementing RWH because there are extra benefits that would be added to the system, such as reduced impact on the environment, research opportunities, and better public perception of government's goals toward sustainability. The cost of implementing a RWH system would be further reduced by including it in the construction of a new building or during a major renovation of an existing building. For all the reasons discussed above, it is highly recommended that RWH systems be installed in newly constructed buildings that have an area greater than 200 m^2 and located in zones of mean annual rainfall greater than 300 mm (Abdulla and Al-Shareef 2009).

The search for a new water source starts with an effort to decrease the present amount of water lost in the distribution system. Equally important is the collection of rainwater in economically feasible cisterns. It must be stressed that rainwater is the only source, which is easy to obtain individually and with minimum cost. The only thing a person needs is the roof of the house to collect the rainwater and a place to store it. In arid and semi-arid countries, where water supply to domestic sector is not based on demand but rather on a rotation system which in many times fails, people have to find other alternatives sources of water. In such case, water harvesting is an important source for drinking and other domestic usages.

Today there is a rapid increase of interest in rainwater harvesting and storage as a potential water supply to meet part of the urban, rural, and agricultural water demand. In this chapter, rainwater harvesting concepts and technologies have been reviewed and discussed. This chapter provides specific recommendations on the most appropriate methods and technologies of rainwater harvesting that should be adapted in different climatic zones. Both domestic and agricultural techniques of rainwater harvesting are addressed in this chapter. Tangible figures on quantities of collected water, water quality and its health and environmental impacts, appropriate designs of water harvesting systems, and cost–benefit analysis have been provided.

1.2 Concept of Rainwater Harvesting

Definition: Rainwater harvesting is a technology used for collecting and storing rainwater from rooftops, land surfaces, road surfaces, or rock catchments using simple techniques such as pots, tanks, and cisterns, as well as more complex techniques such as underground check dams (Appan 1999; Prinz 1995; Zhu et al. 2004). Currier (1973) has defined water harvesting as "the process of collecting natural precipitation from prepared watersheds for beneficial use." In the *Handbook on Water Harvesting*, Frazier and Myers (1983) have defined the term "water harvesting" as the process of collecting and storing water from an area that have been treated to increase precipitation runoff. In scientific terms, rainwater harvesting refers to the collection and storage of rainwater and also other activities aimed at harvesting surface and groundwater: prevention of losses through evaporation and seepage, other hydrological studies and engineering interventions aimed at conservation, and efficient utilization of the limited water endowment of physiographic unit as a watershed (Agrawal and Narain 1997). The basic principle of rainwater harvesting is to "*Capture rain where it falls.*"

Rainwater harvesting can also mean collecting rainwater before it infiltrates into the ground and becomes underground water. Harvested rainwater is a renewable source of clean water that is ideal for domestic and landscape uses. Water harvesting systems provide flexible solutions that can effectively meet the needs of new and existing water demand for domestic and agricultural uses.

All water harvesting systems must have the following three components (see Figure 1.1):

(i) the catchment area, which is the part of land that contributes some or all of its share of rainwater to another area outside its boundaries. The catchment area can be small as few square meters or as large as several square kilometers. It can be agricultural, rocky, or marginal land or even a rooftop or paved road.
(ii) The storage facility, which holds surface runoff. Water can be stored in a surface reservoir, subsurface reservoirs (such as cisterns), in the soil profile (such as soil moisture), or a groundwater aquifer.
(iii) The target area, which is where the harvested water is used. It can be used for agricultural production in which the target is plants or livestock, in the domestic sector for drinking and cleaning, or it can be used for industrial purposes and other enterprises.

Figure 1.1 Components of a water harvesting system.

1.3 Technologies of Rainwater Harvesting

According to Oweis et al. (2004), there are several classifications of water-harvesting technologies, mostly based on the type of use or storage, but the method most widely used is based on the size of the catchment. Prinz (2011) presented wider classifications of water harvesting systems as shown in Figure 1.2. Also, Ngigi (2003) indicated that rainwater harvesting can fall into the following categories: in situ water conservation, conservation tillage, and runoff farming (i.e. storage systems for supplemental irrigation and direct runoff application, flood diversion and spreading systems, small external catchment systems, micro-catchment systems). Apart from in situ water conservation, normally rainwater harvesting systems have runoff producing areas, runoff collection structures, and storage facilities.

In situ water conservation technologies aim at conserving the rainfall where it falls in the cropped area or pasture. The in situ technology consists of making storage available in areas where the water is going to be utilized. Some water conservation methods such as mulching, deep tillage, contour farming, and ridging are often referred to as in situ rainwater harvesting techniques (Habitu and Mahoo 1999; Abu-Zreig and Tamimi 2011). The purpose of these methods is to ensure that rainwater is held long enough on the cropped area to allow more water infiltration into the soil. Deep tillage is a water conservation technique that improves soil moisture capacity by increasing soil porosity. In addition, runoff is reduced through increased roughness at the soil surface, which in turns increases the time available for water to infiltrate into the soil. The primary importance of such technologies is to reduce in-field runoff, increase the amount of water available within the root zone, and reduce soil erosion (Abu-Zreig et al. 2000). In situ water conservation practices are simple and cheap to apply. They include practices like; mulching, ridging, bench terraces, and addition of manure (FAO 2002). On not-so steep slopes, ridging, bench terraces, contour bunds, and small stone barriers can be used in order to slow down or prevent runoff so that rainwater sinks into the ground. On steeper hills, terracing can be applied, though it is quite labor-intensive.

Conservation tillage is a tillage system that conserves soil, water, and energy resources through the reduction of tillage intensity and retention of crop retention. It is also known as any tillage practice where about 30% mulch or crop residue cover is left in the field throughout the year, with the major objective to reduce soil and water loss (Dinnes 2004; Ngigi 2003). Conservation tillage increases infiltration and the water-holding capacity of the soil. The practice also saves labor due to reduced traction needs.

Runoff farming, a water harvesting technique, is the diversion of rainwater from a collecting area to a cropping area, thereby increasing the quantity water of available for crop growth. Runoff farming involves collecting runoff, generated either within the field or from external catchments, and applying the water either directly in the field or storing it for future use. Runoff farming involves technologies for storage of runoff for supplemental irrigation. In many dry parts of the world, simple and cheap structures (e.g. earth dams, farm ponds, and underground tanks) have been developed for storage of rainwater for supplemental irrigation (FAO 2002). Water loss from the tanks and ponds through seepage and evaporation reduces the value of this technology for rainwater harvesting. Runoff farming can be considered a rudimentary form of irrigated agriculture. The major differences are that with runoff farming the water is provided by natural runoff from catchment areas rather than artificial irrigation systems; water is stored in the soil of cultivated land rather than reservoirs, ponds, and cisterns; and the farmers have no control over timing since runoff can only be harvested when it rains. As a result, several innovations like lining tanks with plastic papers and cementing have been tried. However, such measures also imply additional costs to a farmer. Another technology for rainwater harvesting involves diversion of runoff and direct application in the cropland/garden. Under this technique, the soil profile acts as the reservoir. Direct runoff application systems include small external

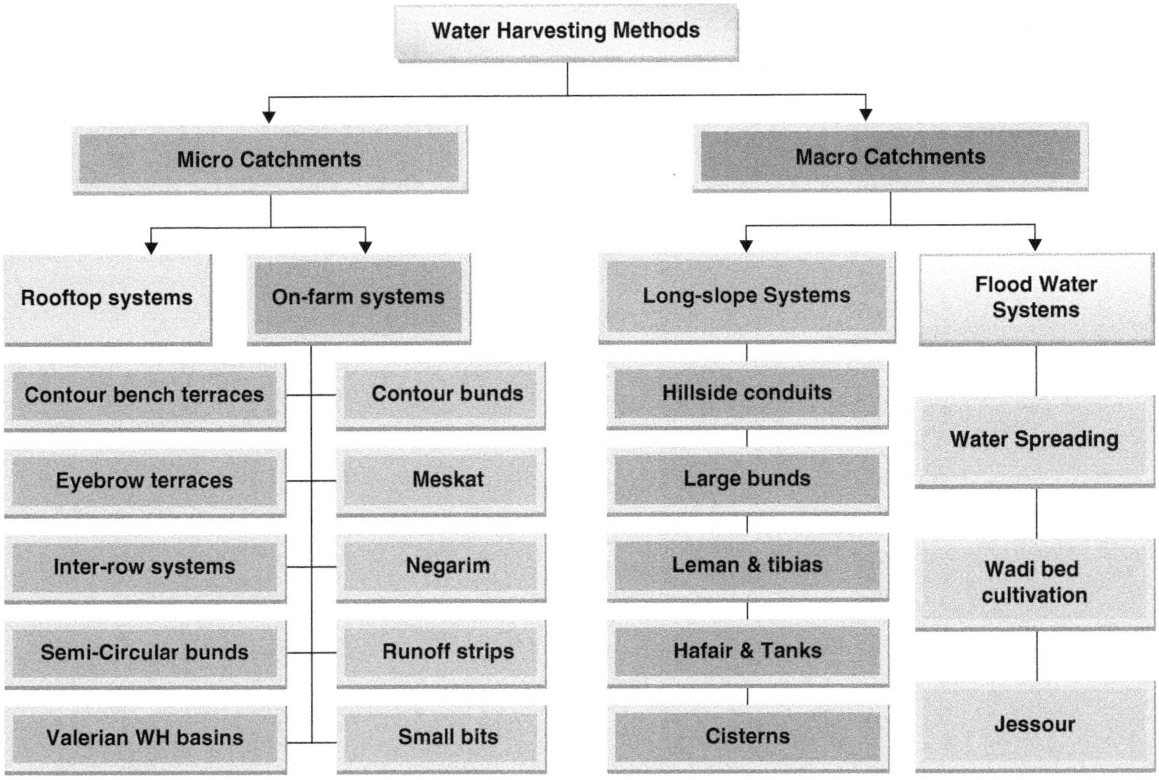

Figure 1.2 Rainwater harvesting technologies. Source: After Prinz 2011.

catchment systems, where small-scale runoff is diverted from roadsides and foot paths, and spread into the garden through a series of cut-off drains, contour bunds, ditches, and trenches (Ngigi 2003). In many cases, shrubs of various types and grass like Napier are planted on the lower sides of the rainwater harvesting structures to stabilize them. Another type of direct runoff application is micro-catchment systems, which involves generation of runoff within a field and concentration of the water on a single crop (i.e. fruit trees), or a garden established along a contour. In rainwater harvesting under micro-catchment systems, the crop land is subdivided into micro-catchments that supply runoff to single crops or a group of crops. Techniques under micro-catchment include moisture retention terraces, contour bunds, infiltration trenches/ditches, semi-circular earth bunds, circular depressions, etc. (e.g. Ngigi 2003). The improved traditional planting pits (zaï) widely used to rehabilitate degraded land in Burkina Faso also follow micro-catchment techniques Figure 1.3.

In this chapter, the classifications presented by Oweis et al. (2004) and Prinz (2011) will be described.

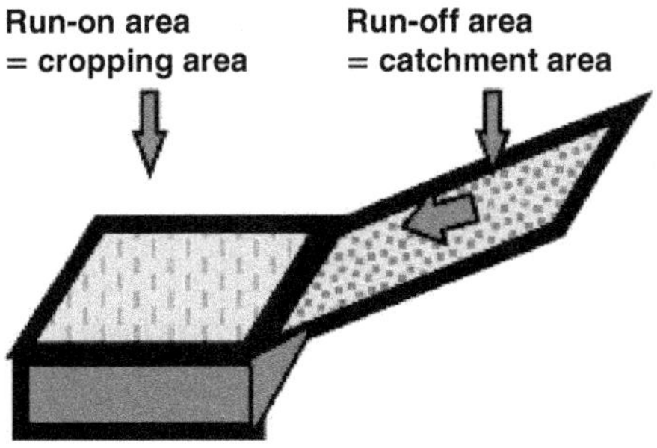

Figure 1.3 Definition of runoff farming. Source: Adapted with courtesy, FAO 1991.

1.3.1 Micro-Catchment Systems

Micro-catchment method is a method of collecting surface runoff from a small catchment area and storing it in the root zone of an adjacent infiltration basin (Prinz and Singh 2006; Boers et al. 1986). Also, micro-catchment can be defined as a specially contoured area with slopes and berms designed to increase runoff from rain and concentrate it in a planting basin where it infiltrates and is effectively "stored" in the soil profile. The water is available to plants but protected from evaporation (Shanan and Tadmor 1979 and Li et al. 2004)

1.3.1.1 Rooftop System

Rooftop rainwater harvesting is the technique through which rainwater is captured from roof catchments and stored in tanks/reservoirs/groundwater aquifers. It consists of conserving rooftop rainwater in urban and rural areas and utilizing it to augment groundwater storage by artificial recharge. It requires connecting the outlet point from the rooftop to divert collected water to an existing well/tube well/bore well or a specially designed well. Rooftop rainwater harvesting systems can be used to direct rainwater that falls on the roof of a building into containers or tanks. These tanks are usually elevated so that when the tap is opened, water flows at a high pressure. This method of rainwater harvesting is good because the accumulated water is mostly clean and usually requires no further treatment to make it fit for human use. This technique addressed by many researchers such as Zaizen et al. 1999; Kahinda et al. 2007; and Chisi et al. 2006.

1.3.1.2 On-Farm Systems

Contour bunds: This system consists of small trash, earth, or stone embankments, constructed along the contour lines (Garcia et al. 2002). The embankments trap the water flow behind the bunds allowing deeper infiltration into the soil. The height of the bund determines the net storage of the structure. This method is a traditional low-cost method of soil conservation suitable for sloping land; it promotes water retention and helps prevent erosion. Contour bunds are constructed in relatively low rainfall areas, having an annual rainfall or less than 600 mm, particularly in areas having light textured soils. For rolling and flatter lands having slopes from 2 to 6%, contour bunding is practiced, in red soils.

Semi-circular bunds: These are constructed in series in staggered formation. Runoff water is collected within the hoop from the area above it and impounded by the depth decided by the height of the bund and the position of the tips (Prinz 1995). Excess water is discharged around the tips and is intercepted by the second row and so on. Normally the semi-circles are of 4–12 m radius with height of 30 cm, base width of 80 cm, side slopes 1:1.5, and crest width of 20 cm. The percentage of enclosed area which is cultivated depends on the rainfall regime of the area. Basic requirements of the semi-circular bunds are:

(i) Ground slope must be less than 3%,
(ii) Soil depth, at least 1 m,
(iii) Average annual rainfall of at least 100 mm.

Meskat-type system: The meskat-type system is a type of micro-catchment system in which the catchment area diverts runoff water directly on to a cultivated area at the bottom of slope (Rosegrant et al. 2002). In this system instead of having CA (Catchment Area) and CB (Cropped basin) alternating like the previous methods, here the field is divided into two distinct parts, the CA and CB, whereby the CB is immediately below the CA.

1.3.2 Macro-Catchment Systems

Macro-catchment water harvesting, also called harvesting from external catchment, is a case where runoff from hill slope catchment is conveyed to the cropping area located at hill foot on flat terrain (Prinz and Singh 2006; Mzirai and Tumbo 2010). Macro-catchment rainwater harvesting includes the collection of water from large areas substantially far away from the cropped areas.

Strip catchment tillage: Strip-tillage is defined as less than full-width tillage of varying intensity that is conducted parallel to the row direction. This involves tilling strips of land along crop rows and leaving appropriate sections of the inter-row space uncultivated to release runoff. It is normally used where the slopes are gentle and the runoff from the uncultivated parts added water to the cropped strips. The catchment:basin area ratios (CBAR) used are normally less than or equal to 2:1. The system can be used for nearly all types of crops and is easy to mechanize.

External (Macro) catchment RWH: This is a system that involves the collection of runoff from large areas which are at an appreciable distance from where it is being used (Gowing et al. 1999). This is sometimes used with intermediate storage of water outside the CB for later use as supplementary irrigation. It is difficult to differentiate this system from conventional irrigation systems but in this chapter the system is called RWH as long as the water for harvesting is not available beyond the rainy season. This system involves harvesting of water from catchments of areas ranging from 0.1 ha to thousands of hectares either located near the cropped basin or long distances away. The catchment areas usually have slopes ranging from 5 to 50%, while the harvested water is used on cropped areas which are either terraced or on flat lands. When the catchment is large and located at a significant distance from the cropped area, the runoff water is conveyed through structures of diversion and distribution networks. The most important systems are described in this section.

There are many ways in which rainwater can be harvested. Some of these methods are very effective and can aid in the collection of a lot of water – even for commercial activities – while others are only suitable for harvesting water meant for domestic use. Every system has its merits and demerits. These are the common methods of rainwater harvesting:

- **Surface catchment systems:** Surface water is simply water that accumulates on the ground's surface. When rainwater falls on the surface of the earth (rock outcrops/slopes, concrete surfaces, plastic sheets, or treated ground surfaces), it usually flows down slopes as it moves toward a point of depression where the moving water can collect. Surface water collection systems enable the collection of surface runoff before it flows to other locations. Examples of such systems include rivers, ponds, and wells. Drainage pipes can be used to direct water into these systems. Water can then be fetched from these sources and used for other purposes such as domestic and livestock consumption. Water quality could be acceptable to beneficiaries. Safe water for human consumption can be assured with proper O&M and simple disinfection techniques if needed. This method is recommended for arid and semi-arid regions.

- **Small-scale Dams:** These are barriers that are designed to trap water. Also, these structures may be a temporary structure constructed with locally available materials (Tiessen et al. 2011). These consist of storage structure (earth dams, concrete dams, or simple excavated ponds, etc.), and structures to extracting water such as horizontal intake pipes. Rainwater or surface runoff can accumulate directly in them or drainage systems can be created to direct water into them. Water collected in these structures is mostly used for irrigation purposes and livestock or treated and then distributed for domestic use. They can also be used to harvest a lot of water because of the way in which they are modeled. Unlike ponds, measures are put in place to reduce the amount of water draining into the ground. Water quality could be acceptable to users and normally consumed without any further treatment. Safe water for human consumption can be assured with proper water extraction structures and simple disinfection methods if needed. These structures are highly recommended in arid and semi-arid region (rainfall between 100 and 500 mm).
- **Underground Tanks:** The systems are mainly comprised of a guttering system, underground storage tank, pump, and walled structure. These are also ideal for collecting rainwater. They are constructed by digging into the ground and creating a space which is then cemented to reduce water infiltration. The top is also sealed, and water is obtained through pipes directed into the tank. To get water out, pumps are used. Underground tanks are wonderful for harvesting rainwater because the rate of evaporation is reduced since they are located underground where sunlight does not really penetrate.
- **Rainsaucer:** Sometimes one can decide to collect rainwater directly as it falls from the sky by using a rainsaucer (RainSaucers 2009). These look like upside-down umbrellas or big funnels. Some are usually attached to a pipe so that the collected water is directed elsewhere. Some people also do a little improvisation by placing the collecting container underground with only the rainsaucer above the ground. It is a simple method yet effective.
- **Water Collection Reservoirs:** Water collected through this method is not really clean and may be contaminated. However, it can still be used for crop irrigation. Such rainwater is harvested from roads and pavements.
- **Barrage:** A barrage is a dam that has several openings which can be closed or opened to control the quantity of water that passes through it. It is usually large and can be used to collect a lot of water.
- **Slopes:** Rainwater tends to collect at the bottom of slopes when it flows on the ground. When it rains heavily, water levels can rise to the hilltop. This is a simple and natural way to harvest rainwater.
- **Trenches:** This is another great way to harvest rainwater for irrigation. When it rains, the water is directed to the farm using trenches. It is one of the traditional methods of rainwater harvesting that is still very much in use today.
- **Rain Barrels:** These are also used for rainwater harvesting. They are specifically designed for this purpose and can be purchased from retail stores. Rain barrels are used for harvesting rainwater that falls on rooftops.

1.4 Advantages and Disadvantages of Rainwater Harvesting

1.4.1 Advantages of Roof Rainwater Harvesting (RRWH)

Roof Rainwater Harvesting (RRWH) technologies are simple to install and operate. Rainwater harvesting is a technology used for collecting and storing rainwater from rooftops, land surfaces, road surfaces, or rock catchments using simple techniques such as pots, tanks, and cisterns as well as more complex techniques such as underground check dams (Appan 1999; Prinz 1995; Zhu et al. 2004). Rainwater harvesting provides the long-term answers to the problem of water scarcity. Rainwater harvesting offers an ideal solution in areas where there is sufficient rain but inadequate ground water supply, and surface water resources are either lacking or are insufficient. Harvested rainwater is a renewable source of clean water that is ideal for domestic and landscape uses. Water harvesting systems provide flexible solutions that can effectively meet the needs of new and existing, as well as of small and large, sites. The greater attraction of the rainwater harvesting system is the low cost compared to other water supply systems, and they are accessible and easily maintained at a household level (Abdulla and Al-Shareef 2009). Harvesting rainwater has a long-term impact on the local water resources by reducing demands for surface and groundwater withdrawals. Also, harvesting rainwater protects the integrity of local water resources by reducing nonpoint source pollution. Including rainwater harvesting in national water supply plans offers an alternative and sustainable water source while protecting the local environment.

Arid and semi-arid countries suffer deficits in water resources. These countries are the most water-stressed countries. The total renewable freshwater resources of these countries are less than the demand required by different sectors. The current water availability in these countries is less than the poverty level of 1000 m^3 per capita per year. The scarcity of water resources in these countries seems to be dictated by climatic conditions, such as aridity and abundance of high solar radiation, and by population pressure which grew at a location distant from water resources. Urban development and increasing demand are putting stress on existing water resources. Attention is now focusing on alternative water sources, such as rainwater harvesting systems, as supplementary water sources with multi-purpose functions. The increasing pressure on the available resources represent the challenge for scientists, engineers, and policy makers because the entire development of the country in different fields depends on the availability of this vital resource (Abdulla and Al-Shareef 2009).

For many years, RWH has been a highly accepted and adopted practice in many places in the world. The practice of RWH systems is highly related to some cultural values that were gained and transferred from one generation to another. The preliminary cost–benefit analysis suggests that implementing a rainwater collection system is not economically feasible at low water price. But this should not prevent implementing RWH because there are extra benefits that would be added to the system, such as reduced impact on the environment, research opportunities, and better public perception of government's goals toward sustainability. The cost of implementing a RWH system would be further reduced by including it in the construction of a new building or during a major renovation of an existing building. For all the reasons discussed above, it is highly recommended that RWH systems should be installed in newly constructed buildings that have an area greater than 200 m^2 and located in zones of mean annual rainfall greater than 300 mm (Abdulla and Al-Shareef 2009).

Due to the variable topographic features of different locations in the world, the distribution of rainfall varies considerably with location. Rainfall amounts vary from 100 mm in arid countries to more than 1000 mm in humid and semi-humid countries.

Search for a new water source starts with an effort to decrease the present amount of water lost in the distribution system. Equally important is the collection of rainwater in economically feasible cisterns. It must be stressed that rainwater is the only source, which is easy to obtain individually and with minimum cost. The only thing which a person needs, is the roof of the house to collect the rainwater and a place to store it. In arid and semi-arid countries where water supply to domestic sectors is not based on demand but rather on a rotation system which in many times fails, people have to find other alternatives sources of water. In such case, water harvesting is an important source for drinking and other domestic usages (Okhravi et al. 2015)

Since WH provides water at the point of consumption, owners (either single-home owners or commercial building owners) have full control of their systems, which greatly reduces operation and maintenance problems. Running costs are also almost negligible. Water collected from roof catchments is usually of acceptable quality for domestic purposes. In addition, rooftop catchments are usually best for satisfactory domestic water requirements because they are close to the dwelling and isolated from many sources of contamination. Rainwater may be utilized for drinking and cooking, for which high water quality is required, or for other domestic purposes, such as washing, bathing, toilet flushing, and landscape irrigation after filtration and disinfection. As it is collected using existing structures not specially constructed for the purpose, rainwater harvesting has few negative environmental impacts compared to other water supply project technologies. In summary, the main advantages of RRWH are:

- Saves water and energy.
- Water for domestic use: Rainwater harvesting is beneficial because it provides a source of water for domestic use. The collected water can be used for house cleaning purposes, washing laundry, and for cooking. When treated, rainwater is good for drinking. It is an easy way of obtaining water for use in the home.
- Sources of energy are not needed to operate the system.
- Water for industrial use: Industries can also harvest rainwater for use in some of their processes. Rainwater meant for industrial use is normally harvested in large scale. Such companies can construct their own dams or have underground tanks to store rainwater.
- Supplementary water source: Many areas experience water shortages during summer due to lack of rain and as a result of the high rate of evaporation. It can be difficult to get a reliable source of water during these periods. Those who sell water may also increase their prices because of the high demand and short supply. Harvesting rainwater is therefore seen as a way of preparing for the sunny days when water is scarce.
- Quality of rainwater can be used as a primary source for specific uses and so reduce the water bill.
- Does not come into contact with soil and rocks where it dissolves salts and minerals. It is soft and can

significantly reduce the quantity of detergents and soaps needed for cleaning.

- Very good for areas that are not served with water.
- Relatively limited technical knowledge is required.
- It uses local construction materials and labor.
- The owner user can easily maintain the system.
- Decreases local erosion and flooding caused by runoff from impervious cover such as pavement and roofs.
- The RRWH is usually found to be socially and environmentally acceptable.

1.4.2 Disadvantages of RRWH

Disadvantages of rainwater harvesting technologies are mainly due to the limited supply and uncertainty of rainfall. Adoption of this technology requires a "bottom-up" approach rather than the more usual "top-down" approach employed in other water resource development projects. This may make rainwater harvesting less attractive to some governmental agencies tasked with providing water supplies in developing countries, but the mobilization of local government and NGO resources can serve the same basic role in the development of rainwater-based schemes as water resource development agencies in the larger, more traditional public water supply schemes.

The main disadvantages are:

- The high initial cost of building the permanent storage facilities; the primary expense is the storage tank.
- The quantity of rainwater available depends on rainfall; for long periods of drought it is necessary to store excessively large volumes of water.
- The mineral-free water is tasteless and could cause nutritional deficiencies; people prefer to drink water rich in minerals.

1.5 Feasibility of Rainwater Harvesting across Different Climatic Zones

There are three important questions that should be asked when undertaking any project in a developing country: Does the community need it? Does the community want it? And can it be done? For a rainwater harvesting system, these translate into assessment of the physical, social, and technical aspects.

1.5.1 Physical Feasibility

The feasibility of rainwater harvesting is highly dependent upon the amount and intensity of rainfall. Other variables, such as catchment area and type of catchment surface, can usually be adjusted according to household needs. As rainfall is usually unevenly distributed throughout the year, rainwater collection methods can serve as only supplementary sources of household water. The viability of rainwater harvesting systems is also a function of the quantity and quality of water available from other sources; household size and per capita water requirements; and budget available.

Physical feasibility depends on the amount of rainfall in the area, the duration of dry periods, and the availability of other water sources. The rainfall pattern over the year plays a key role on determining whether RWH can compete with other water supply systems. Various questions arise: For the given location, does it rain and how often? Does the amount of rainfall per month or per season warrant the usefulness of a rainwater harvesting system? There is no recommended minimum amount of annual rainfall that can be used a guide to the implementation of RWH. As a general rule, rainfall should be over 50 mm/month for at least half the year or 300 mm/year to make RWH physically feasible. This rule may not seem applicable for arid and semi-arid regions. Another source recommends 400 mm per year (United Nations Environmental Programme 1997) for feasible RWH. Past experiences and reviewing of existing RWH in arid and semi-arid regions indicated that RWH technology is adopted in areas having annual rainfall around 100 mm. Not all rainfall storms can generate surface runoff, even on impervious surfaces such as roofed buildings made from cement and concrete. For example, storms of less than 5 mm of rainfall will not generate surface runoff.

Due to the variable topographic features, the distribution of rainfall varies considerably with location. In most countries, the rainy season extends for about six months (for example, in the Middle East and North Africa [MENA] region it extends from October to May), with the peak of precipitation taking place during January and February. Rainfall data in this region indicate that there is a significant amount of rainwater that can be harvested. The highest rates of precipitation are commenced over the highlands which receive the long-term annual average.

1.5.2 Technical Aspects

The technical assessment seeks to answer the question "Can it be done?" by taking into consideration the resources required for the implementation of the system, by determining expected supply and demand for water based on gathered data, and, where applicable, by taking into consideration previously attempted projects and their reception by the people. In summary, the construction of a RWH system is determined by several critical technical factors:

- Use of impermeable roofing material such as concrete surfaces, tiles, etc.
- Availability of sufficient roof area
- Water consumption rate (number of users and types of uses) and storage capacity required
- Availability of other water sources, either groundwater or surface water that can be used when stored rainwater runs out
- Availability of laborers with technical building skills
- Availability of required, suitable local construction material

1.5.3 Social Aspects

The social assessment goes on to answer the whys of the physical assessment. For example, why is one source of water more preferred than another? Is a water source located in an area by choice or by circumstance? Why do people not practice rainwater harvesting? Is there a real need for better water provision?

People may have the need for an improved water supply, but there are several reasons the people may not be receptive to the idea of a RWH system. Depending on the kind of system presented, the technology may be above the education level of the people. There may be other priorities, depending on the season. It may not be considered an immediate need, or there may already be multiple sources of water, each with its own specified purpose. There may be traditional RWH systems already in place.

Cultural perceptions and religious views regarding the use of water, as well as traditional preferences for its location, taste, smell, or color are all important and need to be taken into consideration. It is those very traditions and social roles that will determine the successful implementation and use of a rainwater harvesting system. It is important to know the people, to be aware of their concerns, and to encourage their participation in every step of the rainwater harvesting process. It has been shown that the more a community is involved, the more potential for a successful project. These and likely other factors not mentioned here can positively or negatively affect a RWH system (Abdulla and Al-Shareef 2009). This practice of RWH systems is highly related to some cultural values that were gained and transferred from one generation to another.

The following social aspects should be considered when designing a household-based system:

- There should be a real felt need in the family for better water provision;
- The design should be affordable and cost effective;
- The family or community should be enthusiastic and fully involved;
- In the case of multi-story (apartment) building, it is difficult to implement such a system because the ownership of roof is not well identified.

1.5.4 Financial Aspects

The financial circumstances may also influence the design of a RWH structure. However, one should realize that financial reasons can hardly be a restriction for building a RWH system. Almost every house or building has a suitable roof, but guttering and the water storage do require some investments. The water storage tank usually represents the biggest capital investment element of the RWH system and therefore requires careful design to provide optimal storage capacity while keeping the costs as low as possible.

Installing a water harvesting system at the household level can cost anywhere from $400 USD to more than $2000 USD. It is difficult to make an exact estimate of cost because it varies widely depending on the availability of existing structures like pipes, tanks, and other materials. Actual cost depends on the final design and size of the tank. The cost would be comparatively less if the system was incorporated during construction of the building.

1.6 Roof Rainwater Harvesting System Components

A rainwater harvesting system consists of six basic components: a collection area (roof), a conveyance system, and a cistern or storage tank, with filtration, delivery system, and treatment. Figure 1.1 shows a schematic of a rooftop catchment system.

1.6.1 Catchment Area

The catchment of a rainwater harvesting system is the surface upon which the rain falls; the surface has to be appropriately sloped, preferably in the direction of the storage facility or the recharge area. The collection area in most cases is the roof of a house or a building; however, catchment areas may include driveways or swales in yards. Rainwater harvested from catchment surfaces along the ground should only be used for irrigation because of the increased risk of contamination. The effective roof area and the material used in constructing the roof influence the efficiency of collection and the water quality. Smoother, cleaner, and more impervious roofing materials are preferred; they contribute to better water quality and greater quantity. Tiled roofs or roofs sheeted with corrugated mild steel, etc., are preferable since they are the easiest to use and give the cleanest water (Abdulla and Al-Shareef 2009). Cement and tiled roofs are the most common roofs used in many parts of the world (Abdulla and Al-Shareef 2009). These types of roofs have good durability and provide good quality water. Composite asphalt and some painted roofs are recommended only for non-potable water use because they could leach toxic materials into rainwater as it touches the roof surface. Roofs that have lead materials should be prohibited because acidic rain may cause contamination of collected water from these roofs. Regardless of the roof material, many designers assume about a 20% loss of annual rainfall. These losses are due to roofing material texture, evaporation, losses occurring in gutters and storage tanks, and inefficiencies in the collection process (Abdulla and Al-Shareef 2009).

1.6.2 Conveyance System

A conveyance system usually consists of gutters or pipes that deliver rainwater falling on the rooftop to cisterns. Gutters or pipes must be properly sized, sloped and installed in order to maximize the quantity of harvested water. The most common materials of gutters are galvanized steel, fiberglass, plastic, stainless steel, copper, cast iron, and UPVC (unplasticized polyvinyl chloride). The gutters and downpipes are usually installed in the wall of the building, and sometimes the downpipes are fitted inside the wall during construction. The size of the gutters depends on the area of the roof and the rainfall amount; the size of the gutters used ranges between 5 and 10 cm diameter (Abdulla and Al-Shareef 2009).The diameter of the gutters can be determined using the Rational Equation ($Q = CIA$), where Q is the discharge in m^3/sec, C is the runoff coefficient (0.8), A is the roof area in m^2, and I is the rainfall intensity in mm/hr and can be obtained using the Intensity-Duration-Frequency (IDF) curves. Both drainpipes and roof surfaces should be constructed of chemically inert materials such as wood, plastic, aluminum, or fiberglass, in order to avoid adverse effects on water quality. Leaf screens are important to keep leaves and other debris from entering the system; the gutter should have a continuous leaf screen made of 0.4 mm fine mesh installed along their entire length.

1.6.3 Storage Tank

The water is ultimately stored in a storage tank or cistern. For a long time, people have been building cisterns to collect and store rainfall from roofs of their houses. There are many options for the construction of these tanks with respect to the shape (cylindrical, rectangular, and square), the size, and the material of construction (brickwork, stonework, cement bricks, plain cement concrete, and reinforced cement concrete). Storage tanks may be constructed as part of the building, or may be built as a separate unit located some distance away from the building. Concrete tanks are the most commonly used tanks in many countries; they can be built above or below ground (Abdulla and Al-Shareef 2009). They are usually made on site and are durable and long lasting. Above-ground tanks make it easy to detect cracks and leaks; water can be extracted via gravity and/or pumps; they can be raised off ground to increase water pressure; they are easy to drain for cleaning; and usually cost less than below-ground tanks (Abdulla and Al-Shareef 2009). But they take up space, they are subject to weather conditions, and require anchoring to the ground for when the tank has lees water. Below-ground tanks can save space, but they are more difficult to extract water from – they usually need a pump; it is hard to detect leaks or problems; they are difficult to drain for cleaning; there is a risk of contamination from septic tanks or floodwaters; they can be damaged by tree roots; if the access point is left uncovered, there is a risk of children, adults, and animals drowning or contaminating the water; and they usually have a large excavation costs. In addition, they can sometimes crack, especially when they are below ground in clay soil. They're good for preventing algal growth (light cannot penetrate) and they keep water cool. The storage tank represents the major cost in the system. Some storage tank (cistern) considerations are:

- A cistern should be durable and watertight;
- It should be close to the water supply and demand source;
- Select maximum height location to avoid pumping costs and extract water by gravity;
- Keep away from contaminant source of a septic tank at least 30 m;
- Make it possible to add water from other sources such as tanker water in arid seasons (near entrance preferred);
- A smooth clean interior surface is needed;
- Joints must be sealed with nontoxic waterproof material;
- They should have a cover to prevent evaporation and mosquito breeding and algae growth from contact with sunlight;
- Should not present an excessive danger to users falling in or by the tank failing in a dangerous way;
- Should provide water of a quality commensurate with its intended use – water that is used for drinking requires particular care;
- The tank should be covered to prevent entry of light, and sealed against intrusion by mosquitoes and small creatures;

- The tank should be ventilated to prevent anaerobic decomposition of any matter that is washed in. Ideally, it should also:
  - Be affordable;
  - Be easy and cheap to maintain in good condition;
  - Have a means by which water can easily enter and easily be withdrawn (into the normal household receptacle used in the area);
  - Have some arrangement to satisfactorily handle tank overflow;
  - Be easy to clean or "self-cleaning"

1.6.4 First Flush

Contaminants washed from a roof are usually concentrated in the first part of the runoff. Such contaminants contain various impurities such as bird droppings and dust. After this initial runoff has washed the roof, the collected water can be considered safe. This process is called the first-flush diversion. The purpose of the first-flush diversion is to collect and disposal of the first flush of water from a roof, especially where the collected rainwater is to be used for human consumption. First-flush devices ensure a certain degree of water quality in harvested rainwater. The first-flush volume is assumed to be about 40 l for each 100 m^2 of roof area. Many first-flush devices are simply and cleverly designed. Such devices include tipping buckets that dump when water reaches a certain level. The most simple of these systems consists of a stand pipe and gutter downspout located ahead of the downspout from the gutter to the cistern; the gutter downspout and top of the pipe are fitted and sealed so water will not flow out the top, and once the pipe has filled, the rest of the water flows to the downspout connected to the cistern (Figure 1.4).

1.7 Calculation of Potential Harvested Water

Water harvesting yields can be calculated for roof areas ranging from 100 m^2 to more 1000 m^2. These roof areas cover residential buildings (single houses, villas, and apartments) and public, commercial, and industrial roofs. Monthly rainfall data can be obtained from the country's Ministry of Water or Meteorological Department for a particular location. The volume of rainwater that could be harvested from each roof for each rainfall zone can be calculated considering the annual rainfall data, the total roof area, and a runoff coefficient (efficiency) of 0.8. Runoff coefficient is the factor which accounts for the fact that all the rainfall falling on a catchment cannot be collected. Such a runoff coefficient indicates a loss of 20% of the rainwater that is discarded for roof cleaning and evaporation. The runoff coefficient, or efficiency, takes into account the losses from the collection surface. Recommended values of 0.8–0.85 are often used for the runoff coefficient, however it may be as high as 0.9 or as low as 0.24, depending on the surface material and other factors which may reduce the efficiency (Gould and Nissen-Peterson 1999). These factors include evaporation, clogging, leakage, infiltration, overspill, and retention. A smooth, clean, impervious surface yields better water quality and greater quantity (Texas Water Development Board 1997). The runoff coefficient is also a measure of the performance of the gutters and downspouts, as this is where most system losses tend to occur (Gould and Nissen-Peterson 1999). Table 1.1 shows the runoff coefficient for various surfaces of rooftops.

Thus, the volume of rainwater that could be harvested from each roof was determined by using Eq. (1.1).

$$VR = R \times A \times C \tag{1.1}$$

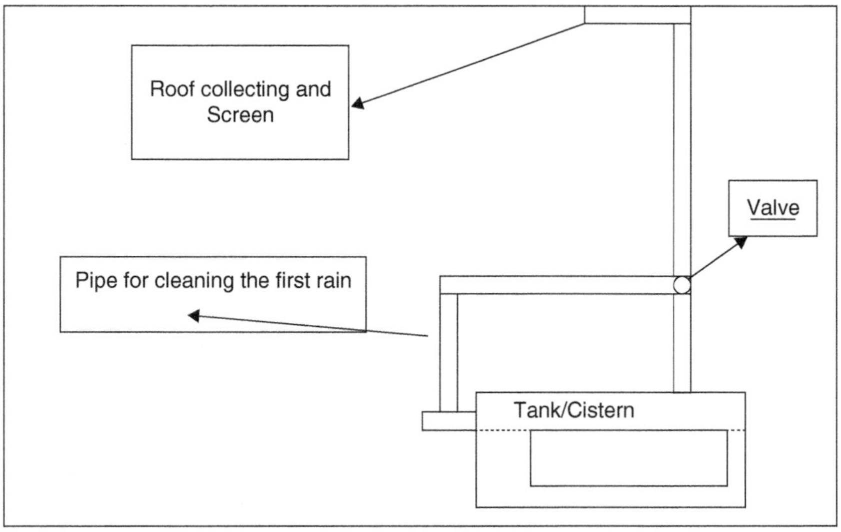

Figure 1.4 Rainwater harvesting system.

Table 1.1 Runoff coefficient for various surfaces.

Types of catchments	Runoff Coeff.
Roof Catchments	
- Tiles	0.8–0.9
- Corrugated metal sheets	0.7–0.9
Ground surface coverings	
- Concrete	0.6–0.8
- Bricks Pavement	0.5–0.6
Untreated ground catchment	
- Soil on slopes less than 10%	0.0–0.3
- Rocky natural catchments	0.2–0.5

Source: Pecey and Cullis (1989)

Where:

VR = Annual volume of rainwater that could be harvested (m^3),
R = Average annual rainfall in each rainfall zone (m/y),
A = Roof area (m^2),
C = Runoff coefficient (non-dimensional).

1.8 Water Quality and its Health and Environmental Impacts

Rainwater itself is of excellent quality; it has very little contamination, even in urban or industrial areas, so it is clear, soft, and tastes good. Contaminants can however be introduced into the system after the water has fallen onto a surface. Studies show that RRWH yields harvested waters with contaminants in levels acceptable by international drinking water standards (Kahinda et al. 2007; Zhu et al. 2004; Abdulla and Al-Shareef 2009) and is thus thought to be a superior option when considering domestic water supply, in particular potable water. Among these studies is one carried out by Abdulla and Al-Shareef (2009). Their analysis of samples of harvested rainwater from residential roofs indicated that the measured inorganic compounds generally matched the WHO standards for drinking water. On the other hand, fecal coliform, which is an important bacteriological parameter, exceeded the limits for drinking water. To be effective, high water quality must be both available during collection and maintained until the water is consumed. In the past, this aspect has not been adequately addressed.

Accounts of serious illness linked to rainwater supplies are few, suggesting that rainwater harvesting technologies are effective sources of water supply for many household purposes. It would appear that the potential for slight contamination of roof runoff from occasional bird droppings does not represent a major health risk; nevertheless, placing taps at least 10 cm above the base of the rainwater storage tanks allows any debris entering the tank to settle on the bottom, where it will not affect the quality of the stored water, provided it remains undisturbed. Ideally, storage tanks should be cleaned annually, and sieves should be fitted to the gutters and downpipes to further minimize particulate contamination. A coarse sieve should be fitted in the gutter where the downpipe is located. Such sieves are available made of plastic coated steel-wire or plastic, and may be wedged on top and/or inside gutter and near the downpipe. It is also possible to fit a fine sieve within the downpipe itself, but this must be removable for cleaning.

As a result, the following research priorities were identified:

(i) Investigation of the effects on the physicochemical quality of stored rainwater of roofing material and construction materials of containers;
(ii) Investigation of the effect on health of use of roofing materials made of asbestos, asphalt, and various metals;
(iii) Use of alternative roofing materials to improve the taste and color of stored rainwater;
(iv) Investigation of the effect of screening devices in preventing mosquitoes in the stored water;
(v) Investigation of the change in bacterial contamination levels of stored rainwater.

1.9 System Operation and Maintenance

Operation and maintenance are simple and depend mainly on cleaning the harvesting system annually before the start of the major rainfall season. Maintenance is generally limited to the annual cleaning of the tank and regular inspection of the gutters and downpipes. Maintenance typically consists of the removal of dirt, leaves, and other accumulated materials. However, cracks in the storage tanks can create major problems and should be repaired immediately. In the case of ground and rock catchments, additional care is required to avoid damage and contamination by people and animals, and proper fencing is required as well.

It is recommended that a tank be flushed at least once a year to remove all silt accumulation from the previous year. The sediment which builds up on the bottom of tanks should be cleaned when significant buildup occurs; if your tank does not provide a bottom clean-out valve then the tank usually must be drained in order to clean it, and all piping systems should be flushed and cleaned.

Roof catchments should also be cleaned regularly to remove dust, leaves, and bird droppings to maintain the quality of the product water; if you need to clean your catchment surface (e.g. roof and gutters) you should be careful that the water used for cleaning does not go into your rainwater tank.

1.10 Conclusion

The practice of harvesting rainwater is an old tradition adopted in many parts of the world and still used as a means for dealing with water problems of today, as well as a new technology that is growing in popularity. Rainfall water harvesting is a source of water and is very critical for the growth of crops and farming. This technology has been adapted in arid and semi-arid areas, rural and urban areas, and can serve as a primary or supplementary water source. The flexibility and the many benefits associated with rainwater harvesting make it a welcomed, widely accepted, and increasingly promoted alternative for the water demands of today. The collection of rainwater is today one of the easiest and cheapest methods of providing a good water supply to urban and rural communities in arid and semi-arid zones. The first reason for this is that does not require mobilizing vast quantities of resources and importing materials and expertise involved as in the planning and building large dams and reservoirs. A small RWH and storage system relies and builds upon local skills and experiences in construction, water consumption rate, and rainfall patterns.

As water shortages become more serious in many regions of the world, rainwater-harvesting systems will become more essential. Building codes must be updated to include these important water-conserving systems. If water-efficient technologies are incorporated at the code level, the resulting cumulative effects on water conservation across the country will be significant, as market opportunities open up for new products, customers demand more water-efficient designs, and designers face less resistance to incorporating efficient systems in buildings. Until they are included in the plumbing codes, it is up to the plumbing engineer to specify them and work with building officials in order to help conserve our increasingly scarce water resources.

Benefit–cost analyses of two roof harvesting systems (pear and concrete tanks) have been applied in this report. The unit price that is assigned for the water harvesting plays an important role in deciding whether the proposed system is economically feasible or not. The results revealed that pear-shaped tanks will be economically feasible in all rainfall zones when a \$1.0 USD m^{-3} is assigned as a cost of harvested water. Concrete tanks will not be economically feasible at this cost. The construction and installation costs of rooftop rainwater harvesting could cost as little as \$2000 USD for concrete tank sizes less than 20 $m^3$and might go up to \$6000 USD for tank size around 100 m^3. In general, the overall system cost typically rises with increasing tank volume, but system cost per m^3 of storage falls.

References

Abdelkhaleq, R. and Ahmad, I. (2007). Rainwater harvesting in an ancient civilization in Jordan. *Water Science and Technology* 7: 85–93.

Abdulla, F.A., and Al-Shareef A.W. (2009). Roof rainwater harvesting systems for household water supply in Jordan Desalination, 243: 195–207.

Abu-Zreig, M. and Tamimi, A. (2011). Field evaluation of sand-ditch water harvesting technique in Jordan. *Agricultural Water Management* 98: 1291–1296.

Abu-Zreig, M., Attom, M., and Nissreen, H. (2000). Rainfall harvesting using sand ditches in Jordan. *Agricultural Water Management* 46: 183–192.

Agrawal, A. and Narain, S. (eds.) (1997). *Dying wisdom, The Rise, Fall and Potential of Indians Traditional Water Harvesting System*. New Delhi: Centre for Science and Environment.

Appan, A.A. (1999). Dual-mode system for harnessing roof water for nonpotable uses. *Urban Water* 4: 317–321.

Boers, T.M., Zondervan, K., and Ben-Asher, J. (1986). Microcatchment water harvesting for arid zone development. *Agricultural Water Management* 12: 21–39.

Chisi, E., Montibeller, A., and Schmidt, R.W. (2006). Potential for potable water savings by using rainwater: an analysis over 62 cities in southern Brazil. *Building and Environment* 41: 204–210.

Currier, W.F. (1973). *Water Harvesting by Trick Tanks, Rain Traps and Guzzlers, in Water-Animal Relations Symposium*. Idaho, USA: Twin-Hall.

Dinnes, D.L. (2004). *Assessment of Practices to Reduce Nitrogen and Potassium Non-point Source Pollution of Iowa's Surface Waters*. Des Moines, IA, USA: Iowa Dept. of National resources.

FAO (1991). Introduction. In: *Water Harvesting: A Manual for the Design and Construction of Water Harvesting Schemes for Plant Production* (eds. W. Critchley and K. Siegert), 133. Rome: UN Food and Agriculture Organization.

FAO (2002). *Crops and drops: Making the best use of water for agriculture*. Rome: Food and Agriculture Organisation (FAO).

Frazier, G.N. and Myers, L.E. (1983). *Handbook of Water Harvesting*, 45. USA: U.S. Dept. of Agriculture, Agriculture Research Service, Agriculture Handbook No. 600.

Garcia, J.N., Gerrits, R.V., and Cramb, R.A. (2002). Adoption and maintenance of contour bunds and hedgerowsin a dynamic environment- experiences in the Philippine urplands mountain. *Research and Development* 22 (1): 10–13.

Gorokhovich, Y., Mays, L., and Lee, L. (2011). A survey of ancient Minoan water technologies. *Water Science and Technology: Water Supply* 11: 388–399.

Gould, J. and Nissen-Peterson, E. (1999). *Rainwater Catchment Systems for Domestic Supply*. England: IT Publications Ltd.

Gowing, J.W., Mahoo, H.F., Mzirai, O.B., and Hatibu, N. (1999). Review of rainwater harvesting techniques and evidence for their use in semi-arid Tanzania. *Tanzania Journal of Agricultural Science* 2 (2): 171–180.

Habitu, N. and Mahoo, H. (1999). Rainwater harvesting technologies for agricultural production: a case for Dodomia, Tanzania. In: *Conservation Tillage with Animal Traction. A Resource Book of Animal Traction Network for Eastern and Southern Africa (ATNESA)* (eds. P.G. Kambutho and T.E. Simalenga). Harare, Zimbabwe: French Cooperation.

Kahinda, M., Taigbenu, A.E., and Boroto, J.R. (2007). Domestic rainwater harvesting to improve water supply in rural South Africa. *Physics and Chemistry of the Earth* 32: 1050–1057.

Li, X.-Y., Zhong-Kui, X., and Yan, X.-K. (2004). Runoff characteristics of artificial catchment materials for rainwater harvesting in the semiarid regions of China. *Agricultural Water Management* 65: 211–224.

Mzirai, O.B. and Tumbo, S.D. (2010). Macro-catchment rainwater harvesting systems: challenges and opportunities to access runoff. *Journal of Animal and Plant Sciences* 7 (2): 789–800. http://www.biosciences.elewa.org/JAPS.

Ngigi, S.N. (2003). What is the limit of up-scaling harvesting in a water basing? *Physics and Chemistry of the Earth* 28: 943–956.

Okhravi, S., Eslamian, S., Eslamian, F., and Tarkesh-Esfahani, S. (2014). Indigenous knowledge as a supportive tool for sustainable development and utilisation of rainwater harvesting systems. *Journal of Flood Engineering* 5 (1/2): 39–50.

Okhravi, S., Eslamian, S., and Adamowski, J. (2015). Water Reuse in Rainwater Harvesting. In: *Urban Water Reuse Handbook* (ed. S. Eslamian), 779–796. Taylor and Francis/CRC Group.

Oleson, J.P. (1995). *The Origins and Design of Nabataean Water Supply System. Studies in the Historyand Archaeology of Jordan*, vol. V. Amman, Jordan: Department of Antiquities.

Oweis, T., Hachum, A., and Bruggeman, A. (eds.) (2004). *Indigenous water harvesting Systems in West Asia and North Africa*. Aleppo, Syria: ICARDA.

Pecey, A. and Cullis, A. (1989). *Harvesting; the Collection of Rainfall and Runoff in Rural Areas*. London: Intermediate Technology Publication.

Prinz, D. (1995). Water harvesting in the Mediterranean environment—its past role and future prospect. In: *Water Resources Management in the Mediterranean under Drought of Water Shortage Conditions*, Internat. Symp., (ed. N. Tsiourtis), 135–144. Nicosia, Cyprus: A.A. Blakema.

Prinz, D. (2011). The concept, components and Methods of Rainwater Harvesting, 2nd Arab Forum, "Living with water Scarcity" Cairo, Egypt.

Prinz, D. and Singh, A.K. (2006). Technological potential for improvement of water harvesting. In: *Water Management: Concepts and Cases* (eds. P. Nair and D. Kumar), 123–135. Hyderabad: ICFAI University Press. ISBN: 81-314-0061-1.

RainSaucers. (2009). RainSaucers. Retrieved September 16, 2013, from http://rainsaucers.com.

Rosegrant, M.W., Cai, X., and Cline, S.A. (2002). *World Water and Food to 2025: Dealing with Scarcity*. Washington, DC, USA: International Food Policy Research Institute.

Salameh, E. and Bannayan, H. (1993). *Water Resources of Jordan, Present Status and Future Potential*. Amman, Jordan: Friedrich Ebert Stiftung.

Shanan, L. and Tadmor, N.H. (1979). *Microcatchment System for Arid Zone Development*, 99. Jerusalem. Israel: Hebrew University.

Texas Water Development Board (TWDB) (1997). Texas Guide to Rainwater Harvesting, 2nd ed. Retrieved December 2004, USA, from http://www.twdb.state.tx.us/publications/reports/rainharv.pdf

Tiessen, K.H.D., Elliott, J.A., Stainton, M. et al. (2011). The effectiveness of small-scale headwater storage dams and reservoirs on stream water quality and quantity in the Canadian Prairies. *Journal of Soil and Water Conservation* 66 (3): 185–171.

de Vries, B. (1997). *The Oxford Encyclopedia of Archaeology in the Near East*, 276–279. London/Umm el-Jimal: Oxford University Press.

Zaizen, M., Urakawa, T., Matsumoto, Y., and Takai, H. (1999). The collection of rainwater from dome stadiums in Japan. *Urban Water* 4: 335–359.

Zhu, K., Zhang, L., Hart, W. et al. (2004). Quality issues in harvested rainwater in arid and semi-arid loess plateau of Northern China. *Journal of Arid Environments* 57: 487–507.

2

Rainwater Harvesting: Recent Developments and Contemporary Measures

Aline Pires Veról[1,5], Marcelo Gomes Miguez[2,3,4], Elaine Garrido Vazquez[3], Fernanda Rocha Thomaz[2], Bruna Peres Battemarco[2], and Assed Naked Haddad[4,5]

[1] *Programa de Pós-Graduação em Arquitetura, Faculdade de Arquitetura e Urbanismo, Universidade Federal do Rio de Janeiro, Rio de Janeiro, Brazil*
[2] *Programa de Engenharia Civil, COPPE, Universidade Federal do Rio de Janeiro, Rio de Janeiro, Brazil*
[3] *Programa de Engenharia Urbana, Escola Politécnica, Universidade Federal do Rio de Janeiro, Rio de Janeiro, Brazil*
[4] *Programa de Engenharia Ambiental, Escola Politécnica, Universidade Federal do Rio de Janeiro, Rio de Janeiro, Brazil*
[5] *Programa de Pós-Graduação em Engenharia Civil, Universidade Federal Fluminense, Niterói, Brazil*

2.1 Introduction

The process of city and population growth in recent decades has raised several concerns about the sustainability of this growth and the possibility of dangerous environmental degradation and depletion of natural resources. In June 1972, the United Nations Conference on the Human Environment was held, bringing together representatives from more than 110 countries in Stockholm, Sweden (United Nations General Assembly 1972). At this meeting, known as the Stockholm Conference, a new concept of development was proposed, aiming at a more intelligent and streamlined process with less waste production. The sustainable development concept was fostered after the United Nations Conference on Environment and Development, also known as Rio 92, held in Rio de Janeiro in 1992 (UNCED 1992).

The establishment of Agenda 21 (UNCED 1992) set out objectives to promote the sustainable development of human settlements, including the provision of adequate housing for all; improvement of the management of human settlements; promotion of sustainable land use planning and management; and provision of environmental infrastructure, such as that related to water, sewage, drainage, and solid waste management. These objectives are strongly linked to a rational and balanced urbanization process to ensure a sustainably built environment.

The 2030 Agenda for Sustainable Development (UN General Assembly 2015) includes Goal 6, namely to "Ensure availability and sustainable management of water and sanitation for all." This goal recognizes that many people still lack access to safe water supplies and sanitation facilities. According to Goal 6, increasing water efficiency and improving water management are crucial to maintaining the balance of the competing and growing water demands from diverse sectors and users.

In this respect, the United Nations General Assembly (Resolution 71/222) proclaimed the period from 2018 to 2028 as the International Decade for Action, "Water for Sustainable Development" (the "Decade"), to further improve cooperation, partnership, and capacity development in response to the ambitious 2030 Agenda (UN General Assembly 2016).

Much is still being discussed about sustainability and how to reach sustainable results, since it is hard to define in absolute terms. The first attempts to move toward sustainable cities came with large-scale proposals for saving natural resources and space (maintaining preserved areas), using the compact city concept, and mixed land use, optimizing services, reducing displacements, and consuming less energy and resources (Miguez et al. 2018).

Any city, however, is a set of interconnected buildings, open spaces, and infrastructure networks. In a systemic view, a building, as the primary cell of urbanization, is essential to the overall performance of urban systems, but this insight is still little explored. The discussion of individual buildings and the entire urban space often occur separately, with particularities of the different scales of each of these contexts.

However, there are several interrelated issues. A "building" uses water and produces sewage. While solid waste is a function of the habits and the consumption of the inhabitants of a city (and therefore of the inhabitants of a building), the generation of superficial flows depends

Handbook of Water Harvesting and Conservation: Basic Concepts and Fundamentals, First Edition. Edited by Saeid Eslamian and Faezeh Eslamian.

on how the building occupies the lot, what is its type of cover, the level of imperviousness, and finally, how many and which compensatory measures are introduced to avoid direct runoff and thus maintain the natural water cycle functions. For instance, on-site detention (OSD) tanks can recover part of the surface retention which has been lost by surface regularization; permeable pavements can restore part of the infiltration which was lost by surface imperviousness; and green roofs can favor evapotranspiration, retain water, and function as a retention reservoir (Woods-Ballard et al. 2015). These actions help to reduce and control flooding in the urban environment, which contains the buildings that are the cause of increased runoff. They can also offer opportunities to use stormwater as a resource.

Likewise, because of the water scarcity risks, the rational use of water is now considered more prominently in the design of hydro-sanitary installations, for example, by using water-saving appliances and fixtures. Among the main advantages, water saving is beneficial since what is not used from the drinking water system remains available for other uses, with a consequently lower water bill to be paid by the consumer (Englehardt et al. 2016). Besides this, a smaller quantity of sewage is produced.

In addition to the rational use of water, sustainable buildings can also employ alternative technologies, which make previously discarded water usable for non-potable purposes. In this case, water reuse and rainwater harvesting can be cited (Okhravi et al. 2015). Reuse is based on the treatment of the sanitary sewage coming from fixtures and appliances, such as wash basins, showers, and washing machines (Juan et al. 2016). On the other hand, rainwater harvesting consists of collecting and storing rainwater, also for non-potable uses, such as washing floors and watering gardens, among others (Campisano et al. 2017a). According to Okhravi et al. (2014), rainwater harvesting is a valuable alternative to overcome the growing water shortage.

Thus, alternatives to optimize water consumption and minimize effluent generation are crucial to reduce vulnerability and increase resilience of urban systems. Ultimately, water savings also revert to energy savings, since a considerable portion of the cost of treatment and distribution of drinking water refers to energy expenditures (Dai et al. 2018). By consuming less energy, on a local scale (in the lot), there is a large-scale chain reaction that benefits the environment and the city itself.

Therefore, taking care of a building's hydraulic and sanitary systems should not only be a concern of the direct user, who can enjoy the rational use of water and obtain economic advantages. There is a significant reduction in the urban scale, contributing to the collective good, to the proper functioning of the city, and pointing to a sustainable development path.

In most cases, the performance of a rainwater harvesting system is estimated from historical rainfall data. However, climate change may affect rainfall patterns, which will affect the performance of the harvesting system. For example, Haque et al. (2016) evaluated the impacts of climate change on the performance of a residential rainwater harvesting system in the Greater Sydney region, Australia, based on the estimated future rainfall data. The results indicated that a given tank size in the study areas would not be able to supply the expected volume of water in the future. According to Alamdari et al. (2018), rainwater harvesting systems designed for current conditions are expected to become less effective in the future for water supply purposes due to climate change for most locations in the western, southern, and central U.S. Thus, the study estimated the necessary additional storage to avoid future reductions in water supply.

This chapter discusses aspects of sustainability related to the buildings' hydraulic and sanitary systems, presenting recent developments and contemporary measures that move toward more sustainable projects. In this sense, the meaning and importance of these alternative technologies are shown, as well as their interaction with urban space and the possibility of supporting the discussion of sustainable development also in the urban setting and hydrographic basin.

2.2 Water Resource Management

In many developing countries, accelerated socioeconomic growth combined with climate change brings uncertainties to water resource planning, where conflicts of interest make the decision-making process a significant challenge (Bhave et al. 2018; Thissen et al. 2017).

Water resource planning should consider the compatibility among various uses and integration with urban planning and land use, which are directly responsible for the change in water availability and water quality. Thus, planning must define criteria for the establishment of public policies that sustainably guide economic development and land use planning.

As the population grows and its income increases, there is also an increase in consumption patterns due to greater demand for products and services with higher water consumption. In the absence of efficient planning, water resources are most likely to be a source of tension among the various water users. This tension reaches the maximum level during periods of drought.

2.2.1 Water Supply

Water is a vital resource for health preservation, food production, and conservation of the environment. The natural water cycle involves the continuous movement of water in the atmosphere, on land (surface, soil, and rock) and in oceans.

The world's total water supply is about 1386 million cubic kilometers, and over 97% is saline. Of the available water, about 2.5% is fresh water, 68.7% of which is in glaciers, 30.1% in groundwater, and only 1.2% surface water, such as rivers, lakes, and lagoons. Rivers are the most used source (USGS 2019).

Although this is a small fraction, surface water still has large volume, around 91 300 km^3 (Gleick 1993; USGS 2019). However, its distribution is not uniform and it is often not present in adequate quantities where the highest population concentrations and demands are located (Pedde et al. 2013).

Thus, problems of water scarcity are related, first, to the unequal geographic distribution, so there are places where physical scarcity is an underlying problem associated with the unfavorable environment.

Nevertheless, the poor management of available resources and excessive pollution (and consequent degradation of water reservoirs) also lead to water scarcity. This fact, together with population and urban growth, are key factors that reduce water availability.

According to UNESCO (2019), more than 2 billion people live in countries that experience high water stress. In 22 countries, water stress, defined as the ratio between freshwater abstractions and total renewable freshwater resources, is above 70% (United Nations 2018). These data show that these countries face critical conditions of water stress. The continuous increase in water stress in several countries has impacts on the sustainability of the resource and causes conflicts among the different uses (UNESCO World Water Assessment Programme 2019).

Another essential aspect is that the water stress figures presented above represent annual averages and do not show the brief periods of acute water scarcity. Approximately two-thirds of the population experience severe water shortages for at least one month of the year (Mekonnen and Hoekstra 2016), for example.

2.2.2 Water Demands

Water is an essential input for world economies. According to UNESCO (2016), half of the global workforce is employed in eight industries that depend on water: agriculture, forestry, fisheries, energy, resource-intensive manufacturing, recycling, building, and transport.

Water uses in the world have been growing at around 1% per year since 1980, and, by 2050, demands will increase by 20–30% relative to the current uses. This growth is largely due to the water demands of developing countries, although the per capita use of these countries is lower than that of the developed countries. The most significant growth is expected to occur in the industrial and domestic sectors, but agriculture will continue to be the largest water user (UNESCO World Water Assessment Programme 2019).

In Brazil, a developing country, there has been an increase in water demand estimated at 80% of the total withdrawn in the last two decades. Water demand by 2030 is expected to increase by 24%. The evolution of water use is directly related to the country's economic development and urbanization process (ANA – Agência Nacional de Águas 2018).

The increase in world population and the improvement of living standards promotes changing patterns of consumption and consequent expansion of irrigated agriculture. These factors are the main drivers of the global increase in water demand (Mekonnen and Hoekstra 2016).

Even though it is such a valuable resource, 663 million people in the world still lack access to safe drinking water sources. The challenge is enormous and is becoming very complex, as population and economic growth are pushing the limits of available water resources. In some situations, water scarcity already restricts economic growth. Even where access to safe water supply and sanitation is available, supply has been characterized for decades by inefficient management and low service levels (WBG 2019).

2.2.3 Water Scarcity

Water scarcity results from overexploitation of water resources and arises when average demand is higher than long-term renewable availability (Sayers et al. 2016). Long droughts can lead to an imbalance between the available supply and the demand for water, resulting in water scarcity in a region. In this situation, conflicts involving water become frequent, especially in metropolitan areas and where large economic enterprises are located.

In recent decades, drought events have increased greatly in intensity and frequency, affecting many regions in the world with economic, environmental, and social impacts.

Economic impacts are those that generate additional costs for individuals and businesses. As an example of economic impact, the United States drought between 1976 and 1977 caused an estimated loss of US$2.663 billion

(Sayers et al. 2016). Howitt et al. (2015) estimated a direct loss in agriculture of US$1.84 billion in the 2015 California drought, as well as the loss of 10 000 seasonal jobs.

The social impacts of drought affect the health and safety of mankind. These impacts include (NDMC 2019): loss of human life, reduced earnings, anxiety or depression due to economic losses during droughts, health problems related to reduced flows and critical conditions of water quality, diseases caused by dust allergies, migration of the population from the city to the countryside or vice versa, etc. An example of social impact occurred in Bengal, India, in 1942, when a severe and widespread drought caused famine, killing about 1.5 million people (Sayers et al. 2016).

Droughts also affect the environment in different ways, where the damage may be temporary or persistent. In certain situations, the damage remains forever (NDMC 2019). An example of a large-scale environmental impact occurred in Canada, where an increase in tree mortality in the Boreal Forest was observed after a series of regional droughts occurring in the period from 1963 to 2008 (Peng et al. 2011).

It is estimated that in Europe, from 1976 to 2006, drought costs were 100 billion euros (Tsakiris 2017). Several countries, including Brazil, have suffered the consequences of droughts, like the ones in the period between 2013 and 2016. Although droughts in Brazil are mostly associated with the Northeast region, these events were also associated with other regions of the country, including the Southeast region, where the climate is usually wet. Cities like São Paulo, Belo Horizonte, and Rio de Janeiro have a high concentration of population and supply sources are increasingly pressed by land use changes and other water uses. These cities were also affected by the drought of 2013–2016.

According to Sayers et al. (2016), the risk of drought in a region is determined by two main components: hazard and consequence. Hazard is a potential threat of damage or loss, formed by the combination of atmospheric processes and hydrological responses, which cause reduction or total loss of water in lakes, rivers, reservoirs, and the soil. The consequences reflect the level of exposure and vulnerability of the system to the possible environmental, social, and economic impacts of drought. This component may also include resilience, which is the ability of adaptation and/or recovery of the system. The concept of adaptability is implicit in this definition; that is, a system that can change or reorganize its behavior without external help aiming to reduce susceptibility to harm has little vulnerability. Therefore, implementing measures for system recovery reduces vulnerability and reflects effective drought risk management.

Since water is a scarce resource, efforts are needed to promote the rational use and conservation of water resources. There is a global concern related to the need to rationalize the consumption of water resources, in order to consolidate contemporary principles of water management. In this context, the search for alternatives to optimize water consumption, as well as minimize the generation of effluents, is a relevant issue to reduce vulnerability and increase system resilience.

The concept of efficient water use encompasses the implementation of technological, institutional, and educational actions for water savings, in addition to focusing on the maintenance and improvement of the quality of this resource. From a technological point of view, it is possible to present new solutions for the design of hydraulic and sanitary systems, aiming at the rationalization of demand and the consequent minimization of consumption. Among these actions are:

- Use of water-saving appliances and fixtures, which act directly to reduce wastewater;
- Individualized measurement, which can be an important tool to raise awareness of rational water use;
- Search for losses in the system, with appropriate maintenance programs;
- Evaluation of the use of alternative water sources (such as use of rainwater or reuse of gray water) to meet the needs of less demanding uses, and to protect the primary sources of water, encouraging water self-sufficiency with increasing local supply.

The rational use of water is a key action for the sustainable development of public water systems. Measures for the conservation and rational use of water can be classified according to different points of view. Schematically, the main classes of measures relate to: function (structural or non-structural); character (active or passive); and interest group (supply or demand management). The structural measures modify the technical characteristics of the systems permanently, while the non-structural ones act on the operation of the system and are reversible. The active or passive character of a measure refers to the possibility of control (or not) by the user. Concerning the interest group, the management framework is divided into supply management (with alternative sources and new sources) or demand (optimization of water consumption), reducing consumption needs.

A building supplied with potable water, with optimized water consumption and appropriate management system, should be capable of maintaining or improving water quality according to each specific type of use, besides generating greater water savings and economic gains.

2.2.4 Regulatory Framework

Several countries have laws and regulations governing rainwater harvesting and provide the necessary requirements and guidelines for its use.

In the USA, according to NCSL (2018), each state has its own legislation to allow, define, and clarify when, where and how the use of rainwater can occur. Currently, there are 18 states and territories with their own legislation and programs: Arizona, Arkansas, California, Colorado, Hawaii, Illinois, Nevada, New Jersey, North Carolina, Ohio, Oklahoma, Oregon, Rhode Island, Texas, Utah, Virginia, Washington, and U.S. Virgin Islands. For example, Texas and Ohio have enacted several laws regulating rainwater harvesting and allow this framework for potable uses, which are frequently excluded from other U.S. states, and in some countries' laws and regulations. Rhode Island, Texas, and Virginia offer tax credits or exemptions on the purchase of rainwater harvesting equipment.

In Europe, the European standard (EN 16491 2018) specifies the requirements for the design, sizing, installation, identification, commissioning, and maintenance of rainwater harvesting systems for non-potable uses.

Interest in rainwater harvesting in Portugal has increased since the creation of a nongovernmental association in 2007 called the National Association for Quality in Building Installations (ANQIP). This association promotes quality and efficiency in the water supply and drainage fittings and fixtures of buildings (Silva-Afonso and Rodrigues 2008; Silva-Afonso and Pimentel-Rodrigues 2011). The technical specification ANQIP-ETA 0701 (ANQIP 2009) establishes technical criteria for the design of rainwater harvesting systems in buildings for non-potable uses.

In recent years, people have been collecting and storing rainwater for household use in the U.K. Despite that, modern systems have only been introduced more recently (Campisano et al. 2017a). In 2006, the government published a document describing sustainable practices in residential buildings, namely the Code for Sustainable Homes. This code was reviewed over time (Department of Communities and Local Government 2010) and on 22 April 2015 the government withdrew it, consolidating some standards into building regulations (Teston et al. 2018). The U.K. has standard BS 8515:2009+A1:2013 – "Rainwater Harvesting Systems-Code of Practice" and standard BS 8595:2013 – "Code of Practice for the Selection of Water Reuse Systems," containing guidelines on how to prepare designs and install rainwater systems (Woods-Ballard et al. 2015).

In Australia, rainwater has been utilized as a clean source of drinking water since the early years of European occupation. Some jurisdictions discouraged the use of rainwater tanks by making it mandatory for householders to pay for the centralized system regardless of being connected. However, some customers still kept rainwater tanks for drinking (Beatty and McLindin 2012). In 2004, the government released a document called "Guidance on use of rainwater tanks," compiling guidelines for rainwater harvesting practices, which was revised in 2010 (enHealth 2010). Four Australian states have mandatory requirements for rainwater tanks: New South Wales (BASIX Building Sustainability Index), Victoria (Five Star Building Code), Queensland (Queensland Development Code), and South Australia (South Australian Building Code) (Marsden Jacob Associates 2009).

In Brazil, the standard NBR 15527 (ABNT 2019) provides guidelines for rainwater harvesting systems on the roof of buildings for non-potable uses. This standard was updated in January 2019, but did not provide requisites for potable uses, which would be an important tool for the Brazilian semiarid region. A bill was proposed in the lower house of Congress (Chamber of Deputies) with the aim of making rainwater harvesting mandatory in buildings larger than 200 m^2 (Bill of Law 1750/2015). Nevertheless, little progress has been made toward enactment of this bill since its introduction. There are many municipal laws and regulations in Brazil, but the number is still small in relation to the number of municipalities in the country. Some cities require implementation by their own regulations, depending on the size of buildings (Teston et al. 2018).

2.2.5 Recent Developments

The water efficiency in buildings has been systematized based on different concepts. For example, the construction of an efficient water cycle system in buildings can be summarized in the 5R Principle, which encompasses five steps: reduce consumption; reduce loss and waste; reuse water; recycle water; and resort to alternative sources. The demand for efficient water use has already been recognized in Portugal as a national priority in the National Program for Efficient Water Use (PNUEA). In this regard, to reduce consumption (first R), ANQIP has created a certification and labeling model for water efficiency for products. The use of rainwater is included in the fifth R (resort to alternative sources) and ANQIP has already established a technical specification for this, namely ETA 0701 (Silva-Afonso and Rodrigues 2008; Silva-Afonso et al. 2011; Silva-Afonso and Pimentel-Rodrigues 2011).

Another concept that has been utilized in some countries is the water-energy nexus, which is related with all the components of the 5R principle. For example, the reduction of water consumption (first R) will lead to a decrease of energy consumption. The concept of water-energy nexus and the net zero water buildings (NZWB) are described below.

2.2.5.1 Water-Energy Nexus

Population growth driven by economic development is directly related to increased demand for natural resources, such as water and energy. External global factors, such as climate change, may also affect the availability of these resources. In this sense, the conservation of water and energy is a key element for sustainable development.

This concern about water and energy security led to the water-energy nexus concept. According to Zhang et al. (2018), nexus can be interpreted as interactions among different sectors, or as a new approach to investigate multi-meaning linkage systems in different contexts. It refers to the search for integrated management through intersectorial coordination as a way to reduce unexpected sectorial tradeoffs and to promote the sustainability and resilience of each sector and the system as a whole (Zhang et al. 2018).

Water and energy are closely related. Several energy production processes require water, such as the extraction and processing of raw materials, electricity generation, cooling of thermal plants, waste treatment, and maintenance of energy-generation facilities (Li et al. 2019). Likewise, the collection, treatment, and distribution of water also require energy. There is also the need for a large amount of energy for the collection, treatment, and disposal of wastewaters (DeNooyer et al. 2016). Thus, analyzing the nexus between water and energy can improve the understanding of quantitative relationships to guide actions and formulate policies to optimize results and minimize risks (Dai et al. 2018). It is noteworthy that the interdependences between water and energy are recognized at different levels, ranging from the city level to the global level, as shown in Figure 2.1.

The concept of water-energy nexus has been implemented in many developed countries such as Australia, the U.K., and the U.S. Radcliffe (2018) analyzed water and energy policies over the past 20 years in Australia, highlighting the crises the country has experienced in each sector and discussing how local governments have acted to respond to the difficulties. There is growing concern about finding solutions to achieve greater efficiency of the water and energy sectors after crises, and how the perception of the nexus between sectors has become increasingly clear, although policies developed for each have been slow to recognize the cause and effect relationship between them. Many measures to promote sector efficiency have been developed in Australia. On the one hand, water utilities

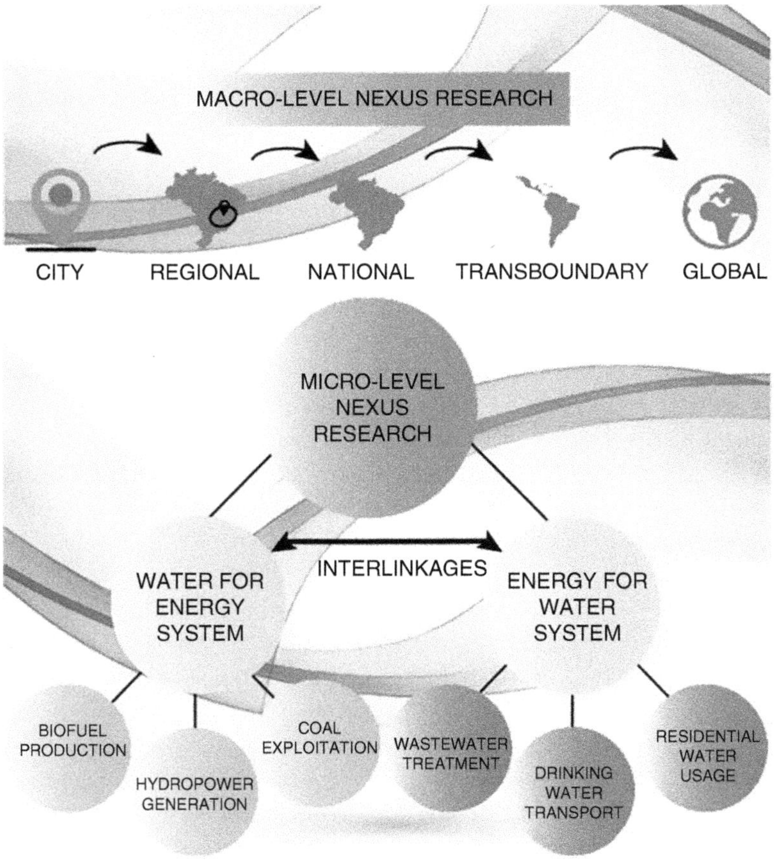

Figure 2.1 Water-energy nexus. Source: Adapted from Dai et al. 2018.

have directed efforts to generate electricity from wastewater treatment plants, supplemented with co-digestion of other organic wastes, to offset their energy needs. On the other hand, a survey conducted in the country about the operational characteristics of water and energy of the main power plants identified the cooling of thermal plants with recycled water, seawater, or even dry cooling. On the residential scale, Radcliffe (2018) pointed out that domestic water consumption in Australia is the second-largest in the world after the U.S., with a value of 320 l per inhabitant per day, and that 30% of energy consumption, approximately, is used to heat water in Australian households. With the context of water and energy crises in the country, some policies and guidelines have been developed, including encouraging the search for alternative sources of water, such as water recycling in buildings, and minimizing water consumption.

An alternative source of water is rainwater harvesting. In the context of the water-energy nexus, Talebpour et al. (2014) conducted a combined analysis of water and electricity consumption in a rainwater tank system for each end-use, such as toilets, washing machines, and irrigation, in 19 residential buildings in Queensland, Australia, using high-resolution water and electricity measurement technologies. The authors concluded that the use of variable speed pumps, although generally more expensive than fixed speed pumps, is more suitable for home applications, since they can adapt to highly variable flow rates, resulting in lower overall energy intensity for the household. The study results showed their potential to provide technical input for building code specifications for stormwater tank systems, particularly the selection of pumps suitable for the purpose (Talebpour et al. 2014).

Konadu et al. (2017), in the report "UK water-energy nexus under climate change: Key issues and priorities," in turn, discussed the main issues associated with water and energy interdependences in the U.K., highlighting the top priorities for research and modeling. The authors cited some sustainable approaches that have been employed by Thames Water to improve water resource availability, while at the same time improving energy efficiency and managing carbon intensity. For example, managing the short-term supply and demand balance at Swindon by promoting behavioral change, and the development of the London main ring, which allows water to flow in any direction under gravity, resulting in large savings, are cited as cases where expensive pumping is reduced. The report also shows that locally applicable technical innovation represents an alternative in the search for the nexus between water and energy, such as the use of efficient turbines in water mains and the use of heat exchangers to recover wastewater heat. The importance of supporting innovation in the joint management of the water and energy sectors, which can lead to new forms of recovery and distributed energy production, is highlighted.

In an analysis of London, the authors found that the relationship between water and energy systems is higher in end use. This can be seen from the data presented by the authors: more than half of total end use of non-transported energy and two-thirds of natural gas use in London is allocated to the city's domestic sector, with about 20% of this energy used for water-related purposes. Thus, the conservation at the service level can save more water together with energy than that used to generate or process this energy. Initiatives aimed at reducing water consumption appear as alternatives, such as rainwater harvesting, which initially used more energy to supply non-potable water, but over the years has been moving toward using low or zero operating energy (Konadu et al. 2017).

In the U.S., the water-energy nexus concept is widespread, even in government spheres. The Department of Energy (DOE), for example, has analyzed the interdependence between the country's water and energy sectors, providing data and listing several challenges and opportunities regarding the application of the water-energy nexus concept (Bauer et al. 2014). According to the report, thermoelectric use currently accounts for over 40% of freshwater withdrawals and 4% of freshwater consumption in the country. In this context, water-stressed regions like Texas and the Southwest have stricter water requirements for power generators. Ohio and Missouri, in their drought plans, classify water use into essential categories, including water use by power plants, which is unrestricted or less restricted during droughts. In California, the California Energy Commission (CEC), responsible for energy policy and planning, recognizes the relationships between energy and water, analyzes permits against water needs and impacts, and assesses how proposed water use can affect other users in the area and the state's overall water supply, prioritizing zero-liquid discharge technologies such as dry cooling (Bauer et al. 2014).

Bauer et al. (2014) also highlighted some opportunities to improve the efficiency of the water and energy sectors, considering their connection. About water for energy, the authors suggest more efficient and less costly alternatives for cooling thermal power plants, as well as options for recovering waste heat, thus reducing the need for cooling plants, as well as improving the quality of the water used in industrial processes and replacing fresh water in energy production. Concerning energy for water, the authors cite the Net-Zero Municipal Wastewater Treatment idea, where advances in the energy efficiency of urban waste treatment and recovery processes can lead to treatment systems that produce as much or more energy than they consume.

Alternatively, energy collection simultaneously with the water supply operation in gravity water distribution systems can be cited. As mentioned by Konadu et al. (2017), an example of success is Lucid Energy Inc.'s LucidPipe™ Power System in Portland, Oregon. It is a solution that enables the production of clean, reliable, and cost-effective electricity from gravity-fed water pipes and effluent streams.

Bringing the water-energy nexus concept to buildings, Chang et al. (2011) conducted a cost-benefit optimization analysis to formulate optimal design strategies for a typical Florida household, integrating green roof with gray water reuse and rainwater harvesting, and achieved some degree of energy savings. The authors found that the additional benefit of these systems may outweigh the additional cost due to water conservation within a specified period. They also concluded that when there is no risk, a green roof would be a viable option for 40–70% of the roofs in Florida, which encourages green building policies in support of water and energy savings.

Thus, it can be seen that a good local example is the use of green roofs as a water conservation measure, integrated with energy saving. Countries such as Japan, the U.S., Australia, and Canada have encouraged the installation of green roofs in the construction of new buildings and the adaptation of old buildings (Vijayaraghavan 2016). As further pointed out by Vijayaraghavan (2016), green roofs have several benefits, such as rainwater retention and consequent delay in peak flow, improved water and air quality, decreased building energy consumption, and reduction of the heat island effect in urban centers. Chang et al. (2011) also cited the contribution of rainwater harvesting for non-potable purposes through water storage if connected to a reservoir.

As for rainwater retention, from research of different sources and locations, Dietz (2007) concluded that green roofs can reduce runoff volume by 60–70%. Hua-peng et al. (2013) found that green roofs perform better in a storm event with late peak intensity. Regarding the reduction of energy consumption, green roofs improve the thermal performance of buildings. During the summer, due to the evapotranspiration effect of plants and the evaporation of soil moisture, the green roof can help with cooling. In the winter, its insulation property limits heat from escaping (Chang et al. 2011). Niachou et al. (2001), in an analysis of the thermal properties and energy performance of green roofs, found a 2–48% reduction in energy used for cooling depending on the type of roof insulation (well insulated, moderately insulated, or not insulated).

Importantly, green roofs can influence water quality positively and negatively. Vijayaraghavan (2016) highlighted some studies on the subject, such as Rowe (2011) and Moran et al. (2003). The first study found the possibility of a positive effect on water quality with the use of green roofs, from a review of research articles. The second study identified higher nutrient concentrations in green roof runoff compared to those of rainwater and control roof runoff. In this study, the authors monitored two green roofs in North Carolina for nine months.

So, it can be said that in buildings, saving water as a result of good practices and water-saving appliances also produces energy savings.

2.2.5.2 Net-Zero Water Buildings

According to the Environmental Protection Agency (EPA), the concept of NZWB indicates that a given resource can be consumed as much as it is produced locally, without generating external dependence, reaching a balance between demand and availability, more sustainably (USEPA 2018). The most advanced studies in the literature are in the area of energy, which was the precursor of the net zero concept, with the definition of buildings with net zero energy balance. However, the concepts of NZWB and net-zero waste buildings have already advanced. The concepts apply to water conservation, reduction of energy use and elimination of solid waste generation, contributing to the improvement of the environment, providing economic benefits as well as helping communities to become more sustainable and resilient (Joustra and Yeh, 2015a). It is important to consider this concept early in the design phase, to facilitate overcoming challenges involved, to maximize consumption reduction and adapt to the local climate.

The aims of a NZWB are to minimize the total amount of water consumed, maximize alternative water sources, minimize wastewater discharge into the environment, and return water to its source. In summary, this is an innovative concept that would make the building fully responsible for its water generation as well as for the treatment of all the waste produced.

Joustra and Yeh (2015b) mentioned that the NZWB assessment can be considered at various hydrological levels: that of the building itself and that of the connection with the urban system, where public water and sewage networks rely on natural water sources to supply water and subsequently dispose of effluents.

In a traditional system (Figure 2.2a), drinking water is supplied by the utility and all sewage generated, both gray and black water, is discharged into the sewer system, which carries it to a treatment plant, while rainwater is carried away by the public stormwater system, without

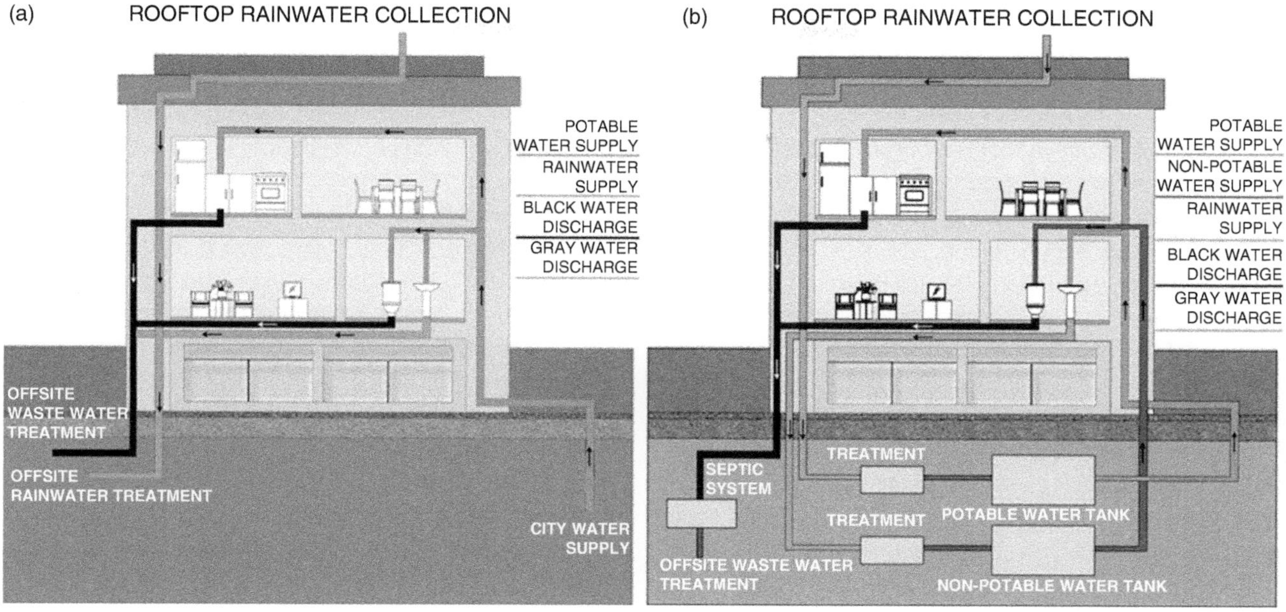

Figure 2.2 (a) Traditional building; (b) Net-Zero water building. Source: Adapted from Silva et al. 2019.

any possible use. The NZWB proposal (Figure 2.2b) determines that rainwater is treated on-site and transformed into potable water. The alternative for non-potable uses is the mixture of rainwater with gray water, which is the less contaminated wastewater from washbasins, showers, bathtubs, and washing machines, while black water is pretreated for disposal in the public network.

If the building is not located within the water basin or aquifer of the original water source, it will be unlikely to return the water to that source (Energy 2019). In such cases, however, water will benefit another basin, similar to a transposition effect. The benefits remain and the reuse decreases the water stress at the source.

It should be noted that new structures should be designed to lower the need for water and energy, to reduce the need for application of reuse techniques.

For a building to be classified as a NZWB and receive some certification, it is necessary for the calculated water balance between demand and reuse to be equal to zero. Although the NZWB idea is a conceptually interesting goal, especially in situations of water scarcity, it is important to emphasize the difficulty of achieving a truly zero balance, since existing standards usually prohibit the use of rainwater for drinking purposes. It is also important to have a safe water supply in terms of quantity and quality, to improve public health. Therefore, this concept, as a reference for a project, can be adapted to situations where minimum consumption is the goal, aiming to rationalize this resource as much as possible and reduce external dependence. This is a more feasible proposal on a large scale, not just as a demonstrative initiative in special buildings.

2.3 Water Management at the Building Scale

Water is the central element of three plumbing systems: drinking water, wastewater, and rainwater. The use of rainwater refers to a fourth system, which when employed, is complementary to the previous ones, collecting rainwater, offering a complementary resource to the drinking water system (non-potable), and possibly generating additional wastewater.

The use of rainwater is a millenarian practice by many societies for agriculture, drinking (when properly treated), and, more recently, for less noble purposes such as washing of clothes, sidewalks, and cars. In this sense, rain is an alternative source of water for non-potable purposes, increasing supply and consequently reducing scarcity. Additionally, the use of rainwater can also contribute to mitigating urban floods at the watershed scale, surpassing the scale of individual buildings, if this initiative is generalized. Hence, there are several positive aspects of rainwater harvesting: reduction in demand from the public water supply, with a direct impact on the amount paid monthly to the utility; conservation of drinking water reserves; runoff control; and prevention of floods.

However, it should be noted that the two main functions of rainwater harvesting systems refer to conflicting objectives: (i) runoff control and flood mitigation; and (ii) use of rainfall water as a resource for alternative supply of buildings. So, to control runoff and prevent flooding, reservoirs must be low, while to use water as a resource, reservoirs must be full. If the two objectives coexist, it becomes necessary to separate the reservoirs that will fulfill each of the two functions. The implementation of a rainwater harvesting system to supply water therefore involves analysis of possible additional supply (depending on local environmental conditions and climate) and demand (depending on consumption habits). Its viability depends on several factors, such as the local average precipitation; the surface collection area and material; the water quality; the elaboration of designs for reservoirs and their complementary systems; the identification of water uses; and the establishment of the type of treatment to be applied to rainwater, according to the quality standard required for each final use. For example, according to Woods-Ballard et al. (2015), in the U.K., rainwater can also be collected from large areas, such as large roofs or parking lots with paved surface. In these cases, rainwater harvesting systems may contribute to reduce the runoff volume, as well as being used as a supply of water for domestic, commercial, industrial and/or institutional properties, considering the required treatment.

The roof is the basic surface for capturing rainwater in buildings, and it must be properly designed, with rain gutters and vertical pipes, which discharge into a storage tank. The stored water must then be treated before being available for use in a building system (separated from the water supply system). After being used, this water should be sent to the public sewage system.

Concerning flood control, it is necessary to define a certain storage volume, depending on the impervious area, or to define a maximum allowed outflow to the system, based on the land occupation before urbanization. Therefore, the core calculation element of rainwater harvesting systems is the capacity of the reservoir. This calculation is usually done by a simplified balance of inputs (defined by the rainfall regime) and outputs (defined by the expected consumption).

2.3.1 Design of a Rainwater Harvesting System

The population that will benefit from the project and the demand for rainwater in each case are the main elements the designer should take into consideration before proposing a rainwater harvesting project. Local hydrological characteristics should also be considered from analysis of the historical rainfall series.

The system may be simpler or more complex, but generally follows the setup presented in Figure 2.3, which illustrates a rainwater harvesting system. Rain falls on the roof used to collect rainwater and through interconnected systems of gutters and pipes, goes to the storage tank. Finally, the water is stored in a reservoir, from which the distribution will occur. The system may also contain water treatment devices, such as first-flush disposers, filters, chlorinators, and ozonators.

The following text presents the elements of a rainwater harvesting system, considering design and sizing aspects with maintenance recommendations.

2.3.1.1 Collection Surface (or Roof Surface)

In a rainwater harvesting project, the collection surface (or the roof surface) should be adapted so that its gutters lead the water to the same final destination: the storage tank. One has to study the location of the rainwater reservoir and the feasibility of directing water to it. Besides this, grates and screens need to be provided for the removal of debris, as shown in Figure 2.4.

The quality of the collected water will be a function of local conditions (air quality, presence of trees, animals), the collecting surface material, and the climate. The adequate material must be nontoxic so as not to contaminate rainwater with toxic substances. Of the most common materials, Brown et al. (2005) recommended metal, concrete, or ceramic tiles. The same authors emphasize that metal roofs have a smooth texture, which retains neither water nor particles, but when they suffer wear, they can contaminate water with metals. Concrete and ceramic roofs, unlike metal ones, retain water that is lost by evaporation and also particles, because they are very porous. Still, they are suitable, especially if coated with nontoxic paint, to eliminate the problem of porosity. Additionally, the use of waterproofing products such as liquid silicone-based agents can also be considered, for example for ceramic tiles.

About the collection surface, the horizontal flat area projected by the impervious surface where the rainwater is collected is considered for design purposes. There is a difference in the volume of precipitated water and what will be collected depending on the roof material, determined by applying the runoff coefficient. Losses represented by this coefficient refer to the water retained by the different roof surface types (as in pitched roofs with tiles, for example) and to the evaporated water. Table 2.1 presents some runoff coefficients for different type of collection surfaces.

2.3.1.2 Gutters and Pipes

The gutters used to collect rainwater from the roof and the pipes, both vertical and horizontal, are responsible for conducting it to the storage tank. Just like the roof, the gutters and pipes should be made of nontoxic materials such as PVC, galvanized steel, lacquered aluminum, and copper alloys.

The rainwater harvesting system may have a device for the disposal of the first-flush water. There is no consensus on the volume to be discarded, but according to Cunliffe

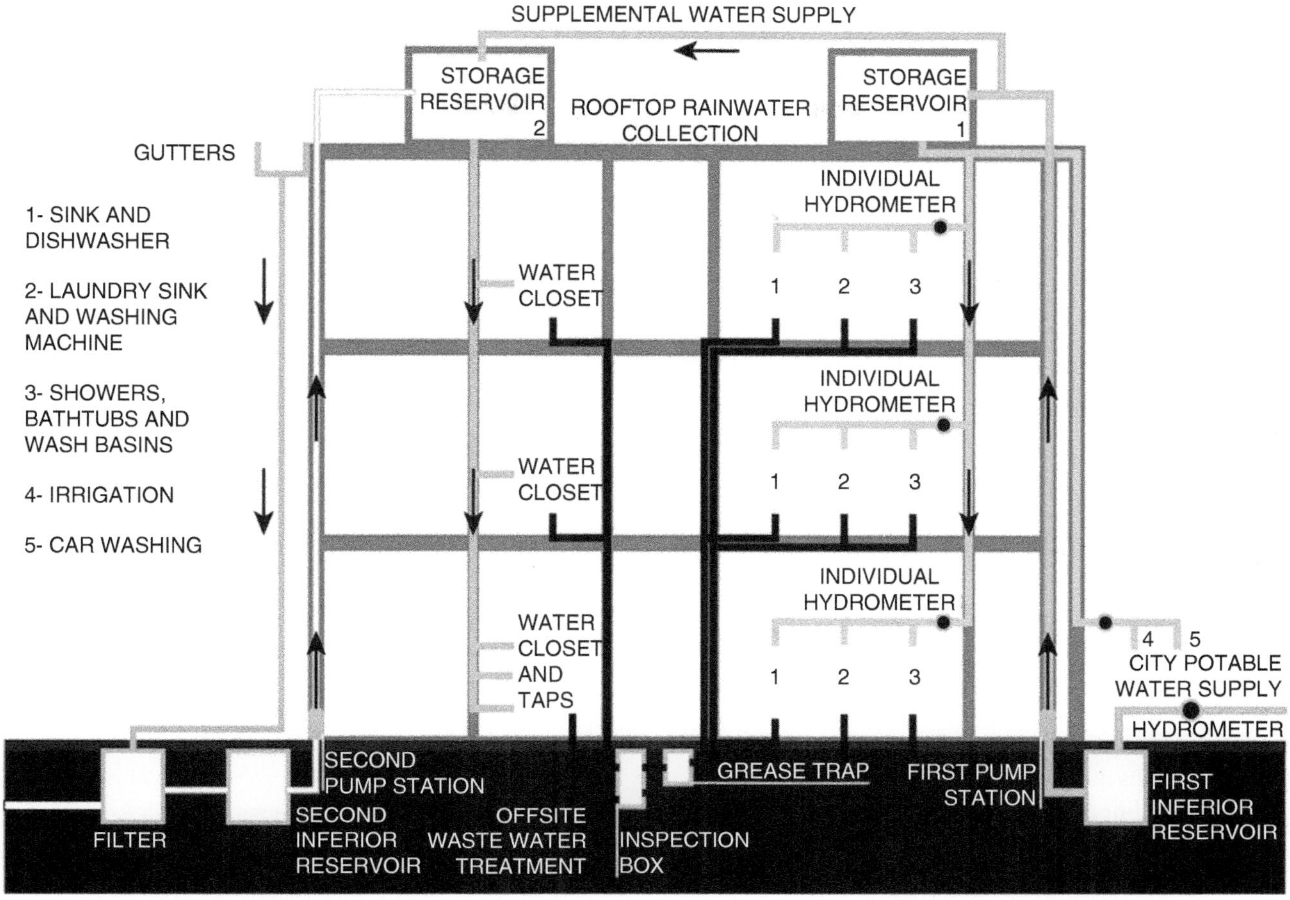

Figure 2.3 Diagram of a rainwater harvesting system.

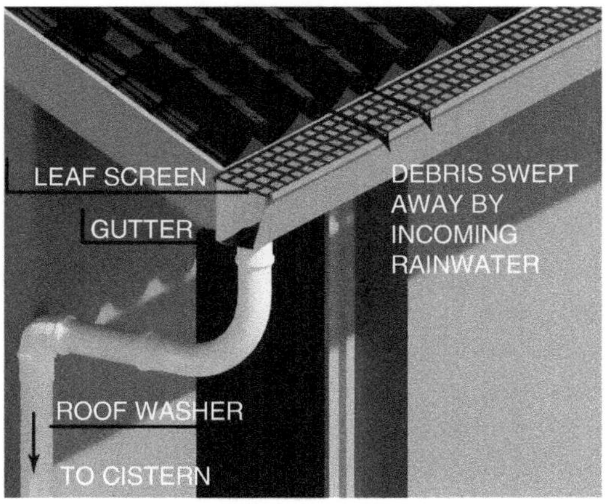

Figure 2.4 Diagram of a gutter with a leaf screen.

Table 2.1 Runoff coefficients for different surface types of roofing from BS 8515:2009 + A1:2013 (Woods-Ballard et al. 2015).

Surface Type	Runoff Coefficient
Pitched roof with profiled metal sheeting	0.95
Pitched roof with tiles	0.90
Flat roof without gravel	0.80
Flat roof with gravel	0.60
Green roof (intensive)[a)]	0.30
Green roof (extensive)[a)]	0.60

a) Green roof runoff yield is particularly uncertain and varies with season. There may also be negative coloration impacts.

(1998), this value is between 20 and 25 l for each 100 m^2 of the collection area. In the absence of more precise measurements, therefore, 2 mm (2 l m^{-2} of roof) of the initial precipitation is usually considered. A simple way of discarding the first flush is by the use of an isolation valve, diverting it from the reservoir in the first minutes, as shown in Figure 2.5a. However, it is better when this device is automatic, such as shown in Figure 2.5b.

2.3.1.3 Storage Tanks (Reservoirs)

The storage of rainwater usually takes place in two reservoirs: the first one receives the water from the rooftop and,

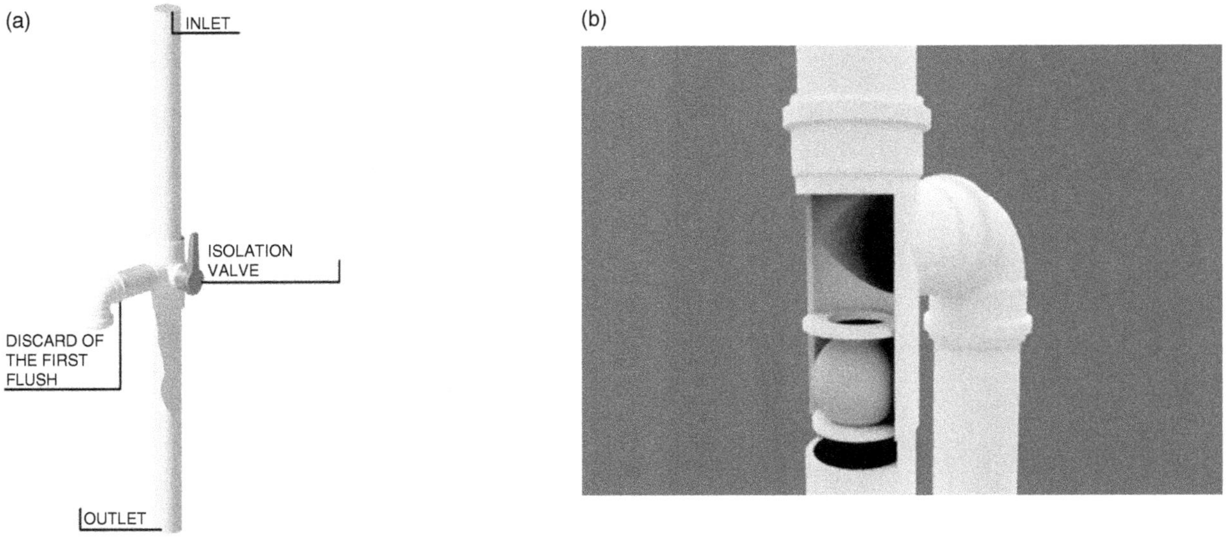

Figure 2.5 (a) Diagram of a simple first flush device; (b) Diagram of an automatic first flush device. Source: Adapted from Silva et al. 2019.

after some treatment this water is sent to the second one, which is responsible for the water supply. They may both be located on the ground floor or the second reservoir may be on the roof. In this case, it may be necessary to pressurize this system, but as a result, the water will be available to reach consumption points by gravity.

The reservoir is considered the most costly element of the rainwater harvesting system. Thus, the choice of the best option (among reservoir types) is crucial, which should be based on technical, economic, and environmental criteria, to ensure a feasible system. The reservoirs can be of different types – they can be buried, half-buried, on the ground, or elevated. They can also be made of various materials, such as polyethylene, PVC, fiberglass, stainless steel, or concrete and masonry, when they are properly waterproofed.

The stored water should be protected against the direct incidence of sunlight and heat (to prevent the proliferation of algae and other microorganisms) as well as from the entrance of animals.

Some devices should be considered in the design of the rainwater harvesting reservoir, just as for traditional supply systems, such as providing an overflow, a cover for the reservoir, and washout taps, as shown in Figure 2.6.

Reservoirs should be cleaned and disinfected with a sodium hypochlorite solution at least once a year. The unusable volume of rainwater may be discharged into storm drains, into a public roadway, or be totally or partially infiltrated, provided there is no danger of groundwater contamination, at the discretion of the competent local authority. There should be no interconnection between the rainwater harvesting system and the public water supply system, although water from the public supply system can also be used to fill the rainwater storage reservoir.

Reservoir design should maximize the use of rainwater. Therefore, it is necessary to characterize the local precipitation, usually using monthly averages (mm/month), and evaluate the availability and size of the collection area, as well as the available space within the property for the installation of the reservoir. Based on water consumption of the devices that will use rainwater after the system installation, one can estimate the demand.

The volume of the reservoirs can be defined according to several existing methods at the designer's discretion. The most common methods are the Rippl Method (Rippl 1883), German Practical Method (DIN 1989-1:2001-10 1989), and Australian Practical Method (enHealth 2010). The existence of these methods leads to great variability of results for reservoir volumes, due to the conceptual difference of the methods and their levels of simplification. While some of these methods take into account the average monthly rainfall distribution and the user demand, such as the Rippl Method, others are simpler, as in the case of the German Practical Method, which only considers empirical estimates based on parameters such as the collection area and the annual average usable precipitation.

The widely used Rippl Method (Mass Diagram Method) is based on the dimensioning of the rainwater reservoir by the maximum accumulated difference between the collected volume and the demanded volume (defined by monthly consumption of the building, which can be constant or variable). The use of daily values is indicated, although monthly values are also acceptable, in the

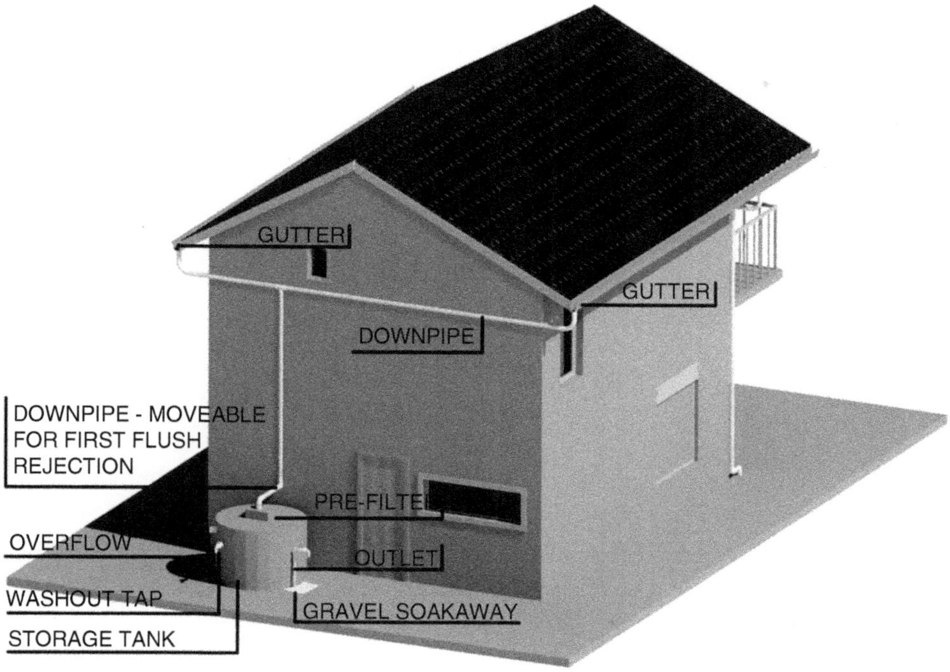

Figure 2.6 Diagram of a rainwater harvesting reservoir and its devices.

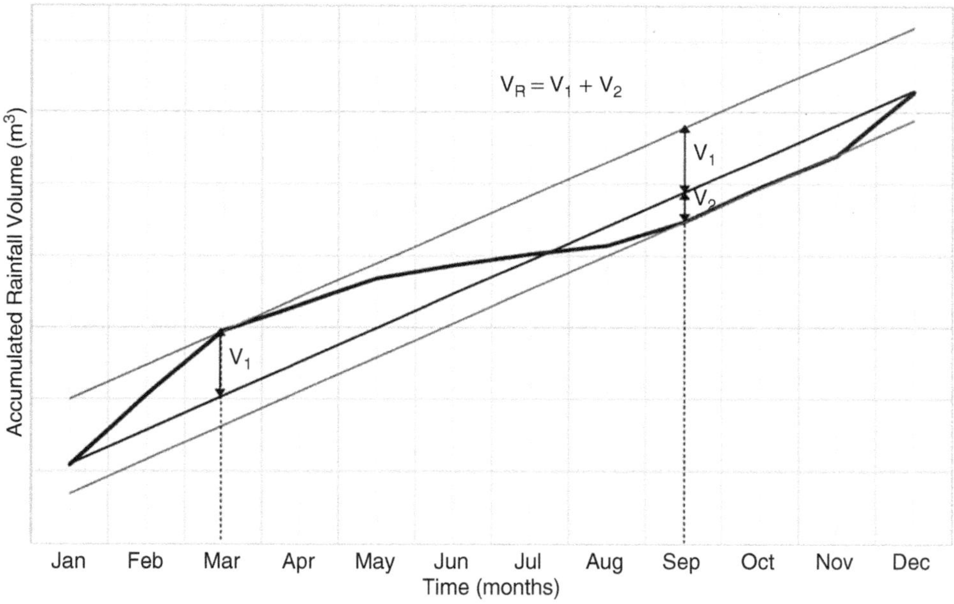

Figure 2.7 Rippl method (Mass Diagram Method).

absence of daily ones. Alternatively, the graphical method can be used, in which the cumulative rainfall volumes and accumulated local demand in one year are recorded on a graph. Parallel lines to the accumulated demands are drawn along the rainfall volume curve, touching the top and bottom points of the graph. The reservoir volume is obtained by the difference (vertical distance) between the two parallels (Figure 2.7).

The Simulation Method consists of preliminarily setting the volume of the reservoir and then verifying the percentage of consumption that will be served. Therefore, by trial and error, the volume is adjusted to satisfy the demand. In this method, the reservoir must be considered full at the beginning of the time t count, and historical data should be representative for future conditions. The continuity equation is applied to the reservoir at each time step.

The German Practical Method considers only the annual volume of the useable precipitation or the annual volume of consumption. The volume of the reservoir is defined as the lowest value between 6% of the annual volume of consumption or 6% of the annual volume of precipitation that can be used. The Australian Practical Method uses continuous analysis through the months, evaluating the remaining volume of the previous month, the total useable rainfall of the month, and the monthly demand. The reservoir volume is calculated by trial and error until the reservoir volume values are optimized, with reliable results.

Amorin and Pereira (2008) presented a comparative study of some design methods for rainwater storage reservoirs, in which they conclude that the practical methods (less complex and easy to apply) are more suitable for application in single-family homes or small establishments, while the more complex methods (Rippl Method, for example) are more suitable for larger projects. Despite this indication, there are no restrictions on the application of any method to different types of buildings. The designer must decide on the best option, given the particularities of each case.

The document "The SUDS Manual" (Woods-Ballard et al. 2015) presents a graphical representation, based on the U.K. framework BS 8515:2009+A1:2013, which allows to obtain the capacity of the storage tank based on an given number of occupants, the annual rainfall depth, and the size of the roof of the property (plan area). Figure 2.8 illustrates this approach. In this figure, the horizontal lines for each property occupancy define the maximum storage that is appropriate for that property, due to the limit in the demand. It is important to mention that the figure was produced considering a roof runoff coefficient of 80%, a water demand per person of 50 l/c/d and 18 days of storage. Thus, it varies from place to place according to the local rainfall and water demand per capita.

2.3.1.3.1 *Example of Reservoir Design*

This example considers a school building, with 1609 m^2 of surrounding garden area and 1008 m^2 of a soccer field area, located in the area covered by the rain gauge station of Ilha do Governador, Rio de Janeiro, Brazil. The building has a roof area of 1066.6 m^2 and an impervious area of 1000 m^2.

Considering the proposed uses for rainwater accepted by Brazilian law, and the possible uses applicable to the school building, this example suggests the use of rainwater for irrigation of the garden and soccer field and also for car washing. Table 2.2 shows different combinations of these possible uses for rainwater, and Table 2.3 presents the forecast of demand for non-potable water for the proposed purposes.

Some adopted premises for the calculation of the volume of the storage reservoirs:

- Irrigation of the gardens and the soccer field will only happen when it is not raining.
- Thus, we counted the average number of dry days in each month of the year, for the considered rain gauge station, and calculated the associated monthly historical average (1997–2013), as presented by Table 2.4.
- Also, for the irrigation of gardens and soccer field, the calculation of the consumption was performed by multiplying the average dry days by the forecast demand/day (Table 2.5).

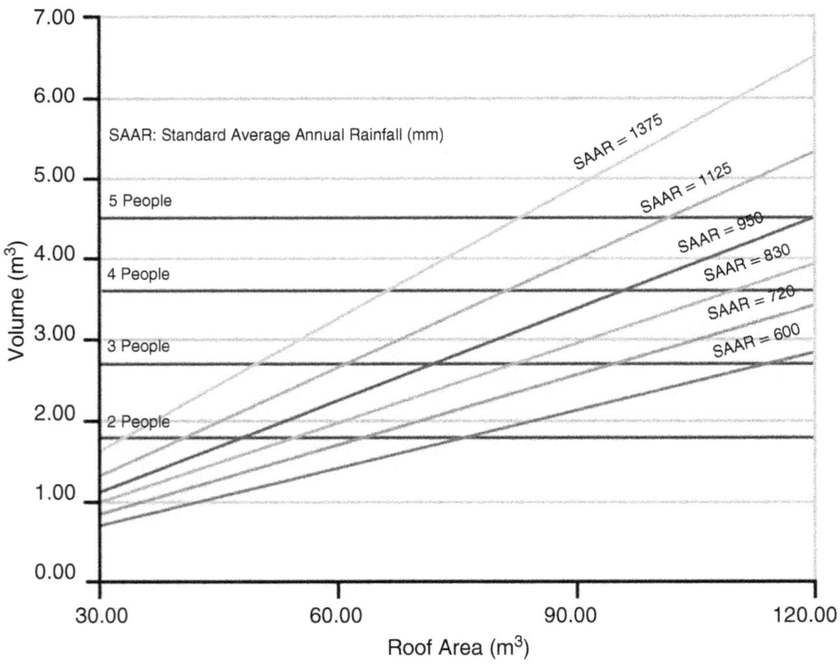

Figure 2.8 A way of dimensioning storage tanks, considering various occupancy rates (Woods-Ballard et al. 2015).

Table 2.2 Alternatives considered for rainwater harvesting in the school building.

Alternative	Description
Alternative 1	Garden Irrigation
Alternative 2	Soccer Field Irrigation
Alternative 3	Car Washing
Alternative 4	Garden Irrigation + Soccer Field Irrigation
Alternative 5	Garden Irrigation + Car Washing
Alternative 6	Garden Irrigation + Car Washing + Soccer Field Irrigation

Table 2.3 Forecast of demand for non-potable water use.

Purpose	Ratio (L/d)	Amount	Demand (L/d)	Demand (m^3/d)	Demand (m^3/month)
Garden Irrigation	1.5	1609 m^2	2413.5	2.4	72
Soccer Field Irrigation	1.5	1008 m^2	1512.0	1.5	45
Car Washing	50	4 unities	200.0	0.2	6

Table 2.4 Average dry days and average historical rainfall – Ilha do Governador Rain Gauge Station.

Month	Average dry days (days)	Average Historical Rainfall (mm)
January	15	206.68
February	18	103.98
March	18	139.25
April	20	89.84
May	22	55.91
June	24	32.25
July	24	40.26
August	25	26.29
September	20	52.03
October	18	87.98
November	15	138.49
December	16	168.50

- The monthly water consumption (m^3) for car washing considered washing 4 cars per day in a typical month of 30 days.
- The rainwater collection area is the building roof area.
- We multiplied the roof area by the monthly average rainfall to obtain the available rainwater volume per month (Table 2.6).

Table 2.5 Consumption (m^3) for irrigation considering the demand and the average dry days.

Month	Dry days	Consumption (m^3)	
		Garden irrigation	Soccer field irrigation
January	15	36.20	22.68
February	18	43.44	27.22
March	18	43.44	27.22
April	20	48.27	30.24
May	22	53.10	33.26
June	24	57.92	36.29
July	24	57.92	36.29
August	25	60.34	37.80
September	20	48.27	30.24
October	18	43.44	27.22
November	15	36.20	22.68
December	16	38.62	24.19

Table 2.6 Available rainwater volume (m^3).

Month	Monthly average rainfall (mm)	Roof area (m^2)	Rain volume on the roof (m^3)
January	206.68	1066.6	220.44
February	103.98	1066.6	110.90
March	139.25	1066.6	148.52
April	89.84	1066.6	95.82
May	55.91	1066.6	59.64
June	32.25	1066.6	34.40
July	40.26	1066.6	42.94
August	26.29	1066.6	28.04
September	52.03	1066.6	55.49
October	87.98	1066.6	93.83
November	138.49	1066.6	147.72
December	168.50	1066.6	179.72

- Using the Analytical Method of Rippl, the calculation of the water deficit was done by the difference between the available rainfall volume on the roof and the consumption, accumulating the successive negative values.

Table 2.7 presents the results for each alternative.

Table 2.8 presents the results obtained for the reservoir in each alternative, which considered the sum of the monthly deficits in each case. It is possible to infer that it is practically unfeasible to construct a reservoir to supply the demand combination of the garden and soccer field irrigation (alternatives 4 and 6), since the calculated volume

Table 2.7 Calculation of reservoir volume – Alternatives 1–6.

Month	Rainfall available volume	Alternative 1		Alternative 2		Alternative 3		Alternative 4		Alternative 5		Alternative 6	
		Consumption (m^3)	Difference (m^3)	Consumption (m^3)	Difference (m^3)	Consumption (m^3)	Difference (m^3)	Consumption (m^3)	Difference (m^3)	Consumption (m^3)	Difference (m^3)	Consumption (m^3)	Difference (m^3)
January	220.44	36.20	184.24	22.68	197.76	6.00	214.44	58.88	161.56	42.20	178.24	64.88	155.56
February	110.90	43.44	67.46	27.22	83.68	6.00	104.90	70.66	40.24	49.44	61.46	76.66	34.24
March	148.52	43.44	105.08	27.22	121.30	6.00	142.52	70.66	77.86	49.44	99.08	76.66	71.86
April	95.82	48.27	47.55	30.24	65.58	6.00	89.82	78.51	17.31	54.27	41.55	84.51	11.31
May	59.64	53.10	6.54	33.26	26.38	6.00	53.64	86.36	−26.72	59.10	0.54	92.36	−32.72
June	34.40	57.92	−23.52	36.29	−1.89	6.00	28.40	94.21	−59.81	63.92	−29.52	100.21	−65.81
July	42.94	57.92	−14.98	36.29	6.65	6.00	36.94	94.21	−51.27	63.92	−20.98	100.21	−57.27
August	28.04	60.34	−32.30	37.80	−9.76	6.00	22.04	98.14	−70.10	66.34	−38.30	104.14	−76.10
September	55.49	48.27	7.22	30.24	25.25	6.00	49.49	78.51	−23.02	54.27	1.22	84.51	−29.02
October	93.83	43.44	50.39	27.22	66.61	6.00	87.83	70.66	23.17	49.44	44.39	76.66	17.17
November	147.72	36.20	111.52	22.68	125.04	6.00	141.72	58.88	88.84	42.20	105.52	64.88	82.84
December	179.72	38.62	141.10	24.19	155.53	6.00	173.72	62.81	116.91	44.62	135.10	68.81	110.91

Table 2.8 Reservoir volume – summary of the alternatives for rainwater harvesting.

Alternative	Description	Reservoir volume (m^3)
1	Garden Irrigation	70.81
2	Soccer Field Irrigation	9.76
3	Car Washing	0
4	Garden Irrigation + Soccer Field Irrigation	230.92
5	Garden Irrigation + Car Washing	88.81
6	Garden Irrigation + Car Washing + Soccer Field Irrigation	260.92

leads to the design of a large reservoir (more than 200 m^3). The other alternatives that consider the garden irrigation (alternatives 1 and 5) lead to the design of reservoirs with lower volume, but still of large size (more than 70 m^3).

The volume to supply the soccer field irrigation (alternative 2) is much smaller than that calculated in the other alternatives (9.57 m^3). This reservoir could have dimensions of 3 m (width) × 4 m (length) × 1 m (depth) and be positioned in the garden area, for example.

From this analysis, it is possible to conclude that the best use of the rainwater collected on the roof of this school building depends not only on the rainwater availability but also on the investment capacity and the space available to introduce the reservoir and adapt the building systems.

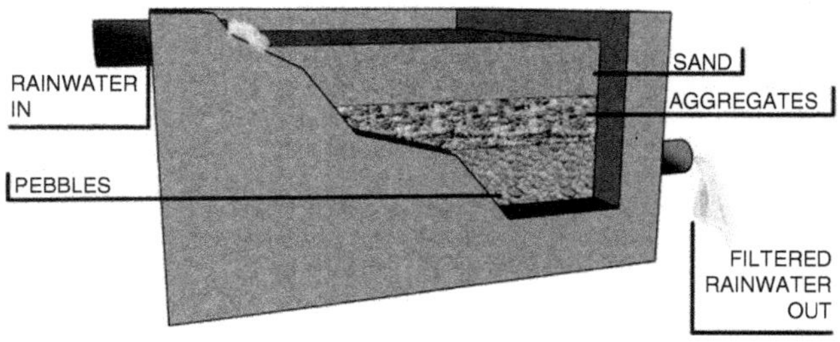

Figure 2.9 Diagram of a sand gravel filter.

2.3.1.4 Rainwater Treatment Systems

The rainwater treatment system is specified according to the rainwater quality and can vary according to the type of the building, purpose of the system, quality of the water demanded, technical and economic viability, and also social acceptance.

The treatment begins with the separation and disposal of solids. Then, the rainwater can go through a filtration process with possible decontamination (which may already occur in the reservoir). In the filtration process, all or at least part of the sediments have been removed.

There are different types of filters; they can hold from large solids to particles of the order of 10^{-9} m. In addition to the filtration capacity, the filters differ in the position occupied in the system and the material used in the process. A simple example is the sand gravel filter, constructed of brick and filleted by pebbles, gravel, and sand (Figure 2.9).

In this kind of filter, water is purified by going through different layers, which are separated from each other by ar wire mesh.

2.3.1.5 Rainwater Pumping Station

The pumping station is responsible for raising the water from the lower reservoir to the upper reservoir, and it consists of pumps and pipes. It should have at least two independent pumps, as a way to keep the system working in the case of failure of one of them. The designer must observe the recommendations of the minimum suction speed and the adequate pump selection for each case. Figure 2.10 presents a diagram of a general rainwater treatment system.

2.3.1.6 Water Supply System (Water Pipes)

It is necessary to adapt the existing building plumbing system to make the non-potable uses of rainwater feasible. The non-potable water system cannot have any connection with the drinking water system. It is necessary to provide a set of pipes, reservoirs, and other components intended exclusively for collecting, storing, treating, and distributing rainwater. Non-potable water outlets in this system should be properly identified to prevent misuse. Additionally, restricted access taps can also be considered as they allow the use only by authorized persons. These taps are installed mainly in outdoor areas around the building, such as garden areas, grass fields, and parking (car wash) areas.

2.3.2 Source Control Systems

The steps described in Section 2.3.1 are intended to increase the overall supply of water resources by harnessing unused water, proposing benefits to the community in terms of reducing floods, increasing the supply of treated water, and reducing demand on the public water supply system. This section has a direct relation with the previous one since the introduction of a rainwater harvesting system also promotes the reduction of outflow, which is capable of acting for urban flood control in a transversal action in the context of public sanitation.

Currently, urban drainage projects should foster an integrated view of the river basin, rather than the traditional way when the solution adopted was to convey the flow more effectively downstream. In this context, water storage devices can be designed in a distributed way in the watershed, as well as the deceleration of flow and infiltration of water into the soil, among others, to restore the original flow patterns of the watershed. The use of on-site tanks is of a measure for distributed rainwater storage, through small reservoirs located inside urban lots. The water collected in the reservoirs can return to the minor drainage system after the rainfall peak, or it can be used by the landowner, especially in non-potable uses such as garden watering and car washing (Figure 2.11). More information can be obtained in Miguez et al. (2011, 2013).

Based on the previous discussion, this section proposes the development of integrated urban drainage projects

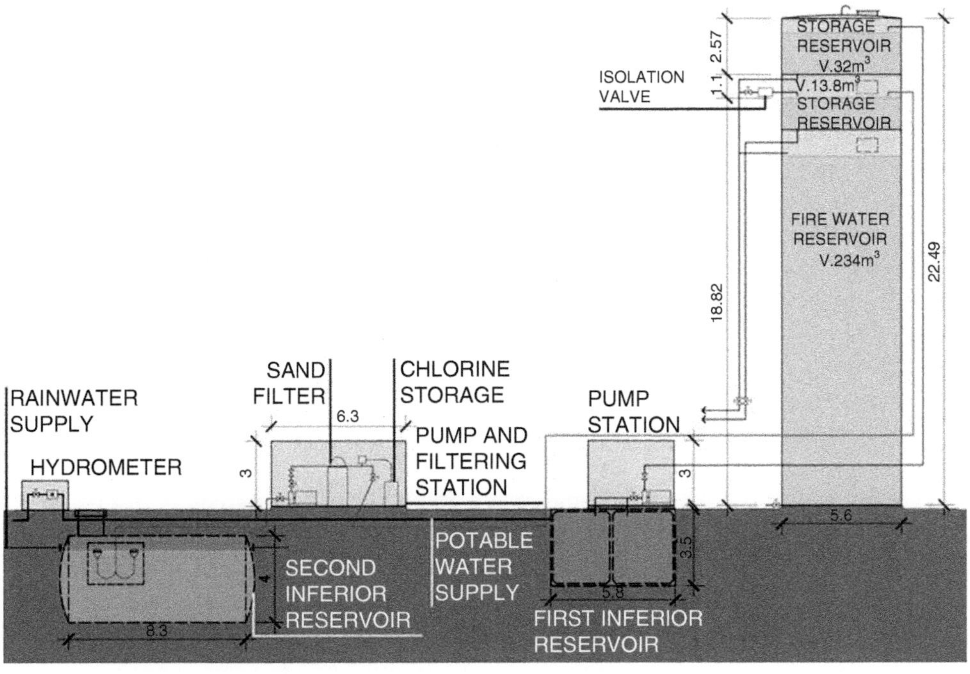

Figure 2.10 Diagram of a general rainwater treatment system.

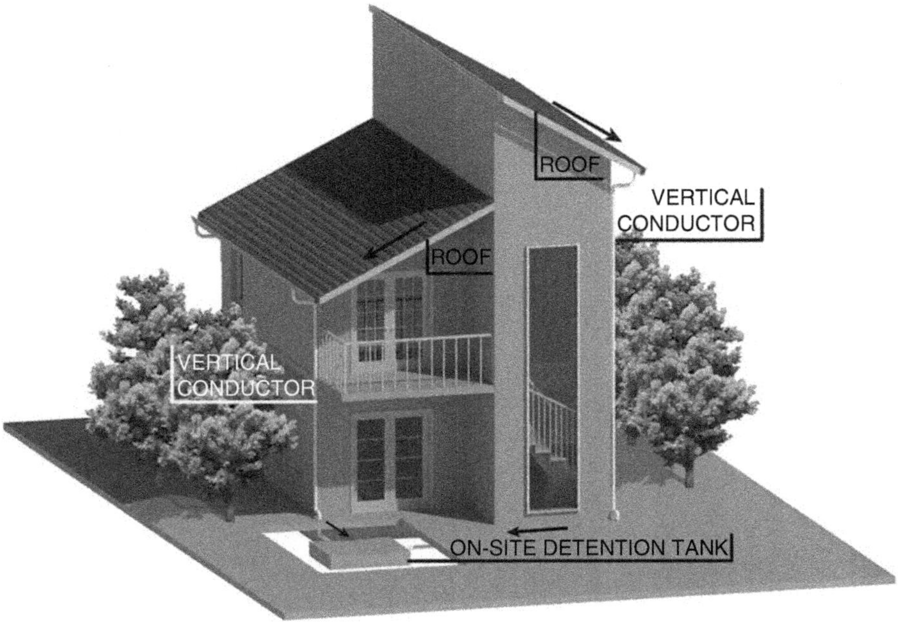

Figure 2.11 Example of an on-site detention tank.

and flood control actions, especially in developing countries, where the investment capacity is lower, based on the following premises, focusing on the use of on-site reservoirs:

- the watershed must be treated as an integrated system and must be organized in terms of areas for applying the OSD measures, to reorganize the general flow pattern, including the definition of mean damping efficiencies expected;
- to reach the desired efficiency, it can be useful to standardize a typical reservoir by sub-region, according to the proposed zoning in an integrated flood control project, including forecasting the time evolution of urbanization patterns;
- the site owner must make a financial contribution to the problem solution – the public authorities should treat the flows in public areas and complement the mitigation project with other measures, structural or non-structural, to compensate for the full non-recovery of the on-site discharges before urbanization;
- when possible, secure and effective on-site reservoirs should be used with multiple purposes, bringing complementary benefits to the owner, to encourage implementation, and more importantly its maintenance in optimal operational conditions over time.

2.4 Analysis of Payback of Rainwater Harvesting Systems

Teston et al. (2018) analyzed many studies that assessed the economic viability of installing rainwater harvesting systems in residential buildings in Brazil based on the payback of investment. According to this study, systems with a payback up to around 10 years were considered feasible. They concluded that demand for non-potable water directly the influences viability of a rainwater harvesting systems, but this depends on the consumption, since there is a minimum rate charged for a volume of up to $10\,m^3$. Thus, even if the consumption of potable water is less than this threshold, the water bill will not vary. Beyond this limit, the rate charged influences the feasibility, whereby the higher the tariff, the lower the payback period (Teston et al. 2018).

Silva et al. (2014) examined the technical feasibility of domestic rainwater harvesting systems for two cities in Portugal, Porto and Almada, with particular weather and water use patterns. The analysis was carried out for a single-family residence containing two bedrooms, a living room, one bathroom, and a kitchen, where the water use pattern was monitored. The water savings efficiency and payback period were estimated for both locations. They concluded that, for an optimum rainwater tank, the water savings potential was similar for Porto and Almada, despite the differences in the average annual precipitation. In addition, the analysis showed an influence of water rates on the economic viability of rainwater harvesting systems.

Some results of the payback analysis of Teston et al. (2018) and Silva et al. (2014) are summarized in Table 2.9.

Campisano et al. (2017b) evaluated the benefits of large-scale installation of domestic rainwater harvesting systems in multi-story buildings. The method developed combines a regression model for water-saving estimation with geospatial analysis tools for semi-automatic collection

Table 2.9 Payback periods for the implementation of rainwater harvesting (Teston et al. 2018; Silva et al. 2014).

References	Feature	Payback (years)	Economic feasibility
(Ghisi and Ferreira 2007)	Multi-family building block A (16 flats and 2.25 residents/flat)	2.4	Feasible
City: Florianópolis-Brazil	Multi-family building block B (17 flats and 2.67 residents/flat)	5	Feasible
	Multi-family building block C (16 flats and 2.33 residents/flat)	–	No saving
(Carvalho 2010)	Residence of 200 m^2 with 4 inhabitants	5.3	Feasible
City: Londrina – Brazil			
(Cruz and Blanco 2017)	Single-family residence (8 residents)	24	Infeasible
City: Rio Branco, Brazil	Single-family residence (5 residents)	27.1	Infeasible
(Cruz and Blanco 2017)	Single-family residence (8 residents)	17.9	Infeasible
City: Belém, Brazil	Single-family residence (5 residents)	20.1	Infeasible
(Silva et al. 2014)	Single-family residence: Tank (200 l) and collection area (103.4 m^2)	23.7	
City: Almada – Portugal	Single-family residence: Tank (1000 l) and collection area (103.4 m^2)	14.7	
(Silva et al. 2014)	Single-family residence: Tank (200 l) and collection area (103.4 m^2)	26.8	
City: Porto – Portugal	Single-family residence: Tank (1000 l) and collection area (103.4 m^2)	16.1	

of spatial information at the building/household level. The study was applied to the historic town of Lipari, the largest municipality of the homonymous Aeolian Islands. The results showed a potential for high yearly water savings (between 30% and 50%), with return on investment in less than 15 years for about 50% of the installed systems.

2.5 Conclusion

This chapter discusses the use of rainwater in buildings, in line with the NZWB concept. The components of this system, which should be seen as complementary to the established ones, such as drinking water and sanitary sewage systems, are presented, as well as a brief case study.

It can be seen that small roof areas may be the main limiting factor for rainwater capture, leading the designer to discard possible alternatives for use when this demands a much larger collection area than the existing one. On the other hand, if the collection area is large, the size of reservoirs may become unfeasible for the site. It is also necessary to consider the impacts of a rainwater reservoir on the building structure and its influence on the sustainability of the project. Large reservoirs have little versatility when it comes to their implementation in existing buildings that do not have enough free space.

Other actions, such as the use of water-saving devices, the treatment and reuse of gray water, should also be combined in the project. Low-use appliances should be considered to reduce the project flow, saving drinking water, and reducing wastewater. Besides acting to reduce consumption, the gray water reuse system would also affect the amount of sewage discharged in the public system and its quality.

Based on the study carried out and presented in this chapter, the following general guidelines are proposed for the architecture design considering rainwater harvesting:

- it is necessary to have a collection area sufficient to collect large amounts of rainwater;
- it is required to have a rooftop to install a second upper reservoir, to make the rainwater harvested available for multiple non-potable uses;
- it is advisable to consider the uses of rainwater and the water demands of the building in the early design process;
- space must be provided on the ground floor (or underground) to house the water tank for treatment;
- consideration should be given to the exchange or use of water-saving devices, making rainwater harvesting more effective; and
- it is necessary to study the buildings on a case-by-case basis to better implement the rainwater harvesting system according to local characteristics.

Acknowledgment

This work was supported by the Coordenação de Aperfeiçoamento de Pessoal de Nível Superior- Brasil (CAPES) [Finance Code 001; 1681776] and by Conselho Nacional de Desenvolvimento Científico e Tecnológico (CNPq)[303240/2017-2].

We thank the undergraduate student Verônica Bahiense, from the School of Architecture and Urbanism of Federal University of Rio de Janeiro, for adapting some figures for this chapter.

References

ABNT. (2019). *NBR 15527: Aproveitamento de água de chuva de coberturas para fins não potáveis - Requisitos*. Brasil.

Alamdari, N., Sample, D.J., Liu, J., and Ross, A. (2018). Assessing climate change impacts on the reliability of rainwater harvesting systems. *Resources, Conservation and Recycling* 132: 178–189.

de Amorin, S.V. and Pereira, D. J. de A. (2008). Estudo comparativo dos métodos de dimensionamento para reservatórios utilizados em aproveitamento de água pluvial. *Ambiente Construído* 8 (2): 53–66.

ANA - Agência Nacional de Águas. (2018). *Conjuntura dos Recursos Hídricos no Brasil 2018 - Informe Anual*.

ANQIP. (2009). "Especificação Técnica ANQIP ETA 0701: Sistemas de Aproveitamento de Águas Pluviais em Edifícios (SAAP)." 1–26.

Bauer, D., Philbrick, M., Vallario, B. et al. (2014). *The Water-Energy Nexus: Challenges and Opportunities*. USA: U.S Department of Energy.

Beatty, R., and McLindin, M. (2012). "Rainwater Harvesting and Urban Design in Australia." *Proceedings of the Water Environment Federation*, 6435–6447.

Bhave, A.G., Conway, D., Dessai, S., and Stainforth, D.A. (2018). Water resource planning under future climate and socioeconomic uncertainty in the Cauvery River basin in Karnataka, India. *Water Resources Research* 54 (2): 708–728.

Brown, C., Gerston, J., and Colley, S. (2005). *The Texas Manual on Rainwater Harvesting*, 2e. Texas Water Development Board.

Campisano, A., Butler, D., Ward, S. et al. (2017a). Urban rainwater harvesting systems : research , implementation and future perspectives. *Water Research* 115 (15): 195–209.

Campisano, A., D'Amico, G., and Modica, C. (2017b). Water saving and cost analysis of large-scale implementation of domestic rain water harvesting in minor Mediterranean islands. *Water (Switzerland)* 9 (916): 1–14.

Carvalho, R.S. (2010). *Potencial Econômico do Aproveitamento de Águas Pluviais: Análise da Implantação de um Sistema Para a Região Urbana de Londrina [Economic Potential of the Use of Rainwater: Analysis of the Implementation of a System for the Urban Region of Londrina]*. Apucarana, Brazil: Paraná Federal University.

Chang, N.B., Rivera, B.J., and Wanielista, M.P. (2011). Optimal design for water conservation and energy savings using green roofs in a green building under mixed uncertainties. *Journal of Cleaner Production* 19 (11): 1180–1188.

da Cruz, W.M. and Blanco, C.J.C. (2017). Aproveitamento De Água Pluvial Para Fins Não Potáveis Residenciais Em Rio Branco – Ac. *Perspectivas Online* 17 (07): 12–24.

Cunliffe, D.a. (1998). Guidance on the use of rainwater tanks. *National Environmental Health Forum Monographs Water* 3 (5210): 1–29.

Dai, J., Wu, S., Han, G. et al. (2018). Water-energy nexus: A review of methods and tools for macro-assessment. *Applied Energy* 210 (August): 393–408.

DeNooyer, T.A., Peschel, J.M., Zhang, Z., and Stillwell, A.S. (2016). Integrating water resources and power generation: the energy-water nexus in Illinois. *Applied Energy* 162: 363–371.

Department of Communities and Local Government (2010). *Code for Sustainable Homes: A Cost Review. Communities and Local Government Publications*. London, UK: Communities and Local Government Publications.

Dietz, M.E. (2007). Low impact development practices: A review of current research and recommendations for future directions. *Water, Air, & Soil Pollution* 186: 351–363.

DIN 1989–1:2001–10. (1989). *Rainwater harvesting systems — Part 1 : Planning , installation , operation and maintenance*. Germany, 1–34.

EN 16491 (2018). *European Standard EN 16941: On-Site Non-potable Water Systems - Part 1: Systems for the Use of Rainwater*. Brussels, Belgium: British Standards Institution.

Energy, O. of E. E. & R. (2019). "Net Zero Water Building Strategies." https://www.energy.gov/eere/femp/net-zero-water-building-strategies> (Sep. 6, 2019).

Englehardt, J.D., Wu, T., Bloetscher, F. et al. (2016). Net-zero water management: achieving energy-positive municipal water supply. *Environmental Science: Water Research & Technology*.

enHealth (2010). *Guidance on Use of Rainwater Tanks*. Australia: Commonwealth Copyright Administration.

Ghisi, E. and Ferreira, D.F. (2007). Potential for potable water savings by using rainwater and greywater in a multi-storey residential building in southern Brazul. *Building and Environment* 42: 2512–2522.

Gleick, P.H. (1993). "Water in crisis: a guide to the world's fresh water resources." International Affairs.

Haque, M.M., Rahman, A., and Samali, B. (2016). Evaluation of climate change impacts on rainwater harvesting. *Journal of Cleaner Production* 137: 60–69.

Howitt, R., Duncan, M., and Medellín-azuara, J. (2015). *Economic Analysis of the 2015 Drought for California Agriculture*. Davis, California, California, USA: Center for Watershed Sciences/*University of California*.

Hua-peng, Q., Zhuo-xi, L., and Guangtao, F. (2013). The effects of low impact development on urban flooding under different rainfall characteristics. *Journal of Environmental Management* 129: 577–585.

Joustra, C.M. and Yeh, D.H. (2015a). Framework for net-zero and net-positive building water cycle management. *Building Research and Information*.

Joustra, C., and Yeh, D. (2015b). "Decision support modeling for net-zero water buildings." Proceedings - Winter Simulation Conference.

Juan, Y., Chen, Y., and Lin, J. (2016). "Greywater Reuse System Design and Economic Analysis for Residential Buildings in Taiwan."

Konadu, D., Fenner, R., Richards, K., and Allwood, J. (2017). *UK Water-Energy Nexus under climate change: Key issues and priorities for modelling and research About the Whole Systems Energy Modelling Consortium (wholeSEM)*. Cambridge, UK.

Li, X., Yang, L., Zheng, H. et al. (2019). City-level water-energy nexus in Beijing-Tianjin-Hebei region. *Applied Energy* 235: 827–834.

Marsden Jacob Associates. (2009). *The cost effectiveness of residential rainwater tanks in Perth. Financial and Economic Consultants ABN*.

Mekonnen, M.M. and Hoekstra, A.Y. (2016). Sustainability: four billion people facing severe water scarcity. *Science Advances* 2 (2): 1–6.

Miguez, M.G., Veról, A.P., Mascarenhas, F.C.B., et al. (2011). "Compensatory techniques on urban drainage for flood control with the aid of mathematical modelling: A case study in Rio de Janeiro city." *WIT Transactions on the Built Environment*.

Miguez, M.G., Verol, A.P., and Santos, R.B. (2013). *Alternative solutions for sustainable urban drainage systems integrating areas of irregular urban growth with the formal city. Applied Mechanics and Materials*.

Miguez, M.G., Veról, A.P., Rêgo, A.Q.d.S.F., and Lourenço, I.B. (2018). *Urban Agglomeration and Supporting Capacity: The Role of Open Spaces within Urban Drainage Systems as a Structuring Condition for Urban Growth." Urban Agglomeration*. InTech.

Moran, A., Hunt, B., and Jennings, G. (2003). A North Carolina field study to evaluate green roof runoff quantity, runoff quality, and plant growth. *World Water & Environmental Resources Congress* 2003: 1–10.

NCSL. (2018). "State Rainwater Harvesting Laws and Legislation." *The National Conference of State Legislatures, USA*, http://www.ncsl.org/research/environment-and-natural-resources/rainwater-harvesting.aspx> (Jun. 15, 2019).

NDMC. (2019). "How does Drought Affect our Lives?" *National Drought Mitigation Center, USA*, https://drought.unl.edu/Education/DroughtforKids.aspx> (Jun. 15, 2019).

Niachou, A., Papakonstantinou, K., Santamouris, M. et al. (2001). Analysis of the green roof thermal properties and investigation of its energy performance. *Energy and Buildings* 33: 719–729.

Okhravi, S., Eslamian, S., Eslamian, F., and Tarkesh-Esfahani, S. (2014). Indigenous knowledge as a supportive tool for sustainable development and utilisation of rainwater harvesting systems. *Journal of Flood Engineering* 5 (1/2): 39–50.

Okhravi, S., Eslamian, S., and Adamowski, J. (2015). Water reuse in rainwater harvesting. In: *Water Reuse in Rainwater Harvesting, Urban Water Reuse Handbook* (ed. S. Eslamian), 779–796. Taylor and Francis, CRC Group.

Pedde, S., Kroeze, C., and Rodrigues, L. N. (2013). "Escassez Hídrica na América do Sul: Situação Atual e Perspectivas Futuras." Bento Gonçalves - RS, 1–8.

Peng, C., Ma, Z., Lei, X. et al. (2011). A drought-induced pervasive increase in tree mortality across Canada's boreal forests. *Nature Climate Change* 1 (9): 467–471.

Radcliffe, J.C. (2018). The water energy nexus in Australia – the outcome of two crises. *Water-Energy Nexus* 1: 66–85.

Rippl, W. (1883). Capacity of storage reservoirs for water supply. *Minutes of Proceedings, Institution of Civil Engineers* 71: 270–278.

Rowe, D.B. (2011). Green roofs as a means of pollution abatement. *Environmental Pollution* 159 (8–9): 2100–2110.

Sayers, P.B., Yuanyuan, L., Moncrieff, C. et al. (2016). *Drought Risk Management: A Strategic Approach*. Paris, France: UNESCO.

Silva, C.M., Sousa, V., and Carvalho, N.V. (2014). Evaluation of rainwater harvesting in Portugal: application to single-family residences. *Resources, Conservation and Recycling* 94: 21–34.

Silva, A.O., Argôlo, A.L.L., Dutra e Mello, L.B.M., et al. (2019). "Aproveitamento de Água de Chuva em Edificações Multifamiliares Considerando o Conceito de Net Zero Water Buildings." *Euroelecs 2019 - III encuentro latinoamericano y europeo sobre edificaciones y comunidades sostenibles*, R. C. Spahn, ed., ANTAC y EcoUrbano, Santa Fé y Paraná, 636–647.

Silva-Afonso, A. and Pimentel-Rodrigues, C. (2011). The importance of water efficiency in buildings in Mediterranean countries: the Portuguese experience. *International Journal of Systems Applications, Engineering and Development* 5 (1): 17–24.

Silva-Afonso, A., Rodrigues, C. (2008). "Water efficiency of products and buildings: The implementation of certification and labelling measures in Portugal." *The 34 th International Symposium on CIB W062 Water Supply and Drainage for Buildings*, Hong Kong, China, 230–240.

Silva-Afonso, A., Rodrigues, F., and Pimentel-Rodrigues, C. (2011). "Water efficiency in buildings: Assessment of its impact on energy efficiency and reducing GHG emissions."

Recent Researches in Energy and Environment - 6th IASME / WSEAS International Conference on Energy and Environment, EE'11, 191–195.

Talebpour, M.R., Sahin, O., Siems, R., and Stewart, R.A. (2014). Water and energy nexus of residential rainwater tanks at an end use level: case of Australia. *Energy and Buildings* 80: 195–207.

Teston, A., Geraldi, M.S., Colasio, B.M., and Ghisi, E. (2018). Rainwater harvesting in buildings in Brazil: A literature review. *Water (Switzerland)* 10 (471): 1–25.

Thissen, W., Kwakkel, J., Mens, M. et al. (2017). Dealing with uncertainties in fresh water supply: experiences in the Netherlands. *Water Resources Management, Water Resources Management* 31 (2): 703–725.

Tsakiris, G. (2017). Drought risk assessment and management. *Water Resources Management.*

UN General Assembly. (2015). *The 2030 Agenda for Sustainable Development.* 1–35.

UN General Assembly. (2016). *International Decade for Action, "Water for Sustainable Development", 2018–2028.* 1–5.

UNCED. (1992). "Earth Summit'92. The UN Conference on Environment and Development - Agenda 21." *Reproduction.*

UNESCO World Water Assessment Programme (2016). *The United Nations World Water Development Report 2016: Water and Job.* Paris, France: United Nations Educational, Scientific and Cultural Organization.

UNESCO World Water Assessment Programme. (2019). The United Nations World Water Development Report 2019: Leaving no One behind. Paris, France.

United Nations (2018). *The Sustainable Development Goals Report-2018.* New York: United Nations.

United Nations General Assembly (1972). *Declaration of the United Nations Conference on the Human Environment.* United Nations.

USEPA, U.S.E.P.A. (2018). "Net Zero Concepts and Definitions." https://www.epa.gov/water-research/net-zero-concepts-and-definitions> (Mar. 28, 2018).

USGS. (2019). "The Fundamentals of the Water Cycle."

Vijayaraghavan, K. (2016). Green roofs: A critical review on the role of components, benefits, limitations and trends. *Renewable and Sustainable Energy Reviews* 57: 740–752.

WBG. (2019). "Water Supply." *The world Bank Group.*

Woods-Ballard, B., Wilson, S., Udale-Clark, H. et al. (2015). *The SUDS Manual.* CIRIA, London, UK: The SuDS Manual.

Zhang, C., Chen, X., Li, Y. et al. (2018). Water-energy-food nexus: concepts, questions and methodologies. *Journal of Cleaner Production* 195: 625–639.

3

Standards for Rainwater Catchment Design

Sisuru Sendanayake[1] *and Saeid Eslamian*[2]

[1] *Department of Civil Engineering, South Asian Institute of Technology and Medicine, Sri Lanka*
[2] *Department of Water Engineering, College of Agriculture, Isfahan University of Technology, Isfahan, Iran, 8415683111*

3.1 Introduction

Ever since the beginning of civilization mankind has depended on water to fulfill their basic needs. For centuries, humans have used and abused the planet's water resources so much so that water has become scarce and is a precious commodity. As has happened time and again, mankind has turned their sights to the heavens, as some ancient civilizations and isolated societies in arid and semi-arid areas have done since time immemorial: looking for rain to be harvested.

Thus, in the last few decades, there has been a mounting interest in rainwater harvesting throughout the world, for both irrigation and domestic use, resulting in extensive research being carried out on the subject (McMahon and Mein 1978; Edwards and Keller 1984; Latham and Schiller 1987; Yaziz et al. 1989; Forster 1991; Butler 1993; Vale and Vale 1995; Fewkes 1999; Hermann and Schmida 1999; Gould and Niesson-Peterson 1999; Zobrist et al. 2000; Zhu et al. 2004; Roebuck and Ashley 2006; Ghisi et al. 2007; Brown 2009; Sendanayake et al. 2014; Hamdy and Eslamian 2015).

> Rainwater harvesting means the systematic collection of water that rains the earth, and the proper use of it.

Rainwater Harvesting (RWH) introduces a means of transforming the existing water cycle through decentralization, where the owner of the system is becoming both the producer and consumer, thus being responsible for individual water security. In addition, RWH reduces the load on the public supply, mitigates overflow quantities and thus the strain on local drainage systems, and the harvested water can be used for non-potable purposes for which treated water is not required.

In general, the practice of RWH is heavily dependent on the local climatic conditions and socio-economic scenario of the user community, as well as the availability/non-availability of centralized public/private water supply. For isolated communities where public water supply is not available, RWH is practiced more by rule of thumb with the intention of collecting the maximum amount of incident precipitation during the rainy season, to survive the subsequent dry period, subject to the limitations of catchment surface and the maximum possible volume of the storage tank that the user can manage to install. In many of these situations the catchment surface is the roof of a dwelling of surface area 25–50 m^2 made of asbestos cement, clay or GI sheets, thus limited in surface area, while the storage tank made out of reinforced cement concrete (RCC), ferro-cement, steel, or polymer of varying volumes, with RCC/ferro-cement tanks typically made up to 5 m^3. The storage tanks of varying shapes, from pumpkin-shaped ferro-cement tanks to barrels made of polymer, can be placed either above ground or below ground depending on ease of maintenance and aesthetic considerations (Thomas and Martinson 2007). The conveyance system usually consists of gutters and pipes that deliver rainwater falling on the roof and in addition, any practical RWH systems will have sub-systems for filtering in collected rainwater, such as first-flushing (FF) devices, pumps to draw-off collected rainwater from the storage facility, and sediment-discharging and cleaning mechanisms for successful operation. Typically, therefore, when RWH is the sole means of water supply for a dwelling, the system sizing is at the total discretion of the user who has to take into consideration all relevant factors which would affect the collection of a sufficient quantity of water for the dry spell depending on the daily usage and for what purpose the harvested rainwater be used. If the collection

Handbook of Water Harvesting and Conservation: Basic Concepts and Fundamentals, First Edition. Edited by Saeid Eslamian and Faezeh Eslamian.

is for potable water, the daily per capita minimum requirement is generally considered to be 2.5 l while for general purposes use the daily per capita consumption is taken as 50 l (Hermann and Schmida 1999; Thomas and Martinson 2007). The above figures again depend on the usage and cultural practices of a particular community, as well as the environmental factors in the locality.

RWH is also considered as a viable supplementary source of water, particularly in tropical countries where adequate rainfall depths are available spread through the year. Conventional Rooftop Rainwater Harvesting (RTRWH) systems consist of a catchment area, gutters and pipes as the conveyance, and a storage tank. From the perspective of builders, designers, and regulatory bodies, as well as the end users, it is important therefore to specify standards for the design of the individual components as well as for the use and maintenance of the system. Thus, in the following sections design standards for the catchment surface, conveyance system, and the storage tanks are discussed in detail and the standards required in pre-treatment and distribution systems are discussed in general. As the vast majority of RWH systems are standalone supplementary water supply units managed by end users, the discussion is limited to the RTRWH scenario.

In designing a RTRWH unit it is important to pay attention to the configuration of the system and its components with relation to the building it serves, as well as the topography of the site, and to make sure the systems confirms with the local plumbing codes. It is also important to note whether the system is to be used as a supplementary source to the reticulated main water supply. Topography of the site is of particular importance as pumping would be required to convey the collected rainwater in the storage tank into the end user points if the tank is placed at a lower elevation, such as in an underground position. However, if the harvested rainwater is to be supplied to end user points through gravity, the tank should be placed at a lower elevation as close as possible to the catchment surface though the supply pressure could be low.

In the following sections, design standards and features of the main components, i.e. the catchment surface, conveyance system, storage tank, and the distribution system are discussed in detail.

3.2 Catchment Surface

The collection area in most cases is the roof of a house or a building. Typical material for roofing includes corrugated iron sheet, asbestos-cement sheet, tiles, or thatch made from a variety of organic materials if thatched tightly. The effective roof area and the material used in constructing the roof influence the efficiency of collection and water quality. All catchment surfaces must be made of non-toxic material. Painted surfaces should be avoided if possible, or, if the use of paint is unavoidable, only non-toxic paint should be used. Lead-, chromium-, or zinc-based paints are not suitable for catchment surfaces due to the presence of heavy metals. Overhanging vegetation should also be avoided. Steep galvanized iron roofs have been found to be relatively efficient rainwater collectors, while flat concrete roofs are very inefficient (Edwards and Keller 1984). However, roofs covered with corrugated galvanized mild steel are found to be easiest to use and give the cleanest water (Zobrist et al. 2000). GI sheets also have the potential to kill bacteria as a result of maintaining high temperatures when exposed to sun.

Rooftop catchment efficiencies range from 70–90%. These losses are due to roofing material texture, evaporation, losses occurring in gutters and storage tanks and inefficiencies in the collection process. It has been estimated that 1 cm of rain on 100 m^2 of roof yield 10 000 l. More commonly, rooftop catchment yield is estimated to be 75% of actual rainfall on the catchment area, after accounting for losses due to evaporation during periods when short, light showers are interspersed with periods of prolonged sunshine (Edwards and Keller 1984), though occasionally, run-off coefficient for hard roofs in humid tropics is taken as 0.85 (Sendanayake 2016a, b, c, d). Typical run-off coefficients are taken as below; 0.9 for GI sheets, 0.6–0.9 for glazed tiles, 0.8–0.9 for Aluminum sheets, 0.6–0.7 for flat cement roofs, 0.8 for Asbestos-Cement and 0.2 for thatched/organic roofs (Lanka Rainwater Harvesting Forum 2004; Sendanayake et al. 2014).

Asbestos roofs, apart from relatively lower collection efficiency of 0.8 due to their rougher surface texture, could promote the growth of coliforms from bird and animal droppings. More seriously, asbestos fibers can come loose if the sheet is damaged, having the potential for human ingestion causing cancer in gastrointestinal tract and pulmonary fibrosis. However, asbestos is not uncommon in most domestic supplies, with concentrations in rivers and lakes around 1 million fibers per liter. The US Environmental Protection Agency (EPA) in 1992 has set drinking water standards for asbestos at 7 million fibers per liter for fibers longer than 10 μm. However, research has indicated that slow sand and gravel filters can remove up to 90% of asbestos fibers and other particulate matter (RHIC network priority).

Microscopically, the coarser surfaces of tiled or asbestos cement roofs allow for higher depositions and entrapment of pollutants from the atmosphere compared to the relatively smoother galvanized iron roofs. High-intensity rain, which tropical countries often experience during the monsoon periods, is more efficient in removing pollutants due to the greater amount of energy present in the rain drops upon impact with the roof surface (Yaziz et al. 1989).

Roofs painted with lead-based paints should not be used to collect rainwater for drinking due to potential leaking in the cases of rainwater having low pH values. Therefore, unpainted and uncoated roof surfaces are the best options to provide drinking water (Cunliffe 2004).

Many materials are used to provide an impervious surface for roofing purposes. It is the general practice to make use of roofs, whether sloping or flat, to collect incident precipitation before conveying the collection to suitably located storage tanks/cisterns. Commonly, clay tiles, asbestos cement sheets, and metal sheets (galvanized iron, zinc alum) are used for sloping roofs, while RCC slabs with waterproofing are used for flat roofs.

In catchment design, two main considerations are the percentage loss of incident rainfall and the possible contamination of the collection due to external and controllable factors.

The collection area in most cases is the roof of a house or a building. The effective roof area and the material used in constructing the roof influence the efficiency of collection and water quality. All catchment surfaces must be made of non-toxic material. Painted surfaces should be avoided if possible, or, if the use of paint is unavoidable, only non-toxic paint should be used. Lead-, chromium-, or zinc-based paints are not suitable for catchment surfaces due to presence of heavy metals. Overhanging vegetation should also be avoided. Steep galvanized iron roofs have been found to be relatively efficient rainwater collectors, while flat concrete roofs are very inefficient (Edwards and Keller 1984).

3.2.1 Collection Efficiency

Coefficient of collection efficiency (C_f) is defined as the percentage of catchment that can be effectively collected from a given impervious surface which is generally based on the roofing material.

Rainfall loss during collection occurs due to absorption by the roofing material and wind effects around the roof. The rainfall loss was modeled using an initial depression storage loss (L) with a run-off coefficient (C_f).

The model is of the general form:

$$\mathrm{Qt} = \sum_{t=1}^{T} Qt = \left(\sum_{t=1}^{T} RtACf \right) - L \tag{3.1}$$

Where,

Qt is the rainwater run-off during rainfall event, t;
T is the duration of rainfall event, t (min);
L is the depression storage loss (L);
C_f is the run-off coefficient;
R_t is the rainfall during rainfall event, t (mm);
A is the effective roof area.

It is noted that L can also be expressed in mm by dividing the depression loss by collection area. L can also be used to accommodate the first-flush volume in a rain event which contributes to storage loss. However, in general the amount of rainwater collected is not found to be significantly affected by wind speed and direction, and therefore the sensitivity of water-saving efficiency (WSE) of rainwater collection sizing (RCS) models to depression storage loss is found to be minimal. Therefore, in calculating

optimum storage capacities, the depression storage loss was set to zero and the sensitivity of the RCS model is investigated using constant proportional losses or run-off coefficients.

Typically, the run-off coefficient C_f is estimated at 0.8–1.0 for clay, asbestos cement, and metal roof surfaces.

3.2.2 Pollutants on the Catchment Surface

Stored rainwater quality should relate to its end use. Some people, who depend on rainwater as their only source of water supply, use it for all household purposes, from drinking and cooking to washing and other domestic uses. Others, who have access to both rainwater and public water supply, use rainwater selectively, for gardening or flushing toilets, and use the public water supply for drinking and other purposes. These varying attitudes are related to the level of education of the uses as well as to their traditional preferences.

In general, the quality of roof run-off is acceptable to supply non-drinking quality domestic uses. Pollutant additions to roof run-off include organic matter, inert solids, fecal deposits from animals and birds, trace amounts of some metals, and even complex organic compounds (Forster 1991).

Factors such as the type of roof material, antecedent dry period (atmospheric deposition), and surrounding environmental conditions (proximity of strong pollution sources, such as motorways or industrial areas) have been shown to influence concentrations of heavy metals in roof run-off (Yaziz et al. 1989), (Forster 1991).

Roof material represents a problem when the roof itself consists of heavy metals, a situation that can be avoided by using an appropriate kind of roof material.

Recent research in roof water quality and health implications of using harvested rainwater have shown that exposure to UV, heat, and desiccation on the rooftop

Table 3.1 Summary of roof run-off water quality (Australian Drinking Water Guidelines, Australian Bureau of Statistics 2005).

	pH	BOD (mg/l)	COD (mg/l)	TOC (mg/l)	Turb (NTU)	TS (mg/l)	SS (mg/l)
Roof Run-off	5.2–7.9	7–24	44–120	6–13	10–56	60–379	3–281
Stored Run-off	6–8.2	3	6–151		1–23	33–421	0–19

Source: Australian Drinking Water Guidelines, Australian Bureau of Statistics 2005.

destroy many bacteria, while wind removes some heavy metals accumulated from atmospheric fallout (Spinks et al. 2003).

The results from a roof run-off quality investigation of four rainwater installations in Hamburg, Germany, reveals that levels of copper, lead, and zinc were well below the standard of drinking water of the World Health Organization (WHO) (Yaziz et al. 1989), (Forster 1991). A comparison of water quality from roof run-off and stored run-off obtained in Australia is given in Table 3.1 In this case the stored run-off is the reticulated water supply to households.

Studies of roof run-off water show that potential pathogenic microorganisms as well as other harmful minerals and heavy metals are present in minute scale only, and are well below standards set by authorities.

In the guidelines given by Nova Scotia Department of Health in Canada (1992) for "Health and Welfare Canada" (Vale and Vale 1995), "the quality of water draining from roofs and gutters of acceptable materials are well within the authorized limits." Similar studies have been carried out in Australia (Coombes 2002) and Germany (Hermann and Schmida 1999) that conclusively prove roof run-off does not pose a health risk. It is also highly unlikely that human waste or a significant amount of animal feces will contaminate the roof catchment (Vale and Vale 1995).

The quality of roof run-off is improved in a rainwater tank by several processes. Biofilms absorb heavy metals, organisms, and pathogens from water. Many bacteria conglomerate in a macro layer on the water surface, whereas many of the heavy metals and other contaminants precipitate out the water column, and settle at the bottom of the tank (Spinks et al. 2003) Water quality in rainwater tanks varies considerably from the water surface to the point of supply near the base of the tank. It is found that the quality of rainwater at the outlet point for delivery from a rainwater tank was significantly better than at the water surface (Coombes 2002).

3.3 Conveyance System

A conveyance system usually consists of gutters or pipes that deliver rainwater falling on rooftop to tanks or other storage vessels. These should be properly supported and sufficiently strong to carry and keep loaded water during the heaviest rain.

Gutters both intercept and transport roof run-off. Increasing the gutter's gradient allows its size and cost to be reduced but also may reduce the fraction of the run-off intercepted (Sendanayake 2016a, b, c, d). Water losses caused by occasional high-intensity rain overshooting gutters are generally acceptable in RWH since for most roofs the actual rainwater collection is taken as about 80% of the product of average rainfall depth and projected roof collection area. The sizes of the gutters depend upon the area of the roof and the rainfall amounts, and are typically in the range of 20–50 cm diameter. To prevent the loss of collection during high-intensity rain events, splash guards can be used.

As a rule of thumb, in humid climates, the gutter cross-section is taken as 1 cm^2 for every 1 m^2 of catchment surface with a roof coefficient of 0.9. Typically, gutters are installed with steeper gradient than 1 : 100 which would increase the water flow by 10–20% (Lanka Rainwater Harvesting Forum 2004).

It is important that the conveyance system to be constructed of chemically inert materials such as plastic, aluminum, or fiberglass in order to avoid adverse effect on water quality.

A conveyance system usually consists of gutters or pipes that deliver rainwater falling on rooftop to tanks or other storage vessels. These should be properly supported and sufficiently strong to carry and keep loaded water during the heaviest rain.

Common roof run-off conveyance materials include aluminum, polyvinylchloride (PVC) pipe, vinyl, and galvanized steel with round bottom gutters and round downspouts. Lead should not be used as gutter and downspout solder since rainwater being slightly acidic can dissolve the lead and contaminate the collected rainwater.

To ensure that debris, decomposing leaves, and other foreign materials do not enter the storage tank, filtering devices and first-flush devices are incorporated to the conveyance system.

Figure 3.1 A typical mesh filter.

3.3.1 Filtering Devices in RWH Systems

Filters are used to filter out the debris that comes with the rooftop water and prevent them being added to the storage tank. These are of two broad types:

(i) *Mesh Filters*
A wire mesh fixed at the mouth of or on the down pipe to prevent leaves and debris from entering the system. While preventing larger objects these filters alone are not sufficient to obtain a reasonable quality rainwater collection. Also mesh filters tend to corrode over time unless the wires are plastic coated. A typical mesh filter is shown in Figure 3.1.

(ii) *First-Flush (FF) devices*
A first-flush (FF) device is a valve that ensures the run-off from the earliest rains is flushed out and does not enter the system. The first flush of run-off water that occurs at the beginning of a storm event has been reported to contain a high proportion of the pollutant load (Fewkes 1999). The main cause of this phenomenon is the deposition and the accumulation of pollutant material to the roof during dry periods. The longer the dry period, the greater the probability of a higher pollutant load in the first flush. It is relatively straightforward to install a device for diverting the first flush away from the collection system (Forster 1991).

The sizing of the FF devices can follow a simple equation relating to the collection area and estimated pollution load on the roof.

$$\text{Flush Volume (L)} = \text{Roof Area (m}^2) \times \text{Pollution Factor} \times 100 \qquad (3.2)$$

Pollution factors are 0.0005, for nil to light pollution, and 0.001–0.002, for heavily polluted sites. This corresponds to 1–to 2 mm of initial rainfall (Zobrist et al. 2000).

As a rule of thumb, the first 1 mm rainfall on a catchment area is to be released through the FF device.

FF devices have a slow release valve which allows the captured water to slowly drain to the garden or storm water outlet and thereby empty and reset for the next rain event. The concept is to flush the contaminants from the roof and gutter into the device which then closes mechanically when full, allowing the remaining roof water to flow into the tank. The release of the FF water commences immediately and studies show that this release rate can be significant to the efficiency of the storage system (Miller et al. 2003). A typical first-flush device is shown in Figure 3.2a.

(a)

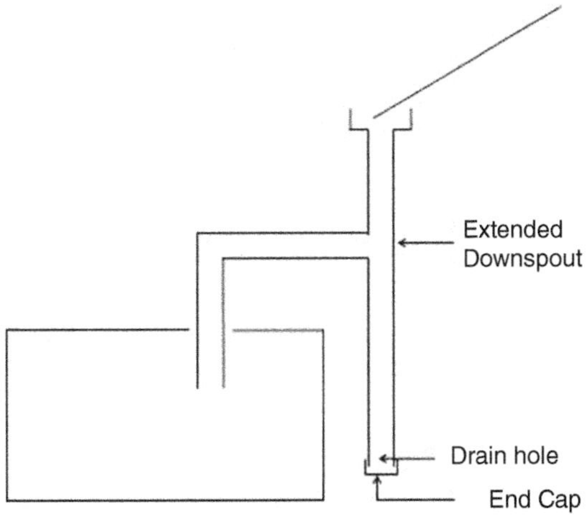

(b)

Figure 3.2 (a) A typical first flush device. (b) Schematic drawing for FF device.

3.4 Storage Tank

A storage tank or recharge tank can be stationed above ground, partly underground, or fully underground depending on the design and spatial arrangements and can be made of RCC, ferro-cement, masonry, plastic (polyethylene), or metal (galvanized iron) sheets. In general, the tanks should be made of water safe non-toxic material with screened openings and adequate foundation to support full tank. While the above-ground tanks are preferably opaque they should be UV and impact resistant. Underground tanks on the other hand should be fully accessible for entry to perform maintenance and repair while capable of supporting overlying soil or other loads. Underground tanks should be positioned a minimum 3 m away from the building foundations and should be strong enough to withstand soil pressure. If high-density polyethylene (HDPE) tanks are used, they should be placed in an RCC enclosure to prevent collapsing when empty. Also positioning of the underground tank in relation to the water table at site is an important consideration. In all tanks, backflow prevention through valve or air gap is essential if the system is hooked up to a municipal backup water supply.

All rainwater tank designs should include as a minimum requirement:

- A coarse inlet filter;
- An overflow pipe;
- A manhole, sump, and drain to facilitate cleaning;
- An extraction system that does not contaminate the water (a tap or a pump);
- Additional features might include:
  - A device to indicate the amount of water in the tank.

Of particular importance is that the overflow must be provided with capacity equal to or greater than inflow capacity and to ensure that it is sufficient to drain the tank while maintaining freeboard. Overflow must also be screened to prevent rodents/insects entering the tank.

3.4.1 Sizing of the Storage Tank

The storage device is the highest cost component of a typical RWH system and therefore its accurate sizing determines the cost of the overall system. The sizing of storage tanks is well covered in the RWH literature. There are a number of different methods used for sizing the tank, from the simple demand based sizing to computer models but sizing the storage for a given collection area (A), rainfall depth (R), and demand (D) using graphically presented correlations less sophisticated but accurate and practical solutions.

3.4.1.1 General Methods of Determining the Tank Capacities of RTRWHS

Two simple methods of determining tank capacities in a typical RTRWH system have been employed for general use. They are:

- Demand-Side Approach
- Supply-Side Approach

3.4.1.1.1 Demand Side Approach

This simple approach assumes sufficient rainfall and catchment area. Calculation of the required tank capacity is as follows:

If consumption of water per capita per day = C

Number of people per household = n

And the longest average dry period = t

Then, the daily consumption = Cn

$$\text{Storage requirement} = \text{Cnt} \tag{3.3}$$

3.4.1.1.2 Supply-Side Approach

In this approach a suitable catchment area with appropriate capture efficiency is determined to optimize the available tank capacity.

$$\text{Supply S (m}^3) = \text{Catchment Area (m}^2) \times \text{Rainfall (m)} \times \text{Run-off coefficient } (C_f) \tag{3.4}$$

If rooftop rainwater harvesting is to be practiced on a large scale, such as in a centralized water supply, or as a system catering to a particular need, for example using collected rainwater for WC flushing only or used as a supplementary system to the main supply, then a more scientific approach is needed to satisfy various parameters to obtain optimum sizes and maximum collection efficiencies. Such systems can be used to compare costs against conventional reticulated water supply systems and to determine cost and energy savings as well as beneficial ecological effects.

3.4.1.2 Sizing Based on Supply (Mass Balance Method or Rainfall Mass Curve Analysis)

This simple method helps determine the storage capacity by balancing the rainwater supply and demand for a specific catchment in a specific geographical location. For the calculation, first a bar graph for cumulative mean monthly roof run-off has to be plotted for the 12 months of the year, then on the same chart, the cumulative rainwater demand is plotted. The plot starts with the first month of the rainy season after a dry period. In the case of the dry zone of Sri Lanka, for example, starting month is October. The capacity of storage is calculated as the greatest excess volume of water over the cumulative water use at any time.

3.4.1.3 Sizing Based on Computer Models

Computer-based programs, developed incorporating the behavioral algorithm of a RWH system, can be used to determine tank sizes accurately for a given set of system parameters. Such models can predict the performance of a RWH system with fluctuating rainfall when long-term monthly rainfall figures are available for a given geographical location. The accuracy can be further increased if long-term daily rainfall data can be obtained which would be particularly important in areas where rainfall is more evenly distributed and more sensitive calculations are necessary. "Rain Cycle" software which allows modeling the tank volume through continuous daily water balance of supply and demand through the year and SimTanka (http://www.geocities.com/RainForest/Canopy/4805) are some examples.

"Rain Cycle" software can be used, which model the tank volume through a continuous daily water supply and demand throughout the year. An optimum volume is chosen when the increase in capacity does not represent significant gains in water collection. WSE of a RWH system can be calculated using the hydraulic computer model Rain Cycle. Rain Cycle is a deterministic mass-balance hydraulic system based on the Yield After Spillage (YAS) algorithm.

3.4.1.4 Sizing Based on Design Charts

For Sri Lanka, a country-specific set of graphs, numbering 23, called Design Charts for RWH, has been developed for corresponding locations to size rainwater tanks by Eng. Mansur in 1998, using over 120 years of monthly rainfall data. For a particular location, daily demands are plotted against plan roof area for a series of storage capacities, with the readings having 95% reliability. The roof coefficient is taken as 0.8 and the graphs can be used to estimate storage capacity for a given demand and catchment area. Graphs are available for locations where weather stations are situated, and therefore are not spatially independent. Sizing of a tank at a particular location therefore needs to refer to the graph for the nearest specific location.

3.4.2 Advanced Methods of Determining Optimum Tank Capacities of RTRWH Systems

Three general types of reservoir sizing models are identified (McMahon and Mein 1978), namely:

- Critical period model
- Moran model
- Behavioral model

3.4.2.1 Critical Period Model

This method identifies and uses sequences of flows where demand exceeds supply to determine the storage capacity. The sequence of flows or time series used in this method is usually derived from historical data. This method is an improved version of previously mentioned "demand-side approach" to determine tank capacities.

Temporal and spatial fluctuations of rainfall data, compounded by climate changes due to global warming, severely limit generalized use of the method over many locations. Further, it is apparent that high rainfall variations affect the overall system efficiency to a great extent.

3.4.2.2 Moran Model

Moran related methods are a development of theory of storage. A system of simultaneous equations is used with this method to relate to reservoir capacity, demand, and supply. The analysis is based upon queuing theory. Moran model also display similar limitations as discussed in Critical period model, affecting the overall system efficiency.

Therefore, a more advanced model, which can readily accommodate temporal and spatial fluctuations in rainfall, is required, and the resultant graphs developed depicting system efficiency can be used as a powerful design tool to determine optimum tank capacities.

3.4.2.3 Behavioral Models

Behavioral models simulate the operation of the reservoir with respect to time by routing simulated mass flows through an algorithm which describes the operation of the reservoir.

The operation of the rainwater collection will usually be simulated over a period of years. The input data, which is in time series form, are used to simulate the mass flow through the model and will be based upon a time interval of either a minute, hour, day, or month. A behavioral model has been used to simulate the performance of rainwater collectors and incorporate the spatial variations of rainfall into the model by using rainfall time series from five different locations and temporal fluctuations in rainfall by using two behavioral models each with different time intervals (Fewkes 1999).

3.4.3 Investigating the Performance of RTRWH System Using the Behavioral Model

Behavioral models have been used by other researchers (Latham and Schiller 1987), to investigate the performance of rainwater stores.

The generic configuration of a rainwater collection system is illustrated in Figure 3.3.

Where

R_t is Rainfall in time t;
D_t is the Demand (time t);

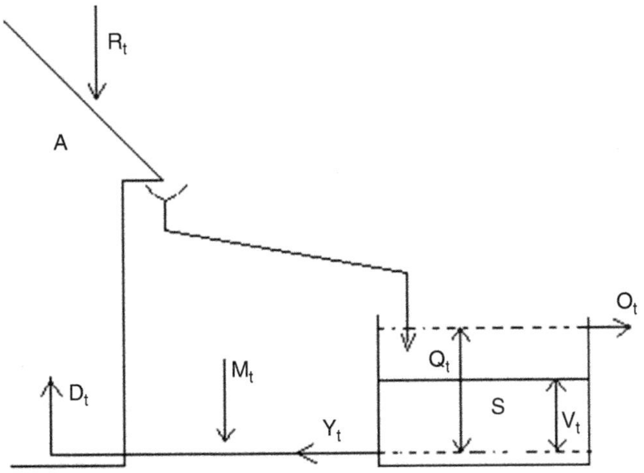

Figure 3.3 Generic configuration of a rainwater collection system.

Y_t is the Yield (time t);
A is the roof area;
S is the storage volume;
Q_t is the roof run-off (t);
O_t is the overflow.

The algorithm that is used to describe the behavioral model is the YAS (yield after spillage) operating rule.

3.4.3.1 Yield after Spillage (YAS) Operating Model

YAS operating rule is;

$$Y_t = \min\{D_t t; V_{t-1}\} \tag{3.5}$$

$$V_t = \min\{V_{t-1} + Q_t - Y_t; S - Y_t\} \tag{3.6}$$

Where

R_t is the rainfall (m) during time interval, t;
Q_t is the rainwater run-off (m^3) during time interval, t;
V_t is the volume in store (m^3) during time interval, t;
Y_t is the yield from store (m^3) during time interval, t;
D_t is the Demand (m^3) during time interval, t;
S is the Store capacity (m^3);
A is the roof area (m^2).

The YAS operating rule assigns the yield as either the volume of rainwater in storage from the preceding time interval or the demand in the current time interval, whichever is the smaller. The rainwater run-off in the current time interval is then added to the volume of rainwater in storage from the preceding time interval with any excess spilling via the overflow and then subtracts the yield.

3.4.3.2 Predicting the Performance of the RTRWH System Using the Behavioral Model

Using the YAS algorithm and a monthly time interval, the reliability or performance of the rainwater store can be expressed using either a time or volume basis. In either case, a reliability or performance of 100% indicates complete security in provision of service water.

The accuracy of behavioral models for the sizing of rainwater collection systems using both different time intervals and reservoir operating algorithms applied to a comprehensive range of operational conditions. The preliminary analysis of their study indicated that the hourly YAS model could be used as a standard of comparison against which other models could be compared and calibrated.

The YAS reservoir operating algorithm was found to give a conservative estimate of system performance irrespective of the model time interval and therefore is preferred for design purposes compared to the YBS (yield before spillage) operating algorithm.

Components of a rainwater collector sizing model are depicted in Figure 3.4.

In developing the system performance curves, two models were used to incorporate the temporal fluctuations of rainfall.

The first model uses a daily time interval, which ignores fluctuations with a time scale less than a day, to predict system performance for different combinations of roof area, demand, storage volume, and rainfall level. A set of curves is produced which enable the performance of rainwater collection systems to be predicted in different locations. The main limitation of this approach is the requirement of daily rainfall time series, which can be both costly and difficult to manipulate.

The second method of modeling uses a larger time interval of one month resulting in a more compact model and economic data set. However, the coarser monthly time interval does not take into account rainfall fluctuations with a time scale less than one month, which may result in an inaccurate prediction of system performance.

The poor resolution of the monthly interval model compared to the daily model is countered by the introduction of a parameter, referred to as the storage operating parameter. The short time scale fluctuations of the daily model are in effect replicated in the monthly model by the storage operating parameter. Values of the parameter are selected so that the monthly model mimics the system performance predicted by the corresponding model using a daily time interval. This approach provides a simple and versatile method of modeling the performance of rainwater collectors which takes into account temporal fluctuations in rainfall.

The performance of the rainwater collection system is described by its WSE (Fewkes 1999).

WSE is a measure of how much main water has been conserved in comparison to the overall demand, and is given

Figure 3.4 Components of a rainwater collector sizing model.

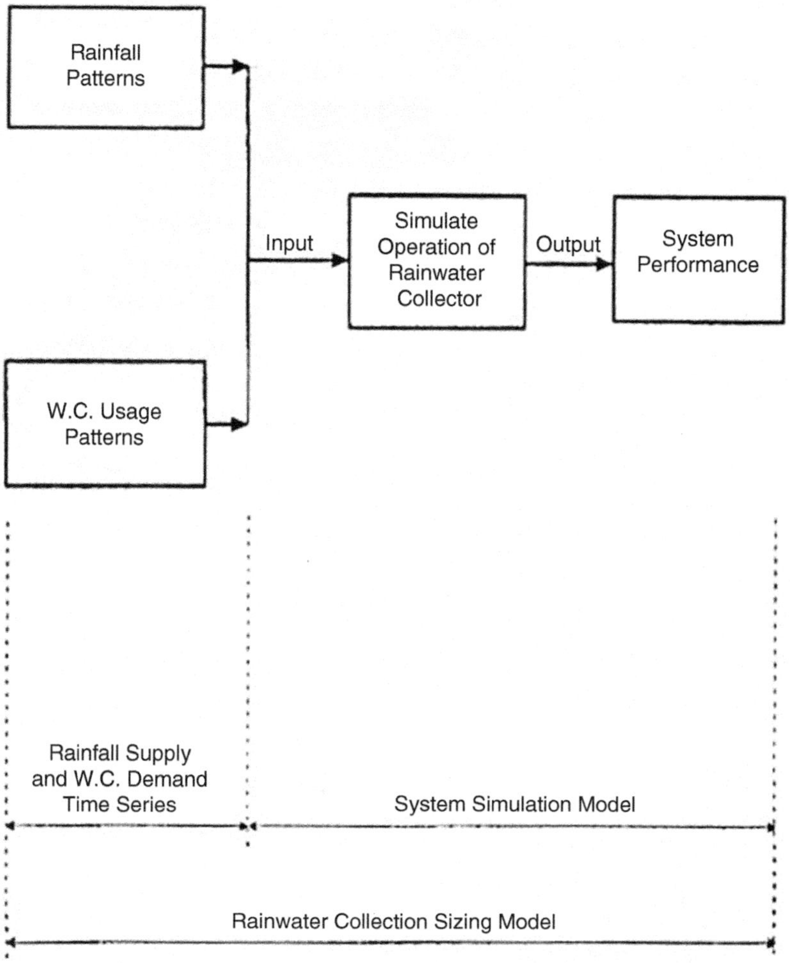

by:

$$WSE = \frac{\sum_{t=1-}^{t=T} Yt}{\sum_{t=1}^{t=T} Dt} x\,100\% \quad (3.7)$$

where, Y_t is the yield from storage facility (m^3) during time interval, t, D_t is the demand (m^3) during the time interval, t. T is the total time under consideration.

In the study conducted (Fewkes 1999), the demand component of the models was limited to WC usage which accounts for approximately 30% of potable household water usage in the UK and was assumed to occur at a constant daily or monthly rate. This assumption was reasonable because the demand time series generated by WC usage did not exhibit excessive daily or monthly variance.

In studies conducted elsewhere, including in Sri Lanka, it is reported that the service water usage is relatively constant and depends on the lifestyle of the users at a particular geographic location. However, if the demand from other domestic appliances such as washing machines was considered, the demand pattern would not be constant and the demand time series required.

The detailed analysis undertaken enabled constraints to be proposed for the application of hourly, daily, and monthly models expressed in terms of storage fraction. It was recommended that hourly models should be used for sizing small stores with a storage fraction below or equal to 0.01. Daily models can be applied to systems with storage fraction within the range 0.01–0.125. Monthly models were only recommended for use with storage fractions in excess of 0.125. Generally, daily models can be used to predict the performance of all stores except small stores with a storage fraction less than or equal to 0.01.

3.4.3.3 Generic Curves for System Performance of a RTRWH System

A generic set of curves using a YAS daily time interval model has been developed (Fewkes 1999), for a range of storage and demand fractions. Different combinations of roof area, store capacity, and demand were expressed in terms of two dimensionless ratios, namely the demand fraction and storage fraction.

The demand fraction is given by D/AR, where D is the annual demand (in m^3), A is the roof area (in m^2), and R is the annual rainfall (in m). The storage fraction is given by S/AR, where S is the store capacity (in m^3). The above fractions can be used to predict the performance of rainwater collectors within a particular geographical area. The performance of the rainwater collection system is described by its WSE (Dixon et al. 1999).

It was observed that the WSE curves at each demand fraction ratio for different sites are of close proximity to each other, suggesting system performance could be adequately represented by a set of average or generic curves. The average WSE of a rainwater collector at demand fractions of 0.25, 0.50, 0.75, 1.00, 1.25, 1.50, 1.75, and 2.00 each with a storage fraction range of 0.005–0.40 is illustrated in Figure 3.5.

Three important factors are considered in developing the generic curves for their effects on performance of the curves. They are:

- Effect of demand pattern
- Effect of roof run-off coefficient (C_f)
- Variation in rainfall data

The generic curves of WSE were plotted against different storage fractions (S/AR) for a given demand fraction (D/AR). In doing so, demand is assumed to be a constant for a particular situation and in the case of WC flushing appears to hold true. It was observed that for a period of 12 months, that WC flushing water demand remained at a fairly consistent level from day to day (Fewkes 1999).

Demand patterns which exhibit significant daily variance will require more precise modeling. Therefore, the generic curves for WSE can be fairly accurately used where demand can be assumed to be a constant.

Rainfall loss during collection occurs due to absorption by the roofing material and wind effects around the roof. The rainfall loss was modeled using an initial depression storage loss (L) with a run-off coefficient (C_f) (Fewkes 1999). However, as the wind speed, etc., does not significantly impact the effective roof collection the initial depression loss can be made equal to zero and the roof collection coefficient can be taken as in Eq. (3.1).

The generic curves for WSE were developed for a particular set of rainfall data. The model was simulated with rainfall data collected in five sites where average annual rainfall varies from 620 to 1600 mm/year (Forster 1991). The performance curves predicted for each site were found to be close together, almost coalescing into a single curve. The modeled performance of rainwater collectors at various demand fractions, except when D/AR is closer to 1.00 when slight sensitivity is shown, appears therefore to be relatively insensitive to fluctuations in daily rainfall patterns experienced at each location. Therefore, for the given rainfall range, the generic curves developed can be adequately used to suggest system performance for given demand and storage fractions.

3.4.3.4 Sample Calculation for Sizing Storage of a RWH System

It is observed that the harvested rainwater can be utilized for WC flushing and cleaning purposes where the amount of water used is approximately 40% of the total water usage. However, such requirements need the delivery of collected rainwater to utility points to be at a sufficient pressure to be used at any given time. One possible energy-efficient arrangement is to position the storage tank at an elevation near the capture area (at roof level) so that the collected water can be fed to utility points through gravity. However, when the tank size increases, the space and strength requirements to support the tank will be beyond the meaningful utilization of harvested rainwater. Further, due to limited availability of ground space in urban multi-story buildings, positioning of a larger storage tank above ground will not be feasible and the entire quantity of harvested rainwater will have to be pumped up to utility points.

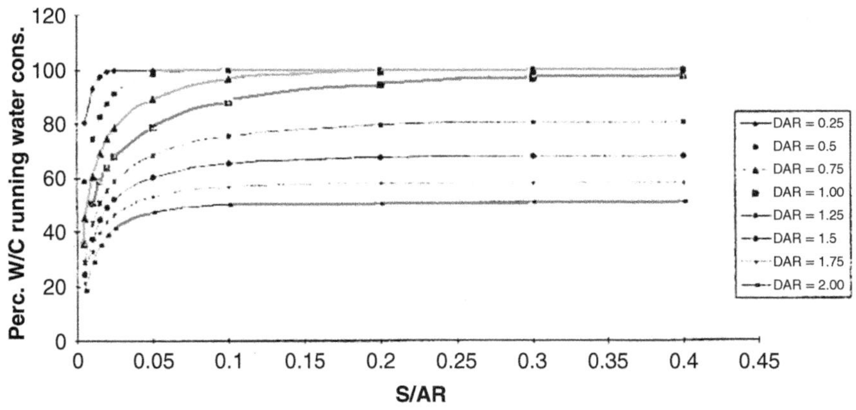

Figure 3.5 Generic curves for Water Saving Efficiency (WSE) (Fewkes 1999).

Therefore, typical sizes of storage tanks will have to be studied to make the model more practical.

Considering a typical household in the wet zone of Sri Lanka, where the annual rainfall is the highest (1500–6000 mm), with a capture area of 50 m^2, the daily water usage for four occupants can be taken as 800 l (at per capita demand of 200 l).

If harvested rainwater is utilized only for WC flushing and cleaning,

Then the demand for harvested rainwater is 800 40% = 320 l/day (116.8 m^3/year),

As the minimum annual rainfall in the wet zone, $R_{min\text{-}wet}$ = 1500 mm,

The value for D/AR can be calculated as D/AR = 1.56.

(It should be noted that the minimum rainfall values are selected as a safety factor for performance reliability.)

From the WSE curves (Figure 3.5), the maximum possible WSE that can be achieved is found to be 65% and the corresponding value for S/AR = 0.15, giving an optimum storage size (S) of 11.25 m^3. Even when the capture area is doubled (100 m^2), it would still give a value of 1.5 m^3 as the storage capacity for the same WSE of 65%. If, however, a WSE of 95% is desired, then the optimum storage capacity (S) will be 15 m^3. Therefore, if a reasonably high and economically acceptable WSE is to be employed (typically over 80%), then a higher value for the optimum tank size (S) is to be expected. Moreover, as the minimum annual rainfall figure (R_{min}) tends to be smaller for the intermediate and dry zones, higher tank capacities are required if the WSE to be achieved above 80%.

It can be observed that in order to provide running water facilities, the storage tank has to be placed at a higher elevation – which is not feasible due to volumes concerned. While such bigger tanks can be accommodated in rural single-story houses with abundant ground space, urban multistory houses which need running water will need a different model to use RWH effectively and meaningfully.

It has been shown that the generic curves for WSE can be used to determine the optimum storage capacities for a given demand and for a desired WSE (Fewkes 1999). The curves are validated for Sri Lanka by Sendanayake et al. (2014). These minimum annual rainfall figures defining the boundary of the domain in which the generalized curves hold true are below the minimum annual rainfall figures in the dry zone of Sri Lanka. As such, the curves given in Figure 3.5 can be used for RWH model system sizing in any region of the country and can be accepted as universal within Sri Lanka. However, as the sizing applications move toward drier regions, unless the capture area is significantly increased D/AR tends to increase, thus falling into regions of lower WSE of the curves. To maximize the WSE for the given D/AR value, S/AR values will have to be chosen beyond the 0.15 range, indicating bigger storage tanks. A similar scenario can be seen when the demand (D) for harvested rainwater increases, even in the wet zone.

3.4.3.5 Use of Reference Maps to Find the Effective Combinations of Roof Area and Storage Capacity

The main parameters affecting the performance of Roof Rainwater Harvesting (RRWH) systems, other than the storage capacity, are the roof catchment area, collection efficiency, rainfall depth, and the daily demand for service water; many methods based on simulation, graphical, mass balance, and statistical models have been developed to determine the supply reliability or the WSE, defined as the percentage yield for a given demand. However, one major drawback for the developers and engineers who are involved in the design stage of RTRWH systems is the dearth of accurate rainfall data for a given location, and if the most effective combinations of roof area and storage capacity can be graphically presented for a series of WSE values for the main climatic regions with known annual rainfall depths, it would be of significant use. For WSE values of 80%, 85%, 90%, and 95% for the given constant daily demand of 400 l in selected locations, a series of graphs are developed relating the roof catchment areas and the corresponding storage capacities (Figure 3.6a). For an example using the graphs, a map is developed for the easy reference of optimum storage capacity required for typical roof catchment areas of 200 m^2 and daily demands of 400 l in an average household in Sri Lanka (Figure 3.6b) (Sendanayake et al. 2016a, b, c, d).

3.4.4 Positioning of the Storage Tank

Various methods of positioning bigger storage tanks, which can be used to provide running water to utility points and the corresponding plumbing configurations possible for typical households, having RTRWH systems supplementing the service water provided by either main supply or from a well/bore hole, are presented below. The practical water supply situations in both single and two-story household situations are looked at in five scenarios:

a) *The storage tank at ground level, and draw-off through pressure-operated pump (PP)*
   Collected rainwater is fed to a separate pipeline, feeding WC end user points, at a higher pressure than the mains. A level sensor operates the pressure pump, to prevent the pump running dry. The system can be used in multi-story situations, but no energy saving is possible. A 5000 l tank connected to a roof area of a minimum 45 m^2 is recommended. A schematic diagram is shown in Figure 3.7.

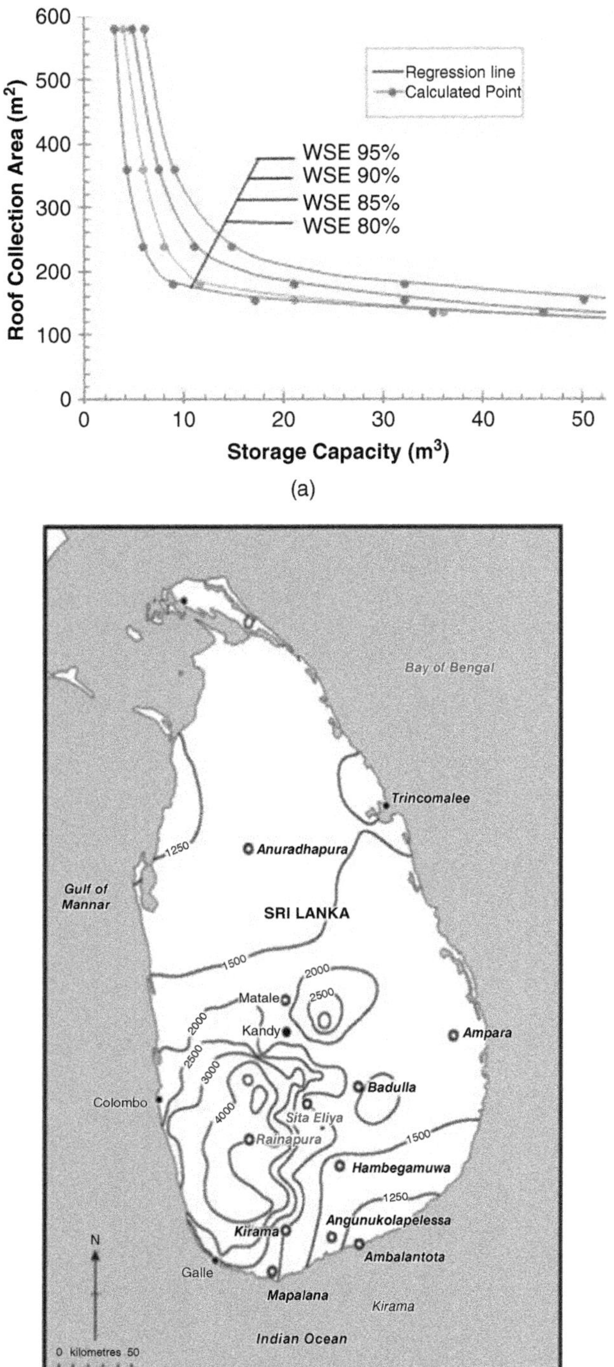

Figure 3.6 (a) Sizing curves for annual average rainfall of 1250 mm (Sendanayake 2016a, b, c, d). (b) Annual average rainfall map for Sri Lanka (Sendanayake 2016a, b, c, d).

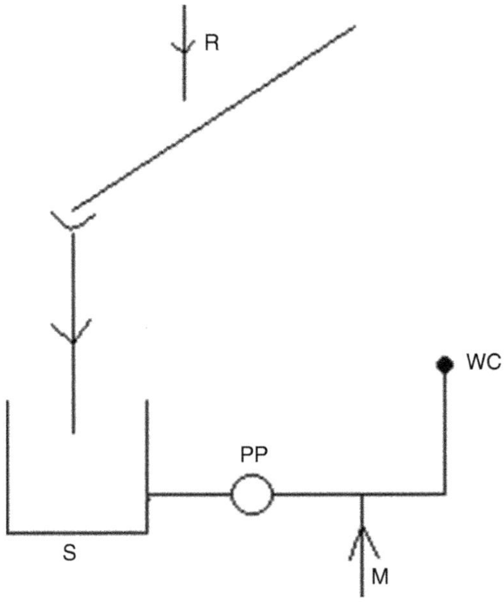

Figure 3.7 Plumbing configuration for RTRWHS – Scenario (a).

b) *The storage tank mounted on eave of multi-story house*
   Rainwater is supplied through gravity; hence no energy consumption occurs. However, supply of water to upper stories is not possible due to lack of head. Since the tank is mounted on the eave, space restrictions could occur. Also, a strength analysis of the eave for its load-bearing capacity is required.
   It should be noted that if the capture area is >200 m², a smaller tank of 2000 l can be utilized, so that the eave can support the additional weight since the tank size is smaller compared to that for a smaller capture area. A schematic diagram is shown in Figure 3.8.
c) *Rainwater pumped from storage facility to an Overhead Tank*
   In this situation an extra energy input is required to pump collected rainwater to the OHT (overhead tank). Therefore, the overall system efficiency could be low. A level sensor to operate the pump P1 fixed in the OHT could improve the efficiency in water saving. This system is suitable for locations where ground water levels drop seasonally. A 5000 l capacity tank connected to a filtering system in between the rainwater tank and the OHT is recommended. A schematic diagram is shown in Figure 3.9.
d) *Rainwater collected in split sump*
   To mitigate the unreliability of mains water supply, many households utilize underground sumps. By partitioning the sump so that one part receives roof collection while the other part receives the mains supply, savings can be made on service water. A 5000 l capacity tank connected to a minimum roof area of 45 m² is recommended for WC flushing water requirement.

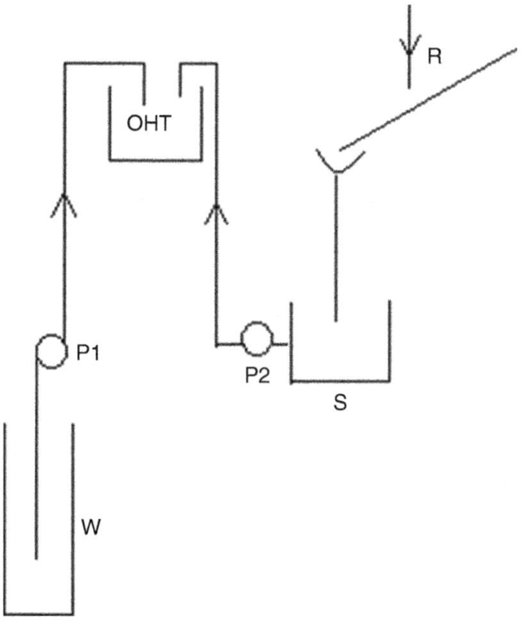

Figure 3.8 Plumbing configuration for RTRWHS – Scenario (b).

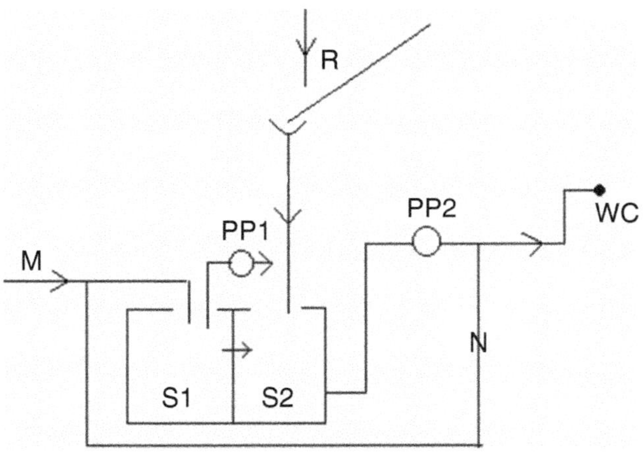

Figure 3.9 Plumbing configuration for RTRWHS – Scenario (c).

e) Rainwater collected in sump with draw-off through filtration
Employing a series of filters such as carbon and sediment filters and a UV sterilizer, drinking quality water can be obtained from the collected rainwater. It can be envisaged that, by selecting suitable storage capacities and collection surfaces, substantial water-saving efficiencies can be achieved. A 10 000 l tank connected to a minimum roof area of 200 m^2 is recommended for this configuration. However, a higher capacity tank will ensure water security even in prolonged drought situations. A schematic diagram is shown in Figure 3.10.
Except in scenario (b), in all other scenarios the requirement of a pump to provide the harvested rainwater

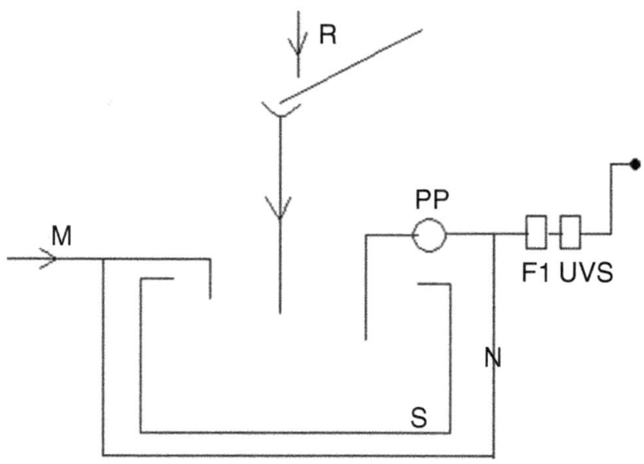

Figure 3.10 Plumbing configuration for RTRWHS – Scenario (d).

either to an overhead tank or directly to the utility points can be observed. Such arrangements while preserving water utilizes energy to transfer the entire quantity of collected rainwater and as such cannot be considered as energy efficient or as promoting the principles of sustainable development for built environments.

3.4.5 Cascading Multi Tank Model

In the following paragraphs a RWH model is introduced with the new concept of decentralizing the storage capacity where the roof collection cascades down through storage tanks located at different floor levels (Sendanayake and Jayasinghe 2016a, b).

In any RWH situation, the storage tank has to be placed at a lower elevation than the collection area, thereby facilitating the flow of collected rainwater into the tank under gravity. However, the retention volume required for improved WSE levels poses a problem in space requirements in built-up areas, besides the bigger problem of pumping back the harvested rainwater into service points for the system to be on par with the centralized systems as far as the user convenience is concerned. Such a system will negate the positive contribution of RWH on sustainability principles by consuming energy in pumping. In order to minimize the energy requirement in transferring collected rainwater, a cascading multi tank rain water harvesting (CMTRWH) model is proposed and analyzed as shown in Figure 3.11.

In the model, a number of smaller volume tanks are positioned at each floor level, with the top-most storage tank just below the collection area, and a bigger volume tank at ground level. Rainwater is fed first to the upper tank, the overflow of which will cascade down to the lower tanks,

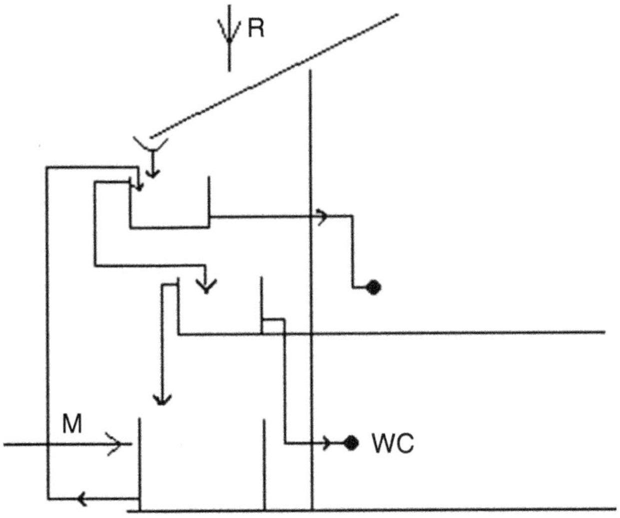

Figure 3.11 CMTRWH system for a two-story house.

finally ending up in the parent tank at ground level. Supply to each floor is from individual smaller capacity tanks by gravity floor and make-up water is pumped from the parent tank to the top-most storage tank as and when required. Essentially the concept of CMTRWH model attempts to distribute the storage capacity of the RWH system at various floor levels so that the requirement for pumping is minimized for the same or marginally improved overall WSE.

In developing an algorithm for the operation of a CMTRWH system, the following are assumed to be valid:

- The height differences between each floor level are a constant.
- The water usage at any given floor level remains constant for a given set of operating parameters.
- No loss of water occurring in system operation, i.e. in cascading down or pumping up of collected rainwater.
- All tanks installed at floor levels other than the ground level are taken as equal capacity.

Advantages and limitations of CMTRWH systems:
Advantages are the following:

- Possibility of gravity feeding the total usage to service points.
- If pumping is required for higher demands, the reduced energy utilization.
- Lower spatial and strength demand on the building structure.
- Reduced adverse impact on the aesthetic appearance of the building envelops.

Limitations are the following:

- Reduced supply pressure at user points.
- Requirement of additional storage tanks for upper floor levels.

In detached domestic dwellings of diffuse settings, rainwater is harvested off the roof and if it is to be supplied to service points under gravity, a structure is needed to elevate the storage tank. The alternative is to position the storage tank at or below ground level and to pump the collected rainwater to a header tank from which allowing gravity feed to service points. Recent research has found that by distributing the storage capacity vertically, where the main storage tank is at or below ground level and a smaller capacity tank feeding the service points is at an elevated position receiving the roof run-off, with the excess cascading down to the main storage, a significant savings on pumping energy is possible (Sendanayake and Jayasinghe 2016a, b). It is noted that the distributed capacity system, identified as the cascading multi-tank system (Sendanayake and Jayasinghe 2016a, b) requires a minimum of 1 m^3 capacity feeder tank to maintain reasonable supply reliability, whereas the simpler header tank system can have smaller capacity feeder storage comparatively. However, in typical housing units, feeder tanks of both types of systems can be assumed to be positioned below the roof without additional structural strengthening (Figures 3.12 and 3.13).

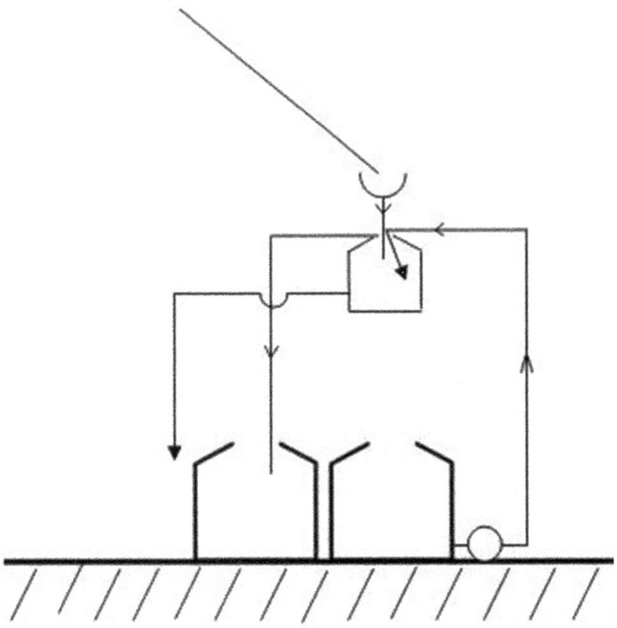

Figure 3.12 Feeder tank for Cascading Multi Tank RWH System.

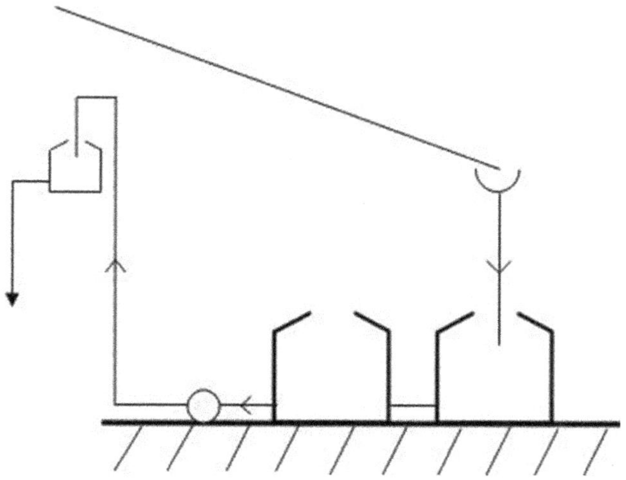

Figure 3.13 Feeder tank to supply harvested RW under gravity.

3.4.6 Tank Materials and Life Cycle Energy (LCE) of Tanks

Many types of materials are used in storage tank construction, of which the most common are RCC, fiberglass, HDPE, ferro-cement, and aluminized steel.

While RCC and aluminized tanks can be used either in above- or below-ground scenarios, HDPE and fiberglass tanks, which are not strong enough to bear the soil pressure when empty in below-ground positioning, are recommended only for above-ground installations. Ferro-cement tanks on the other hand, though cheaper in construction compared to RCC tanks, have a relatively shorter lifespan with the potential to crack and leak, particularly when open to roots of nearby vegetation, hence they are more suitable for above ground installations.

Another method of comparing the suitability of storage tanks for RTRWH systems in the light of sustainability is the life cycle analysis (LCA). The life cycle energy (LCE) of each type of tank can be compared to ascertain the tank which will have the durability as well as the potential to keep in line with the sustainable principles.

LCE is a tool to quantify the embodied energy in a system or product from construction, through use to disposal, during its useful lifetime. In the process, it is important to normalize the calculated values for the differences in durability of the components of the product or system. For this purpose, a functional unit is defined for the study as that of 1 m^3 of rainwater used per capita per year. It has been found that the LCE of RCC tanks are comparatively lower than tanks made of other commonly available materials such as HDPE, fiberglass, and ferro-cement (Sendanayake 2016a, b, c, d). Significantly, RCC compares favorably in life cycle cost (LCC) calculations as well, mainly due to its longer life span of 50 years or more (Sendanayake 2016a, b, c, d).

3.5 Pre-treatment of Roof Collection

Rainwater in its original form is devoid of impurities, hardness, and turbidity, but gets contaminated depending on the cleanliness of the catchment surface and the local air quality. Due to occasional bird droppings and animal excreta, usually the harvested rainwater is known to contain bacterial pathogens such as coliforms but unless the surrounding air is highly polluted with industry and vehicle emissions there is no likelihood of Cd, Pb, and As contamination. It is also reported that fluoride content in harvested rainwater is well below the allowable safe limit (Australian Bureau of Statistics 2005) while pH value is around 5.9 indicating a slight acidity. However, if RCC tanks are used for storing of rainwater, the pH value is found to be increased due to calcium leaching from cement paste in concrete, thus improving the taste (Vale and Vale 1995). The high pH value however will decrease with time when the tank gets flushed off several times during the course of maintenance.

While it is required to clean the catchment surface, guttering, and storage tank to minimize the dissolved organic content (DOC) in the harvested water, most of the contaminants can be removed by utilizing a first-flush (FF) device where the initial collection corresponding to the first 1 mm of rain after a dry spell is collected into an extension of the downspout leading to the tank (Sendanayake 2016a, b, c, d). This should be in addition to preventive measures such as gutter screening to minimize biological materials such as leaves entering the tank increasing the DOC. Further, it is recommended that the draw-off be located so that there is no disturbance of the sediments at the bottom of the tank.

It is reported that contaminants up to 95% can be removed by sending the draw-off into a slow sand filter of 1 m depth (Sendanayake 2016a, b, c, d) and the rest of the bacterial pathogens destroyed by boiling or UV sterilization. For rainwater systems in general, it is recommended that the draw-off goes through a series of filters starting with 50-μm sediment filters, going down in size to 10 or 5 μm filters, followed by a carbon filter after appropriate chemical disinfection. However, in rural areas, considering the high cost of filters it is advisable to send the draw-off through a clean cloth to remove most of the suspended particles. Besides, finer filters may not suit the situation as the water pressure supplied through gravity from a tank located at the roof level is low. The same can be said about RO systems, which need higher pressure besides being expensive. In addition to the inherent problem of RO systems, the production of waste water may complicate the supply issues, taking into account the water scarcity in the region. It is also reported that the RO method does not remove more than 10% of ionic impurities, including ionized fluoride and calcium compounds. Typically, disinfection such as chlorination, ozonization, or UV light is used to destroy microorganisms. Though chlorination is recommended where doses of 10 g/1000 l to be added to rainwater so that the amount of free chlorine exceeds 0.4–0.5 mg/l (Sendanayake 2016a, b, c, d), it has also been revealed that chlorine reacting with DOCs and producing by products such as tri halo methanes (THMs) that could be carcinogenic. As the required chlorine dosage will depend on the quantity of water to be treated, pH, and the temperature, it is not recommended as a practice.

If UV sterilization is to be used, despite its cost, it is important that water has to be fed through a filtration mechanism to avoid bacterial pathogens casting shadows in the flowing water, allowing live organisms to pass through unharmed.

3.6 Distribution System and Related Regulations

RWH is highly encouraged primarily as a measure of water security in already water-stressed regions, as well as in areas where potential scarcity is anticipated. The main feature of RWH is the availability of the source of water at the same location of use in a decentralized setting where the user wields authority over both supply and demand. Many countries have introduced legislation therefore to monitor and regulate issues related to decentralized use of water, particularly health risks posed by the consumption of harvested rainwater.

Quality of harvested rainwater depends on the cleanliness of the catchment and the local air quality. Therefore, bacterial pathogens from animal and bird droppings are common, and if the catchment is in a built-up area, heavy metals such as lead (Pb) from vehicle emissions can also be found in collected rainwater, thus posing a grave risk to people's health. Besides, rainwater is naturally acidic with a low pH value, thus of a corrosive nature. Pb can also enter the harvested rainwater from solder joints of metal guttering and higher-than acceptable limit zinc (Zn) contents are possible from galvanized iron (GI) roofs. Even though simpler methods, such as discarding the first 1 mm of rain after a dry spell using a first-flush (FF) device, is reported to remove 90% of the pathogens and other contaminants, harvested rainwater still cannot be considered safe for potable use as per World Health Organization (WHO) standards. Filtering methods such as slow sand filtering, disinfection through chlorination, activated carbon filtering, UV sterilization, and pasteurizing can bring the collected rainwater to be on par with treated reticulated supply. Still, without a central authority monitoring the operations and therefore quality, the risk persists.

Although in many situations, this aspect is not sufficiently covered in relevant legislation, several countries do not allow harvested rainwater to be used for potable uses. Ensuring water quality standards and public health concerns are met, provisions exist to make sure that harvested rainwater does not mix with the potable water supply by way of a check valve. However, the lack of a mechanism that this requirement is adhered to is noticeable, jeopardizing the public health. In France for example, a home owner is obliged to inform authorities when using water other than that provided by a public service. Further, when a household is supplied by mains water as well as rainwater, a disconnector or air gap between the two supply inlets is mandatory and supplying a home simultaneously with public water and stored rainwater is prohibited. In the United States of America (USA), most states allow RWH only for non-potable use. The exceptions are the US Virgin Islands, and the states of Ohio and Texas, who allow the use of rainwater for potable uses but only after informing the authorities, so that the state public supply is not held liable in case of a health issue arising from the consumption of rainwater. In the state of Ohio, the Ohio Department of Health regulates private RWH systems. Many states in Australia actively promote RWH at individual dwellings, but for non-potable use such as toilet flushing and garden watering. In India on the other hand, though RWH is made mandatory for new constructions, in many states no regulations are imposed on whether collected rainwater is to be used exclusively for non-potable purposes.

The distribution system usually includes an appropriately sized pump producing sufficient pressure for all intended end user points, while all pipes marked/color coded for potable or non- potable use.

3.7 Conclusion

Rainwater harvesting is perceived as an acceptable method of supplementing the reticulated service water supply, particularly in water-stressed locations. Mostly used as standalone types for the provision of service water for secondary purposes such as laundry, the biggest drawback in the RWH system is its rainwater storage tank occupying a considerable space depending on the daily demand for water, roof collection area, and the average rainfall depth. The storage tank is also the main cost component of the system and attempts have been made to optimize the storage tank capacity for a desired WSE with the aim of reducing the size of the tank, which would also allow the tank to be placed just below the roof level to feed the service points through gravity. However, due to practical and aesthetic considerations, in many standalone RTRWH systems the storage tanks are placed at various other positions, including below the ground. Once the optimum sizing of the tank is determined and the positioning with respect to the building is secured, the catchment surface and the effective area required can be determined and a suitable conveyance system and distribution system can be arranged. In essence, in all the above steps, adherence to specified design standards is important for the optimum and effective advantage to the end user. Therefore, in this chapter attention has been made to the selection of catchment surface, calculation of the effective rainfall collection after allowing for rainwater losses, pre-treatment of rainwater, and standard sizing of the guttering and downspouts for minimization of losses. Attention has also been focused on catchment surface material and possible contamination from external sources.

Design standards for the positioning of the storage tank were discussed with the outlining of sizing methods to minimize the cost and space requirements. Quality issues in using harvested rainwater, post treatment, and maintenance have also been touched upon in discussing the distribution of collected rainwater.

References

Australian Bureau of Statistics; Water Account Australia, (2005) Australian Government.

Brown, R. (2009). Practitioner perceptions of social and institutional barriers to advancing a diverse water source approach in Australia. *Water Resources Development* 25 (1): 15–28.

Butler, D. (1993). The influence of dwelling occupancy and day of the week on domesic appliance eastewater discharges. *Building and Environment* 28 (1): 73–90.

Coombes, P. (2002) "Rainwater tanks revisited. New opportunities for urban water cycle management." PhD Thesis, Department of Civil Engineering, University of New Castle, Australia.

Cunliffe, D., (2004) "Guidance on use of rainwater tanks," National Environmental Health Forum Monograph", Australia Health Council, Water Series No. 3, pp 29.

Dixon, A., Butler, D., and Fewkes, A. (1999). Computer simulation of domestic water reuse systems: investigating greywater and rainwater in combination. *Water Science and Technology* 21 (2): 99–106.

Edwards, D., Keller, K. (1984) "A workshop design for rainwater catchment systems (Training Guide) & appendix; "Rainwater Harvesting for Domestic Water Supply in Developing Countries," La Paz, UNESCO/ROSTLAC, UNICEF.

Fewkes, A. (1999). Modeling the performance of rainwater collection systems towards a generalized approach. *Urban Water* 1: 323–333.

Forster, J. (1991). Roof run-off pollution. In: *Proceedings of Second European Junior Scientist Workshop* (eds. M. Grottker and W. Schilling), 145–158. Kastienbaum, Switzerland.

Ghisi, E., Lapolli Bressan, D., and Martini, M. (2007). Rainwater tank capacity and potential for potable water savings by using rainwater in the residential sector of Southeastern Brazil. *Building and Environment* 42: 1654–1666.

Gould, J. and Niesson-Peterson, E. (1999). *Rainwater Catchment Systems for Domestic Supply; Design, Construction and Implementation*. London, U.K.: Intermediate Technology.

Hamdy, A. and Eslamian, S. (2015). Sustainable reuse and recycling of treated urban wastewater. In: *Urban Water Reuse Handbook*, Ch. 80 (ed. S. Eslamian), 1039–1054. Taylor and Francis, CRC Group.

Hermann, T. and Schmida, U. (1999). Rainwater utilization in Germany: efficiency, dimensioning, hydraulic and environmental aspects. *Urban Water* 1: 307–316.

Lanka Rainwater Harvesting Forum (2004) "Rainwater Harvesting in urban areas." Pilot Project on Rainwater Harvesting Technology for Urban Households, GTZ Project.

Latham, B. and Schiller, E.J. (1987). A comparison of commonly used hydrologic design methods for rainwater collectors. *Water Resources Development* 3: 165–170.

McMahon, T.A. and Mein, R.G. (1978). *Reservoir Capacity and Yield, Developments in Water Science*. Amsterdam: Elsvier.

Miller, G., Yu, B., Gardner, T. (2003) "Rainfall catch efficiency for domestic water supply," 28th International Hydrology and water resources Symposium, Wollongong, NSW, Australia, Institution of Engineers, Australia.

Roebuck, R.M. and Ashley, R.M. (2006). Predicting the hydraulic and lifecycle cost performance of rainwater harvesting systems using a computer based modeling tool. In: *Proceeding if the 4th International Conference on Water Sensitive Urban Design*, Apr. 2-7, 699–706. Melbourne.

Sendanayake, S. (2016a). *Rainwater Harvesting for Urban Living*. South Asian Institute of Technology and Medicine. ISBN: 978-955-43389-0-6.

Sendanayake, S. (2016b). Life cycle analysis of ferro-cement rainwater tanks in Sri Lanka: a comparison with RCC and

HDPE tanks. *International Journal of Advances in Engineering Research* 12 (2): 30–42.

Sendanayake, S. (2016c). Life cycle analysis of sub-surface brick-dome rainwater tanks in Sri Lanka: a comparison with below ground RCC tanks. *International Journal of Scientific Research and Innovative Technology* 3 (5): 92–101.

Sendanayake, S. (2016d). Optimum storage capacities for rooftop rainwater harvesting systems in Sri Lanka; development of a reference map. *Imperial Journal of Interdisciplinary Research (IJIR)* 2 (10): 1722–1727.

Sendanayake, S. and Jayasinghe, M.T.R. (2016a). Multi tank model for energy efficient rainwater harvesting. *Ideal Journal of Engineering and Applied Sciences* 2 (2): 52–55.

Sendanayake, S. and Jayasinghe, M.T.R. (2016b). Cost effective cascading multi-tank rainwater harvesting systems for multi-storey buildings. *International Journal of Advances in Engineering Research* 11 (5): 49–55.

Sendanayake, S., Miguntanna, N.P., and Jayasinghe, M.T.R. (2014). Validation of design methodology for rainwater harvesting for tropical climates. *Asian Journal of Water, Environment and Pollution* 11 (1): 87–93.

Spinks, A.T., Dunstan, R.H., Coombes, P..Kuczera, G. A. (2003) "Water quality. treatment processes and health implications of using harvested rainwater in the urban environment." In: 28th Hydrology and Water Resources symposium, Institution of Civil Engineers, Wollongong, Australia,

Thomas TH, Martinson DB (2007) Roofwater harvesting: a handbook for practitioners; Technical paper series, no 49, IRC International Water and Sanitation Center Delft, The Netherlands, 160 pp.

Vale, R., Vale, B. (1995) "Water and sewage treatment in the autonomous house"; Plumbing magazine,; NSW, Australia, p. 17–19.

Yaziz, M.I., Gunting, H., Sapari, N., and Ghazali, A.W. (1989). Variations in rainwater quality from roof catchments. *Water Research* 23: 761–365.

Zhu, K., Zhang, L., Hart, W. et al. (2004). Quality issues in harvested rainwater in arid and semi arid loess plateau of Nothern China. *Journal of Arid Environments* 57 (4): 487–505.

Zobrist, J., Muller, S.R., Amman, A. et al. (2000). Quality of roof run-off for ground water infiltration. *Water Resources* 34 (5): 1455–1462.

4

Water Security Using Rainwater Harvesting

Adebayo Eludoyin[1], Oyenike Eludoyin[2], Tanimola Martins[3], Mayowa Oyinloye[1], and Saeid Eslamian[4]

[1] *Department of Geography, Obafemi Awolowo University, Ile-Ife, Nigeria*
[2] *Department of Geography and Planning Sciences, Adekunle Ajasin University, Akungba-Akoko, Ondo State, Nigeria*
[3] *Medical School, University of Exeter,, UK*
[4] *Department of Water Engineering, College of Agriculture, Isfahan University of Technology, Isfahan, Iran*

4.1 Introduction

The Institute for Water, Environment & Health of the United Nations (UNU-IWEH 2013) defined water security as the capacity of a population to safeguard sustainable access to adequate quantities of acceptable quality water for: (i) sustaining livelihoods, human well-being, and socioeconomic development; (ii) ensuring protection against water-borne pollution and water-related disasters; and (iii) for preserving ecosystems in a climate of peace and political stability. Water security is important for food and energy security, and it is considered a national goal in many countries (He et al. 2014; Hoekstra et al. 2018). An important way of meeting the challenge of water security is sustainable water supply. Water supply is, however, limited, hence many countries declared various security targets, given their level of water security types and level (UNEP/GRID-Arendal 2009).

Water scarcity is a well-known problem in West Asian and North African regions, where critical levels of scarcity and water-related issues have been reported (Oweis and Hachum 2006). Figure 4.1 shows an illustration of concerns for food security in parts of Africa (Rockström and Falkenmark 2015) owing to water scarcity, which is likely to worsen due to issues relating to climate extremes. Whereas arid regions on the continent are emphasized in Figure 4.1, problems with water availability are common; and in areas that are not impacted by aridity, poor water technology and policy are challenges that hinder adequate freshwater supply. Freshwater scarcity is also prominent in many rainfed tropical regions of South America and Africa, mainly due to low water storage capacity, low infiltration, larger interannual and annual fluctuations of precipitation, and high evaporation demand (Sivanappan 2006; Roa-García 2009). It is therefore important that different means of water conservation and sustainable water use techniques be explored to meet the challenges of water scarcity, and inform future development of policies on water sustainability.

In this chapter, an exploratory review of the position of rainwater harvesting in meeting water security challenges is presented. For simplicity, the chapter is divided into five main subsections. Section 4.1 examines the concept of rainwater harvesting, focusing largely on micro-catchment harvesting systems.

Sections 4.3 and 4.4 focus on the importance of rainwater harvesting and quality assessment of harvested rainwater. In Section 4.4, authors examine the problems associated with rainwater harvesting. The remaining sections include a summary of the chapter and the authors' concluding remarks.

4.2 Concept of Rainwater Harvesting

Rainfall is the primary source of water into the hydrological cycle, and it is rainfall that mobilizes the availability of water for harvesting. Appan (1999) described rainwater harvesting systems as:

> *"... old as the mountains. The standard adage – as in all water supply scheme is – store water (in a reservoir) during the rainy season so that you can use it when you need it most during the summer. In other words, 'Save for the dry day!' The principle, methods of construction, usage and maintenance are all available. And, most important of all, there are many financial models to suit developing and developed countries. What is needed most is the*

Handbook of Water Harvesting and Conservation: Basic Concepts and Fundamentals, First Edition. Edited by Saeid Eslamian and Faezeh Eslamian.

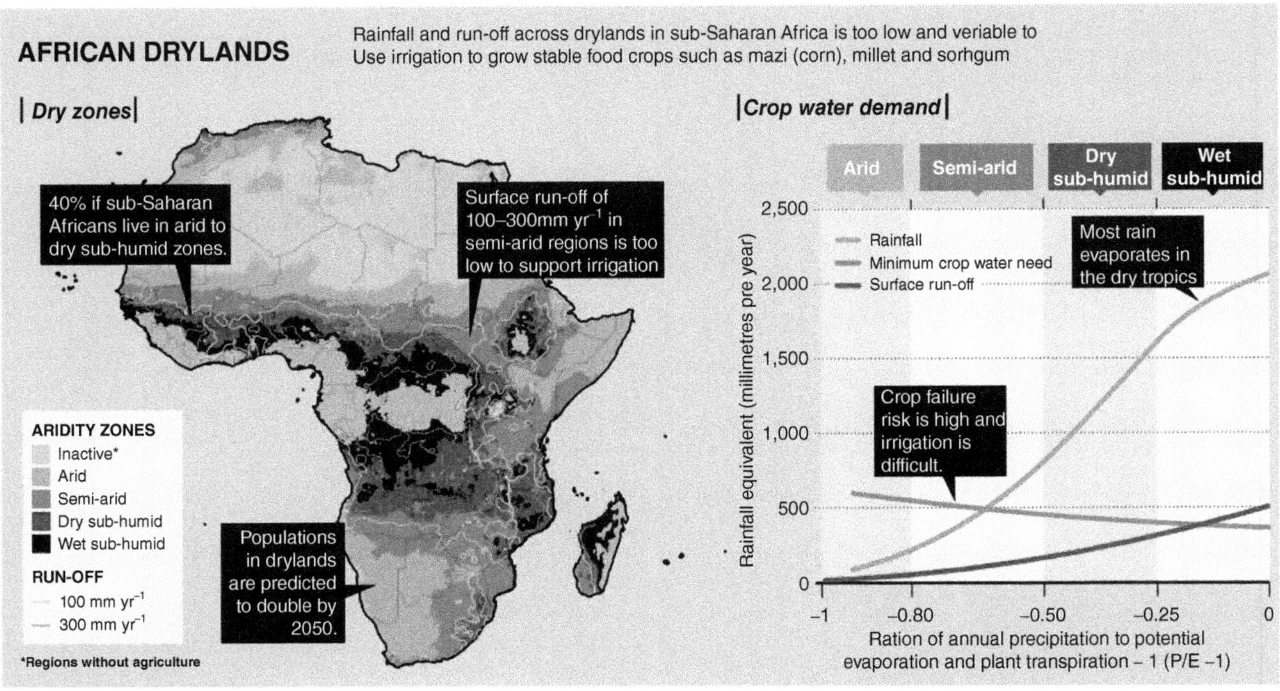

Figure 4.1 Concerns for food shortage, consequent on increasing rate of dryness and water scarcity in Africa.

moral acceptance of the technology and the political will to implement the system".

Appan (1999)'s view of rainfall harvesting is in line with many authors' perception on the age of the practice and its purposes. There is no consensus on the exact period when the idea of rainwater harvesting started, probably because civilization began at different time across locations and ages. Scholars who did present information on the probable periods included Dwivedi and Bhadauria (2009) and Mays et al. (2013) that argued that although the practice probably started about 6000 years ago in China, 4000 years ago in Africa, and around 2000 BCE in Israel and some other parts of the Middle East, it developed through progress in technologies and engineering methods across different societies.

An evaluation of the review by Prinz (1998) indicated that research on water harvesting became prominent in parts of the USA, Asia, Middle East, and Australia from 1950. In sub-Saharan Africa, many water harvesting projects were based on individual indigenous efforts, except for few programs such as the contour stonebunds (which later combined with the indigenous *zai* system) at the Yatenga Province of Burkina Faso, which became useful to about 8000 ha, in over 400 villages (Critchley et al. 1994).

In Ethiopia, the Sudan, and Botswana, small check-dams made of earth (*haffirs*) were used to harvest moderate overland flow passing down slight slopes. In the Americas, Mexico showed evidence of indigenous water harvesting with traditional check-dam, earthen-walled basins that harvest water diverted from seasonal creeks (*arroyos*), a system that is still used in the cultivation of crops in the region (Barrow 2016). Many western Australian communities involved topography modification in the form of catchment treatment (*roaded catchments*) that is made up of steep, bare, and compacted earth, surveyed at a gradient that allows runoff to occur without causing erosion of the intervening channels (Frasier 1994). In general, although a wide range of acceptance of water harvesting as a major approach to combating water challenge is common, it is not so with the use of harvested water – a situation that may be regarded as response to concerns for its quality. For example, in the case of rooftop rainwater harvesting, there is no consensus about the quality of the harvested water as studies from different environment showed that the quality may be good or acceptable (Adeniyi and Olabanji 2005; Melidis et al. 2007; Uba and Aghogho 2000) to severely polluted (Chang et al. 2004; Gromaire et al. 2002), despite that it is being consumed without treatment in some communities, including many rural communities in Nigeria.

4.3 Rainwater Collection Systems

Rainwater harvesting involves the collection, storage, and use of rainwater, either as the primary source or part of conjugate water sources for potable and non-potable purposes (Fewkes 2006). The rainwater collection approach

Table 4.1 Description of selected rainwater harvesting systems (Biazin et al. 2012).

Rainwater harvesting system	Type/structure and countries where method is practiced in the sub-Saharan Africa	Description
Micro-catchment rainwater harvesting systems	Pitting (Zai pits, Ngoro pits, trenches, tassa pits, etc.) (practiced in Burkina Faso, Mali, Tanzania, Kenya, Somalia, Uganda, Ethiopia, Zimbabwe, and South Africa)	Zai pits: A grid of planting pits is dug across plots that could be less permeable or rock-hard; organic matter is sometimes added to the bottom of the pits; Ngoro pits: A series of regular traditional pits, 1.5 m^2 by 0.1–0.5 m deep with the crops grown on the ridges around the pits; Trenches: pits are made along the contour sometimes with a bund downslope either staggered or continuous to check the velocity of runoff, conserve moisture and increase ground water recharge.
	Contouring (stone/soil bunds, hedgerows, vegetation barriers). Practiced in Kenya, Ethiopia, Tanzania, Burkina Faso, and South Africa	Stone and soil bunds: A stone or sometimes earthen bank of 0.50–0.75 cm height is piled on a foundation along the contour in a cultivated hill-slope, sometimes stabilized with grasses or other fodder plant species; Hedge rows: Within individual cropland plots, strips of land are marked out on the contour and left unplowed in order to form permanent, cross-slope barriers of naturally established grasses and herbs. Alternatively, Shrubs are planted along the contour.
	Terracing (Fanya Juu, Semi-circular and hillside terraces). Practised in Kenya, Ethiopia, and Tanzania	Bunds in association with a ditch, along the contour or on a gentle lateral gradient are constructed in different forms. The Fanya Juu terraces are different from many other terrace types in that the embankment is put in the upslope position.
	Micro-basins (Negarims, half-moons, and eyebrows). Practiced in Ethiopia, Kenya, Tanzania, Uganda, Burkina Faso, Mali, and Niger	Different shapes of small basins, surrounded by low earth bunds are formed to enable the runoff to infiltrate at the lowest point, where the plants are grown. The differences between the different structures is basically in their shapes, *Negarims* (diamond), *Halfmoons* (semi-circular), etc.
Macro-catchment rainwater harvesting systems	Traditional open ponds (storage capacity of 30–50 m^3). Practiced mainly in East Africa (Kenya, Ethiopia, Tanzania, Somalia)	Runoff collected from cultivated hill slopes, natural watercourses, footpaths or cattle tracks is stored in unplastered and open ponds. The stored water usually suffers from losses due to seepage and evaporation.
	Cisterns (30–200 m^3 storage capacity). Practiced mainly in East (Kenya, Ethiopia, Tanzania, and Uganda) and South (Zimbabwe, Botswana) African countries.	Runoff collected from bare lands, cultivated hill slopes, or road catchments is guided and stored in underground storage tanks. The cisterns have plastered walls and covered surfaces. In most cases, settling basins are attached in front of the inlet to reduce sedimentation and otherwise, regular cleaning is required.
	Earthen dams (microdams). Obtained in Tanzania, Ethiopia, Botswana, and Burkina Faso.	Larger sized rainwater storage systems such as *ndivas* in Tanzania and micro-dams in Ethiopia are communally constructed around foots of hill slopes to store the runoff from ephemeral or perennial rivers. The reservoirs are neither plastered at their walls nor covered on their surfaces. The water is mostly used for supplemental irrigation communally and for cattle.
	Sand dams. Popular with East Africa (Kenya, Ethiopia)	Dams constructed to store part of the natural flow in seasonal rivers. The sand carried by the river will settle upstream of the dam and gradually fill the streambed. Hence, the sand will reduce evaporation and contamination of the water in the sand body behind the dam.
	Ephemeral stream diversions and spate irrigation. Mainly practiced in East Africa (Eritrea, Ethiopia, Tanzania)	Ephemeral streams from uplands are diverted from their beds at an *agim* (temporary diversion structure) to irrigate adjacent crop fields downstream usually before planting.

(continued)

Table 4.1 (Continued)

Rainwater harvesting system	Type/structure and countries where method is practiced in the sub-Saharan Africa	Description
In-situ rainwater harvesting systems	Ridging (occurred in many parts of the sub-Saharan Africa)	Basins that are wider than the traditional furrows are created either by manual hoeing or during tillage using a modified plowing instrument. They can be designed to be tied every 3–6 m distance for holding water and facilitating infiltration in low and erratic rainfall areas.
	Mulching (in Western and Eastern Africa)	The use of both crop residues and material from non-cultivated areas, including stones, aimed at covering the soil. This improves infiltration of water into the soil and prevents evaporation out of the soil.
	Furrowing and pothoeing (in Western and Eastern Africa)	Different furrowing techniques are used before and after planting to conserve soil moisture in areas where oxen plowing and hand-hoeing practices are common. In the Sahel, small shallow holes are dug manually at correct intervals and the seeds are covered with soil; Two weeks after the emergence of the crop they add fertilizer about 10 cm from the plant.
	Conservation tillage. Practiced in many parts of sub-Saharan Africa (South Africa, Kenya, Tanzania, and Ethiopia)	It encompasses a wide range of tillage techniques ranging from non-inversion plowing and reduced tillage to ripping and sub-soiling in sub-Saharan Africa.

includes macro-catchment, micro-catchment, and in-situ harvesting systems (Boers and Ben-Asher 1982; Biazin et al. 2012) (Table 4.1).

Macro-catchment rainwater harvesting systems refer to the collection of runoff farming water from a catchment area, using water channels, dams, or diversion systems, and storing it in a surface reservoir or in the root-zone of a farmed area for direct use (Boers and Ben-Asher 1982). Becker and Robson (2010) described macro-catchment harvesting systems as those which collect runoff in larger catchments such as hillsides with long slopes; also known as runoff harvesting or long-slope runoff farming systems, because of its dominant application in agriculture. The systems often require construction of elaborate structures and labor-intensive maintenance. An example of the macro-catchment water harvesting system is the hillslope conduit system that is constructed to supply agricultural fields with water in a way that the runoff is conveyed to where is required within the system. Practices of rainwater harvesting systems include *Hafirs* and *Tabias* or *Limas* in Sudan and Iberian Peninsula, respectively, and the *Aljibe* systems in Spain.

A macro-catchment water harvesting system also involves floodwater harvesting that may occur "within stream bed" (i.e. spate irrigation; involving blocking water flow to inundate the valley bottom of the entire flood plain) or as diversion (forcing the water to leave its natural course and conveying it to a preferred unit within the catchment). The practice often requires construction of elaborate hydraulic structures like large dams or dikes and distribution facilities. These include the Great Dam in Ma'rib, Yemen, Harbaqa dam in Syria, and *Jessour* system in Tunisia.

The micro-catchment harvesting system, on the other hand, involves collecting surface runoff from the contributing area, which is typically over a flow distance of less than 100 m, and storing it for consumptive use in the root-zone of an adjacent infiltration basin that may be planted with a single tree, bush, or with annual crops (Boers and Ben-Asher 1982). Micro-catchment rainwater harvesting systems are designed to collect runoff from a relatively small catchment area, mostly 10–500 sq. m, with the ratio of collection catchment to the cultivated target area typically varying between 2:1 and 10:1. They are also easily controlled by the farmer, and this makes them easier to adapt and replicate (Biazin et al. 2012). Micro-catchment harvesting system includes contour catchment water harvesting, desert strip farming, contour bench farming, and runoff-based pitcher farming. The in-situ system includes primarily of a rooftop rainwater harvesting system, which is often practiced in households in many rural and semi-urban settlements in sub-Saharan Africa to augment domestic water supplies. There have been concerns about the quality of the water and its suitability for direct consumption in recent years. Nonetheless, the in-situ system provides a major source of domestic water in the majority of homes in tropical Africa, where pipe-borne water supply is lacking, especially in the wet season.

4.4 Rainwater Storage

Approaches for harvesting or storing harvested rainwater within the catchment system may vary with technology and targeted usage (agricultural or domestic purpose). The approaches can be classified into surface runoff and rooftop rainwater harvesting methods. The surface runoff rainwater harvesting collects overland flow during a rain event into a tank below the surface (Figure 4.2). Gunnell and Krishnamurthy (2003) described the use of tanks (aboveground or underground) in India and Sri Lanka in agricultural fields, especially in paddy rice fields, as "a structuring node of the agricultural landscape where water is concentrated, stored, and redistributed for farming during the dry season." The practice involves storing rainwater in tanks within the farmlands in *Ayacut* wells from where the water is calculatedly released to recharge the water table. *Ayacut* wells are considered effective for water conservation, and are crucial as a risk-management device in years of rainfall deficit. Another example of runoff harvesting is the "Vallerani system" (named after its inventor, Italian agronomist Venanzio Vallerani) in Jordan. This method involves the use of a special plough to build contour ridges and semi-circular water harvesting catchments at user-selected intervals, up to 7500 micro-catchments per day (Abdulla and Al-Shareef 2009).

The second general class are the rooftop rainwater harvesting. Rooftop rainwater harvesting involves capturing, diverting and recharging or storing of rainwater from the roof catchments of domestic houses or commercial buildings into storage tanks. The main objective of rooftop rainwater harvesting is to make water available for future domestic use, to improve the quality of ground water, among other functions. The rainwater storage system (especially tanks) varies with culture and technological development. Common storage materials include barrels, buckets, and different models of storage tanks that can be placed under the downspout of guttering, such that rain from rooftop is diverted into it.

Table 4.2 Common estimates of runoff coefficients from roof types (Ferrany et al. 2011).

Roof type		RC
Roofs (in general)		0.7–0.95
Sloping roofs	Concrete/asphalt	0.9
	Metal	0.81–0.95
	Aluminum	0.7
Flat roofs	Bituminous	0.7
	Gravel	0.8–0.85
	Level	0.81

Rooftops are the first candidate for rainwater harvesting in urban areas, probably because they make the first contact with falling water, and subsequently get channeled into a waiting container, placed on the ground surface. The rainwater harvesting potential of a roof (in liters per year) can be estimated using Eq. (4.1) (Farreny et al. 2011):

$$R \times CA \times RC \tag{4.1}$$

Where

R = Rainfall (mm year^{-1}),
CA = Catchment Area (m^2),
RC = Runoff Coefficient (RC).

The RC is a dimensionless value that estimates the portion of rainfall that becomes runoff, taking into account losses due to spillage, leakage, catchment surface wetting, and evaporation (Mahmoud et al. 2014; Eludoyin 2013). It is, therefore, a useful tool for predicting the potential water running off a surface, which can be conveyed to a rainwater storage system. Liaw and Tsai (2004) described typical rooftop rainwater harvesting system (in Taiwan) as comprising three basic sub-systems: a catchment system (the roof), a delivery system (filters and gutters), and a storage system. Consequently, they mathematically determined the optimum storage capacity for selected roof types using RC and other relevant inherent variables (including the quantity of rainfall, the area of the roof, and the calculated rainwater yield), such that water supply reliability isoquant (U, a line joining locations of same quantity of water supply) is given as a function of the storage capacity (SC) and roof area (RA) (Eq. (4.2)):

$$U = \mathrm{f}(SC, RA) \tag{4.2}$$

Ferrany et al. (2011) provided an estimate of RC for different roof types. Table 4.2 shows typical range of values for RC, but there no consensus on RC values for different roof types under diverse environmental climatic conditions. The values may vary with the degree of imperviousness, infiltration capacity of the drainage surface, and frequency of storms.

They are also capable of being influenced by climatic variables (including size and intensity of rain event, antecedent moisture, prevailing winds) and architectural characteristics of the surface area (including slope, roof material, surface depressions, infiltration, roughness).

4.5 Importance of Rainwater Harvesting

The practice of rainwater harvesting has been recognized as a global practice to meet the increasing water security risks (Hoekstra et al. 2018). It is a centuries-old water

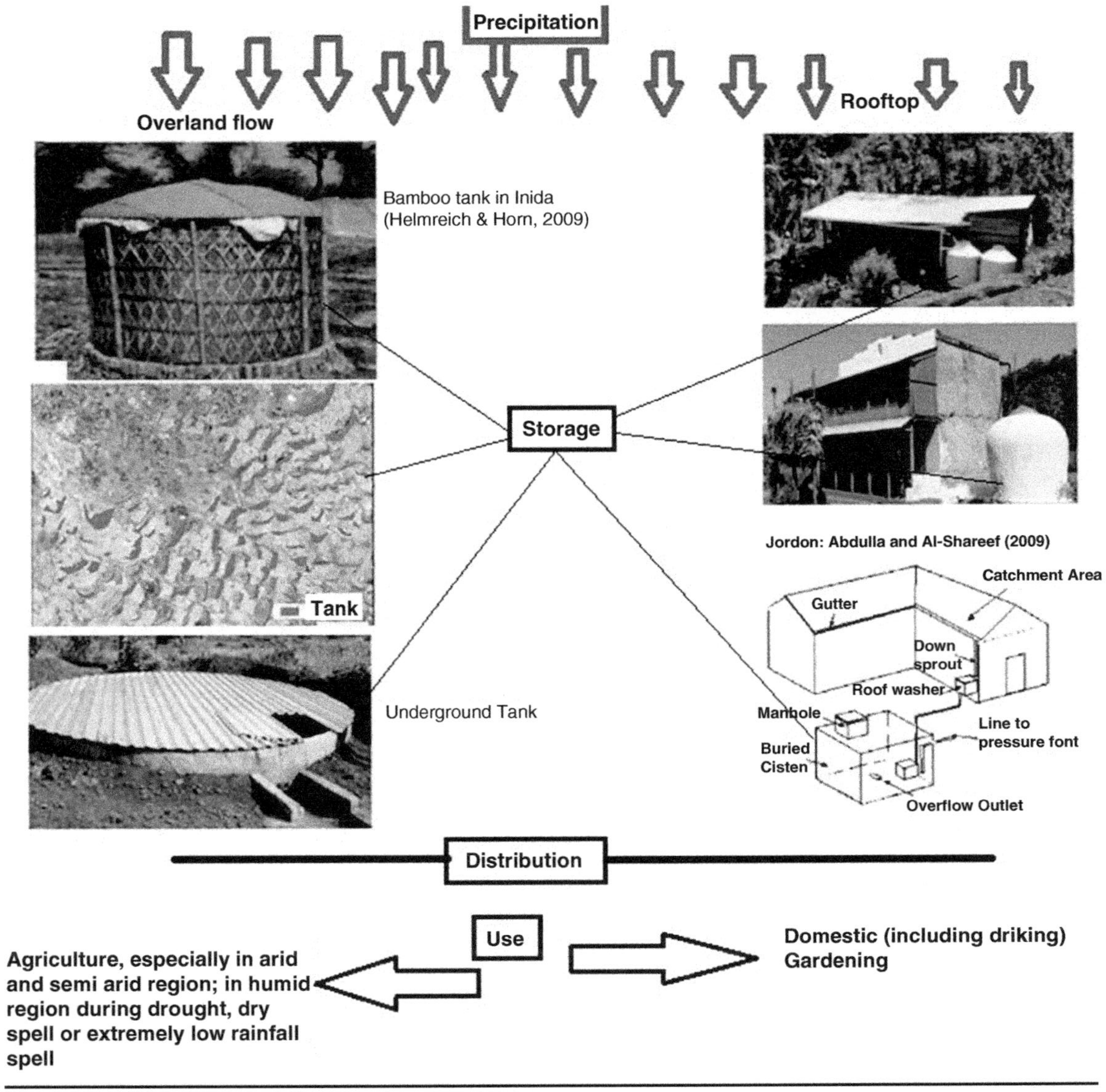

Figure 4.2 Conceptualized components of rainwater harvesting and examples of storage methods.

supply technology that plays a major role in meeting the increasing demand for water (owing to population increase and urbanization) and coping with the effects of climate change and variability (Rahman and Hajani 2017). Rainwater harvesting contributes to the achievement of global Sustainable Development Goals: ensuring availability and sustainable management of water and sanitation for all (Sridhar et al. 2001).

A number of studies have examined the importance of rainwater harvesting in different parts of the world. For example, Van der sterren et al. (2012) and Hanson and Vogel (2014), among others, showed that rainwater harvesting is generally used in urban areas as an alternative water supply means for non-potable purposes (e.g., toilet flushing, laundry, irrigation, and car washing), and for controlling stormwater. Studies from developing countries, including Nigeria, have shown that the practice is adequate for control of water-related diseases and meeting daily water demand (Imteaz et al. 2012). According to Sridhar et al. (2001):

> *"In Igbole-Elerin (Nigeria), water shortage is severe and the communities were suffering from guineaworm epidemic during 1960s. A principal…of Igbo-Elerin Grammar School saw this problem and organized the rainwater harvesting from all the roofs of the school in a systematic way. He started collecting from 1960 and continued regularly. The harvested rainwater was led through a surface tank into 20 underground wells of about 4 m deep. All the school children numbering over 1000 were ordered to use only that source of water and never to go to ponds. He also advised them to take some water for their homes. Subsequent principals (and some teachers) maintained this tradition up to the mid-1980s. In a year's time, the guineaworm was under control. Having noticed this incident, the community came together, collected adequate funds and went for a borehole to augment the water demand of the community. While the other villages suffered from guineaworm, this area with 34 villages stood as a model in eradicating the age-old disease. Till to date the rain harvesting structures are well maintained and community stands as a model for harvesting the rainwater for a larger population and in the eradication of an age-old disease, guineaworm".*

In a peri-urban community of Guatemala, Elgert et al. (2016) demonstrated how rainwater harvesting systems have been developed to serve as a potential buffer against water scarcity and insecurity by supplementing or replacing municipal water services meant for geophysical, sociopolitical, or economic purposes, all of which have limited capacity. The system, implemented for 12 households, was designed to improve water quantity, quality, and access. Elgert et al. (2016) remarked that:

> *"an even greater positive impact on drinking water, as better quantity combined with quality means that families rarely, if ever, drink water from the public tap."*

Furthermore, Padre (2003) presented case studies of farmers in Asian sub-continent (especially in India, Taiwan, Thailand, and China) whose farms were resuscitated with sensible harvesting and storage of rainwater. Such was the success of the method that in Thailand, the government (in 1980s) initiated a program that involved making millions of tank-like "cement jars" for household to store rainwater that flow off the roof of their houses. Similarly, Rahman et al. (2003) reported on the acceptability of rainwater harvesting systems by rural communities in Bangladesh. They showed that although, the people initially rejected the use of rainwater for drinking and cooking due to perceived staleness and pollution of rainwater, they were nonetheless conscious of its economic value, especially in the dry season. Indeed, an estimated 85% of Bangladeshi people under the rainwater harvesting program used harvested rainwater for drinking and cooking purposes (Rahman et al. 2003).

In general, rainwater harvesting is considered very useful for both agricultural and domestic purposes, and as a substitute to public water supply in many regions in developing countries with poor water infrastructure reach. Vohland and Barry (2009), among others, argued that rainwater harvesting is important in the context of agricultural production in African drylands as they provide opportunity to stabilize agricultural landscapes in semiarid regions, thereby making them more productive and more resilient toward climate change. Rainwater harvesting serves as an important means of restoring degraded cultivated and/or natural grazing lands. According to Adeniyi and Olabanji (2005), "it is not uncommon for schools in rural areas to use rainwater sources in place of distilled or deionized water, while food and water vendors use it to produce packaged water (commonly called 'pure water' in Nigeria) for sale to the public".

Rainwater harvesting is recommended by the United Nations Convention to Combat Desertification, describing it as a "strategic tool for combating drought mitigation and desertification" (Sharma and Smakhtin 2006; Reinstädtler et al. 2017). In a Southern Indian state, Bitterman et al. (2016) found that the many small-scale farmers harvested rainwater in small reservoirs for irrigation. They showed that such practices led to an increasing productivity and moderating of agricultural production, thereby alleviating poverty, and providing ecosystem services. Rainwater harvesting is also known to improve resilience of residents to climate extreme-impacted water shortages (Folke et al. 2004). Rainwater harvesting practices are also useful in restoring degraded areas, and ensuring water security in water-scarce environments. Box 4.1 contains a list of the importance of rainwater harvesting in restoring degraded environments and ensuring water security in water-scarce environments.

Box 4.1 Significance of Rainwater Harvesting (Ruhela et al. 2004)

- Provision of relatively high-quality water, soft and low in minerals at low cost;
- direct capturing of rainwater which significantly reduces reliance on water from dams/reservoirs and canal systems. This is capable of exerting less pressure on the storage capacity at macro level, and thereby reduce the need to expand dams or build new ones;
- promotion of self-sufficiency and fostering appreciation of water as a resource. It also promotes water conservation;
- ability to reduce local erosion and flooding from impervious cover in built up areas because local rainfall is diverted into collection tanks in rainwater harvesting; and
- promotion of the capability to encourage households and institutions to be well equipped with onsite and decentralized water supply of reliable quality.

4.6 Quality Assessment of Harvested Rainwater

Studies have revealed increased interest in the monitoring of rooftop water quality as urban users of the water source have also increased in recent times (Skarżyńska et al. 2007). This is probably because rooftop runoff is often regarded as unpolluted or, at least, it presents relatively good quality standards compared to the rainwater from surface catchment areas (Ferrany et al. 2011), and because rooftop runoff is usually harvested for domestic and consumptive (drinking) purposes. Fellany et al. (2011) noted that research relating to the quality of rooftop harvested rainwater have been carried out in other parts of the world (in East Asia [Appan 2000; Kim et al. 2005], Europe [Albrechtsen 2002; Gromaire et al. 2002; Moilleron et al. 2002], United States of America [Chang et al. 2004; Van Metre and Mahler 2003] and Oceania [e.g. Evans et al. 2006; Kus et al. 2010; Magyar et al. 2007; Simmons et al. 2001]). Existing studies also indicate that there is not yet a consensus on the quality assessment of roof runoff water. For example, whereas studies (e.g. Adeniyi and Olabanji 2005; Melidis et al. 2007; Uba and Aghogho 2000) found the harvested water to be within good or acceptable limits of recognized standard guidelines, other studies (Chang et al. 2004; Gromaire et al. 2002; Simmons et al. 2001) found the water to be polluted and unfit for consumption.

Factors which influence the quality of harvested rooftop rainwater, include the type and hygiene of roof type and certain environmental conditions (Avila and Alarcon 1999). The environmental conditions comprise the natural climatic conditions and anthropogenic factors including urbanization, industrialization, and transportation (Kus et al. 2010). For examples, the neutralization effect that is associated with atmospheric concentration of certain compounds has been known to account for higher level of acidity in rains in many European countries than countries in Africa (Evans et al. 2006). Also, bird excrement, together with moss and lichens on the roofs, can cause an increase in ammonium as well as phosphorus levels (Gobel et al. 2007).

4.7 Problems Associated with Rainwater Harvesting

Studies have argued that even though rainwater harvesting is a helpful technique to cope with water scarcity and stress in many regions, there are some problems hindering its planning, implementation, and integration with available water management policies. For example, Vohland and Barry (2009) and Helmreich and Horn (2009) argued that the technology for rainfall collection is often inadequate to meet the requirements of many regions in the developing countries. They may also be expensive for rural dwellers in many regions. Furthermore, there is a lack of acceptance, motivation, and participatory policy makers-users in many regions in developing countries. In Nigeria, for instance, rainwater harvesting is practiced at very small, mostly individual scale. A major hindrance to the development of the scheme in Nigeria is the problem with maintenance; and this exemplifies the situation in many developing countries in the world, suggesting that major water conservation ingenuity may only work for a few years except where there is institutional support.

4.8 Conclusion

This review described the concept of rainwater harvesting in terms of micro, macro, and rooftop approaches to collection, storage, management, and use of rainwater from existing knowledge. It also explored the importance and quality assessment of and presented the problems that are associated with rainwater harvesting in different regions of the world, especially among developing countries, as well as arid and semi-arid countries. In terms of the quality assessment, this review agrees with other scholars that proposed that the physical and chemical properties of rainwater can be superior to sources of surface and groundwater sources that may have been subjected to contamination, especially where rainwater harvesting is being practiced with institutionalized supports. It is recommended that further study should be carried on rainwater harvesting practices in regions with water scarcity-related problems, particularly the potentials, sustenance, and quality of harvested rainwater. Indigenous knowledge among people in the regions practicing it can be harnessed, assessed, and supported with appropriate technology, and institutional and structural policies in regions with poor water supply problems.

References

Abdulla, F.A. and Al-Shareef, A.W. (2009). Roof rainwater harvesting systems for household water supply in Jordan. *Desalination* 243 (1–3): 195–207.

Adeniyi, I.F. and Olabanji, I.O. (2005). The physico-chemical and bacteriological quality of rainwater collected over different roofing materials in Ile-Ife, southwestern Nigeria. *Chemistry and Ecology* 21 (3): 149–166.

Albrechtsen, H.J. (2002). Microbiological investigations of rainwater and graywater collected for toilet flushing. *Water Science and Technology* 46 (6–7): 311–316.

Appan, A. (1999). Trends in water demands and the role of rainwater catchment systems in the next millennium. In 9° *Conferência Internacional sobre Sistemas de Captação de Água de Chuva*. Anais,. Petrolina–PE.

Appan, A. (2000). A dual-mode system for harnessing roofwater for non-potable uses. *Urban Water* 1 (4): 317–321.

Avila, A. and Alarcón, M. (1999). Relationship between precipitation chemistry and meteorological situations at a rural site in NE Spain. *Atmospheric Environment* 33 (11): 1663–1677.

Barrow, C.J. (2016). *Water Resources and Agricultural Development in the Tropics*. USA: Routledge.

Becker, A. and Robson, B.J. (2010). Riverine macroinvertebrate assemblages up to 8 years after riparian restoration in a semi-rural catchment in Victoria, Australia. *Marine and Freshwater Research* 60 (12): 1309–1316.

Biazin, B., Sterk, G., Temesgen, M. et al. (2012). Rainwater harvesting and management in rainfed agricultural systems in sub-Saharan Africa–a review. *Physics and Chemistry of the Earth, Parts A/B/C* 47: 139–151.

Bitterman, P., Tate, E., Van Meter, K.J., and Basu, N.B. (2016). Water security and rainwater harvesting: a conceptual framework and candidate indicators. *Applied Geography* 76: 75–84.

Boers, T.M. and Ben-Asher, J. (1982). A review of rainwater harvesting. *Agricultural Water Management* 5 (2): 145–158.

Chang, M., McBroom, M.W., and Beasley, R.S. (2004). Roofing as a source of nonpoint water pollution. *Journal of Environmental Management* 73 (4): 307–315.

Critchley, W.R.S., Reij, C., and Willcocks, T.J. (1994). Indigenous soil and water conservation: a review of the state of knowledge and prospects for building on traditions. *Land Degradation and Development* 5 (4): 293–314.

Dwivedi, A.K. and Bhadauria, S.S. (2009). Domestic Rooftop Water Harvesting – a Case Study. *Groundwater* 4 (6): 14–27.

Elgert, L., Austin, P., and Picchione, K. (2016). Improving water security through rainwater harvesting: a case from Guatemala and the potential for expanding coverage. *International Journal of Water Resources Development* 32 (5): 765–780.

Eludoyin, A.O. (2013) Modelling the Hysteretic Patterns of Solute Concentration-Discharge Relationships and their Significance for Hydrological Pathways at the Farm-Scale, Doctoral Thesis, University of Exeter / Rothamsted Research-North Wyke Farm Platform, United Kingdom, 272 pp, http://hdl.handle.net/10871/15374

Evans, C.A., Coombes, P.J., and Dunstan, R.H. (2006). Wind, rain and bacteria: the effect of weather on the microbial composition of roof-harvested rainwater. *Water Research* 40 (1): 37–44.

Farreny, R., Morales-Pinzon, T., Guisasola, A. et al. (2011). Roof selection for rainwater harvesting: quantity and quality assessments in Spain. *Water Research* 45 (10): 3245–3254.

Fewkes, A. (2006). The technology, design and utility of rainwater catchment systems. *Water Demand Management*, pp. 27–51.

Folke, C., Carpenter, S., Walker, B. et al. (2004). Regime shifts, resilience, and biodiversity in ecosystem management. *Annual Review of Ecology, Evolution, and Systematisc* 35: 557–581.

Frasier, G. (1994). Water harvesting/runoff farming systems for agricultural production. In: *Water Harvesting for Improved Agricultural Production*, 57–71. Rome: FAO.

Göbel, P., Dierkes, C., and Coldewey, W.G. (2007). Storm water runoff concentration matrix for urban areas. *Journal of Contaminant Hydrology* 91 (1–2): 26–42.

Gromaire, M.C., Chebbo, G., and Constant, A. (2002). Impact of zinc roofing on urban runoff pollutant loads: the case of Paris. *Water Science and Technology* 45 (7): 113–122.

Gunnell, Y. and Krishnamurthy, A. (2003). Past and present status of runoff harvesting systems in dryland peninsular India: a critical review. *Ambio: A Journal of the Human Environment* 32 (4): 320–324.

Hanson, L.S. and Vogel, R.M. (2014). Generalized storage–reliability–yield relationships for rainwater harvesting systems. *Environmental Research Letters* 9 (7): 075007–075011.

He, C., Zhang, L., Zhang, X., and Eslamian, S. (2014). Water security: concept, measurement, and operationalization. In: *Handbook of Engineering Hydrology: Environmental Hydrology and Water Management*, Ch. 28,, vol. 3 (ed. S. Eslamian), 545–554. USA: Francis and Taylor/CRC Group.

Helmreich, B. and Horn, H. (2009). Opportunities in rainwater harvesting. *Desalination* 248 (1–3): 118–124.

Hoekstra, A.Y., Buurman, J., and van Ginkel, K.C. (2018). Urban water security: A review. *Environmental Research Letters* 13 (5): 053002.

Imteaz, M.A., Adeboye, O.B., Rayburg, S., and Shanableh, A. (2012). Rainwater harvesting potential for Southwest Nigeria using daily water balance model. *Resources, Conservation and Recycling* 62: 51–55.

Kim, R.H., Lee, S., and Kim, J.O. (2005). Application of a metal membrane for rainwater utilization: filtration characteristics and membrane fouling. *Desalination* 177 (1–3): 121–132.

Kus, B., Kandasamy, J., Vigneswaran, S., and Shon, H.K. (2010). Analysis of first flush to improve the water quality in rainwater tanks. *Water Science and Technology* 61 (2): 421–428.

Liaw, C.H. and Tsai, Y.L. (2004). Optimum storage volume of rooftop rain water harvesting systems for domestic use. *Journal of the American Water Resources Association (JAWRA)* 40 (4): 901–912.

Magyar, M.I., Mitchell, V.G., Ladson, A.R., and Diaper, C. (2007). An investigation of rainwater tanks quality and sediment dynamics. *Water Science and Technology* 56 (9): 21–28.

Mahmoud, S.H., Mohammad, F.S., and Alazba, A.A. (2014). Determination of potential runoff coefficient for Al-Baha region, Saudi Arabia using GIS. *Arabian Journal of Geosciences* 7 (5): 2041–2057.

Mays, L., Antoniou, G.P., and Angelakis, A.N. (2013). History of water cisterns: Legacies and lessons. *Water* 5: 19161940.

Melidis, P., Akratos, C.S., Tsihrintzis, V.A., and Trikilidou, E. (2007). Characterization of rain and roof drainage water quality in Xanthi, Greece. *Environmental Monitoring and Assessment* 127 (1–3): 15–27.

Moilleron, R., Gonzalez, A., Chebbo, G., and Thévenot, D.R. (2002). Determination of aliphatic hydrocarbons in urban runoff samples from the "Le Marais" experimental catchment in Paris Centre. *Water Research* 36 (5): 1275–1285.

Oweis, T. and Hachum, A. (2006). Water harvesting and supplemental irrigation for improved water productivity of dry farming systems in West Asia and North Africa. *Agricultural Water Management* 80 (1–3): 57–73.

Padre, S. (2003). Rain water harvesting. *Appropriate Technology* 30 (3): 43–48.

Prinz, D. (1998). *Rainwater harvesting for dry land agriculture-Developing a methodology based on remote sensing and GIS*, In Proceeding XIII. International Congress Agricultural Engineering, 02.-06. 02. 1998. Morocco: Rabat.

Rahman, A. and Hajani, E. (2017). Rainwater harvesting in arid regions of Australia. In: *Handbook of Drought and Water Scarcity, Vol. 2: Environmental Impacts and Analysis of Drought and Water Scarcity* (eds. S. Eslamian and F. Eslamian). USA: Francis and Taylor.

Rahman, M.H., Rahman, M.M., Watanabe, C. and Yamamoto, K. (2003) Arsenic contamination of groundwater in Bangladesh and its remedial measures. In: Arsenic Contamination in Groundwater-Technical and Policy Dimensions. Proceedings of the UNU-NIES International Workshop United Nations University, Tokyo, Japan. pp. 9–21.

Reinstädtler, S., Islam, S.N., and Eslamian, S. (2017). Drought Management for Landscape and Rural Security, Ch. 8. In: *Handbook of Drought and Water Scarcity, Vol. 3: Management of Drought and Water Scarcity* (eds. S. Eslamian and F. Eslamian), 195–234. USA: Francis and Taylor/CRC Press.

Roa-García, C., (2009) *Wetland and water dynamics in small headwater catchments of the Andes*, Doctoral dissertation, University of British Columbia, Canada.

Rockström, J. and Falkenmark, M. (2015). Increase water harvesting in Africa. *Nature* 519: 283–285.

Ruhela, M., Bhutiani, R., and Anand, A. (2004). Rain water harvesting. *Environment Conservation Journal* 5 (1–3): 21–34.

Sharma, B.R. and Smakhtin, V.U. (2006). *Potential of Water Harvesting as a Strategic Tool for Drought Mitigation, International Water Management Institute (IWMI)*. New Delhi, India: South Asia Liaison Office.

Simmons, G., Hope, V., Lewis, G. et al. (2001). Contamination of potable roof-collected rainwater in Auckland, New Zealand. *Water Research* 35 (6): 1518–1524.

Sivanappan, R.K. (2006) Rain water harvesting, conservation and management strategies for urban and rural sectors. In National Seminar on Rainwater Harvesting and Water Management, 11 (12): 1–2.

Skarżyńska, K., Polkowska, Ż., Namieśnik, J., and Przyjazny, A. (2007). Application of different sampling procedures in studies of composition of various types of runoff waters-A review. *Critical Reviews in Analytical Chemistry* 37 (2): 91–105.

Sridhar, M.K.C., Coker, A.O. and Adegbuyi, S.A., (2001) Rainwater harvesting in Nigeria: Prospects and problems. In *Proceedings at the 10th International Rainwater Catchment Systems Conference*, Mannheim, Germany, 2p

Uba, B.N. and Aghogho, O. (2000). Rainwater quality from different roof catchments in the Port Harcourt district, Rivers state, Nigeria. *Journal of Water Supply: Research and Technology-AQUA* 49 (5): 281–288.

UNEP/GRID-Arendal (2009). *Increased global water stress, Vital Water Graphics 2*. Norway: UNEP.

UNU-IWEH (2013). *Water Security and the Global Water Agenda*. Japan: Tokyo Available online at: collections.unu .edu/eserv/UNU:2651/Water-Security-and-the-Global-WaterAgenda.pdf.

Van der Sterren, M., Rahman, A., and Dennis, G.R. (2012). Implications to stormwater management as a result of lot scale rainwater tank systems: a case study in Western Sydney, Australia. *Water Science and Technology* 65 (8): 1475–1482.

Van Metre, P.C. and Mahler, B.J. (2003). The contribution of particles washed from rooftops to contaminant loading to urban streams. *Chemosphere* 52 (10): 1727–1741.

Vohland, K. and Barry, B. (2009). A review of in situ rainwater harvesting (RWH) practices modifying landscape functions in African drylands. *Agriculture, Ecosystems and Environment* 131 (3–4): 119–127.

Part B

Water Harvesting Resources

5

Single-Family Home and Building Rainwater Harvesting Systems

Duygu Erten[1,2]

[1] *Department of Civil Engineering, İstanbul Medipol University, 34810 İstanbul, Turkey*
[2] *Vice Director of Climate Change Research Center, 34810 İstanbul, Turkey*

5.1 Introduction

Among the most pressing environmental challenges facing us are extreme weather events and temperatures; accelerating biodiversity loss; pollution of air, soil, and water; failures of climate-change mitigation and adaptation; and transition risks as we move to a low-carbon future. Water demand issues and climate change risks are happening here and right now. This societal risk has interconnectedness to all environmental risks and even to involuntary migration.

The negative impacts of climate change on freshwater systems will most likely outweigh its benefits. Current projections show that crucial changes in the temporal and spatial distributing of water resources and the frequency and intensity of water-related disasters rise significantly with increasing greenhouse gas emissions. Exploitation of new data sources, better models, and more powerful data analysis methods, as well as the design of adaptive management strategies, can help respond effectively to changing and uncertain conditions (IPCC 2007).

Water is an important factor of production, contributing both directly and indirectly to economic activity across all sectors and regions of the global economy. Water scarcity may therefore go beyond having important consequences for people, society and ecological systems but may also pose a threat to economic growth (Distefano and Kelly 2017). Optimizing water management systems like rainwater harvesting (RWH) and utilization is an environmentally sound solution for single-family residences. "Rainwater" by definition is precipitation that is collected from relatively clean, above-ground surfaces – usually rooftops. Because rooftop rainwater has minimal contamination, it is an ideal source for harvesting.

RWH is a process of collecting and storing rainwater that falls on a catchment surface (typically a roof, although almost any external surface could be suitable) for use, independent from or supplemental to the mains (utility) water supply.

5.2 Historical Development of RWH and Utilization

RWH has been practiced for more than 4000 years, owing to the temporal and spatial variability of rainfall (Panigrahi 2017). From the time of early civilizations, people in arid and semi-arid regions have relied on harvesting rainfall runoff and storing the water in cisterns. For example, these early systems were used in the Mediterranean region to collect runoff from hillsides, open yards, and roofs mainly for domestic purposes (Angelakis 2013). Evidence of roof catchment systems date back to early Roman times. Roman villas and even whole cities were designed to take advantage of rainwater as the principal water source for drinking and domestic purposes since at least 2000 BCE. In the Negev desert in Israel, tanks for storing runoff from hillsides for both domestic and agricultural purposes have allowed habitation and cultivation in areas with as little as 100 mm of rain per year. The earliest known evidence of the use of the technology in Africa comes from northern Egypt, where tanks ranging from 200 to 2000 m^3 have been used for at least 2000 years – many are still operational today. The technology also has a long history in Asia, where rainwater collection practices have been traced back almost 2000 years in Thailand. The small-scale collection of rainwater from the eaves of roofs or via simple gutters into traditional jars and pots has been practiced in Africa and Asia for thousands of years. In many remote rural areas, this is still the method used today.

Anatolia is a territory where many ancient civilizations were developed. In particular, the richness in the historical

Handbook of Water Harvesting and Conservation: Basic Concepts and Fundamentals, First Edition. Edited by Saeid Eslamian and Faezeh Eslamian.

water structures is much more than other regions (Bildirici and Bildirici 1997). A great variety of hydraulic structures were implemented during the last four thousand years on Anatolian land, which was at the crossroads of civilizations, making Turkey one of the foremost open-air museums of the world with respect to ancient water works (Hamza 2017). The world's largest rainwater cistern is the Yerebatan Sarayi in Istanbul, Turkey. This was constructed during the rule of Caesar Justinian (527–565 CE). It measures 140 by 70 m and has a capacity of 80 000 cubic metres.

On Crete during the Venetian period, many water cisterns and fountain houses were constructed in both the towns and the countryside. In several Venetian cities and villages (e.g. in the Pediada region), which were densely populated and rich in water, significant water supply systems – expressed mainly in water cisterns and fountain houses – were constructed (Panagiotakis 2006). In general, the Venetian accomplishments in hydraulics are worth noting, such as the construction and operation of aqueducts, cisterns, wells, fountains, baths, toilets, and harbors. Many of these technologies were developed and used in the famous castles constructed during that period. Thus, several cisterns have been found in Venetian Rethymnon, on the island of Gramboussa, and in the Viannos Vigla castle. Also, small cisterns have been located in several villages in the area of Vamos, Chania, such as those in the village of Gavalochori, and those located with the wells at Agios Pavlos and Paleloni. Later evidence from the Venetian period suggest the existence of more than 500 cisterns in the city of Iraklion after c. 1500 AD (Spanakis 1981).

The history of RWH in Asia can be traced back to about the ninth or tenth century and the small-scale collection of rainwater from roofs and simple brush dam constructions in the rural areas of South and Southeast Asia. Rainwater collection from the eaves of roofs or via simple gutters into traditional jars and pots has been traced back almost 2000 years in Thailand (Prempridi and Chatuthasry 1982). RWH has long been used in the Loess Plateau regions of China. More recently, however, about 40 000 well storage tanks, in a variety of different forms, were constructed between 1970 and 1974 using a technology which stores rainwater and stormwater runoff in ponds of various sizes. A thin layer of red clay is generally laid on the bottom of the ponds to minimize seepage losses. Trees, planted at the edges of the ponds, help to minimize evaporative losses from the ponds (UNEP 1982).

The historical examples of cistern technologies researched by Mays et al. (2013) may even have importance for today's water engineering. Some lessons learned include:

(i) Throughout history, cisterns have been an essential part of water supply technology for human survival and well-being to obtain water resource sustainability;
(ii) A combination and balance of smaller-scale measures (such as cisterns for water harvesting systems) and the large-scale water supply projects (such as cisterns for storage of aqueduct flows) were used by many ancient civilizations;
(iii) The ancient water technologies of cisterns should be considered not as historical artifacts, but as potential models for sustainable water technologies for the present and the future;
(iv) Ancient water technologies such as cisterns were characterized by simplicity, ease of operation, and the requirement of no complex controls, making them more sustainable (Mays 2013). Nevertheless, the successful design and operation of some of these systems were massive achievements in engineering;
(v) Cisterns were used by ancient civilizations for water resource sustainability and have been used ever since, though their importance to modern-day water supply purposes have vanished somewhat in developed parts of the world, despite having continued in many developing parts of the world (Mays et al. 2013);
(vi) The ancients considered water security as one of the critical aspects of the design and construction of their water supply systems.

5.3 Pros and Cons of RWH Systems

A RWH system can alleviate the impact of stormwater runoff, save potable water, reduce energy use, and can contribute to groundwater preservation. (Younos 2011). Figure 5.1 shows components of a typical RWH system. The most attractive part of single-family home RWH systems is the low upfront costs. The rainwater does not

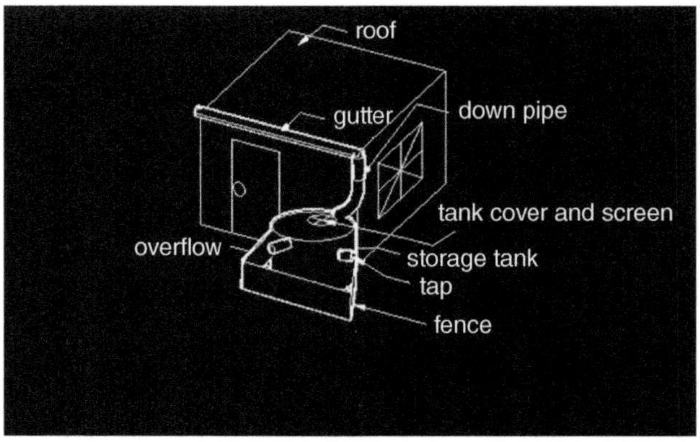

Figure 5.1 Components of a typical RWH system.

cost anything. The construction of rainwater catchment and storage are very simple, low cost and needs little maintenance. The water storage tanks come in different materials depending on the brands, but they are generally lightweight and designed for shallow digs. There is no need for long pipe installation since most systems are on the roof, mounted on the wall, or installed next to the home underground. An added benefit is having free water to irrigate the landscape, which can solve some drainage problems on the property.

Rainwater generally is of higher quality than most traditional, and many improved, water sources found in the developing world. Contrary to popular beliefs, rather than becoming stale with extended storage, rainwater quality often improves as bacteria and pathogens gradually die off (Wirojanagud et al. 1989). Water harvesting helps to preserve the existing surface water resources because capturing runoff reduces local flooding, the potential for water pollution is reduced. Harvesting rainwater can reduce utility produced water consumption by up to 50% so it offers protection against rising water prices. Rainwater pH is ideal for plants because there is no artificial water softening chemicals in harvested rainwater. Rainwater can provide an important source of drinking water with appropriate treatment as well as a useful source of water for blending with other water sources to reduce the levels of contaminants of health concern, such as arsenic and fluoride.

There are some disadvantages of single-family home RWH systems. Some systems are not reliable due to uncertainty of rainfall. Even within the same country, the volume of rainfalls may vary radically. Erratic rainfall patterns may prevent a majority of people from making an initial investment in RWH systems. Because the responsibility of the maintenance and operation of the system relies solely on the homeowner, the system is not attractive to some homeowners due to risks involved in harvesting rainwater. One other reason why single-family RWH systems are not so widely used can be the risk of reducing the homeowner's water bill, which will affect the budgets of municipalities and private water companies. Based on the research completed for this chapter, it is obvious that there are no established policies for RWH in the regions where they are most needed.

5.3.1 Economics of RWH

According to Hammerstrom and Pushard (2016), the benefits of investing in rainwater systems can best be explained with a traditional economic payback approach. In certain areas of the US and the world, catching rainwater for household use is the only cost-effective solution. In locations like central Texas, parts of Northern Mexico, and arid rocky areas where no municipal water system is available and where well drilling can be prohibitively expensive, harvesting rainwater can be a smart economic choice. For homeowners with monthly water bills of $200 or more, short-term, positive returns can usually be demonstrated, and long-term returns may also be realized through increase in property values. Higher water rates and new, tiered rate systems (i.e. rate systems that are not flat fees per water used, but tiered so the larger the amount of water used, the higher the cost per gallon) are frequently making RWH systems economically viable. Water rates can be expected to continue to increase faster than the rate of inflation, so in more areas RWH will make great economic sense. These increases are being driven by population growth and demand, diminishing freshwater supplies, and greater costs to process water to make it drinkable. Increased costs will spur more conservation, but it will also make the option of RWH more viable.

Hammerstrom and Pushard (2016), provided a calculation tool that can determine if a RWH investment will pay for itself (i.e. the Payback Period). This is a simple comparison tool that weighs one expenditure against another. Calculating the Payback Period is a simple method that evaluate the investment on a RWH system, given local water rates.

$$\textbf{Payback Period} = \text{Amount to be Invested}/ \\ \times \text{Estimated Annual Savings}$$

An installed active rainwater system (i.e. with pumps and associated components that are more expensive than passive systems) will typically cost about $3–$6 per gallon of system size, depending on system specifics (above or below ground storage tank, amount of piping required, number of pumps, etc.), not including gutters. The payback will vary widely for specific local rainfall amount, tank size and other features, water rates, installation costs, etc.

A study by Sweeney and Pate (2017), investigated the performance and economics of a rainwater collection system and underground cistern in a single-family residential setting, focusing on rainfall, rainwater collection, and life cycle costs. Their study reports on an expanded investigation of the hydraulic and economic performance of residential RWH systems, including model details, model sensitivity analysis, and a Monte Carlo simulation. Daily water demand averages of $1.39\,m^3$/day were used in the simulation with a 30% variation. Hydraulic inputs, outputs, and economics were evaluated over a 50-year life cycle. Sensitivity and stochastic analyses were performed with below average, average, and above average estimates of input parameters to assess contributions, and to assess a probabilistic interpretation of the lifecycle results. Using 30-year rainfall normals and Monte Carlo stochastic methods, there was a 45–48% probability of 73% of the annual demand being harvested.

5.3.2 Cisterns as Flood Mitigation/Control Systems

A prototype analysis done for determining the "Stormwater Retention and Water Supply Benefits of Cisterns" by Canfield and Shipek show that cisterns have considerable flood control benefit, but that the benefit depends to some extent on soil type. The analysis shows that cisterns can provide significant flood control benefits even under particularly problematic situations such as impermeable cover over more than half the site.

5.3.3 Types of RWH Systems

There are many types of single-family home rainwater harvesting systems. The evolution of RWH systems to increase water use efficiency and the continuous effort to preserve the environment for sustainable development is proven by the many innovative technologies developed for water harvesting.

5.3.4 Water Harvesting: Water Collection Source

RWH using ground or land surface catchment areas can be a simple way of collecting rainwater. Compared to rooftop catchment techniques, ground catchment techniques provide more opportunity for collecting water from a larger surface area. By retaining the flows (including flood flows) of small creeks and streams in small storage reservoirs (on surface or underground) created by low cost (e.g. earthen) dams, this technology can meet water demands during dry periods. There is a possibility of high rates of water loss due to infiltration into the ground, and because of the often marginal quality of the water collected, this technique is mainly suitable for storing water for agricultural purposes (UNEP 2002).

The following site design measures implemented as part of the project design help RWH (PAMC Section 16.11 Stormwater Pollution Prevention Plan):

(i) Direct roof runoff into cisterns or rain barrels for reuse.
(ii) Direct roof runoff onto vegetated areas.
(iii) Direct runoff from walkways, driveways, and patios onto vegetated areas.
(iv) Construct walkways, driveways, and patios with permeable surfaces.

Although it is technically possible to harvest runoff from walkways, driveways, and patios, it is more difficult since a subterranean cistern or a pump is usually needed to move the water into an above-ground rain barrel or cistern. They typically contain contaminants that are undesirable in a rainwater catchment system. Due to these complexities, it is more common to harvest rainwater from rooftops and many different systems are developed for this purpose.

5.3.5 RWH System: System Components

The categorization of RWH systems depends on factors like the size and nature of the catchment areas and whether the systems are in urban or rural settings. The most common harvesting system is the Roof Water Collection system.

The typical main units are (Figures 5.1 and 5.2):

Rainwater conductors. Leaders and gutters or an internally piped roof drainage system that conveys the storm water from the roof to the cistern.

Cistern. A storage tank that allows large particulate matter to settle out of the water. Assuming that RWH has been determined to be feasible, two kinds of techniques–statistical and graphical methods–have been developed to aid in determining the size of the storage tanks. These methods are applicable for rooftop catchment systems only, and detail guidelines for design of these storage tanks can be found in Gould (1992), Pacey and Cullis (1989), Sojka et al. (2016).

Figure 5.2 A typical contemporary RWH systems. Source: https://www.kingspan.com/gb/engb/products/wastewater-management/rainwater-harvesting.

In order to get an idea of the efficiency of having a RWH tank at home, some simple simulations show on a day-to-day basis the evolution of the rainwater contained in the tank depending on the rainfall and the household's water consumption. As the aim of these simulations are to get an overall idea of the behavior of tanks in different situations and not to get precise continuous numbers for tank water quantity, a number of simplifying hypothesis have been made. First of all, a daily time step has been selected in accordance to the available rainfall data. This excludes subtleties such as daily water consumption peaks and rainfalls of high intensity and short length of time, but gives a clear view of the tank's water reaction to exterior events on a larger time frame. Also, the daily household's water consumption has been set as constant, and the values used were the one calculated previously of 77 l per person per day, and another one of 50 l per person to represent a household with more reasonable consumption along with exclusively water saving devices. Furthermore, the rain catchment efficiency was set at 0.8, which includes roof and gutter spillage as well as loss through the first-flush diverter equipped with drain holes of a diameter of 1.25 mm (Gardner et al. 2004).

Overflow from cistern. A pipe that takes overflow from the cistern to the storm drainage system.

Pumping system. Provides the pressure required at the fixture most distant from the tank.

Disinfection system. Various filtration and disinfection systems can be used.

Potable water makeup. Makeup water provided to the tank during dry seasons. Appropriate backflow prevention is required.

Filters. Continuous R&D done on RWH resulted in patented filters rising in the market. Different models of filters are found in the market to capture rainwater from varied rooftops.

5.3.6 Rooftop Material

Rooftop catchment, rainwater storage tanks can provide good quality water, clean enough for drinking, as long as the rooftop is clean, impervious, and made from non-toxic materials (lead paints and asbestos roofing materials should be avoided), and located away from overhanging trees, since birds and animals in the trees may defecate on the roof. If the roof material is asphalt or wooden shingles, the harvested rainwater should only be used for non-edible landscapes, unless the water is treated first. Petroleum or other chemicals from these roofing materials can leach into the rainwater. Roofs with cement, clay, or metal surfaces are ideal for harvesting water for a wide variety of uses. While the bacteriological quality of rainwater collected from ground catchments is poor, that from properly maintained rooftop catchment systems, equipped with storage tanks having good covers and taps, is generally suitable for drinking, and frequently meets World Health Organization (WHO) drinking water standards. Notwithstanding, such water generally is of higher quality than most traditional, and many improved, water sources found in the developing world.

5.3.7 Roof Washers

There are several other very important construction features that will help ensure good quality cistern water. A lot of dirt and dust collects on the roof-catchment surface between rainstorms. This debris can include particles of lead and other atmospheric pollutants as well as bird droppings. These contaminants will enter the cistern along with the roof water unless steps are taken to prevent contamination. The use of roof washers and roof-water filters can reduce the amount of these contaminants entering the system.

The first water to come off the roof at the beginning of a rainstorm is the most contaminated. The degree of contamination will depend on several things including the length of time since the last rainfall, proximity of the catchment to a highway or other local source of airborne pollution, and the local bird population. Also, certain types of materials are preferable for the catchment surface, as will be detailed later.

A roof washer is a mechanism that diverts this initial, highly contaminated roof water away from the cistern. Once the catchment surface has been washed off by an adequate amount of rainfall, the roof water is once again routed to the cistern for storage. Usually the first 0.01 in. of rainfall is considered to be adequate to remove most of the dust and dirt from the surface of the catchment. In this way, only the cleanest roof water is collected in the cistern, whereas the contaminated roof wash is discharged to waste.

There are several ways of accomplishing this. The roof water can be diverted manually through a series of valves within the spouting system, or automatic roof washers may be fabricated by the cistern owner or purchased from commercial distributors.

5.3.8 Maintenance

Rainwater is acidic and therefore corrosive. Unless steps are taken to neutralize this water, it will corrode household distribution systems, adding toxic metals such as lead and cadmium to the tapwater. Corrosion processes are very complex chemical reactions that involve many different factors. Employing the recommendations to completely eliminate corrosion within the cistern system or reducing it to tolerable levels is mandatory. Minimizing the amount of corroded metals in the finished tapwater is the goal.

Maintenance is generally limited to the annual cleaning of the tank and regular inspection of the gutters and downpipes. Maintenance typically consists of the removal of dirt, leaves, and other accumulated materials. Such cleaning should take place annually before the start of the major rainfall season. However, cracks in the storage tanks can create major problems and should be repaired immediately. In the case of ground and rock catchments, additional care is required to avoid damage and contamination by people and animals, and proper fencing is required.

A neutralizing agent needs to be added directly to the cistern, and there is one simple thing that you should do before using the tapwater for drinking or cooking purposes. You should always allow the cold water to run for about a minute before using it for drinking or cooking. This will flush the "stale" water from the supply line, leaving you with tapwater of acceptable quality. This practice is especially important after a tap has gone unused for several hours, or overnight. Rather than just letting the water run down the drain during this procedure, you may use it for purposes other than drinking or cooking.

5.3.9 Smart Rainwater Systems

The smart systems collect rainwater from a building or site and supply filtered rainwater to the landscape drip irrigation system. The system can save tens of thousands of gallons of water a year. The Smart Rainwater System is easy to install and can be used for renovation or new construction projects. The structure of the system is shown in Figure 5.3.

The wall-mounted rainwater manifold has a rainwater controller as seen on Figure 5.4. This rainwater controller automates the entire rainwater system and operates the pump, backflushes the filter, tracks water usage, and displays system information and alarms. The pressure sensor maintains a constant water pressure of 45–65 psi to the irrigation system. Rainwater is filtered through a 100 μm filter at the manifold. The filter is automatically backflushed based upon how many gallons has passed through the filter. Filterflush water is sent to the landscape. If the rainwater tank is empty, the rainwater controller opens the backup valve to partially fill the rainwater tank to maintain a minimum water level for the pump. The water meters track both greywater and backup water use and displays the information on the controller LCD screen. The flow rate of the irrigation system is also displayed on the LCD screen. The system has low maintenance requirements.

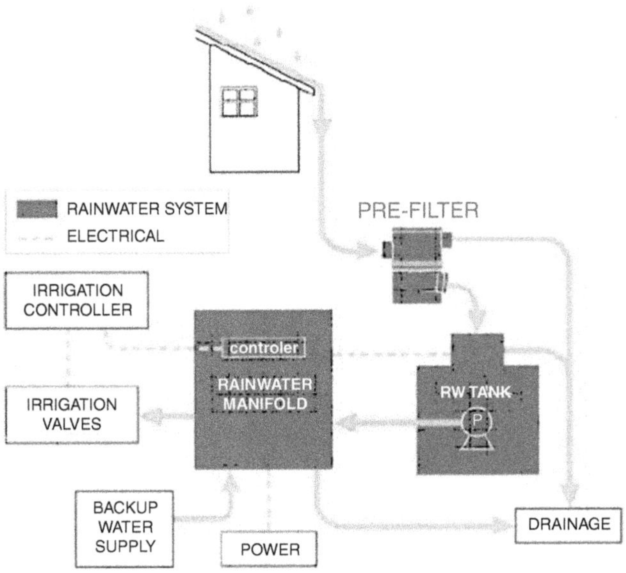

Figure 5.3 Smart rainwater systems. Source: https://www.kingspan.com/gb/engb/products/wastewater-management/rainwater-harvesting.

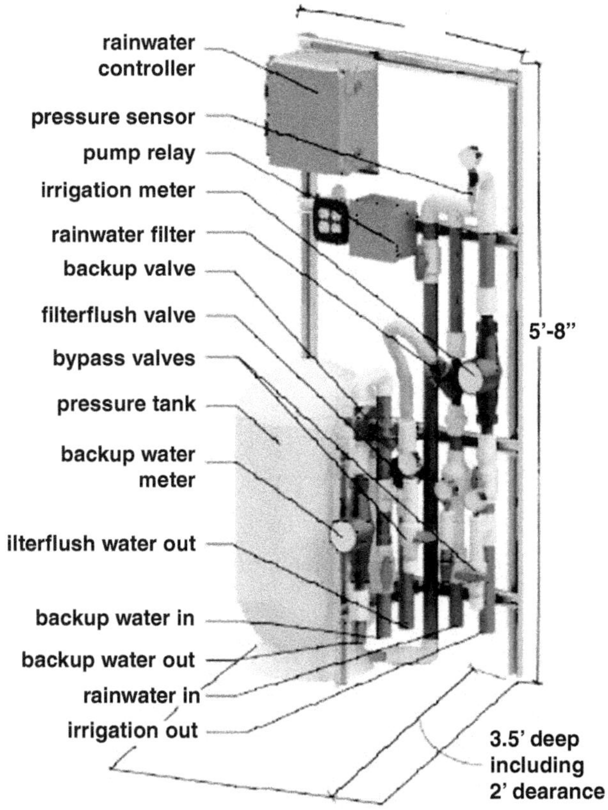

Figure 5.4 Wall mounted rainwater manifold. Source: http://www.water-lab.net.

5.3.10 RWH Systems with Solar Electric Pump

One of the innovative technologies is the rain collection system with a solar panel to power an electric pressure pumpThe RainPerfect pump provides an eco-friendly and cost effective way to power the water out of your rain barrel. The pump is powered by the included solar panel which is easy to install and works with most standard rain barrels. The RainPerfect doesn't require electricity and provides plenty of pressure through an ordinary garden hose to run most low pressure sprinklers, water your garden, or wash your car.

5.3.11 Water Harvesting from Air

A water generator engineering design is created for water harvesting from the air. This system, which is applicable to single-family homes, can serve as a reference design for scalable adoption and adaptation by local communities. Aside from using readily available components, the prototype is to run on DC to enable compatibility with solar powered systems; the prototype thus offers a viable clean water system in places where public water and grid power systems are not in place.

The water-generating device for obtaining potable water from ambient air inside or outside a structure or dwelling has ducts for bringing this supply of ambient air to the device and for releasing the air back outside the device after it has been processed. There is an air filter for filtering the air prior to processing of the air. The air filter includes a one-time sensing element which renders it unusable when removed from the generator. A condenser is provided for extracting water vapor in the air brought thereto by the ducts. Within the ducts there is a fan or blower to move air from outside the device through the condenser and to return the air back outside the device after it has traversed the condenser. Between the condenser and the collection point, there is an immediate temporary holding reservoir, or plumbing to this reservoir, which contains an ultraviolet light to kill existing microorganisms, as well as a pump to transport the water through a subsequent water filter, a second exposure to ultraviolet light, and into the ultimate internal or external water storage unit. The water filter also includes a one-time sensing element which renders it unusable when removed. An internal container is positioned to receive and collect the water after it leaves the second exposure to ultraviolet light, and there is a water sensor below the top of the internal container for shutting down the device when the container is full of water. Prior to this internal container, which is removable and may be reusable, is a valve for diverting this water instead to external water storage units. A switch is provided for automatically deactivating the device until the nonreusable air filter element is replaced after a predetermined pressure drop in the air after the air filter compared to that of the air before the air filter. A timer is provided for deactivating the device until a nonreusable water filter element is replaced after a predetermined number of hours of operation. A sensor is provided for deactivating the device when the UV light(s) fail to operate. A thermostat and humidistat can be set in conjunction with each other by a user to minimize energy consumption and maximize water yield. There is also a manual override switch to these conditional settings. First, second, and third indicators provide a signal when the air filter element, the water filter element, and the UV light element(s) are in need of being replaced. A fourth indicator provides a signal when the internal water container is full. It also demonstrates a hybrid system, coupling rainwater collection with water-from-air condensation, designed for off-grid operation. For the portable air-water generator, a United States Patent numbered 5 106 512 has been issued for a condenser that extracts water vapor from air. The air filter is installed before the condenser and a water filter is also built into the system to clean the water after it is being condensed. This concept has started an array of products in line with water generation from air, including water generation coupled with water purification (Reidy 1992). A recent patented design of "A portable, potable water recovery and dispensing apparatus" contains the main parts of a water-generating device and the additional features of UV disinfection, solid-core carbon filter, and mineralizing cartridge, and a choice of hot or cold dispensed water (Xziex 2008). A recently launched water generator product in the US can produce 19.5 l of water at conditions of 70% relative humidity and 25 °C temperature. It has the following systems in one package: hot and cold water dispensing, air and water filtration, ultraviolet disinfecting with automatic shutdown, water level indicator, water leak alarm, and an easy-to-read LCD function display. All these products operate on AC wall-plug system and cost about $300–$350 excluding tax and freight charges.

The Solar-Powered Atmospheric Water Generation and Purification (SAWGAP) system adapts the basic water generation and purification features of the previously mentioned devices, explores the possibility of using readily available parts for low-cost community production, and runs on DC power. A custom DC–AC inverter was also developed and built into the system to enable UV treatment in off-grid solar power locations (Cabacungan et al. 2009).

5.4 Current Practices Around the World

The idea of RWH is not only putting together a tank and a pipe to collect rainwater from a roof. The reality is that RWH is becoming a viable alternative for supplying houses with water. There are many countries such as Germany and Australia where RWH is the norm. Due to the green building movement, RWH systems are becoming more popular in the US and other countries.

Singapore, which has limited land resources and a rising demand for water, is on the lookout for alternative sources and innovative methods of harvesting water. Almost 86% of Singapore's population lives in high-rise buildings. A light roofing is placed on the roofs to act as catchment. Collected roof water is kept in separate cisterns on the roofs for non-potable uses. A recent study of an urban residential area of about 742 ha used a model to determine the optimal storage volume of the rooftop cisterns, taking into consideration non-potable water demand and actual rainfall at 15-minute intervals. This study demonstrated an effective saving of 4% of the water used, the volume of which did not have to be pumped from the ground floor. As a result of savings in terms of water, energy costs, and deferred capital, the cost of collected roof water was calculated to be $0.96 against the previous cost of $1.17 per cubic meter (UNEP 2002).

St. Thomas, US Virgin Islands, is an island city where annual rainfall is in the range of 1020–1520 mm. A rainwater utilization system is a mandatory requirement for a residential building permit in St. Thomas. A single-family house must have a catchment area of 112 m^2 and a storage tank with 45 m^3 capacity. There are no restrictions on the types of rooftop and water collection system construction materials. Many of the homes on St. Thomas are constructed so that at least part of the roof collects rainwater and transports it to storage tanks located within or below the house. Water quality test of samples collected from the rainwater utilization systems in St. Thomas found that contamination from fecal coliform and Hg concentration was higher than EPA water quality standards, which limits the use of this water to non-potable applications unless adequate treatment is provided (UNEP 2002).

RWH is an accepted freshwater augmentation technology in Asia. Singapore is currently committing to infrastructure that is designed to capture every drop of rain that falls. The city of Melbourne, Australia, is piloting a project to capture rainwater from roofs and store it in tanks to be used for toilet flushing and lawn watering. Storing rainwater from rooftop runoff in jars is an appropriate and inexpensive means of obtaining high-quality drinking water in Thailand. The jars come in various capacities, from 100 to 3,000 liters and are equipped with lid, faucet, and drain. The most popular size is 2,000 liters and holds sufficient rainwater for a six-person household during the dry season, lasting up to six months. The results of the program are good with 10 million rainwater jars constructed in just over a 5-year period. Rainwater jars have been successful in the Northeast Thailand because the technology is simple, inexpensive, and understood by a majority of the rural population. Among other factors are the acceptance of rainwater in this region, traditional use of rainwater for drinking, common usage of traditional earthen vessels for rainwater collection for domestic use, relatively cheap cost of the technology, access to water at each house, and the unpalatability of ground water due to high salinity and hardness. The role of the government in the supply and installation of rainwater jars is now over and this role has been taken over by the private sector. However, the success of the Thai jar program has reached international recognition, and other countries are pursuing similar technologies. (Areerachakul 2013).

5.5 Health Risks of Roof-Collected Rainwater

Due to contamination of surface water as well as groundwater, and the occurrence of saltwater intrusion in freshwater aquifers, the quality of water from RWH systems is generally better than that of the above alternatives (Gould and Nissen-Petersen 1999).

If the water is to be used for potable uses then there are chances of water contamination due to not only biological means but also xenobiotics from airborne pollution. Contaminated air can come from a variety of sources and these contaminants are often partitioned onto aerosols, and could be precipitated out of the air via rain. The issue of contamination, and subsequent treatment or use of specific treatments, are important concerns the user should have.

In the US, filtration and disinfection of rainwater for potable water use is commonplace. Several different disinfection processes are used, such as:

Particulate filter(s) with an ultraviolet disinfection unit: Filters must be changed and the unit must be cleaned periodically.

Ozone: The ozone acts as a powerful oxidizing agent that reduces color, eliminates foul odors, and reduces total organic carbon in water. It is disinfectant and is effective against both parasites and viruses. It is produced by passing an electrical current through air or oxygen. It has a very short half-life in water and therefore must be efficiently introduced into the water.

Chlorine: Automatic self-dosing systems are available. A chlorine pump injects chlorine into the water as it enters the building. Appropriate contact time is critical to kill bacteria. Chlorination is often used in combination with two technologies – ultraviolet and ozonation – because a free chlorine residual can be maintained in the distribution system, providing ongoing treatment for viruses.

UV Light: Ultraviolet light is more effective against parasites like Giardia and Cryptosporidium than it is against viruses. UV radiation is effective only in relatively clear water; it is important to have a filter installed upstream to remove turbidity. UV systems do not maintain a disinfection residual in the water, avoiding the chlorine taste disliked by many people. However, there is increased risk for bacterial regrowth in the plumbing system.

Polishing Processes: Additional treatment processes may be applied after the disinfection step. These include corrosion control and activated carbon filtration at the tap. Rainwater lacks the buffering capacity that mineral salts provide to ground and surface water. The low pH of pure rainwater (8% lead), it may be advisable to install granular activated carbon (GAC) filters at drinking water tapes to remove heavy metals. This is unlikely to be a problem in homes constructed after 1988. Any POU (point-of-use) filter that is installed should be certified by ANSI/NSF Standard 53 to reduce the target contaminant (Siegel 2015).

5.6 Guides, Policy, and Incentives

Water harvesting is increasingly important in urban water management strategies globally and many governments are requiring sustainable drainage of surface water to be included in developments that require planning approval, or have drainage implications. Although RWH is an ancient practice, it has never been directly addressed by national plumbing codes. This problem has been recently solved by updates to necessary codes. The IAPMO (International Association of Plumbing and Mechanical Officials Green Plumbing and Mechanical Code Supplement 2012), Uniform Plumbing Code, and ARCSA/ASPE Rainwater Catchment Plumbing Engineering Design Standard (2018) now combine to provide the necessary tools for inspectors, installers, and system designers to safely and confidently embrace and utilize rainwater. With code obstacles eliminated, it appears that rainwater collection and use are poised to become mainstream as jurisdictions around the world turn to rainwater reuse to address intensifying water crises (IAPMO Green Plumbing and Mechanical Code Supplement 2012). The next hurdle to overcome is the lack of incentives. Financial incentives and tax exemptions encourage the installation of RWH systems. Like the Texas Legislature, many states in the US have passed bills, and some local taxing entities have adopted rules that provide tax exemptions for RWH systems. A few public utilities have implemented rebate programs and rain barrel distribution events that encourage RWH by residential, commercial, and industrial customers. In addition to financial incentives, performance contracting provisions in state code can be used to encourage installation of RWH systems.

The following guides can be used internationally and many countries like Canada and India have their own local guides:

ARCSA – American Rainwater Catchment Systems Association, whose mission it is to provide resources and information on rainwater collection, promote the advancement of rainwater conservation, and work with state, county and other local governmental units in promoting rainwater catchment, has several guides. https://www.arcsa.org/store/ViewProduct.aspx?id=3642480

ERCSA – European Rainwater Catchment Systems Association Guide (http://www.ercsa.eu)

EPA – Managing Wet Weather with Green Infrastructure (https://www.epa.gov/sites/production/files/2015-10/documents/gi_munichandbook_harvesting.pdf)

USGBC – LEED V4 NC Water Efficiency Credits (www.usgbc.org)

ASLA – Sustainable Sites (http://www.sustainablesites.org/resources)

The following codes have been introduced to the market on RWH:

ASHRAE/USGBC/ASPE/AWWA Standard 191 – Standards for the efficient use of water in building, site, and mechanical systems. Covers all uses of water within a site and a building.

IGCC – International Green Construction Code, Chapter 7 Rainwater Collection and Distribution Systems Allows ANSI/ASHRAE/USGBC IES Standard 189.1 as an option (http://spc191.ashraepcs.org)

IAPMO (2012)– International Association of Plumbing Mechanical Officials, (https://www.standardsportal.org/usa_en/trade_associations/iapmo.aspx#Standards)

2010 Green Plumbing and Mechanical Code Supplement covers all aspects of a potable and non-potable rainwater catchment system and is recommended to be usedwith all codes. (http://www.iapmo.org/Documents/2012GreenPlumbingMechanicalCodeSupplement.pdf)

2012 IAPMO Uniform Plumbing Code (UPC), (http://iapmomembership.org/index.php?page=shop.product_detailsandflypage=flypage_iapmo.tplandproduct_id=155andcategory_id=27andoption=com_virtuemartandItemid=3andvmcchk=1andItemid=3andredirected=1andItemid=3)

CSI – Construction Specification Institute RWH Systems and Components, Gutters and Downspouts, Domestic water Filtration https://www.edmca.com/media/35207/masterformat-2016.pdf

ARCSA and ASPE – American Rainwater Catchment Systems Association and American Society of Plumbing Engineers Standards for designers on all components of a RWH system. https://www.aspe.org/sites/default/files/webfm/pdfs/ARCSA_ASPE_Draft4_for_public_review.pdf

NSF International Protocol P151 – Health effects from rainwater catchment system components. Additional standards from NSF and ANSI include ANSI Standard 14, 42, 53, 55, 60, and 61 are developed. https://www.nsf.org/newsroom_pdf/water_rainwater_catchment.pdf

The plumbing industry now has authoritative tools at its disposal to standardize the safe and reliable use of rainwater for potable and non-potable applications. These tools are used as references worldwide through green building certification systems like Leadership in Energy and Environmental Design (LEED).

RWH provisions were introduced in the GPMCS and the UPC by the IAPMO Green Technical Committee (GTC) as part of a broader effort to reduce the energy and water consumption of plumbing and mechanical systems while ensuring that these systems are safe and reliable. The GTC is comprised of the broadest group of expert stakeholders ever assembled to develop sustainable plumbing and mechanical requirements. ASPE and ARCSA are well represented on the GTC and played a critical role in the development of the first model code provisions for RWH.

UPC and the International Plumbing Code (IPC) do not directly address RWH systems in either the potable water or stormwater section of the code. Some jurisdictions enforce Appendix J of the 2015 UPC reclaimed water code for RWH systems. It is important to note that RWH systems are not reclaimed water systems. Many local jurisdictions provide guidance in the design of rainwater-harvesting systems. This is an area where the code-writing bodies must quickly adapt to the realities of the building industry and the water shortage crisis throughout the world.

The city of San José, California, revised the "Post Construction Urban Runoff Management" policy in 2011 (City of San Jose, Annual Stormwater Report 2016). Beginning on December 1, 2012, for "Large Detached Single Family Home Projects: Site Design Measures are required. Detached single family home projects, which are not part of a larger plan of development must create or replace 2500 square feet or more of impervious surface, and are required to incorporate one or more site design measures in accordance with provision C.3.i of the MRP (Municipal Regional Stormwater Permit). These measures include the use of permeable surfaces to construct driveways, walkways and patios, directing runoff to vegetated areas, or into cisterns or rain barrels."

RWH System Incentives in Central Texas: listed below are some of the local incentives to promote RWH (https://www.watercache.com/rebates). These offers are available to residents of the respective cities or counties:

- City of Austin offers rebates up to $5000 for systems of all sizes, not to exceed 50% of the system cost.
- City of Sunset Valley offers rebates up to $3500 for systems larger than 300 gal.
- City of San Marcos offers rebates up to $5000 for systems of all sizes, not to exceed 50% of the system cost.
- Hays County offers a property tax reduction by your rainwater collection system construction costs. If you qualify for any of the rebates listed above, they will provide you with all of the necessary documentation for the rebate program. A sample application for RWH Incentive Program for single family homes is presented in the application form presented in the form that follows.

All equipment used solely for rainwater collection in the State of Texas is exempt from sales tax. The city of Austin has a Green Building Program where auxiliary water systems such as rainwater collection systems can provide up to 10 points. The intent of the credit is to reduce potable water use and associated costs for treating and pumping. Requirements meet one of the following for RWH:

- Rainwater is sole source of potable water; 20 000 gal minimum storage; back-up well allowed (10 points)
- >5001 gal storage (5 points)
- 1001–5000 gal (4 points)
- 501–1000 gal storage (3 points)
- 110–500 gal storage (1 point)

Preventing cross connections between auxiliary and drinking water systems is important to protect the health and safety of public water system users. City, state, and federal regulations apply to auxiliary water sources used with drinking water service because they may not meet drinking water standards.

Mumbai, India: The state government has made RWH mandatory for all buildings that are being constructed on plots that are more than 1000 m^2 in size.

New Delhi, India: Since June 2001, the Ministry of Urban affairs and Poverty Alleviation has made RWH mandatory in all new buildings with a roof area of more than 100 sq. m and in all plots with an area of more than 1000 sq. m, that are being developed (New Delhi 2001). Furthermore, the Central Ground Water Authority (CGWA) has made RWH mandatory in all institutions and residential colonies in notified areas (South and southwest Delhi and adjoining areas like Faridabad, Gurgaon, and Ghaziabad). This is also applicable to all the buildings in notified areas that have tube wells.

Sample Form

Hays County Development Services P.O Box 1006 San Marcos TX 78667-1006 2171 Yarrington Road San Marcos TX 78666 512-393-2150 / 512-493-1915 fax

Hays CountyApplication for RWH Incentive Program

Name: ____

Site Address: ________________________________

Legal Description:____________________________

Daytime Phone:

____

Construction Start Date: ___________________

Estimated Completion Date: ________________

Square Footage of Collection Area: ____________ SQ FT

Total Storage Capacity: ____________Gallons

Planned use:

Potable Non-potable Distribution Method:

Gravity Flow Pumped

Disinfection: Yes ________No____________

Method: ________________________________

Do you authorize release of your address and system capacity for fire department mapping and emergency use?
Yes No

Additional Application Requirements:

- Project Summary
- Site Plan with detailed system design
- Detailed Cost Estimation: ____$

(List capital outlays) or Firm

Project Bid: ____________________$

Completion Information:

Date of Completion:

Completion Verification:

County or Central Appraisal District Official

Date of Development Application: _____Permit#:_______

Haryana, India: Haryana Urban Development Authority (HUDA 2019) has made RWH mandatory in all new buildings irrespective of roof area.

Himachal Pradesh, India: All commercial and institutional buildings, tourist and industrial complexes, hotels, etc., existing or upcoming, and having a plinth area of more than 1000 m^2 will have rainwater storage facilities commensurate with the size of roof area. No objection certificates, required under different statutes, will not be issued to the owners of the buildings – unless they produce satisfactory proof of compliance of the new law. Toilet flush systems will have to be connected with the rainwater storage tank. It has been recommended that the buildings will have rainwater storage facility commensurate with the size of roof in the open and set back area of the plot at the rate of 0.24 cft. Per sq m of the roof area (CSE 2019).

Bangalore, India: In order to conserve water and ensure ground water recharge, the Karnataka government in February 2009 announced that buildings constructed in the city will have to compulsorily adopt rainwater harvesting facility. Residential sites that exceed an area of 2400 ft^2 (40×60 ft), shall create rain harvesting facility according to the new law (CSE 2019).

In Australia, RWH is retained for subsequent use such as irrigation and toilet flushing. RWH potential from the roof catchments is investigated based on the rainfall data at 15 different locations in the arid regions of Australia (Rahman et al. 2017). Ten different rainwater tank sizes and three different combinations of water uses are considered. It is found that a 20 kl tank can provide a reliability of 59–98% for toilet and laundry use depending on the location within the Australian arid regions. For irrigation and combined use (toilet, laundry, and irrigation), estimated reliability values are smaller than 30%. At the current Australian water price, RWH system is not financially viable in the Australian arid regions as the benefit–cost ratio is much smaller than 1.0. The harvested rainwater in the arid regions is four to eight times more expensive than the current water price in Australian cities.

Victoria, Australia: Since July 2005, new houses and apartments in Victoria must be built to meet the energy efficiency and water management requirements of the five Star standard, which requires either a rainwater tank for toilet flushing, or a solar hot water system.

South Australia: New homes are required to have a rainwater tank plumbed into the house.

Sydney and New South Wales, Australia: The BASIX (Building And Sustainability Index) building regulations call for a 40% reduction in mains water usage. In order to meet the BASIX target for water conservation, a typical single-dwelling design must include a rainwater tank or alternative water supply for outdoor water use and toilet flushing and/or laundry, among other water conservation devices.

Gold Coast, Australia: Construction of 3000-l (800-gal) rainwater tank is mandatory in the Pimpama Coomera Master Plan area of Gold Coast. This is for all homes and business centers connected to the Class A+ recycled water system (those approved for development after 29 August 2005). The tank should be plumbed to their cold-water washing machine and outdoor faucets.

Queensland, Australia: Residents can get a rebate of up to $1500 for the purchase and installation of home rainwater storages.

Germany: Rain taxes in Germany are a great example of internalizing externalities for a more fair system. Fees are collected for the amount of impervious surface cover on a property that generates runoff directed to the local storm sewer. That means that the more the rainwater is caught and conserved, the less rainwater runs off and is added to the storm drains. Less runoff allows for smaller storm sewers, which, in turn, saves construction and maintenance costs at the site. Thus there is a large incentive to convert impervious pavement/roof into a porous surface. The construction of a RWH system does not require a building approval but it is advisable to report it to the local public health office as well as the local water supplier in Germany. Some regulations and standards (especially DIN 1989) should be taken into consideration during construction and maintenance of a RWH system (Kloss 2008).

Various levels of governmental and community involvement in the development of RWH technologies in different parts of Asia were also noted. In Thailand and the Philippines, both governmental and household-based initiatives played key roles in expanding the use of this technology, especially in water-scarce areas such as northeast Thailand (UNEP-IETC 2002). Compiled by the South Pacific Applied Geoscience Commission (SOPAC) for the United Nations Environment Programme (UNEP) in conjunction with the Tonga Community Development Trust (TCDT) and funded by The Swedish Development Agency (SIDA), a training manual is prepared for teaching single-family RWH. This is a practical guide to be used by extension workers who are involved in RWH projects or programs. It is designed to help them support community members to build technical knowledge and skills related to the maintenance and repair of RWH systems, as well as address critical social and community issues (SOPAC 2004).

In Turkey, on June 23rd, 2017, the Regulation of Water Harvesting Collection, Storage and Discharge was revised and put into force by the Ministry of Environment and Settlements in Turkey detailing the RWH systems (RWHCSD) (2017).

5.7 Green Building Certification Systems and RWH

Green building rating systems are a type of building certification system that rate or reward relative levels of compliance or performance with specific environmental goals and requirements. Rating systems and certification systems are frequently used interchangeably.

Green building rating and certification systems require an integrated design process to create projects that are environmentally responsible and resource-efficient throughout a building's life cycle: from siting to design, construction, operation, maintenance, renovation, and demolition. While the philosophy, approach, and certification method vary across these the systems, a common objective is that projects awarded or certified within these programs are designed to reduce the overall impact of the built environment on human health and the natural environment. Green building rating systems exist to address every project type from single-family houses and commercial buildings to entire neighborhoods according to Whole Building Design Guide (WBDG) of National Institute for Building Sciences in the US (2016). All systems have credits for water efficiency and rainwater collection can provide several points.

RWH scores highly in the UK-based BREEAM (2016) and USA-based LEED v3.1 rating systems, which are used worldwide. The LEED Green Building Rating System was devised as a voluntary, consensus-based national standard for developing high-performance, sustainable buildings.

Projects seeking LEED certification earn points across nine categories addressing sustainability issues, such as sustainable sites and water efficiency. Depending on the number of points earned, projects are awarded one of four levels of LEED certification: Certified, Silver, Gold, and Platinum. LEED has several different systems for different functions of the buildings.

Rainwater management is under Sustainable Sites credit and awards 3 points. This credit applies to LEED Homes. The intent of the credit is to reduce rainwater runoff volume from the site. HOMES Projects that must comply with local requirements of the National Pollutant Discharge Elimination System (NPDES) must follow Case 2 (Ref. EPA)

Case 1. Low Impact Development. Use low-impact development (LID) techniques to minimize the amount of stormwater that leaves the site. Examples of acceptable techniques include the following:

- planting areas with native or adapted plant material (e.g. trees, shrubs);
- installing a vegetated roof;

Table 5.1 Points for permeable area, as percentage of total lot area.

Percentage (%)	Points
50–64	1
65–79	2
80	> 3

Source: *Whole Building Design Guide* 2016

- using permeable paving, consisting of porous above-ground materials (e.g. open pavers, engineered products), a base layer designed to drain water away from the home, and (often) a 6-in. deep (150 mm) subbase; and
- installing permanent infiltration or collection features (e.g. vegetated swale, rain garden, rainwater cistern) that can handle 100% of the runoff from a two-year, 24-hour storm. Single-family home projects may use Tables 5.1 or 5.2 to determine points. To determine compliance for single-family homes, the percentage of the lot area needs to be calculated, including the area under roof, which is permeable or can direct water to an on-site catchment or infiltration feature.

As an alternative approach to determining compliance for single-family homes only, credit is given for reducing the total impermeable area compared to the ENERGY STAR reference home, as listed in Table 5.2.

Thresholds for total impermeable area are then calculated according to the values in Table 5.3, column 1.

Case 2. NPDES: Projects Using LID and green infrastructure to replicate natural site hydrology, manage on-site the runoff from the developed site for the percentile regional or local rainfall events listed in Table 5.4. Use daily rainfall data and the methodology in the US Environmental Protection Agency's Technical Guidance on Implementing the Stormwater Runoff Requirements for Federal Projects, under Section 438 of the Energy Independence and Security Act, to determine the percentile amount.

Water harvesting touches on many of the principles embodied in LEED V4 certification. Harvesting efforts can earn to 15 points in LEED V4. These include:

- Conserving water
- Reducing energy consumption

Table 5.3 Points for reducing total impermeable area (square feet).

Impermeable area (square feet) points	Impermeable area (square feet) points
Points Reference home size * 1	1
Reference home size * 0.66	2
Reference home size * 0.33	3

aust.

Table 5.4 Points for on-site management of water from rainfall events.

Percentile rainfall event	Points
95th	2
98th	3

- Reducing the depletion of natural resources and materials
- Creating a sustainable site
- Use of innovative design

On average, LEED-certified buildings use 30% less water than conventional buildings, which translates to about 1–3 million gallons of water saved per year. Reducing the amount of water that needs to be treated by municipal waste water treatment facilities also reduces pumping and process energy required to these systems.

LEED V4 certification promotes on-site storage and use of rainwater and greywater through water harvesting to lower water consumption cost, as it reduces the impact on storm drainage and municipal treatment systems. Water harvesting efforts can earn a significant number of LEED points across several categories. LEED V4 points that can be received by water conservation are:

Water Efficiency Credits:

Credit 1: Outdoor Water Use Reduction (2 points)

Option 1: No irrigation required. Show that the landscape does not require a permanent irrigation system beyond a maximum two-year establishment period.

Option 2: Reduced irrigation. Reduce the project's landscape water requirement by at least 30% from the calculated baseline for the site's peaking watering month.

Table 5.2 Conditioned floor area of reference home, by number of bedrooms.

	1	2	3	4	5	6	7	8 or more
Floor area (square feet)	1000	1600	2200	2800	3400	4000	4600	+600 ft^2 per additional bedroom
Floor area (square meters)	93	148	204	260	315	371	426	+55.6 m^2 per additional bedrooms

Credit 2: Indoor Water Use Reduction (6 points)

Earn 1 point for a 25% reduction in indoor water usage, 2 points for a 30% reduction, 3 points for 35%, etc., all the way up to 6 points for a 50% reduction.

Credit 3: Cooling Tower Water Use (2 points)

For 2 points, maximum number of cycles achieved without exceeding any filtration levels or affecting operation of condenser water system (up to maximum of 10 cycles).

For 3 points, achieve a minimum 10 cycles by increasing the level of treatment in condenser or make-up water, OR achieve the number of cycles for 1 point and use a minimum 20% recycled non-potable water.

Credit 4: Water Metering (1 point)

Install permanent water meters for the following water subsystems, as applicable to the project:

Irrigation. Meter water systems serving at least 80% of the irrigated landscaped area. Calculate the percentage of irrigated landscape area served as the total metered irrigated landscape area divided by the total irrigated landscape area.

Indoor plumbing fixtures and fittings. Meter water systems serving at least 80% of the indoor fixtures and fitting described in WE Prerequisite Indoor Water Use Reduction, either directly or by deducting all other measured water use from the measured total water consumption of the building and grounds.

Sustainable Sites Credits:

Credit 4: Rainwater Management (3 points)

Implement the following strategies to reduce the annual volume of rainwater runoff from the existing site's baseline condition:

Use LID practices to capture and treat water from 25% of the impervious surfaces from 1.2″ (30 mm) of rainfall.

Rainwater collection system capturing and reusing 25% of the runoff from impervious surfaces.

Integrative Process Credit:

Beginning in pre-design and continuing throughout the design phases, identify and use opportunities to achieve synergies across disciplines and building systems described below. Use the analyses to inform the owner's project requirements (OPR), basis of design (BOD), design documents, and construction documents. Worth one (1) point.

Other than the LEED system, some of the other systems that give credits for RWH are as follows:

DGNB system (2018) ENV2.2_Drinking Water and Waste Water give credit for gray water usage.

BREEAM International 2016 gives points if there is a harvesting amount of 25–50%.

BS 8515:2009 "RWH Systems, Code of Practice," establishes standards for the installation, testing, and maintenance of RWH systems for non-potable applications. It includes standards for filtration, for the manufacture and installation of storage tanks, and a series of approaches for calculating the sizes of tanks.

The Code for Sustainable Homes 2010 (used in the UK) is an environmental assessment rating method for new homes which assesses environmental performance in a two-stage process (design stage and post-construction stage) using objective criteria and verification has been used in UK until 2015. The results of the Code assessment are recorded on a certificate assigned to the dwelling. The Code for Sustainable Homes has provisions to restrict surface water runoff for new build developments and encourages fitting underground tanks to new-build homes to collect rainwater. The credits given are as follows:

5.7.1 Code for Sustainable Homes/BREEAM Support/Points Awarded

Credit Maximum	Achievable (%)	Credit (%)	Description
CSH Wat 1	7.5	1.5	Credits for reduced water consumption
CSH Wat 2	1.5	1.5	Credits for rainwater use for landscaping
BREEAM Wat 1	3.0		3.0%

Minimum requirement for "GOOD" rating and higher.

If the base case is above 5.5 m^3/person/year, all 3 credits could be achievable.

5.8 Conclusion

Single-family RWH systems are one of the major solutions to a pressing water scarcity problem. There are a remarkable amount of ways developed since the 1990s which make harvesting an irresistible option to make water available to single homes. But still, the usage of these systems is not widespread due to lack of incentives and liabilities due to maintenance and health problems. In order to meet upcoming challenges such as climate change, rapid growth, shrinking cities, and water scarcity, water infrastructure needs to be more flexible, adaptable, and sustainable (Sitzenfrei et al. 2013). Sustainable urban drainage systems and water-sensitive urban design solutions are needed for minimizing the hydrological impacts of urban development on the environment (Okhravi et al. 2015).

As water shortages become more prevalent globally, rainwater-harvesting systems will become more essential. The codes must be updated to include these important water-conserving systems. If water-efficient technologies are incorporated at the code level, the resulting cumulative effects on water conservation across the world will be significant, as market opportunities open up for new products, customers demand more water-efficient designs, and designers face less resistance to incorporating efficient systems in buildings.

It is a major defect both socially and economically today to use potable water for areas than can handle the usage of harvested water. The usage of RWH systems in regions that have reasonable rain volume will bring multiple benefits to the user. The wisdom of several ancient civilizations must be put into practice so that water management becomes decentralized and subsequently increases water availability. The codes must be updated to include these important water-conserving systems.

References

Angelakis, A.N. (2013) Evolution of RWH and Use in Crete, Hellas through the Millennia. In Proceedings of the 13th International Conference on Environmental Science and Technology (CEST2013), Athens, Greece, 5–7 September 2013.

ARCSA/ASPE (2018) Rainwater Catchment Plumbing Engineering Design Standard, USA.

Areerachakul, N. (2013). Overviews of RWH and utilization in Thailand: Bangsaiy municipality. *World Academy of Science, Engineering and Technology International Journal of Environmental and Ecological Engineering* 7 (7): 488.

Bildirici, M., Bildirici, Ö. (1997) Historical Water Structures in Konya and surroundings, Türkiye İnşaat Mühendisliği 14. Teknik Kongresi (İzmir, İnşaat Mühendisleri Odası), pp. 1115–1128.

Cabacungan, P. and Cheng C., et al., (2009) Solar-Powered Atmospheric Water Generation and Purification System. 4th International Conference on Humanoid, Nanotechnology, Information technology, Communication and Control, Environment and Management, At Century Park Hotel, Manila.

Center for Science and Environment, (2019) New Delhi, India https://www.cseindia.org/cses-work-on-draft-legal-framework-for-conservation-and-protection-of-inland-wetlands-in-south-asia-5263. (1.1.2019) Accessed on 4.6.2020

City of San José Environmental Services (2016), City of San José Stormwater Management Annual Report, Department of Watershed Protection Division Stormwater Management Section, USA.

Code for Sustainable Homes, (2010) Technical Report, Nov. 2010 ISBN 978 1 85946 331 4.

DGNB, (2018) http://www.dgnb-system.de/de/system/kriterien/neubau_gebaeude (1.2.2018).

Distefano, T. and Kelly, S. (2017). Are we in deep water? Water scarcity and its limits to economic growth. *Journal of Ecological Economics* 142 (C): 130–147.

Gardner, T., Baisden, J. and G. Millar (2004) Rainwater first flush devices – are they effective? In: Proceedings of the Sustainable Water in the Urban Environment Conference. Brisbane, Australia.

Gould, J.E. (1992) Rainwater Catchment Systems for Household Water Supply, Environmental Sanitation Reviews, No. 32, ENSIC, Asian Institute of Technology, Bangkok, Thailand.

Gould, J. and Nissen-Petersen, E. (1999). *Rainwater Catchment Systems for Domestic Supply: Design, Construction and Implementation*. London: Intermediate Technology Publications. 335 pp.

Hammerstrom J. and Pushard D., (2016) Is RWH a Good Investment? Published at online portal: http://HarvestH2o.com

Hamza Ö, (2017) Traditional Water Harvesting In Anatolia; Water Harvesting Prospects For Turkey, World Water Diplomacy and Science News- 2017-10002, ENISSN: 42016–10005.

HUDA: (2019) https://www.cseindia.org/laws-and-policy--1161 (19.09.2019).

IAPMO (2012) Green Plumbing and Mechanical Code Supplement, Uniform Plumbing Code, USA https://www.iapmo.org/green-plumbing-and-mechanical-code-supplement/.

IPPC (2007) 4th Assessment Report, Impacts, Adaptation and Vulnerability, Climate Change, Contribution of Working Group II to the Fourth Assessment Report of the Intergovernmental Panel on Climate Change, 2007, Parry M.L., Canziani O.F., Palutikof J.P., van der Linden P.J. and Hanson C.E. (eds), Cambridge University Press, Cambridge, United Kingdom and New York, NY, USA.

Kloss, C. (2008). *Managing Wet Weather with Green Ionfrastructure Municipal Handbook: RWH Policies*. USA: EPA.

Mays, L.W. (2013). Use of cisterns during antiquity in the Mediterranean region for water resources sustainability. Water Sci. Technol. *Water Supply* 2013 (13): 735–742.

Mays, L., Antoniou, G.P., and Angelakis, A.N. (2013). History of water cisterns: legacies and lessons. *Water* 5 https://doi.org/10.3390/w5041916.

New Delhi (2001) http://www.rainwaterharvesting.org/policy/img/Notice.jpg (1.1.2018).

Okhravi, S., Eslamian, S., and Adamowski, J. (2015). Water Reuse in RWH. In: *Urban Water Reuse Handbook*, Ch. 59 (ed. S. Eslamian), 779–796. Taylor and Francis, CRC Group.

Pacey, A. and Cullis, A. (1989). *RWH: The Collection of Rainfall and Runoff in Rural Areas*. London: WBC Print Ltd.

Panagiotakis, N. (2006). Water Management in the Pediada Region in Central Crete through time. In: *Proceedings of 1st IWA International Symposium on Water and Wastewater Technologies in Ancient Civilizations* (eds. A.N. Angelakis and D. Koutsoyiannis), 271–280. Iraklion, Greece, 28–30 October 2006: National Foundation for Agricultural Research, Institute of Iraklion.

Panigrahi, S. (2017). Roof-top RWH system for official/multistoried building with reference to malda district, WB. *International Journal of Engineering Research and Applications*, ISSN: 2248-9622 7 (1), (Part -1): 111–115.

Prempridi, T. and Chatuthasry, C. (1982). Past and present use of ponds as rainwater storage in Thailand. In: *Proc. of 2nd International Conference on Rainwater Cistern Systems*, 1–21. St. Thomas, United States Virgin Islands, C2.

Rahman, A., Hajani, E., and Eslamian, S. (2017). RWH in arid regions of Australia. In: *Handbook of Drought and Water Scarcity, Vol. 2: Environmental Impacts and Analysis of Drought and Water Scarcity* (eds. S. Eslamian and F. Eslamian), 489–500. USA: Francis and Taylor, CRC Press.

Regulation of Water Harvesting Collection, Storage and Discharge of Turkey (RWHCSD) (2017) www.resmigazete.gov.tr/eskiler/2017/06/20170623-8.htm

Reidy J.J., (1992) "Potable water generator," United States Patent 5149446.

Siegel M (2015), Water treatment in RWH Systems for Potable Use, Water Resources Action Project, Inc.

Sitzenfrei, R., Möderl, M., and Wolfgang, R. (2013). Assessing the impact of transitions from centralised to decentralised water solutions on existing infrastructures – Integrated city-scale analysis with VIBe. *Water Research* 47 (20): 7251–7263. https://doi.org/10.1016/j.watres.2013.10.038, ISSN 0043-1354.

Sojka, S., Younos, T., and Crawford, D. (2016). Modern urban RWH systems: design, case studies, and impacts. In: *The Handbook of Environmental Chemistry*, vol. 47, 351, 209–234. Springer Publishers. ISBN: 978-3-319-29335-6.

SOPAC Joint Contribution report, 178, (2004) Harvesting the heavens: A manual for participatory training in RWH, Prepared for printing by the South Pacific Applied Geoscience Commission (SOPAC) Suva, Fiji Islands

Spanakis, S. (1981) The Water Supply of Iraklion, 828–1939 (in Hellenic); The Technical Chambsweeneyer of Hellas: Iraklion, Greece.

Sweeney, J.F. and Pate, M.B. (2017). Lifecycle analysis of a single-family residential RWH system in a subtropical, metropolitan environment. *Journal of Sustainable Water in the Built Environment* 3 (4): 324–333.

UNEP (2002), RWH and utilisation: an environmentally soundly approach for sustainable urban water an introductory guide for decision makers. http://www.unep.or.jp/Ietc/Publications/Urban/UrbanEnv-2/index.as

UNEP [United Nations Environment Programme] (1982). *Rain and Storm Water Harvesting in Rural Areas*. Dublin: Tycooly International Publishing Ltd.

UNEP-DTIE-IETC/Sumida City Government/People for Promoting Rainwater Utilisation (2002), Sumida City Government.

Whole Building Design Guide (WBDG), National Institute for Building Sciences, (2016), Green Building Standards and Certification Systems, USA.

Wirojanagud W., Hovichitr P., Mungkarndeee P., et al. (1989), Evaluation of Rainwater quality; heavy Metals and Pathogens. IDRC, Canada.

Xziex – Water Generator From Air, (2008) Patented design of a portable, potable water recovery and dispensing apparatus, USA.

Younos, T. (2011). Paradigm shift: holistic approach for water management in urban environments. *Frontiers in Earth Science* 5 (4): 421–427. https://doi.org/10.1007/s11707-011-0209-7, Higher Education Press and Springer-Verlag Berlin Heidelberg.

6

Water Harvesting in Farmlands

Elena Bresci and Giulio Castelli

Department of Agriculture, Food, Environment, and Forestry (DAGRI), Università degli Studi di Firenze, Firenze, Italy

6.1 Introduction

Water Harvesting (WH) represents one of the most important strategies for coping with water shortage in arid and semi-arid regions, where water scarcity has physical and meteorological causes, and where political, social, and economic reasons may hamper a safe access to water.

In drought-prone farmlands, WH represents a fundamental source of water for sustaining agricultural production and increase productivity (Rockstrom et al. 2002; Motsi et al. 2004; Okhravi et al. 2015; Zahraei et al. 2017). It can represent also a viable option to restore degraded agro-ecosystems, that can both provide renovated ecosystem services to farmers and produce animal fodder in restored grazing lands (Castelli et al. 2017; Oweis 2017).

In order to fully develop the potential of implementing WH to support farming (and agro-pastoral) systems in drought-prone areas of the world, researchers, professionals, and practitioners should develop a complete knowledge of all the possibilities connected to these techniques. WH is also often rooted in the historical knowledge of rural populations, and, sometimes, these traditional, site-specific, techniques need only to be rediscovered and inserted in water management strategies.

To do so, this chapter firstly explores the definitions of WH, identifying four main sub-categories (Section 6.2). The fundamental structure of a WH system and the physical and management strategies are then presented. Section 6.3 presents the main features of floodwater harvesting (FWH) and, with a case study example from Ethiopia, discuss the importance to involve local communities in designing and managing the system. Section 6.4 discusses macro-catchment water harvesting systems (MWH), and, with a case study in Kenya, presents an example of planning multiple sand dams combining Geographical Information System (GIS) approach and indigenous knowledge. Section 6.5 analyses micro-catchment water harvesting (mWH), presenting a case study from Ethiopian Highlands where mWH triggered upstream–downstream issues connected with water allocation. Section 6.6 presents the main applications of rooftop water harvesting (RWH) in farmlands, and, with an example from rural Guatemala, explains the need for a regular monitoring and maintenance of WH infrastructure. In Section 6.7, the reasons for integrating soil fertilization with WH are explained and discussed. Section 6.8 reports a short resume of the main features regarding planning, designing, managing, and maintaining WH systems in farmlands, discussing upstream–downstream issues that may occur at watershed scale, including also the diverse strategies that can be taken in consideration for FWH, MWH, mWH, and RWH. Future developments are also presented. While not representing a review, the study drafts on multiple experiences carried out around the globe about water harvesting in farmlands. While adopting the most common definition for WH, it should be acknowledged that other definitions and subdivisions of WH techniques can be found in the literature.

6.2 Water Harvesting: Definitions

Definitions and classifications of WH have been proposed by many authors over the years and they are available in literature. An updated overview of definitions can be found in Mekdaschi Studer, and Liniger (2013). Furthermore, in old references, the term "water harvesting" has been used more frequently than "rainwater harvesting" (Myers 1975; Boers 1994). Very often the two forms have been interchangeably used (Biazin et al. 2012), as "the collection and storage of

Handbook of Water Harvesting and Conservation: Basic Concepts and Fundamentals, First Edition. Edited by Saeid Eslamian and Faezeh Eslamian.

any form of water either from runoff or creek flow for irrigation use" (Boers and Ben-Asher 1982; Critchley and Siegert 1991; Siegert 1994; Nasr 1999; Oweis et al. 1999; Falkenmark et al. 2001).

In the last 30 years, attention in Sub-Saharan Africa (SSA), the Middle East, and Southeast Asia has been put on developing and applying techniques for collecting, storing, and using precipitation for agricultural purposes (Rockstrom 2000; Rockstrom et al. 2002; Oweis et al. 2004; Humphreys and Bayot 2008). In agricultural uses, the supplemental irrigation for rainfed grown crops, the provision of water for livestock, fodder, and tree production and, less frequently, water supply for fish and ponds are included.

Beyond the definition, the overall aim of WH is to collect runoff or rainwater from areas of excess or where it is not going to be used, store it, and make it available and utilizable where and when there is water need. Water availability is then increased, allowing to overcome lack of water in space and time. Through WH, more water is then available for domestic, livestock, and agricultural uses, bridging drought spells and dry seasons through the collection and the storage of water when it is available.

More recently, the concept has been extended to encompass in situ techniques and appropriate land management practices which enhance infiltration and reduce surface runoff and soil evaporation (Rockstrom et al. 2002; Temesgen 2007). WH interventions can be seen as part of the sustainable land and water management system, defined as the use of land resources, including soil, water, animals, and plants, for the production of goods to meet changing human needs, while simultaneously ensuring the long-term productive potential of these resources and ensuring their environmental functions (Mekdaschi Studer and Liniger 2013).

An updated all-embracing definition of WH technologies can be found in Ouessar (2012) as: "The collective term for a wide variety of low-cost interventions which are primarily or secondarily intended to collect natural water resources which otherwise would have escaped from human reach, and buffer them through storage and/or recharge on or below the soil surface. The effect is increased retention of water in the landscape, enabling management, and use of water for multiple purposes. Water harvesting technologies can operate either as independent units, or require embedding in a larger system of environmental management interventions, or require specific natural conditions."

A WH system is composed by the following elements: a catchment or collection area, the runoff conveyance system, a storage component, and an application area. In some cases, the components are adjacent to each other; in other cases they are connected by a conveyance system. The storage and application areas may also be the same, typically where water is concentrated in the soil for direct use by plants. The catchment or collection area is where rain is harvested in the form of runoff. The catchment may be as small as a few square meters or as large as several square kilometers. It may be a rooftop, a paved road, compacted surfaces, rocky areas, open rangelands, cultivated or uncultivated land, and natural slopes. The conveyance system is where runoff is transported through gutters, pipes (in case of RWH), or overland, rill, gully, or channel flow and either diverted on cultivated fields (where water is going to increase the soil water content and being stored in it) or into storage facilities. Unlike other authors, this chapter adopts the recommendation of Oweis (2017), who include in WH techniques only those ones that involved the transformation of rainfall in runoff, and the direct harvesting of runoff. Techniques such as terraces are then not considered as WH, since they function as soil and water conservation structures after rainfall infiltration, but not runoff. Similarly, structures like Qanats or underground dams, which harvest groundwater flows and not runoff, will not be considered.

The storage component is where the harvested runoff water is going to be stored until it is used by people, animals, or plants. Water may be stored in the soil profile as soil moisture, above ground (jars, ponds, or reservoirs), underground (cisterns), or as groundwater (near-surface aquifers) (Oweis et al. 2012). When concentrated runoff is directly diverted to fields, the application area is the same as the storage area, as plants directly uptake the accumulated water in the soil. The different storage systems can keep the water until it is used either adjacent to the storage facilities or further away. The application area or target is where the harvested water is put into use either for domestic consumption (drinking and other household uses), for livestock consumption, or agricultural use (including supplementary irrigation) (Mekdaschi Studer and Liniger 2013). The two common criteria used to classify WH systems are the catchment type and size, and the method of water storage. The classification of WH based on catchment type (Table 6.1) considers four groups: FWH, MWH, mWH, and RWH (Mekdaschi Studer and Liniger 2013). This categorization considers the size of catchment and takes account of storage methods and end use. It integrates the classifications used by Critchley and Siegert (1991), Oweis et al. (2012), and Tuinhof et al. (2012).

6.3 Floodwater Harvesting in Farmlands

FWH is defined as the harvesting of floodwater from ephemeral catchment streams and riverbeds (Mekdaschi

Table 6.1 Overview of water harvesting systems.

	Floodwater harvesting (FWH)	Macro-catchment water harvesting (MWH)	Micro-catchment water harvesting (mWH)	Rooftop water harvesting (RWH)
Catchment: application area ratio	100 : 1–10 000 : 1	10 : 1–100 : 1	1 : 1–10 : 1	—
Catchment area	2–50 km^2	0.1–200 ha	10–1000 m^2	—
Catchment type	Ephemeral river catchment	Hillsides, pasture land, forests or roads, and settlements	Generally bare, with sealed, crusted, and compacted soils	Rooftop
Source type	Temporary channel flow	Overland flow or rill flow	Sheet and rill flow	Sheet flow

Source: Modified from Mekdaschi Studer and Liniger (2013).

Studer and Liniger 2013). The rainfall accumulation catchment can be several kilometers long, and precipitation can occur far away from the harvesting site. In these systems, the dry river bed is the flow conveyance system, while floodwater can be either be diverted or harvested within the river bed (Van Steenbergen et al. 2010; Mekdaschi Studer and Liniger 2013). Water is usually applied before the planting season, and seeds are planted after floodwater recession from fields, taking advantage of soil moisture stored in the soil. Minor floods, after the main rainy season, represent an additional source of water that can be used as supplemental irrigation during the growing phase.

FWH can be then subdivided in two groups (Mekdaschi Studer and Liniger 2013):

- Floodwater diversion systems: in which floodwater naturally overflows or is forced to leave the riverbed and then conveyed to cultivated fields.
- FWH within the streambed: where floodwater is directly dammed within the riverbed, allowing on-site infiltration, and then the riverbed itself cropping.

A list of main FWH techniques is reported in Table 6.2.

FWH is widely applied in hyper-arid to semi-arid farmlands across the world, where potential evapotranspiration largely exceeds available rainfall. Irrigation schemes are influenced by the hydrology typical of wadi (ephemeral rivers) catchments. Rainfall occurs in high mountain ridges and it is often scattered and unpredictable, both in frequency and intensity. At management level, FWH systems are characterized by elevated unpredictability in water availability, since both extremely dry or water abundant year may occur, and are risk-prone, since most violent floods can damage the irrigation infrastructures and flooding human settlements and villages.

To cope with these characteristics, farmers have identified diverse strategies, documented around the world. Firstly, to cope with the unpredictability of interannual water availability, farmers living in flood-based agricultural systems rely on multiple livelihood sources, like pastoralism of off-season works, to maintain their livelihood in dry years (Van Steenbergen et al. 2010). Water rights are

Table 6.2 Classification of main FWH techniques.

Name	Type	Description	References
Spate irrigation	Floodwater diversion	Deviation of floodwater in fields adjacent to riverbed, made with weirs and/or diversion bunds that work as discharge separators. After the diversion, water is conveyed with primary canals to irrigated areas.	Tesfai and Stroosnijder (2001), Komakech et al. (2011), Van Steenbergen et al. (2011), and Zimmerer (2011)
Flow recession farming	Floodwater diversion	Crop cultivation over areas naturally flooded, after flood recession.	Saarnak (2003) and Motsumi et al. (2012).
Water spreading weirs	Floodwater diversion	Spreading of floodwater over a large area with a masonry weir built perpendicular to flow direction.	Mekdaschi Studer and Liniger (2013)
Jessour	Floodwater harvesting in streambed	Agricultural terraces built within a wadi riverbed	Gabriels et al. (2005), Abdelli et al. (2012), and Adham et al. (2016)

Source: Mekdaschi Studer and Liniger 2013.

reactive, describing acceptable practices in a given situation, rather than prescribing quantifiable water amount to be delivered to individual farmers. As an example, in spate irrigation systems, farmers are allowed to impound floodwater for a certain time over their field, before breaking the field bund leaving water flowing to downstream areas (Mehari et al. 2011). To cope with frequent destructive floods, farmers groups are usually characterized by a high level of cooperation, in order to rebuild damaged structures during the dry season (Van Steenbergen et al. 2010). Flood-based farming areas are also characterized by human settlements exposed to flood hazard, given the proximity of farmers villages to flooding area. Recent studies have pointed out how development project on flood-based farming system should also take into account flood mitigation strategies for settlements (Castelli and Bresci 2017).

In addition to this, it is widely recognized that FWH systems depend and are supported by a strong indigenous knowledge, gained by local population in centuries of practice (Mehari et al. 2005; Motsumi et al. 2012). Incorporating local knowledge in modernization and planning projects is thus vital for successful implementation of development initiatives.

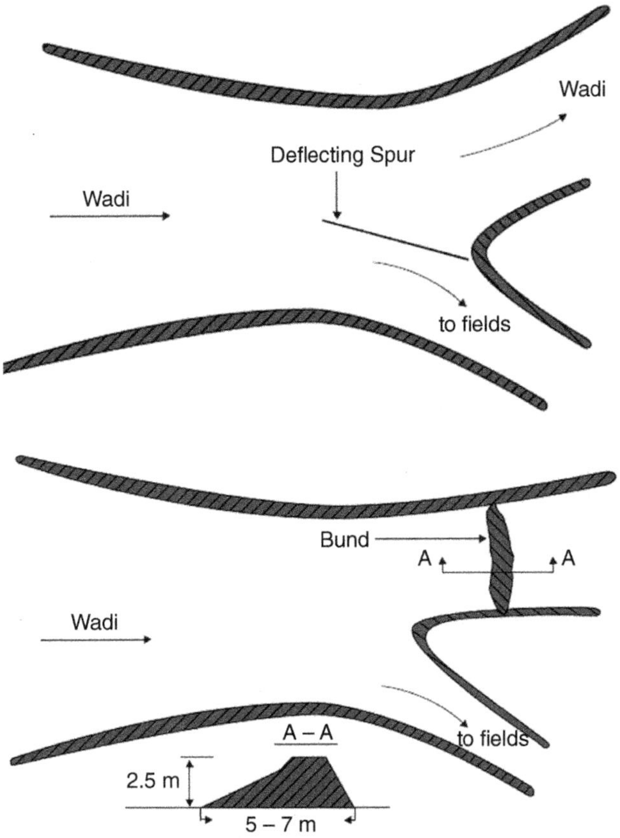

Figure 6.1 Types of spate irrigation bunds. Top: spur-type diversion; bottom: bund-type diversion. Source: Van Steenbergen et al. (2010).

6.3.1 Case Study: Spate Irrigation Systems in Raya Valley

Among many FWH techniques, spate irrigation represents one of the most ancient and diffused. The technique is based on the diversion of floodwater from ephemeral rivers (wadi) for multiple uses, including groundwater recharge, with bunds and weirs made in earth, brushwood, or concrete (Van Steenbergen et al. 2010; Hashemi et al. 2013).

Diversion can be of two main types: Spur-type diversions are bunds built parallel to water flows, that deviate floodwater toward the main irrigation canal; while bund-type diversions are realized with a bund that raises water level during flood events, forcing water to enter in an irrigation canal situated on the riverbank (Figure 6.1).

Spate irrigation was originally born in Yemen 5000 years ago and today it covers around three million hectares of irrigated land in arid and semi-arid regions of Asia, Africa, and Latin America (Tesfai and Stroosnijder 2001; Van Steenbergen et al. 2010, 2011; Zimmerer 2011; Hashemi et al. 2013). In arid and semi-arid areas of Ethiopia, spate irrigation has been practiced since centuries. In particular, in Raya Valley, a dryland plateau located in the northern Tigray Region, the technique is widely diffused, with many development projects realized and research undergone (Hagos et al. 2014; Fenta et al. 2015; Demssie 2016; Castelli and Bresci 2017).

6.3.1.1 Modernization of Spate Irrigation in Raya Valley

Raya Valley is characterized by fertile soils and moderate water availability, representing one of the main niches for the development of agricultural production in arid Tigray Region (COTWRD 2005; Ayenew et al. 2013). Traditional spate systems are based on the regular reconstruction of diversion headworks washed away by high flooding event. Headworks are made by poor materials, such as earth, stones, and brush woods. In order to improve agricultural productivity, Ethiopian government and development agencies developed several modernization projects in the last 20 years, by implementing new schemes based on concrete structures.

However, despite the high effort, most modernization projects failed because the main characteristics of the wadi environment were not considered, adopting rather a design strategy typical of perennial rivers (Embaye et al. 2013; Flood Based Livelihood Network Foundation 2016; Castelli and Bresci 2017). The main problems involved the destruction of diversion structures and heavy siltation. After a

Figure 6.2 Hybrid diversion headwork, Oda spate irrigation system, Raya Valley. Source: Flood Based Livelihood Network Foundation (2016).

long series of failed projects, rural farmers' expertise began to be considered within the structure design, leading to the realization of "hybrid" diversion structures (Flood Based Livelihood Network Foundation 2016), with an elevated diversion angle of 135°, like spur-type diversion style, with an open off-take (Figure 6.2). The new implemented systems also include a multiple intake scheme, rather than a single intake. This solution is based on traditional spate irrigation schemes, and, as shown by Castelli and Bresci (2017), increases the resilience of the system: if one modernized intake is damaged and stops working, the spate irrigation scheme cannot convey water to farmers' field, while if there are structural failures in any of the multiple intakes, the system can keep working.

Recent studies also highlighted that farmers in Raya Valley autonomously developed technical solutions to cope with extreme conditions, typical of flash floods in wadi environments (Mehari et al. 2005; Castelli and Bresci 2017). One of the most interesting examples is represented by uphill diversion channels found the traditional spate system of Harosha (Castelli and Bresci 2017), that are conceived for reducing sedimentation in the scheme. Floodwater, that has a high velocity, is slowed down in the uphill channel, allowing the occurrence of most of sedimentation in the first tenth of meters of the channel. Farmers then dredge this first portion, while heavy siltation of cropped field is avoided.

6.3.1.2 Water Rights and Regulation of Raya Valley Spate Irrigation Systems

As for many FWH techniques, water rights in Raya Valley spate systems are reactive, rather than prescribing precise amounts of water to be withdrawn. If there is enough water, all tertiary canals are flooded at the same time, if floodwater is scarce, or comes from a minor flood, each canal is flooded alone. To establish an irrigation order, it is typical to organize irrigation turns in the basis of the hours of contribution of each single farmer. Women and old farmers are exempted from the law, and women can get water first if they contribute regardless the number of hours of work. These rules help to establish a sustainable management of the cyclic reconstruction of the headworks and to increase the cooperation within the farmers' community (Castelli and Bresci 2017).

6.4 Macro-Catchment Water Harvesting in Farmlands

MWH is classified as a technique of harvesting of runoff water with structures that are located at the outlet of a catchment area characterized by compacted surfaces, such as roads, rocky areas, open rangelands, cultivated and uncultivated land, and natural slopes (Mekdaschi Studer and Liniger 2013). The collection catchment may range from 2 to 200 ha. The subdivision between MWH and FWH is often unclear. According to Mekdaschi Studer and Liniger (2013), in the present work FWH is intended as the collection of floodwater from ephemeral rivers (wadi) characterized by flash-flood hydrology, while MWH collects sheet flow and runoff from hillslope and minor streams.

MWH encompasses water storage in the aquifer, in the soil, and in open water surfaces. The main techniques include large semi-circular or trapezoidal bunds (earth or stone), road water harvesting and open surface water storage in dams, ponds, and pans, sand dams, and horizontal and injection wells.

6.4.1 Case Study: Sand Dams in Kenya

A significant measure to climate change adaptation is represented by the local water storage, increasingly seen as a method for ensuring water availability and, then, food security to rural and urban populations, especially in developing countries (Kashyap 2004). In particular, for arid and semi-arid areas outside the reaches of perennial rivers and where there is no or little groundwater available, the need for increasing storage capacity and consequently water security, is of a primary importance. In the drier areas of SSA, water conservation is of high priority. The possibility of collecting water in the rainy season and storing for the dry season, or even from wet years to dry years, is highly important. People living in areas characterized by extremely variable and erratic rainfall may experience both droughts and floods, with a consequently insecure livelihood (Ertsen and Hut 2009).

Water supply solutions in rural areas require low-cost systems that can be constructed, operated, and easily maintained, with a very high degree of local community involvement (Lasage et al. 2008). Water harvesting solutions for water storage have been applied and used since ancient times in arid and semi-arid regions, such as in the Middle East. Indigenous knowledge makes easy the development and maintenance of such interventions. Therefore, they can be seen as a valid adaptation strategy to climate variability and change.

A well-known example of local "community-based" adaptation to drought can be found in southeast Kenya, in Kitui District, 150 km east from Nairobi. In this area, more than 700 small-scale dams, so-called "sand dams," have been realized over the last 50 years, with the majority in the last 20 years, to reduce the vulnerability of the local communities to droughts. Limited scientific research has been done on this subject, however results show positive, sustainable, environmental, and social impacts that could increase adaptive capacity to climate change conditions (Lasage et al. 2008; Pauw et al. 2008; Quilis et al. 2009). It is, then, also possible to think to the upscaling of sand dams as an appropriate technology for drylands in other regions (Ryan and Elsner 2016).

Sand dams are concrete structures, realized across ephemeral sand-river beds to harvest water for supporting multiple uses, including water for human consumption, small-scale irrigation, and livestock watering (Foster and Tuinhof 2004; Hut et al. 2008). Once the dam construction is finished, the first rains will fill the upstream area of the dam with water, and silt and sand will fill both upstream and downstream portions. The coarser sand, having the highest settling velocity, will deposit upstream of the dam starting the filling operation. The new stratum of deposited sand is able to provide additional water storage capacity, retained in the void sand pores, then creating an artificial aquifer. Material in suspension, with smaller grain sizes, such as silt, will pass over the top of the dam wall and continue to move downstream.

Sand dams may be considered fully functioning within a period of five to eight years after the end of construction (Ryan and Elsner 2016) and, once filled, dams can retain up to 25–40% of their total volume as water (Maddrell and Neal 2012).

The potentiality of a sand dam filling operation is a function of the availability of sand to be transported along the ephemeral river and to be trapped by the wall dam, the longitudinal slope of the wadi, which allows construction of weirs, the hydrogeology to suit storage structures, and the presence of a population willing to collaborate in the dam construction and make use of the water (Nissen-Petersen 1997).

The presence of sand dams increase the volume of groundwater available for abstraction, prolonging the period in which groundwater is available for withdrawal (Nissen-Petersen 1982).

The stored water in the aquifer can be extracted by using wells, pumps, or digging a scooping hole, directly in the riverbed or banks in the upstream regions of the dam (Quilis et al. 2009). Great advantages by storing the water in the sand may be seen in terms of both quantity and quality, because stored water is protected from high evaporation losses and contamination from animals (Guiraud 1989; Tuinhof and Heederik 2003). In addition, as water moves through the artificial sand aquifer, it is also filtered, and biological threats like bacteria can be reduced (Huisman and Wood 1974) and fewer mosquitos are present in the area, because of a lack of surface water. Water is then safe for drinking (Avis 2014).

A sand dam construction induces land cover changes due to the additional water presence and its movement as a result of the new aquifer formation and its recharge. Attention in the design should be posed so that the natural flow of the ephemeral river is not altered, and to avoid erosion downstream of the dam due to a reduction of sediment transportation.

The dam also obstructs groundwater naturally flowing through the permeable riverbed. This creates higher upstream groundwater levels that subsequently move the water into the adjacent riverbanks, inducing groundwater levels raising. The subsurface groundwater flows and seasonal rains recharge the groundwater aquifer (Hoogmoed 2007).

The section where the dam is going to be constructed should be identified in a long straight stretch of the river rather than before a bend. A central spillway in the wall dam will release the excess water to the riverbed and wing walls should be able to keep the flood waters remaining in the river and not going around the dam, causing erosion and eventually undercutting the dam walls.

Land management practices, like terracing and plantings, are in general associated with sand dam construction in the upstream portion of the catchment to control erosion and transport of silt material. In fact, silt reduces the dam's ability to store water. Small rills are blocked to prevent soil erosion with plantings, sandbags, or smaller sand dams (Maddrell and Neal 2012).

6.4.1.1 GIS and Local Knowledge for Selecting Best Sites for Sand Dam Constructions in Kenya

Sand dams are recognized as an example of adaptive response to drought in rural areas, but they are still currently only promoted by a small number of national and international non-governmental organizations (NGOs)

(Ryan and Elsner 2016). In Kitui, East Sub County (Kenya), construction of sand dams has been carried out since the 1990s by a local NGO named Sahelian Solutions (SASOL), with a community-based approach. The process includes community mobilization to plan and agree on modalities of intervention, such as material availability, labor, modalities of accessing the groundwater, etc. The realization of a sand dam requires then operating at larger scale, such as watershed scale rather than planning singular interventions at a local level. This makes it necessary to address issues, like land ownership for the site where the sand dam is constructed, as well as involvement of local institutions.

In selecting suitable sites for sand dam construction, the SASOL involves the community, through meetings organized by the local administration, informing the community on the sand dam construction project, focusing mainly on the benefits offered by the sand dam implementation. The criteria required for choosing a suitable site are mainly based on the local knowledge about the physical characteristics of the area. Specifically, the best areas are considered the ones with prolonged accumulation of water after rainy seasons, areas where agricultural activities are mostly carried out after the rains due to availability of water, as well as areas with vegetation presence, like fig trees. Furthermore, the selection of the areas for the construction of the sand dam depends on the willingness of the community to contribute with labor both for mobilizing local materials and dam construction, as well as on land owners to allow access to the facilities to be used by the communities and undertake horticultural activities and livestock watering. Furthermore, local construction materials like stones, sand, and water should be available nearby (SASOL Foundation 2004). Based on local knowledge, the community identifies some sites, and then SASOL proceeds with the verification of the suitability based on a technical assessment to ensure the sites meet the conditions, such as the presence of a stony catchment (source of sand), a sandy riverbed, two high and strong river banks, a maximum width of 25 m, and waterholes/scoop holes. When the community and SASOL agree on the necessity of dam construction, a site committee (or dam committee) is established, to coordinate community involvement in the building process. The members of this committee are selected by the community. In general, 20 families are involved in dam work construction. Within this committee agreement on site selection, rules and inputs of work are made. SASOL facilitates the site selection and the construction works since they have a technical expertise. During the process of dam construction, SASOL representatives are present in the field, to support the community with technical knowledge. The construction starts with digging a ditch in the riverbed to reach the bedrock. This ditch is then filled with rock material and a mortar used, with the wall dam rising 1–4 m above the surface, depending on the site conditions. The work for dam construction is done by roughly 15 persons, taken from the community. The construction can take from three to six months, depending on the local conditions. In general, the water users are committed in the maintenance of the dams (Beimers et al. 2001; Mutiso 2003; Aerts and Lasage 2005; Borst and de Haas 2006).

As a result of dam construction, the average walking distance for accessing water per capita has been highly reduced and then, additional economic activities have been able to start, contributing to the diversification and stabilization of the income base for families. In general, the increased availability of water boosted agricultural production of the region and has significantly helped communities in adapting to unfavorable climate conditions (Lasage et al. 2006).

In the literature, few examples of methodologies for the selection of the best sites for the construction of sand dams and underground structures in general can be found, while most of the studies (El-Awar et al. 2000; Al-Adamat et al. 2010; Ziadat et al. 2012) focus on surface water harvesting interventions. In 2015, the University of Firenze (GESAAF Department) started to work on the identification of suitable sites for sand dam construction using GIS as an expeditious method for the first identification of potential sites to be then discussed and approved by communities. Local knowledge plays a predominant role in selecting the sites on the basis of a suitability map produced through GIS techniques.

The procedure makes use of reliable data, such as soil and topographic suitability, land cover/land use, and runoff generating potential. Hydrologic modeling, remote sensing, and GIS techniques are applied to select a suitable site. In selecting a suitable site, decision rules, which specify how to combine a set of criterion maps according to some preferences based on evaluation criteria, are used. The decision rule is the weighted linear combination (WLC) operator. WLC involves standardization of the suitability maps, assigning the weights of relative importance to the suitability maps, and then combining the weights to obtain an overall suitability score. To assign the weights to the previously mentioned factors, analytic hierarchy process (AHP) is used and the sites with the highest score are selected. The weights of the themes and their features were assigned and normalized using the AHP. A suitability map for sand dams' potential sites has been obtained and then validated by comparing map with the position of 20 existing sand dams constructed by SASOL between 2015 and 2016. The proposed GIS method has been validated, and the obtained suitability map can be considered for future construction of sand dams as well as in allocating new projects.

Further tests will be done on areas where sand dam construction is going to take place (Mburu 2016).

6.5 Micro-Catchment Water Harvesting in Farmlands

mWH is characterized by small catchment areas almost equal to the application area. Multiple small structures are built in a regular pattern along a hillside to block and store in the soil the rainfall falling in the upstream area. Structures include traditional WH technologies, such as Negarim and Half-moons (Mekdaschi Studer and Liniger 2013), and more simple forms of water impounding, such as trenches, infiltration pits, and contour ridges (Oweis 2017). Regular patterns can be obtained also through mechanization of WH, as for example with the so-called Vallerani system (Figure 6.3), developed by Italian agronomist Venanzio Vallerani, that allows to create multiple mWH basins along a hillslope through a modified tractor plough (the "Delfino" system) (Oweis 2017).

(a)

(b)

Figure 6.3 (a) Ploughing operations with Vallerani System, (b) mWH ponds obtained with Vallerani Delfino Plough. Source: Photo by W. Critchley, from Mekdaschi Studer and Liniger (2013).

6.5.1 Case Study: Multiple Micro Catchment Systems in Ethiopia

Ethiopian highlands have been the subject of a large effort of land rehabilitation, including widespread adoption of mWH, also called in situ WH (Grum et al. 2017). However, it is still uncertain if in such concentration of water conservation systems, the upstream portions of ephemeral catchment can hamper downstream farmlands. Catchments of northern Ethiopia, in fact, are characterized by steep slopes and hilly landscapes in upstream portions, while the downstream part is characterized by alluvial lowlands where ephemeral rivers (wadi) end in alluvial fans. Spate irrigation systems are distributed along the lowland wadi channels (see Section 6.3.1), relying on flash floods and on residual milder discharges generated by scarce and ephemeral rainfall in the upstream part of the catchment. Thus, in arid areas, the intensification of WH in the upstream part of a catchment can thus reduce water availability for downstream farmers (Dile et al. 2016).

The project "Harnessing Floods to Enhance Livelihoods and Ecosystem Services" funded by the Water and Land Ecosystem research program of CGIAR aimed to analyze upstream–downstream dynamics in two Tigray Region, Oda catchments, with few interventions of soil and water conservation, and Guguf catchment, with intense soil and water conservation, including mWH structures like contour bunds and trenches. In this particular case, mWH was not only used to provide water for crops, but also to restore the degraded landscape to provide multiple benefits to farmlands. The project adopted an ecosystem services-based approach, to quantify impacts on both water availability and other services provided by the catchment to upstream and downstream farmers. A joint study carried out by the University of Florence (Italy) and Mekelle University (Ethiopia) (Castelli et al. 2017) revealed how, in the Ethiopian context, mWH adoption slightly decreased the water availability for downstream part of the Guguf catchment, but enhanced ecosystem services at the full watershed scale, including the reduction of erosion, restoration of grazing lands, increase of soil moisture, while these benefits were not visible in the Oda catchment. However, the same study emphasized how the increased water availability for downstream farmers, given by extensive water conservation at catchment scale and leading to a more stable flow in the riverbed, allowed them to build an irrigation system in the upstream area of the Guguf catchment. It was reported that this latter implementation may lead to emerging conflicts for water allocation at the catchment scale. Thus, this representative case study shows that widespread mWH adoption for increasing water availability is not a magic black box, and, especially when implemented at a large scale, should be adequately monitored to avoid inequalities.

6.6 Rooftop Water Harvesting in Farmlands

RWH is a subsection of rainwater harvesting. Rain falls on roofs and then runs off from it. It is then collected directly from the surface where it falls (Thomas and Martinson 2007). RWH is a low-tech system, a simple and affordable tool that can be easily conformed both to urban and rural areas (Worm and van Hattum 2006), significantly increasing the adaptability of smallholder farming systems to extreme weather events and to climate changes, by providing a more stable access to water resources. In many arid areas of the world, it can be considered the only source of water supply (Bailey et al. 2017).

The collected runoff is extremely variable since it is dependent on rainfall. In years with low rainfall, the runoff flow would be very limited. However, since the collected runoff is channeled into a collection tank, water can be taken from it whenever it is needed, even some time after the last rainfall. In addition, as the tank is generally located directly next to the building whose roof is utilized as collecting surface, the stored water is used as water supply for the building uses, reducing the needs of transporting the water somewhere far from it. In rural areas, roof systems are mainly utilized for household use (Gur and Spuhler 2010).

A roof water harvesting system consists of a suitable roof as collecting catchment area, a filter, a storage tank, and a supply facility (Mun and Han 2012). To be considered suitable, a roof should be made of tiles, metal, or plastic sheets; grass and palm leaf are not suitable. Gutters are usually placed at the lower roof to concentrate and redirect rainwater into the collection tank.

More commonly, rainwater is harvested from roofs of individual houses. There are also examples of collecting it from roof in institutions, such as a school. In this case, problems related to management may arise, mainly related with who owns the water or who is responsible for the system management (Thomas and Martinson 2007).

Many examples of roof water harvesting can be found in the literature. Das et al. (2017) showed the results of roof water harvesting use in the northeastern region of India, where the mean annual rainfall is higher than 2000 mm. Despite the rainfall amount during post and pre-monsoon season, water scarcity is experienced and agricultural activities are limited during summer and winter seasons. The harvested rainfall water was used by farmers for diversification activities such as raising crops (broccoli, maize, tomato, French bean, etc.) and livestock growing (poultry and pig), in addition to domestic use. The farmers with no roof water harvesting were only able to use rainfed cropping.

6.6.1 Case Study: Rooftop Water Harvesting in Guatemala

Areas within the region of Corridor Seco, consisting of seven departments in Guatemala and extending into El Salvador, Honduras, and Nicaragua, suffer famines and droughts on an almost annual basis. Erratic and unreliable annual rainfall (around 940 mm y^{-1}, but concentrated in summer) with extreme events and a long dry period, caused, in recent years, conditions of food and water scarcity. In 2001, a 40-day drought caused the loss of corn and bean crops in the municipalities of Camotán, Olopa, and Jocotán, in the department of Chiquimula, Guatemala, also causing 48 deaths. Furthermore, in 2017, between 150 000 and 400 000 families were at risk of famine and drought (Wirtz 2017).

In 2013, the project "Accesso alla risorsa idrica con tecniche appropriate e sostenibili nelle comunità rurali guatemalteche del Municipio de Jocotán del Guatemala per garantire la sovranità alimentare e combattere la denutrizione infantile" – "Improving water access with appropriate and sustainable techniques in the Guatemalan rural communities of the Municipio de Jocotán of Guatemala to guarantee food sovereignty and combat child malnutrition" – installed 34 household RWH systems in the municipality of Camotán (Laurita et al. 2018). These structures were realized to facilitate water access for rural families, since the few springs able to satisfy their water needs were about to dry up due to climate change effects. The collected water was utilized for irrigating family gardens, realized on terraced areas, aiming at favoring the diet improvement with crop diversification. EMAS (Escuela Móvil de Aguas y Saneamiento) pumps were utilized for pumping water from the underground reservoir, realized excavating and made impermeable with the use of a plastic film of 3 mm thickness (Bresci et al. 2013). The first results showed promising perspectives for what concerns the acquisition of consciousness of the benefits provided by a varied and balanced diet and the awareness of the role played by family horticulture, not only for self-sustenance, but also as a means for land reappropriation to cope with large-scale agro-business activities (Laurita et al., 20 189). A participatory design methodology has been applied in 2017 during a two-week field activity, to identify and propose improvements in the roof water harvesting collection and storage system. The participatory analysis showed that the main problem was related to the storage system, in terms of the reduced amount of stored water due to water losses for infiltration through the broken plastic film. Breakages in the plastic coating were originated both by rodents' attacks and irregularities in the surface of the reservoir. Additional problems were caused by the presence of insects and snakes within the cistern. The analysis of the results of the project showed an improvement in the diet variety, the increase of seed production, the availability of enough water for domestic use and horticultural production, which allowed local communities to sell vegetables in the local market, and then livelihood improvement.

The investigation work at the site was carried out by experts from GESAAF department (University of Florence, Italy), from the Asociación Santiago Jocotán (Guatemala), and the local users from Dos Quebradas and Lantinquin area in May 2017. The main outcomes of the process were the realization of a more resistant cistern, with a regular and raised edge to avoid the intrusion of rodents, snakes, and insects. A pyramid trunk-shaped excavation was realized with base dimensions of 2.7 × 2 m and a depth of 0.50 m. In order to increase the volume of the cistern and to have a uniform edge for the cistern closing, with another polyethylene mesh, the maximum level was raised by constructing an adobe wall along the edges of the excavation. Adobe were realized by families in loco, using local clay and straw. The excavation was then plastered with clay and coated with a polyethylene sheet. The choice of adobe and clay coating instead of the initially requested cement one helped avoid rodent-related problems with a more sustainable and environmentally appropriate approach, also making it possible to use locally sourced materials.

After the field activities, 23 modified cisterns were realized by the two communities, 13 in Lantinquin and 10 in Dos Quebradas, respectively, and they were all functioning after the first rainy season.

6.7 Water Harvesting and Fertilization

WH can sustain and support agricultural activities in arid farmlands, by increasing agricultural production, supporting grazing land development, and enhancing water ecosystem services at catchment scale. However, practitioners and development agencies should be aware that, in many cases water is not the only limiting factor to agricultural development. Fox and Rockstrom (2000) firstly showed how, in a rural region of Burkina Faso, sorghum harvest increased by 40% with additional irrigation from WH, by 180% with WH irrigation and soil fertilization, and by 132% with only improved fertilization. In this sense, fertilization can have a major impact for increasing agricultural production, but still WH represents a vital factor reducing the risk of total crop failure, since interventions aiming to restore soil fertility alone showed high production in wet years, but low or no production in dry years (Rockstrom et al. 2002; Andersson et al. 2013).

As Rockstrom et al. (2002) showed, in arid and semi-arid farmlands, uncertainty is the major factor affecting farmers' decision. Thus, it is difficult that farmers invest in soil fertilization, even if they know the potential production increase, because of the risk of crop failure given by limited water availability. Upgrading agricultural farming systems in drylands requires a "blended" approach, both increasing water conservation for overcoming dry spells with WH, and by supporting agricultural production with soil fertility management (Rockstrom et al. 2002).

6.8 Conclusions and Future Perspectives

This chapter analyzes the main WH applications at the farmland level, discussing the most relevant issues at planning and operational level. First, a classification of WH techniques based on the catchment size is proposed as the clearest to frame the different application at farmland level, including: FWH, MWH, mWH, and RWH.

From a planning point of view, GIS techniques of WH siting can facilitate the decision making for large-scale WH implementation, especially for MWH adoption at regional or at catchment scale. In doing so, it is vital that water managers and land planners consider local knowledge, especially when data are scarcely available.

For WH designing and management, the incorporation of local wisdom in the design of the structures represents a fundamental factor for successful implementation, especially when dealing with extreme hydrological phenomena, such as wadi flows in FWH. Local farmers, in fact, have a remarkable knowledge of these phenomena, that often is lacking by the side of planners and designers.

Maintenance is a key factor. WH interventions are often conceived for harsh conditions, where arid and semi-arid climates hinder economic and social development. The implementation of a WH project without dedicated monitoring, follow-up, and community mobilization can show good results after one year, and then can fail a few years after, due to the insufficient inputs for maintenance, or too-complicated technological realization. Again, communities' involvement is a key factor in securing the long-term sustainability of the project, as shown by the example on RWH.

When WH interventions are implemented, they impact water allocation at catchment scale. As every water storage intervention, they may trigger upstream–downstream conflicts, and this effect is particularly under-rated when dealing with many small mWH interventions. At catchment level, diverse farmlands system may coexist, relying on different fluxes of the water cycle, and pre- and post-assessments of WH impacts are essential to evaluate trade-offs, and avoiding conflicts.

Last but not least, it should be noted that soil fertilization management represents an important and often neglected factor for WH adoption in farmlands. WH alone can stabilize yields, avoiding the risk of a complete crop failure if extended dry spells occurs, but it is only through fertilization that crop production can be increased.

Considering future perspectives, flood-based farming areas are quite common in arid and semi-arid areas, but still some critical elements are present. Human settlements, located very close to the flooded areas, are always at risk of being flooded and flushed away. Recent studies showed the necessity of implementing flood mitigation strategies for human settlements. In addition to this, FWH systems are dependent upon and supported by a strong indigenous knowledge that should be incorporated in modernization and planning projects for guaranteeing the successful implementation of such development initiatives. Another issue to be solved is related to the approach used for designing the modernization of spate irrigation intake structures. Problems of being flushed away and heavy siltation are resulting from the application of hydrologic and hydraulic formulas derived for perennial rivers to ephemeral rivers. Research should focus more on the investigation of the hydrology of ephemeral rivers. Regarding sand dams siting, the proposed approach based on best siting can reduce the time for selecting the best places due to a preliminary identification of potential sites based on technical variables, but the final identification should be made at the sites with the involvement of local people. More studies should be made on the verification of the potential site maps identified with the sand dam best siting procedure and those identified with the common procedure mainly based on indigenous knowledge approach.

The proposed case study on mWH intervention adoption for increasing water availability showed rising conflicts for water allocation at catchment scale. It is then necessary to consider not only the benefit coming from an increase of water availability, but also the setting of regulation on water uses and the need of monitoring action to avoid inequalities.

For RWH, future perspectives of research should refer to the analysis of most cost-effective sizing of cisterns, especially considering application in large cities, focusing also on RWH long-term reliability. In rural context, attention should be given to identifying the best management and monitoring strategies, and to the analysis of health implications, connected to the possible presence of insects induced by standing water within human settlements.

References

Abdelli, F., Ouessar, M., and Khatteli, H. (2012). Methodology of identification of existing structures and potential sites for Jessour. *Revue des Sciences de l'Eau* 25 (3): 237–254.

Adham, A., Riksen, M., Ouessar, M., and Ritsema, J.C. (2016). A methodology to assess and evaluate rainwater harvesting techniques in (semi-) arid regions. *Water* 8 (5): 198.

Aerts, J.C.J.H. and Lasage, R. (2005). *Mission Report Kenya*. Amsterdam, The Netherlands: Institute for Environmental Studies.

Al-Adamat, R., Diabat, A., and Shatnawi, G. (2010). Combining GIS with multicriteria decision making for siting water harvesting ponds in Northern Jordan. *Journal of Arid Environments* 74 (11): 1471–1477.

Andersson, J.C.M., Zehnder, A.J.B., Wehrli, B. et al. (2013). Improving crop yield and water productivity by ecological sanitation and water harvesting in South Africa. *Environmental Science and Technology* 47 (9): 4341–4348.

Avis, O. (2014) A microbial analysis of water in Sand Dams and associated abstraction methods. Masters thesis, London School of Hygiene and Tropical Medicine, U.K.

Ayenew, T., GebreEgziabher, M., Kebede, S., and Mamo, S. (2013). Integrated assessment of hydrogeology and water quality for groundwater-based irrigation development in the Raya Valley, northern Ethiopia. *Water International* 38: 480–492.

Bailey, R.T., Beikmann, A., Kottermair, M. et al. (2017). Sustainability of rainwater catchment systems for small island communities. *Journal of Hydrology* 557: 137–146.

Beimers, P.B., van Eick, A.J., Lam, K.S., and Roos, B. (2001) Building sand-storage dams, SASOL Foundation Kitui District, Kenya, Practical work report, Delft University of Technology, the Netherlands.

Biazin, B., Sterk, G., Temesgen, M. et al. (2012). Rainwater harvesting and management in rainfed agricultural systems in Sub-Saharan Africa – a review. *Physics and Chemistry of the Earth* 47–48: 139–151.

Boers, Th.M. (1994) Rainwater Harvesting in Arid and Semiarid Zones – International Institute for Land Reclamation and Improvement. Publication No. 55.

Boers, T.M. and Ben-Asher, J. (1982). A review of rainwater harvesting. *Agricultural Water Management* 5: 145–158.

Borst L, and de Haas, S.A. (2006) Hydrology of sand storage dams. A case study in the Kiindu catchment, Kitui District, Kenya. M.Sc. Thesis, Vrije Universiteit, Vrije Universiteit, Amsterdam, The Netherlands.

Bresci, E., Giacomin, A., and Preti, F. (2013). Experiences of improving water access in rural areas in Guatemala. *Journal of Agricultural Engineering* XLIV: 856–860.

Castelli, G. and Bresci, E. (2017). Participatory rural appraisal for diagnostic analysis of spate irrigation systems in Raya Valley, Ethiopia. *Journal of Agriculture and Rural Development in the Tropics and Subtropics* 118: 129–139.

Castelli, G., Minelli, A., Tefera, M. et al. (2017). Impacts of rainwater harvesting and rainwater management on upstream – downstream agricultural ecosystem services in two catchments of southern Tigray, Ethiopia. *Chemical Engineering Transactions* 58: 685–690.

COTWRD (2005) Engineering Feasibility Report of Gereb Fokissa Spate Irrigation Project: Head Work and Infrastructure reports. Mekelle, Ehiopia.

Critchley, W. and Siegert, K. (1991). *Water Harvesting. A Manual for the Design and Construction of Water Harvesting Schemes for Plant Production*. Rome, Italy: FAO.

Das, A., Singh, R.K., Ramkrushna, G.I. et al. (2017). Roof water harvesting in hills-innovations and livelihood improvement. *Current Science* 113 (2): 1–8.

Demissie, B. (2016). Ephemeral stream dynamics, land cover changes and climate variability in the marginal grabens of the northern Ethiopian Rift Valley the problem statement and study context. *Afrika Focus* 29: 93–100.

Dile, Y.T., Karlberg, L., Daggupati, P. et al. (2016). Assessing the implications of water harvesting intensification on upstream – downstream ecosystem services: a case study in the Lake Tana basin. *Science of the Total Environment* 542: 22–35.

El-Awar, F.A., Makke, M.K., Zurayk, R.A., and Mohtar, R.H. (2000). A hydro-spatial hierarchical method for siting water harvesting reservoirs in dry areas. *Applied Engineering in Agriculture* 16 (4): 395–404.

Embaye, T.G., Mezgebu, A., and Yazew, E.H. (2013) Developments in the Design and Construction of Modern Spate Systems in Tigray, Ethiopia.

Ertsen, M. and Hut, R. (2009). Two waterfalls do not hear each other. Sand-storage dams, science and sustainable development in Kenya. *Physics and Chemistry of the Earth* 34: 14–22.

Falkenmark, M., Fox, P., Persson, G., and Rockstrom, J. (2001) Water Harvesting for Upgrading of Rainfed Agriculture: Problem Analysis and Research Needs. Stockholm International Water Institute (SIWI) Report 11, Stockholm, Sweden.

Fenta, A.A., Kifle, A., Gebreyohannes, T., and Hailu, G. (2015). Spatial analysis of groundwater potential using remote sensing and GIS-based multi-criteria evaluation in Raya Valley, northern Ethiopia. *Hydrogeology Journal* 23: 195–206.

Flood Based Livelihood Network Foundation (2016) Practical Note 29 - Improvements in the Design of Flood Diversion Structures. The Netherlands.

Foster, S., and Tuinhof, A. (2004) Sustainable groundwater management. Lessons from practice. Case Profile Collection Number 5. Brazil.

Fox, P. and Rockstrom, J. (2000). Water-harvesting for supplementary irrigation of cereal crops to overcome intra-seasonal dry-spells in the Sahel. *Physics and Chemistry of the Earth, Part B: Hydrology, Oceans and Atmosphere* 25: 289–296.

Gabriels, D., Schiettecatte, W., Ouessar, M. et al. (2005). Soil hydrology for ecohydrology in drylands: examples from Tunisia, China and Cape Verde Islands. A review. *Ecohydrology and Hydrobiology* 5 (3): 183–193.

Grum, B., Assefa, D., Hessel, R. et al. (2017). Improving on-site water availability by combining in-situ water harvesting techniques in semi-arid Northern Ethiopia. *Agricultural Water Management* 193: 153–162.

Guiraud, R. (1989). Les Barrages d'Inféroflux Leur Intérêt pour l'Afrique Saharienne et Sahélienne [Interflux dams: their advantage for Saharan and Sahelian Africa]. In: *Proceedings of the Sahel Forum, the State-of-the-Art of Hydrology and Hydrogeology in the Arid and Semi-Arid Areas of Africa, Ouagadougou, Burkina Faso, 18–23 February 1989*, 320–329. Urbana, IL: International Water Resources Association.

Gur, E and Spuhler, D. Rainwater harvesting (rural). SSWM [online] (2010). http://www.sswm.info/category/implementation-tools/water-sources/hardwar.

Hagos, E.Y., Gebru T., G/Egziabher T., et al. (2014) Spate Irrigation Network Overview Paper 13: Spate Irrigation Systems in Raya Valley, Ethiopia.

Hashemi, H., Berndtsson, R., and Persson, M. (2013). Floodwater harvesting for artificial recharge and spate irrigation in arid area. *Journal of Water Management and Research* 69: 93–100.

Hoogmoed, M. (2007) Analyses of impacts of a sand storage dam on groundwater flow and storage. Groundwater flow modelling in Kitui District Kenya. MSc. Thesis, Vrije Universiteit, Amsterdam, The Netherlands.

Huisman, L. and Wood, W.E. (1974). *Slow Sand Filtration*. Illinois, USA: World Health Organization.

Humphreys, E. and Bayot, R.S. (2008). Increasing the productivity and sustainability of rainfed cropping systems of poor smallholder farmers. In: *Proceedings of the CGIAR Challenge Proram on Water and Food International Workshop on Rainfed Cropping Systems, Tamale, Ghana, 22–25 September, 2008, The CGIAR Challenge Program on Water and Food*. Sri Lanka: Colombo, 311 pp. http://46.20.115.203/download/cis/93596.pdf.

Hut, R., Ertsen, M., Joeman, N. et al. (2008). Effects of sand storage dams on groundwater levels with examples from Kenya. *Physics and Chemistry of the Earth* 33 (1): 56–66.

Kashyap, A. (2004). Water governance: learning by developing adaptive capacity to incorporate climate variability and change. *Water Science and Technology* 29 (7): 141–146.

Komakech, H.C., Mul, M.L., Van Der Zaag, P., and Rwehumbiza, F.B.R. (2011). Water allocation and management in an emerging spate irrigation system in Makanya catchment, Tanzania. *Agricultural Water Management* 98: 1719–1726.

Lasage, R., Mutiso, S., Mutiso, G.C.M. et al. (2006). Adaptation to droughts: developing community based sand dams in Kitui, Kenya. *Geophysical Research Abstracts* 8: 01596.

Lasage, R., Aerts, J., Mutiso, G.-C.M., and de Vries, A. (2008). Potential for community based adaptation to droughts: sand dams in Kitui, Kenya. *Physics and Chemistry of the Earth* 33: 67–73.

Laurita, B., Castelli, G., Giacomin, A. et al. (2018). Participatory design of rooftop water harvesting systems for smallholder farmers' in the corridor seco region. *JUNCO – Journal of UNiversities and International Development COoperation* 2 (1): 456–465. http://www.ojs.unito.it/index.php/junco/issue.

Maddrell, S. and Neal, I. (2012). *Sand Dams: A Practical Guide*. London: Excellent Development. http://www.excellentdevelopment.com/resources/publications.

Mburu, S.M. (2016) Integration of GIS techniques with local knowledge in selection of suitable sites for sand dam construction: a case study of Kitui east sub-county, Kenya. Master degree thesis dissertation. University of Firenze, Italy.

Mehari, A., Schultz, B., and Depeweg, H. (2005). Where indigenous water management practices overcome failures of structures: the Wadi Laba spate irrigation system in Eritrea. *Irrigation and Drainage* 54 (1): 1–14.

Mehari, A., Van Steenbergen, F., and Schultz, B. (2011). Modernization of spate irrigated agriculture: a new approach. *Irrigation and Drainage* 60: 163–173.

Mekdaschi Studer, R. and Liniger, H. (2013). *Water Harvesting: Guidelines to Good Practice*. Bern/Amsterdam, the Netherlands/Wageningen, the Netherlands/Rome, Italy: Centre for Development and Environment (CDE)/Rainwater Harvesting Implementation Network (RAIN)/MetaMeta/The International Fund for Agricultural Development (IFAD).

Motsi, K.E., Chuma, E., and Mukamuri, B.B. (2004). Rainwater harvesting for sustainable agriculture in communal lands of Zimbabwe. *Physics and Chemistry of the Earth* 29: 1069–1073.

Motsumi, S., Magole, L., and Kgathi, D. (2012). Indigenous knowledge and land use policy: implications for livelihoods of flood recession farming communities in the Okavango Delta, Botswana. *Physics and Chemistry of the Earth, Parts A/B/C* 50–52: 185–195.

Mun, J.S. and Han, M.Y. (2012). Design and operational parameters of a rooftop rainwater harvesting system: definition, sensitivity and verification. *Journal of Environmental Management* 93: 147–153.

Mutiso, S.M. (2003). *Community Groundwater Management Based on Integrated Sand Dam Development*. Nairobi: Sasol Foundation.

Myers, L.E. (1975). Water harvesting 2000 BC to 1974 AD. In: *Proceedings of the Water Harvesting Symposium, Phoenix, Arizona, ARS W-22* (ed. G.W. Frazier), 1–7. USDA.

Nasr, M. (1999) Assessing desertification and water harvesting in the Middle East and North Africa: discussion papers on development policy. No. 10. ZEF–Bonn, Germany.

Nissen-Petersen, E. (1982). *Rain Catchments and Water Supply in Rural Africa: A Manual*. Bungay, Suffolk, UK: Hodder and Stoughton.

Nissen-Petersen, E. (1997). *Groundwater Dams in Sand-Rivers*. Myanmar: UNCHS/UNDP.

Okhravi, S., Eslamian, S., and Adamowski, J. (2015). Water reuse in rainwater harvesting. In: *Urban Water Reuse Handbook* (ed. S. Eslamian), 787–803. USA: Taylor and Francis, CRC Group.

Ouessar, M., Hessel, R., Sghaier, M., and Ritsema, C. (2012) Report on Water Harvesting inventory, History and Success Stories. WAHARA Report series, the Netherlands-Tunisia.

Oweis, T.Y. (2017). Rainwater harvesting for restoring degraded dry agro-pastoral ecosystems: a conceptual review of opportunities and constraints in a changing climate. *Environmental Reviews* 25: 135–149.

Oweis, T., Hachum, A., and Kijne, J. (1999) Water harvesting and supplemental irrigation for improved water use efficiency in the dry areas. SWIM Paper 7, International Water Management Institute, Colombo, Sri Lanka.

Oweis, T., Hachum, A., and Bruggeman, A. (eds.) (2004). *Indigenous Water Harvesting Systems in West Asia and North Africa*. Aleppo, Syria: International Center for Agricultural Research in the Dry Areas (ICARDA).

Oweis, T.Y., Prinz, D., and Hachum, A.Y. (2012). *Water Harvesting for Agriculture in the DA*. Balkema, Leiden, the Netherlands: ICARDA, CRC Press.

Pauw, W.P., Mutiso, S., Mutiso, G., et al. (2008) An Assessment of the social and economic effects of the Kitui Sand Dams community based adaptation to climate change. SASOL Foundation and IVM Institute for Environmental studies Vrije University, Amsterdam, The Netherlands.

Quilis, R.O., Hoogmoed, M., Ertsen, M. et al. (2009). Measuring and modeling hydrological processes of sand-storage dams on different spatial scales. *Physics and Chemistry of the Earth* 34: 289–298.

Rockstrom, J. (2000). Water resources management in smallholder farms in Eastern and Southern Africa: an overview. *Physics and Chemistry of the Earth, Part B: Hydrology, Oceans and Atmosphere* 25: 275–283.

Rockstrom, J., Barron, J., and Fox, P. (2002). Rainwater management for increased productivity among small-holder farmers in drought prone environments. *Physics and Chemistry of the Earth* 27: 949–959.

Ryan, C. and Elsner, P. (2016). The potential for sand dams to increase the adaptive capacity of East African drylands to climate change. *Regional Environmentak Change* 16: 2087–2096.

Saarnak, N.L. (2003). Flood recession agriculture in the Senegal River Valley. *Geografisk Tidsskrift-Danish Journal of Geography* 103: 99–113.

SASOL Foundation (2004) Kitui sand dams: Construction and Operation. Kenya.

Siegert, K. (1994). Introduction to water harvesting: some basic principles for planning, design and monitoring: water harvesting for improved agricultural production. In: *Proceedings of the FAO Expert Consultation, Cairo, Egypt, November 1993*, 9–22. Rome, Italy: FAO.

Temesgen, M. (2007) Conservation Tillage Systems and Water Productivity Implications for Smallholder Farmers in Semi-arid Ethiopia. PhD thesis. Taylor and Francis/Balkema, Leiden, The Netherlands.

Tesfai, M. and Stroosnijder, L. (2001). The Eritrean spate irrigation system. *Agricultural Water Management* 48: 51–60.

Thomas, T.H. and Martinson, D.B. (2007). *Roofwater Harvesting. A Handbook for Practitioners*. Delft, The Netherlands: IRC International Water and Sanitation Centre (Technical Paper Series; No. 49).

Tuinhof, A., and Heederik, J.P. (2003) Management of Aquifer Recharge and Subsurface Storage – Making Better Use of Our Largest Reservoir, NNC-IAH publication No. 4, Netherlands National Committee – International of Our Largest Reservoir, NNC-IAH Publication No. 4, Netherlands National Committee.

Tuinhof, A., van Steenbergen, F., Vos, P., and Tolk, L. (2012). *Profit from Storage: The Costs and Benefits of Water Buffering*. Wageningen, The Netherlands: 3R Water Secretariat.

Van Steenbergen, F., Lawrence, P., Mehari, A. et al. (2010). *Guidelines on Spate Irrigation*. Rome, Italy: FAO.

Van Steenbergen, F., Mehari, A., Alemehayu, T. et al. (2011). Status and potential of spate irrigation in Ethiopia. *Water Resources Management* 25: 1899–1913.

Wirtz, T. (2017) Climate change and food security in Guatemala: The first in a series of reports – Global Americans (https://theglobalamericans.org/2017/05/climate-change-food-security-guatemala-first-series-reports) (Retrieved 25/01/2018).

Worm, J. and van Hattum, T. (2006). *Rainwater Harvesting for Domestic Use*. Wageningen, the Netherlands: Agromisa Foundation and CTA.

Zahraei, A., Saadati, S., and Eslamian, S. (2017). Irrigation deficit: farmlands. In: *Handbook of Drought and Water Scarcity: Management of Drought and Water Scarcity*, vol. 3 (eds. S. Eslamian and F. Eslamian), 345–359. Taylor and Francis, CRC Press, US.

Ziadat, F., Bruggeman, A., Oweis, T. et al. (2012). A participatory GIS approach for assessing land suitability for rainwater harvesting in an arid rangeland environment. *Arid Land Research and Management* 26 (4): 297–311.

Zimmerer, K.S. (2011). The landscape technology of spate irrigation amid development changes: assembling the links to resources, livelihoods, and agrobiodiversity-food in the Bolivian Andes. *Global Environmental Change* 21: 917–934.

7

Rainwater Harvesting for Livestock

Billy Kniffen

Biological and Agricultural Engineering Department, Texas A&M University Water Texas AgriLife Extension Service Resource Associate, College Station, Texas, USA

7.1 Introduction

Livestock, like all living things, require water for survival. A sufficient amount of clean drinking water is the first and most important ingredient in animal husbandry for milk and meat production. "Water Is Life" has been quoted as the key ingredient for discovering life and living things. Access to water determines the production of food and water available for humans, forage, and livestock. The timing and amount of rainfall determine seasons of abundance of water and forage or areas not usable for livestock. Then there are droughts, flooding, climate conditions, and animal adaptability to those conditions that dictate survival and productivity. Most locations in the world have seasons of rainfall and seasons without rain. Harvesting and storing that rainfall when it comes, whether in the soil, aquifers, lakes, ponds or containers, and having access to water is a daily requirement for livestock and forage production. Although free access to water is critical for survival, capturing water on the land to grow forage is also an important component of livestock production and land health. Forage or grain production requires water and water that runs off from a range site, pasture, or field does not help the plants to grow. Runoff also carries with it topsoil, nutrients, organic matter, and seed. Management of land to harvest and store rainfall for aquifer recharge, plant growth, and reducing runoff production requires management as well. Desertification is taking place globally due to overgrazing and mismanagement of our land. Just as animals need water, our land needs water stored in rainy seasons to support life during dry seasons.

Harvesting rainwater for livestock requires planning and capital to cover storage and component costs of the system. Storage location, sizing conveyance components, pump if required, and delivery method must be determined to effectively harvest the water for livestock. It could include the cost of guttering a roof or expanding the impervious surface to meet the demand for livestock. The amount of storage required depends on animal species, rainfall patterns, and amounts. Storage may meet only seasonal needs, supplement other supply sources, provide emergency water, or in some cases meet all required demand for livestock.

The quality of the water stored for livestock needs to be such that there are no odor or taste issues along with carried pathogens that may affect animal health. Proper screening of water prior to storage is the most critical requirement to ensure proper water quality. Too much organic matter from dust, debris, or bird or rodent droppings entering the storage tank can cause the water to turn anaerobic or septic, causing poor consumption and health issues. Water needs to be covered or sealed and protected from mosquitoes, rodents falling in and drowning, or other contaminants ruining water quality.

The water supply is finite and the supply may run out, leaving livestock without water. Piping and delivery supplies can break or leak, causing water supplies to be lost. Maintenance and constant monitoring are required to ensure livestock always have a good supply of quality water to drink. Backup plans and alternative sources of water must always be available. Although not part of this discussion, the harvesting of rainwater from livestock facilities will reduce runoff and water saturating pens and livestock loafing areas. This will reduce the amount of manure contamination entering waterways and ponds.

7.2 Rainfall Harvesting on the Land

The soil, soil moisture, and plants are like a bank. If you continually draw your money out of a bank without putting any back you will wind up broke. The same is true with our land. Overharvesting forage will reduce forage production and eventually kill desirable plants. Plants

Handbook of Water Harvesting and Conservation: Basic Concepts and Fundamentals, First Edition. Edited by Saeid Eslamian and Faezeh Eslamian.

take the energy from the sun through their leaves and convert it into the growth of more leaves, roots, and seed production. Without leaves, plants have no way to capture the sun's gift. Overgrazing forces plants to reduce all processes except leaf growth. Continued grazing will reduce or stop root growth and replacement and roots die. This leaves the plant vulnerable during extended dry periods to death of our best native forage plants and replacement by less desirable, quicker maturing plants and eventual bare ground. A livestock producer must practice good livestock grazing practices before all the plant is consumed. Plant leaves are required to continue the collection of sunlight and provide growth and eventual litter cover on the ground to slow water movement across the land. "Graze half and leave half" has been encouraged by range specialists to leave enough of the plant so it can fully recover. *Goals of maintaining the health and productivity of rangeland should supersede other management goals for long-term ranch benefits*. The higher the productivity potential of a land site and individual plants, the more water it requires. For recovery from grazing, plants require healthy roots and leaves (money in the bank) to intercept and infiltrate rain drops. No leaves – no plant survival. It is hard to leave a site with forage still on it or rotate to another site so plants can fully recover before being grazed again, but it is a requirement of proper land/range management.

There are four basic principles of grazing management:

- Proper use – stocking rate
- Proper season of use – seasons of growth
- Proper distribution – more uniform grazing
- Proper kind and class of animal – diet components of grazing animals

Plant protects the soil, slows raindrops down, encourages water infiltration, and makes it hard for water to runoff. As overgrazing continues, forage production decreases, stored energy, plant root depth and volume, and soil organic matter decreases. The plant community will change from high-producing plants to plants that can survive on less intercepted rainfall and harsher grazing conditions, and it does it by producing less forage on a shorter growing season and replacing forage with shallow roots, quick-maturing seed heads, and less leaves that do not have the same ability as a healthy range site. Desertification takes place and it depletes the land of its water-holding ability and usefulness in forage and livestock production. Proper grazing management is the first component of rainfall capture for livestock.

The land's productivity level varies depending on plant species, rainfall, soil fertility, usable and unreachable acres due to dense brush, rocks, and fenced out areas. The number of animals a land will support while still managing the health and productivity of the land is a criterion that a producer must consider for long-term productivity of forage and land health. The amount of rainfall that stays on the land and assists in plant growth is called effective rainfall. It is determined by seeing how much of the rainfall stays vs. the amount that runs off. A healthy plant community is most important in improving that amount that is retained on the land. Again – the goal of maintaining the health and productivity of rangeland should supersede other management goals for long-term ranch benefits. That means you need to continually monitor rainfall, plant growth, and animal numbers to keep the numbers of animals in balance with growing conditions. In severe droughts it might even require total destocking to maintain land health.

7.3 Animal Water Requirements

Rainfall amounts, patterns, and water availability vary all across the globe. The selection of particular animal species and breeds within a species must reflect those particular conditions. Heavy producing dairy females require much more water than those only nursing a single offspring. Larger animals require more food and water for body maintenance and production than smaller animals of the same species. Animal units is a term to reflect the forage requirements for one cow weighing 1000 lb. In comparison it takes six goats and five sheep to equal one animal unit. That too can be adjusted for body weight of given animals. A 1400-lb cow has the equivalent of 1.4 animal units and a 500-lb calf equals ½ animal unit.

Cattle – 1000-lb cow – 1 animal unit
Horse – 1.1 animal units
Goat – 0.18 animal units (6 to make 1 animal unit)
Sheep – 0.2 animal units (5 to make 1 animal unit)

Note: One animal unit is based on 26 lb of forage (oven dry weight) per day.

There are many variables affecting water demand requirements that must be considered in livestock production, such as travel time and difficulty of terrain to reach a water source or desired grazing land. The more distance and time spent by an animal to reach food and water, the less time available and more energy required for body maintenance and less is available for growth or milk. Keeping water sources closer to available forage, the less energy is spent just in maintaining body health and condition. Hot, dry conditions increase respiration and body water loss which requires more water consumption. And the reverse is true for humid cool climates. The forage moisture level too is a source of water. Moisture levels can

Figure 7.1 Livestock water with rainwater requires a roof, conveyance system, storage container, and supply source for livestock. This system uses a float to keep the water available for the sheep.

be as high as 80% or as low as 5% or less depending on the growth stage of the plant or if it is dead and dried out. All these affect daily water requirements. But again, limiting animal water consumption limits an animal's production ability as well as its health and wellbeing (Figure 7.1).

One must understand animal water requirements and demands based on temperature, forage moisture content, and animal activity or class of animal. Hot dry conditions demand more water for livestock and in cool conditions there is less water demand and while lush green forage has a high-water content and alternatively, dry mature forage has little available moisture to supplement water requirements. Distance to water and feed sources, terrain, and energy required for body maintenance also affect water consumption and these all need to be considered when providing supplemental water needs. Other factors like high temperatures increase animal respiration and thus increased water demand (Figure 7.2).

Water needs to be easily accessible. The results of improper water placement affect a number of things, including:

- Improper grazing patterns. Animals stay closer to a water source and over graze close to it instead of utilizing the whole pasture.
- lower weight gain or milk produced. If animals have to spend time traveling, they have less time to eat and thus may gain less weight, produce less milk, and in the case of nursing females, decrease weight or body condition due to lost forage intake.
- Offspring too will suffer, burn more energy traveling, and as they get older also have lower weight gains.

Different species and classes of animals consume different amounts of water. As a rule of thumb they will consume about 1–2 gal per 100 lb body weight. The amount will vary depending on animal size and weight, activity,

Figure 7.2 Water quality affects livestock health and performance. Small shallow ponds allow animals to wade into them and defecate, reducing water intake.

Table 7.1 Typical water demand for farm animal species.

Animal	Water consumption, typical	
	(Gallons per Day)	(Liters per day)
Chickens/100	6	23
Cow, Dry	15	57
Milking Cows	35	130
Dairy Calves (1–4 mo)	2.4	9
Dairy Heifers (5–24 mo)	6.6	25
Dry Cows	9.3	41
Hog	4	15
Horse, Steer	12	45
Pig, feeder	1.1–2	5–9
Sheep	2	7.5
Turkeys/100	20	75

Source: Meehan et al. (2015).

milk production, and environmental conditions and species/breed adaptation. Cattle will demand 7–18 gal of water while sheep and goats may require 1–4 gal d^{-1}. Dry cows require 40 l d^{-1} vs. milking cows, which could be 100–150 l d^{-1}. Swine consume 10–20 l d^{-1}. Table 7.1 provides some typical species' water demand (Kniffen and Machen 2007).

7.4 Harvested Rainfall as a Source for Livestock

The source of water will vary from ranch to ranch depending on location, availability of rivers, streams, springs, or stock ponds and impounded water. Groundwater availability and depth to water makes a difference on the equipment needed to pump and access water. Distance to electricity or use of solar pumps or windmills all make a difference in the amount of water which needs to be stored and have available for livestock.

Rainwater may be a primary, secondary, or emergency water source for livestock. It can be a support for a pond, well, stream, or other water source. Small numbers of livestock or poultry may be watered solely by rainwater.

7.5 Requirements for Harvesting Rainwater for Livestock

Rainwater harvesting requires a number of components and each needs to be designed or evaluated to ensure that the rainfall reaches the storage container in sufficient quantity and quality to meet the requirements of the livestock. Each component is designed and sized based on roof area, rain intensity, elevation, and storage capacity.

A roof or impervious area is needed to harvest rainfall from. This can be a barn, shed, farmhouse, or a roof built primarily for harvesting precipitation; a rock ledge, tarp tied between trees, posts, or some other secure anchor; or even from the trunk of a tree. The larger the catchment surface, the more water that can be collected. There is the potential to collect 0.623 gal of water per square foot of roof (25.37 Liters per square meter) (Audrey 2015). The surface area is measured in length times the width of a structure's drip line – not by the total length of the roof material. A roof with lots of slope to it will have more length than a flat roof but the same amount of water will be captured by both. Multiply the length times width to get the number of square feet of roof. A 10 by 20 ft roof has the potential of collecting almost 125 gal of water with a 1 in. rainfall event. ($10 \times 20 \times 0.623 = 125$ gal). The roofing or surface material will affect the efficiency and actual percentage of runoff harvested. Metal roofing will absorb less moisture than wooden shingles or a sod roof although some water may evaporate from a metal roof on an extra hot day. There are estimates on the percent efficiency of a roofing surface but most metal, composition shingles are around 95% efficient. The amount will decrease in high winds, hard rains, or blowing snow, or sites where trees or other structures block some of the precipitation or it may splash more water off the roof or past the gutters. Wooden shingles or thatched roofs may be less efficient. A roof that is 95% efficient in capturing the precipitation is equivalent to approximately .6 gal for every ($0.623 \times 0.95 = 0.6$) square foot of roof ($10 \times 20 \times 0.6 = 120$ gal). Surfaces used from natural areas like packed clay soils, rock ledges, or other natural impervious surfaces may also be considered though the exact calculation for runoff will vary depending on the absorption, evaporation, or spillage of water you are trying to capture. Roofs designed strictly or primarily for harvesting rain, snow, or even dew or fog have been constructed in some locations to increase potential runoff or expand the collection surface in a particular location. Even certain species of trees channel water toward their trunk and a wrap around a tree can divert water into a container. Water is heavy and the collection surface must be strong enough to hold the weight of the water running off. The flatter the surface the slower water runs off and conversely the more slope, the faster it runs off and the lighter the load bearing ability it has to be. Even a vertical surface will have the ability to harvest some water.

Measuring the size of guttering and conveyance piping is determined by the roof size and rain intensity for a given location. There are tables available to guide the sizing of conveyance piping. Rain intensity may be measured in inches, millimeters, feet, or meters. There are tables that give rain intensity for many cities in inches or millimeters per hour and that number is converted to gallons/millimeters per minute per square foot/meter of roof area. An area where the rain intensity for a one-hour event with a return of 100 years could be 4 in h^{-1}; that number is converted to 0.042 gal per minute per square foot. A 1000-square foot roof multiplied by 0.042 equals a flow rate of 42 gal per minute. Pipes must then be large enough to handle that flow vertically and/or horizontally. Water flows faster and a pipe can handle more water when it moves vertically vs. horizontally. Sizing pipes to insure water reaches the collection tank without spillage is important.

Water running off a roof surface must be directed toward a storage container. Gutters typically block and divert water from the roof toward a conveyance pipe, although canals, roof drains, or diversion piping may be standard in certain locations. Gutters are most often made from galvanized or painted tin, aluminum, vinyl, copper, or wood and bamboo. Gutters should be sloped to drain toward an opening or downspout that will transport the water to the storage container. The more slope, the faster the gutter will drain and the more water it can carry. Also, gutters must drain dry to prevent mosquitoes and potential disease hazards in the standing water. The size of the gutter depends on:

(i) Size of roof area being captured from (square foot or meter of roof being diverted into one downspout);
(ii) Slope of the gutter;
(iii) Rain intensity at a given location.

Rain intensity varies all across the globe. Areas with huge rain events may have rains that may be as high as 4–5 or more inches per hour. Other locations may only have their most intense rain events that amount to 1 in or less per hour. Areas in the US closest to the Gulf of Mexico experience frequent heavy downpours in the 4.5–5 in h^{-1} category. Locations on the west coast may actually receive more annual rainfall but these rain events have much smaller and slower falling raindrops. This rain intensity must be considered in sizing gutters to ensure the precipitation is harvested and does not overwhelm the gutter. Gutter size may be reduced by adding more downspouts or increasing the slope of a gutter.

Downspouts and or conveyance piping transport the water from the gutter to the ground or divert it to a storage container. There are three methods of getting water from a gutter to a storage container.

The first is called a "dry line." This is most preferred. It is where water is directed from the gutter directly into the top of a storage container. The size of this pipe transporting water from the gutter to the storage container again depends on the flow rate of the water entering the pipe (gallons per minute at peak rain intensity), slope of the pipe and inside diameter of the pipe. There are charts called "Horizontal Conveyance" that provide the information similar to sizing gutters (Figure 7.3).

An alternative to the dry system is the "wet system." This is used when the storage container cannot be located next to or close to the roof area. In livestock pens, the container may best be placed outside the pen and barn area. It is most common in warmer climates or protected from freezing. Here piping is dropped below ground and channeled to a desired location, and piping then goes back up and enters the tank at a point lower than the gutter or sealed pipe. As in sizing gutters, rain intensity, roof area, distance, and size of pipe affect the elevation difference between the gutter and entry point going into the storage container.

The third method is where there is not enough elevation difference between the roof surface and the entry point of the storage container. A "lift station" or pump is required to push the water to a higher location than the gutter or in a wet system, the elevation difference is nor there and the water must be pumped into the storage container. Pump selection then must be considered.

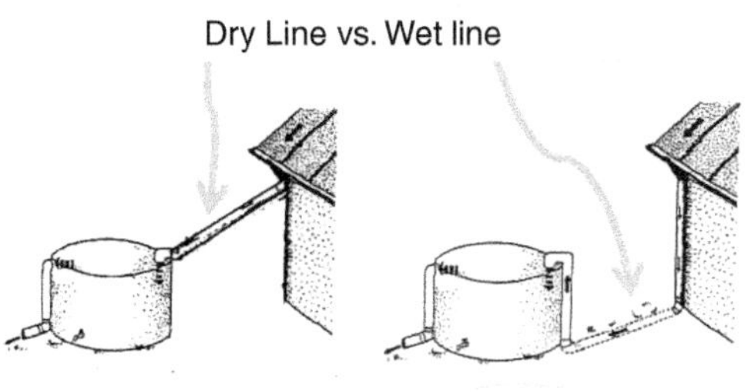

Figure 7.3 Comparison of dry and wet conveyance. Water will remain in the conveyance piping of a wet system if a drain is not included.

Screening of water prior to entering a stored container is the most critical part of a rainwater harvesting system. Too much organic matter in the storage tank will turn the stored water anaerobic or septic. Livestock pens traditionally have lots of dried manure and organic dust in the area and particles of manure end up on the roof surface. This manure, as well as leaves or pollen, could be washed into the storage tank during a rain event and begin to deteriorate. In the decaying process it pulls oxygen out of the water. The removal of oxygen and the introduction of rotting material creates an ideal habitat for anaerobic bacteria to proliferate. These are what we consider "bad bacteria" as they live in the guts of animals and are the main group of pathogens and make humans and animals sick. This also causes the water to have a foul odor, taste, and appearance and makes it unsuitable for livestock consumption. Care must be taken in the collection process to restrict the amount of organic material from entering the collection tank. This can be done with either a manufactured prescreen or a homemade product. There are a number of manufacturers that offer screens in a large price range. Australia and New Zealand produce a number of products, and more sophisticated products are manufactured in Europe. However, sand and gravel prefilters can be made and have been used for many years. Larger ones – a meter square or larger – could be made with cinder blocks, cement, or plastic containers. Smaller ones could be made with thick plastic containers like plastic barrels cut to the desired volume or smaller buckets. The size is dictated by the flow rate of water coming off the roof. Water would enter the top and be screened by a layer of pea gravel and coarse sand. A screen in the bottom holds the media in place and a drainpipe allows water to go through the media and be channeled into the storage tank. A plate with holes in it is placed above the sand and gravel to spread out the water so it penetrates evenly through the media. This type of prefilter may require hand removal of leaves but the top layer of sand will trap the contaminants and aerobic bacteria will help break down the trapped organic material. The top layer of sand/gravel may be removed and replaced annually if required.

The first flush is an optional device to mechanically divert the first water running off a collection surface prior to entering a prefilter or collection tank. A certain amount of runoff first fills a diversion container before proceeding to the storage tank. Once it is full water passes by and is prescreened before entering the storage tank. This water drains out the bottom of the diversion container slowly (preferably within 24 hours).

The amount of water diverted depends on the size of roof area and degree of contamination. In simple terms:

(i) Low contamination – 1 gal per 100 square foot of roof surface;
(ii) Medium contamination – 2 gal per 100 square foot of surface;
(iii) Heavy contamination – 3 or more gallons per 100 square foot of roof surface.

The first flush is often a source of contamination itself if it is not checked and maintained. Too often the drain is plugged up leaving a container of rotting debris that will enter the collection tank during the next rain event. The first flush should have a small hole in the bottom as part of a larger plug that can be unscrewed to drain the first flush if it becomes plugged. Commercially made products with automatic drains are available. But the first flush should be checked often to ensure it has not sealed up and has not drained. Water in these devices is also more susceptible to freezing in colder climates. However, these devices allow for water to be diverted during high pollen season or contaminate load times to prevent any water from entering the collection tank by opening the drain completely and closing it during times of desired rain capture. Look elsewhere in this book for more information on first flushes.

The prefilter may be before the first flush to prevent larger particles being diverted into the first flush, or after the first flush. Prior to the first flush is preferred, but the author used a screen basket in the lid of his collection tank requiring the first flush to be placed before the screen and it has worked effectively for over 15 years.

Stored containers are the most expensive part of a collection system. There are a number of commercially available products on the market and usually the price goes down as the container size increases. Durability and life of the container is important to reduce the cost of the system. Polyethylene containers are limited to their size and multiple containers may be connected together to obtain the desired storage capacity. Corrugated metal containers set on a sand or fine gravel bed with a vinyl liner holding the water allows construction to be completed on site and may increase overall holding capacity upwards of over 50 000 gal. Containers made from steel, stainless steel, fiberglass, concrete, and other materials are also available. More common in recent years is the diversion of rainwater from an existing livestock barn or shed into an existing concrete or metal storage tank supported by a windmill or solar pump. There are times when the windmill may not keep up with demand or conditions do not support the pumps and the rainwater may fill that void. In the US there is additional concern with runoff and the water from around livestock pens having an increased load of manure running into streams and lakes. Capturing this water and diverting it reduces that load before it hits the ground.

The placement of the storage container and components must be away from livestock activity or fenced off to protect components from being damaged. Curious animals will step on valves and piping causing breakage or leaks. Livestock may also lay down on the shaded side of the container during the day forcing undue strain on a near empty container and fighting animals may bump and move or damage the container. But they need to be placed where the roof water will enter the storage container by gravity and close to livestock pens so livestock can have access to it. The container takes up space and needs to be placed so it does not obstruct animal movement or work in and around livestock working facilities and vehicle movement.

Containers that allow light to penetrate through or have an open top will have algae growing in it. This organic material may plug up piping, filters, and water control devices. Water may need to be taken out, preferably at least 4 in. from the bottom or from a floating extractor to reduce the contaminants from entering the pipes and clogging the system. Translucent containers may be painted, placed in full shade, or covered to reduce algae growth. Also, the color of the container may affect the temperature of the water inside it. The darker the color the warmer the water inside. Black may be preferred in cold climates and white or reflective silver may be preferred in hotter climates.

Piping water from the storage container to pens or pastures may need longer distances and require digging and laying piping below frost lines and sometimes rocky areas. The further the distance, the slower the flow due to friction, so piping needs to be larger when moving longer distances (Figure 7.4).

Pumps may be required if moving to watering troughs at an elevation higher than water in the container or water needs to be under pressure or moved long distances.

Figure 7.4 Water may have to be pumped to meet the elevation or distance requirements to keep a trough filled.

However, most storage containers supply water to watering devices at the same level and gravity pressure to supply water to a float device set on the side or in the center of a water trough is all that is needed to meet demand. Again, there is a limited supply of water, and malfunctioning watering devices that leak can quickly drain a storage container.

Containers with a closed or sealed top should be both mosquito tight and not allow sunlight in or penetrate through the walls of the containers. Sunlight in a container encourages algae growth which may clog pipes and as it deteriorates cause the water to become aerobic or septic. The container is the most expensive part of the system and the water demand of livestock will require a larger stored container. There are calculators to help determine the size of container to install. The calculator on the Texas A&M University Rainwater Harvesting website (Kniffen et al. 2012) is one that can help us calculate the supply and demand to install the best size of container for the roof area and rainfall pattern.

7.6 Distribution of Water for Livestock

Distribution of the water may be by gravity and allowing the weight and height of the water in the container to push water through lines and to water troughs. Water troughs may be filled by hand but using automatic floats that keep the water level constant is more common. Floats may need to be protected from livestock and wildlife which may damage piping and the watering devices. If water troughs are at a level higher than the water storage tank a pump will be required to move water to a higher elevation or long distances. Solar pumps may be used to either push water to another storage container at the elevation necessary to provide water for the livestock on a continual basis or an electrical pump may be required to provide a constant supply of water for livestock. In any case the livestock producer must be vigilant to ensure livestock never run out of water where rainwater may be in a limited supply.

Watering containers need to be large enough to meet the demand for a herd of livestock or poultry but small enough to reduce evaporation and keep water fresh and cool. The size of water troughs or pans will depend on the size of flock or herd to insure each has the space required so all animals have access to water.

The water may be exposed to sunlight and moss or algae may grow on the sides of the trough and may become an issue, but dust, debris or even birds or animals may fall into the water trough and drown. Adding a wooden board, rocks, or metal near the edge to help animals escape will help maintain water quality. Maintaining water quality for livestock is important as it is a limited amount of water and cannot be wasted by dumping and refilling if not necessary (Figure 7.5).

Figure 7.5 Escape access for wildlife or small livestock must be added to prevent them from drowning.

7.7 Rainwater System Maintenance

The rainwater system requires constant monitoring to ensure the finite amount of water is meeting the demand and the system does not run out of water. Livestock must have water on a daily basis. Pipes and components may be hit and cracked or broken by livestock movement. Busted or leaking pipes may drain a system in a short time. Observation and maintenance is a daily chore. Containers and distribution piping need to be protected from livestock bumping or trampling components and piping. Piping needs to be protected from freezing so livestock have access to water daily. And, piping on top of the ground may heat up the water, making it less desirable for animals to consume and also may cause the PVC or plastic pipe to deteriorate unless it is protected from sunlight. Painting exposed components with a latex paint will extend the life of PVC pipe and keep it from becoming brittle and less flexible.

7.8 Conclusion

Rainwater can be used in a livestock management plan, especially where there are extreme rainfall patterns and a lack of water during certain times of year. It can be utilized in pens, traps, or pastures on a temporary basis where other water supplies are not available. Where sources are limited, harvested rainwater may extend an existing source. Once again, looking at average daily water consumption by species, it is easy to see that a large water supply may be needed for larger species and high numbers of animals. It is expensive and difficult for harvested rainwater to meet the total needs of large numbers of livestock but may best serve to reduce the demand on other water supplies, be a supplemental source, or an emergency source of water.

The deciding factors to consider when considering capturing water off a barn for livestock include:

(i) What species or kind of livestock to be watered? This may include cattle vs. goats and yearling calves vs. nursing cows.
(ii) How many will be watered with your source of water?
(iii) Where you live, your annual and monthly average of rain, how long a dry spell might last, and do you have a back-up plan;
(iv) Other animals may use this as a watering source. If wildlife frequents this as a water source, they must be added in.

Planning and designing a rainwater system to effectively divert water from a rooftop into a storage container is critical or water will be lost. The most expensive part of the system is the storage container. Knowing the needed amount and rainfall will help ensure the storage will meet the intended need. Prefiltration is the most critical part of a system to ensure water being delivered is of acceptable quality to meet animal needs. And finally, maintenance must be a daily chore to ensure there is water available, leaks are stopped, and the system is working effectively. Harvested rainwater is a valuable resource and has and will continue to be used to support livestock production.

References

Audrey, A., 2015. Rainwater Harvesting Manual. American Rainwater Catchment Systems Association, Tempe, AZ, USA. https://www.engineeringtoolbox.com/farm-use-animals-water-consumption-d_1588.htm.

Kniffen, B., Machen, R. 2007. Rainwater Harvesting: Livestock, Texas AgriLife Extension Service. Bulletin E450. USA.

Kniffen, B., Clayton, B., Kingman, D., Jaber, F. 2012. Rainwater Harvesting: System Planning, Texas AgriLife Extension Service. B6240. USA.

Meehan, M., Stokka, G., Mostrom, M. 2015. Livestock Water Requirements, North Dakota State University. Bulletin AS1763. USA.

8

Road Water Harvesting

Negin Sadeghi[1] and Saeid Eslamian[2]

[1] *Department of Architecture, Isfahan (Khorasgan) Branch, Islamic Azad University, Isfahan, 81595-158, Iran*
[2] *Department of Water Engineering, College of Agriculture, Isfahan University of Technology, Isfahan, Iran*

8.1 Introduction

The basic right for clean and potable water availability defines it as a great issue for all aspects of life development (including socioeconomic). Nowadays the growth of population, distribution of living districts, and increase in water demand lead to improvement in water supply systems. On the other hand, water shortage is a contemporary problem. As many sources of water are seasonal or limited, optimizing sources is an important issue in numerous countries.

Water is a fundamental input for a vast range of uses beside ecosystem, and biodiversity including domestic, industrial, agricultural, and recreational uses (Rosegrant et al. 2002; FAO 2003; Sivanappan 2006; FAO 2008). The increase in water demand (for agriculture, industry, and energy production uses) is predicted to grow globally over the coming decades (Connor 2015).

There is worldwide attention to the multi-objective and multi-criteria challenging task of addressing scarcity and environmental problems in the field of water (Biswas and Rahman 2009). In The Global Risks Report-2016, the World Economic Forum stated that a water crisis was the natural world's main risk of the century (FAO 2016). As a multidisciplinary field, water relates to several aspects, such as biophysical, environmental, and socioeconomic conditions, so it is important to monitor water management in order to organize sustainable life conditions for generations. Due to increasing urbanization, and sparse green cover, rainwater as an accessible water source, cost-effective alternative of water reuse, and the current need, water availability is an important issue, so rainwater harvesting (RWH) is a global point of consideration. The constantly increasing world' population, as well as the water demand rate, has direct influence, and limitations in water resource supplies, for example global water demand, has increased sixfold between 1990 and 1995 while the population only doubled, and the demand of the agricultural sector is almost 70% of the total demand (Tilman et al. 2001).

As water and water development are permanently connected with land use management, rural and urban settlements, and agricultural, infrastructural, and industrial development, it is necessary to integrate water management with these sectors in the best interests of the country and its people. Integrated water management is crucial in developing efficient management at the appropriate level. The major limitations for water availability in a big number of countries lies in the uneven distributions and mismatch of the available water resources with agro-ecological and settlement patterns. Moreover, despite sometimes-high aggregate annual rainfall, it falls either too early or too late with a characteristic high intra- and interannual variation in quantity and in terms of the spatial and temporal distributions of the seasonal rainfall (Mati 2006; Okhravi et al. 2014). Many of these prevailing limitations in terms of rainfall distribution and amounts could be effectively addressed if RWH is taken seriously. Also, for countries in which water is either a "zero-sum-game" or where there is abundance of water, the management of these water resources is key in order to make it available to recharge for productive use. RWH practices and their recognition as alternative options to supplement other water sources has been practiced for generations. Among the different ways of water harvesting, water harvesting from roads is key in this undertaking.

Roads have a major impact on the landscapes immediately surrounding them – determining the movement of water, sediment, and dust, among other things. They have a particularly important impact on rain runoff. De facto roads often act either as an embankment or as a drain and thus bring major changes to the natural hydrology.

Handbook of Water Harvesting and Conservation: Basic Concepts and Fundamentals, First Edition. Edited by Saeid Eslamian and Faezeh Eslamian.

These modifications now often have negative impacts: roads cause local floods and waterlogging along the way, whereas the more concentrated discharge from drains and culverts causes erosion and sedimentation. These negative impacts are related to the prevailing practice in road engineering to evacuate water away from the roads as soon as possible rather than making use of the water for beneficial purposes. All this undermines the resilience of roadside communities.

These negative impacts, however, can be turned around into opportunities and roads can be systematically used as instruments for water harvesting. This has been implemented in Tigray since 2014 and in Amhara since 2015. It can generate substantial positive impacts: more secure water supply, better soil moisture, reduced erosion, and respite from harmful damages. It leads to better returns to land and labor and a higher ability of people, households, and communities to deal with and thrive in the face of shocks and stresses. RWH (collecting the water from raindrops, treatment, storage) has been the basic source of water origin for different uses for many years, before using systems of water conveyance. The collection was from the sky directly or from roofs or other surfaces (Patrick 1997; Fewkes 2006). In this chapter, the necessity, structure, function, procedure, and details of water harvesting systems (WHS), especially road WHS systems, are described. Despite a vast range of studies in the field of WHS, concentration on the potentials of road systems (planning, design, management, and maintenance) is limited.

Although the history of WHS is largely undocumented – the first records are of roads built in the fourth or fifth centuries CE in Samangan (Thakht-e Rustam) – the earliest significant signs of RWH date back to over 9000 years ago in southern Jordan (Boers and Ben-Asher 1982). Historic evidence of techniques of RWH has been discovered in several civilizations in countries around the world, like Jordan, Palestine, Syria, Tunisia, and Iraq, especially for supporting seasonal water for agriculture (Fewkes 2006; Bringezu et al. 2014). Some regions like South America, India, Arabian Peninsula, North America, Asia-Pacific, and Europe have adopted RWH in order to reduce the climate change impact on the water supply. Lately, traces of global warming and climate change on natural systems are becoming visible. Rainwater harvesting systems (RWHS) are being used in many countries like Japan, Fiji, Thailand, and the US (Bruckner et al. 2003) (Table 8.1).

Instead of new system of canals, pipes, and sprinklers which aren't very efficient and can even be wasteful, in some ancient countries like India, the tradition of RWH as an earthwork in deserts dates back to 4500 BCE, as in Rajasthan-Based Paar, Talaab, Saza kuva Johad, Pat, Kunds/Kundis, Tankas, and Eri. Different water harvesting techniques are used in Indian administrators due to varying locations and weather across the country (Pandey et al. 2003) (Table 8.2).

Table 8.1 Approaches of some countries as a response to water deficiency.

Alteration of environment			*South America*	*Several studies outcomes indicate climate variability over the continent may lead to reduction of 20 to 40% in rainfall (Brains et al. 1986).*
	Extreme climate events such As drought and flood	*Many commies adopting strategies to conserve the available water resources Including promoting the usage of rainwater harvesting technique for landscaping and agriculture*	*Arabian Peninsular*	*Because of aridity and consequent decay in groundwater levels rainwater was used in order to recharge the groundwater and hold structures used for management, food production and growth of pastures and promoting vegetation together with the conservation of the environment.*
Global weather			*India*	*India has a long precedent on rainwater harvesting systems management. Climate variations, such as large spatial fluctuations in Holocene monsoon and temperature are well-resolved now. Proverbially, it is estimated that winter rainfall may decline by 5 to 25%, may lead to droughts in the dry summer months for forthcoming decades. Traditional village tanks, ponds and earthen mounds exceed more than 1.5 million, harvest rainwater in 660,000 Indian villages yet and persuade growth of vegetation (Kumar et al. 2005).*

8.2 Water Harvesting Systems and Their Characteristics

WHS as a proper method of collecting water can be applied in various ways. The WHS is mostly defined as "the procedure of collecting and managing floodwater and rainwater runoff in order to increase water availability for both agricultural and domestic use to further ecosystem sustenance." The general term of RWH describes the concentration/collection/storage/use of rainwater runoff in alternative domestic consumption and agricultural purposes (Gould and Nissen-Petersen 1999; Kahinda et al. 2008).

RWH has many definitions. Geddes provided one of the oldest definitions of RWH, as quoted by Kahinda et al. 2007: "The act of collecting and storing of any farm waters, runoff or creek flow, in order to irrigation use" Critchley et al. (1991) defined RWH as the act of collecting runoff for productive use. Gould and Nissen-Petersen (1999) said it consists of inducing methods to collect, store, and conserve runoff of local surface for agriculture in arid and semi-arid regions (ASARs) or arid and semi-arid lands (ASAL). The definition in the World Overview of

Table 8.2 India WHS (Kumar et al. 2006).

		Important Techniques Used in Rainwater Harvesting in India
1	*Johads of Rajasthan*	In order to improve percolation and groundwater recharge, Johads are small earthen bunds/round ponds/check dams to capture and conserve rainwater. Since 1984 there are about 3000 johads in 650 villages of Rajasthan. This caused a general rise of the groundwater level. Some even make some dried rivers (like River Arvari) perennial. These are earthen check dams that were meant to collect rainwater. Because of their earthen nature, water percolated easily into these systems. They resulted in a tremendous rise of the groundwater levels.
2	*Kunds/Kundis*	These structures look like a dome-shaped cover and an upturned cup nestling in a saucer that gently slopes toward the center in order to harvest rain. The water can be drawn out with a bucket. The depth and diameter of kunds depend on their use (drinking or domestic water requirements).
3	*Tankas*	Tankas (small tank) are underground tanks, found traditionally in most Bikaner houses, which are built in the main house or in the courtyard. They were circular holes made in the ground, lined with fine polished lime, in which rainwater was collected, sometimes beautifully decorated with tiles, which helped to keep the water cool to use only for drinking. In this way, the people of Bikaner were able to meet their water requirements. The tanka system is also to be found in the pilgrim town of Dwarka where it has been in existence for centuries. It continues to be used in residential areas, temples, and hotels.
4	*Eri*	The kings of the South India had built small and large open tanks called Eris, and ponds and lakes, which formed a complete collection and storage system. If a tank overflowed, the water would flow through a canal to another tank at a lower level. Approximately one- third of the irrigated area of Tamil Nadu is watered by Eris (tanks). Eris have played several important roles in maintaining ecological harmony as flood-control systems, preventing soil erosion and wastage of runoff during periods of heavy rainfall. Assignments of revenue-free lands, called manyams, were made to support village functionaries who undertook to maintain and manage Eris. These allocations ensured eri upkeep through regular desilting and maintenance of sluices, inlets, and irrigation channels. The early British rule witnessed disastrous experiments with the land tenure systems in a quest for larger land revenues. The enormous expropriation of village resources by the state led to the disintegration of the traditional society, its economy, and polity. Allocations for maintenance of Eris could no longer be supported by the village communities, and these extraordinary water harvesting systems began to decline.
5	*Rajasthan-Based Paar*	Paar was a harvesting practice used in the desert areas of Rajasthan. This involved collecting rainwater from the catchment to let it percolate into the soil. The water in the soil was then accessed by masonry construction of dugs (kuis) in the storage area; they are about 5 m to 12 m deep.
6	*Talaab*	These have been popular since the days of the kings. Talaabs are reservoirs – natural (as in Bundelkhand) or man-made (as in Udaipur). These reservoirs were used to meet irrigation and drinking water requirements. These constructions lasted only as long as the monsoon. Post-monsoon, the beds of these water bodies were cultivated with rice.
7	*Saza kuva*	These wells were initiated on a partnership basis. With multiple users, these saza kuvas were primarily used for irrigation. A group of farmers usually had one made amongst themselves.
8	*Pat*	The pats of the Jhabua district of Madhya Pradesh are irrigation panels. These irrigation panels are fed using water that is diverted from fast-moving hill streams.

Conservation Approaches and Technologies (WOCAT) database defines "The floodwater or rainwater runoff collection and management to increase water availability for domestic and agricultural consumption for ecosystem sustenance." (Mekdaschi and Liniger 2013) (Table 8.3).

The main usual WHS components are source, conveyance system, storage, and application area or target, as shown in Figure 8.1.

8.2.1 Rainwater Harvest System

RWHS as a perfect remedy for climate change, and can balance water demand and supply. RWH is a management tool for dynamic storm water, which can provide conservation in the amount of storm water and energy, so has many financial benefits. RWH as a process can help to minimize wastage of rainwater in all scales of homes, towns and even cities. RWH is being used in many countries in order to make use of the rainwater falling on the roofs that is both beneficial to resident's water-use self-sufficiency as well as the environment. The storage of rainwater can be via soil, manufactured dams, tanks, and containers in order to use or recharge aquifers (Brikké and Bredero 2003; Isioye et al. 2012). Rainwater usage can be potable (including cooking, drinking, bathing, and washing) but should be treated in order to remove contaminants, and non-potable (including watering garden, flushing toilets, and cleaning floors), which does not require treatment. The process steps are

Table 8.3 Parts of a RWH System (El-Fadel et al. 2000).

Rainwater Harvesting	Storage Tanks	For that receive a great deal of rain, or for buildings with exceptionally large rain-gathering footprints, larger storage tanks may be required to take advantage of all the runoff. If a large storage tank is present, they will generally be filled using custom-built guttering systems with a single downspout to prevent wasting runoff, while some storage tanks may be runoff. Although some storage tanks may be gravity-fed, especially if they can be placed on an upper story of a building, most rely on an electrical pump to utilize the water they capture.
	In-Ground Storage	Underground storage tanks are popular in areas where the majority of the year's rain comes in a single season. An underground tank is insulated and has an extremely low rate of evaporation. They also have the advantage over surface storage tanks as the water in an underground tank will not freeze if it is buried below the frost line. Underground tanks must be connected to an electric pump since a gravity feed is not a viable option in most cases. Due to their storage capacity and year-round usability, they are a popular choice for drinking water systems.
	Retention Ponds	A large-scale version of rainwater harvesting uses ditches to channel runoff from a wide area into a storage pond. This type of pond is usually mud-bottomed, but may be lined with concrete in some cases. The most common use of a rain-fed retention pond is watering livestock, but the water can also be pumped out to water lawns, crops, or other plants. Pond harvesting is only in viable areas with a great deal of rainfall and soil that is rich in clay, since sandy soils soak up water too quickly to allow for much runoff.

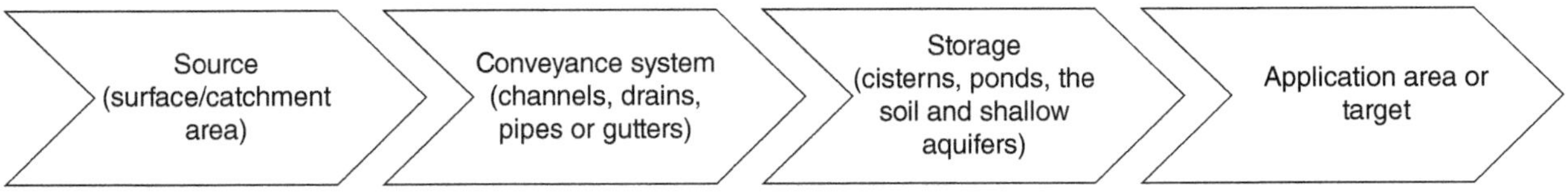

Figure 8.1 Main WHS components.

Table 8.4 Three components of A RWHS for irrigation purposes (Abdulla and Al Shareef, 2009).

Three components of a rainwater harvesting system for irrigation purposes	The supply sources as the catchment (basic components of ideal domestic rainwater-harvesting systems)	The roof (catchment)
		Gutters and downpipes
		Primary screening and first flush diverters
		Storage tanks
	The demand (landscape water requirement)	
	The conveyance system (system that moves the water to the plants)	The pipes
		Water treatment unit

collecting, conveying, and storing water from the range of increased rainfall or snowmelt runoff and involves natural watersheds. Recharging groundwater is required in places with unsuitable quality/quantity of rainfall. Bore wells or trenches can be used as these recharge structures.

Three components of a RWH system for irrigation purposes are: the supply source (the catchment), the demand (landscape water requirement), and the conveyance system, which moves the water to the plants (Abdulla and Al-Shareef 2009). In harvesting systems, the volume of collected rainwater varies by occasion and weather.

Water harvesting can be from various sources: fog and dew harvesting (water in the air), rain-flood water harvesting (overland flow), or groundwater harvesting (groundwater). Collecting the wasting water from raindrops as a sustainable strategy has been the basic source of water origin for different uses for many years, before using systems of water conveyance; the collection was from the sky directly or from roofs as a sustainable strategy to conserve water from climate change. Water, as a main life source, affects daily existence directly or indirectly, in both social and economic aspects. Road design and integrating water harvesting is a recent concept that has attracted a lot of attention (Table 8.4).

Despite several water-harvesting technologies using roof water harvesting (individual or community scale), partial research has been conducted regarding rural or urban road surface runoff harvesting for agricultural use, both rained and irrigated (Figure 8.2).

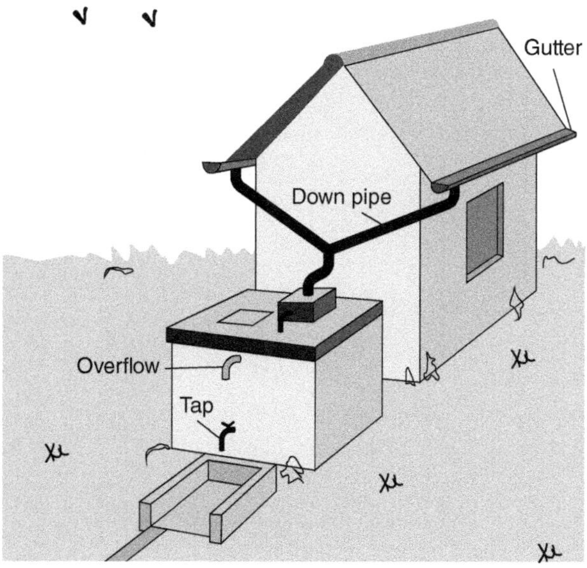

Figure 8.2 RWH system from rooftop (Brikké and Bredero 2003).

Table 8.5 A sample of a simple rainwater harvesting process.

Collection of rainwater, not allowed to flow away Gathering from catchments (roofs and yards)	Placing Mesh filters on terraces Stops external matter like leaves and similar objects form entering the tanks with rain water.	Filter units of tanks Helps filter the rainwater which is accompanied with dust and grime.	The second stage of filtering (with charcoal or sand filters) Removes debris from the water before disposing it to the storage tank or the recharge area.	Accumulation of the rainwater in underground tanks (masonry or plastic tank)which with motors, conduits, or pipes direct the water properly

The optimized WHS should be incorporated designed as an integral part of a site's storm water with multiple uses. The objective of the spreadsheet is to put RWH on a level playing field with other stormwater BMPs (best management practices) so that its use can become more widespread. By creating a common statewide specification, the multiple benefits of RWH can become realized on a broader scale. RWH is known as a sustainable, aesthetic method (Kahinda et al. 2008; Kahinda and Taigbenu 2011; Terêncio et al. 2017). A simple RWH process is shown in Table 8.5.

8.2.2 Necessity and Advantages of WHS

Already, ASALs as the agent of 35% of earth's land, are regularly facing the obstacle of water scarcity, for both drinking and vegetation (Ziadat et al. 2012). The RWHS advantages are pollution reduction, conserving water resources, help to control flooding, and reduction in impact of weather change. So the trend of water managing is to use RWH systems in building requirements of flood conservation (Scholz et al. 2007; Scholz and Sadowski 2009; Robinson and Martin 2010; Yang et al. 2011) (Table 8.6).

Hence it is helpful to use the flexible technologies of WHS which are labor-friendly in construction, operation, and maintenance.

8.2.3 Types of Water Harvesting Systems

An effective technique of increasing the recharge of groundwater is by storing rainwater locally, through roof water harvesting, refilling of dug wells, recharging of hand pumps, construction of percolation pits, trenches around fields, and bunds or dams on small rivulets. The RWHSs are widely various in the scale and complexity. According to the kind of catchment surface used, RWH is classified into infield RWH (IRWH), ex-field RWH (XRWH), and domestic RWH (DRWH). DRWH are systems that collect water from rooftops, courtyards, and compacted or treated surfaces, and store it in RWH tanks for domestic consumption. Consumption systems of IRWH use part of the target side as the catchment side, so long as XRWH systems use an uncultivated side as their catchment side (Figure 8.3).

8.3 Road Water Harvesting

Infrastructure has a substantial impact on watershed hydrology. This impact is often negative, as processes of uncontrolled flooding, erosion, sedimentation, and wetland loss clearly show. Roads may block water, concentrate runoff in limited drainage canals, and affect sub-surface streams. Roads are a main reason for drainage congestion and waterlogging. This trend can be reversed, with roads becoming instruments of beneficial water harvesting and management. Many measures can be implemented to manage water with roads and make roads instruments for food and water security, landscape management, and environmental protection. Road bodies and drainage infrastructure can be used to harvest water in dry areas by guiding water runoff from roads to recharge areas or surface storage, or applying it directly to the land. Roads can also be used to manage water catchments by controlling runoff speed, compartmentalizing and mitigating floods, and influencing sedimentation processes in catchments.

Overall, the implementation of road water management provides three types of benefits. First, road and landscape damage are reduced when flood events strike. Second, economic potential is unlocked by reducing the downtime of

Table 8.6 Main Objectives of WHS (Hulme et al. 1999; Durling et al. 2007; UNEP, 2009).

Main objectives of rainwater harvesting	To manage more efficient use of natural available resources
	Provides a suitable supplement to other water sources/ utility systems, as a relief for other water sources
	To meet the increasing demand of water (environmentally and economically efficient)
	To reduce the runoff which chokes the drains
	To manage water crisis and avoid flood and droughts
	To raise the underground water table
	To reduce groundwater pollution (reducing contaminant into water bodies)
	To reduce soil erosion (better environmental protection)
	To diminish flood and storm drainage load in roads
	To supply domestic freshwater needs (not contaminated with chemicals and industrial refuse)
	To assign water supply buffer during emergency time or natural disasters

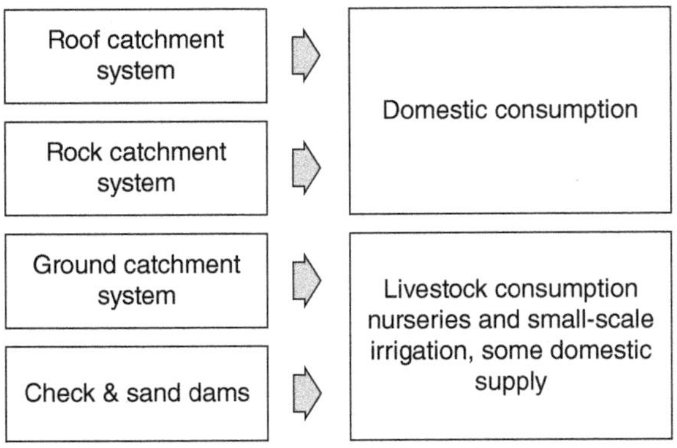

Figure 8.3 Small-scale rain WHS and uses (Gould and Nissen-Petersen 1999).

roads after flood events that would otherwise hamper economic activity in an area. Third, the use of the harvested water from roads has benefits for men and women in rural communities, such as improvements in agricultural productivity and water availability; expanded livelihood opportunities in terms of crop choices, agroforestry, and livestock grazing; improved health and nutrition; and economic empowerment. Table 8.6 summarizes positive environmental impacts derived from road runoff harvesting.

Advantages of road water harvesting are: (Limao and Venables 1999; Sheridan et al. 2006; Machiwal and Singh 2009; Norris 2012; Rahman et al. 2017)

- Less damage to roads – reducing in particular damage from water
- Prevention of water stagnation on roads, thus helping to ease traffic and avoid accidents
- Less damage caused by roads to adjacent landscapes – such as erosion and sedimentation
- Cessation of gully expansion by retaining water, thereby in combination with vegetation also retaining soil
- Substantial benefits in terms of groundwater recharge, increased soil moisture, and water storage – hence increased water availability for agricultural use
- Prevention of flooding on roads and open spaces, and stabilizing floods with road bodies
- Recovery of environment's ability to retain natural resources and increased resilience to withstand, alleviate, and recover from periods of drought
- Increased agricultural production with road runoff to be used for irrigation and livestock, leading to improved food security and nutrition standard for local communities in ASALs
- Increased stability of households and environments to vagaries of climate change, improved short-term coping strategy and long-term sustainability (Table 8.7)

Different WHS techniques can be used for sub-road such as:

(i) Road surface WHS uses road surface as main catchment channel for collecting and diverting water to the destination using the (inward slope) of the cross section or the rolling dips and humps.
(ii) Rolling Dips WHS uses sand and dry stone built check walls, stone paved rolling dips or humps (earth rolling dips) to divert, collect road surface water to the proposed locations.
(iii) Channels WHS uses longitudinal channels parallel to the road alignment made of dry stone to lead the surface water from rolling dips and spillways to the proposed location (Figures 8.4–8.7).

Road designers deal with steps of road survey, design (geometry, design speed, drainage, stream crossing structures), and construction. The road management process

Table 8.7 The potential points of road WHS design (Zeedyk 2006; Van Steenbergen et al. 2011; Kubbinga 2012; Nissen-Petersen 2006; Hwang 1998; Niemczynowicz 1999; Puertas et al. 2014; Baum et al. 2009; Nissen-Petersen 2006).

Planning road alignments	Planned road alignments determine the runoff pattern and determine area to serve by recharge and water harvesting. (Notice to slope that influence runoff speed and Sediment load content Modeling runoff behavior according to topography, structure of soil and local rainfall patterns. In order to feasibly conceive of harvesting water or leading runoff to storage or recharge areas, it is imperative to build roads, with the slopes ranging between 5% and 40%, that are able to drain and maintain, so the runoff can be reused in lower area.
Using road surfaces	Road alignments cause the natural runoff patterns to shift, so the road surface is available to be water harvested directly. The extent of water from the road surface is on the grade, width, and surface of the road
	Runoff collected from the road surface leads to recharge parts or storage ponds using drainage techniques (common methods of road surface drainage are: rolling dips for din roads, lead-off ditches, and flat land drains).
	Rolling dips are a preferred technique in dirt roads.
Cross-drainage and culverts	The cross drains location and multiplicity specifies how and where the runoff is (given our feasibility studies). In a downhill area, check dams (armoring) is needed on the drains leading in front of the road in order to avoid gully formation and stream scouring.
Roadside drains	Roadside drains arc great sources for water besides culverts and cross-drains) The design (one both side) depends on the shape of the road.
Borrow pits	The use of borrow pits is systematically as storage, recharge, or seepage ponds They are built to near the road to fiber materials (sand, gravel, soil). The shape and size of the ponds are related to the aims, soil conditions, and geology in the district (maximize effective storage of round shapes: less evaporation loss of deeper ponds).
Permeability of road foundations	The road foundation is dependent on the road type and supporting traffic, it can support the road structure from water intrusion. Drainage systems if combined with water storage, can be used as a system for water harvest tag.
Fords and flood water spreading weirs	Fords (low causeways, drifts, or Irish budges) are created because of dirt roads crossing dry riverbeds or water streams. Fords can have multiple functions. Crossing road traffic, proxy sand dam as local aquifers to store and retain water.
Spring capture	Roads made in deep cuts of terrain crossing hilly areas may open springs in mountain aquifers. It is significant to estimate the discharge of spring flows. The newly opened springs are available water supply sources especially in semi-arid regions.
Use roadside vegetation	The uses of roadside vegetation as a natural banter against runoff and erosion arc slowing down runoff, capturing sediment, fixing turning pollutants, stabilizing land surrounding the rood, and helping increase infiltration. Rock bunds and side drains can be designed with planted vegetation to aid infiltration.
Managed sand harvesting	One of the main issues of road maintenance costs is due to the harvesting sediment out of runoff-sediment record in road pass minis, drainage systems, and downstream lands. Sand harvesters filter the sand from the sand dam in horizontal layers, not to disturb the riser (tow and later sand traps can be emptied, so choosing construction material for nearby projects can lower maintenance costs).

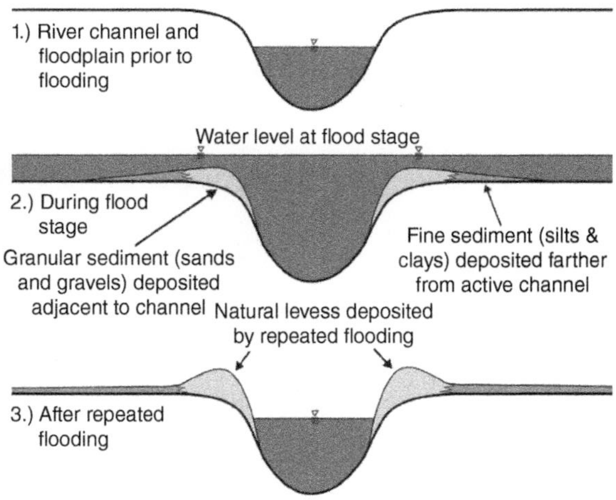

Figure 8.4 Humps and earthen channels.

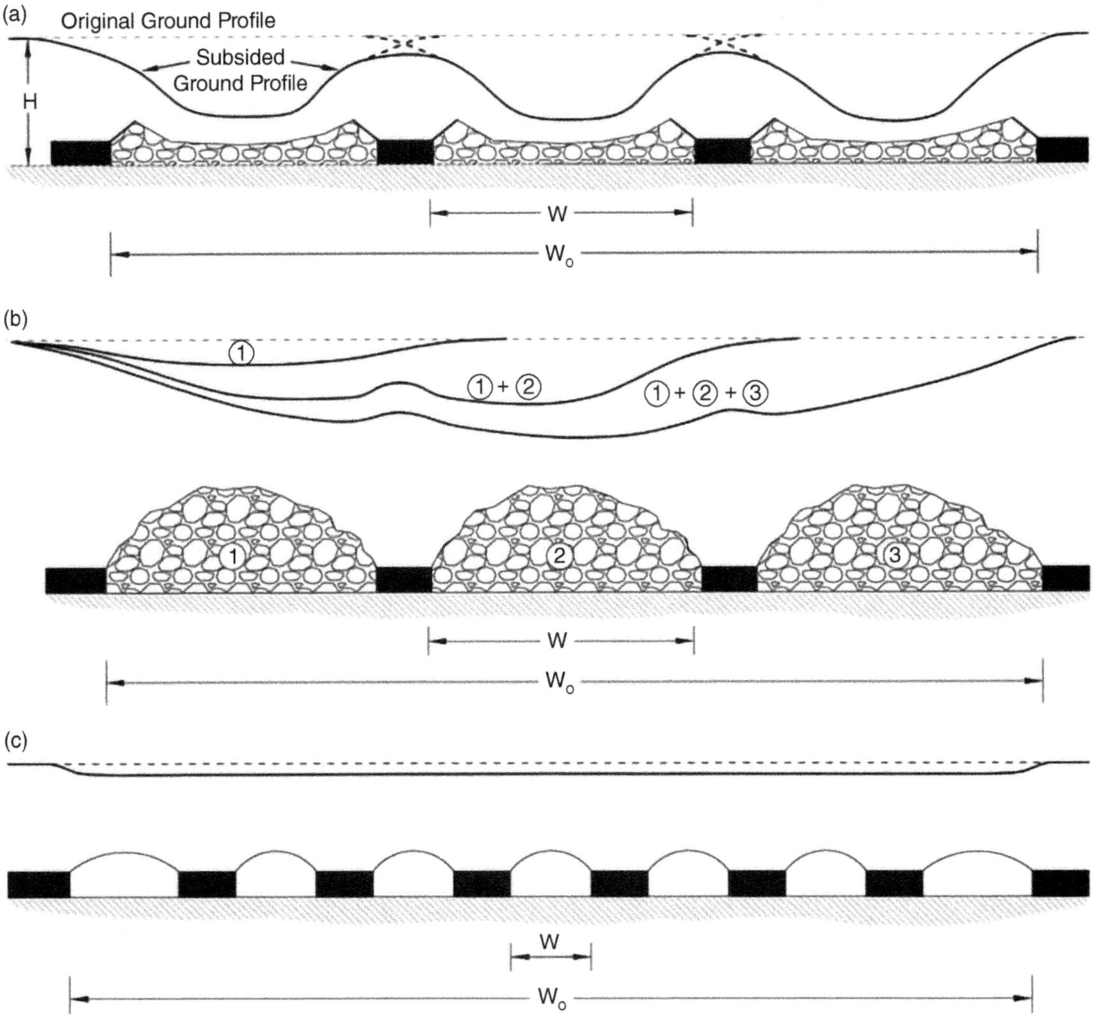

Figure 8.5 Dip stone paved rolling.

Figure 8.6 Earthen channels, cisterns.

Figure 8.7 Longitudinal parallel channels.

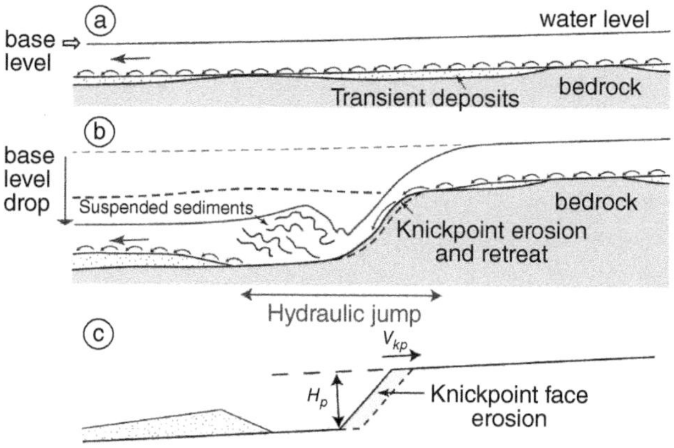

consists of clear information on road planning, locating, designing, and construction. Road location, which minimize costs–distance, affects the system of drainage and erosion due to the landscape and slopes (Figure 8.8).

WHS directs the water falling on roads and upslope of roads into side drains and then culverts so it can be moved safely and avoid damage to the road. Often the missing part of WHS is the storage infrastructure and system of drainage adapted to divert water into it. Roads are potentially suitable tooling for WHS in multiple uses.

The various behavior of subsurface flow depends on typology and permeability of road soils, embankments, subgrades, and pavements. The amount of obtained water from the road surface is due to the road slope or grade, width, surface, and the runoff coefficient. RWS (road water surface) with different road-cross slopes and side drainage design can be used depending on the type of terrain. In flat terrain, it is important to use a crown section. For designing rolling gentle cross slopes (less than 15%) the road should be sloped downhill (at 4–6% slope) without upslope drain. In case of steeper terrain, the road should slope toward the upslope drain and then water can be carried under the road with a culvert (Tague and Band 2001).

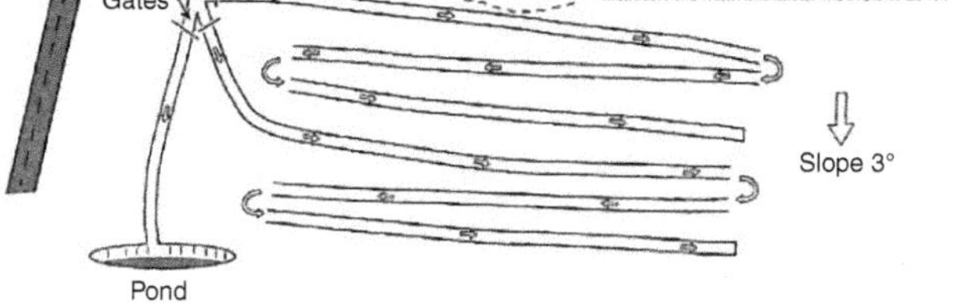

Figure 8.8 Plan of road runoff WHS (Mutunga et al. 2001).

Generated runoff by the road surface should be diverted to recharge areas/storage ponds by drainage techniques. Rolling dips, water bars, and lead-off ditches are the most usual road surface drainage methods.

8.3.1 Rolling Dips

Rolling dip as a collecting surface directs runoff from the roadway ditch across and away from the roadway on the down slope. Rolling dips are the sterling technique in dirt roads (Zeedyk 2006) (Figure 8.9).

Rolling dips should be built ideally perpendicular or with an angle of maximum 25° to the road, constructed in multiples all the way along. It is necessary to add some armoring (broken stones, riprap/gravel) at the bottom of the dip, especially when the excavation beneath the road reaches softer soils. It is significant to protect the outlet of the rolling dip, to avoid erosion. The collected water can be used to recharge areas and reservoirs, or spread over fields (Daniels 2004).

8.3.2 Water Bars

Proper water bars can divert water off roads/rails/landings effectively. However, they are hard to drive over and

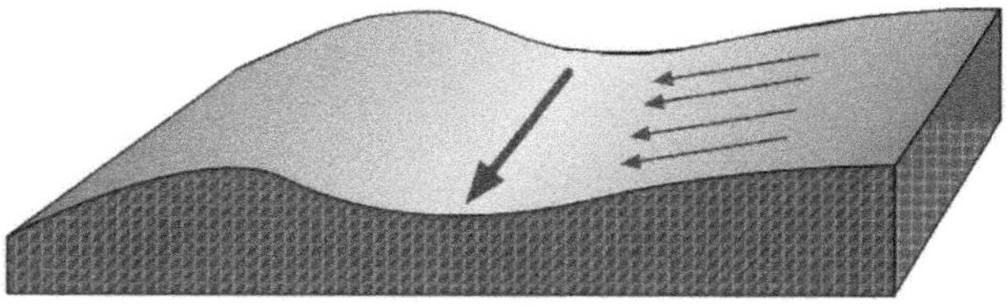

Figure 8.9 A typical rolling dip layout.

difficult to maintain. Also, they are not efficient for active traffic surfaces. Rocky soils are limited in their use. Rolling dips are more desirable for busy roads. Water bars are slender, earthen bunds built across roads. The water is lead away from roads or trails into vegetated areas preventing erosion (Lickwar et al. 1992). In the case of road damage by traffic, cattle, or heavy rainfall, maintenance to rebuild berms are needed.

8.3.3 Side Drains

Water harvesting directly along roads can be used via water from downslope drains. The water from the drain of road may be directly routed to the land in order to recharge structures, small reservoirs or improved structures (Kubbinga 2012). The act of planning regular miter drains, by paving the drain with riprap, planting vegetation, or using scour check can cause erosion limitation. There are some advantages for connecting roadside drainage to recharge and storage systems. They help store peak discharges. If areas have permeable soils, infiltration of drainage system runoff can serve to recharge shallow aquifers. In addition, the recharge areas can be "enhanced" by developing infiltration trenches or micro-basins. In the case that the water is directly applied to the field, techniques of moisture storage are most appropriate; in semi-arid areas, mulching and deep plowing will ensure the availability of water later (Van Steenbergen et al. 2011).

8.3.4 Miter

Miter drains are useful to reduce the amount of water collection inside drains and to lead it sequentially to the side of the road. The perfect angle between the miter drain and the side drain should be 30–45° and the suitable slope is about 2–5%. Laid stones at the end of drain can prevent erosion (Mrema et al. 2012).

8.3.5 Culverts

Culverts are an efficient, cheap way of transferring water across roads. The construction and management process can be done by local labor and material but they aren't proper for areas of high-volume flows. Culverts do not need to slow down terrific at the cross. Designing of side ditches and spacing of culverts depends on terrain and paved/unpaved road. In flat terrain, no culverts are needed. On roads inclined to flooding, concrete drifts are proper (Waterfall 2004).

8.3.6 Gully Prevention and Reclamation

A gully, as an expensive undertaking of good land husbandry in the case of land misuse, may form on one or both sides of the road. In general, gullies are planned due to high runoff volume/peak runoff rate. In some areas, the best method would be to stabilize small or primary gullies. The first, most and effective way of gully control is runoff control; if it can be controlled, it would be suitable to grow vegetation in the gully. Three methods of gully control are: to improve gully catchments in order to reduce and control run-off volume and maximum rates; diversion of gully area runoff water upstream; and to stabilize gullies by structural factors and escort re-vegetation (Begum et al. 2008) (Table 8.8). To control gullies in some areas with large rains, all three methods are essential.

Obtained water can be used in various ways depending on the landscape characteristics and the last use of the harvested water (for crop/grass/tree production, and growing trees of different kinds).

Major considerations of WHS site selection are:

- roadway safety (the site must be highly visible for the driver);
- manage the wet season soil drainage problem in certain areas;
- have access to water sources;
- sites shall not be considered with established animal paths;
- priority of sites with nearby households engaged in farming, etc.;
- priority of sites with access to nursery;
- sites shall not be severely degraded;
- priority of sites with a positive attitude community toward the project.

Roads are usually considered infrastructures delivering transport services, but they can have peculiar effects on the hydrology of construction areas. Due to the major cost factors (largest public investments in developing countries), of road maintenance, the role of water in these surfaces is important in road development with the aim of integrating water harvesting. As for the increasing area of roads

Table 8.8 System of managing road runoff.

Runoff harvesting from roadside	Storage structure	Ponds □ Earth dams □ Cisterns
	Spread over land	Bunds □ Terraces □ Pits □ Micro-basins □ Trenches □ Borrow pits
	Shallow Aquifer Recharge	Spread □ Trenches □ Recharge structures □ Tube recharge □ Borrow pits

globally: 35–80% of damage to paved and unpaved roads is due to water. The role of rivers in the changing surface of hydrology, has major impacts on flooding, runoff, waterlog, sedimentation and erosion, and damages to land and houses.

The win-win case of multi-functional planning, designing and aligning roads, road drainage control, and road surface and crossing control as inclusive, technical, and social development has some advantages and can prevent the following: land degradation, the jeopardizing of road embankments, increased sign maintenance costs, and reduction of road sign durability. This systematic combination of water harvesting and road building systems can optimize the use of roads (mobility, growth) and minimize drainage and surface services besides recharging water resources, so turning a threat into an opportunity (Table 8.9).

Roads can cause major impact on climate resilience as they define the way water moves across the land; roads can also prevent water flow or centralize waters as drains. Thus, it is possible for water from roads to be used for various productive uses to the rural communities' benefit. Water management with roads can cause incensement in the resilience of communities to droughts, and minimize road damages. The main factors of road design are terrain, soils, climate, and road class.

Table 8.9 Impacts of unplanned roads on local water management.

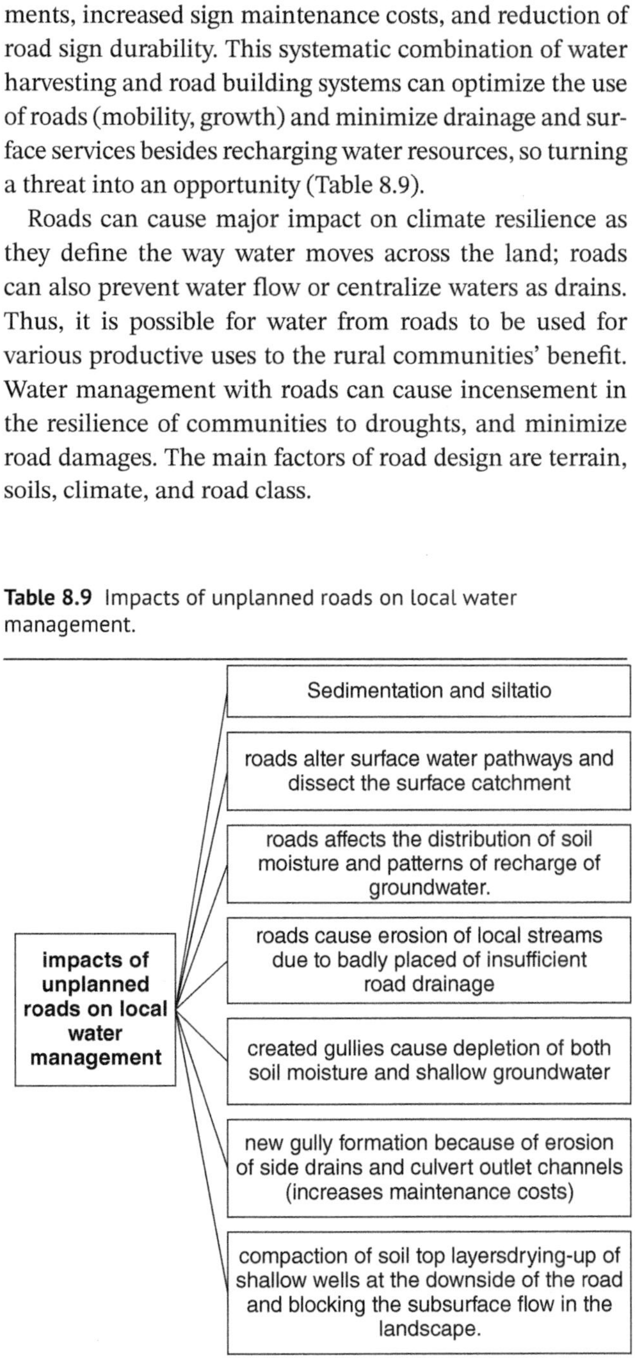

8.3.6.1 Terrain

Slope is a critical factor for designing roads. It can be classified in four groups:

- flat (0–10 five-meter contours per km): the natural slope is below 3%. Construction and maintenance of roads on steeper slopes is more expensive and more prone to damage and landslides of road water.
- rolling (11–25 five-meter contours per km): the natural slope is between 3% and 25%.
- mountainous (26–50 five meter contours per km): the natural slope is above 25%
- escarpment (typical gradients greater than those encountered in mountain terrain): as special geological features involve engineering risks (Furniss et al. 1991).

8.3.6.2 Climate

Different climate zones are characterized according to rainfall and evapotranspiration (zones of rainfall-intensity-duration frequency).

8.3.6.3 Soils

The type of soil shows the road and the surrounding land's vulnerability. Low clay content soils are weak and have less internal cohesion. Therefore, they will slide at a lower slope angle and are more apt to erosion.

Road WHS Techniques are micro-catchments, macro-catchments, floodwater harvesting, and storage reservoirs. The road harvesting system (RHS) catchment water harvesting method is leading runoff water from the road and its surrounding catchment toward ditches of roadside and repartition into farmland or holding basins for vegetation or future use (Rockstrom 2000; Rockstrom and Barron 2007; Rockström et al. 2009) Water can be a valuable and vital resource, but also a trouble. Therefore, practical measures for water harvesting and conservation are required, which can support population growth and increasing water demand, as well as recharge of ground water including moisture conservation. As a most effective technique, water management can prevail in a water crisis to ensure management of sustainable water resources under changing environmental conditions (Oweis and Hachum 2006; Glendenning et al. 2012; Jha et al. 2014). There is a promising tool for addressing adaption and water scarcity to climate and socioeconomic change topics, which is artificial recharge of groundwater coupled with RWHS. The best-known technique for harvesting runoff water from roads is storing the harvested water in reservoirs of soil. The process of construction is easy and economical, without any need for special materials; some structures can even be created by farmers (Table 8.10).

Table 8.10 Examples of harvesting rainwater from roads (Nissen-Petersen 2006).

Murram pits	Murram pits (borrow pits) as manmade systems, always situated along roads. They can easily convert into water reservoirs, as they are excavations of some trenches. The material (murram) excavated from them can be used for road construction. Outcome water from murram pits would be drawn out in several ways depending on the water usage. 
Small pans/Large pans	Pans as natural depressions are being constructed without any dam walls around the water reservoirs. Silanga ya ndovu, (elephant dams) is another name of pans. It is believed that elephants scooped out Pans at first. Rainwater running off roads is diverted into pans by excavated trenches or along stone bunds.
	Storage or runoff water from roads in large pans is known as common water sources in some flat lands. The uses of pans are raising fish, watering livestock, and domestic uses.
Ponds	Ponds are tiny-scale earth dams (Charco dams, hillside dams, and valley dams), made by unweaving soil which uses that soil constructing a dam wall at the lower side of the water reservoir. 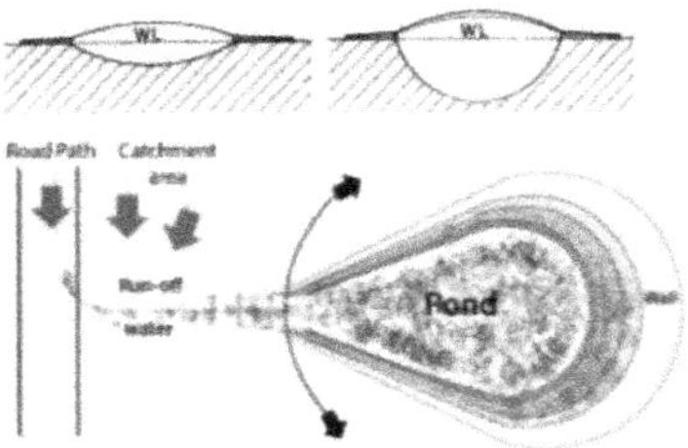 Plan and Cross sectional of Charco dam showing the first and last stage of excavation work.

Managing roads makes saving water, improving vegetation cover, and increasing crop yields conceivable and concurrent and can protect soils from erosion. Also, rainwater runoff from rural roads often creates gullies or causes other damages while finding way into the bush or fields, before it ends up in the sea or in basement aquifers (Nissen-Petersen 2006).

In some roads, gullies from the runoff alongside the constructed roads and surrounding construction are crucial problems, especially during rainy days. It is effective to design systems for collecting this wasted water using culverts leading to close ponds, which can later have used as a potential water harvesting source with economical/environmental benefits. Harvesting runoff from roads can not only be used as additional water source for supplementary irrigation but also minimizes the damage caused by flood on farms along the roadside, as well as on the rural roads, which in turn reduces the cost of maintenance of the road itself for damage that is caused by excess runoff. The aim of water harvesting and conservation techniques is to maximize the available water via "catchment area" to gather water and lead it to "cropping area" and then using soil and biomass for conserving water by reduction of runoff and keeping water at the maximum level. Due to the reliability, RWHS are in 4 types: occasional (water storage for only a few days), intermittent (store water during rainy seasons), partial, and full (water storage for year-round domestic use). Road development hydrological effects are: incensement in erosion of local streams and roadsides drainages, sedimentation/siltation of reservoirs, farmlands, alteration of sub-surface shallow groundwater flows, and waterlogging in the upstream areas.

8.3.7 Inclusive Planning/Water-Friendly Road Design

For roads to truly become multipurpose infrastructure, close collaboration between the agencies responsible for road development and those that promote agriculture and watershed management is required. Moreover, local communities need to be involved in the design phase to indicate local water needs and identify opportunities and constraints for water capture along roads. This will require a different style of working for road engineers, but it may go a long way in reducing water damage to roads – now the single largest cost in road repairs.

Optimizing water harvesting's socioeconomic benefits from roads is the main aim of environmental protection.

The procedure can be useful for landscape and road engineers, managing water and roads through alternative planning, design, and managing. There are some general usual road drainage structure designs (such as culverts, ditches) practiced by engineers, while the specific road characters should be determined (geology, terrain, experience, geometric parameters, need for specific plan and profile design, approach to integrated water resource, hydrologic and hydraulic studies, and cost restraint). The generalized, optimized road planning, design, management, and drainage system lead to a suitable water harvest system. The road WHS should both be supported by government and local society, furthermore being improved technically and institutionally (by taking advantage of previous experience locally and globally) to support and protect the road drainage network which can result in the suitable RWHS, regarding water rights, erosion control, environmental sustainability, and responding to social and gender expectations. Road planning is a process to meet current needs, maximize investments, and minimize impacts on the environment and people. Objectives of road management are to define the road's purpose, standards, use, management, maintenance, and funding, and also applicable best practices for the road. The process of road planning involves all aspects of implementation of the design and fitting the project to the ground, use of national standards, and careful alignment to avoid high costs.

Designing roads is a multi-dimensional process consisting of: access and mobility, environmental, economic, and social affects, justice, and availability of services (Limao and Venables 1999; Adger et al. 2002; Van de Walle and Cratty 2002; Crawford and Semlitsch 2008; Enfors and Gordon 2008; Puertas et al. 2014; Demenge et al. 2015; Van Steenbergen et al. 2018).

Toward development, roads are accompanied by sufficient policies (education, demonstration schemes, and access to credit) and the right conditions must be set for existing markets and employment opportunities (Demenge et al. 2015).

One key to increase livelihoods and beneficial points of roads is water provision, which can have many beneficial effects for urban and rural communities (Ngigi 2006). Water harvesting techniques can affect water availability, agricultural productivity, agro-forestry, improved environmental conservation and protection, social integration, economic empowerment, and new, creative sources of income (Ngigi 2006; Nissen-Petersen 2006; Kubbinga 2012).

There is vast potential to use existing roads to harvest/manage water, via design modifications to existing roads, adapting and implementing new design approaches, and integrating road water harvesting options. In order to use these potentialities of roads and inclusiveness methods of road engineers it is necessary to pay attention to technical knowledge and development of technical guidelines, advocacy at political and decision-making levels, capacity building (for road experts), and mapping of potential areas of intervention. Poorly engineered roads (planning/designing/construction) affect people's property and livelihoods negatively. Also, lack of road maintenance (potholes, ruts, soft-muddy, roadway surfaces, rills, and gullies) can cause problems for users (Figure 8.10).

Water can affect road costs, duration, and maintenance and roads affect water resources negatively or positively. Although there is growing interest in WHS in the last decades, people should be asked to voluntarily take up the responsibility to harvest RWH as an easy and inexpensive affair through traditional or modern technologies, and to see the potential of surface roads that can be used as

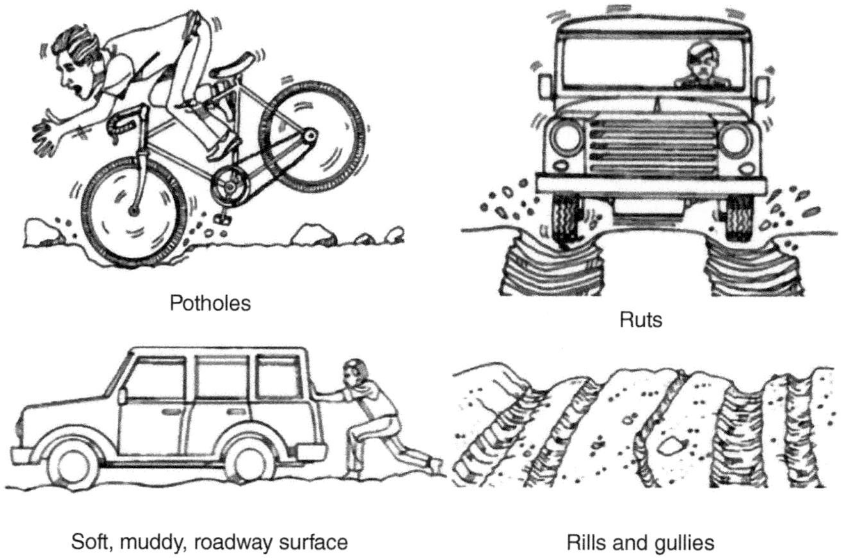

Figure 8.10 Difficulties caused by road damage.

a resilient, flexible tool, especially in arid and semi-arid areas to:

- provide approaches and technologies that are community-based (the concept of ownership, participation, sustainability).
- upgrade the quality and quantity of (ground/surface) water.
- use the adaptable/flexible technology of RWH.
- change from reactive to proactive community engagement approaches (consultative, collaborative, collegial).

8.3.8 Road WHS and Planting

As in an African proverb, "first to planting trees, plant the water," water harvesting as a clean source can be used for small- or large-scale landscapes. There are factors like types of plants, size, age, and distance between serial plants that chip in the benefits of using rainwater systems for irrigating landscape areas. The limited restrictions of water harvesting can be covered by suitable design and planning (Figure 8.11).

Roadside tree planting for environmental mitigation and economic benefits contains any vegetation growing on a roadside and a wide variety of trees, and can be coupled with road water harvesting to provide a vast variety of benefits.

Plants selected to re-vegetate roadsides must be resistant to harsh conditions. Native species have a higher survival rate and are adapted to local conditions. The objectives of plantation (economic or environmental) are important, and height, growth rate, and density characteristics are important considerations. The principles of plant selection and design in a road WHS are shown in (Table 8.11).

A dense plantation provides higher protection over short distances, while a less dense plantation provides less protection but over a greater distance. With a porous plantation, a large part of the airflow goes through it. Pollution is better trapped in porous plantations, because there is more contact with the leaves of the trees and shrubs. To accomplish a good degree of porosity, plantations should be approximately 5–20 m wide, consisting of tall trees with a bush layer underneath. The capture is enhanced by the turbulence in the plantation. This turbulence is caused by the presence of irregularities, such as branches, leaves, and leaf structure. The more irregularities the structure contains, the more dust, wind and pollutants will be trapped.

Benefits of roadside planting are:

- reduce soil erosion (holds soils in place);
- decrease the air pollutants or dust;
- wind break (against invasive weeds);
- control flood (absorb or slow road runoff);
- rectify the water quality (the ability of planting to increase water infiltration);
- increase road stability (lower local water tables by planting to affect the road formation and pavement);
- Safety (reinforcing road alignment, serving as crash barriers, protecting view planes, and reducing wind speeds – CO_2 sequestration);
- create shade and prevent straight sun radiation.

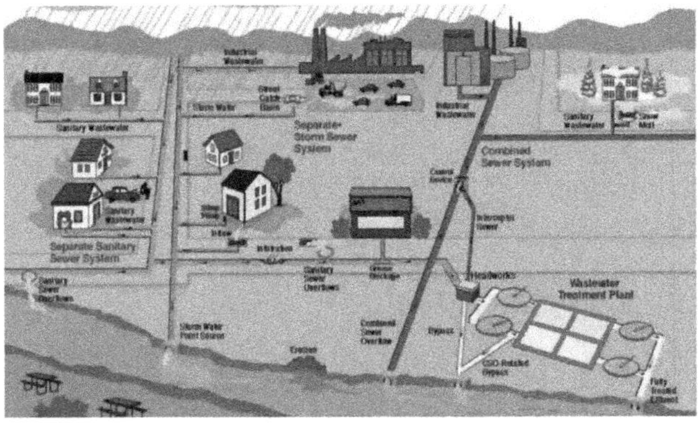

Figure 8.11 Rain WHS landscape.

Table 8.11 Important issues in the design of roadside plantations and plants selection.

Plant selection	Design of roadside plantations
• Evergreen plants or remain green over most of the period of tohe year • More horizontal than vertical extension • Tolerant of seasonal drought and insect and pest • Deep rooted to resist wind power, drought stress, and to avoid damaging the road • Tree species shall permit the growth of other plant species • Tree and grass species shall not be invasive • Tree species shall be fast growing • Productive use and secondary benefits • Grass species shall be capable of covering areas through rhizomes	• Leave room for safe passage of vehicles, animals, and people • Suitable with adjacent land use • Spacing between trees depends on species (for trees average is 3–4 m) • Shrubs can be placed at closely • Trees 5–6 m away from the roads • On curves and crossings 10–12 m away • Plant at the beginning of the main rainy season, 8 weeks before the end of the rainy season

Table 8.12 Planting effects in road design.

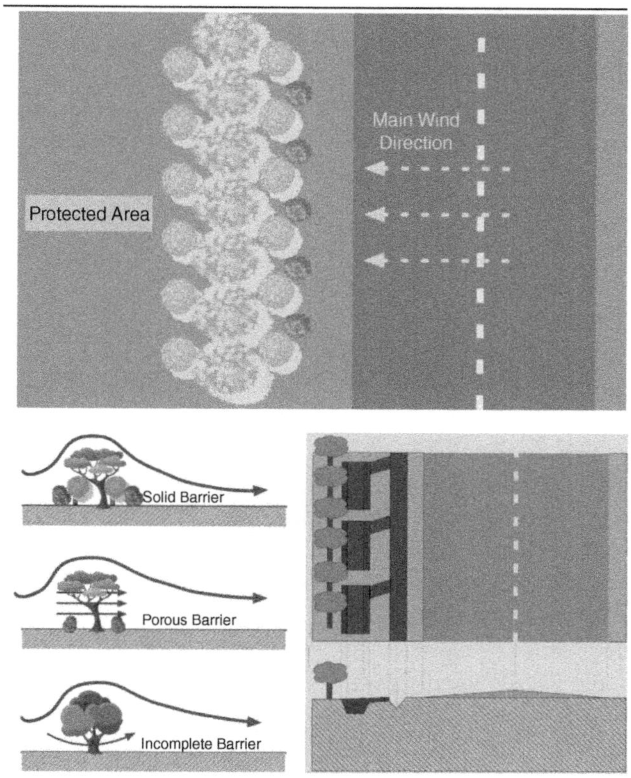

In the way forward, toward the ongoing development and hence the increasing need for water, the issue of road WHS considers a strong need for collaboration of sectors for addressing the sustainability in all aspects (environmental, social, economic). The exact system can divert negative effects of water wasted from roads to positive by managing the appropriate WH system. Therefore, its procedures to plan an integrated set of planning, designing, building, maintenance, and management in association with WHS. The outcome of road WHS is saving the environment, saving the road, providing water security, and improving livelihood in the district (Table 8.12).

When trees are planted with the additional function of wind break and dust barriers, it is necessary to plant several parallel rows. It is preferable to have plantations of two to four rows to protect a larger area (Table 8.13).

8.3.8.1 Site Selection

The outcome of RWH systems relates to identifying suitable sites and technical design (Al-Adamat et al. 2012).

Criteria for site selection are preferably access to water sources, access to nursery, apart from established animal paths, priority of farmer households, not severely degraded, priority and positive attitude of community and individuals to water harvesting, preventing runoff patterns, clear views, and amenity of site road and trees.

Table 8.13 Suitable planting against populations.

Type of pollution	Mechanism	Suitable leaf characteristics
Ozone, nitrogen dioxide	Adsorption	Wide, flat leaves of deciduous trees
Volatile organic compounds (PCB [polychlorinated biphenyls], dioxins, furans)	Adsorption	Thick and greasy wax layer (cuticle) on leaves, particularly conifers
Particulate matter (PM10)	Interception	Pointed shape such as conifer needles, or rough, hairy and sticky leaves

8.4 Conclusion

Road projects can be a potential to endow road communities with additional water and soil resources. Humans have an essential duty to conserve resources, especially water, for future use. In order to do this a prime concept is collecting wastewater from surfaces to recharge underground water surfaces. As various types of channelizing techniques have been used from ancient times in different countries, it seems that with today's development in engineering and technology, we can find more effective ways which lead to advantages for urban/rural areas.

Urbanization as an irreversible, fast-forwarding process causes more and more people to live on localized land, and more reliance on the water supply. Water is an unevenly distributed resource across the world. The main RWH design options are designing road hydraulic structures (culverts, roadside, drainages/ditches, etc.) to manage/harvest water, designing road embankments as water storage dams, promoting river crossings that enhance surface and sub-surface water buffering, integrating road water harvesting as a standard procedure in road design/construction procedures, and using proper guidelines and standards.

Many cities are trying to adopt developed, updated water infrastructure concepts like waterway city, water cycle city, and water sensitive city, in order to solve the water crisis. In the United States, low-impact development (LID) is commonly known; in Europe, it is the integrated water management system; and in China, there are sponge cities, which prioritize lower impact of urban development to natural water system. However, there are small steps that can be done by individual households or local municipalities. Traditional WHS is still current in rural areas, using surface storage frames (irrigation tanks, eri, lakes, ponds, etc.). Due to the lack of open space in urban areas, it is essential to harvest rainwater as ground water. The prevalent event in urban spaces is often the vast exactly paved surfaces causing water runoff, floods, and other problems. To regain and reuse this valuable rainwater it is important to choose suitable WHS.

Rain WHS is an ideal approach, particularly for regions with lacking or insignificant surfaces and supplies of ground water. It can utilize the rainfall runoff (that is going to sewer or storm drains) and reduce flooding. The system infrastructure is common, low-cost, and eco-friendly. There are some challenges in RWH design options such as: lack of awareness at all levels (from the top of the managers up to level implementers), poor integration among stakeholders, resource integration, road harvesters investing in pond digging, road constructors carrying soil to fill borrow pits, technical integration, and lack of proper guidelines for implementing road water harvesting.

References

Abdulla, F. and Al Shareef, A.W. (2009). Roof rainwater harvesting systems for household water supply in Jordan. *Desalination* 243 (1–3): 195–207.

Adger, W.N., Kelly, P.M., Winkels, A. et al. (2002). Migration, remittances, livelihood trajectories, and social resilience. *Ambio: A Journal of the Human Environment* 31 (4): 358–366.

Al-Adamat, R., AlAyyash, S., Al-Amoush, H. et al. (2012). The combination of indigenous knowledge and geo-informatics for water harvesting siting in the Jordanian Badia. *Journal of Geographic Information System* 4: 366–376.

Baum, A.W., Patzek, T., Bender, M. et al. (2009). The visible, sustainable farm: a comprehensive energy analysis of a Midwestern farm. *Critical Reviews in Plant Sciences* 28 (4): 218–239.

Begum, S., Rasul, M.G., and Brown, R.J. (2008). A comparative review of storm water treatment and reuse techniques with a new approach: green gully. *WSEAS Transactions on Environment and Development* 4 (11): 1002–1013.

Biswas, S.K. and Rahman, I. (2009). Environmental toxicity, redox signaling and lung inflammation: the role of glutathione. *Molecular Aspects of Medicine* 30: 60–76.

Boers, T.M. and Ben-Asher, J. (1982). A review of rainwater harvesting. *Agricultural Water Management* 5 (2): 145–158.

Brikké, F. and Bredero, M. (2003). *Linking Technology Choice with Operation and Maintenance in the Context of Community Water Supply and Sanitation: A Reference Document for Planners and Project Staff*. Geneva, Switzerland: WHO and IRC International Water and Sanitation Center.

Bringezu, S., Schütz, H., Pengue, W. et al. (2014). *Assessing Global Land Use: Balancing Consumption with Sustainable Supply*. Nairobi: United Nations Environment Programme.

Bruckner, A.W., Johnson, K.A., and Field, J.D. (2003). Conservation strategies for sea cucumbers: can a CITES appendix II listing promote sustainable international trade. *SPC Bêche-de-mer Information Bulletin* 18: 24–33.

Bruins, H.J., Evenari, M., and Nessler, U. (1986). Rainwater-harvesting agriculture for food production in arid zones: the challenge of the African famine. *Applied Geography* 6: 13–32.

Connor, R. (2015) "The United Nations World Water Development Report 2015: Water for a Sustainable World.

Crawford, J.A. and Semlitsch, R.D. (2008). Post-disturbance effects of even-aged timber harvest on stream salamanders in southern appalachian forests. *Animal Conservation* 11: 369–376.

Critchley, W., Siegert, K., and Chapman, C. (1991). *Water Harvesting*. Rome, Italy: FAO.

Daniels, B. (2004). *Managing Forests for Water Quality: Forest Roads*. NR/FF/010,. Logan, UT: Utah State University.

Demenge, J., Alba, R., Welle, K. et al. (2015). Multifunctional roads: the potential effects of combined roads and water harvesting infrastructure on livelihoods and poverty in Ethiopia. *Journal of Infrastructure Development* 7: 165–180.

Durling, N.E., Catchpole, O.J., Grey, J.B. et al. (2007). Extraction of Phenolics and essential oil from dried sage (*Salvia officinalis*) using ethanol–water mixtures. *Food Chemistry* 101 (4): 1417–1424.

El-Fadel, M., Zeinati, M., and Jamali, D. (2000). Water resources in Lebanon: characterization, water balance and constraints. *International Journal of Water Resources Development* 16: 615–638.

Enfors, E. and Gordon, L.J. (2008). Dealing with drought: the challenge of using water system technologies to break Dryland poverty traps. *Global Environmental Change* 18 (4): 607–616.

FAO (2003) "Food and Agriculture Organization". Diet, Nutrition and the Prevention of Chronic Diseases." Report of a Joint WHO/FAO Expert Consultation: 147–149.

FAO (2008) "Food and Agriculture Organization of the United Nations". Diet, Nutrition and the Prevention of Chronic Diseases." Report of a Joint WHO/FAO Expert Consultation, Vol.15.

FAO (2016) Global tuberculosis report 2016, ISSN: 924156539X.

Fewkes, A. (2006). The technology, design and utility of rainwater catchment systems. *Water Demand Management*: 27–51.

Furniss, M.J., Roelofs, T.D., and Yee, C.S. (1991). Road construction and maintenance. *American Fisheries Society Special Publication* 19: 297–323.

Glendenning, C., van Ogtrop, F.F., Mishra, A.K. et al. (2012). Balancing watershed and local scale impacts of rain water harvesting in India—A review. *Agricultural Water Management* 107: 1–13.

Gould, J. and Nissen-Petersen, E. (1999). *Rainwater Catchment Systems for Domestic Supply*. Intermediate Technology.

Hulme, M., Barrow, E.M., Arnell, N.W., Harrison P.A., Johns T.C., and Downing T.E., (1999). Relative impacts of human-induced climate change and natural climate variability. Nature 397: 688–691.

Hwang, I.H. (1998). Paving method of water-permeable concrete, Google Patents.

Isioye, O.A., Jobin, P., and Youngu, T.T. (2012). An assessment of digital elevation models (DEMs) from different spatial data sources. *Asian Journal of Engineering, Sciences and Technology* 2 (1): 42–51.

Jha, M.K., Chowdary, V.M., Kulkarni, Y., and Mal, B.C. (2014). Rainwater harvesting planning using geospatial techniques and multicriteria decision analysis. *Resources, Conservation and Recycling* 83: 96–111.

Kahinda, J.M. and Taigbenu, A. (2011). Rainwater harvesting in South Africa: challenges and opportunities. *Physics and Chemistry of the Earth, Parts A/B/C* 36 (14–15): 968–976.

Kahinda, J., Mwenge, T., Akpofure, E., and Boroto, J.R. (2007). Domestic rainwater harvesting to improve water supply in rural South Africa. *Physics and Chemistry of the Earth, Parts A/B/C* 32 (15–18): 1050–1057.

Kahinda, J.M., Lillie, E., Taigbenu, A.E. et al. (2008). Developing suitability maps for rainwater harvesting in South Africa. *Physics and Chemistry of the Earth, Parts A/B/C* 33 (8–13): 788–799.

Kubbinga, B. (2012) "Road Runoff Harvesting in the Drylands of Sub-Saharan Africa: Its Potential for Assisting smallholder farmers in coping with water scarcity and Climate Change, Based on Case Studies in Eastern Province, Kenya" Based on Case Studies in Eastern Province, Kenya [MSc thesis]. Vrije University, Amsterdam, the Netherlands.

Kumar, R., Singh, R.D., and Sharma, K.D. (2005). Water resources of India. *Current Science*: 794–811.

Kumar, M.D., Ghosh, Sh., Patel, A. et al. (2006). Rainwater harvesting in India: some critical issues for basin planning and research. *Land Use and Water Resources Research* 6 (1): 1–17.

Lickwar, P., Hickman, C., and Cubbage, F.W. (1992). Costs of protecting water quality during harvesting on private forestlands in the southeast. *Southern Journal of Applied Forestry* 16 (1): 13–20.

Limao, N., Venables, A. (1999). "Infrastructure Geographical Disadvantage and Costs." World Bank Policy Research Working Paper (2257).

Machiwal, D. and Singh, P.K. (2009). Investigating the water balance of on-farm techniques for improved crop productivity in rain fed systems: a case study of Makanya catchment, Tanzania. *Physics and Chemistry of the Earth, Parts A/B/C* 34 (1–2): 93–98.

Mati, B.M. (2006). *Overview of Water and Soil Nutrient Management under Smallholder Rain-Fed Agriculture in East Africa*. IWMI.

Mekdaschi, R. and Liniger, H. (2013). *Water Harvesting: Guidelines to Good Practice Centre for Development and Environment*. Bern, Germany: Centre for Development and Environment.

Mrema, G.C., Gumbe, L.O., Chepete, H.J., and Agullo, J.O. (2012). *Rural Structures in the Tropics: Design and Development*. Food and Agriculture Organization of the United Nations.

Mutunga, K., Critchley, W., Lameck, P. et al. (2001). *Farmers' Initiatives in Land Husbandry: Promising Technologies for the Drier Areas of East Africa*. Nairobi, Kenya: Regional Land Management Unit, RELMA/Sida.

Ngigi, S.N. (2006). *Hydrological Impacts of Land Use Changes on Water Resources Management and Socio-Economic Development of Upper Ewaso Ng'iro River Basin in Kenya*. UNESCO-IHE, Institute for Water Education.

Niemczynowicz, J. (1999). Urban hydrology and water management–present and future challenges. *Urban Water* 1 (1): 1–14.

Nissen-Petersen, E. (2006). *Water from Dry Riverbeds*. Denmark: ASAL Consultants Limited for the Danish International Development Assistance.

Norris, D. (2012). An analysis of long-term trends, seasonality and short-term dynamics in water quality data from Plynlimon, Wales. *Science of the Total Environment* 434: 186–200.

Okhravi, S., Eslamian, S., Eslamian, F., and Tarkesh-Esfahani, S. (2014). Indigenous knowledge as a supportive tool for sustainable development and utilisation of rainwater harvesting systems. *Journal of Flood Engineering* 5 (1/2): 39–50.

Oweis, T. and Hachum, A. (2006). Water harvesting and supplemental irrigation for improved water productivity of dry farming systems in West Asia and North Africa. *Agricultural Water Management* 80 (1–3): 57–73.

Pandey, D., Narayan, G., Anil, K. et al. (2003). Rainwater harvesting as an adaptation to climate change. *Current Science*: 46–59.

Patrick, E. (1997)."Assessing Water Harvesting Suitability." FAO Corporate Document Repository, Water Report 16, Italy.

Puertas, D., Garcia-Landarte, W., Kifle, M. et al. (2014). Roads for water: the unused potential. *Waterlines* 33 (2): 120–138.

Rahman, A., Hajani, E., and Eslamian, S. (2017). Rainwater harvesting in arid regions of Australia, Ch. 26. In: *Handbook of Drought and Water Scarcity*, Environmental Impacts and Analysis of Drought and Water Scarcity, vol. 2 (eds. S. Eslamian and F. Eslamian), 489–500. USA: Francis and Taylor/CRC Press.

Robinson, J.P. and Martin, S. (2010). ITS use and declining social capital? More cold water from the general social survey (GSS) and the American time-use survey (ATUS). *Social Science Computer Review* 28 (1): 45–63.

Rockstrom, J. (2000). Water resources management in smallholder farms in eastern and southern Africa: an overview. *Physics and Chemistry of the Earth, Part B: Hydrology, Oceans and Atmosphere* 25 (3): 275–283.

Rockström, J. and Barron, J. (2007). Water productivity in rain-fed Systems: overview of challenges and analysis of opportunities in water scarcity prone Savannahs. *Irrigation Science* 25 (3): 299–311.

Rockström, J., Falkenmark, M., Karlberg, L. et al. (2009). Future water availability for global food production: the potential of green water for increasing resilience to global change. *Water Resources Research* 45 (7): 102–118.

Rosegrant, M.W., Cai, X., and Cline, S.A. (2002). *World Water and Food to 2025: Dealing with Scarcity*. Washington, DC: International Food Policy Research Institute.

Scholz, M. and Sadowski, A.J. (2009). Conceptual classification model for sustainable flood retention basins. *Journal of Environmental Management* 90 (1): 624–633.

Scholz, F.G., Bucci, S.J., Goldstein, G. et al. (2007). Biophysical properties and functional Significance of stem water storage tissues in neotropical Savanna trees. *Plant, Cell and Environment* 30 (2): 236–248.

Sheridan, G.J., Noske, P.J., Whipp, R.K., and Wijesinghe, N. (2006). The effect of truck traffic and road water content on sediment delivery from unpaved forest roads. *Hydrological Processes* 20 (8): 1683–1699.

Sivanappan, R.K. (2006). Rain water harvesting, conservation and management strategies for urban and rural sectors. *National Seminar on Rainwater Harvesting and Water Management* 11 (12).

Tague, C. and Band, L. (2001). Simulating the impact of road construction and Forest harvesting on hydrologic response. *Earth Surface Processes and Landforms* 26 (2): 135–151.

Terêncio, D.P.S., Fernandes, L.F.S., Cortes, R.M.V., and Pacheco, F.A.L. (2017). Improved framework model to allocate optimal rainwater harvesting sites in small watersheds for agro-forestry uses. *Journal of Hydrology* 550: 318–330.

Tilman, D., Fargione, J., Wolff, B. et al. (2001). Forecasting agriculturally driven global environmental change. *Science* 292: 281–284.

UNEP, F (2009). *Fiduciary Responsibility: Legal and Practical Aspects of Integrating Environmental, Social and Governance Issues into Institutional Investment*. Geneva, Switzerland: UNEP FI.

Van de Walle, D. and Cratty, D. (2002). *Impact Evaluation of a Rural Road Rehabilitation Project*. Washington, DC: World Bank, USA.

Van Steenbergen, F., Haile, A.M., Alemehayu, T. et al. (2011). Status and potential of spate irrigation in Ethiopia. *Water Resources Management* 25 (7): 1899–1913.

Van Steenbergen, F., Woldearegay, K., Perez, M. et al. (2018). *Roads: Instruments for Rainwater Harvesting, Food Security and Climate Resilience in Arid and Semi-Arid Areas. Rainwater-Smart Agriculture in Arid and Semi-Arid Areas*, 121–144. Springer.

Waterfall, P. (2004). *Harvesting Rainwater for Landscape Use*. University of Arizona Cooperative Extension, pub, No. AZ1344, USA.

Yang, J., Xu, M., Zhang, X. et al. (2011). Life-cycle analysis on biodiesel production from microalgae: water footprint and nutrients balance. *Bioresource Technology* 102 (1): 159–165.

Zeedyk, B. (2006). A Good Road Lies Easy on the Land…Water Harvesting from Low-Standard Rural Roads. The Quivira Coalition, Zeedyk Ecological Consulting, LLC, the Rio Puerco Management Committee – Watershed Initiative, and the New Mexico Environment Department – Surface Water Quality Bureau, USA.

Ziadat, F., Bruggeman, A., Oweis, T. et al. (2012). A participatory GIS approach for assessing land suitability for rainwater harvesting in an arid rangeland environment. *Arid Land Research and Management* 26 (4): 297–311.

Part C

Hydroinformatic and Water Harvesting

9

Application of RS and GIS for Locating Rainwater Harvesting Structure Systems

Dhruvesh Patel[1], Dipak R. Samal[2], Cristina Prieto[3], and Saeid Eslamian[4]

[1] *Department of Civil Engineering, Pandit Deendayal Petroleum University, Gandhinagar, Gujarat, India, 382007*
[2] *Faculty of Technology, CEPT University, Ahmedabad, Gujarat, India, 380009*
[3] *Environmental Hydraulics Institute, IH Cantabria, University of Cantabria, Santander, Spain*
[4] *Department of Water Engineering, Isfahan University of Technology, Isfahan, Iran*

9.1 Introduction

Water is an essential natural resource for sustaining life and the environment (Shinde and Gaikwad 2016). Water plays a vital role not only in fulfilling the basic human need for life and health, but also plays a vital role in socioeconomic development. As rainfall is the primary source of water, it becomes necessary for us to harvest it effectively so we can maximize the storage and minimize the wastage of rainwater (Prasad et al. 2014). Therefore, the natural resources like land and water are planned and managed within a naturally defined region called the watershed. In a watershed-based approach, soil and water conservation measures are considered to be effective when the excess rainfall or runoff is drained to a common outlet and controlled by various morphological parameters related to size, shape, and relief of the watershed (Saptarshi and Raghavendra 2009). The size of the watershed varies from few hectares to several square kilometers. According to the catchment area, the classification of the watershed is given in Table 9.1.

Table 9.1 System of classification of watershed in India.

Category	Number	Size ranges (10^3 ha)
Regions	6	25 000–100 000
Basin	35	3 000–25 000
Catchments	112	1 000–3 000
Sub-catchments	500	200–1 000
Watersheds	3 237	50–20
Sub-watersheds	12 000	10–50
Mini-watersheds	72 000	1–10
Micro-watersheds	400 000	0.5–1

Source: Foundation for Ecology. Source Book of Soil and Water Conservation 2008.

The water harvesting structure is one of the important components of watershed development to conserve soil and water (Patel et al. 2012). The water harvesting structure controls the runoff, recharges groundwater table, increases the vegetation cover, minimizes the reservoir sedimentation, and provides water whenever required, specifically on non-rainy days (Gavit et al. 2018). The various rainwater-harvesting structures viz., check dams, farm ponds, boulder bunds, nalla bunds, percolation tanks, etc., are constructed at an appropriate site to check runoff and soil which as a result store the water for irrigation, domestic, industry, and groundwater recharge (Singh et al. 2009). The identification of potential sites for water harvesting is an important step to increase the land productivity through efficient utilization and management of water resources in the area. The traditional fragmented approach of identification of potential sites for water harvesting is no longer viable and a more holistic approach to water management is needed (Gavit et al. 2018). It involves massive field work and it is an expensive and time-consuming technique. Recently, remote sensing (RS) coupled with geographical information system (GIS) has been proven to be an efficient tool to identify suitable sites for water harvesting structures with minimal field visits. Moreover, digital elevation model (DEM)-based terrain visualization, processing, and quantification of topographic attributes make GIS a powerful tool for locating water harvesting structures. RS is a technique to acquire or gather the information about an object without physical contact. Remote-sensing imagery has many applications in mapping land use land cover (LULC), agriculture, soil mapping, forestry, city planning, archeological investigations, military observation, geomorphological surveying,

Handbook of Water Harvesting and Conservation: Basic Concepts and Fundamentals, First Edition. Edited by Saeid Eslamian and Faezeh Eslamian.

land cover changes, deforestation, vegetation dynamics, water quality dynamics, urban growth, etc. (Sivakumar et al. 2004). GIS is a powerful tool for collecting, storing, retrieving, transforming, and displaying spatial data from the real world. The GIS data model allows the geographic features in real-world locations to be digitally represented and stored in a database so that can be abstractly presented in map (analogy) form and can also be worked with and manipulated to address some problem. GIS techniques are increasingly used for planning, development, and management of natural resources at the regional, national, and international level (Weerasinghe et al. 2011). They have been applied for the assessment and mapping of several water-related environmental challenges such as weather parameters (rainfall and temperature), soil erosion, degradation of land by waterlogging, ground and surface water contamination, and ecosystem changes (Jasrotia et al. 2002; Matouq et al. 2014).

One of the useful applications of GIS is for watershed prioritization, which refers to the ranking of different mini-watersheds according to the order of development. Through prioritization of watersheds, one can conclude which watershed produces more discharge due to excessive amounts of rainfall and erosion (Chowdary et al. 2009; Edet et al. 1998; Javed et al. 2009; Javed et al. 2011). Several researchers have also demonstrated the use of earth observation datasets and GIS for determining the suitable sites for water harvesting structures by overlaying of DEM, soil map, and slope maps (Javed et al. 2009; Javed et al. 2011; Patel et al. 2012; Patel et al. 2015; Rahman et al. 2017).

DEM is an important tool to represent the earth surface in 3D; therefore, it can be utilized to derive different topographical parameters such as slope, aspect, perspective three-dimensional view, hill shading, curvature, flow direction, flow accumulation, catchment area, drainage network, etc. Nowadays, DEM is available with no cost for the whole globe in varied spatial resolution. The Shutter

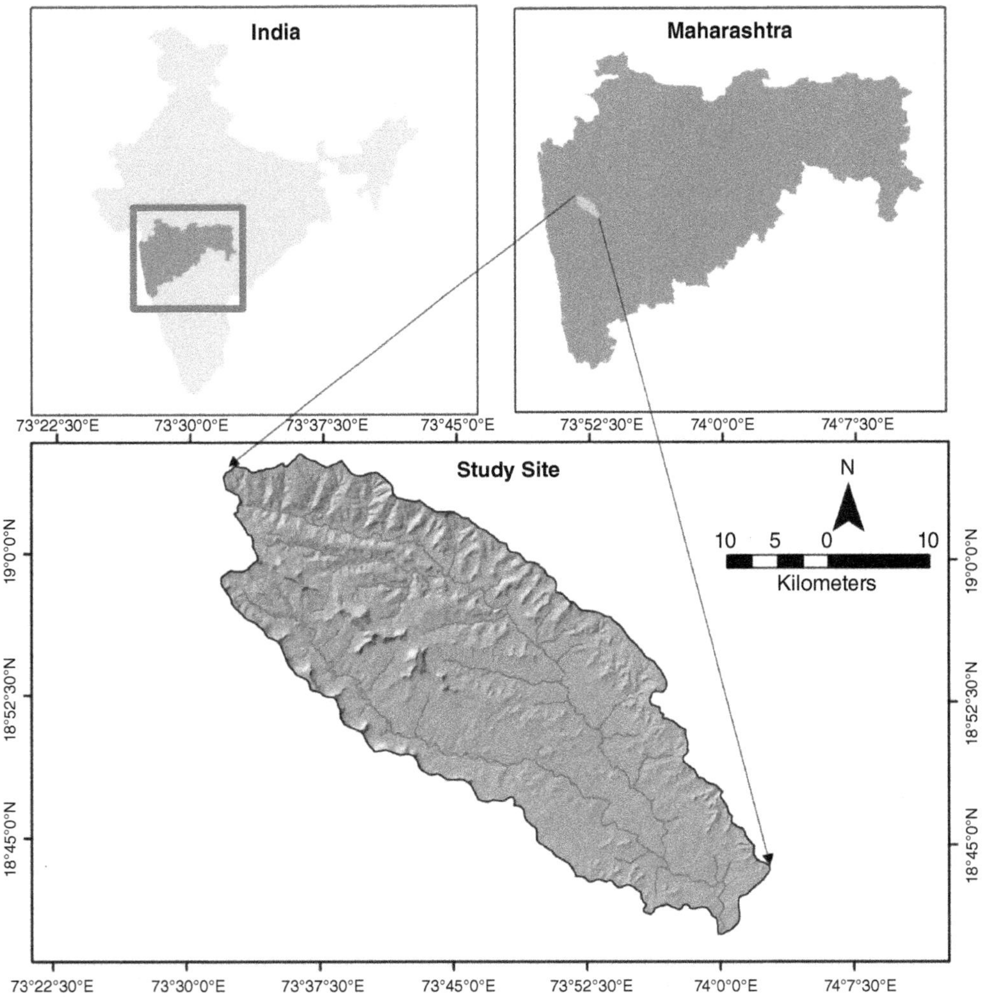

Figure 9.1 Location map of the Upper Bhīma watershed.

Radar Topographical Mission (SRTM) of 1 arch-second interval DEM is one of the freely available global DEMs which has been used in watershed management studies. High-resolution DEM along with GIS is an efficient technique for watershed delineation and terrain visualization. It is an important component to help understand, analyze, and explain the distribution of phenomena on the surface of the earth, and will become increasingly important as volumes of digital spatial data become more unmanageable (Buttenfield and Mackaness 1991). Geovisualization has been characterized as a kind of geoinformation, with emphasis on individuals using interactive visual tools in the search for unknowns (MacEachren and Taylor 1994).

The present study aims to identify suitable sites for various water harvesting structures such as check dams, farm ponds, and boulder bunds in Upper Bhima micro-watersheds using GIS and RS techniques. RS images for the study area were obtained from Earth observation satellites, whereas several spatial analysis tools in GIS are used to prepare the various thematic maps like drainage, LULC, slope, and soil maps of the study area. Morphometric analysis was carried out for deriving the basic, linear, and shape parameters of the basin. The prioritization concept is helpful to understand the morphology of individual watersheds. Finally, Integrated Mission for Sustainable Development (IMSD) guidelines prioritized watershed integrated with soil, slope, and LULC map along with geovisualization concept to be used to locate suitable site for check dams, farm ponds, and boulder bunds. Additionally, an attempt has been made to visualize a site and take the decisions for positioning an appropriate water harvesting structure using geovisualization techniques. In this way, a lot of expenses and labor may be saved involved in costly field visits. These structures directly check the excessive water coming from the watersheds and hence, lead to soil and water conservation.

9.2 Experimental Site

The study watershed lies in the Pune district of Maharashtra state in India. It covers an area of 1132 km^2, mainly covered with wasteland (50.8%), followed by agricultural land (26.3%), forest (17.9%), water bodies (3.8%), and built-up area (1.1%). The watershed lies between north latitudes 18° 40′ and 19° 5′ and east longitudes 73° 31′ and 74° 2′ (Figure 9.1).

It receives an average annual rainfall of about 777 mm and the average annual evapotranspiration is about 450–467 mm year^{-1}. Due to the presence of the Western Ghats, the rainfall pattern changes from west to east. The mean minimum and maximum temperature is about 12 °C and 39 °C, respectively. The Bhīma river originated from the basin and flows downwards to merge in Krishna river. The elevation in the basin ranges from 541 to 1255 above MSL (mean sea level), with mean elevation 705 m. Similarly, the slope ranges from 0 to 63 degrees, with a mean slope of 6 degrees.

9.3 Methodology

The methodology adopted for the present work is shown in Figure 9.2.

9.3.1 Drainage Network

The drainage map was prepared by digitizing drainage from SOI (Survey of India) toposheets as well as updated from satellite imagery (Figure 9.3).

Each stream segments ordered according to Strahler's stream ordering technique. The morphometric parameters for micro-watershed were calculated based on the formula suggested by Horton (1945), Strahler (1964), Schumm (1956), Nooka Ratnam et al. (2005), Thakkar and Dhiman (2007), and Miller (1953) and are given in Table 9.2.

9.3.2 Digital Elevation Model and Slope

The DEM is obtained from SRTM 1 arc 30 m gird interval (https://earthexplorer.usgs.gov). The elevation ranges from 541 to 1255 meter above mean sea level (Figure 9.4). The topography is highly undulated in the western part and reduced to a relatively flatter area toward the watershed outlet. The slope map is generated from DEM using the ArcGIS tool and grouped under different classes (Figure 9.5). Slope values are classified on the basis of the guidelines mentioned in IMSD document. In the study area, slopes are categorized as: flat (0–1), very gentle sloping (1–3%), gently sloping (3–8%), moderately sloping (8–15%), and steep sloping (>15%).

9.3.3 Soil Map

The soil map of the study area was obtained from National Bureau of Soil Science and Land Use Planning, Nagpur (NBSSLUP 1996) at 1:50,000 scale. Based on the association with geomorphic units, the soil types are classified into clayey soil, loamy soil, fine clayey soil, and very fine soil. The soil depth varies from shallow to very shallow (Figure 9.6), with moderate to severe erosion. Several cereal and commercial crops are grown in the region with surface and ground water irrigation.

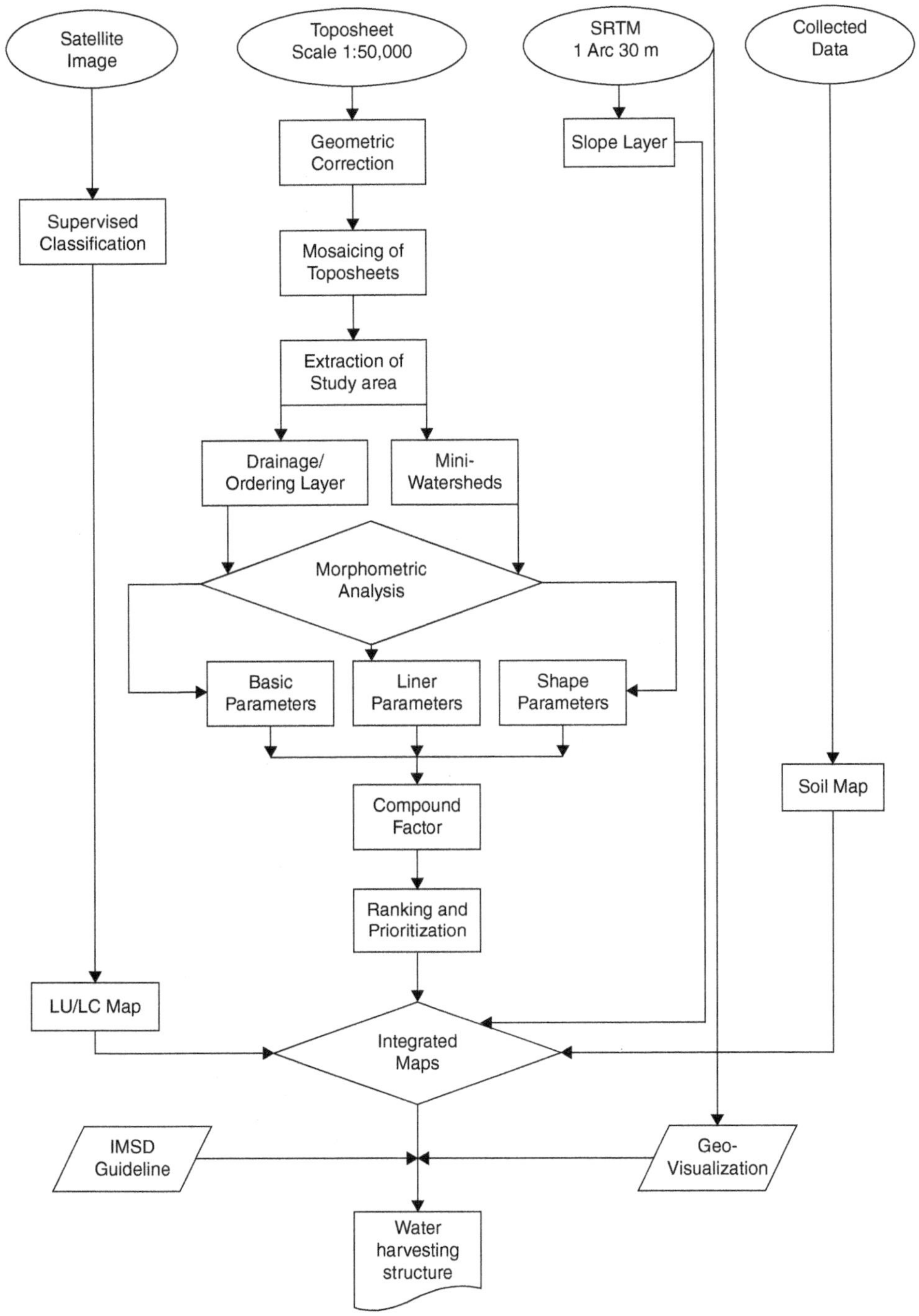

Figure 9.2 Flow chart for locating water harvesting structures.

9.3.4 Land Use and Land Cover (LULC)

Land use refers to a piece of land being used by human beings. The land use map for the study area was prepared by digital classification of earth observation satellite images (Figure 9.7). An object-based image processing technique was applied to classify the area under five major classes such as agricultural land, wasteland, forest, built-up areas, and water bodies (Samal and Gedam 2015). A major portion of the study area is covered by the wasteland category followed by agricultural land, forest, water bodies, and built-up areas. Locating surface water harvesting structures might be useful to store water and utilize water in non-monsoon period. The agricultural land is confined to river banks and forest areas are closely associated with hilly terrain of Western Ghats region. The water bodies refer to water spreading areas of reservoir and rivers in the watershed.

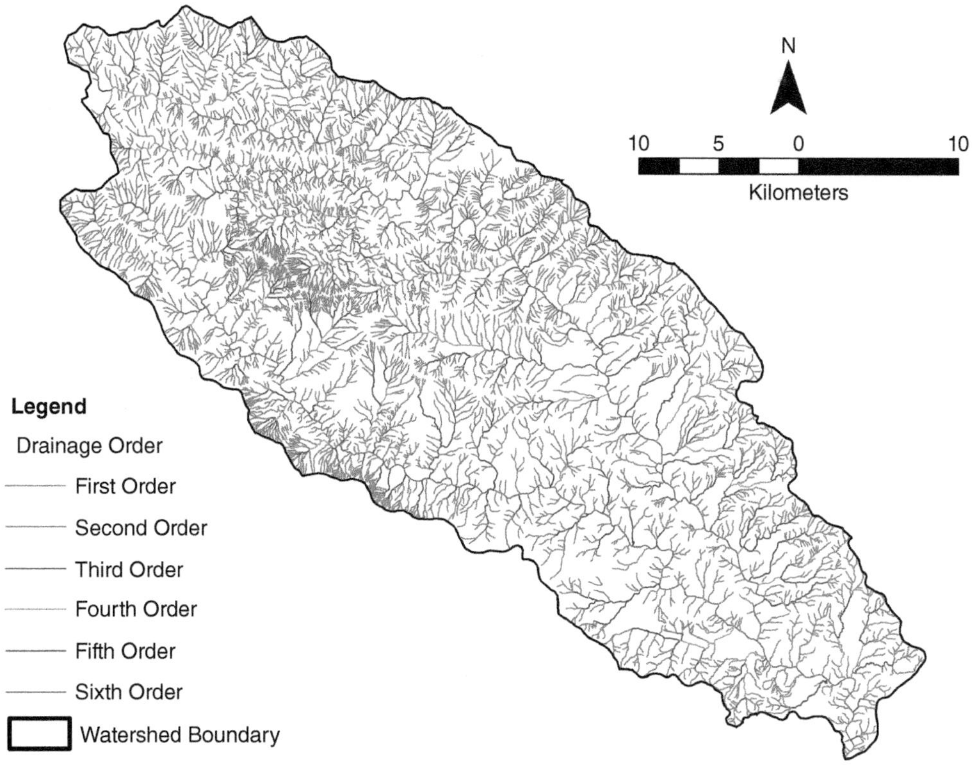

Figure 9.3 Drainage order map.

9.3.5 Morphometric Analysis

The information about basic morphometric parameters such as area (A), perimeter (P), length (L), and number of streams (N) was obtained from mini watershed delineated layer and basin length (L_b) was calculated from stream length, while the bifurcation ratio (R_b) was calculated from the number of stream segments. Other morphometric parameters were calculated using the equations as described in Table 9.2. Liner parameters have a direct relationship with erodibility (Nooka Ratnam et al. 2005); the higher the value, the higher the erodibility. The highest value of the linear parameter was ranked 1, the second highest value ranked 2, and so on. On the contrary, the shape parameters have an inverse relation with linear parameters, so that lower their value more is the erodibility. Thus the lowest value of the shape parameter was rated as rank 1, the second lowest as rank 2, and so on.

A compound factor was then worked out by summing all the ranks of liner parameters as well as shape parameters and then dividing by number of parameters. From the group of these mini watersheds, the highest prioritized rank was assigned to the mini watershed having the lowest compound factor and so on.

9.3.6 Decision Rules for Site Selection of Water Harvesting Structures

Selecting suitable sites for various water harvesting structures as per the guidelines of IMSD are as follows:

1) Check Dams
   The slope should be 1–3%.
   The higher-order stream defined banks with narrow width channel.
   The type of the sand should be sandy/gravel zone.
2) Farm Pond
   The slope should be 1–2%.
   The infiltration rate of soil should be semi-pervious to impervious.
3) Boulder Bund
   The slope should be 2–3%.
   Across 1st and 2nd order stream.
   The infiltration rate of soil should be semi-pervious to pervious.
4) Nala Plug/Bund
   The slope should be 2–5%.
   Across 1st and 3rd order stream.
   The rate of soil erosion is high.
5) Percolation Tank
   The slope should be 2–3%.
   Across 3st and 4nd order stream.

Table 9.2 Formulas for computation of morphometric parameters.

Morphometric parameters	Formula	Reference
Area of the basin	A = Area of the Basin in Km^2	Nooka Ratnam et al. (2005)
Perimeter of basin	P = Perimeter in Km.	Nooka Ratnam et al. (2005)
Total No. of streams	N = No. of streams	Nooka Ratnam et al. (2005)
Total No. of first-order streams	N_1 = Total no. of first-order streams	
Stream order (u)	Hierarchical rank	Strahler (1964)
Basin length (L_b)	$L_b = 1.312*A^{0.568}$ Where, L_b = Length of Basin (km) A = Area of Basin (km^2)	Nooka Ratnam et al. (2005)
Stream length (L)	Length of the stream	Horton (1945)
Bifurcation ratio (R_b)	$R_b = N_u/N_{u+1}$ Where, R_b = Bifurcation Ratio N_u = Total number of stream segment of order 'u' N_{u+1} = Number of segment of next higher order	Schumm (1956)
Drainage density (D_d)	$D_d = L_u/A$ Where, D_d = Drainage density L_u = Total stream length of all order A = Area of the basin	Horton (1945)
Stream frequency (F_u)	$F_u = N_u/A$ Where, F_u = Total number of streams of all order A = Area of the Basin (km^2)	Horton (1945)
Texture ratio (T)	$T = N_u/P$ Where, N_u = Total number of streams of all orders P = Perimeter (km)	Horton (1945)
Length of overland flow (L_g)	$L_g = 1/D*2$ Where, L_g = Length of the Overland Flow D = Drainage density	Horton (1945)
Form factor (R_f)	$R_f = A/Lb^2$ Where, R_f = Form Factor A = Area of the basin (km^2) Lb^2 = Square of the basin length	Horton (1945)
Shape factor (B_s)	$R_f = Lb^2/ A$ Where, B_s = Shape Factor A = Area of the basin (km^2) Lb^2 = Square of the basin length	Nooka Ratnam et al. (2005)
Elongation ratio (R_e)	$R_e = (2/Lb)*(A/P)^{½}$ Where, R_e = Elongation Ratio Lb = Length of basin (km) A = Area of the basin (km^2)	Schumm (1956)
Compactness constant (C_c)	$C_c = 0.2821P/P^2$ Where, C_c = Compactness Ratio P = Perimeter of the basin (km)	Horton (1945)
Circularity ratio (R_c)	$R_c = 4\pi A/P^2$ Where, R_c = Circularity Ratio A = Area of the basin (km^2) P = Perimeter (km)	Miller (1953)

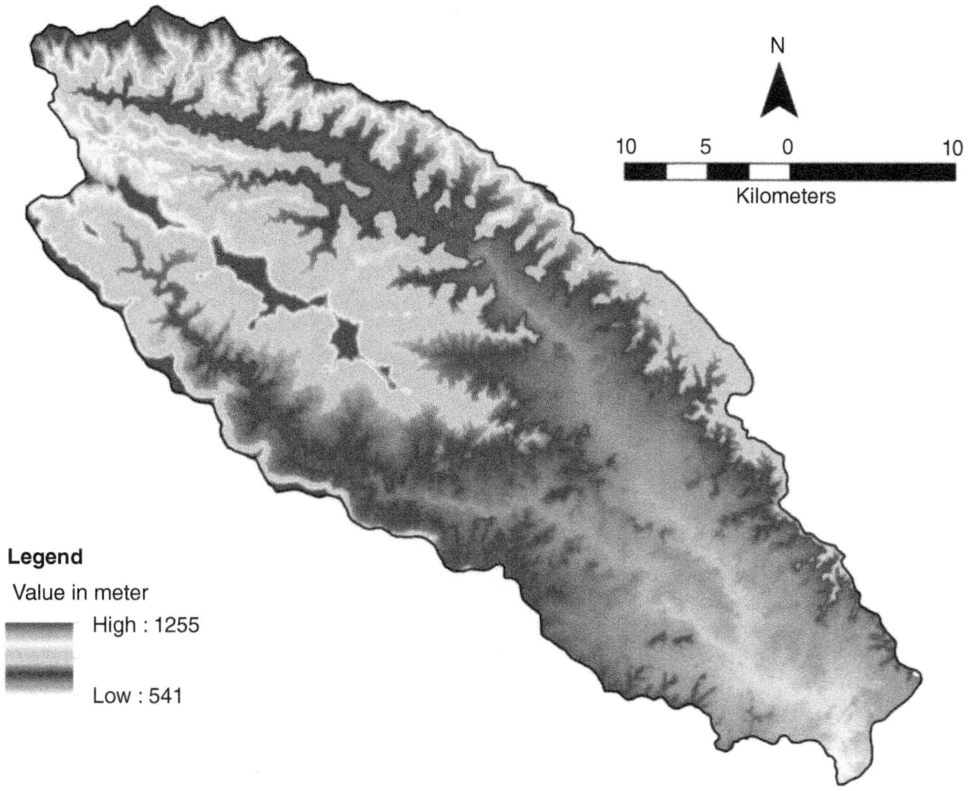

Figure 9.4 SRTM 1 arc-second interval DEM of Upper Bhīma watershed.

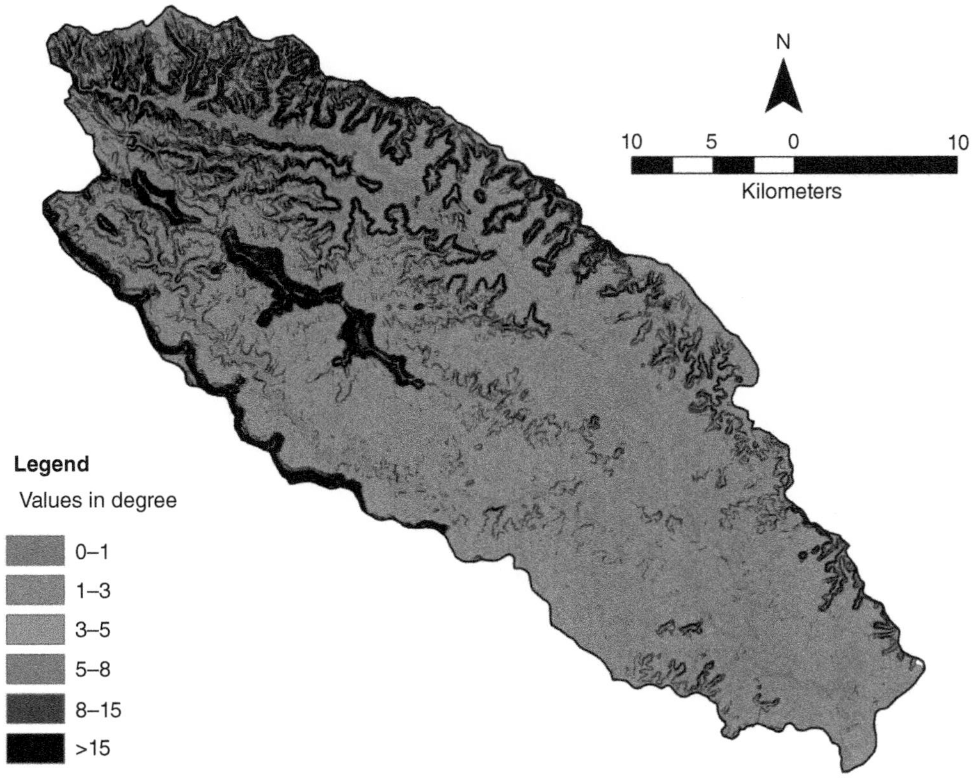

Figure 9.5 Slope map derived from SRTM DEM.

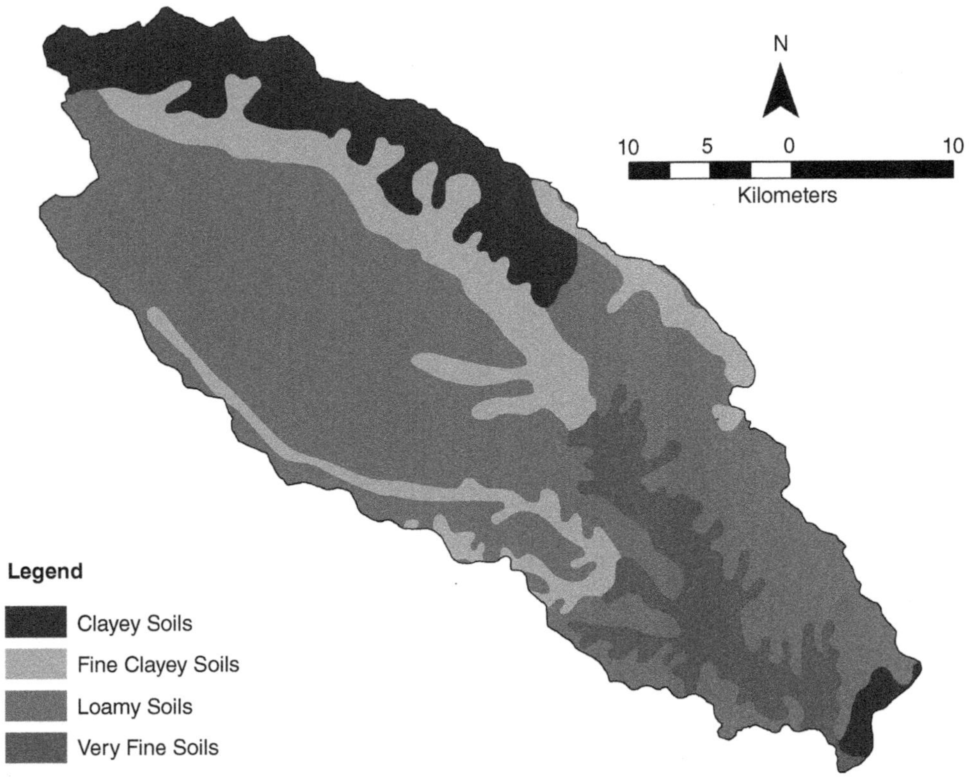

Figure 9.6 Soil map for study site. Source: NBSS&LUP, Nagpur, India.

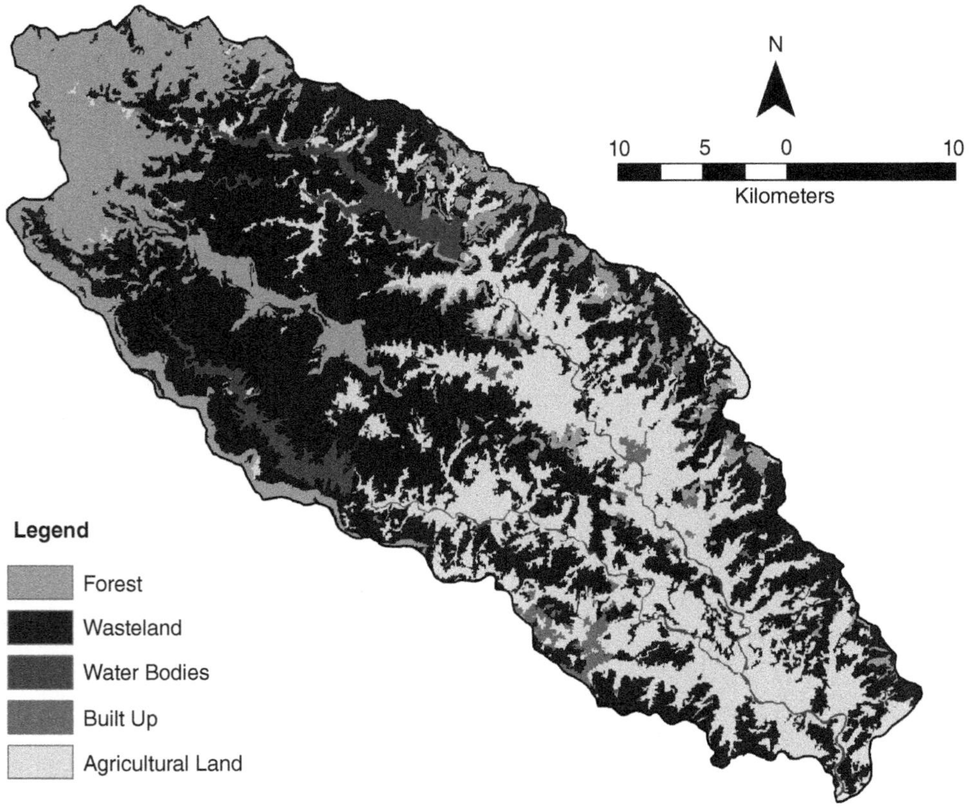

Figure 9.7 Land use land cover map.

The infiltration rate of soil should be semi-pervious to pervious.

To find suitable locations, the site was visualized through the fly tool of ArcScene 10.1 as per the priority assigned. This means higher-ranked watersheds were visualized first. Thus, we can visualize each mini watershed without field visit and find the best feasibility for positioning a water harvesting structure by overlaying of DEM, soil map, and slope map along with the guidelines of IMSD.

9.4 Results and Discussions

The importance of the morphometric parameters and results obtained by morphometric analysis are discussed here. These parameters were calculated using the formulas tabulated in Table 9.2, and the results are in Tables 9.3–9.5.

9.4.1 Basic Parameters

Basic parameters include watershed area, perimeter, stream length, stream order, and basin length.

9.4.1.1 Area (A) and Perimeter (P)

The drainage area (A) is probably the single most important watershed characteristic for hydrologic design and reflects the volume of water that can be generated from rainfall. The present result shows that watershed no. 5 covers the maximum area of 156.92 km^2 while watershed no. 26 has a minimum area of 7.56 km^2. The basin perimeter can be defined as the length of the line that defines the surface divide of the basin. In the present case, the maximum value as 64.09 km is found in mini-watershed no. 1, whereas the minimum of 26.87 km in sub-watershed no. 04.

9.4.1.2 Total Length of Streams (L)

Addition of the lengths of all streams, in a particular order, defines total stream length. The numbers of streams of various orders in a mini-watershed were counted and their lengths measured, shown in Table 9.3. These results help to find the drainage density.

9.4.1.3 Stream Order (u)

The concept of stream order was introduced by (Horton 1945) and (Strahler 1964) to describe the basins in quantitative terms. In this research work, stream order analysis is performed using the Strahler method: where the first-order streams do not consist of any tributaries, and the confluence of the two first-order streams formed the second-order streams and so on. However, as the stream orders increase, the total number of streams of the particular order decreases. This supplemented the study of stream order of the watershed. Among 11 watersheds, no. 8, 2, 7, and 5 are the watersheds having stream no. 809, 730, 313, and 181, respectively, as shown in Table 9.3. In watersheds no. 8 out of 809 streams, 618 are first-order streams, whereas none is having stream order VII.

9.4.1.4 Basin Length (L_b)

Basin length can be defined as the distance measured along the main channel from the watershed outlet to the basin divide. It is an important factor in the estimation of various morphometric analyses and it is proportional to drainage area. As per the result, the basin length of mini-watersheds varies from 23 km (mini-watershed no 1) to 9.5 km (mini-watershed no 6), shown in Table 9.4.

Table 9.3 Stream ordering of study watershed.

	Number of Streams for each stream order						
Watershed no.	**I**	**II**	**III**	**IV**	**V**	**VI**	**Total**
1	532	123	23	6	1	0	685
2	561	135	26	6	2	1	730
3	524	118	25	7	2	1	676
4	152	35	9	1	0	0	197
5	131	38	9	2	1	0	181
6	219	58	15	4	1	1	297
7	239	56	13	4	1	1	313
8	618	148	33	9	1	0	809
9	444	105	28	3	1	0	581
10	202	52	12	1	1	0	268
11	202	43	13	3	0	1	261
Total	3824.00	911.00	206.00	46.00	11.00	5.00	

Table 9.4 Calculated morphometric parameters.

Watershed no.	Basic parameter						Linear parameter					Shape parameter				
	A km^2	P km	L km	N	N1	L_b km	R_1	D_4 Km Km^{-2}	F No Km^{-2}	T	L_b	R_1	B_1	R_1	C_1	R_1
1	140.89	64.09	23	685	532	23	4.88	3.18	1.53	8.30	0.16	0.27	3.75	0.39	0.19	0.43
2	104.48	52.14	19.7	730	561	19	4.17	3.79	1.84	10.76	0.13	0.27	3.71	0.41	0.20	0.48
3	134.76	60.03	16	676	524	15.7	4.06	3.22	1.56	8.73	0.16	0.53	1.90	0.55	0.19	0.47
4	30.44	26.87	10.3	197	152	10.5	5.74	3.57	1.81	5.66	0.14	0.29	3.48	0.57	0.27	0.53
5	47.86	34.96	14.7	181	131	14.5	3.54	2.75	1.38	3.75	0.18	0.22	4.52	0.45	0.24	0.49
6	108.73	54.59	8.6	297	219	9.5	3.85	2.32	1.18	4.01	0.22	1.47	0.68	0.97	0.20	0.46
7	83.98	46.45	10.3	313	239	10.5	3.96	2.70	1.38	5.14	0.19	0.79	1.26	0.75	0.21	0.49
8	149.83	57.72	19.5	809	618	19.5	5.33	3.37	1.60	10.71	0.15	0.39	2.54	0.44	0.18	0.56
9	139.64	58.36	18.3	581	444	18	5.08	2.91	1.43	7.61	0.17	0.42	2.40	0.47	0.18	0.52
10	108.62	52.06	13.4	268	202	13.5	6.74	1.97	1.25	3.88	0.25	0.60	1.65	0.61	0.20	0.50
11	82.99	47.07	9.5	261	202	9.1	4.11	2.52	1.25	4.29	0.20	0.92	1.09	0.82	0.21	0.47

Table 9.5 Calculation of compound factor and prioritized ranks.

Watershed no	R_b	D_4	F_a	T	L_a	R_f	B_4	R_f	C_4	R_f	Compound factor	Ranks
1	5	5	5	4	7	2	10	1	4	1	4.4	1
2	6	1	1	1	11	3	9	2	6	5	4.5	2
3	8	4	4	3	8	7	5	6	3	3	5.1	4
4	2	2	2	6	10	4	8	7	11	10	6.2	6
5	11	7	7	11	5	1	11	4	10	7	7.4	10
6	10	10	11	9	2	11	1	11	7	2	7.4	11
7	9	8	8	7	4	9	3	9	8	6	7.1	8
8	3	3	3	2	9	5	7	3	1	11	4.7	3
9	4	6	6	5	6	6	6	5	2	9	5.5	5
10	11	11	10	10	1	8	4	8	5	8	6.6	7
11	7	9	9	8	3	10	2	10	9	4	7.1	9

9.4.2 Linear Parameters

Linear parameters include bifurcation ratio, drainage density, stream frequency, texture ratio, and length of overland flow.

9.4.2.1 Bifurcation Ration (R_b)

The term bifurcation ratio was first proposed by Schumm (1956). It is defined as the ration of the number of the streams of a particular order to the number of streams of the next higher orders. Generally, only a small range of variation is observed in different regions except in the basin which is denominated by powerful geological control (Strahler 1964). The lithological and geological development of the river basin brings about a change in its value (Strahler 1964). Lower R_b values are the characteristics of structurally less disturbed watersheds without any distortion in drainage pattern (Nag 1998). Table 9.4 shows that of bifurcation ratios (R_b) of Upper Bhīma watersheds, watershed no. 5 has the least bifurcation ratio of 3.54 and no. 10 has a maximum ratio of 6.74. Most of the micro watersheds have their R_b values close to or > 4.00, which indicates the presence of hilly terrain. Values > 5.00 in mini-watershed no. 4 (5.74) and mini-watershed no. 8 (5.33) are controlled by structural characteristics of the micro watersheds. The values in the range of 4.00–5.00 indicate that the stream network in the micro watersheds is greatly influenced by slope and topography in the region than structural characteristics (Samal et al. 2015)

9.4.2.2 Drainage Density (D_d)

It is the most important independent morphometric parameter which is related to other parameters viz. ruggedness number (R_n), length of overland flow (L_g), and constant of channel maintenance (C). It is calculated as the ratio of the total stream lengths of all orders per drainage area. It indicates how closely the stream segments are spaced with each other in a river basin (Langbein 1947). In the regions where topographical structures are eroded by streams, the D_d is an important indicator of the linear characteristics of the land features (Horton 1932). It helps to determine the relationship between the slope gradients and precipitations to measure the rate of runoff in the watershed. It is observed that the D_d of the micro watershed region is related to the number of streams and their distribution, relief, topography, length of streams, rock types, climate, and infiltration capacity (Smith 1950). The character and quantity of the surface runoff is influenced by the amount and type of precipitation. In regions with a very high rainfall, a large amount of rainfall is lost as runoff which results in more drainage lines on the land surface. The rate of surface runoff is affected by amount of vegetation and infiltration capacity of the soils. Hence, it influences the drainage texture of the region. Generally, in the areas of low relief, dense vegetation, and a highly permeable or resistant sub-surface material, a low D_d value is observed. Conversely, in the regions of high relief, sparse vegetation, and impermeable or weak sub-surface material, a high D_d value is measured. Low D_d results in coarse drainage (Miller 1953) texture in the region whereas high D_d results in fine drainage texture. Observations from Chankao (1982) suggest that watersheds with adequate drainage have $D_d > 5$, and with poor drainage have $D_d < 5$. High drainage density (> 13.7) is indicative of impermeable subsoil thereby leading to fine drainage texture, mountainous relief, and sparse vegetation. Low drainage density (<5) indicates highly permeable subsoil and low relief, which leads to coarse drainage texture and thick vegetative cover. Lower values generally tend to occur on gneiss, granite, and schist regions and areas which are highly resistant. The D_d in the mini watersheds range from a lowest 2.32 in mini watersheds no. 6 to 3.79 in mini watersheds no. 2 (Table. 9.4). A high value of the drainage density would indicate a relatively high density of streams and thus a rapid storm response, values of D_d are as shown in Table 9.4.

9.4.2.3 Stream Frequency (F_u)

Stream frequency/channel frequency (F_u) is the total number of stream segments of all order per unit area (Horton 1932). Its value indicates that topographical and rainfall conditions are the factors which control the development and origin of streams in the river basins. A low stream frequency value in the river basin shows that the streams have experienced fewer structural disturbance. The higher-order streams cause a high rate of surface runoff and fast stream flow (Kaliraj et al. 2015). The value of stream frequency ranges from 1.18 to 1.84 for mini watershed no. 6 and 2, as shown in Table 9.4.

9.4.2.4 Texture Ratio (T)

The texture ratio can be defined as the ratio of total number of streams of the first order to the perimeter of the basin (Horton 1945). It is an important parameter to understand the geomorphology of the watershed. It represents the relative space between the drainage lines. According to (Horton 1945), infiltration capacity is the single most important factor which influences drainage texture. Based on the drainage density, (Smith 1950) classified five different drainage textures in the watersheds. If the D_d of the watershed is < 2 it represents very coarse T. D_d in the range of 2 and 4 represents coarse T. D_d between 4 and 6 shows moderate T. D_d in the range of 6 and 8 indicates fine T whereas T > 8 represents very fine T. Generally, over impermeable areas drainage lines are more than permeable areas (Aher et al. 2013). The value of the texture ratio ranges from 3.75 to 10.76 as shown in Table 9.4.

9.4.2.5 Length of Overland Flow (L_o)

It is the length of water over the ground before it gets concentrated into definite stream channels and is equal to half of drainage density (Horton 1945). Length of overland flow relates inversely to the average channel slope. Table 9.4 reveals the length of overland flow for Upper Bhīma watersheds.

9.4.3 Shape Parameters

Shape parameters include form factor, shape factor, elongation ratio, compactness ratio, and circulatory ratio.

9.4.3.1 Form Factor (R_f)

The form factor can be defined as the ratio of the area of the basin to a square of the basin length (Horton 1945). For a perfectly circular basin, its value is generally greater than 0.78 (Strahler 1964). The smaller the value of form factor, the more the basin will be elongated. The basin with high form factors has peak flow of shorter duration, whereas an elongated mini watershed with low form factors has lower peak flow with longer duration. In the present case, the value of form factor is varied between 0.22, for watershed no. 5, and 0.92 for no. 11, as shown in Table 9.4. This value indicates the elongated shape of the basin and having flatter peak flow for a longer duration, which helps to manage the flood easily than those of the circular basin.

9.4.3.2 Shape Factor (B_s)

The shape factor can be defined as the ratio of the square of the basin length to the area of the basin (Horton 1945) and is in inverse proportion with form factor (R_f). Shape factor lies between 0.68 and 4.52 in present work, which indicates the elongated shapes of the basin.

9.4.3.3 Elongation Ratio (R_e)

The elongation ratio is the ratio between the diameters of the circle of the same area as that of the drainage basin to the maximum length of the basin. A circular basin is more efficient in runoff discharge than an elongated basin (Singh et al. 2009). The value of elongated ratio is varied between 0.6 and 1.0 in typical regions of very low relief, whereas values ranging between 0.6 and 0.8 are associated with high relief and steep ground slope (Strahler 1964). The lower value of the elongation ratio indicates that particular mini watershed is more elongate than others. The elongation value can be grouped into three categories, namely circular basin ($R_e > 0.9$), Oval basin (R_e: 0.9–0.8), Less elongated basin ($R_e < 0.7$). In this study, (Table 9.4), major values are less than 0.7 and hence the basins are elongated in shape.

9.4.3.4 Compactness Coefficient (C_c)

Compactness coefficient can be expressed as basin perimeter divided by the circumference of a circle to the same area of the basin. It is quite opposite (inversely related) to the elongation ratio in a basin and is responsible for causing erosion in the basins. The lesser the value of compactness coefficient means the more elongated the shape of basin and less erosion, while a higher value indicates the basin is less elongated and more erosion prone. It was also observed that the values of compactness coefficient exhibit a variation from 0.18 to 0.27, shown in Table 9.4.

9.4.3.5 Circularity Ratio (R_c)

This ratio shows the shape characteristics of the watershed. The R_c is measured from the ratio of the watershed area to the area of the circle having circumference equal to the perimeter of the micro watershed (Kaliraj et al. 2015; Miller 1953). It is mainly concerned with the frequency and length of the streams, LULC, geological structures, relief, climate, and slope of the watershed, which influence its value. A value of $R_c < 0.4$ indicates strongly elongated and homogenous rock with high runoff. A value equal to 1 indicates a circular shaped basin. In this study area, all the mini watersheds have the R_c value of ≤ 4.00 which indicated that all the mini watersheds are perfectly elongated with homogenous rock material catering high runoff (Table 9.4). R_c lies between 0.4–0.5 when the basin shape is more elongated in shape and very permeable with homogeneous materials. As seen in the results, the maximum circularity ratio was observed in mini watershed no. 1 (0.43) while the minimum in mini watershed no 8 (0.56), which reflects the highly elongated shape of the basins.

9.4.4 Compound Factor and Ranking

In this study, a compound factor is used for the prioritization of mini watershed in the Upper Bhīma river basin. For this, both shape parameters and linear parameters are taken into consideration.

Linear parameters are directly correlated to the erosion (the higher the value, the more erodible), whereas as in the case of shape parameters it is vice versa (the lower the value, the more erodible). First of all, individual ranking is assigned to both linear and shape parameters depending on the parameter values and afterwards; finally a compound factor is obtained by summing up all the parameter rankings divided by the number of parameters. From the compound factor, the first rank is assigned to the lower most value, and the last rank to the higher most value. In the present case, watershed no. 1 ranked first (4.4), followed by watershed no. 2 and 8 with second and third respectively, whereas watershed no. 6 has the last rank (7.4) and details are demonstrated in Table 9.5 and Figure 9.8 as well.

9.4.5 Positioning a Water Harvesting Structure

The suitable sites for water harvesting structures were identified with the application of remote sensing and GIS. The watershed boundary map, drainage map, land use map, soil map, and DEM map were prepared using satellite imagery and SOI toposheets of the Upper Bhīma watershed. The five classes of land use/land cover and hydrological soil group map were prepared. The overlay operation of land use map, hydrological soil group map, slope map, and watershed prioritization map was carried out for selecting suitable sites for water harvesting structures and presented through a site suitability map (Figure 9.9).

The suitability of check dam sites can be confirmed as the site is located on third-order drainage and satisfies the conditions of land use, soil type, and slope as per IMSD guidelines. A total of 4 sites are selected for check dams, 8 for farm ponds, and 14 sites for boulder bund construction, however it has been observed that 3 minor dams already existed on proposed site which validates the entire approach. For locating a proposed check dam in WS 6, the study site is visualized through fly-through tool in Arc GIS and the check dam is positioned at an appropriate location without field visit (Figure 9.9). In this way a lot of

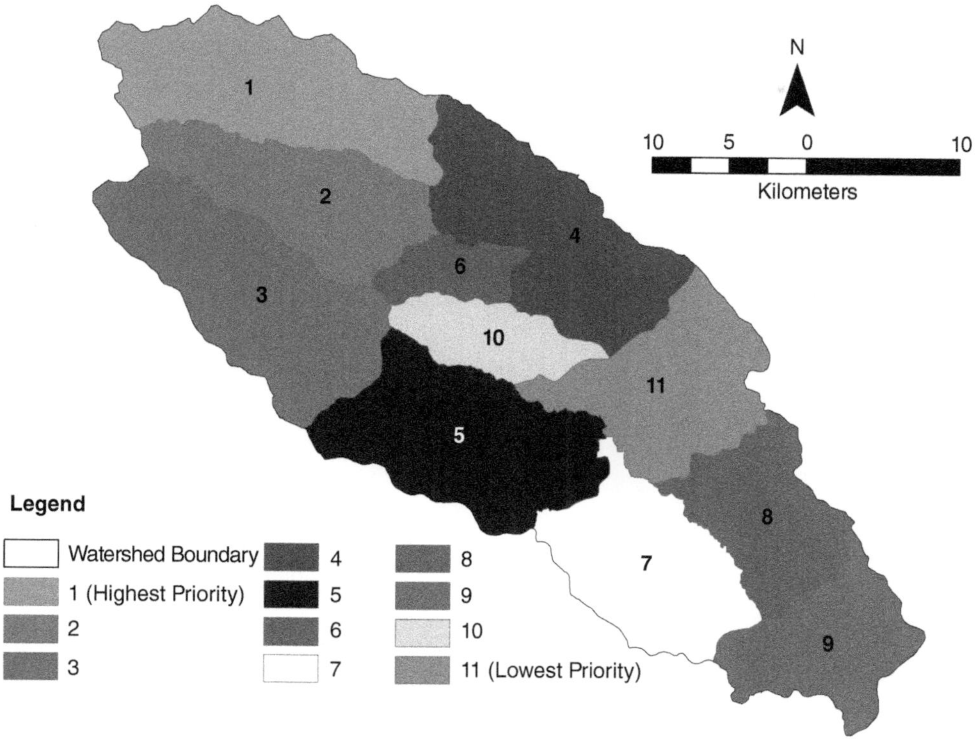

Figure 9.8 Watershed prioritization map.

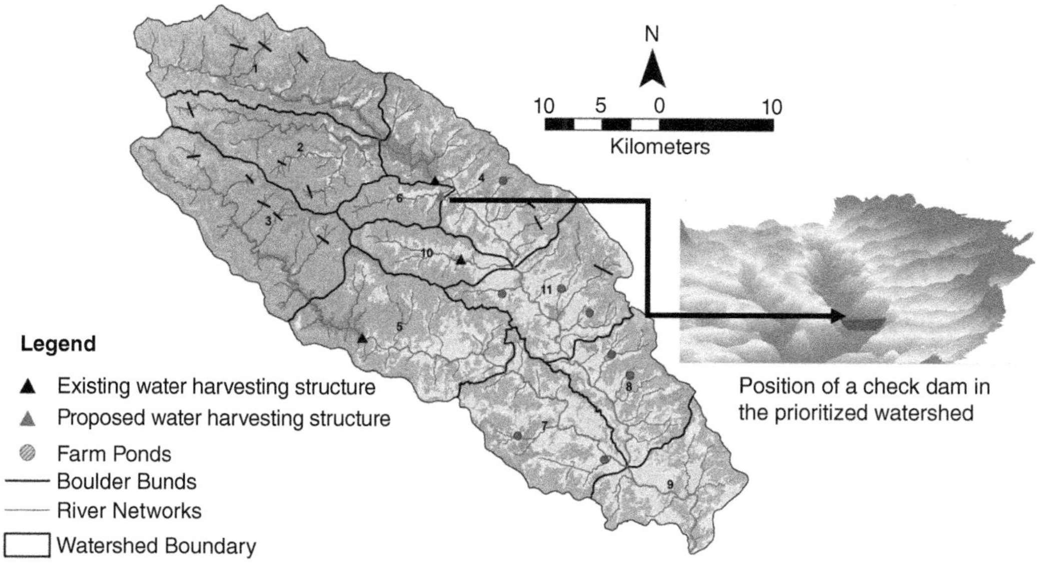

Figure 9.9 Integrated map showing positioning of water harvesting structures.

expenses and labor may be saved involved in costly field visits. These water harvesting structures directly check the excessive water coming from the watersheds and hence, lead to soil and water conservation. The proposed water harvesting structures are also useful for improving the groundwater condition, thus improving crop and fodder production.

9.5 Conclusion

This study reveals that the application of RS, GIS, and geovisualization technique is extremely useful to locate the appropriate water harvesting structures like check dams, farm ponds, and boulder bunds, in mini watershed whereas morphometric analysis is a very useful tool for prioritization of mini watersheds. Water harvesting structures are important to conserve precious natural resources like soil and water, which are depleting day by day at an alarming rate. It leads to soil as well as water conservation and reduces the high runoff and flood potential. Thus, for sustainable development of watersheds, soil and water conservation can be implemented by positioning suitable water harvesting structures in the Upper Bhīma watersheds.

References

Aher, P., Adinarayana, J., and Gorantiwar, S. (2013). Prioritization of watersheds using multi-criteria evaluation through fuzzy analytical hierarchy process. *Agricultural Engineering International: CIGR Journal* 15: 11–18.

Buttenfield, B.P. and Mackaness, W.A. (1991). Visualization. In: *Geographical Information Systems*, vol. 1 (eds. D.J. Maguire, M.F. Goodchild and D.W. Rhind), 427–443. Harlow, Essex, UK: Longman Scientific and Technical.

Chankao, K. (1982). *Principle of Watershed Management Faculty of Forestry*. Bangkok, Thailand: Kasetsart University.

Chowdary, V., Ramakrishnan, D., Srivastava, Y. et al. (2009). Integrated water resource development plan for sustainable management of Mayurakshi watershed, India using remote sensing and GIS. *Water Resources Management* 23: 1581–1602.

Edet, A., Okereke, C., Teme, S., and Esu, E. (1998). Application of remote-sensing data to groundwater exploration: a case study of the Cross River State, southeastern Nigeria. *Hydrogeology Journal* 6: 394–404.

Gavit, B., Purohit, R., Singh, P. et al. (2018). Rainwater harvesting structure site suitability using remote sensing and GIS. In: *Hydrologic Modeling* (eds. V.P. Singh, S. Yadav and R. Narayan Yadava), 331–341. Springer.

Horton, R.E. (1932). Drainage basin characteristics. *Transactions of the American Geophysical Union* 13: 350–361.

Horton, R.E. (1945). Erosional development of streams and their drainage basins; hydrophysical approach to quantitative morphology. *Geological Society of America Bulletin* 56: 275–370.

Jasrotia, A., Dhiman, S., and Aggarwal, S. (2002). Rainfall-runoff and soil erosion modeling using Remote Sensing and GIS technique—a case study of tons watershed. *Journal of the Indian Society of Remote Sensing* 30: 167–180.

Javed, A., Khanday, M.Y., and Ahmed, R. (2009). Prioritization of sub-watersheds based on morphometric and land use analysis using remote sensing and GIS techniques. *Journal of the Indian Society of Remote Sensing* 37: 261–274.

Javed, A., Khanday, M.Y., and Rais, S. (2011). Watershed prioritization using morphometric and land use/land cover parameters: a remote sensing and GIS based approach. *Journal of the Geological Society of India* 78: 63–75.

Kaliraj, S., Chandrasekar, N., and Magesh, N. (2015). Morphometric analysis of the River Thamirabarani sub-basin in Kanyakumari District, South west coast of Tamil Nadu, India, using remote sensing and GIS. *Environmental Earth Sciences* 73: 7375–7401.

Langbein, W.B. (1947). *Topographic Characteristics of Drainage Basins*. Washington, DC, USA: US Government Printing Office.

MacEachren, A.M. and Taylor, D.R.F. (1994). *Visualization in Modern Cartography*, vol. 2. Pergamon Press.

Matouq, M., Al-Bilbisi, H., El-Hasan, T., and Eslamian, S. (2014). GIS applications in a changing climate, handbook of engineering hydrology. In: *Modeling, Climate Change, and Variability*, vol. 2 (ed. S. Eslamian), 297–312. Taylor and Francis.

Miller, V.C. (1953). *Quantitative Geomorphic Study of Drainage Basin Characteristics in the Clinch Mountain Area, Virginia and Tennessee Technical Report*, 389–402. New York: Columbia University, Department of Geology.

Nag, S. (1998). Morphometric analysis using remote sensing techniques in the Chaka sub-basin, Purulia district, West Bengal. *Journal of the Indian Society of Remote Sensing* 26: 69–76.

NBSSLUP (1996). *Maharashtra Soils*. Nagpur, India: National Bureau of Soil Survey and Land Use Planning.

Nooka Ratnam, K., Srivastava, Y., Venkateswara Rao, V. et al. (2005). Check dam positioning by prioritization of micro-watersheds using SYI model and morphometric analysis—remote sensing and GIS perspective. *Journal of the Indian Society of Remote Sensing* 33: 25–38.

Patel, D.P., Dholakia, M.B., Naresh, N., and Srivastava, P.K. (2012). Water harvesting structure positioning by using geo-visualization concept and prioritization of mini-watersheds through morphometric analysis in the Lower Tapi Basin. *Journal of the Indian Society of Remote Sensing* 40: 299–312.

Patel, D.P., Srivastava, P.K., Gupta, M., and Nandhakumar, N. (2015). Decision Support System integrated with Geographic Information System to target restoration actions in watersheds of arid environment: a case study of

Hathmati watershed, Sabarkantha district, Gujarat. *Journal of Earth System Science* 124: 71–86.

Prasad, H.C., Bhalla, P., and Palria, S. (2014). Site suitability analysis of water harvesting structures using remote sensing and GIS-A case study of Pisangan Watershed, Ajmer District, Rajasthan. *The International Archives of Photogrammetry, Remote Sensing and Spatial Information Sciences* 40: 1471.

Rahman, A., Hajani, E., and Eslamian, S. (2017). Rainwater harvesting in arid regions of Australia. In: *Environmental Impacts and Analysis of Drought and Water Scarcity*, Handbook of Drought and Water Scarcity, vol. 2 (eds. S. Eslamian and F. Eslamian), 489–500. Francis and Taylor/CRC Press.

Samal, D.R. and Gedam, S.S. (2015). Monitoring land use changes associated with urbanization: an object based image analysis approach. *European Journal of Remote Sensing* 48: 85–99.

Samal, D.R., Gedam, S.S., and Nagarajan, R. (2015). GIS based drainage morphometry and its influence on hydrology in parts of Western Ghats region, Maharashtra, India. *Geocarto International* 30: 755–778.

Saptarshi, P.G. and Raghavendra, R.K. (2009). GIS-based evaluation of micro-watersheds to ascertain site suitability for water conservation structures. *Journal of the Indian Society of Remote Sensing* 37: 693–704.

Schumm, S.A. (1956). Evolution of drainage systems and slopes in badlands at Perth Amboy, New Jersey. *Geological Society of America Bulletin* 67: 597–646.

Shinde, S.D. and Gaikwad, V. (2016). Application of GIS for mapping rainwater harvesting potential: a case study of Nidhal Village in Satara District, Maharashtra, India. *International Journal of Research in Applied* 4: 141–148.

Singh, J., Singh, D., and Litoria, P. (2009). Selection of suitable sites for water harvesting structures in Soankhad watershed, Punjab using remote sensing and geographical information system (RS&GIS) approach—a case study. *Journal of the Indian Society of Remote Sensing* 37: 21–35.

Sivakumar M, Roy P, Harmsen K, Saha S (2004) Satellite remote sensing and GIS applications in agricultural meteorology. In: Proceedings of the Training Workshop in Dehradun, India. AGM-8, WMO/TD.

Smith, K.G. (1950). Standards for grading texture of erosional topography. *American Journal of Science* 248: 655–668.

Strahler, A.N. (1964). Quantitative geomorphology of drainage basin and channel networks. In: *Handbook of Applied Hydrology* (ed. V. Chow), 439–476. New York: McGraw Hill.

Thakkar, A.K. and Dhiman, S.D. (2007). Morphometric analysis and prioritization of miniwatersheds in Mohr watershed. *Gujarat using remote sensing and GIS techniques. J Indian Soc Remote Sens* 35: 313–321. https://doi.org/10.1007/BF02990787.

Weerasinghe, H., Schneider, U., and Loew, A. (2011). Water harvest-and storage-location assessment model using GIS and remote sensing. *Hydrology and Earth System Sciences Discussions* 8: 3353–3381.

10

Information Technology in Water Harvesting

S. Sreenath Kashyap[1], M.V.V. Prasad Kantipudi[2], Saeid Eslamian[3], Maryam Ghashghaie[4], Nicolas R. Dalezios[5], Ioannis Faraslis[5], and Kaveh Ostad-Ali-Askari[6]

[1] *Department of ECE, Kommuri Pratap Reddy Institute of Technology, Hyderabad, Telangana, India*
[2] *Department of ECE, Sreyas Institute of Engineering and Technology, Hyderabad, Telangana, India*
[3] *Department of Water Engineering, College of Agriculture, Isfahan University of Technology, Isfahan, Iran*
[4] *Department of Water Resources Engineering, Faculty of Agriculture, Bu-Ali Sina University, Hamedan, Iran*
[5] *Department of Civil Engineering, University of Thessaly, Volos, Greece*
[6] *Department of Civil Engineering, Islamic Azad University, Najafabad Branch, Isfahan, Iran*

10.1 Introduction

Water is a precious natural resource that is colorless, tasteless, odorless. Water is one of the constituent substances of Earth's streams, a fluid in any living organism. It is very important for all forms of life. Nearly 70% of the Earth consists of water. A large percentage – 96.5% – is mainly found in seas and oceans, whereas a small portion – nearly 1.7% – is seen in the unique spaces like Antarctica and Greenland. Unfortunately, only 1.7% is available as ground water. In total, only 2.5% is freshwater content, and nearly 98% is ice.

Water also plays a major role in the world's economy. Safe drinking water is an essential component for the livelihood of human beings. Water is also useful for agricultural, industrial, and manufacturing units. Fruitful usage of water without any wastage is the responsibility of an individual.

Harvesting water for future usage should be the prime concern of human beings. Specifically, where surface water is scarce, ground water plays a major role for drinking and irrigation purposes. Groundwater tables are declining very rapidly, mainly due to urbanization. There is an alarming need for replenishment of ground water in urban areas. This can be compensated by using artificial recharge of groundwater.

The variety of Internet of Things (IoT) technologies in water harvesting techniques will lead to a fruitful solution in determining the parameters associated with water harvesting anytime and anywhere.

This chapter is organized as follows: first, water harvesting methods are briefly presented. Then, IoT is briefly described, followed by applications of IoT in water harvesting. The next section covers the assessment of the available subsurface resources using IoT. Finally, IoT devices for efficient agricultural/irrigation usage are presented.

10.2 Water Harvesting Methods

10.2.1 Basin Method

This method consists of two to four basins beside the river (Figure 10.1). The river flow is diverted into these basins, and after filling them the water enters in the river (Tucsonaz.gov 2018). Because of the water storage capacity, the water infiltrates into subsurface layers.

10.2.2 Stream Channel Method

In this method, the stream flow of the river is channelized and increased such that the infiltration time is increased. The flow control can be enhanced by constructing various obstruction structures (Figure 10.2) (Dbstephens.com 2018).

10.2.3 Ditch and Furrow Method

Water is distributed in a series of ditches and furrows, which are shallow and flat bottomed (Figure 10.3). The width range of the ditch is appropriately 0.3–1.8 m (User 2018).

Handbook of Water Harvesting and Conservation: Basic Concepts and Fundamentals, First Edition. Edited by Saeid Eslamian and Faezeh Eslamian.

Figure 10.1 Basin method.

Figure 10.2 Stream channel method.

Figure 10.3 Ditch and furrow method.

10.2.4 Flooding Method

This method is applicable to flat topography areas. In this method the water is supplied to the open places, through which this water infiltrates into subsurface layers (Figure 10.4) (The Hans India 2016).

10.2.5 Irrigation Method

In the absence of agricultural fields, water is supplied to crop land through canals, thus, leading to an increase of the groundwater levels (Figure 10.5) (Indiamart Business Video 2014).

10.2.6 Pit Method

This is the most effective method for recharging water in urban areas. A pit is excavated up to permeable formation for ground water recharge. This pit consists of a layer of gravel, charcoal, and sand. This method is applicable to hard rock areas and clay soils. Figure 10.6 shows a pit

Figure 10.4 Flooding method.

Figure 10.5 Irrigation method.

Figure 10.6 Pit method.

consisting of a 2 m width, 1 m depth for effective infiltration of water. The size of the pit depends on the recharge area (Shramajeevi 2015).

10.2.7 Recharge Well Method

Artificially injecting water into the well is known as the recharge well method. In this method, a bore hole is drilled up to the groundwater zone. This hole consists of a 0.25 nm mesh, which filters dust and other suspended salts. This method is applicable to areas where the groundwater table is deep (Figure 10.7). This method is mainly useful for urban areas, where water scarcity is becoming an issue (Sravan Kumar 2014).

The choice of a specific method depends upon the topography, soil type, rainfall, and geology of the area. There are benefits of the artificial recharge method, such as: reduction in the pumping cost; increased agriculture production; water quality improvement; storage of water in subsurface layers during the rainy periods; prevention of sea water intrusion in costal aquifers; and renovation of wastewater and reduction of flood flows (Rao et al. 2011).

Figure 10.7 Recharge well method.

10.3 The Internet of Things (IoT)

In this era of the twenty-first century, the advancement in the field of science and technology is appreciable. The advancement is the technology, which has resulted in the development of handheld devices, which are user-friendly and capable of providing services at all times and anywhere through the IoT. The IoT is a network of sophisticated devices, which constitute a database and enable users to exchange data and store the data for future purposes. The IoT allows the objects to sense the information or signals and can be controlled remotely by the user without any interference through which the user can save the time and can access information or devices from anywhere.

10.3.1 Applications of the IoT in Water Harvesting

The IoT technology and devices can be used for multidisciplinary applications. There are several parameters of water harvesting that can be measured using IoT-based applications, such as: estimation of the soil moisture content; determining the quality of groundwater; rate of infiltration in the soil; delineation of aquifer boundaries and estimation of storability of aquifer; estimation of the depth of aquifer from the surface of the earth; and identification of sites for artificial recharge structures. This is only a small portion of the potential applications.

10.3.1.1 Estimation of the Soil Moisture Content

The amount of water content present in the soil is known as moisture content, which is a very important factor for determining the rate of movement of water into the ground. The moisture content of the soil depends on the recent exposure of rain, specific retention, porosity, and temperature of the area.

Estimation of moisture content can be conducted by the residual weighted method, which involves the weight of the soil moisture (W_1). After drying in the oven, the weight of the soil becomes (W_2) and the difference of the W_2 and W_1 gives the moisture content of the soil (Sharma and Baranwal 2003).

10.3.1.2 Determining the Quality of Groundwater

Water consists of physical, chemical, and biological characteristics (Vafakhah and Eslamian 2017). Depending upon the resistivity values, it is possible to determine the quality of groundwater. Fresh water acts as a conductor, showing less resistivity value for resistivity values ranging between 0 and 10-Ω meters. If the water consists of any impurities, like dissolved solids, salts, and seawater, then the resistivity value ranges between 10 and 100-Ω meters.

10.3.1.3 Rate of Infiltration in the Soil

The water table from the aquifer inclines and declines (Kouhestani et al. 2016). Due to water harvesting, the ground can be estimated through a water level indicator, which gives a sound when it reaches the water. To measure the groundwater table, an automatic water level indicator is used, which indicates the depth of water present in the aquifer through sound.

10.3.1.4 Delineation of Aquifer Boundaries and Estimation of Storability of Aquifer

Multiple sensors are placed in and around the aquifer and, based on the temperature and resistivity, the delineation of aquifer boundaries can be determined.

10.3.1.5 Depth of Aquifer from the Surface of the Earth

In this method two electrodes are used acting as a current electrode, and two electrodes acting as a potential electrode. When the current is passing through the current electrodes, the potential difference is calculated by the potential electrodes, and based on these values, the resistivity of the material is identified. Based on the resistivity values, the depth of the aquifer is determined. If the resistivity is in the range of 0–10-Ω meters, then it is considered to be groundwater.

10.3.1.6 Identification of Sites for Artificial Recharge Structures

The location for artificial recharge structures depends upon the type of soil, geology, and topography of the area. With the use of the sensors it is feasible to identify the zones of less resistivity. These zones are suitable for recharging water harvesting (Dhakate et al. 2008).

10.4 Assessing the Available Subsurface Resources Using the IoT

In many countries the decline of subsurface water signifies the depletion of the resources. It is necessary to assess the amount of water content in the subsurface layers for the usage of water for future purposes. A case study is presented about the geophysical investigations in the Medak district of Telangana State in India (Rao and Thangarajan 1992).

Close to 72 electrodes are connected by a multi-core cable (Figure 10.8). The current and potential electrode pairs are connected to the earth resistivity meter, which provides the output current. The resistivity against the depth is along the survey line data and is collected by automatic profiling along the line until the electrode is reached. The spacing is increased by the minimum electrode separation and the process is repeated for depth investigations. The data is transformed to resistivity, which also depends on the type of the electrode used. Once converted, the data is modeled using finite element and least squares inversion methods in order to calculate a true resistivity versus depth pseudo section.

The entire area under the study is being divided into various regions:

- Raipol: In this region, based on the imaging results, it is found that the resistivity increased with depth (Figure 10.9). It is also observed that the subsurface layer consists of saprolite and granite.
  However, there is possibility for deep potential aquifer zones (Figure 10.10).
- Bhumpali: In this region, based on the imaging results, it can be noted that the resistivity decreases significantly

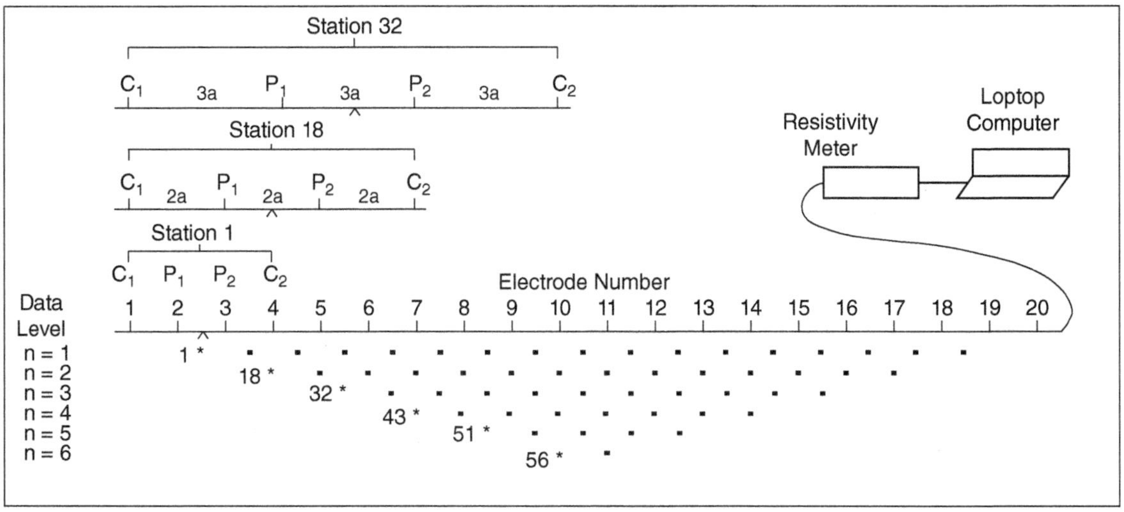

Figure 10.8 Sequence of measurements to build up a pseudo section.

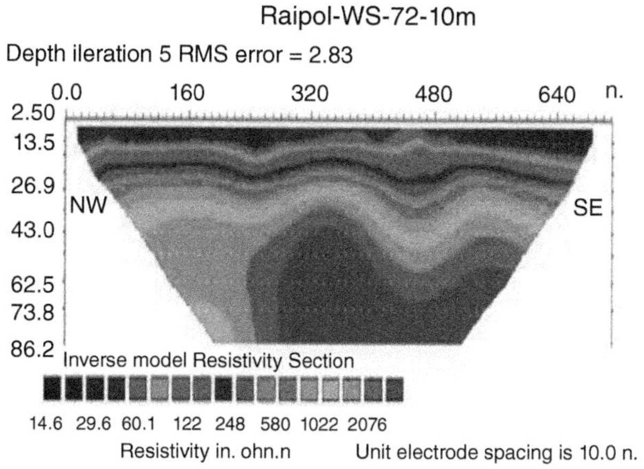

Figure 10.9 Resistivity imaging of Raipol village.

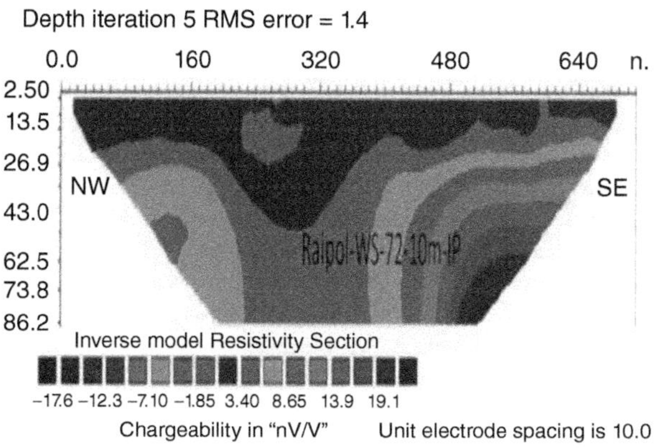

Figure 10.10 Chargeability imaging of Raipol village.

with depth (Figure 10.11). It is also observed the subsurface layer is a deep fracture zone (Figure 10.12).

The chargeability shows that the resistive zones are of high chargeability which is mainly due to the presence of oxidized minerals.

- Dubbak: The 2D imaging reveals that the subsurface layers consist of the sharp resistivity and contrast between the weathered and basement zones (Figure 10.13).

  The chargeability imaging figure shows that the low chargeability is due to the presence of fresh water (Figure 10.14).
- Gollapalli: The 2D imaging figure shows the multiple profile of resistivity. The fresh granite basement is noticed in the bottom-most layer, which is of a depth greater than 500 m (Figure 10.15).
- Sangapur: This area is the border area of two districts. The 2D imaging results show the low resistivity values in the northeastern part and high resistivity values in the southeastern part (Figure 10.16).

  The chargeability figure shows the high resistivity region, which is composed of granite (Figure 10.17).

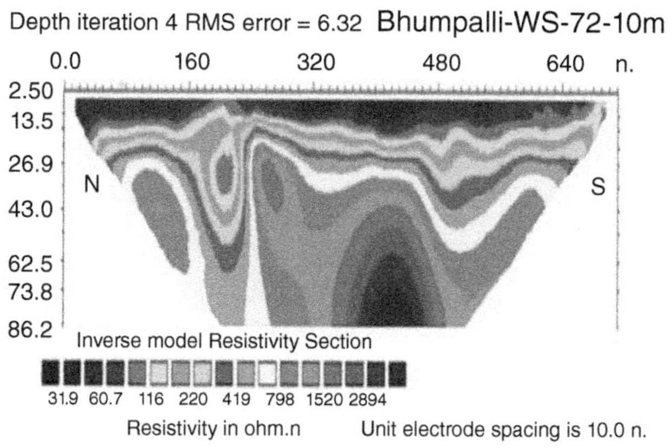

Figure 10.11 Resistivity imaging of Bhumpalli village.

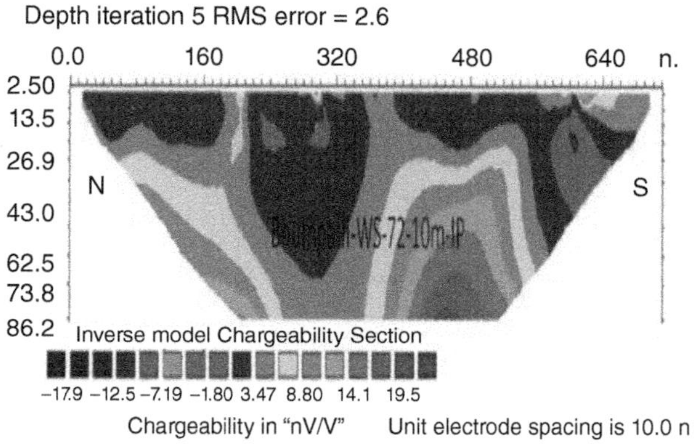

Figure 10.12 Chargeability imaging of Bhumpalli village.

Figure 10.13 Resistivity imaging of Dubbak village.

In this experimental investigation, the 2D resistivity and chargeability survey is carried out. The analysis and observations are to be stored in the database. The data available will be helpful to the user in such a way that the interpretation of data will help to find the topographical overview of the area. This interpretation can be helpful for the user, such that the composition of the subsurface layers can be

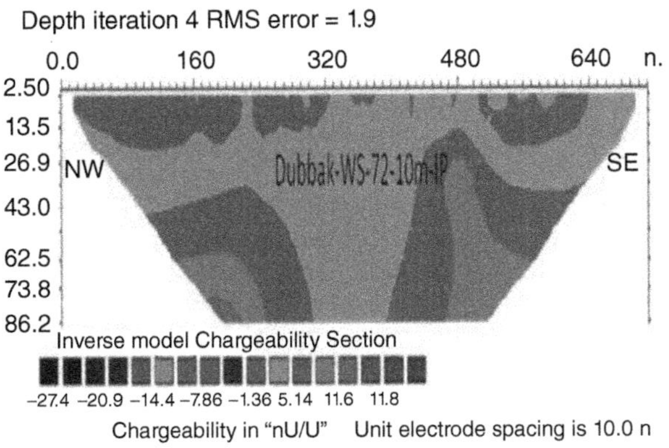

Figure 10.14 Chargeability imaging of Dubbak village.

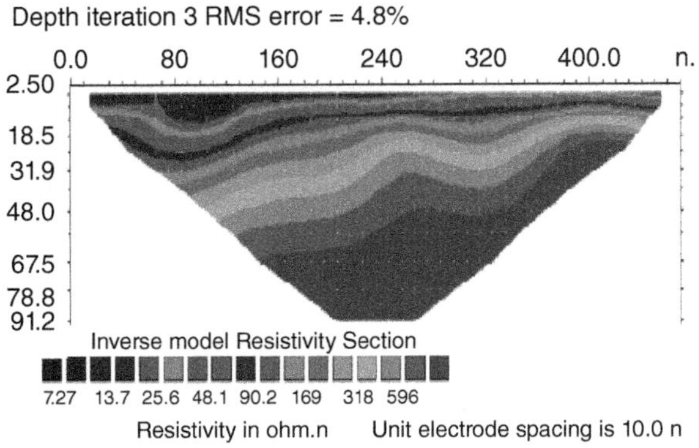

Figure 10.15 Resistivity imaging of Gollapalli village.

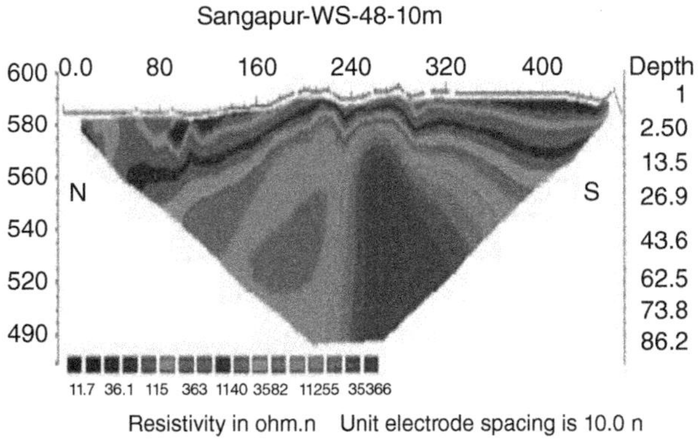

Figure 10.16 Resistivity imaging of Sangapur village.

found, while drilling the bore or developing the artificial recharge structures.

10.5 The IoT Devices for Efficient Agricultural/Irrigation Usage

The IoT devices can be used to determine the atmospheric conditions of a specific region, as well as the soil condition.

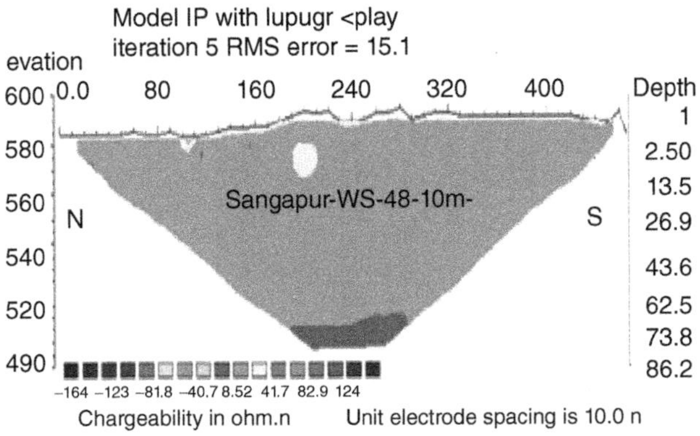

Figure 10.17 Chargeability imaging of Sangapur village.

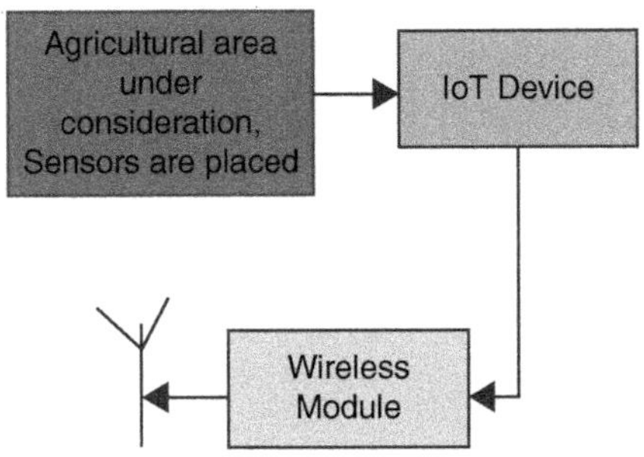

Figure 10.18 IoT-based wireless module associated at the site/agricultural area under consideration.

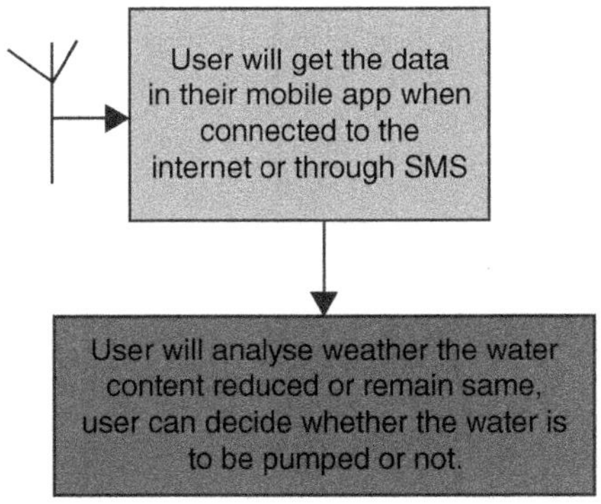

Figure 10.19 Module associated at the site/agricultural area under consideration.

Sensors which can detect the amount of temperature can be used for measuring the soil temperature (Figure 10.18).

The temperature sensor senses the value and gives the signal to the IoT device connected to it. This IoT device transmits the data or the details to the user using wireless technology (Figures 10.18 and 10.19). This will help the user to decide on the requirement of water for the crop at that instant based on the temperature of the soil or the data sent to the user. If the temperature of the soil increases, then it means that the moisture content in the soil is reduced. If the temperature of the soil remains the same or there is minor variation, then the amount of moisture content remains the same or is similar, such that no water is required at that instant of time. With the help of these devices, which will work based on the IoT technology the user can access data and know the condition at any point of time from anywhere. The usage of water can be controlled. This also helps the other users to determine the quality of the soil and to determine the suitability of the soil for that specific crop.

10.6 Conclusions

"Save water, save life" should be the motto of mankind. Efforts are to be made to improvise the groundwater in the subsurface layers. The usage of information technology and IoT devices for water harvesting applications help the current scenario of experimental investigations and analysis at any time and any place. This technology helps the user to find the topographical conditions of the area, to determine the various parameters of the soil, the suitability of the soil for the specified crop, etc., which in turn helps to identify the type of artificial recharge techniques that can be applicable for that area for efficient water harvesting.

The development of suitable devices of the IoT for agricultural and irrigation purposes improves the efficiency of farmers and mankind's work in farming and crops to be cultivated, which indirectly improve the natural resources in that area and also the gross domestic product of the country. The database developed is expected to be useful for the future in understanding and identifying suitable areas and methods for their irrigation and agricultural purposes.

References

Dbstephens.com (2018) Managed Aquifer Recharge [online] Available at: http://www.dbstephens.com/managed-aquifer-recharge.aspx [Accessed 1 Apr. 2018].

Dhakate, R., Singh, V.S., Negi, B.C., and Chandra, S. (2008). Geomorphological and geophysical approach for locating favorable groundwater zones in granitic terrain, Andhra Pradesh, India. *Journal of Environmental Management* 88 (4): 1373–1383.

Indiamart Business Video (2014) Borewell Recharging by Farmland Rainwater Harvesting Systems. [video] Available at: https://www.youtube.com/watch?v=wZ_4oo0jGyY [Accessed 1 Apr. 2018].

Kouhestani, S., Eslamian, S.S., Abedi-Koupai, J., and Besalatpour, A.A. (2016). Projection of climate change impacts on precipitation using soft-computing techniques: a case study in Zayandeh-rud Basin, Iran. *Global and Planetary Change* 144: 158–170.

Rao, V.T. and Thangarajan, M. (1992). Application of electrical imaging for borehole siting in hard rock regions of India. *Journal of the Geological Society of India* 61 (2): 147–158.

Rao, V.T., Prasad, S.Y., and VaraLakshmi, V. (2011). Resistivity 2D imaging at success and failed water wells in granitic and basaltic terrains of Andhra Pradesh. *Journal of Geo Physics, India* XXXI (1 and 2): 105–109.

Sharma, S.P. and Baranwal, V.C. (2003). Delineation of groundwater – bearing fractured zones in a hard rock area integrating very low frequency electromagnetic and resistivity data. *Journal of Applied Geophysics* 57 (2): 155–166.

Shramajeevi, S. (2015) Bore well recharge technique, Project by Centre for World Solidarity: [image] Available at: https://www.youtube.com/watch?v=_j55sdgO9LE [Accessed 1 Apr. 2018].

Sravan Kumar, A. (2014). 2-D resistivity and induced polarization surveys for characterisation of hard rocks in Kudliar watershed, Medak district. In: *Two-Day National Conference on Water, Environment and Society (NCWES-2014)* (ed. M.V.S.S. Giridhar), 50–56. Hyderabad, India: Center for Water Resource Institute.

The Hans India (2016). Recharge pits to bring up groundwater. [online] p.NABARD, Rainwater Harvesting Pits. Available at:http://www.thehansindia.com/posts/index/Telangana/2016-06-23/Recharge-pits-to-bring-up-groundwater/237005 [Accessed 1 Apr. 2018].

Tucsonaz.gov (2018) Central Avra Valley Storage and Recovery Project (CAVSARP) [online] Available at: https://www.tucsonaz.gov/water/cavsarp [Accessed 1 Apr. 2018].

User, S. (2018) Furrow irrigation system [online] Siracoirrigation.com. Available at: http://siracoirrigation.com/uganda/index.php/what-weoffer/irrigation-systems/furrow-irrigation-system [Accessed 1 Apr. 2018].

Vafakhah, M. and Eslamian, S. (2017). Application of intelligent technology in rainfall analysis. In: *Handbook of Drought and Water Scarcity, Vol. 1: Principles of Drought and Water Scarcity* (eds. S. Eslamian and F. Eslamian), 441–460. USA: Francis and Taylor, CRC Press.

11

Global Satellite-Based Precipitation Products

Zhong Liu[1,2], *Dana Ostrenga*[1,3], *Andrey Savtchenko*[1,3], *William Teng*[1,3], *Bruce Vollmer*[1], *Jennifer Wei*[1], *and David Meyer*[1]

[1]*Goddard Earth Sciences Data and Information Services Center, NASA Goddard Space Flight Center, Greenbelt, MD, USA*
[2]*Center for Spatial Information Science and Systems, George Mason University, Fairfax, VA, USA*
[3]*Adnet Systems Inc., Bethesda, MD, USA*

11.1 Introduction

Water is essential for life on Earth. Rainwater harvesting is a process of collecting water from rainfall for different purposes (e.g. home gardening, agriculture irrigation). Scales of rainwater harvesting can vary a lot, ranging from backyard or rooftop to reservoir. Accurate and timely available precipitation data and information are critical for research, planning, and applications of rainwater harvesting and conservation around the world. However, precipitation is notoriously difficult to measure and predict. As a result, precipitation data are not uniformly available in space and time and even when they are available, other issues can exist. For example, data access can be a daunting task to many, making research and applications on rainwater harvesting and conservation even more difficult.

Precipitation data obtained from ground-based gauge measurements have existed for many years. Even today, they still play an important role in research and applications, as well as providing valuable information for ground validation and bias correction on precipitation products derived from other techniques and algorithms. After World War II, ground-based weather radars were developed with new observational techniques, providing a new avenue for remotely and continuously measuring precipitation in three dimensions and greatly expanding temporal sampling and spatial coverage of precipitation systems. However, drawbacks in gauge and weather radar networks exist and both can be costly and difficult for deployment, calibration, and maintenance. In mountainous regions, weather radars also suffer from a beam blockage issue (Maddox et al. 2002). As a result, ground-based precipitation data are still sparsely available around the world, especially in developing countries where resources are limited, as well as in remote and mountainous regions. Furthermore, there are virtually no long-term observations from both gauge and radars over vast oceans that cover nearly 71% of the Earth's surface.

Over the past several decades, satellite-based techniques have been developed. Infrared, passive, and active microwave instruments onboard satellites are used to measure precipitation from space. With a constellation of multiple satellites in low-Earth orbits (LEO) and geostationary Earth orbits (GEO), spatial coverage and temporal sampling of precipitation measurement have been greatly extended, and thereby observations of precipitation are no longer limited at local or regional scale compared to ground-based measurement networks, making global precipitation measurement (GPM) a reality.

Climate change, due primarily to human activities that release greenhouse gases, can greatly influence rainfall patterns (spatial, intensity, duration, etc.) around the world. According to recent research (e.g. Global Change 2018; Hegerl et al. 2015; Groisman et al. 2005; Fischer and Kutti 2016; Allan and Soden 2008), extreme and record-breaking precipitation events can become more frequent in a warming climate, often at the expense of light to moderate rainfall that becomes less frequent. As a result, droughts and floods can occur more frequently in such warming climate scenarios (Ahmed et al. 2013; Parr et al. 2015; Wang et al. 2017). All these changes can affect rainwater harvesting and conservation activities. Therefore, it is essential to understand changes in rainfall intensity, location, type, and duration in such climate scenarios.

Both near-real-time and historical data for global and regional precipitation play an important role in research and applications (Dalezios et al. 2018). However, discovering, accessing, and using satellite-based precipitation data

Handbook of Water Harvesting and Conservation: Basic Concepts and Fundamentals, First Edition. Edited by Saeid Eslamian and Faezeh Eslamian.

can be a daunting task to many, especially for first-timers outside the satellite precipitation community. For example, precipitation datasets can be in different formats and data structures that often require investment in time and resources at various levels. Over the years, there have been many global and regional precipitation products developed by different organizations around the world but identifying the right one suitable for a specific research or application can be equally difficult, especially without in-depth knowledge about algorithms and ground validation results.

In this chapter, a list of commonly used NASA global and regional satellite-based precipitation products for research and applications of rainwater harvesting and conservation are introduced. The NASA Goddard Earth Sciences Data and Information Center (GES DISC) is home to global satellite-based precipitation products from two NASA precipitation satellite missions, the Tropical Rainfall Measuring Mission (TRMM) and the GPM, as well as other global precipitation products. In particular, a suite of the latest precipitation products, the Integrated Multi-satEllite Retrievals for GPM (IMERG), consisting of both near-real-time and research-grade precipitation products (spatial coverage: 60°N–S) at half-hourly temporal and 0.1° grid spatial resolutions, can be suitable for activities of rainwater harvesting and conservation around the world.

User-friendly data services to facilitate data access, visualization, and evaluation are described along with examples. In particular, Giovanni, a Web-based online visualization and analysis tool without downloading data and software, is introduced. Other related ancillary global datasets at GES DISC for rainwater research and applications are introduced as well. Furthermore, precipitation data quality issues are discussed. In recent years, extreme rainfall events have become more frequency and intense (Global Change 2018; Hegerl et al. 2015; Groisman et al. 2005; Fischer and Kutti 2016; Allan and Soden 2008). Such events have impacted many countries, including their economy, infrastructure, and people's daily life. As an example, preliminary results of an ongoing investigation using model and satellite-based products on extreme rainfall events in Maine, located in the northeastern United States, are presented. Seasonal and interannual variability of rainfall in the Amazon watershed to demonstrate our data products and services is also presented.

This chapter is organized as follows: Section 11.2 briefly describes current satellite-based precipitation measurement techniques; Section 11.3 introduces a list of frequently used global and regional near-real-time and research precipitation products at GES DISC; data services and tools for accessing these products are described in Section 11.4; examples of using precipitation products and services are in Section 11.5, followed by the conclusion.

11.2 Precipitation Measurements from Space

On April 1, 1960, the first weather satellite, the Television Infrared Observation Satellite (TIROS), was successfully launched from Cape Canaveral, Florida, and an era of satellite-based Earth's observations began. In particular, the NASA Earth Observing System (EOS) program that was formulated in the 1980s and implemented in the early 1990s consists of a coordinated mission series of polar-orbiting and low inclination satellites for long-term global observations of radiation, clouds, water vapor, and precipitation; the oceans; greenhouse gases; land-surface hydrology and ecosystem processes; glaciers, sea ice, and ice sheets; ozone and stratospheric chemistry; and natural and anthropogenic aerosols (NASA 2018a). As of this writing, there are over 28 active NASA satellite missions currently in space to provide observations to scientific and application users around the world (NASA 2018a).

As stated in the introduction, precipitation is notoriously difficult to measure and predict. Rain gauge method is a traditional way to measure precipitation at a single point or station, suffering from impractical and costly deployment over a large area, especially in remote and mountainous regions, in addition to many other issues. Ground-based weather radars can overcome the single point issue in the rain gauge method, providing continuous and three-dimensional precipitation measurements. However, it requires an adequate deployment of radar network to cover a large region to avoid gaps in coverage. Inter-calibration activities among radars with different frequency bands are also needed. Radars require regular maintenance to prevent unexpected breakdown and loss of service, especially during severe weather events. Furthermore, the beam blockage is an issue in mountainous regions with weather radars (Maddox et al. 2002) and as a result, observational gaps commonly exist.

At present, there are three major techniques used to measure precipitation from satellites. The first one is to infer cloud top temperatures measured from infrared instruments onboard satellites (mainly geostationary satellites due to high sampling rates) to surface rainfall rates. This technique suffers from underestimation of warm rain and overestimation (or false alarms) of cirrus clouds and anvils. For example, orographic rainfall, which is common in tropical regions, can be underestimated due to warm rain processes (AMS 2012). However, its abundance of data makes it a workhorse in many algorithms in order to fill in data gaps that are not available from the rest of techniques. The second technique uses passive microwave (PMW) instruments onboard LEO satellites such as TRMM and GPM to measure precipitation particles and provide improved precipitation estimates compared to the first technique. Over oceans, precipitation estimates retrieved from PMW instruments can rival ground-based weather radars according to studies (Spencer et al. 1989; Kummerow et al. 2001). However, problems exist such as variations of emissivity over land (Spencer et al. 1989; Kummerow et al. 2001) and underestimation of orographic rainfall associated with warm rain processes (AMS 2012). The last technique uses active microwave instruments onboard satellites (e.g. TRMM and GPM) that are also known as space-borne precipitation radars (PRs). Difficulties (e.g. Iguchi et al. 2000) with PR include attenuation correction, complex terrain, minimum detectable signal power, etc. More detailed description of each technique is available in the Web portal of the University of Utah Precipitation Measurement Mission Science (Utah 2018).

Given the challenges and difficulties in different measurement techniques, blended algorithms for precipitation estimates have been developed to utilize strengths of different observational techniques from both space-borne and ground-based instruments. For example, PMW techniques provide accurate precipitation estimates, however, observations are quite limited (Huffman et al. 1997, 2007; Adler et al. 2003). On the other hand, data from less reliable IR measurements are abundant and can be used to fill in gaps from PMW observations (Huffman et al. 1997, 2007; Adler et al. 2003). Products from combined PMW and PR algorithms can provide a baseline for calibration with other precipitation algorithms. Ground observations such as the GPCC (Global Precipitation Climatology Centre) can be used for bias correction. To overcome limited observations from a single satellite, blended methods require a constellation of satellites from domestic and international government agencies to expand spatiotemporal coverage of observations.

11.3 Overview of NASA Satellite-Based Global Precipitation Products and Ancillary Products at GES DISC

NASA Earth Observation (EO) data are managed by the NASA Earth Observing System Data and Information System (EOSDIS) that consists of 12 discipline-based Distributed Active Archive Centers (DAACs) in the United States. The EOSDIS currently hosts ~22 PB of EO data at 12 DAACs and it is expected to grow rapidly over the coming years, to more than 37 (246 PB) PB by 2020 (2025) (NASA 2018b). Such large collection of EO data is an important asset for environmental research and applications around the world, including rainwater harvesting activities. As one of the 12 NASA DAACs, the GES DISC hosts global precipitation products including TRMM and GPM, as well as other satellite missions and reanalysis projects. In addition, the GES DISC also hosts global and regional satellite-based interdisciplinary data products such as solar irradiance, atmospheric composition and dynamics, global modeling, etc. Currently, over 2700 unique data products are archived at GES DISC and available for public distribution.

11.3.1 TRMM and GPM Missions

Launched on November 27, 1997, TRMM (Kummerow et al. 2000) was designed to: (i) advance understanding of the global energy and water cycles by providing distributions of rainfall and latent heating over the global tropics; (ii) understand the mechanisms through which changes in tropical rainfall influence global circulation and to improve ability to model these processes in order to predict global circulations and rainfall variability at monthly and longer timescales; (iii) provide rain and latent heating distributions to improve the initialization of models ranging from 24-hour forecasts to short-range climate variations; (iv) help to understand, to diagnose, and to predict the onset and development of the El Niño, Southern Oscillation, and the propagation of the 30–60-day oscillations in the Tropics; (v) help to understand the effect that rainfall has on the ocean thermohaline circulations and the structure of the upper ocean; (vi) allow cross calibration between TRMM and other sensors with life expectancies beyond that of TRMM itself; (vii) evaluate the diurnal variability of tropical rainfall globally; and (viii) evaluate a space-based system for rainfall measurement.

TRMM (1997–2015) carried five instruments. Table 11.1 lists four precipitation-related instruments including: (i) the first space-borne Precipitation Radar (PR) that operated at Ku band (13.8 GHz) and provided three-dimensional rain distribution over land and ocean surfaces; (ii) the TRMM Microwave Imager (TMI); (iii) the Visible and Infrared Scanner (VIRS); and (iv) the Lightning Imaging Sensor (LIS). TRMM data play an important role for providing baseline measurements for satellite intercalibration to minimize systematic differences among sensors, a necessary step for developing multi-satellite and multi-sensor precipitation products. Its multi-satellite and multi-sensor merged global precipitation product suite, the TRMM Multi-Satellite Precipitation Analysis (TMPA), has been widely used in many disciplines and applications and was the highest cited product suite in the Journal of Hydrometeorology of the American Meteorological Society (AMS 2018a).

Built upon the success of TRMM, GPM (2014–present) was launched on February 27, 2014, as one of NASA's

Table 11.1 TRMM precipitation related instruments.

Instrument name	Description				
	Band frequencies/wavelengths	Spatial resolution		Swath width	
		Pre-boost	Post-boost	Pre-boost	Post-boost
Visible and Infrared Scanner (VIRS)	5 channels (0.63, 1.6, 3.75, 10.8, and 12 μm)	2.2 km	2.4 km	720 km	833 km
TRMM Microwave Imager (TMI)	5 frequencies (10.7, 19.4, 21.3, 37, 85.5 GHz)	4.4 km at 85.5 GHz	5.1 km at 85.5 GHz	760 km	878 km
Precipitation Radar (PR)	13.8 GHz	4.3 km. Vertical: 250 m	5 km. Vertical: 250 m	215 km	247 km
Lightning Imaging Sensor (LIS)	0.7774 μm	3.7 km	4.3 km	580 km	668 km

Table 11.2 GPM precipitation instruments.

Instrument name	Description		
	Band frequencies/wavelengths	Spatial resolution	Swath width
GPM Microwave Imager (GMI)	13 frequencies (10–183 GHz)	5.1 km at 85.5 GHz	885 km
Dual-frequency Precipitation Radar (DPR)	13.8 GHz (KuPR), 33.5 GHz (KaPR)	5 km. Vertical: 250 m (KuPR); 250 m/500 m (KaPR)	245 km (KuPR), 120 km (KaPR)

foundational missions to further advance GPMs and science, in particular in light rain and snow, as well as improve our understanding of precipitation distribution and processes, especially in the polar regions. New instruments and capabilities (Table 11.2) of GPM are: (i) a dual-frequency PR (DPR) that contains a new Ka band frequency (35 GHz) for measuring frozen precipitation and light rain and (ii) new high-frequency channels in the microwave instrument to enhance the capability for measuring precipitation intensity and type through all cloud layers.

The GES DISC is home to the permanent archive of TRMM and GPM data products. The following sections give more details about their commonly used products in research and applications, and data services.

11.3.2 Multi-Satellite and Multi-Sensor Merged Global Precipitation Products

As mentioned earlier, precipitation retrieval algorithms that utilize multi-satellites and multi-sensors (i.e. microwave and geostationary infrared sensors) from a constellation of satellites, or blended algorithms, have been developed to overcome a very limited spatiotemporal coverage from a single satellite and utilize strengths of different observational techniques (Adler et al. 2003; Huffman et al. 2007, 2009, 2010; Huffman and Bolvin 2012, 2013; Joyce et al. 2004; Mahrooghy et al. 2012; Hong et al. 2007; Behrangi et al. 2009; Aonashi et al. 2009). These near-global precipitation products are available uniformly in space and time and are widely used in hydrometeorological research and applications (Liu et al. 2012, 2017).

For example, the TMPA products in Table 11.3 (Huffman 2017; Huffman et al. 2007, 2010; Huffman and Bolvin 2012, 2013), developed by the Mesoscale Atmospheric Processes Laboratory at NASA Goddard Space Flight Center (GSFC), provide precipitation estimates at 3-hourly and monthly temporal resolutions on a 0.25°×0.25° grid available from January 1998 to present. The TMPA consists of two product groups: near-real-time (3B42RT, spatial coverage: 60° N–S) and research-grade (3B42, spatial coverage, 50° N–S). The former is less accurate due to time constraints for additional input products and calibration, but provides quick precipitation estimates suitable for near-real-time monitoring and modeling activities (e.g. Wu et al. 2012). The latter, available approximately two months after observation, is processed with additional input data and calibrated with gauge data, different sensor calibration, and additional post-processing in the algorithm. The resulting products are more accurate and suitable for research (Huffman et al. 2007, 2010). Over the years, the TMPA products have been widely used in various research and applications (e.g. Wu et al. 2012; Bitew et al. 2012; Gourley et al. 2011; Su et al. 2011; Gianotti et al. 2012; Engel et al. 2017).

During the GPM era, the IMERG product suite (Huffman et al. 2017) has several improvements compared

Table 11.3 Summary of variables in TMPA products (Huffman 2017).

3-hourly Near-Real-Time Product (3B42RT)
Calibrated precipitation
Calibrated precipitation error
Satellite source identifier
Uncalibrated precipitation
3-hourly Research Product (3B42)
Multi-satellite precipitation
Multi-satellite precipitation error
Satellite observation time
PMW precipitation
IR precipitation
Satellite source identifier
Monthly Research Product (3B43)
Satellite-gauge precipitation
Satellite-gauge precipitation error
Gauge relative weighting

Daily products are also available and provided by GES DISC as value-added products.

Table 11.4 Summary of variables in IMERG products (Huffman et al. 2017).

Half-hourly Products (IMERG Early, Late, and Final)
Calibrated multi-satellite precipitation
Uncalibrated multi-satellite precipitation
Calibrated multi-satellite precipitation error
PMW precipitation
PMW source identifier
PMW source time
IR precipitation
IR KF weight
Probability of liquid-phase precipitation
Precipitation quality index
Monthly Research Product (IMERG Final)
Satellite-gauge precipitation
Satellite-gauge precipitation error
Gauge relative weighting
Probability of liquid-phase precipitation
Precipitation quality index

Daily products are available and provided by GES DISC as value-added products.

to the TMPA products: (i) the grid spatial resolution is improved from 0.25° to 0.1° to meet increasing demand for high-resolution data; (ii) the temporal resolution is improved from 3-hourly to half-hourly; (iii) the spatial coverage is expanded from 50° N–S to 60° N–S (the full global coverage is planned in future releases); (iv) light rain detection is improved with advanced instruments and algorithms; (v) frozen precipitation is available for the first time; and (vi) quality index is added. The IMERG suite (Table 11.4) contains of three output run products, "Early satellites" (lag time: ~4 hours), "Late satellites" (lag time: ~12 hours), and the final "Satellite-gauge" (lag time: ~3.5 months). Compared to TMPA, additional new input and intermediate variables such as precipitation quality index have been added in IMERG for data quality diagnosis (Tables 11.3 and 11.4). The first two runs only contain half-hourly products and the last both half-hourly and monthly. To facilitate research and applications, the GES DISC has developed their daily products for all three runs (Early, Late, and Final). Detailed comparison between TMPA and IMERG is available (Huffman 1997). The retro-processing of the IMERG product suite back to the TRMM era will be released in the near future to allow users developing baseline products for anomaly detection and other applications. Once these products are available, the TMPA suite will retire (Huffman 2016).

Other similar global and regional precipitation products are also available (IPWG 2018; Huffman et al. 2018). Huffman compiled a list of global precipitation datasets on the website of the International Precipitation Working Group (IPWG 2018). Precipitation datasets are grouped according to data types: (i) satellite combination datasets with gauge data; (ii) satellite combination datasets without gauge data; (iii) single-source datasets; and (iv) precipitation gauge analyses. Group 1 contains a list of datasets based on input data from different satellite sensors and gauges (e.g. TMPA, IMERG). In Group 2, rain gauges are not used as input data. Group 3 consists of datasets from a single satellite sensor type. Group 4 only contains rain gauge datasets.

11.3.3 Global and Regional Land Data Assimilation Products

Global and regional land data assimilation products at GES DISC consist of quality-controlled, and spatially and temporally consistent, land-surface model (LSM) datasets from the best available observations and model output to support modeling activities (NASA 2018c). These products are generated by ingesting satellite- and ground-based observations and using advanced land-surface modeling and data assimilation techniques (NASA 2018c).

The precipitation field is available in the forcing group of the North American Land Data Assimilation System – Phase 2 (NLDAS-2) and is derived from a temporal disaggregation (from daily analysis to hourly intervals) of a gauge-only NOAA (National Oceanic and Atmospheric Administration) CPC (Climate Prediction Center) analysis of USA daily precipitation (Higgins et al. 2000) with an orographic adjustment based on the widely applied PRISM (Parameter-elevation Relationships on Independent Slopes Model) climatology (Daly et al. 1994; NASA 2018c). On the other hand, more precipitation products (NASA 2018c) are used to derive the precipitation field in the forcing group of the Global Land Data Assimilation System (GLDAS) and they consist of: (i) satellite-based observed precipitation products from the Naval Research Laboratory, the NASA Goddard TMPA near-real-time algorithm (3B42RT), and PERSIANN; and (ii) the CPC merged analysis of precipitation (CMAP) using a merged satellite and gauge algorithm.

The GES DISC hosts precipitation products from GLDAS, NLDAS, NCA-LDAS (National Climate Assessment), and FLDAS (Famine Early Warning Systems Network – FEWS NET). All LDAS products contain forcing and model products (NASA 2018c). GLDAS provides global 3-hourly, daily, and monthly land-only products at 0.25° grid resolution from 1948 to 2010. NLDAS covers North America and it contains hourly and monthly products at 0.125° grid resolution from 1979 onwards. NCA-LDAS products are for North America only and available at daily and 0.125° grid resolution from 1979 onwards. FLDAS covers Africa and Central Asia available daily and monthly at 0.1° grid resolution from 2001 onwards.

11.3.4 Modern-Era Retrospective Analysis for Research and Applications, Version 2 (MERRA-2) Products

Modern-Era Retrospective Analysis for Research and Applications, Version 2 (MERRA-2) is developed by the NASA Global Modeling and Assimilation Office (GMAO) at NASA GSFC. MERRA-2 provides global datasets from 1980 onwards and runs a few weeks behind real time (Gelaro et al. 2017). The MERRA project emphasizes historical analyses of the hydrological cycle on a broad range of weather and climate time scales and places the NASA EOS suite of observations in a climate context (Gelaro et al. 2017). Alongside the meteorological data assimilation using a modern satellite database, MERRA-2 includes an interactive analysis of aerosols that feed back into the circulation, uses NASA's observations of stratospheric ozone and temperature (when available), and takes steps toward representing cryogenic processes (Gelaro et al. 2017). Compared to the previous version, new observation types have been added and advanced techniques have been made in the assimilation system that enables assimilation of modern hyperspectral radiance and microwave observations, along with GPS-Radio Occultation datasets (Rienecker et al. 2011; Reichle and Liu 2014; Suarez and Bacmeister 2015; Takacs et al. 2015). MERRA-2 also includes advances in both the Goddard Earth Observing System Model, Version 5 (GEOS-5) and the GSI (Gridpoint Statistical Interpolation) assimilation system and NASA ozone observations after 2005 (Gelaro et al. 2017).

There are two types of precipitation parameters in MERRA-2: (i) precipitation from the atmospheric model and (ii) observation-corrected or bias-corrected precipitation (Reichle and Liu 2014; Bosilovich et al. 2015). Observational data are introduced in the latter parameter due to considerable errors that propagate into land surface hydrological fields and beyond (Rienecker et al. 2011).

Bosilovich et al. (2015) have conducted a general evaluation of MERRA-2 precipitation estimates, including precipitation climatology, interannual variability, diurnal cycle, Madden-Julian Oscillation (MJO) events, global water cycle, and US summertime variability. Major findings (Bosilovich et al. 2015) are: (i) an overestimate of the modeled precipitation in the tropical west Pacific Ocean, the eastern tropical ITCZ (the Intertropical Convergence Zone), and the SPCZ (the South Pacific convergence zone) in DJF and JJA (Bosilovich et al. 2015); (ii) Extreme values of modeled precipitation in the vicinity of high topography in the tropics; (iii) An upward trend in the MERRA-2 time series exists and by contrast no trend is observed in GPCP (the Global Precipitation Climatology Project); (iv) Larger modeled precipitation diurnal cycle (PDC) amplitude is found over the high mountains; (v) The phases of modeled PDC are not well reproduced in several regions such as the US Great Plains; (vi) MJO signal from modeled precipitation is stronger than GPCP; (vii) MERRA-2 can reproduce the observed precipitation and anomalies in US summertime, reasonably well. Although the preliminary evaluation (Bosilovich et al. 2015) provides a basic understanding of the MERRA-2 precipitation products, evaluation for extreme rainfall events is missing and as a result, it is not clear about MERRA-2 precipitation behavior and characteristics in extreme events.

The complete list of MERRA-2 products along with documentation and more is available on the official MERRA-2 website (NASA 2018d). A special collection of research papers, entitled MERRA-2, has been published by the American Meteorological Society (AMS 2018b). The collection includes an overview article of MERRA-2 (Gelaro et al. 2017) and articles about MERRA-2 assessment in different disciplines and regions around the world. To facilitate data access, the GES DISC has developed data services and tools to be described in Section 11.4.

11.3.5 Ancillary Products at GES DISC

In addition to global and regional precipitation products, the GES DISC also archives interdisciplinary products for rainwater related research and applications. As one of 12 NASA discipline-based DAACs, the focus areas at GES DISC are atmospheric composition, water and energy cycles, climate variability, as well as the carbon cycle and ecosystem. Major NASA missions and projects are TRMM, GPM, MERRA, NLDAS, MODIS-Aqua, MODIS-Terra, etc. Users can search and download these datasets with the newly released Web interface to be described in the following section. Users who want datasets in other disciplines that are not available at GES DISC can search the EOSDIS Earthdata portal (NASA 2018e). The portal provides access to datasets archived at all NASA DAACs.

11.4 Data Services

Although a large collection of NASA global satellite data is available for research and applications around the world, many users find it challenging to discover, access, and use NASA satellite remote sensing data (Liu and Acker 2017). Heterogeneous data formats, complex data structures, large-volume data storage, special programming requirements, diverse analytical software options, and other factors often require a significant investment in time and resources, especially for first-time users (Liu and Acker 2017). Over the years, data services have been developed at NASA's EOSDIS DAACs to improve NASA data discovery and access. Due to space limitation, it is difficult to describe all data tools and services at DAACs in one article. Since precipitation datasets for rainwater harvesting are the main focus, only data services at GES DISC are presented.

11.4.1 Point-and-Click Online Tools

Surveys (e.g. Kearns 2017) and experience from user support services at the GES DISC show that non-expert users and those who occasionally use satellite-based products prefer point-and-click data tools in order to obtain graphic and data assessment results. As mentioned earlier, new dataset assessment activities may not be straightforward and can be costly. Point-and-click tools provide fast and easy access to satellite-based data products for all users without the need of coding and downloading data and software. Furthermore, they can be further developed for in-depth data analysis and visualization. Here, a point-and-click online tool developed by GES DISC is introduced: the Giovanni (Figure 11.1).

Giovanni stands for the Geospatial Interactive Online Visualization and Analysis Infrastructure (Giovanni, NASA 2018f), developed by GES DISC to assist a wide range of users with data access and evaluation, as well as with scientific exploration and discovery (Liu and Acker 2017; Acker and Leptoukh 2007). Currently, 8 disciplines, 74 measurements, and over 2000 variables are available in Giovanni. Over 1700 peer-reviewed papers across various Earth science disciplines and other areas were published with acknowledgement of Giovanni.

Giovanni offers both keyword and faceted search capabilities in its Web interface (Figure 11.1a) for searching variables of interest. For example, a search for "gpm precipitation" returns GPM-related variables (Figure 11.1a). Commonly used analytical and plotting capabilities (Table 11.5) are in Giovanni (NASA 2018f; Liu and Acker 2017). Mapping options include time-averaging, animation, accumulation (precipitation), time-averaged overlay of two datasets, and user-defined climatology. For time series, options include of area-averaged, differences, seasonal, and Hovmöller diagrams. Cross-sections include latitude-pressure, longitude-pressure, time-pressure, and vertical profile for 3D datasets from AIRS (Atmospheric Infrared Sounder) and MERRA. For comparison, Giovanni has built-in processing code for datasets that require measurement unit conversion and regridding. Commonly used comparison functions include map and time-series differences, as well as correlation maps and X-Y scatter plots (area-averaged or time-averaged). Zonal means and histogram distributions can also be plotted.

Visualization features (Liu and Acker 2017) include interactive map area adjustment; animation; interactive scatter plots; adjustments of data range; change of color palette; contouring; and scaling (linear or log). The on-the-fly area adjustment feature allows an interactive and detailed examination of a result map without replotting data. Animations are helpful to track evolution of an event or seasonal changes. Interactive scatter plots allow identification of the geolocation of a point of interest in a scatter plot. Adjustments of any of these plots provide custom options to users.

To support increasing socioeconomic and Geographical Information Systems (GIS) activities in Earth sciences, vector shapefiles have been added for countries, states in the United States, land-sea masks, and major watersheds around the world. Available functions for shapefiles are time-averaged and accumulated maps, area-averaged time series, and histogram.

All data files involved in Giovanni processing are listed and available in the lineage page. Available image formats are PNG, GeoTIFF, and KMZ (Keyhole Markup Language) that can be used for applications and software packages; for example, KMZ files can be conveniently imported into Google Earth where a rich collection of overlays is available. ASCII (CSV [comma-separated values]) is available for selected functions (e.g. time series). All input and output data are in NetCDF, which can be handled

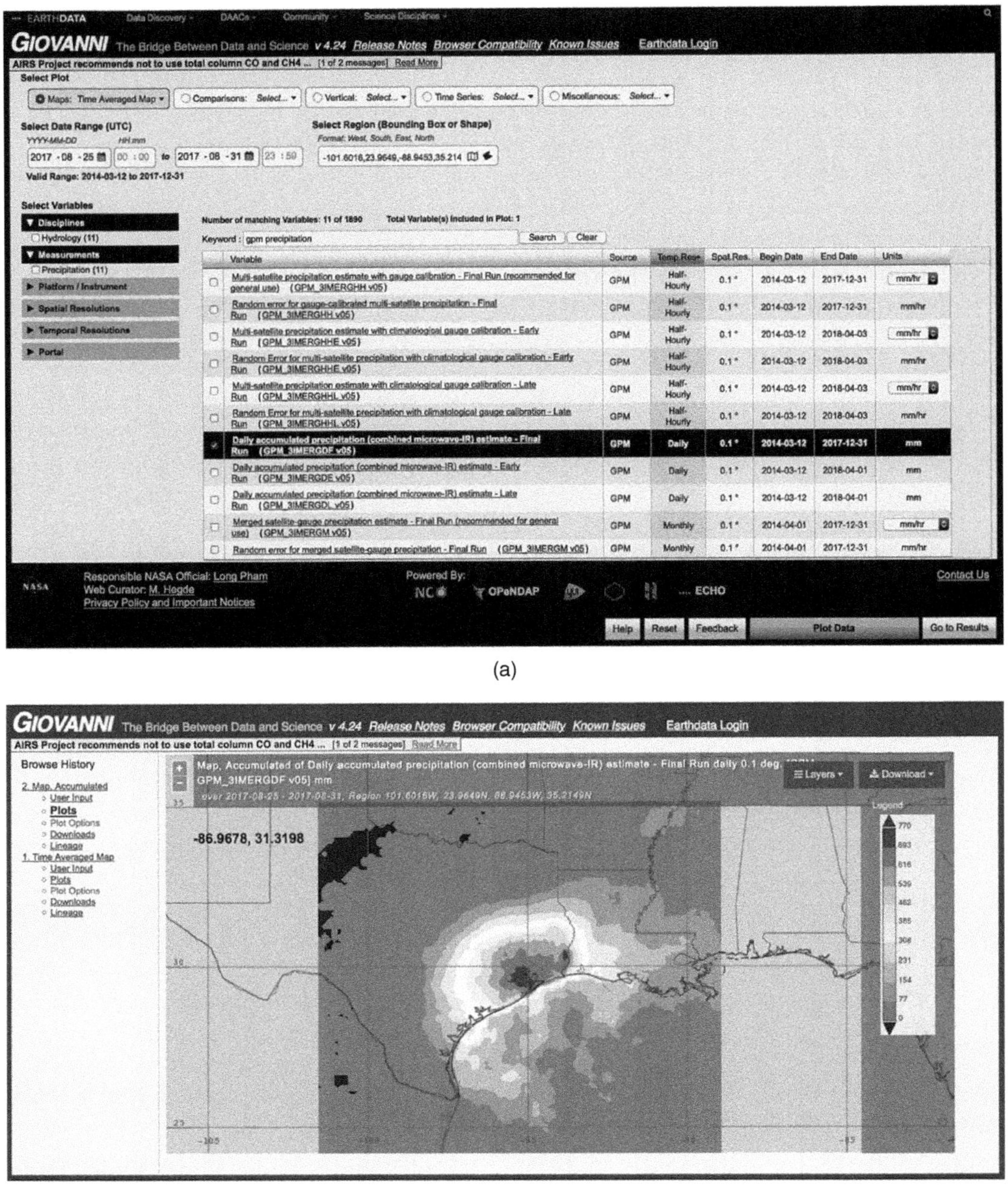

(a)

(b)

Figure 11.1 The NASA GES DISC Giovanni: (a) a screenshot of the Web interface showing "gpm precipitation" in the keyword search box and the search results with the IMERG-Final daily precipitation product selected; (b) the result page of Giovanni showing the accumulated rainfall of Hurricane Harvey during August 25–31, 2017.

by many off-the-shelf software packages. Furthermore, users can bookmark URLs generated by Giovanni processing for reference, documentation, or sharing with other colleagues.

11.4.2 Data Rod Services

Providing long time-series data to the hydrology community can be a challenge (Teng et al. 2016). In hydrology, earth surface features are expressed as discrete spatial objects such as watersheds, river reaches, and point observation sites; and time varying data are contained in time series associated with these spatial objects. Long-time histories of data may be associated with a single point or feature in space. Most remote sensing precipitation products are expressed as continuous spatial fields, with data sequenced in time from one data file to the next (Teng et al. 2016). Hydrology tends to be narrow in space and

Table 11.5 Analytical and plotting capabilities in Giovanni (NASA 2018f; Liu and Acker 2017).

Maps	Comparisons	Vertical	Time Series	Miscellaneous
Time Averaged Map	Map, Correlation	Cross Section, Latitude-Pressure	Hovmöller, Longitude-Averaged	Zonal Mean
Animation	Scatter, Area Averaged (Static)	Cross Section, Longitude-pressure	Hovmöller, Latitude-Averaged	Histogram
Difference of Time Averaged	Scatter (Interactive)	Cross Section, Time-Pressure	Area-Averaged Differences	
Accumulated	Scatter (Static)	Vertical Profile	Area-Averaged	
Time Averaged Overlay Map	Scatter, Time-Averaged (Interactive)		Seasonal	
Monthly and Seasonal Averages				

deep in time, which poses a challenge during the GPM era. For example, to generate a one-year time series, one needs to pull all the 0.1 deg., half-hourly IMERG product, which can be time consuming and not suitable for online data services due to the large volume of data (Teng et al. 2016).

The concept of data rods (Teng et al. 2016; Gallaher and Grant 2012; Rui et al. 2012, 2013) can be applied to this challenge. Teng et al. (2016) proposed two general solutions: (i) retrieve multiple time series for short time periods and stitch the multiple time series into desired single long time series and (ii) reprocess (parameter and spatial subsetting) and archive data as one-time cost approach. The resultant time series files would be geospatially searchable and could be optimally accessed and retrieved by any user at any time (Teng et al. 2016). One drawback for the data rod approach is that a large number of files need to be generated and maintained. For example, for IMERG, the number of half-hourly files will be $1300 \times 3600 = 4\,680\,000$, posting a file management issue. Data rods have been implemented in CUAHSI-HIS (Consortium of Universities for the Advancement of Hydrologic Science, Inc.-Hydrologic Information System) and other hydrologic community tools (Rui et al. 2013) where time series data of NLDAS, GLDAS, MERRA-Land, the TMPA 3-hourly product, etc. can be accessed.

11.4.3 Subsetting and Format Conversion Services

For users who belong to the advanced category and wish to do more complex data processing, it is necessary to directly work on data in their native formats. All datasets at GES DSIC can be searched and downloaded from the newly designed Web interface (NASA 2018g). The interface (Figure 11.2a) provides a search capability for dataset collections, data documentation, product alerts, Frequent Asked Questions (FAQs), glossary, How-to recipes, and more. For example, a search for "imerg" (Figure 11.2a) returns all IMERG-related products (Figure 11.2b). Each product has its own all-in-one dataset landing page that contains information about product summary, dataset citation, documentation, and data access. Product summary lists basic information about the product such as version, format, spatial, and temporal resolutions, etc. Dataset citation provides necessary information for citing the dataset in publications. Documentation section provides related technical documents in case one needs to know details about the product. Data access lists different tools and methods to access data. In short, a dataset landing page serves a one-stop shop for all data related services and information.

Subsetting and format conversion services are essential because spatial coverage of NASA data products is normally global in the Hierarchical Data Format (HDF) that some users are not familiar with and might have difficulties to handle it. A subsetting service is needed for users who plan to do local or regional studies and applications. To facilitate data access, the GES DISC has developed data subsetting services for precipitation and other products. This feature has been included in the GES DISC interface for search and downloads (NASA 2018g). When "Subset/Get Data" icon is clicked on, a list of options is available for product subsetting, format conversion, and more. Figure 11.3 is an example of the Web interface for the IMERG-Final half-hourly product, showing the interface (Figure 11.3a), the date range picker (Figure 11.3b), the spatial subset map (Figure 11.3c), the selection of variables (Figure 11.3d), and the format selection (Figure 11.3e).

11.4.4 Other Web Data Services and Information

NASA satellite-based data products at the GES DISC are also accessible (NASA 2018h) via other Web services and protocols (Table 11.6) including https for data archive, OPeNDAP, WMS (Web Map Service), GDS (GrADS Data

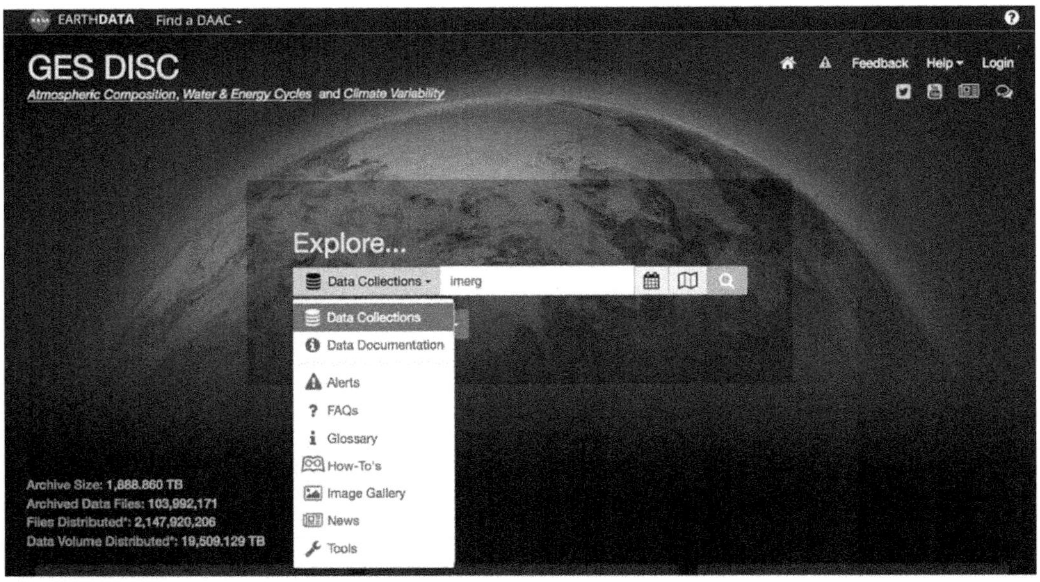

(a)

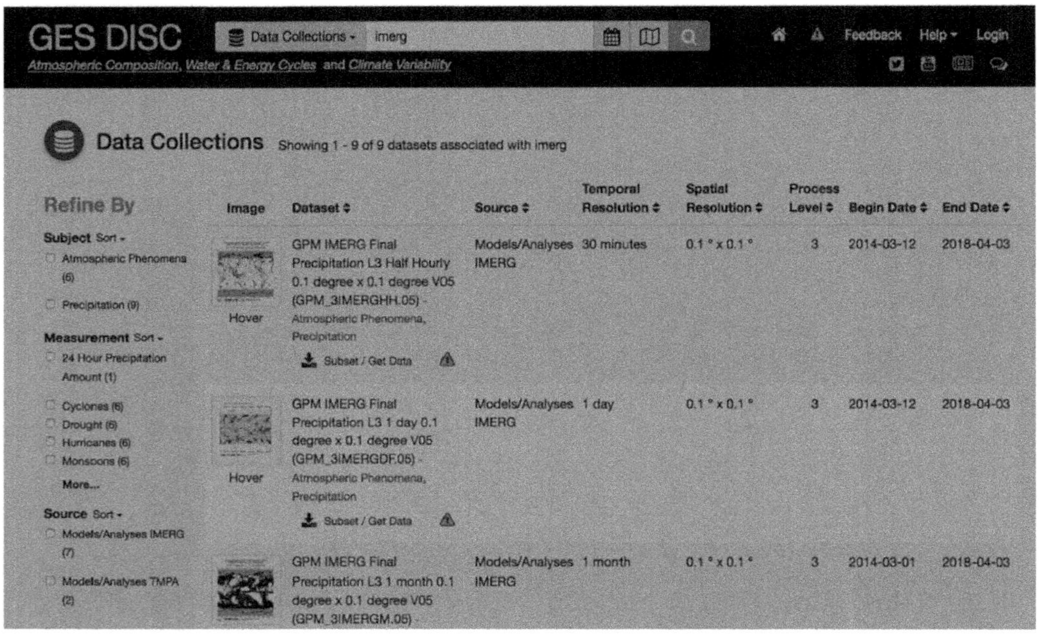

(b)

Figure 11.2 The search and order Web interface at the GES DISC: (a) a screenshot of the GES DISC Web portal showing a keyword search for "imerg" and different search options (e.g. Data Collections, Data Documentation, etc.); (b) the result page from the search in (a) showing available IMERG products.

Server), etc. These protocols support for data downloading activities such as daily operations on the user's side. The https method provides direct access to product archives. OPeNDAP, WMS, GDS, etc. provide remote access to individual variables within datasets in a form usable by many tools and software packages such as IDV, McIDAS-V, Panoply, Ferret, GrADS, etc. OPeNDAP is a framework that simplifies all aspects of scientific data networking and makes local data accessible to remote locations regardless of local storage format (OPeNDAP 2018). With OPeNDAP, users can have options to download TRMM and GPM data in NetCDF, binary, and ASCII. WMS is a standard Web protocol for serving georeferenced map images over the Internet generated by a map server using data from a GIS database and the specifications developed and published by the Open Geospatial Consortium (OGC) in 1999 (WMS

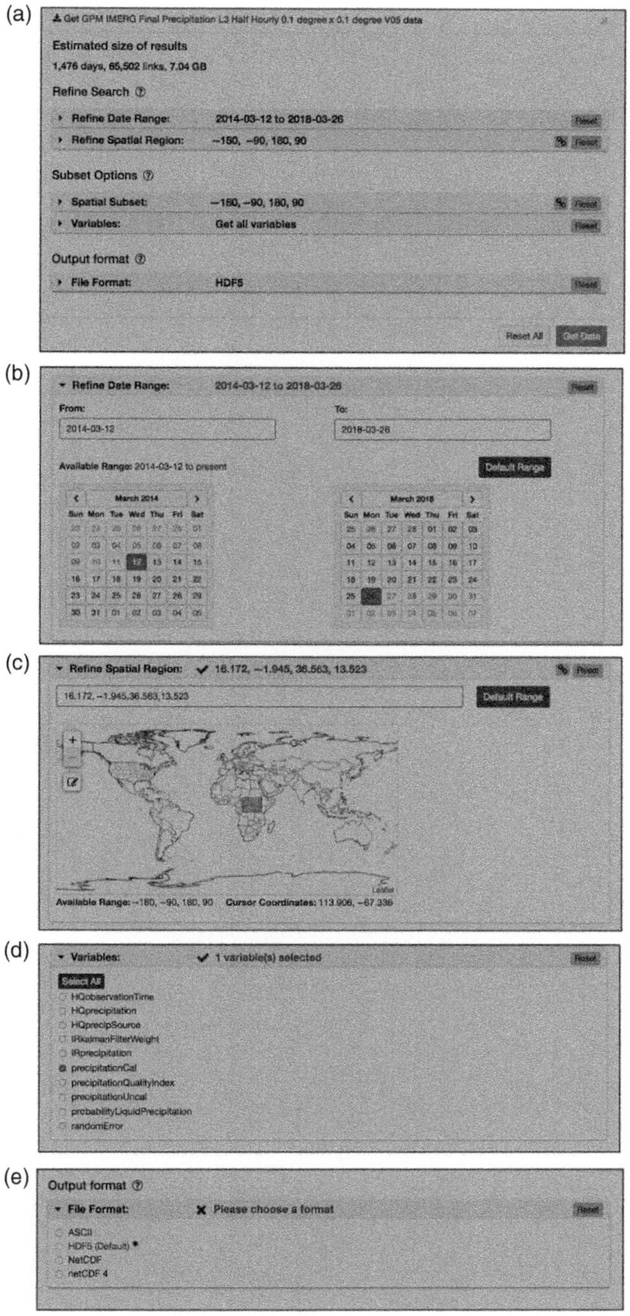

Figure 11.3 Web interface for data subsetting: (a) The interface; (b) Date range picker; (c) Spatial region selection; (d) Variable selection; and (e) Format selection.

2018). The GDS is a stable, secure data server that provides subsetting and analysis services across the internet (GDS 2018). The core of the GDS is OPeNDAP and GrADS, a software framework used for data networking that makes local data accessible to remote locations. With GDS, users can access TRMM TMPA and GPM IMERG products and generate visualizations without downloading data. For example, one can easily generate animations in mp4 using

Table 11.6 Other precipitation data services at GES DISC.

Service	Description
GrADS Data Server	Stable, secure data server that provides subsetting and analysis services across the internet. The core of GDS is OPeNDAP (also known as DODS), a software framework used for data networking that makes local data accessible to remote locations.
OPeNDAP	The Open Source Project for a Network Data Access Protocol (OPeNDAP) provides remote access to individual variables within datasets in a form usable by many tools, such as IDV, McIDAS-V, Panoply, Ferret, and GrADS.
OGC Web Map Service	The Open Geospatial Consortium (OGC) Web Map Service (WMS) provides map depictions over the network via a standard protocol, enabling clients to build customized maps based on data coming from a variety of distributed sources.
THREDDS	The THREDDS Data Server (TDS) is a Web server that uses OPeNDAP, OGC WMS and WCS, HTTP, and other remote data access protocols to provide metadata and data access for datasets.

GDS as remote data access in Panoply. The GES DISC also provide user services including FAQs, data recipes, user forums, email, or phone inquiry, etc.

11.5 Examples

11.5.1 Maps of Seasonal Averages of Precipitation

Long-time averages of precipitation provide information on spatial distribution of precipitation in an area of interest. Average of a product for over 30 years is called climatology based on the definition from the World Meteorological Organization (WMO 2018). Most NASA satellite missions are much shorter than 30 years; however, products from data assimilation projects (e.g. MERRA, NLDAS, and GLDAS) are well over 30 years. Giovanni (Figure 11.1a) has the capability to generate maps of long-time monthly and seasonal averages. The procedure is as follows:

(1) Identify a precipitation variable by using keyword search (Figure 11.1).
(2) Select area/country/state (US only)/watershed.
(3) Select Monthly and Seasonal Averages from Maps.
(4) Select month or season from Select Seasonal Dates and select the time period for averaging.
(5) Click on the "Plot Data" button.

The data for download are available in the output page (Figure 11.1b). User registration is required for downloading data and Giovanni provides the information and steps for facilitating registration.

Figure 11.4 is an example of using a NLDAS-2 monthly precipitation product and Giovanni to generate seasonal climatology (1979–2017) maps in the United States. It is seen that the spatial distribution of precipitation in the United States is highly uneven in all seasons (Figure 11.4). In winter, high precipitation regions are mainly located in both coasts and the regions in between receive less precipitation (Figure 11.4a). Higher precipitation regions are found on the West Coast (e.g. Washington, Oregon, northern California), compared to the East Coast. In summer, most high precipitation regions are along the coastlines of the East Coast, especially in Florida, and in the Midwestern United States (Figure 11.4c). The spatial distribution patterns of precipitation in spring and autumn are similar (Figure 11.4b,d) with high precipitation areas in the upper West Coast and the states along the lower Mississippi river.

Another example is to use the shapefile feature in Giovanni to generate average seasonal maps of precipitation for an irregular area such as Brazil with the TMPA monthly product (3B43). The procedure is similar to the previous example, except the replacements of the variable with 3B43 and the area with Brazil from the country list. After the maps are generated, the data can be downloaded and plotted with Panoply, which is free software from NASA GISS. Since the data format is in NetCDF, it can be directly imported into Panoply (Figure 11.5). Figure 11.5 consists of average seasonal precipitation maps for four austral seasons. It is seen that the spatial distribution patterns for austral spring and summer and are similar with high precipitation regions mainly in northwest Brazil (Figure 11.5b,c). In austral spring, another high precipitation region is found in the southernmost part of Brazil (Figure 11.5b). Similar spatial patterns are found in both austral winter and autumn (Figure 11.5a,d) with high precipitation regions are mainly located at the top and bottom of Brazil, respectively, although sizes of precipitation areas are different. Larger size of higher precipitation is at the top and lower precipitation area at the bottom in austral autumn (Figure 11.5d). Very little precipitation is found in central Brazil during austral winter (Figure 11.5a).

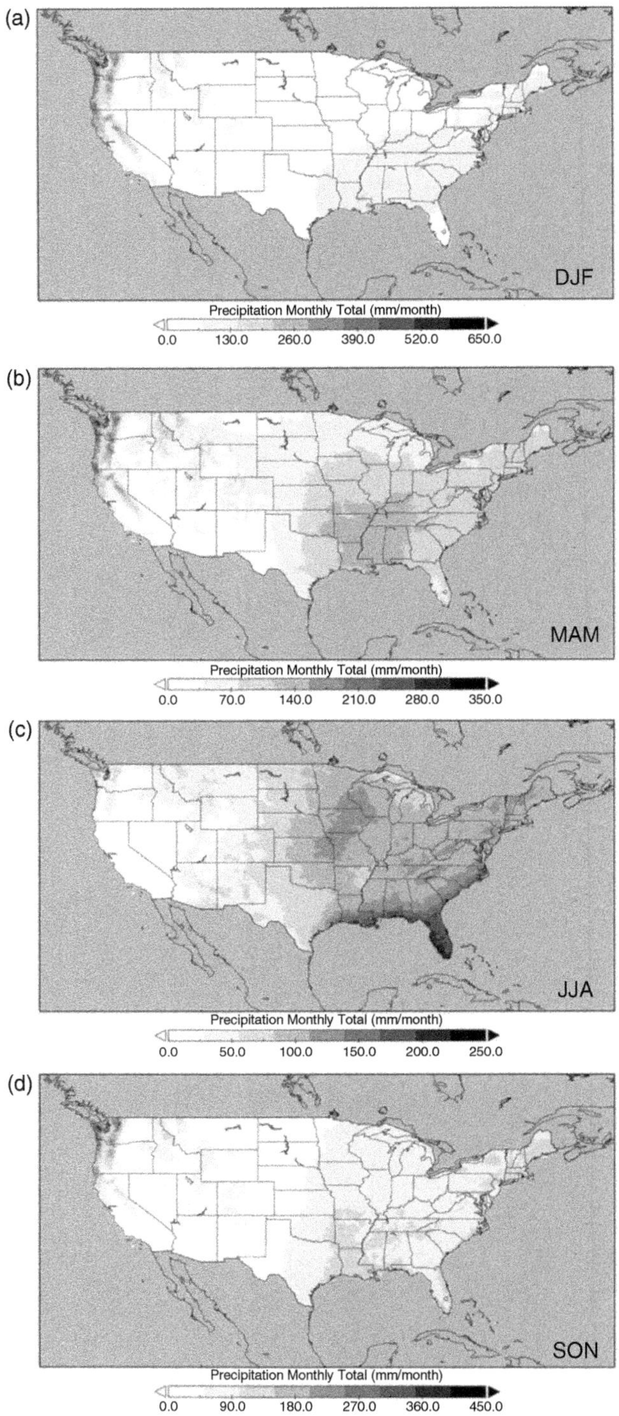

Figure 11.4 Seasonal precipitation climatology (1979–2017) maps in the United States, averaged from the NLDAS-2 monthly precipitation product.

11.5.2 Time Series Analysis of Precipitation in Watersheds

Time series information is essential for rainwater harvesting activities. With Giovanni, one can obtain time series plots and data for a single point, a rectangular area, watersheds, states in the United States, and countries. Meanwhile, users can use Data Rods to obtain time series data in ASCII as mentioned earlier. The procedure of using Giovanni is:

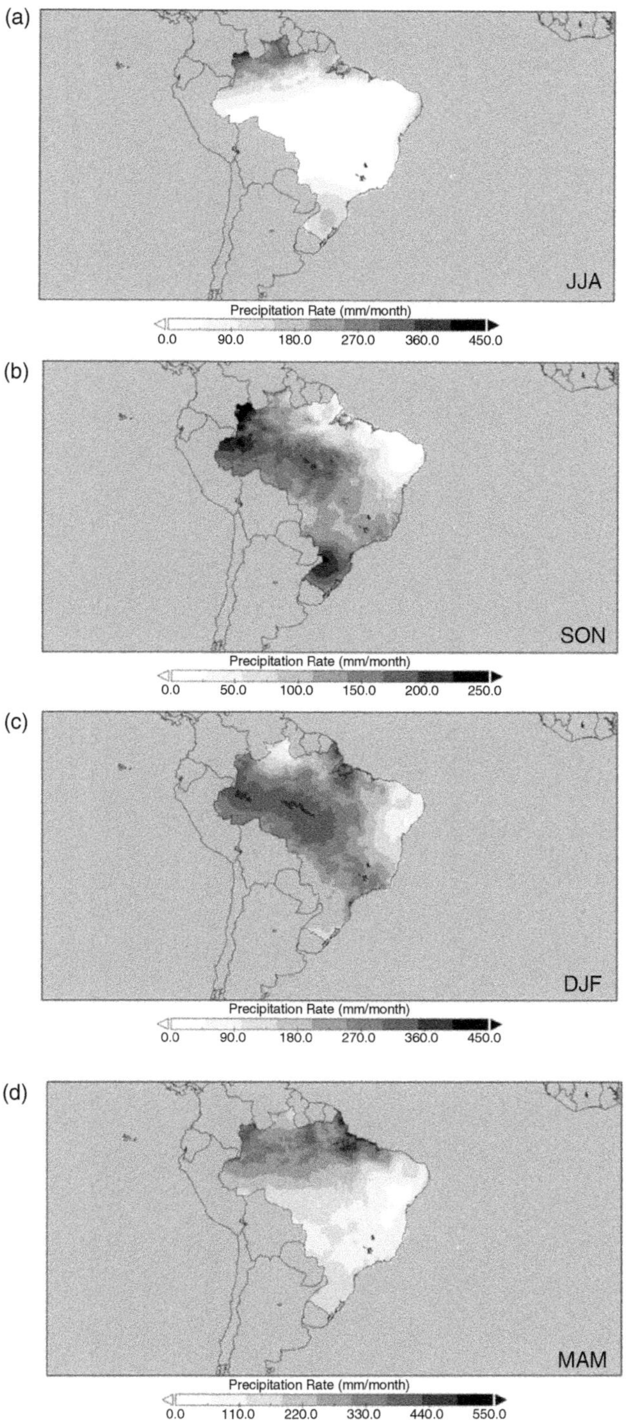

Figure 11.5 Average seasonal precipitation (1998–2017) maps in Brazil from the TMPA 3B43 monthly precipitation product.

(1) Identify a precipitation variable by using keyword search (Figure 11.1a).
(2) For a single point: type in the geolocation information (latitude, longitude) in this format: west, south, east, north. For an area: select area/country/state (US only)/watershed.
(3) Select Area-Averaged or Seasonal.
(4) Select month or season from Select Seasonal Dates and select the time period for time series.
(5) Click on the "Plot Data" button.

For example, using the geolocation (23.133333° N, 113.266667° E) of a megacity, Guangzhou, China, and the TMPA monthly product (3B43), one can use Giovanni to plot its time series from 1998 to 2017. Figure 11.6 shows two plots: (i) the time series of monthly precipitation (Figure 11.6a) and (ii) the time series of seasonal precipitation (Figure 11.6b). Located in the sub-tropical region, the climate in the city of Guangzhou is associated with monsoons that are characterized as high (low) precipitation in summer (winter) as seen in Figure 11.6a,b. The interannual variation of rainfall is quite significant (Figure 11.6a). For example, the highest monthly rainfall in one year can be more than twice of the highest in the other year. Tropical cyclones such as typhoons from the Western Pacific or the South China Sea frequent the region and rainfall from tropical cyclones or typhoons is one of the modulating factors for the interannual variations of precipitation. Figure 11.6b shows the interannual variations of seasonal rainfall. The highest rainfall season is summer, followed by spring, autumn, and winter. The interannual variations from high to low are found in this order, summer, spring, autumn, and winter, as rainfall is modulated by tropical cyclones or typhoons. Again, the data, available in the output page, can be downloaded in CSV (Microsoft Excel) for further analysis.

Likewise, Figure 11.6c contains the seasonal time series plots (1998–2017) for Brazil using 3B43. It is seen that the highest rainfall season is in austral summer, followed by autumn, spring, and winter. Large interannual variations are found in high precipitation seasons in austral summer and autumn compared to the other two low precipitation seasons.

Figure 11.7 consists of time series plots of monthly and seasonal precipitation for the Amazon watershed. Figure 11.7a is a map of the Amazon watershed that covers several countries such as Brazil, Peru, Bolivia, etc. Figure 11.7b is a time series plot of the TMPA monthly product (3B43), showing a typical monsoon precipitation pattern. The seasonal time series plot is in Figure 11.7c. Unlike the seasonal time series plots for Brazil (Figure 11.6c), the rainfall difference between austral summer and autumn is much smaller.

11.5.3 Changes in Precipitation Patterns

As mentioned in the introduction, precipitation patterns can change in a warming climate, as studies reveal. New England, in the United States, is among the regions with

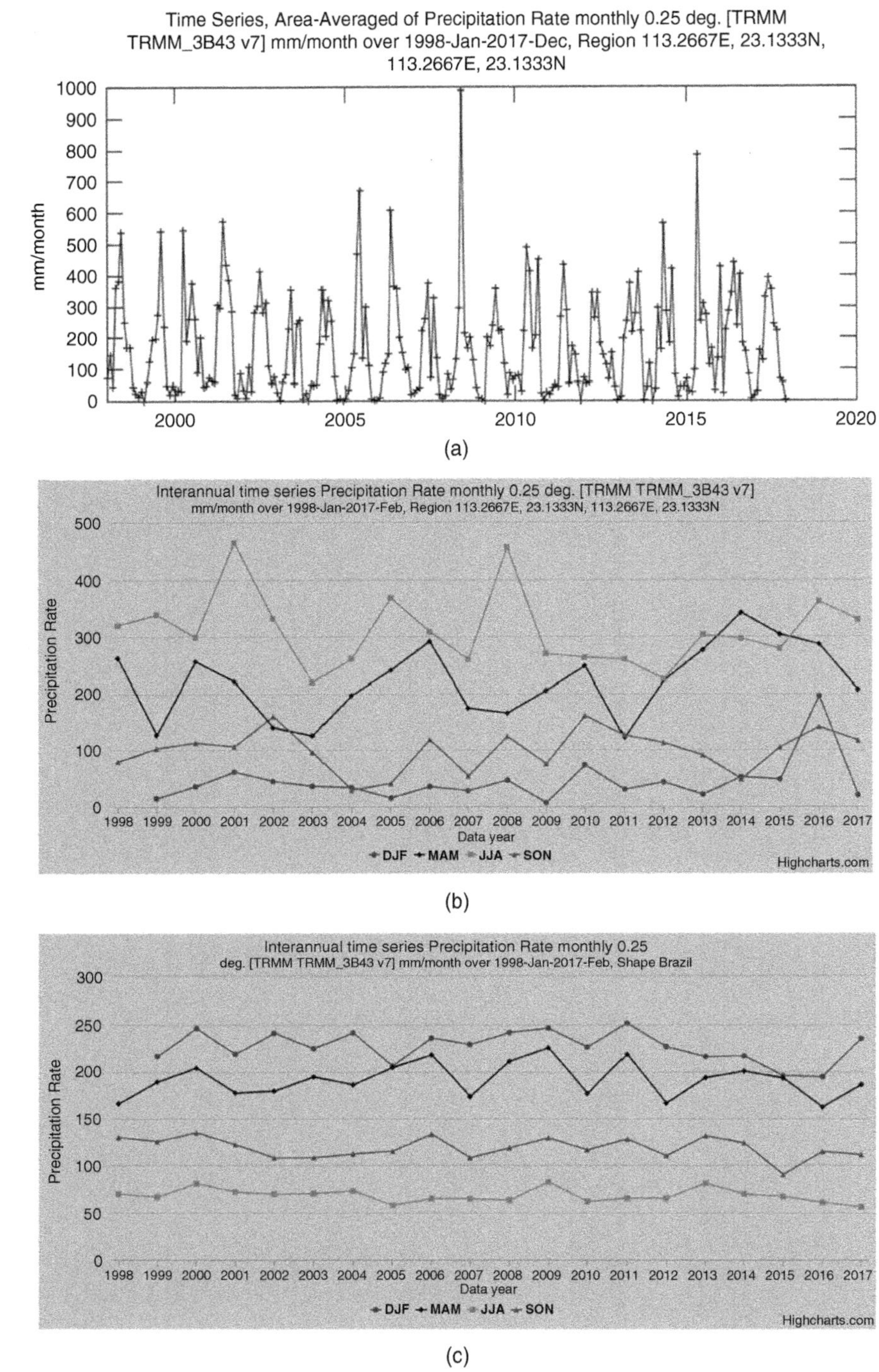

Figure 11.6 Time series of 3B43 precipitation (mm/month) for Guangzhou (113.2667° E, 23.1333° N), China: (a) monthly and (b) seasonal. Time series of 3B43 seasonal precipitation (mm/month) for Brazil (c).

increasing precipitation and extreme precipitation events (e.g. Spierre and Wake 2010; Walsh et al. 2014; Yu et al. 2016). Maine, the largest state in New England, is located in the easternmost region. Precipitation in Maine is examined with the NLDAS-2 hourly and monthly precipitation datasets and other variables (air temperature and surface pressure).

Figure 11.8 contains monthly time series plots of NLDAS-2 seasonal air temperature at 2 m above the ground (Figure 11.8a), NLDAS-2 seasonal precipitation (Figure 11.8b), and 3B43 seasonal precipitation. In Figure 11.8a, it is seen that the monthly air temperature has been rising steadily since 1979, suggesting that Maine is experiencing a slow warming trend. The time series plots of seasonal precipitation (Figure 11.8b) are a bit noisy compared to air temperature (Figure 11.8a), however, it can still be seen that the precipitation in the first part (before 1996) of the seasonal time series plots is slightly lower

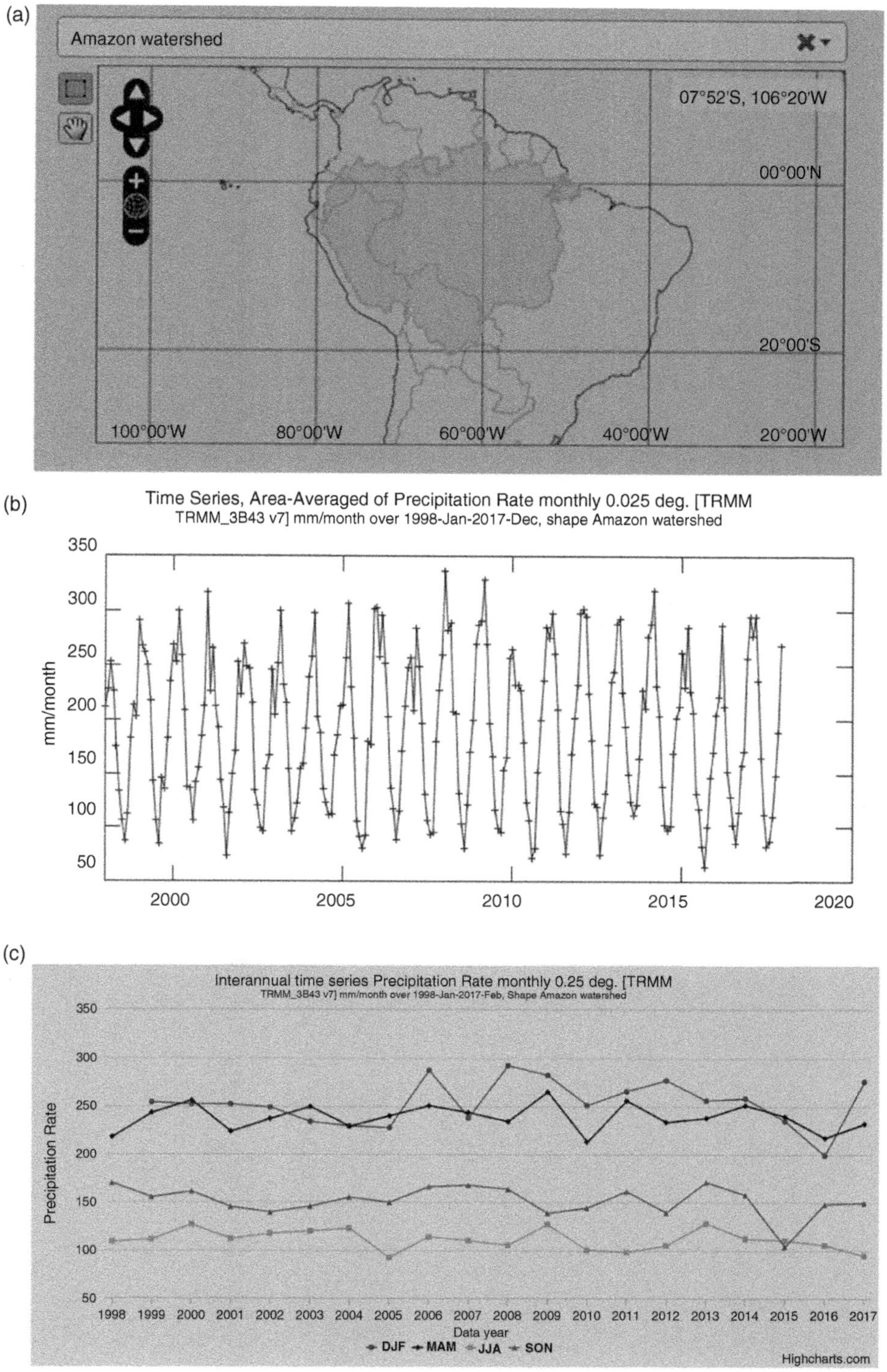

Figure 11.7 Time series of TMPA 3B43 seasonal precipitation (mm/month) for the Amazon watershed: (a) map of Amazon watershed; (b) time series of monthly precipitation; and (c) time series of seasonal precipitation.

than the second part (after 1996), indicating a slow rising trend. Based on the time series plots of air temperature and precipitation, it seems that there is a connection between rising air temperature and precipitation. However, precipitation quality issues have been reported recently (e.g. Ferguson and Mocko 2017) and users need to be careful when using NLDAS-2 precipitation.

Figure 11.8c contains precipitation time series plots similar to NLDAS-2 in Figure 11.8b, except from TMPA 3B43 and a shorter time period (1998–2017). It is seen that 3B43 can capture the main features in the time series plots in Figure 11.8b. For example, two high precipitation seasons that are found in the spring and autumn of 2005 in Figure 11.8b are also found in 3B43 (Figure 11.8c),

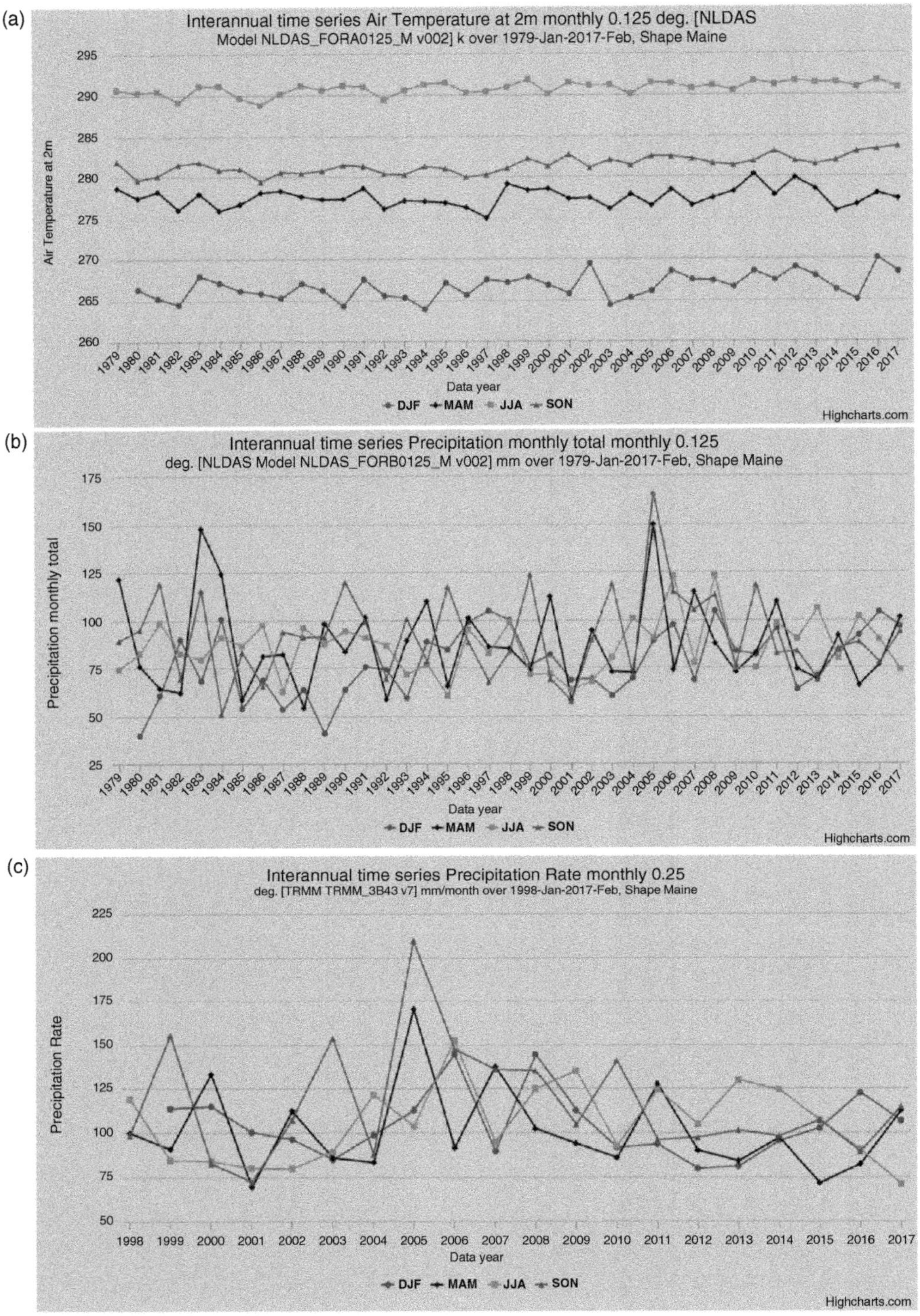

Figure 11.8 Time series plots from NLDAS-2 for Maine, the United States: (a) 2 m monthly air temperature (K); (b) seasonal precipitation in mm/month; and (c) seasonal precipitation in mm/month from 3B43.

although 3B43 shows a high bias issue in precipitation estimates.

To further understand the spatial distribution of precipitation in high precipitation seasons, the autumn in 2005, identified as the highest precipitation season (Figure 11.8b), is selected for further analysis. Figure 11.9a,b show two NLDAS-2 precipitation maps for the autumn of 2005: precipitation total and its anomaly (baseline: 1979–1997). It is seen that high rain areas are mostly concentrated along the coastlines of Maine and precipitation decreases in areas further away from the coastlines (Figure 11.9a,b). To further understand the

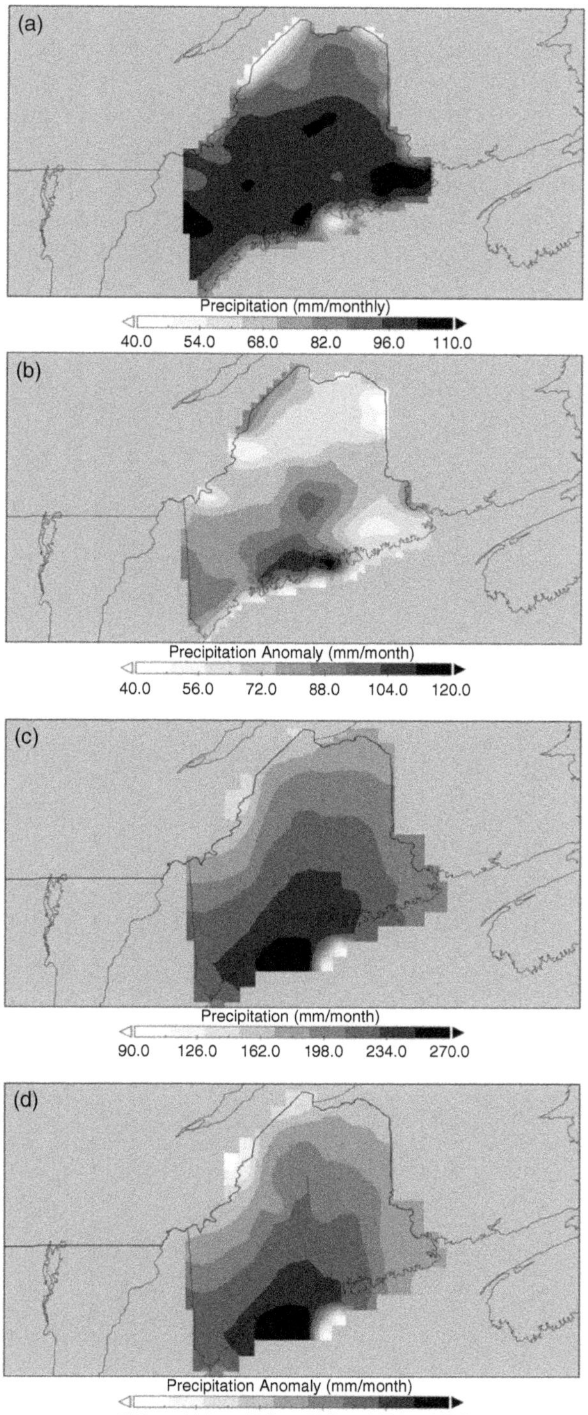

Figure 11.9 NLDAS-2 precipitation (mm/month) for SON, 2005 (a) and its anomaly (climatology: 1979–2017) (b) for Maine. 3B43 precipitation (mm/month) for SON, 2005 (c) and its anomaly (departure from its monthly average: 1998–2017) (d).

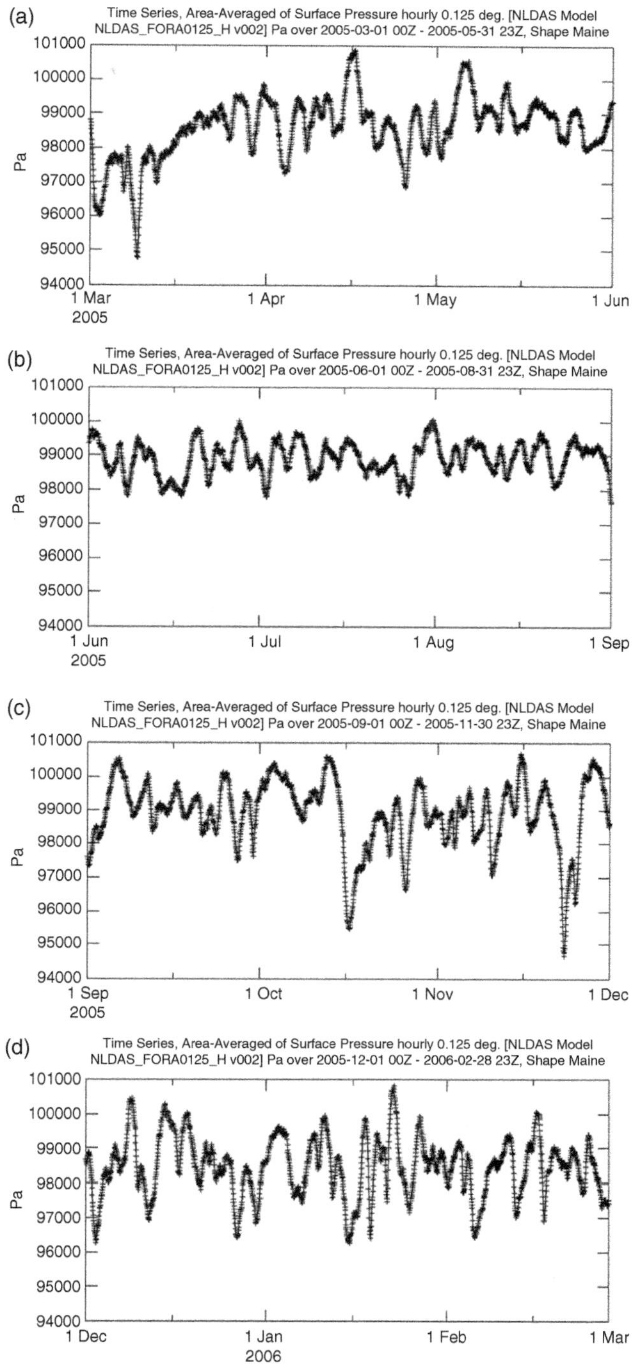

Figure 11.10 Time series of NLDAS-2 hourly surface pressure (in Pa) averaged over Maine in 2005: (a) MAM (March, April, May); (b) JJA (June, July, August); (c) SON (September, October, November); and (d) DJF (December, January, February).

bias issue in precipitation estimates, the similar maps from 3B43 are plotted in Figure 11.9c,d. Although both NLDAS-2 and 3B43, in general, share a lot of similarities in spatial distribution, details are a bit different. For example, the high bias issue found in Figure 11.8 appears in Figure 11.9 as well. Due to the spatial resolution difference between the two products (0.125° for NLDAS-2 vs. 0.25° for 3B43), it is seen that the NLDAS-2 maps provide more details regarding spatial distribution of precipitation. In short, when using a satellite-based precipitation product, users need to understand any potential biases along with other reported issues, which may involve studies to estimate biases with gauge data if no prior knowledge is available.

From Figure 11.8b, it is seen that there is no significant difference in precipitation for all seasons in Maine, despite that air temperature can vary a lot in different seasons (Figure 11.8a). Now the question is what factors cause the above-normal precipitation in the autumn of 2005. From Figure 11.9, the above-normal precipitation areas found along the coastlines suggest that the anomalies may be associated with coastal storms. Previous studies (e.g. Agel

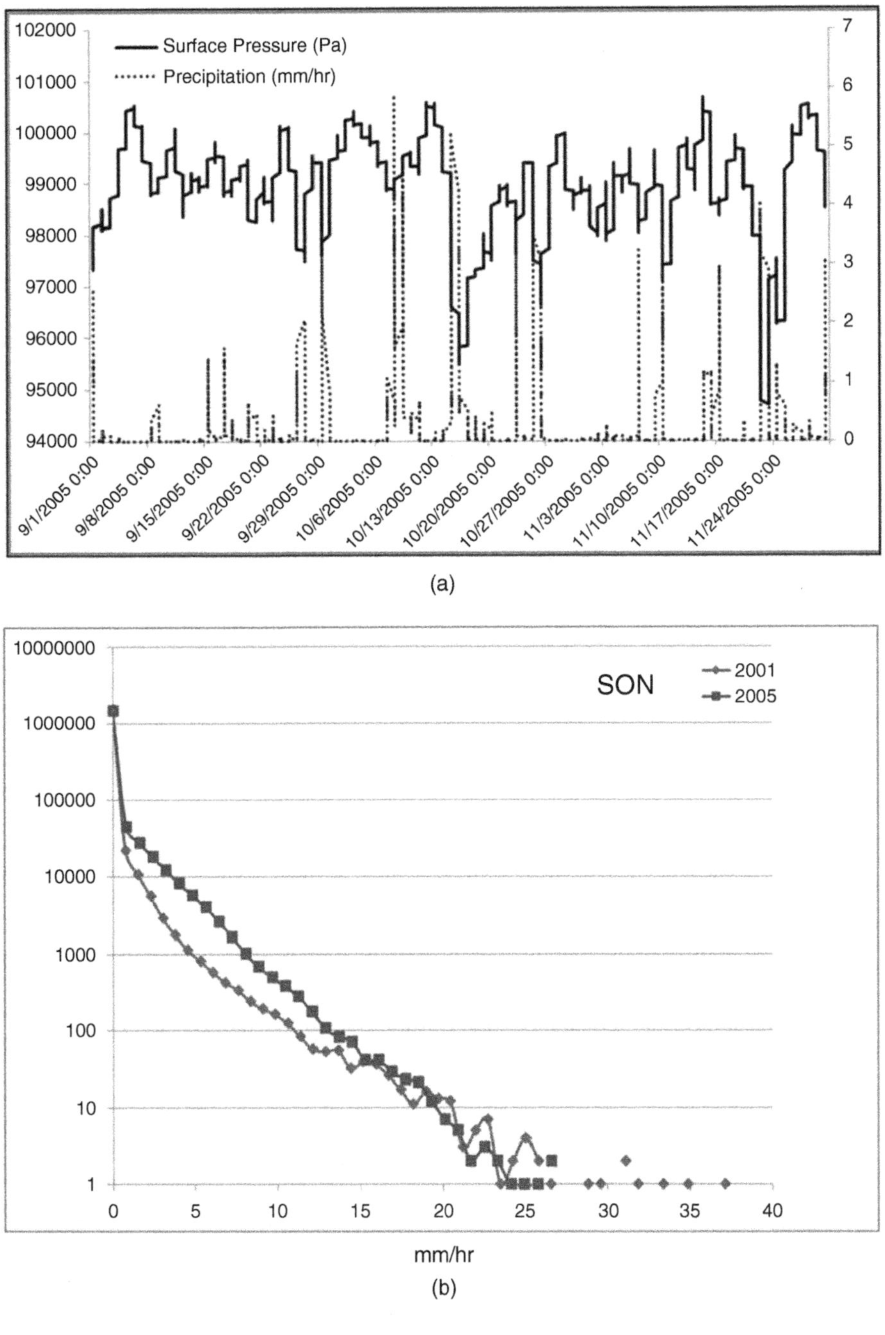

Figure 11.11 Time series of NLDAS-2 hourly surface pressure and precipitation averaged over Maine for SON in 2005 (a). Histograms of NLDAS-2 hourly precipitation rates for SON in 2001 and 2005 (b).

et al. 2015) show that extratropical and tropical storms are the main sources for increasing precipitation and extreme precipitation events in New England.

Figure 11.10 contains the time series plots of NLDAS-2 hourly surface pressure in all seasons of 2005. It is seen that large variations in surface pressure are found in the high precipitation seasons: the spring (Figure 11.10a) and autumn (Figure 11.10c), compared to the two remaining seasons. The largest variations are found in the autumn, followed by the spring, winter, and summer (Figure 11.10). Comparing to two low precipitation seasons (spring and autumn) in 2001 (Figure 11.8b), the variations in surface pressure in the spring and autumn of 2005 are much larger than those in 2001 (not shown). Combining precipitation and surface pressure (Figure 11.11a), the connection between the two becomes more obvious. Precipitation occurs in Maine when surface pressure is dropping. Precipitation rate appears to be related to the magnitude of the pressure dropping, however, in some cases, pressure dropping does not yield significant precipitation compared to other similar situations such as in the early autumn of 2005 (Figure 11.11a).

Histograms of NLDAS-2 hourly precipitation rates (Figure 11.11b) are used to understand precipitation rate distribution in a season. For comparison, the corresponding season in 2001, due to the relatively low monthly precipitation (means: 0.09 mm hr^{-1} [2001], 0.26 mm hr^{-1} [2005]), is selected (Figure 11.8b). In Figure 11.11b, the histograms are generated with all grid points in Maine. Compared to the histograms in 2001 (Figure 11.11b), the following are found for the autumn in 2005: (i) fewer non-precipitation grids; (ii) fewer high precipitation rates; and (iii) increasing counts at the middle and lower precipitation rates (Figure 11.11b).

Overall, this preliminary investigation for Maine shows that: (i) the precipitation characteristics (spatial distribution and intensity) can change, especially along the coastal regions, as air temperature slowly rises; (ii) seasonal precipitation is not that sensitive to seasonal temperature changes; (iii) major spatiotemporal features can be captured by the satellite-based TMPA monthly product (3B43), although a high bias issue exists; (iv) above-normal precipitation in the autumn of 2005 appears to be linked with abnormal (larger) surface pressure perturbations; (v) fewer non-precipitation counts and high precipitation counts; and (vi) more counts in precipitation rates are found in the middle and lower part of the distribution. However, more questions remain to be answered, such as why fewer high precipitation counts are found in high precipitation seasons? Is this because less convective rain is found in high precipitation seasons? Nonetheless, more studies are needed.

11.6 Conclusion

In this chapter, techniques for precipitation measurements from space including infrared, passive, and active microwave instruments, as well as their strengths and weaknesses, have been described. Blended methods that utilize strengths of different measurement techniques and bias correction with ground-based gauge data have also been described. Frequently used global and regional precipitation products that are archived and distributed at GES DISC are introduced, with emphasis on two blended precipitation product suites (TMPA and IMERG) from TRMM and GPM.

The TMPA product suite contains near-real-time and research-grade products. Two near-real-time products are available: the 3-hourly (3B42RT) and its daily product developed by GES DISC as a value-added product. For research-grade products, three are available: 3-hourly, daily, and monthly. Several input and intermediate variables are also available for data quality diagnosis. The TMPA product suite will retire after the retro-processing of IMERG is finished in the near future.

The IMERG suite contains of three output products: "Early satellites" (lag time: ~4 hours), "Late satellites" (lag time: ~12 hours), and the final "Satellite-gauge" (lag time: ~3.5 months). The retro-processing of the IMERG product suite back to the TRMM era will be released in the near future. Similar to TMPA, intermediate variables are also found in IMERG with new ones such as precipitation quality index.

Global and regional land data assimilation system data products include forcing and optimal fields of land surface states and fluxes. They are generated by ingesting satellite- and ground-based observational data products and using advanced land surface modeling and data assimilation techniques. Both forcing data and model results support water resources applications, numerical weather prediction studies, numerous water and energy cycle investigations, and also serve as a foundation for interpreting satellite and ground-based observations.

MERRA-2 is developed by the NASA GMAO at NASA GSFC. MERRA focuses on historical analyses of the hydrological cycle on a broad range of weather and climate time scales and places the NASA EOS suite of observations in a climate context. MERRA-2 provides global data from 1980 onwards and runs a few weeks behind real time. Alongside the meteorological data assimilation using a modern satellite database, MERRA-2 includes an interactive analysis of aerosols that feed back into the circulation, uses NASA's observations of stratospheric ozone and temperature (when available), and takes steps toward representing cryogenic processes.

Other global and regional data products at GES DISC have been briefly introduced, including solar irradiance, atmospheric composition and dynamics, global modeling, etc. Currently, there are over 2700 unique data products archived at GES DISC. More NASA EO products including NASA's EOS and other past NASA satellite mission data are available and distributed by 12 NASA EOSDIS DAACs located throughout the United States. Science disciplines include atmosphere, cryosphere, human dimensions, land, ocean, and calibrated radiance and solar radiance, etc.

Data services at GES DISC to facilitate precipitation data access, evaluation, visualization, and analysis have been presented. Giovanni is a point-and-click Web-based tool. One can access popular precipitation data and other datasets at GES DISC without downloading data and software, making it very suitable for first-time users to evaluate and explore global and regional precipitation from different satellite missions and projects. Other popular products are available in Giovanni. A new GES DISC Web-based search and download interface has been introduced. Users can search over 2700 datasets archived at the GES DISC as well as their documents, FAQs, How-to recipes, etc. A dataset landing page offers a one-stop place for data access, dataset citation, technical documents, and more. Data subsetting and format conversion are essential for conducting research and applications by reducing data volume and making data more suitable for processing. A user-friendly subsetting tool is described. Advanced users can use open-access methods (e.g. OPeNDAP, GDS) to remotely access precipitation data at GES DISC.

Several examples are given to demonstrate basic data access with Giovanni such as time-averaged maps and time series plots for regular and irregular shapes. An in-depth analysis on precipitation pattern change in Maine, with datasets from NLDAS-2 and TRMM, has also been presented. Although the analysis is pretty preliminary, users can quickly learn basic characteristics of precipitation in a changing environment, which can further stimulate more investigation activities.

As of this writing, NASA EOSDIS hosts ~22 PB of Earth Observation (EO) data at 12 DAACs and it is expected to grow to more than 37 (246 PB) PB by 2020 (2025). Such large EO dataset collection is an important asset for environmental research and applications around the world. Over the years, NASA EOSDIS has developed data services and tools to facilitate data discovery and access at 12 discipline-oriented DAACs. For complex environmental issues, it often requires a multi-disciplinary approach which needs an information system that can integrate all these data products archived at 12 DAACs as well as other related data products from users. Therefore, it is essential to provide a one-stop shop for data and services by removing obstacles in data discovery, access, interoperability, etc. Giovanni is an example that makes multi-discipline data variables available in one place and efforts are being carried out to include additional datasets from other DAACs in Giovanni.

The ongoing NASA EOSDIS Cloud Evolution Project will have the potential for scaling up data services and developing a new information system that supports multidisciplinary data products and services. Moving from discipline-oriented to multidiscipline-oriented data services is not a simple task, which will involve in a team of data scientists and software engineers from NASA and other organizations as well as from end users or stakeholders due to many obstacles to be overcome such as data formats, data volume, data structures, terminology in different disciplines, etc. Nonetheless, still a lot of work needs to be done to develop better information systems and services for efficiently solving problems, not only in rainwater harvesting but also in other areas.

Acknowledgments

The team effort of all past and current members at the GES DISC for their contributions to the development of data services and tools such as Giovanni is recognized. Thanks extend to dataset algorithm developers and many users for their feedback and suggestions. The GES DISC is funded by NASA's Science Mission Directorate (SMD).

References

Acker, J.G. and Leptoukh, G. (2007). Online analysis enhances use of NASA earth science data. *Eos, Transactions, American Geophysical Union* 88 (2): 14–17.

Adler, R.F., Huffman, G.J., Chang, A. et al. (2003). The version 2 global precipitation climatology project (GPCP) monthly precipitation analysis (1979-present). *Journal of Hydrometeorology* 4: 1147–1167.

Agel, L., Barlow, M., Qian, J.-H. et al. (2015). Climatology of daily precipitation and extreme precipitation events in the Northeast United States. *Journal of Hydrometeorology* 16: 2537–2557. https://doi.org/10.1175/JHM-D-14-0147.1.

Ahmed, K.F., Wang, G., Silander, J. et al. (2013). Statistical downscaling and bias correction of climate model outputs for climate change impact assessment in the U.S. Northeast. *Global Planet Change* 100: 320–332.

Allan, R. and Soden, B. (2008). Atmospheric warming and amplification of precipitation extremes. *Science* 321: 1481–1484.

AMS (2012) Warm rain process, in Glossary of Meteorology. Available online, http://glossary.ametsoc.org/wiki/Warm_rain_process.

AMS (2018a) Journal of Hydrometeorology - Most Cited. Available online, https://journals.ametsoc.org/toc/hydr/current

AMS (2018b) A special collection on Modern-Era Retrospective analysis for Research and Applications version 2 (MERRA-2). Available online: https://journals.ametsoc.org/topic/merra-2.

Aonashi, K., Awaka, J., Hirose, M. et al. (2009). GSMaP passive, microwave precipitation retrieval algorithm: algorithm description and validation. *Japan Meteorological Society* 87A: 119–136.

Behrangi, A., Hsu, K.-L., Imam, B. et al. (2009). PERSIANN-MSA: a precipitation estimation method from satellite-based multispectral analysis. *Journal of Hydrometeorology* 10: 1414–1429. http://dx.doi.org/10.1175/2009JHM1139.1.

Bitew, M.M., Gebremichael, M., Ghebremichael, L.T., and Bayissa, Y.A. (2012). Evaluation of high-resolution satellite rainfall products through streamflow simulation in a hydrological modeling of a small mountainous watershed in Ethiopia. *Journal of Hydrometeorology* 13: 338–350. http://dx.doi.org/10.1175/2011JHM1292.1.

Bosilovich, M., S. Akella, L. Coy, et al. (2015) Techincal Report Series on Global Modeling and Data Assimilation, Vol. 43, R. Koster, Editor. Available online: http://gmao.gsfc.nasa.gov/pubs/tm/docs/Bosilovich803.pdf.

Dalezios, N.R., Dercas, N., and Eslamian, S.S. (2018). Water scarcity management: part 2: satellite-based composite drought analysis. *International Journal of Global Environmental Issues* 17 (2/3): 262–295.

Daly, C., Neilson, R.P., and Phillips, D.L. (1994). A statistical-topographic model for mapping climatological precipitation over mountainous terrain. *Journal of Applied Meteorology* 33: 140–158. https://doi.org/10.1175/1520-0450(1994)033<0140:ASTMFM>2.0.CO;2.

Engel, T., Fink, A.H., Knippertz, P. et al. (2017). Extreme precipitation in the West African cities of Dakar and Ouagadougou-atmospheric dynamics and implications for flood risk assessments. *Journal of Hydrometeorology* 18: 2937–2957.

Ferguson, C.R. and Mocko, D.M. (2017). Diagnosing an artificial trend in NLDAS-2 afternoon precipitation. *Journal of Hydrometeorology* 18 (4): 1051–1070.

Fischer, E.M. and Kutti, R. (2016). Observed heavy precipitation increase confirms theory and early models. *Nature Climate Change* 6: 986–991.

Gallaher, D. and G. Grant (2012) Data Rods: High speed, time-series analysis of massive cryospheric data sets using pure object databases, Geoscience and Remote Sensing Symposium (IGRASS), 22–27 July 2012. Available: http://ieeexplore.ieee.org/xpl/articleDetails.jsp?reload=true&arnumber=6352413&contentType=Conference+Publications

GDS (2018) The GrADS Data Server (GDS). Available online, http://cola.gmu.edu/grads/gds.php.

Gelaro, R., McCarty, W., Suárez, M.J. et al. (2017). The modern-era retrospective analysis for research and applications, version 2 (MERRA-2). *Journal of Climate* 30: 5419–5454. https://doi.org/10.1175/JCLI-D-16-0758.1.

Gianotti, R.L., Zhang, D., and Eltahir, E.A.B. (2012). Assessment of the regional climate model version 3 over the maritime continent using different cumulus parameterization and land surface schemes. *J. Climate* 25: 638–656. http://dx.doi.org/10.1175/JCLI-D-11-00025.1.

Global Change (2018) National Climate Assessment, Available online, https://nca2014.globalchange.gov.

Gourley, J.J., Hong, Y., Flamig, Z.L. et al. (2011). Hydrologic evaluation of rainfall estimates from radar, satellite, gauge, and combinations on Ft. Cobb Basin, Oklahoma. *Journal of Hydrometeorology* 12: 973–988. http://dx.doi.org/10.1175/2011JHM1287.1.

Groisman, P., Knight, R.W., Easterling, D.R. et al. (2005). Trends in intense precipitation in the climate record. *Journal of Climate* 18: 1326–1350.

Hegerl, G.E., Black, E., Allan, R.P. et al. (2015). Challenges in quantifying changes in the global water cycle. *Bulletin of American Meteorological Society* 96: 1097–1115.

Higgins, R.W., W. Shi, E. Yarosh, and R. Joyce (2000) Improved United States precipitation quality control system and analysis. NCEP/Climate Prediction Center Atlas No. 7.

Hong, Y., Gochis, D., Cheng, J. et al. (2007). Evaluation of PERSIANN-CCS rainfall measurement using the NAME event rain gauge network. *Journal of Hydrometeorology* 8: 469–482. http://dx.doi.org/10.1175/JHM574.1.

Huffman, G.J. (1997). Estimates of root-mean-square random error contained in finite sets of estimated precipitation. *J. Appl. Meteor.* 36: 1191–1201.

Huffman, G.J., Adler, R.F., Arkin, P. et al. (1997). The global precipitation climatology project (GPCP) combined precipitation dataset. *Bulletin of American Meteorological Society* 78: 5–20.

Huffman, G.J. (2016) The Transition in Multi-Satellite Products from TRMM to GPM (TMPA to IMERG). Available online, https://pmm.nasa.gov/sites/default/files/document_files/TMPA-to-IMERG_transition_161025.pdf.

Huffman, G.J. and D.T. Bolvin (2012) Real-Time TRMM Multi-Satellite Precipitation Analysis Data Set

Documentation. Available online: ftp://trmmopen.gsfc.nasa.gov/pub/merged/V7Documents/3B4XRT_doc_V7.pdf.

Huffman, G.J. and D.T. Bolvin (2013) TRMM and Other Data Precipitation Data Set Documentation. Available online: ftp://meso-a.gsfc.nasa.gov/pub/trmmdocs/3B42_3B43_doc.pdf.

Huffman, G.J., Adler, R.F., Bolvin, D.T. et al. (2007). The TRMM multi-satellite precipitation analysis: quasi-global, multi-year, combined-sensor precipitation estimates at fine scale. *Journal of Hydrometeorology* 8 (1): 38–55.

Huffman, G.J., Adler, R.F., Bolvin, D.T., and Gu, G. (2009). Improving the global precipitation record: GPCP version 2.1. *Geophysical Research Letters* 36: L17808. https://doi.org/10.1029/2009GL040000.

Huffman, G.J., Adler, R.F., Bolvin, D.T., and Nelkin, E.J. (2010). The TRMM multi - satellite precipitation analysis (TAMPA). Chapter 1. In: *Satellite Rainfall Applications for Surface Hydrology* (eds. F. Hossain and M. Gebremichael), 3–22. Springer Verlag. ISBN: 978-90-481-2914-0.

Huffman, G.J., D. Bolvin, D. Braithwaite, et al. (2017) IMERG Algorithm Theoretical Basis Document (ATBD), Available online: https://pmm.nasa.gov/sites/default/files/document_files/IMERG_ATBD_V4.6.pdf

Huffman, G.J., D.T. Bolvin, E.J.Nelkin (2018) Integrated Multi-satellitE Retrievals for GPM (IMERG) Technical Documentation. Available online: https://pmm.nasa.gov/sites/default/files/document_files/IMERG_doc_180207.pdf.

Iguchi, T., Kozu, T., Meneghini, R. et al. (2000). Rain-profiling algorithm for the TRMM precipitation radar. *Journal of Applied Meteorology* 39: 2038–2052.

IPWG (2018) Global precipitation datasets. Available online: http://ipwg.isac.cnr.it/data/datasets.html.

Joyce, R.J., Janowiak, J.E., Arkin, P.A., and Xie, P. (2004). CMORPH: a method that produces global precipitation estimates from passive microwave and infrared data at high spatial and temporal resolution. *Journal of Hydrometeorology* 5: 487–503.

Kearns, E. (2017) Improving Access to Open Data through NOAA's Big Data Project. Available online: https://bigdatawg.nist.gov/Day2_08_NIST_Big_Data-Kearns.pdf.

Kummerow, C., Simpson, J., Thiele, O. et al. (2000). The status of the tropical rainfall measuring Mission (TRMM) after two years in orbit. *Journal of Applied Meteorology.* 39: 1965–1982. https://doi.org/10.1175/1520-0450(2001)040<1965:TSOTTR>2.0.CO;2.

Kummerow, C., Hong, Y., Olson, W.S. et al. (2001). The evolution of the Goddard Profiling Algorithm (GPROF) for rainfall estimation from passive microwave sensors. *Journal of Applied Meteorology* 40: 1801–1820.

Liu, Z. and Acker, J. (2017). Giovanni: the bridge between data and science. *Eos* 98 https://doi.org/10.1029/2017EO079299. Published on 24 August 2017.

Liu, Z., Ostrenga, D., Teng, W., and Kempler, S. (2012). Tropical rainfall measuring mission (TRMM) precipitation data and services for research and applications. *Bulletin of the American Meteorological Society* http://dx.doi.org/10.1175/BAMS-D-11-00152.1.

Liu, Z., Ostrenga, D., Vollmer, B. et al. (2017). Global precipitation measurement Mission products and services at the NASA GES DISC. *Bulletin of the American Meteorological Society* 98 (3): 437–444. https://doi.org/10.1175/bams-d-16-0023.1.

Maddox, R.A., Zhang, J., Gourley, J.J., and Howard, K.W. (2002). Weather radar coverage over the contiguous United States. *Weather Forecasting* 17: 927–934.

Mahrooghy, M., Anantharaj, V.G., Younan, N.H. et al. (2012). On an enhanced PERSIANN-CCS algorithm for precipitation estimation. *Journal of Atmospheric and Oceanic Technology* 29: 922–932. http://dx.doi.org/10.1175/JTECH-D-11-00146.1.

NASA (2018a) NASA's Earth Observing System Project Science Office. Available online, https://eospso.gsfc.nasa.gov.

NASA (2018b) NASA Earth Science Data and Information System Project. Available online, https://earthdata.nasa.gov/about/esdis-project.

NASA, 2018c, Land Data Assimilation Systems. Available online, https://ldas.gsfc.nasa.gov/index.php.

NASA (2018d) Modern-Era Retrospective analysis for Research and Applications, Version 2, Available online, https://gmao.gsfc.nasa.gov/reanalysis/MERRA-2.

NASA (2018e) Earthdata Search. Available online, https://search.earthdata.nasa.gov/search.

NASA (2018f) NASA GES DISC Giovanni. Available online, https://giovanni.gsfc.nasa.gov.

NASA (2018g) NASA GES DISC Homepage. Available online: https://disc.gsfc.nasa.gov.

NASA (2018h) Web Services at GES DIC. Available online, https://disc.gsfc.nasa.gov/information/tools?title=OPeNDAP%20and%20GDS.

OPeNDAP (2018) OPeNDAP – Advanced Software for Remote Data Retrieval. Available online, https://www.opendap.org

Parr, D.T., Wang, G.L., and Bjerklie, D. (2015). Integrating remote sensing data on evapotranspiration and leaf area index with hydrological modeling: impacts on model performance and future predictions. *Journal of Hydrometeorology* 16: 2086–2100.

Reichle, R.H., and Q. Liu (2014) Observation-Corrected Precipitation Estimates in GEOS-5. NASA/TM–2014-104606, Vol. 35.

Rienecker, M.M., Suarez, M.J., Gelaro, R. et al. (2011). MERRA: NASA's modern-era retrospective analysis for research and applications. *Journal of Climate* 24: 3624–3648. https://doi.org/10.1175/JCLI-D-11-00015.1.

Rui, H., B. Teng, R. Strub, and B. Vollmer (2012) Data reorganization for optimal time series data access, analysis, and visualization, AGU Fall Meeting, San Francisco, CA, Dec. 3–7, 2012.

Rui, H., R. Strub, W.L. Teng, et al. (2013) Enhancing access to and use of NASA earth sciences data via CUAHSI-HIS (Hydrologic Information System) and other hydrologic community tools, AGU Fall Meeting, San Francisco, CA, Dec. 9–13, 2013.

Spencer, R.W., Goodman, H.G., and Hood, R.E. (1989). Precipitation retrieval over land and ocean with the SSM/I: identification and characteristics of the scattering signal. *Journal of Atmospheric and Oceanic Technology* 6: 254–273.

Spierre S.G. and C. Wake (2010) Trends in Extreme Precipitation Events for the Northeastern United Sttes (1948–2007). Available online: https://sustainableunh.unh.edu/sites/sustainableunh.unh.edu/files/media/2010_northeastextremeprecip.pdf.

Su, F., Gao, H., Huffman, G.J., and Lettenmaier, D.P. (2011). Potential utility of the real-time TMPA-RT precipitation estimates in Streamflow prediction. *Journal of Hydrometeorology* 12: 444–455. http://dx.doi.org/10.1175/2010JHM1353.122.

Suarez, M. and Bacmeister, J. (2015). Development of the GEOS-5 atmospheric general circulation model: evolution from MERRA to MERRA2. *Geoscientific Model Development* 8: 1339–1356. https://doi.org/10.5194/gmd-8-1339-2015.

Takacs, L.L., M. Suarez, and R. Todling (2015) Maintaining Atmospheric Mass and Water Balance Within Reanalysis. NASA/TM–2014-104606, Vol. 37.

Teng, W., Rui, H., Strub, R., and Vollmer, B. (2016). Optimal reorganization of NASA earth science data for enhanced accessibility and usability for the hydrology community. *Journal of American Water Resources Association (JAWRA)*: 1–11. https://doi.org/10.1111/1752-1688.12405.

Utah (2018) Methods to Measure Rainfall, University of Utah PMM Science. Available online, http://trmm.chpc.utah.edu.

Walsh, J., Wuebbles, D., Hayhoe, K. et al. (2014). Ch. 2: our changing climate. Climate change impacts in the United States: the third national climate assessment. In: *U.S. Global Change Research Program* (eds. J.M. Melillo, T.C. Richmond and G.W. Yohe), 19–67. https://doi.org/10.7930/J0KW5CXT.

Wang, G. et al. (2017). The peak structure and future changes of the relationships between extreme precipitation and temperature. *Nature Climate Change volume* 7: 268–274. https://doi.org/10.1038/nclimate3239.

WMO (2018) What is climate? Available online: http://www.wmo.int/pages/prog/wcp/ccl/faqs.php.

WMS (2018) Web Map Service. Available online, https://en.wikipedia.org/wiki/Web_Map_Service.

Wu, H., Adler, R.F., Hong, Y. et al. (2012). Evaluation of global flood detection using satellite-based rainfall and a hydrologic model. *Journal of Hydrometeorology* 13: 1268–1284.

Yu, L., Zhong, S., Pei, L. et al. (2016). Contribution of largescale circulation anomalies to changes in extreme precipitation frequency in the United States. *Environmental Research Letters* 11 (4): 044003.

12

Risk Analysis of Water Harvesting Systems

Maria Do Céu Almeida[1], *Nelson Carriço*[2], *João Santos*[3], *and Saeid Eslamian*[4]

[1] *Department of Hydraulics and Environment, National Laboratory of Civil Engineering, 1700-066 Lisboa, Portugal*
[2] *Barreiro School of Technology, Polytechnic Institute of Setubal, 2839-001 Lavradio, Portugal*
[3] *School of Management and Technology, Polytechnic Institute of Beja, 7800-295 Beja, Portugal*
[4] *Department of Water Engineering, College of Agriculture, Isfahan University of Technology, Isfahan, Iran*

12.1 Introduction

Rainwater harvesting systems (RWHS) can result in important benefits, but there are a number of potential negative risks. Clear understanding and management of risks by users, regulating authorities, water utilities, and society in general, are essential to ensure quality, safety, and return on investment. Especially where water is scarce, alternative sources of water need to be identified and case-specific solutions used.

Climate dynamics impose a more responsible use of resources and consideration of a portfolio of solutions to increase the resilience of urban systems (Muller 2007). The challenge resides on how to deal with existing and expected levels of risk and to take advantage of opportunities. The risks associated with RWHS depend on several factors and duly consideration is required to ensure that risks associated with this type of solutions are low.

Therefore, a generic and systematic approach to manage risk is required. This approach should include identification of risks and opportunities related to alternative actions and incorporate uncertainties. Given the effects of climate dynamics together with socioeconomic changes, RWHS as an adaptation measure should address all system components, reliability as water source, the interactions of people with RWHS components, and the types of water uses.

Rainwater can be used for urban, industrial, and agricultural uses, with different quality requirements, including different potable or non-potable uses as well as indoor (e.g. cooling process, boiler water, flushing, washing, drinking water) or outdoor (e.g. sanitary sewer flushing, irrigation, vehicle or building washing, street cleaning, firefighting, ornamental, and recreational water features) uses. Whatever the purpose, an underlying risk is associated with the local precipitation regime versus the storage capacity available. This risk is analyzed in the section on rainwater collection reliability as water source.

Several approaches are available to undertake risk analysis of RWHS. Whatever the methodology used, it is essential that concepts and terminology used are clear to avoid misunderstandings. For this chapter, the following section introduces a brief overview of concepts and terminology adopted. Subsequently, an overview of common approaches to risk management are presented. The following sections focus on specific aspects for the application of these approaches to RWHS, with emphasis on risk identification, specific risk treatment action, and monitoring.

12.2 Concepts and Terminology

The concepts and terminology on risk assessment and management vary depending on the type of application and the context. Given the existing differences an effort was made to harmonize and integrate recent developments in risk management standards (main source documents used are the ISO Guide 73:2009 and ISO 31000:2018). Table 12.1 presents the definitions adopted within this chapter, to clarify the meaning as used by the authors.

12.3 General Approaches to Risk Management Applicable to RWHS

Several authors have investigated risks associated with RWHS in different applications (Campling et al. 2008; Fewtrell et al. 2008; Gwenzi et al. 2015; Lye 2002; Sadhu et al. 2014). Points of view considered include:

Handbook of Water Harvesting and Conservation: Basic Concepts and Fundamentals, First Edition. Edited by Saeid Eslamian and Faezeh Eslamian.

Table 12.1 Risk management definitions adopted.

Expression	Definition
Consequence	Result of an event affecting the objectives. A consequence can have positive or negative effects on objectives.
Control	Course of action with the purpose to modify risk. Controls include, but are not limited to, any process, policy, device, practice, or other conditions. Despite intended to modify risk, effectiveness can differ from the expected.
Event	Occurrence or change of a particular set of circumstances. An event can have one or several associated occurrences. An event has a degree of uncertainty associated.
Exposure	Extent to which an organization or individual is subject to an event.
Hazard	Source of potential harm. A hazard can be a risk source.
Hazardous event	A situation that can cause harm, e.g. leading to the presence or release of a hazard. The hazardous event is part of the event pathway starting at the risk source.
Likelihood	Chance of something happening, described using general terms or mathematically such as a probability or a frequency over a given time period. In some languages probability is used with the same broad meaning.
Resilience	Adaptive capacity of an organization, a system or a person in a complex and changing environment.
Risk	Effect of uncertainty on objectives. An effect is a deviation from the expected. It can be positive, negative, or both, and can address, create, or result in opportunities and consequences. Objectives can have different aspects and categories and can be applied at different levels. Risk is usually expressed in terms of risk sources, potential events, their consequences, and their likelihood.
Risk analysis	Process to comprehend the nature of risk and to determine the level of risk.
Risk assessment	Overall process including risk identification, analysis, and evaluation.
Risk evaluation	Comparing the results of risk analysis against risk criteria to determine whether the level of risk is acceptable or tolerable.
Risk factor	Something that can have an effect on the risk level, by changing the probability or the consequences of an event. Risk factors are often causes or causal factors that can be acted upon using risk reduction measures. Typically, three main categories are considered, namely human factors, environmental factors, and equipment/infrastructure factors.
Risk identification	Process of finding, recognizing, and describing risks. Risk identification involves the identification of risk sources, events, their causes, and their potential consequences. It can involve using historical data, theoretical analysis, informed and expert opinions, and stakeholder's needs.
Risk management	Coordinated activities to direct and control an organization with regard to risk.
Risk source	Element which alone or in combination has the intrinsic potential to give rise to risk.
Risk treatment	Process to modify risk. Risk treatment can involve: (i) avoiding the risk by deciding not to start or continue with the activity giving rise to the risk; (ii) removing the risk source; (iii) changing the likelihood; (iv) changing the consequences; (v) sharing the risk with another party or parties; (vi) retaining the risk by informed decision.
Vulnerability	Degree of susceptibility of something, which if exposed to any risk source or hazardous event, can lead to an event with consequences.

- public health and safety – adequate water quantity and quality (e.g. chemical or microbiological) to ensure basic needs and quality of life, exposure to water related hazards (e.g. flooding, contaminated surface water);
- social – water availability and affordability;
- economic – costs of water, required investments;
- environmental impacts – overuse of resources (e.g. water, energy), pollution by overland flows and untreated combined sewage, greenhouse gases production (GHG);

Available approaches to assess these types of risks include the Risk Management Framework (RMF) (ISO, 2018), the Water Safety Plan (WSP) approach (Davison et al. 2005; Bartram et al. 2009), the Hazard Analysis and Critical Control Point (HACCP) (Codex 2003), and the ISO 22000 (2005) family of International Standards, which addresses food safety management, among others. The general principles of these approaches have a clear parallel since broadly all include identification of risks and relevant factors, evaluation of magnitude of risks, monitoring actions, and a continuous improvement type of process.

The RMF is a broad approach applicable to any field and to any type of risk. This general systematic approach is aimed at development and implementation of risk management to control, prevent, or reduce negative

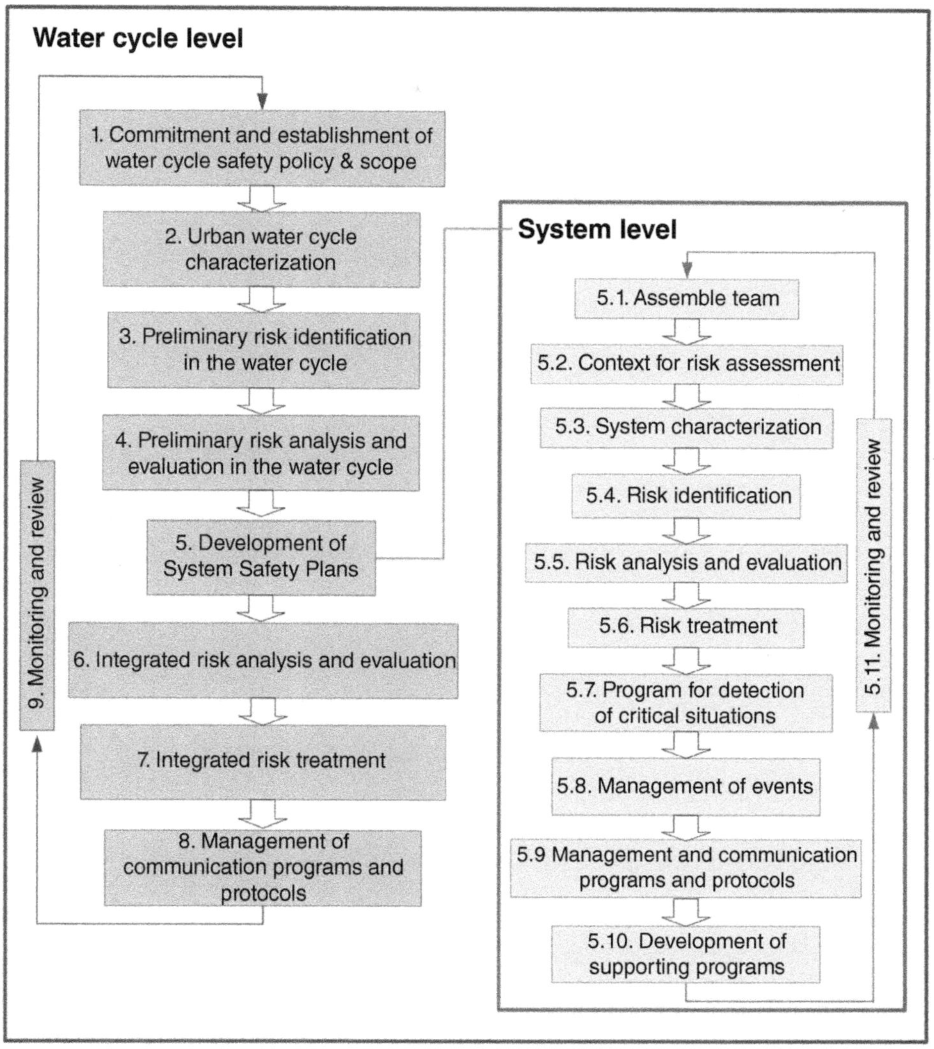

Figure 12.1 Water cycle safety plan approach (Almeida et al. 2014)

consequences (e.g. loss of life, illness, injury, damage to property, environmental impacts) allowing at the same time identification of opportunities (ISO, 2018). The RMF incorporates a risk management process (RMP) for the effective implementation of risk management principles at all relevant levels and functions of the organization.

The RMP usage covers a wide range of activities and fields. Application and adaptation of the RMP to manage drinking water quality is, for instance, within the scope of the Australian Drinking Water Guidelines (ADWG) (NHMRC 2004). Other specific approaches have been developed and applied to water supply systems (WSS) aiming at improving the safety of drinking water. One of the best known is the WSP (Davison et al. 2005; Bartram et al. 2009) developed within the scope of the World Health Organization (WHO) guidelines for drinking-water quality (WHO 2005) to contribute to reducing risk to the consumer's health. The WSP evolved from the HACCP method (Codex 2003; Havelaar 1994), which is a multi-barrier approach conceived to assure food safety and safety for consumption of food products reaching the consumer, today widely applied to the food industry.

The ISO 22000 (2005) was a development integrating the principles of HACCP and specifying requirements for a food safety management system (Blanc 2006). This standard goes beyond HACCP as it adds in further recommendations, and is auditable and suitable for certification purposes (ISO 2005). Drinking water as a product in the food chain can benefit from approaches as HACCP and ISO 22000 (2005) that have been applied in the water industry (Almeida et al. 2014).

Both approaches recommend water utilities to follow risk assessment and risk management by implementing WSP, aiming at consistently ensuring the safety and acceptability of drinking water considering health-based targets. This approach incorporates not only end-product testing but also process control from source to tap (WHO 2009, 2017). WSP is specific to WSS and focuses on risks related to human health. A WSP can be used to identify, assess, and treat risks associated with RWHS, especially when uses require drinking-water quality.

As emphasized by the WHO (2017), application of the WSP concept at household or small-size facilities may not be practicable. However, regulators, health authorities, or water authorities can use it as a basis to establish recommendations and good practices to users as well as control actions. The water cycle safety plan (WCSP) concept, a comprehensive risk-based approach combining the WSP with RMF in a two-level decision-making process (integrated and per subsystems) applied to the urban water cycle, allows contribution of different stakeholders into a common approach and planning for action (Almeida et al. 2014). It includes the RWHS as one of the subsystems and can provide the overall framework to responsible authorities to deal with risks associated with water cycle (Figure 12.1).

A central part of most methodologies is risk analysis, which combines the steps of risk identification and estimation. The criteria and methods should be defined when the context is established. There are several techniques or methods that can be used to analyze risks being, generally, grouped in qualitative (e.g. fault trees, risk matrix), or quantitative (e.g. quantitative microbial risk assessment [QMRA] or failure mode and effects analysis [FMEA]). Qualitative methods define likelihood, consequence, and level of risk through ordinal scales, such as high, medium, and low, and combine the likelihood and consequences levels to obtain a level of risk through qualitative criteria. Quantitative methods define likelihood, consequence, and level of risk numerically in specific units (ISO 2018). The ISO/IEC 31010:2009 standard lists about 30 techniques and some of their attributes, and presents about one-page summary of each technique. These techniques may be applied to different steps of the risk assessment process such as for risk identification, risk analysis (likelihood, consequence or risk level) or risk evaluation.

A systematic procedure to undertake risk identification and identify the relevant events is essential to facilitate risk analysis and to ensure the effectiveness of risk management. Risk identification can be a complicated task since multiple risk sources and risk factors interplay and event pathways can lead to many different events, each potentially escalating to different consequences. A good understanding of concepts and methods used is essential to the success of this task. Failure to identify risks or events corresponds to the implicit acceptance of the risks not analyzed.

Within the several techniques available, the FMEA has been widely used in different applications. The FMEA allows identifying reasonably foreseen failures and predict its effects of the various parts of a system. This well-established technique is used to identify failure modes (defects or errors), mechanisms, and their effects, and can be used as a qualitative or quantitative technique that prioritizes failures according to severity, frequency, and detectability (ISO 2018). For instance, Sadhu et al. (2014) adopted FMEA in a health risk analysis associated with domestic rooftop water harvesting systems where reported failure effects can be the presence of fluoride, total dissolve solid (TDS), nitrate, and other chemicals present beyond their permissible limit in potable water. The severity (S) can be judged based on given parameter in a scale from 0 (no effect) to 10 (critical/hazardous). An occurrence (O) ranking from 1 (no known occurrences on similar products or processes) to 10 (very high, failure is almost inevitable) can be used to quantify how frequently a failure mode occurs. Detectability (D) refers to degree of difficulty in detecting failures, this parameter can also be rated from 1 (fault will be caught on test) to 10 (fault will be passed to beneficiaries undetected) (Sadhu et al. 2014). A risk priority number (RPN) should be calculated for the entire design or process considering the product of the three above parameters (i.e. $RPN = S \times O \times D$). The highest RPNs should get highest priority for corrective measures.

QMRA is a "formal, quantitative risk assessment approach that combines scientific knowledge about the presence and nature of pathogens, their potential fate and transport in the water cycle, the routes of exposure of humans and the health effects that may result from this exposure, as well as the effect of natural and engineered barriers and hygiene measures" (WHO 2016). QMRA typically follows the steps of: (i) problem formulation (or hazard assessment, Seidu et al. 2007), (ii) exposure assessment, (iii) health effects assessment, (iv) and risk characterization (WHO 2016). These authors recognized the potential of QMRA but emphasize the need for data inputs to capture the uncertainties and variability of the effects of waterborne pathogens in health, and the limitations in dose-response relationships characterization for a wide range of water-related pathogenic microorganisms. QMRA demands higher technical knowledge and resources when compared to other approaches. Therefore, application is limited by organizations' capacity and is not suitable for the small-scale individual RWHS. However, responsible authorities can use QMRA for general risk management of these solutions and to establish recommendations and good practices to users as well as control actions, as mentioned previously for WSP.

Customized risk-based approaches are found in the literature. DH (2013) proposes a methodology for non-drinking applications in multi-residential, commercial, and community facilities to develop a plan for rainwater supply management, integrating a RMP (Table 12.2).

Table 12.2 Key steps for developing a RWHS management plan.

Steps	Description
Step 1: Commitment to responsible use and management of rainwater	Organizational commitment to the application of preventative measures and nomination of a responsible for the rainwater supply system to ensure the responsible use and management of the water supply. This role is essential to ensure adequate maintenance, operation, and routine monitoring of the system to ensure rainwater of appropriate quality.
Step 2: Detailed description of the RWHS	A detailed description and map of the rainwater supply system is important to help understand how the system works.
Step 3: Identify hazards and risk assessment and treatment	Completion of a thorough risk assessment of the RWHS. A good understanding of the RWHS is required in order to identify all potential risks to the rainwater uses and possible sources of contamination. The identified risks need to be managed and adequately addressed by measures such as routine maintenance of the system and if required, appropriate treatment. Factors to consider when identifying hazards: possible sources of contamination and associated control measures; source water contamination sources; storage and risk of contamination inputs; treatment failure; pipework (cross-connection risks and biofilm growth).
Step 4: RWHS operation, monitoring and maintenance procedures documentation	Document standard operating procedures for RWHS, including procedures to follow for regular operation, monitoring, and maintenance. Tailor maintenance activities to suit the system. Monitoring activities planning to set up and record keeping procedures.
Step 5: Incident management plan	Plan how to respond if an event occurs, include contingency and emergency plans and who to notify.
Step 6: Review	Periodic (annual) review to ensure the management plan is effective and up to date.

Source: Adapted from DH (2013).

12.4 Supporting Risk Management for RWHS

As mentioned before, one of the first steps in any RMP is risk identification. The identification process is more comprehensive if including not only hazard identification but also risk sources, risk factors, and events. The resulting set of events identified and characterized must include all reasonable foreseen risks and ways leading to their materialization. For each event, evaluation of risk levels allows subsequent evaluation and decision on risk treatment needs (Almeida et al. 2013). During this process questions may arise, such as:

- What can go wrong for users if a RWHS is used as a potable source of water?
- How can the usage of a RWHS for irrigation or other non-potable uses endanger public health?
- Which safety issues can arise with recommended operation and maintenance of RWHS?
- Which negative impacts can result to the environment?

During the identification process, it is necessary to consider the type of system and its components, the foreseeable uses and associated requirements, the local climate and associated variability, and positive and negative consequences that can derive from this alternative source of water.

An important requirement for hazard identification and risk analysis is to detail the RWHS components. In general, a RWHS includes different component subsystems such as rainwater collection, storage, treatment, distribution, and uses as depicted in Figure 12.2.

The widespread use of a RWHS is hindered by uncertainty over quantity, quality, and perceptions of health risk (Evans et al. 2007). As emphasized by the WHO (2017), rainwater harvesting is widely used at a household level and increasingly also at community scale, not only as an additional source of drinking water but, in some circumstances, for mixing with other sources to reduce the levels of hazardous substances to health such as arsenic and fluoride.

Several countries in the world are encouraging the use of RWHS for domestic purposes as a potable and non-potable alternative source of water in regions where water is scarce (Ahmed et al. 2010). The use of a RWHS as an alternative WSS has many environmental, social, economic, and climate benefits and consequences.

Climate change may affect existing hazards, risk sources, and risk factors or may lead to new ones (Almeida et al. 2014). It is very important that the users recognize the main hazards and risk factors associated to a RWHS.

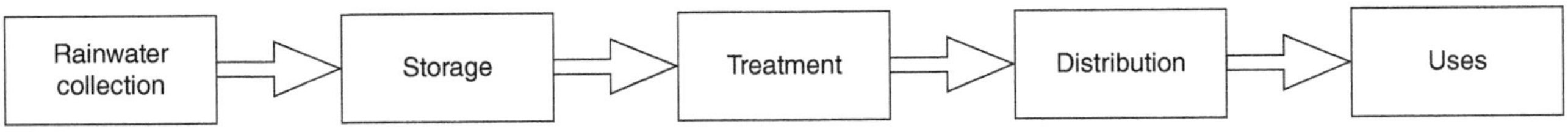

Figure 12.2 Component subsystems of a typical RWHS.

12.5 Hazards and Exposure Modes

Generally, rainwater is considered of good quality but can become contaminated if it absorbs airborne pollutants and contaminants from surfaces and component subsystems (WHO 2016). Rainwater can be collected in different ways and over a wide range of surfaces, such as from building roofs, paved surfaces, and green spaces in urban areas. The sources from which rainwater is collected determine its water quality characteristics as well as factors deteriorating water quality during the pathway until point of use. Contamination of rainwater reaching collection surfaces can occur due to various types of debris accumulated, leaching of RWHS component materials, poor maintenance in water storage, and poor hygiene at the point of use. A possible source of enteric pathogens includes fecal material deposited on roofs and other surfaces by birds, lizards, and small animals or from dead animals deposited on surfaces (Coombes 2006). Collected water is stored in tanks, cisterns, or barrels and, depending on water quality and intended uses, some type of treatment can be applied.

Exposure analysis should be undertaken to evaluate how a hazard might reach a vulnerable target and associated intensity or in what amount, or in broader terms, cause any type of consequence. It often involves a pathway analysis, which considers the different routes the hazard might take, the barriers that might prevent it from reaching the target and the factors that might influence the level of exposure (WHO 2016). For instance, in considering the risk of infection associated with the exposure to potential pathogens from roof-harvested rainwater used as potable or non-potable water the exposure analysis would consider the pathogen number in the source water (i.e. tank rainwater) and the amount of volume ingested or inhaled by a person (Ahmed et al. 2010). If a RWHS is designed for non-potable uses, exposure mechanisms should include inhalation of aerosolized rainwater (e.g. from sprinklers used for irrigation), dermal exposure, and unintentional ingestion (Struck 2011; WHO 2016). Exposure to hazards during operations and maintenance of RWHS, often of occupational nature, can be relevant and result in consequences such as drowning and severe injuries, for instance, from falls at height (Fewtrell et al. 2008). Table 12.3 presents examples of relevant exposure mechanisms and corresponding hazard type associated with use of RWHS.

Table 12.3 Exposure mechanisms and hazards for risks associated with public health, safety, and social.

Risk class	Exposure to hazard	Hazard type
Public health and safety	Consumer /user • Intentional ingestion • Unintentional ingestion (e.g. water intended for non-potable uses) • Aerosol inhalation • Skin contact • Insect bites	• Presence of microbial pathogens in water • Presence of chemical contaminants in water • Vector-borne pathogens (e.g. transmitted by mosquitoes breading in stored water)
	Public/occupational • Inhalation • Skin contact • Maintenance activities (e.g. during cleaning operation)	• Presence of microbial pathogens in water • Presence of chemical contaminants in water • Presence of toxic chemicals in the atmosphere of locations where public or workers might have access to height
Social	Low water availability and affordability	• Extended periods without supply

The number of studies focusing on risk to health and potable uses is significant. Conversely, studies looking at risks associated with non-potable uses are fewer. Non-potable uses include irrigation, fountains, toilet or urinal flushing, cleaning, washing, laundry, heating and cooling systems, industrial processes, and firefighting (DH 2013; Struck 2011). A relevant aspect is the possible cross-connection between potable and non-potable piping, which was reported in several studies, and the evidence of non-potable water allowing introduction of a range of pathogens in households and other facilities, which would normally not occur (Struck 2011). Lim et al. (2015) investigated the viral health risk associated with a generic urban stormwater harvesting solution and its application in three non-potable uses – namely toilet flushing, showering, and food-crop irrigation – using QMRA, concluding that food-crop irrigation presented the higher risk, followed by showering and toilet flushing. The two latter uses were found to be within WHO recommended disease burdens.

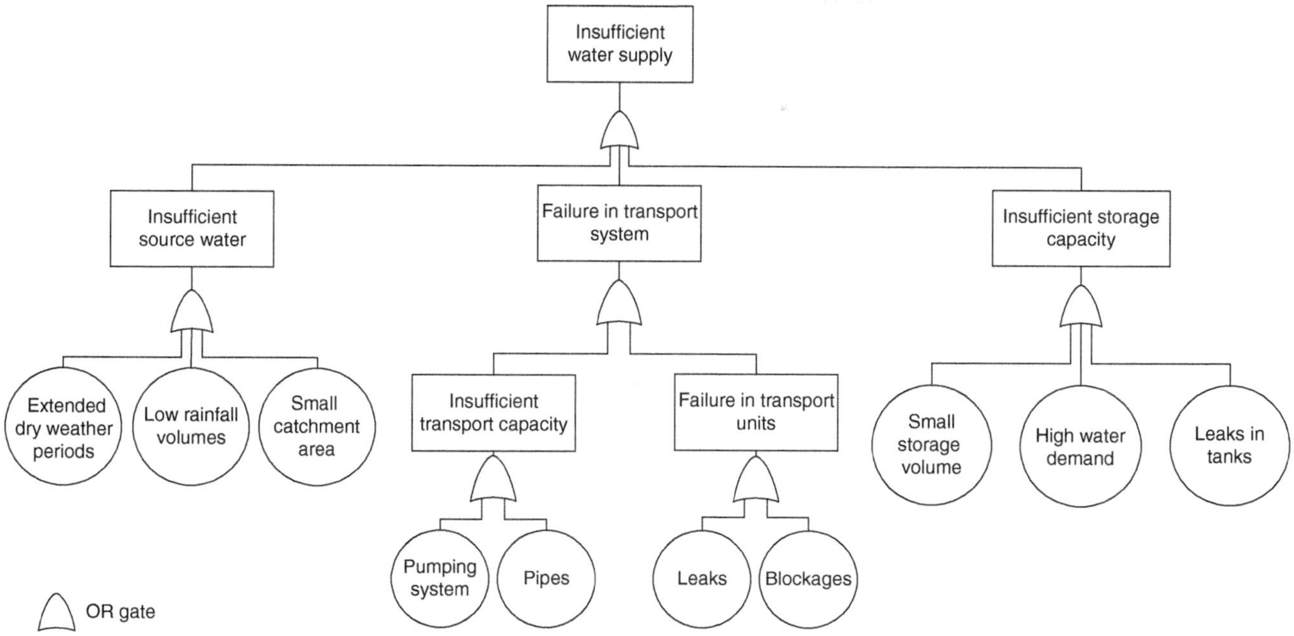

Figure 12.3 Example of qualitative fault tree for hazardous event insufficient water supply.

The identification of typical risk events can be supported by techniques such as the fault trees or event trees. A fault tree is a logical diagram showing the relation between system failures (i.e. a specific hazardous event) and failures of the components of the system. An event tree analysis is carried out by posing a number of questions where the answer is either "yes" or "no" (Aven 2008). In Figure 12.3 and Figure 12.4 there are two examples of simplified fault trees for the hazardous events of insufficient water supply and presence of microbial pathogens in water used for irrigation.

Definition of the typical events in each case requires consideration of the requirements for the different uses and parametric values for water quality. Therefore, risks linked to potable and non-potable uses must be defined in alignment with indicator parameter values to support acceptability to consumers or users and for operational monitoring. These parameters should meet water quality standards specified by public health or other regulatory authorities.

12.6 Rainwater Collection Reliability as Water Source

Reliability can be defined as "the ability of an item to perform a required function, under given environmental and operational conditions and for a stated period of time" (Rausand and Hoykand 2004). For a RWHS an essential function is to supply water for the demand. Assessment of RWHS performance as water source should include metrics allowing quantification of the reliability for this function (Ursino 2016). The satisfaction of a target demand is generally the basis for the design of any RWHS, from small rooftop tanks to large storage facilities, with assumptions that determine the underlying risk of failure even if not explicitly calculated. Variability on both demand and meteorological conditions adds uncertainty to the estimations of the reliability as water source. Climate dynamics is also a source of uncertainty adding to current situation uncertainty levels.

In general, the reliability of a RWHS depends on three main issues: the precipitation, the characteristics of the system, and the water demand. The precipitation is determined by local climate and associated variability, both influencing rainfall depth, intensity, and seasonality, and factors such as temperature and wind. The characteristics of the system influencing supply reliability include catchment area, storage tank volume, and water losses in the system due to processes such as evaporation, leakage, and overflows. Relevant water demand characteristics include average demand flow, demand patterns in time (daily patterns and seasonality), and changes in uses. All these issues are interrelated, with rainfall regime strongly affecting the overall functioning of the system (Hanson and Vogel 2014), but the effect of precipitation on systems' performance depends on storage volumes and water demand.

In climate research, scenarios of plausible descriptions of how the future may evolve with respect to a range of variables including socioeconomic change, technological

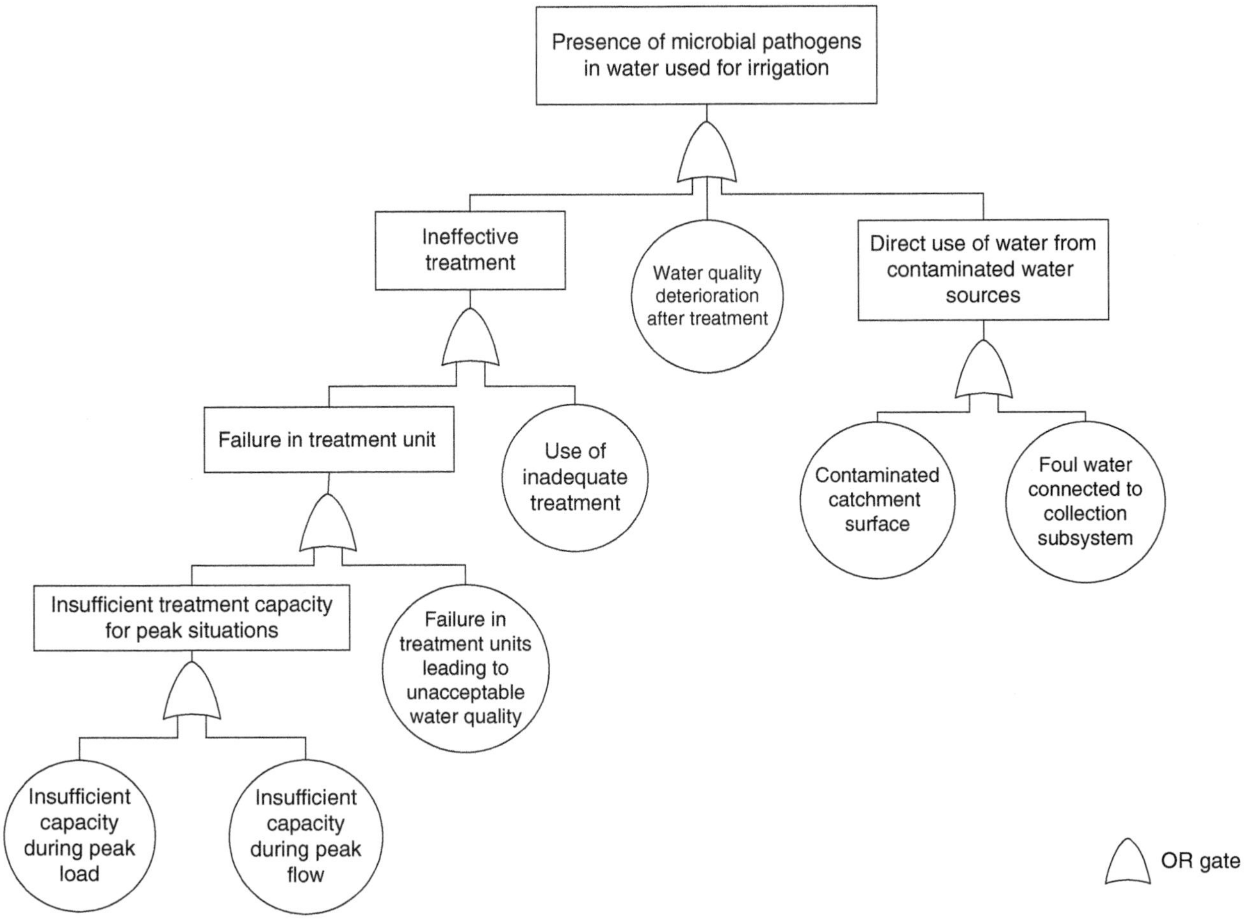

Figure 12.4 Example of qualitative fault tree for hazardous event presence of microbial pathogens in water.

change, energy and land use, and emissions of GHG and air pollutants are used (van Vuuren et al. 2011). The representative concentration pathways (RCPs) were developed with the purpose of providing information on possible development trajectories of greenhouse gas concentration and emissions pathways designed to support research on impacts and potential policy responses to climate change (Rihai et al. 2011). There are four RCPs described in literature: one mitigation scenario leading to a very low forcing level (RCP2.6), two medium stabilization scenarios (RCP4.5/RCP6), and one very high baseline emission scenario (RCP8.5) (van Vuuren et al. 2011).

Estimated effects of climate changes (considering the RCP8.5 scenario) in precipitation are not uniform (IPCC 2014): high latitudes and the equatorial Pacific are likely to experience an increase in annual mean precipitation; in many mid-latitude and subtropical dry regions, mean precipitation is likely to decrease; in many mid-latitude wet regions, mean precipitation is likely to increase. As emphasized by IPCC (2014), "Climate change will amplify existing risks and create new risks for natural and human systems. Risks are unevenly distributed and are generally greater for disadvantaged people and communities in countries at all levels of development."

Climate dynamics is expected to reduce performance of RWHS, especially in the regions where estimated precipitation trends are for reduction in yearly rainfall and in aggravation of dry season conditions, including increasing number of days without precipitation (Haque et al. 2016; Kahinda et al. 2010). Climate change will produce beneficial impacts in some regions and harmful ones in others. Within the broader management of water, adaptation measures such as increasing the use of RWHS in regions of water scarcity can play a role in improving health and resilience as alternative water source to reduce stress on traditional supply practices. In regions where rainfall increases, the use of RWHS can be an adaption measure for flood mitigation in urban areas.

McMahon et al. (2006) listed usual system reliability metrics considering the case of reservoirs in rivers, namely time-based reliability (proportion of intervals during a pre-defined period that the system can meet the target demand) and volumetric reliability (volume of water demand supplied divided by the total target demand during a pre-defined period). The volumetric reliability is less restrictive than time based because it takes into account even the time step where the demand is partially met. For a RWHS, the time step (hour, day, month, or year) for analysis or simulations used in the water balance strongly influences the results' accuracy. Calculation of reliability can be carried out using measured data (often not available) or from water balance simulations.

Measures of RWHS reliability are often used but not always coherent in terms of terminology and concept. Haque et al. (2016), in a study, used the time-based reliability as described above but names the volumetric reliability as "water savings." Assuming stable water demand, results show a positive impact of climate change scenarios on time-based reliability for wet months and strong negative effect for dry months. Male and Kennedy (2006) also used time-based reliability to evaluate the impact of storage volume for a fixed catchment area, with time step of one week, concluding that daily analysis could provide more realistic results. Kisakye et al. (2018) adopted time-based reliability and concludes that effects of climate change scenarios are more significant for dry seasons. Palla et al. (2011) analyzed the performance of RWHS for residential use in Italy based on comparison of total daily rainfall and total daily demand. These authors compare available rainwater with demand directly independent of storage volume (named water-saving efficiency by the authors) for rainfall data series from 33 to 167 years in three locations (Genoa, Florence, and Catania) with three different precipitation and water demand regimes, obtaining yearly water-saving efficiency values from 0.4 to 2.34. Differences found between cities are not significant and the most relevant parameter is the variation in water demand level.

Mehrabadi et al. (2013) adopted the volumetric reliability as the basis to study efficiency and effectiveness of RWHS for non-potable uses in Iran and to analyze the effect of rainfall regime, storage volume, roof area, and water demand. Ghisi et al. (2009) evaluated the efficiency of a RWHS for washing vehicles in Brasilia (Brazil) based on 30 years of daily rain observed data. Using volumetric reliability, both authors concluded for different rainfall regimes, the trade-off between catchment area, water demand, and tank capacity has significant impact on reliability, with results varying from values as low as 7.4–73% and are all important design variables.

In summary, rainwater collection reliability is a key aspect in risk analysis and management. Reliability metrics can add value to time-based analysis, in terms of water balance and systems' efficiency and effectiveness being complementary. Uncertainties need to be taken into account by either using probabilistic approaches or considering typical scenarios. RWHS limitations in terms of reliability as water source can be reduced if combined with other alternative sources such as greywater reuse (Li et al. 2010; Zavala et al. 2016).

12.7 Specific Risk Treatment Actions

Risk treatment is the process and the implementation of measures or actions to modify risk, and includes the identification of control measures for all significant risks in order to prevent or reduce the occurrence or to minimize the consequences of hazardous events (Aven 2008; Almeida et al. 2014). These measures can act on risks in different ways (Almeida et al. 2011):

- reducing the likelihood, by removing the risk source, acting on relevant risk factors or causes;
- reducing the consequences, considering all potential dimensions of the consequence; and
- avoiding the risk by deciding not to use a RWHS that originates the risk.

Retaining the risk by informed decision is a possible course of action, but it can hardly be considered as a risk reduction measure (Almeida et al. 2014).

Frequently, types of measures are associated with risk factors, often divided into human, environment, and equipment. Different types of risk reduction measures (RRM) can be considered. Almeida et al. (2011) resumed types of measures reported in the literature as follows:

- Barriers – any physical impediment or containment method that tends to confine or restrict a potentially damaging condition, reducing the probability of events, or containment of event after its occurrence, thus reducing consequences.
- Redundancy – Additional, identical, and redundant components in a system introduced to decrease the likelihood of failure of subsystems.
- Increase components or systems reliability – substitution of critical elements by more reliable ones, structural modifications of the systems, or changes to the safety systems logics.
- Increase components or systems effectiveness – substitution or improvement of system elements with more efficient ones, including upgrading of technology.
- Maintenance – adequate preventive or corrective maintenance activities can reduce failure rates and consequently the likelihood of events.
- Control systems – detection of failure states, existence of unsafe conditions, by means of monitoring, testing, or inspection, and actions to change the state of systems.

- Avoidance of a risk – measures that involve deciding not to start or continue with the activity that gives rise to the risk, including not initiating or discontinuing an activity (e.g. rainwater use for potable use) or a technical process (not using a specific technical process).
- Adaptation of user and public behavior – changes in behavior of system users or public in general allowing the risk reduction by decreasing the probability or the consequence of an event.

For some risks, multiple measures can be identified and used individually or in combination (e.g. "multiple barriers") to accomplish more effective risk reduction. For each measure, appropriate actions should also be described since they are relevant in terms not only of implementation effort but also for effectiveness and efficiency of implementation.

Assessment of alternative RRM should be carried out using appropriate criteria and metrics to balance costs and risks of implementation against expected benefits. Aspects to consider in the assessment of each RRM include the level of risk to be controlled; effectiveness (achievement of the desired reduction in risk); efficiency (achievement of the desired effect with the least resource consumption); sustainability (effectiveness under various future scenarios); cost of implementation; side effects (e.g. some RRM may create secondary risks); and legal and regulatory viability, if existing.

Several engineering tools are available, not specific to risk, for mathematical modeling, failure analysis, or to support multiple criteria decision making, which allow detailed analysis of the potential effects of the measures. After assessment, RRM alternatives should be prioritized using the selected criteria and a decision made on which RRM to implement.

An important issue to take into consideration, when selecting the RRM, is the level of monitoring and control on aspects influencing risk. For instance, while controlling access of mosquitoes to stored water is feasible, controlling climate variables, such as temperature, is out of reach. Monitoring the RWHS is an essential part of risk treatment because it shows whether the risk control measures are working properly or not (DH 2013).

For households or small-scale applications, the outcome of a risk assessment by responsible authorities can result in good practices guidance. Examples provided by the WHO (2017), applicable to both potable and to non-potable applications, include:

- guidance for well-designed RWHS with catchments less prone to accumulation of debris: covered storage tanks, adequate treatment, placement of wire meshes or inlet filters placed over the top of downpipes to prevent leaves and other debris from entering storage tanks, and careful selection of materials that do not leach contaminants; a system to divert the contaminated first flow of rainwater from roof surfaces is necessary and automatic devices that prevent the first flush of runoff from being collected in storage are recommended, and a mechanism to empty the storage to facilitate cleaning;
- cleaning and maintenance guidance to ensure good hygiene at point of use; regular cleaning of storage tanks to reduce deterioration of water quality; regular cleaning of catchment surfaces and gutters should be undertaken to minimize the accumulation of debris;

Opportunities are often identified when analyzing or designing a RWHS and can be considered within risk analysis. A RWHS can be used as a stormwater control measure (Steffen et al. 2013). Collecting and using rainwater where it falls reduces discharges to the sewer system since less rainwater enters to the network during a precipitation event (Melville-Shreeve et al. 2016).

12.8 Process Control and Monitoring

A key outcome of risk management is the identification of control measures and monitoring to detect critical situations and alert for the need to act to avoid emergence of risk events. Therefore, procedures should be set up in each case to guide periodic assessment of specific risk factors and risk sources, as well as be the basis for establishing a control and monitoring plan. The controls can be selected to prevent or eliminate a hazard or reduce it to an acceptable level, to prevent a hazardous event, or, should this event be already on going, to reduce the magnitude of consequences.

Critical limits for variables to be controlled and the corresponding monitoring program should also be established to detect loss of control in time to make adjustments through corrective actions (Almeida et al. 2014).

It is important to have control measures to ensure rainwater quality and eliminate contamination pathways to safeguard public health (Gwenzi et al. 2015). A higher water quality requirement for use needs more stringent control measures. If the risk of contamination causing harm to rainwater users is deemed to be a significant risk, after risk assessment, then the quality of the rainwater should be tested (DH 2013). Combining sanitary inspection with laboratory analytical results enables prioritization of mitigation/control measures (Gwenzi et al. 2015). A sanitary inspection consists of an on-site examination and evaluation by qualified professionals that typically make use of standardized "sanitary inspection forms" containing a systematic checklist of a limited number of specific questions on actual or potential danger to the health and well-being of the user (WHO 2016). Control measures for RWHS may include (DH 2013):

- roof catchment protection and maintenance; using correctly designed and maintained roof catchments is a key step to protecting rainwater from contamination;
- correct material selection and installation of the rainwater storage, distribution, and plumbing;
- treatment, such as filtration and disinfection, where deemed necessary;
- regular inspection and maintenance of the supply system (pipes, tanks, pumps, and other elements).

Monitoring the RWHS is an essential part of a multiple barrier approach. The results of monitoring shows whether the risk control measures are working properly (DH 2013). For example:

- if the RWHS has a filtration system it should be monitored and maintained accordingly with the manufacturer's advice;
- the chlorine injection units should be checked to ensure they are fully operational and have an adequate supply of chlorine.

12.9 Conclusion

Benefits of using RWHS can be important but negative risks need to be dealt with. To ensure quality, safety, and return on investments, combined action between regulatory authorities, householders, water utilities, managers, and users is essential. Looking at the water cycle in a comprehensive way reveals the potential of RWHS to reduce several potential risks, including insufficient water supply, but also, for instance, those related to flooding. Consideration of risks associated with RWHS and adoption of appropriate measures allows us to ensure acceptable risks associated with these types of solutions.

A generic and systematic approach to manage risk is required. The approach should include identification of risks and opportunities related to alternative actions and incorporate uncertainties. Given the effects of climate dynamics together with socioeconomic changes, RWHS as an adaptation measure should address all system components, reliability as water source, the interactions of people with RWHS components, and the types of water uses.

In the vast literature on risk analysis different concepts and terms are found depending on the type of application and the context. Therefore, in Section 12.2 the concepts and terminology on risk assessment and management used in this chapter are presented.

Different approaches on how to determine risks such as WSP, RMP, and HACCP can be used, but all of them have some common steps. The first step in any risk management or risk analysis process is risk identification. This step consists of finding, recognizing, and describing risks and involves the identification of risk sources, events, their causes and their potential consequences. Several techniques exist for carrying out such an identification process such as FMEA or QMRA.

After risk identification, risks should be analyzed and evaluated. Risk analysis is the process of comprehending the nature of risk and determining the level of risk, while evaluation is the process in which the results of risk analysis against risk criteria are compared to determine whether the level of risk is acceptable or tolerable.

The following step is risk treatment, which is the process and the implementation of measures or actions to modify risk and includes the identification of control measures for all significant risks to prevent or reduce the occurrence or to minimize the consequences of hazardous events.

Finally, control measures and monitoring to detect critical situations and alert for the need to act to avoid emergence of risk events should be carried out.

References

Ahmed, W., Vieritz, A., Goonetilleke, A., and Gardner, T. (2010). Health risk from the use of roof-harvested rainwater in Southeast Queensland, Australia, as potable or non-potable water, determined using quantitative microbial risk assessment. *Applied and Environmental Microbiology* 76 (22): 7382–7391.

Almeida, M.C., Strehl C., Leitão, J.P. and Mälzer, H. (2011) Risk reduction measures. Supporting document for RRDB structure. Deliverable D 2.4.1. PREPARED. Report number 2011.025. PREPARED project. 38 p.

Almeida, M.C., Vieira, P. and Smeets, P. (2013) Water cycle safety plan framework. Deliverable D 2.1.4. PREPARED. Report number 2013.025. PREPARED project. 103 p.

Almeida, M.C., Vieira, P., and Smeets, P. (2014). Extending the water safety plan concept to the urban water cycle. *Water Policy* 16 (2): 298–322. https://doi.org/10.2166/wp.2013.137.

Aven, T. (2008). *Risk Analysis: Assessing Uncertainties Beyond Expected Values and Probabilities*. Wiley. ISBN: 978-0-470-51736-9.

Bartram, J., Corrales, L., Davison, A. et al. (2009). *Water Safety Plan Manual: Step-by-Step Risk Management for Drinking-Water Suppliers*. Geneva/London: WHO/IWA.

Blanc, D. (2006) ISO 22000. From intent to implementation. *ISO Management Systems*. 7-11.

Campling, P., De Nocker, L., Schiettecatte, W., et al. (2008) *Assessment of alternative water supply options. Final summary report* (extended version). EU – DG-Environment.

Codex (2003) *Hazard and critical control point system and guidelines for its application. Annex to the recommended international code of practice – General principle of food hygiene*. Codex Alimentarius Commission.

Coombes, P. (2006). *Guidance on the Use of Rainwater Harvesting Systems*. Callaghan, Australia: University of Newcastle.

Davison, A., Howard, G., Stevens, M. et al. (2005). *Water Safety Plans: Managing Drinking-Water Quality from Catchment to Consumer*. Geneva, Switzerland: World Health Organization.

DH (2013) *Rainwater use in urban communities. Guidelines for non-drinking applications in multi-residential, commercial and community facilities*. Department of Health, State of Victoria: Melbourne. www.health.vic.gov.au/water/alternative/rainwater>.

Evans, C.A., Coombes, P.J., Dunstan, H. et al. (2007). *Roof Harvested rainwater – indicator organisms, water quality and risk assessment*. Environmental Science.

Fewtrell, L., Kay, D., and McDonald, A. (2008). Rainwater harvesting – an HIA of rainwater harvesting in the UK. In: *Health Impact Assessment for Sustainable Water Management* (eds. L. Fewtrell and D. Kay), 45–68. London, UK: IWA Publishing. ISBN: 9781843391333.

Ghisi, E., Tavares, D., and Rocha, V.L. (2009). Rainwater harvesting in petrol stations in Brasília: potential for potable water savings and investment feasibility analysis. *Resources, Conservation and Recycling* 54 (2): 79–85. https://www.sciencedirect.com/science/article/pii/S0921344909001360.

Gwenzi, W., Dunjana, N., Pisa, C. et al. (2015). Water quality and public health risks associated with roof rainwater harvesting systems for potable supply: review and perspectives. *Sustainability of Water Quality and Ecology* 6 (2015): 107–118. https://doi.org/10.1016/j.swaqe.2015.01.006.

Hanson, L.S. and Vogel, R.M. (2014). Generalized storage–reliability–yield relationships for rainwater harvesting systems. *Environmental Research Letters* 9 (7): 075007. http://stacks.iop.org/1748-9326/9/i=7/a=075007?key=crossref.aa58bfd970e76eb73c220c4273d6d8e5.

Haque, M., Rahman, A., and Samali, B. (2016). Evaluation of climate change impacts on rainwater harvesting. *Journal of Cleaner Production* 137 (2016): 60–69. https://doi.org/10.1016/j.jclepro.2016.07.038.

Havelaar, A.H. (1994). Application of HACCP to drinking water supply. *Food Control* 5 (3): 145–152.

IPCC (2014). *Climate Change 2014: Synthesis Report. Contribution of Working Groups I, II and III to the Fifth Assessment Report of the Intergovernmental Panel on Climate Change* (eds. Core Writing Team, R.K. Pachauri and L.A. Meyer), 151. Geneva, Switzerland: IPCC.

ISO (2005) ISO 22000:2005 Food safety management systems. Requirements for any organization in the food chain. International Standards Organization.

ISO (2018) *ISO/IEC 31010:2018 Risk management. Risk assessment techniques*. International Standards Organization.

Kahinda, J., Taigbenu, A.E., and Boroto, R.J. (2010). Domestic rainwater harvesting as an adaptation measure to climate change in South Africa. *Physics and Chemistry of the Earth* 35 (13–14): 742–751. https://doi.org/10.1016/j.pce.2010.07.004.

Kisakye, V., Akurut, M., and Van der Bruggen, V. (2018). Effect of climate change on reliability of rainwaterharvesting systems for Kabarole District, Uganda. *Water* 10 (1): 1–34. https://doi.org/10.3390/w10010071.

Li, Z., Boyle, F., and Reynolds, A. (2010). Rainwater harvesting and greywater treatment systems for domestic application in Ireland. *Desalination* 260 (1–3): 1–8. https://doi.org/10.1016/j.desal.2010.05.035.

Lim, K.-Y., Hamilton, A., and Jiang, S. (2015). Assessment of public health risk associated with viral contamination in harvested urban stormwater for domestic applications. *Science of the Total Environment* 523 (2015): 95–108. https://doi.org/10.1016/j.scitotenv.2015.03.077.

Lye, D. (2002). Health risks associated with consumption of untreated water from household roof catchment systems. *Journal of the American Water Resources Association* 38 (5): 1301–1306.

Male, J.W. and Kennedy, M.S. (2006). Reliability of rainwater harvesting. *Eco-Architecture: Harmonisation between Architecture and Nature* 86: 391–400. https://doi.org/10.2495/ARC060391.

McMahon, T.A., Adeloye, A.J., and Zhou, S.-L. (2006). Understanding performance measures of reservoirs. *Journal of Hydrology* 324 (1–4): 359–382. https://doi.org/10.1016/j.jhydrol.2005.09.030.

Mehrabadi, M.H., Saghafian, B., and Fashi, F. (2013). Assessment of residential rainwater harvesting efficiency for meeting non-potable water demands in three climate conditions. *Resources, Conservation and Recycling* 73: 86–93. https://doi.org/10.1016/j.resconrec.2013.01.015.

Melville-Shreeve, P., Ward, S., and Butler, D. (2016). Rainwater harvesting typologies for UK houses: a multi criteria analysis of system configurations. *Water* 8 (4): 129. https://doi.org/10.3390/w8040129.

Muller, M. (2007). Adapting to climate change: water management for urbanresilience. *Environment and Urbanization* 19 (1): 99–113.

NHMRC (2004). *Australian Drinking Water Guidelines.* Australian Government: National Health and Medical Research Council.

Palla, A., Gnecco, I., and Lanza, L.G. (2011). Non-dimensional design parameters and performance assessment of rainwater harvesting systems. *Journal of Hydrology* 401 (1–2): 65–76.

Rausand, M. and Hoykand, A. (2004). *System Reliability Theory. Models, Statistical Methods and Applications.* USA: Wiley. ISBN: 0-471-47133-X.

Rihai, K., Rao, S., Krey, V. et al. (2011). RCP 8.5—a scenario of comparatively high greenhouse gas emissions. *Climatic Change* 109: 33–57. https://doi.org/10.1007/s10584-011-0149-y.

Sadhu, G., Chaudhary, M., and Chakravarty, N. (2014). Health risk assessment of domestic roof water harvesting using failure mode and effects analysis: evidence from Nagaur, Rajasthan, India. *Human and Ecological Risk Assessment: An International Journal* 20 (3): 668–688. https://doi.org/10.1080/10807039.2013.810941.

Seidu, A., Heistad, A., Lindholm, O. et al. (2007). Integrating quantitative microbial risk assessment into health risk management of water supply systems in Norway. *Vann* 4 (2007): 329–336. https://vannforeningen.no/wp-content/uploads/2015/06/2007_364874.pdf>.

Steffen, J., Jensen, M., Pomeroy, C.A., and Burian, S.J. (2013). Water supply and stormwater management benefits of residential rainwater harvesting in U.S. cities. *Journal of the American Water Resources Association* 49 (4): 810–824. https://doi.org/10.1111/jawr.12038.

Struck, S. (2011). *Rainwater Harvesting for Non-potable Use and Evidence of Risk Posed to Human Health*, 17. British Columbia Centre for Disease Control. http://www.ccnse.ca/sites/default/files/BCCDC-Rainwater_Harvesting_Oct_2011.pdf.

Ursino, N. (2016). Risk analysis approach to rainwater harvesting systems. *Water* 8 (8): 7. https://doi.org/10.3390/w8080337.

van Vuuren, D.P., Edmonds, J., Kainuma, M. et al. (2011). The representative concentration pathways: an overview. *Climatic Change* 109: 5–31. https://doi.org/10.1007/s10584-011-0148-z.

WHO (2005). *Water Safety Plans. Managing Drinking-Water Quality from Catchment to Consumer*. Geneva, Switzerland: World Health Organization.

WHO (2009). *Water Safety Plan Manual. Step-by-Step Risk Management for Drinking-Water Suppliers.* Geneva, Switzerland: World Health Organization.

WHO (2016). *Quantitative Microbial Risk Assessment: Application for Water Safety Management*. Geneva, Switzerland: World Health Organization.

WHO (2017). *Guidelines for Drinking-Water Quality*. Geneva, Switzerland: World Health Organization.

Zavala, M., Veja, R., and Miranda, R. (2016). Potential of rainwater harvesting and greywater reuse for water consumption reduction and wastewater minimization. *Water* 8 (6): 18. https://doi.org/10.3390/w8060264.

Part D

Hydrological Aspects of Water Harvesting

13

Return Period Determination for Rainwater Harvesting System Design

Sandeep Samantaray[1], Dillip K. Ghose[1], and Saeid Eslamian[2]

[1]*Department of Civil Engineering, NIT, Silchar, 788010, Assam, India*
[2]*Department of Water Engineering, College of Agriculture, Isfahan University of Technology, Isfahan, Iran*

13.1 Introduction

Rainwater harvesting (RWH) is the accession and accumulation of rainwater from reprocess to runoff. Rainwater is harvested from reservoirs, rivers, or roofs and transported to a deep pit. Uses for rainwater include water for livestock, irrigation, gardens, and domestic purposes. Groundwater recharge is another process for harvesting water. In urban areas, for water harvesting, combined sanitary and stormwater transfer provisions must be designed to include a system for stormwater detention. The systems which collect rainwater and conserves safely are assimilated by the existing system. Unfortunately, many conservation systems require rainwater capture and detention do not allow for harvesting. RWH is an exercise of detention and retention of rainwater to meet up prospective requirements. Various types of RWH system are presented in Figure 13.1.

RWH consigns to compilation with store room of rainfall, which aims for yielding surface and groundwater, by preventing the evaporation and seepage loss in a watershed (Agarwal and Narain 1997). RWH is an efficient scheme for budding sustainable water resources in metropolitan landscapes. RWH mainly comprises assortment and staging confined to rain in watersheds. Water resources are limited, variable, and required for utilization (Choudhary and Aneja 1991). The functioning of the RWH structures and the design prospective inserts a new layer of difficulty to capital projects, leading to complex infrastructures (Nam and Tatum 2006). The RWH structures are included to improve intensive, iterative, and complex water harvesting structures (Kobet et al. 1999; Reed and Gordon 2000; Kashyap et al. 2003; Magent et al. 2005; Horman et al. 2006). RWH is a prime method to trim down the problem of water scarcity by providing an alternative foundation of water conservation in a normal environment (Oweis et al. 1999; TRHEC 2006). Harvesting rainwater through different technology is the single optimal solution for growing demand in the environment (Athavale 2003; White 2007; Ibraimo and Munguambe 2007; Grady and Younos 2008; Jothiprakash and Sathe 2009; Partzsch 2009). There is an important untapped potential to produce additional water supplies in Texas through RWH for urban and suburban areas (Krishna 2003; TRHEC 2006). In East Asia, the United States, and Australia, rainwater conservation is an ample method for water harvesting (NCLS 2012; White 2007, 2010; NRDC and LIDC 2011; Han and Park 2007; Gaston 2010). RWH helps in many ways to conserve water (Levarios 2007; Gaston 2010). RWH is a complex process (Chatfield and Coombes 2007; Ibraimo and Munguambe 2007; Jothiprakash and Sathe 2009; Angrill et al. 2012; City of Bellingham 2012; Mun and Han 2012). RWH is more effective than other centralized systems (Partzsch 2009; City of Tucson 2009; City of Bellingham 2012). They are segmented into two types of design practices: passive and active mode of conservation systems. Passive methods of RWH are designed to direct water to an assigned area without storing it in an impermanent suppression system. Passive systems do not need piping or metering in terms of water supply to support them (City of Tucson 2009; Gaston 2010; City of Bellingham 2012). Design of RWH structures are an elemental component of complex water resource methodology which includes the proficiency largely united with the design (WBDG 2012; Ku and Mills 2010). RWH appears as a practical selection for water protection (Mun and Han 2012). RWH systems are exceptional as the technology offers dual benefits, water supply, and stormwater organization (Guo and Baetz 2007; Zhang et al. 2009; Fewkes and Warm 2000; Basinger et al. 2010; Ahmed et al. 2011). There is a growing attention in RWH to allay the amount and use of stormwater runoff, from local to global range (Burns et al. 2015; Kim and Furumai 2012). The

Handbook of Water Harvesting and Conservation: Basic Concepts and Fundamentals, First Edition. Edited by Saeid Eslamian and Faezeh Eslamian.

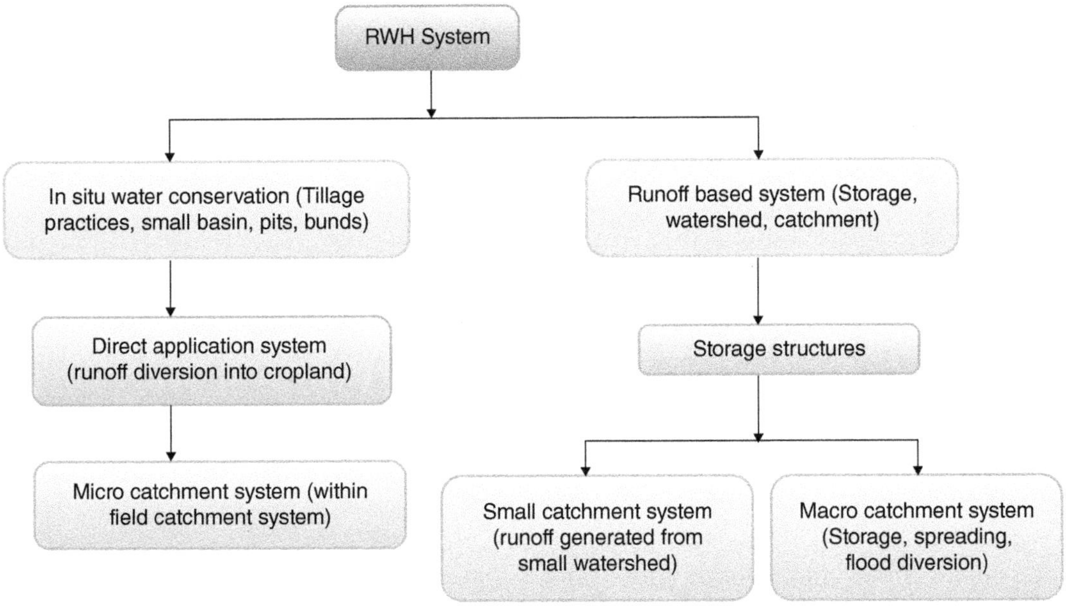

Figure 13.1 Types of RWH system.

RWH methods escalate additional water demands to the system with volume and frequency analysis. The optimum requirement for plant growth and automation with human intervention depends upon water capacity for irrigation (De Busk 2013; De Busk et al. 2013; Jones and Hunt 2010). Reduction of water availability for future use can jeopardize water potential to utilize in a system. The RWH system involves the methodology for enhancing the benefits of water management. Preservation of the water conservation system includes passive and active mechanisms. The passive mechanism includes dual purposes for dividing a storage tank into two branches: detention and retention storage volume (Brodie 2008; Reidy 2010; Herrmann and Schmida 1999). They described various hydrological effects on environment toward soil contamination, river low-flow, urban water quality, and desertification (Rahman et al. 2017). Implementation of indigenous knowledge by farmers is discussed to recognize potential sites for RWH (Okhravi et al. 2014). Rapid urbanization and industrialization have direct impact on the rainfall of any region due to climatic and topographic changes. A sustainable and eco-friendly cost effective method is necessary for measuring water harvesting and water use efficiency. At primitive sites of mining areas RWH is an important activity to conserve rainwater. Non-potable purposes of water usage through water harvesting are utilized in the form of:

(i) gardening;
(ii) dust suppression through sprinkling;
(iii) toilet flushing;
(iv) independence from groundwater;
(v) measures of mitigating adverse effect of long-term draw down of groundwater like change in groundwater flow pattern, land subsidence, and permanent lowering of water table.

The objectives of the study include (i) interpretation of return period with frequency analysis; (ii) design of specific site for ecological sustainability of RWH; (iii) to predict runoff over short-term and long-term return periods; and (iv) to compare various methods for designing return period.

The objective of the study is to evaluate runoff using various statistical methods for long term return period.

13.2 Study Area

The area under present study is a circular area of 10 km radius, keeping Ramchandrapur village in the centre as a buffer zone established in the district of Ganjam, Odisha, India. The study area falls under Survey of India Topo sheet No.74 A/15 and 74 A/16 and 74 E/3. It is bounded by 19°13′48″ to 19°24′50″ N latitudes and 84°50′56″ to 85°02′36″ E longitudes. The site is under the Chatrapur Subdivision and hardly 2.5 km from coast. The site near the village Ramchandrapur is accessible by a narrow road of 2 km from NH 5, i.e. the Chennai–Kolkata national highway. The site is a predominantly undulated land with gentle slopes. The area comes under Seismic Zone-III as per IS 1983. The meteorological data, rainfall, and temperature are used to evaluate the return period from 1993 to 2016 to accesses the capacity of RWH.

The study area is characterized by a tropical monsoon climate having three district seasons in a year: winter, summer, and rainy seasons. Winter commences from the month of November and continues up to February. Mid-December to Mid-January is coldest period of the year when the temperature drops down to around 14.70 °C (mean daily minimum temperature). The maximum temperature (mean) during the winter season varies from 14.70 to 24.57 °C. The basis of the rainfall depends upon the southwest monsoon, which contributes 75–80% of the total annual rainfall. The normal rainfall in this area is 1029 mm.

Water level is an index of water harvesting below ground called groundwater harvesting. Evaluation of the status of the water level in the buffer zone due to availability of water harvested annually with moderately intensive well inventory has been carried out during the pre-monsoon period. The sites of groundwater observation wells are depicted in Figure 13.2. Ramachandrapur (star mark) is considered the study area. The wells of GWS&I, Govt. of Orissa and CGWB, Govt. of India have been taken into account in the buffer zone before preparing depth to water level and fluctuation maps.

13.2.1 Water Level Fluctuation

Seasonal fluctuation of water level occurs between pre monsoon and post monsoon. The perusal of the data reveals that water level fluctuation is in the order of 4.5–5.0 m around the projected study area. The maximum and minimum fluctuation during the pre- and post-monsoon period is observed between 5.36 and 2.02 m, respectively. Water level fluctuation is shown in Figure 13.3.

The study area is situated in Rishkulya River basin. The main course of the Rushikulya River surges from west to east, and is situated in the north of the buffer zone (7.2 km) and then takes a turn toward the south and thus having flow from north–west to south–east in the buffer zone. Rainwater transports and intermingled with the Bay of Bengal. The topography of the Rushikulya River basin is undulating in nature having low to moderate gradient and dense vegetation. The other small river or nala which runs in the area is Khari nala, which flows from north–west to south–east of the catchment. The area is a canal command area having a network of canals originating from Rushikulya River. The drainage pattern is mostly dendritic. Besides these, there

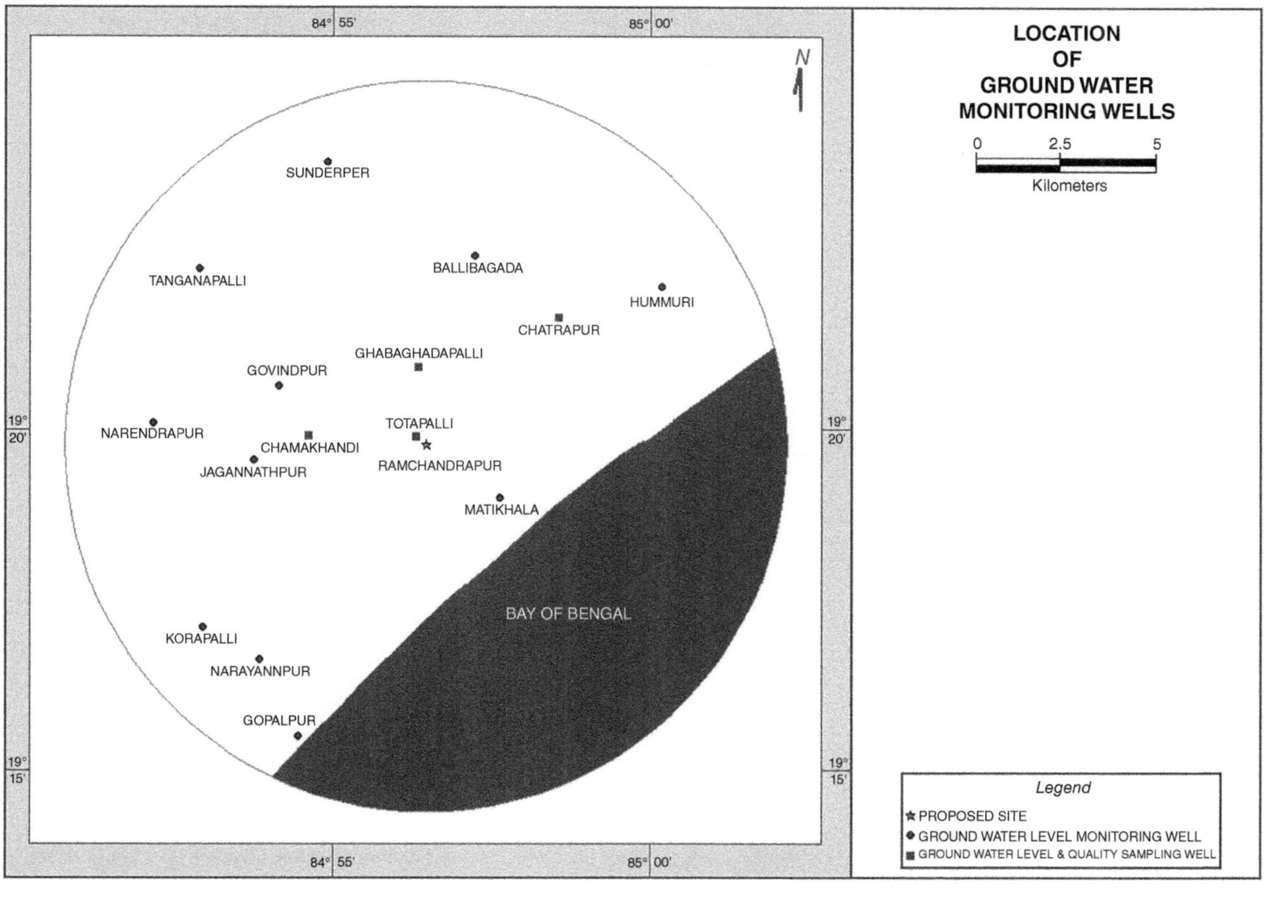

Figure 13.2 Groundwater monitoring well.

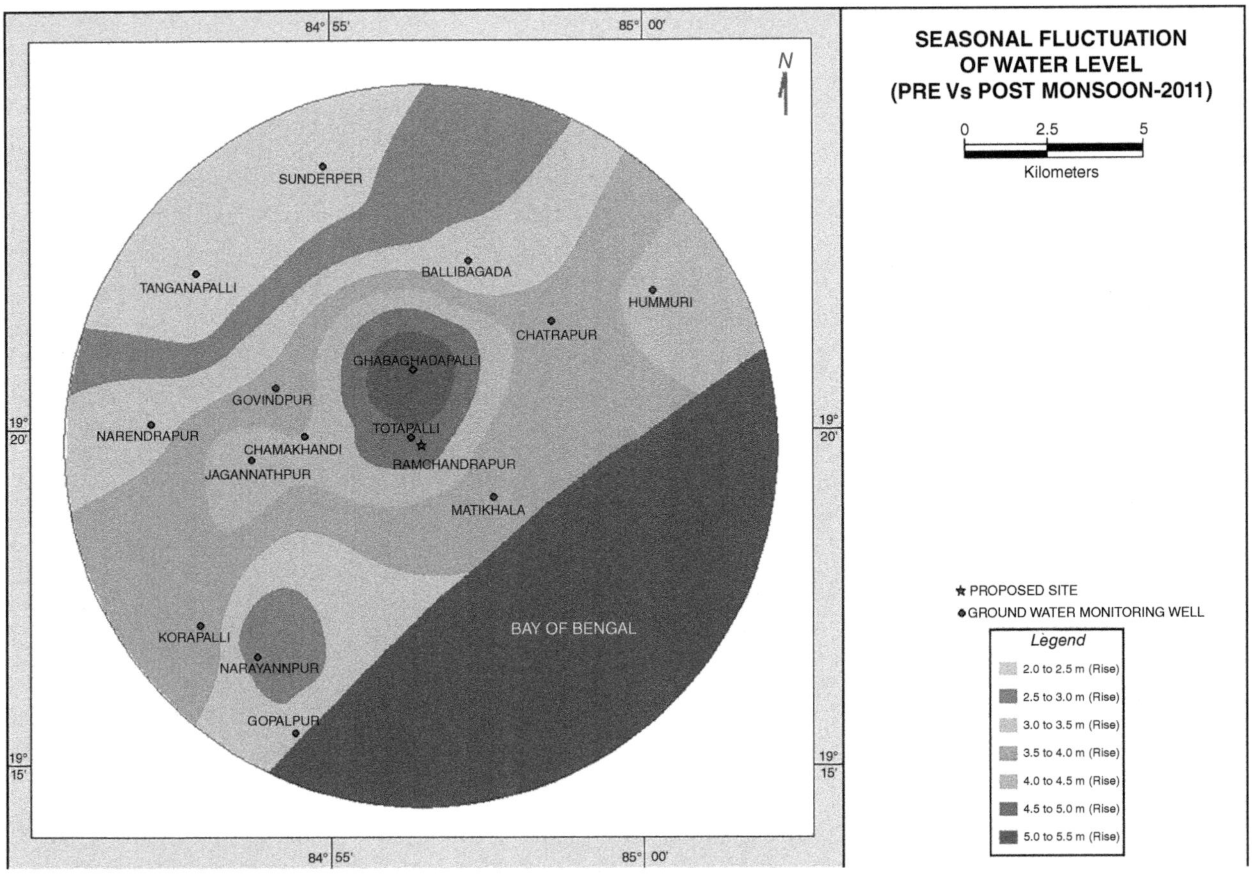

Figure 13.3 Seasonal fluctuation of water level.

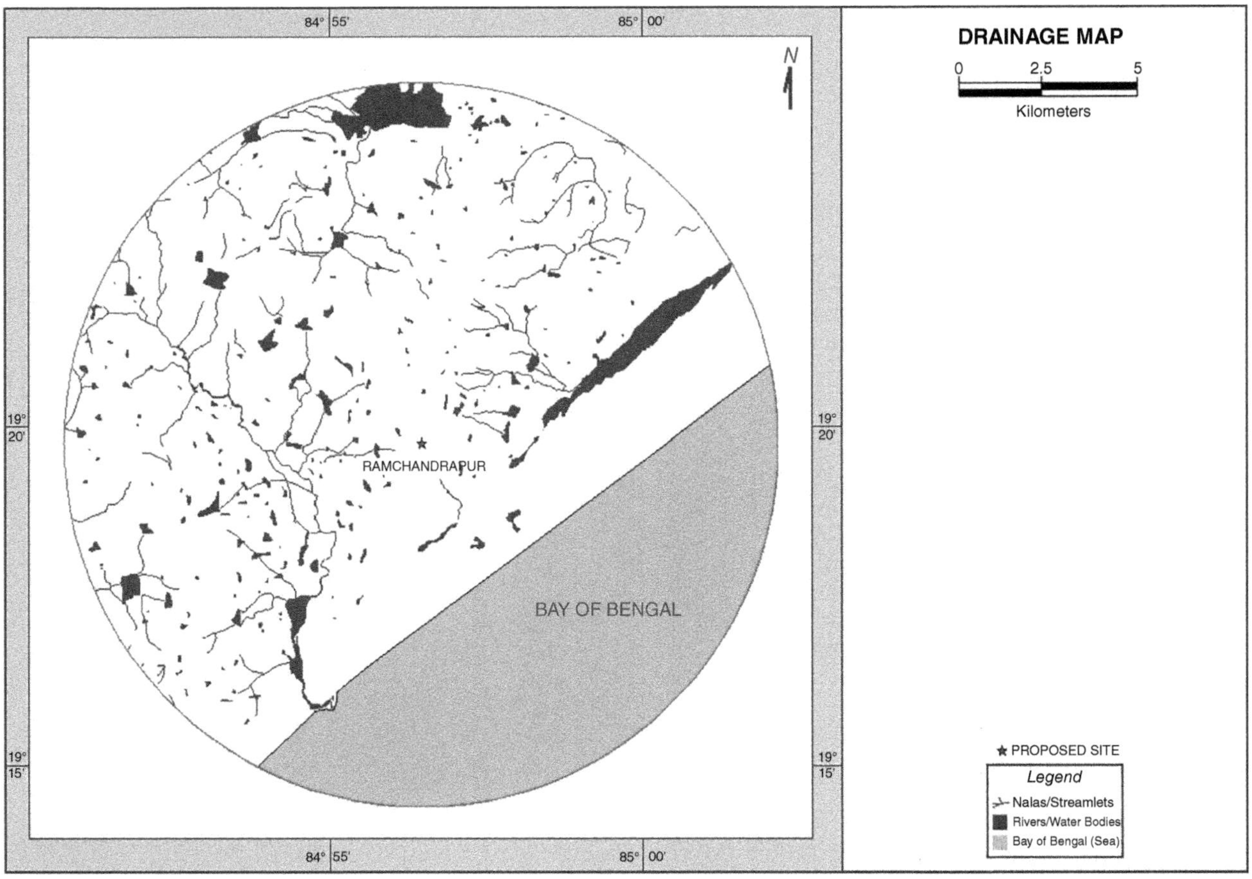

Figure 13.4 Drainage map.

are numerous small water tanks available in the area, which is a traditional source of water storage (Figure 13.4).

13.3 Overview of Rainwater Harvesting

The process of collecting and concentrating runoff water from runoff area into run-on area is called water harvesting. The collected water is directly applied to the cropping area and then stored in the soil profile for immediate use by the crop. Runoff is diverted from water reservoir for future uses in the form of domestic, livestock, aqua culture, and irrigation. Some part of collected water is used for recharge of ground water. In the catchment water drawn is larger than the command area for practical uses. Catchment area to command area ratio is inversely related to the amount of rainfall and intensity of rainfall. Classification of water harvesting is shown in Figure 13.5.

13.3.1 Different Types of Water Harvesting Techniques

13.3.1.1 Rooftop Water Harvesting (RTWH)

The productive utilization for rainwater falling on rooftops of structures is known as Rooftop Water Harvesting (RTWH). The rooftops are usually impervious and occupy considerable land area in urban regions. Municipal water supply is inadequate, inefficient, and unreliable in different cities. In this situation, collecting runoff from rooftops of individual structures is a step forward for storing water. Later, the storage water is found to be useful in numerous circumstances. Inadequacy of water and cost of supply is observed in many industries. Many organizations in urban areas utilize RTWH systems. For efficient and economic storage, new technologies have to be worked out for designing storage structures. Water collected from rooftops is also used for recharging ground water. Rainfall characteristics like intensity, duration, and the nature of the rainfall season are required to design a RTWH system.

13.3.1.2 Micro-Catchment System of Rainwater Harvesting (MiCSRWH)

In this system, the catchment is a small area for any productive purpose. The catchment length is between 1 and 30 m. The ratio area of catchment to area of cultivation is 5 : 1. Runoff storage depends upon the characteristics of the soil profile.

13.3.1.3 Macro-Catchment System of Rainwater Harvesting (MaCSRWH)

A macro-catchment system of rainwater harvesting (MaCSRWH) system is designed for bigger catchments, where overland flow and rill flow occurs. The catchment is 30–200 m long ratio of catchment to cultivated area is in the range 3 : 1.

13.3.1.4 Floodwater Harvesting (FWH)

Floodwater harvesting (FWH) is used for larger catchments. In the drainage system water is harvested during flow. The area of the catchments is several kilometers long. The ratio of catchment area to command area is greater than 10 : 1. Flood harvesting occurs (i) using storage structures and (ii) spreading technique.

13.3.1.5 Storage Structure Systems

Small storage structures are put together across the drainage to accumulate runoff known as check dams and nala bunds. These structures are used for arresting erosion from the catchment and prevent the deepening and widening of guiles. In general, check dams are recommended

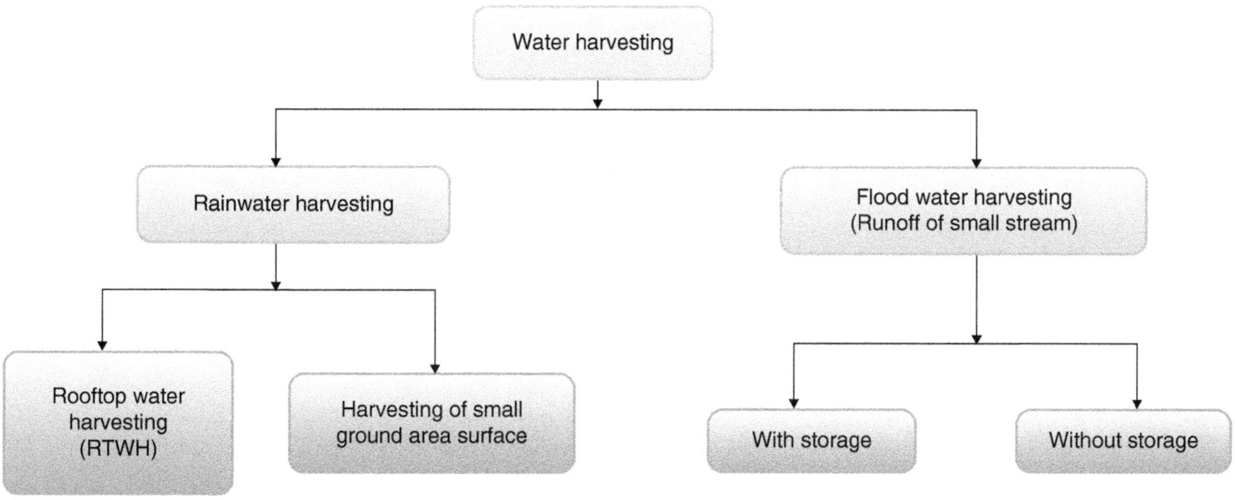

Figure 13.5 Classification of water harvesting.

where fluctuation of water table is high, and the watercourse is influent. Formations similar to nala bunds but in larger dimension are known as percolation tanks.

13.3.1.6 Spreading of Water

Diversion of runoff across the drainage will flow to adjacent watershed. In the water course appropriate bunds cause spreading of water over the command area and the harvesting through spreading of water by maintaining the soil characteristics and terrain slope in the catchment.

13.4 Methodology

13.4.1 Evaluation of Return Period

Mainly hydraulic engineering applications concerned with measurement of precipitation, runoff, and probability of occurrence of rainfall play a vital role, and principally work on probability, empirical, and statistical methods. The probability occurrence of an event of random variable whose magnitude is equal to or in excess of specified magnitude is called P, that is probability. The recurrence interval or return period is defined as $T = {}^{1}/{}_{P}$. The probability analysis is either made by empirical or analytical methods. The following empirical formula by different methods used to calculate probability as given in the next sections.

13.4.2 Design of Water Harvesting Structures

13.4.2.1 Design Approach

The important components, which are to be evaluated for RWH, are:

- Hydrogeology of the area including nature and extent of aquifer, soil cover, topography, depth to water table, and quality of ground water;
- Area contributing to runoff, i.e. total area and land use pattern, whether industrial, residential, or green belts and general built-up pattern of the area;
- Hydro meteorological characters, viz. rainfall duration, general pattern, and intensity of rainfall.

13.4.2.2 Estimation of Runoff Rate

The peak runoff rate should be assessed accurately for designing the recharge structure and may be assessed by following formula.

$$\text{Peak rate of runoff} = \text{Catchment area} \times \text{Runoff Coefficient} \times \text{Rainfall Intensity} \quad (13.1)$$

13.4.2.3 Estimation of Runoff Volume

Runoff volume is estimated using the USDA-SCS (soil conservation service) Curve Number technique. The storage capacity or potential retention is related to the Curve Number by the following relationship:

$$S = \left(\frac{100}{CN}\right) - 1 \quad (13.2)$$

Where,

S = potential retention parameter (mm);
CN = weighted curve number representing the hydrological soil cover, which is a function of antecedent moisture condition and hydrologic soil group.

The SCS curve number is a purpose of the ability of soils to allow infiltration of water with respect to land use/land cover (LU/LC) and antecedent soil moisture condition (AMC). Based on the USDA, SCS soils are distributed into four hydrologic soil groups – group A, B, C, and D – with respect to rate of runoff and final infiltration.

AMC is considered low when there is little prior rainfall and high when there has been considerable preceding rainfall to the modeled rainfall event. For modeling purposes, AMC II in the watershed is essentially an average moisture condition. Runoff curve numbers from LU/LC and soil type are taken for the average condition, AMC II, dry condition, AMC I, or wet condition, AMC III, and the equivalent curve number can be computed with Eqs. (13.2) and (13.3). The curve number value is recognized in the case of AMC II (USDA 1986) (Tables 13.1 and 13.2). The following equations are used in the cases of AMC-I and AMC-III

$$CN(I) = \frac{CN(II)}{2.282 - 0.0128\, CN\,(II)} \quad (13.3)$$

$$CN(III) = \frac{CN(II)}{0.427 + 0.00573\, CN\,(II)} \quad (13.4)$$

Where

$CN(I)$ is the curve number for dry condition,
$CN(II)$ is the curve number for normal condition,
$CN(III)$ is the curve number for wet condition.

$$CN_w = \sum CN_i * {}^{A_i}/{}_{A} \quad (13.5)$$

Where

CN_w is the weighted curve number,
CN_i is the curve number from 1 to any number N,
A_i is the area with curve number $CN_{i,}$
A is the total area of the watershed

Table 13.1 Soil conservation service classification.

Hydrologic soil (HSG)	Soil textures	Runoff potential	Water transmission	Final infiltration
Group A	Deep, well drained sand and gravels	Low	High rate	7.5
Group B	Moderately deep, moderately well drained	Moderate	Moderate rate	3.8–7.5
Group C	Clay loams, Shallow sandy loam, soils with moderate to fine textures	Moderate	Moderate rate	1.3–3.8
Group D	Clay soils that swell significantly when wet	High	Low rate	<1.3

(Source: USDA-SCS 1974).

Table 13.2 Group of antecedent soil moisture losses (AMC).

AMC group	Soil characteristics	Five day antecedent rainfall in mm	
		Dormant season	Growing season
I	Wet condition	<13	<36
II	Average condition	13–28	36–53
III	Heavy rainfalls	>28	>53

Table 13.3 Runoff coefficients vis-a-vis type of area.

SL. no.	Type of area	Standard values of runoff coefficients		Adopted values of runoff coefficients
		Min	Max	
1	Rooftop	0.75	0.95	0.75
2	Paved area	0.50	0.85	0.50
3	Bare ground	0.10	0.20	0.10
4	Green area	0.05	0.10	0.05

13.4.2.4 Runoff Coefficients

Runoff coefficient is the ratio of runoff to rainfall. It plays an important role in assessing the runoff availability and it depends upon catchment characteristics. Based on the standard values depending on the local condition, the runoff coefficients adopted for the study area is given in table below, which are used for estimating the runoff (Table 13.3).

13.4.2.5 Normal Distribution Method

1) From the daily data, the actual discharge (Q) for yearly, monsoon season, non-monsoon season (pre-monsoon and post-monsoon) on yearly basis have been calculated.
2) Then the standard mean (μ) and standard deviation (σ) of the calculated discharges are computed for both monsoon and non-monsoon data.
3) For the required return period (T), the probability factor (P) is evaluated in percentage. The conversion formulae used to evaluate the probability is given as

$$P = {}^{1}/_{T} \tag{13.6}$$

4) From the standard normal distribution table, by interpolation, the frequency factor (K_t) is computed based on the different return periods, where frequency factor equal to the standard normal deviate (z).
5) Finally, the predicted discharge (Q_p) is found using the standard normal distribution formula for the different return periods for the respective seasons:

$$Q_p = \mu + K_t \sigma \tag{13.7}$$

Where,
Q_p = predicted discharge
μ = standard mean
σ = standard deviation
K_t = frequency factor

13.4.2.6 Gumbel Distribution Method

1) From the given data, the actual discharge (Q) for yearly, monsoon season, non-monsoon season (pre-monsoon and post-monsoon) have been evaluated.
2) Then the standard means (μ) and standard deviation (σ) formulae have been used to standardize the calculated discharges for the respective seasons.
3) For the required return period (T), the reduced variate (Y_t) has been assessed by using the formula:

$$Y_t = -[\ln \ln(\frac{T}{T-1})] \tag{13.8}$$

4) Then the reduced mean (Y_n) and reduced standard deviation (S_n) has been determined from the Gumbel distribution table for the given sample size (N).

5) Then the frequency factor (K_t) is estimated using the formula:

$$K_t = \frac{(Y_t - Y_n)}{S_n} \tag{13.9}$$

K_t = frequency factor
Y_t = reduced Variate
Y_n = reduced mean
S_n = reduced Standard deviation

6) Finally, the predicted discharge (Q_p) is computed using the standard normal distribution formula for the different return periods for the respective seasons:

$$Q_p = \mu + K_t\sigma \tag{13.10}$$

Where
Q_p = predicted discharge
μ = standard mean
σ = standard deviation
K_t = frequency factor

13.4.2.7 Extreme Value Type-I Distribution

1) From the given data, the actual discharge (Q) for yearly, monsoon season, non-monsoon season (pre-monsoon and post-monsoon) has been weighed.
2) Then the standard mean (μ) and standard deviation (σ) are gauged from the calculated discharges for the respective seasons.
3) The frequency factor is calculated using the formula:

$$K_t = -\frac{\sqrt{6}}{\pi}\{0.5772 + \ln\ln(^T/_{T-1})\} \tag{13.11}$$

Where T = return period

4) Finally, we calculated the predicted discharge (Q_p) by using the standard normal distribution formula for the different return periods for the respective seasons:

$$Q_p = \mu + K_t\sigma \tag{13.12}$$

Where,
Q_p = predicted discharge
μ = standard mean
σ = standard deviation
K_t = frequency factor

13.4.2.8 Log Pearson Type-III Distribution

1) From the given data, we calculate the actual discharge (Q) for yearly, monsoon season, non-monsoon season (pre-monsoon and post-monsoon).
2) Then calculate the natural logarithm of the actual discharges (Z) and find out its standard logarithmic mean (μ) and standard logarithmic deviation (σ) of the calculated discharges for the respective seasons.

$$Z = \log_{10}Q \tag{13.13}$$

3) Then the coefficient of skewness (C_s) is calculated using the logarithmic discharges (Z) and for the required return period (T), we calculated the probability (P) in percentage ($P = {}^1/_T$)
4) From the standard normal distribution table, by interpolation, we calculated the standard normal deviate (z).
5) The frequency factor depends on return period and coefficient of skewness
6) When Cs = 0, the frequency factor is equal to the standard normal deviate z and is calculated as in case of normal deviation.
7) When Cs ≠ 0, the frequency factor (K_t) by using the formula (By Kite):

$$K_t = Z + (Z^2 - 1)k + \frac{1}{3}(z^3 - 6z)k^2$$
$$-(z^2 - 1)k^3 + zk^4 + \frac{1}{3}k^5 \tag{13.14}$$

Where z = normal deviate

$$K = \frac{\mathbf{Cs}}{\mathbf{6}} \tag{13.15}$$

K_t = frequency factor

8) Now predicted logarithmic discharge is calculated by using formula:

$$Q_p = \mu + K_t\sigma \tag{13.16}$$

Where,
q_p = predicted logarithmic discharge
μ = standard logarithmic mean
σ = standard logarithmic deviation

9) Finally, the predicted discharge (Q_p) is calculated by taking antilog of q_p

$$Q_p = antilog\,(q_p) \tag{13.17}$$

Table 13.4 Estimation of runoff coefficient.

Type of runoff catchment area	Effective area (m²)	Runoff coefficients
Rooftop rainwater harvesting		
Rooftop area	1 200	0.75
Stormwater harvesting		
Roads + paved areas	28 328	0.5
Green area	169 968	0.05
Barren	28 328.02	0.1
Miscellaneous	338 736	0.5
Total	566 560	

Table 13.5 Estimation of water harvesting capacity.

Year	Rainfall (cm)	Rainfall (m)	Roads + paved areas	Green area	Barren area	Miscellaneous area	Water harvesting capacity
1993	129.47	1.2947	18 338.13	11 002.88	3 667.629	219 280.7	252 289.4
1994	179.09	1.7909	25 366.31	15 219.78	5 073.265	303 321.2	348 980.5
1995	140.04	1.4004	19 835.27	11 901.16	3 967.056	237 182.9	272 886.4
1996	72.28	0.7228	10 237.74	6 142.644	2 047.549	122 419.2	140 847.1
1997	126.63	1.2663	17 936.01	10 761.61	3 587.206	214 472.4	246 757.2
1998	90.57	0.9057	12 828.33	7 697.001	2 565.669	153 396.6	176 487.6
1999	94.95	0.9495	13 448.72	8 069.231	2 689.745	160 814.9	185 022.6
2000	81.35	0.8135	11 522.41	6 913.448	2 304.484	137 780.9	158 521.2
2001	160.44	1.6044	22 724.72	13 634.83	4 544.948	271 734	312 638.5
2002	83.1	0.831	11 770.28	7 062.17	2 354.058	140 744.8	161 931.3
2003	166.98	1.6698	23 651.05	14 190.63	4 730.213	282 810.7	325 382.6
2004	103.56	1.0356	14 668.24	8 800.943	2 933.65	175 397.5	201 800.3
2005	114.67	1.1467	16 241.86	9 745.115	3 248.374	194 214.3	223 449.6
2006	171.87	1.7187	24 343.67	14 606.2	4 868.737	291 092.8	334 911.4
2007	138.46	1.3846	19 611.47	11 766.88	3 922.298	234 506.9	269 807.6
2008	164.1	1.641	23 243.12	13 945.87	4 648.628	277 932.9	319 770.5
2009	144.26	1.4426	20 432.99	12 259.79	4 086.6	244 330.3	281 109.7
2010	108.4	1.084	15 353.78	9 212.266	3 070.757	183 594.9	211 231.7
2011	115.46	1.1546	16 353.75	9 812.253	3 270.753	195 552.3	224 989.1
2012	126.78	1.2678	17 957.12	10 774.27	3 591.426	214 724.8	247 047.6
2013	151.73	1.5173	21 491.04	12 894.62	4 298.21	256 982.1	295 665.9
2014	157.77	1.5777	22 346.54	13 407.93	4 469.312	267 211.9	307 435.7
2015	94.13	0.9413	13 332.57	7 999.544	2 666.517	159 426.1	183 424.7
2016	100	1	14 164	8 498.4	2 832.802	169 368	194 863.2

13.5 Results and Discussions

Calculated runoff coefficient using SCS-CN method for Roads + Paved Areas, Green Area, Barren, and Miscellaneous areas are presented in Table 13.4 with effective area and runoff coefficient.

The value arrived at through both the methods do not vary considerably. However, it is better if the minimum value is taken into consideration. The calculation for water harvesting capacity of the study area during 1993–2016 using different nature of area of the catchment is presented in Table 13.5 using Eq. (13.1).

Return period is calculated for 10 years interval from 10 successive years to 30 successive years and an interval of 5 years for checking the variation for two successive 5-year periods, that is from 30 to 35 years and 35 to 40 years capacity of RWH in the study area and so on up to 75-year return period as shown in Table 13.6 for Normal distribution Method, Table 13.7 for Gumbel method, Table 13.8 for Extreme method, and Table 13.9 for Log Pearson type III method.

Table 13.6 Capacity of water harvesting using normal distribution method.

T (y)	P = 1/T (%)	Kt	Q (cumec)
10	10	1.28	324 258
20	5	1.65	347 201.6
30	3.33	1.84	358 983.5
35	2.86	1.9	362 704.1
40	2.5	1.96	366 424.7
50	2	2.05	372 005.5
60	1.67	2.13	376 966.3
70	1.43	2.19	380 686.9
75	1.34	2.22	382 547.2

By comparing different methods for capacity of RWH it is found that the harvesting capacity increases in every distribution method with respect to increasing return period.

Table 13.7 Capacity of water harvesting using Gumbel method.

T (yr)	Yt	Sn	Yn	Kt	Q (cumec)
10	2.25	1.0628	0.5236	1.624 388	345 613.5
20	2.97	1.0628	0.5236	2.301 844	387 622.4
30	3.38	1.0628	0.5236	2.687 618	411 544.1
35	3.54	1.0628	0.5236	2.838 163	420 879.4
40	3.67	1.0628	0.5236	2.960 482	428 464.4
50	3.9	1.0628	0.5236	3.176 891	441 883.9
60	4.08	1.0628	0.5236	3.346 255	452 386.1
70	4.24	1.0628	0.5236	3.496 801	461 721.4
75	4.31	1.0628	0.5236	3.562 665	465 805.6

Table 13.8 Capacity of water harvesting using Extreme value type-I method.

T (yr)	X	Y	Kt	Q (cumec)
10	0.105 361	−2.25037	1.305 071	325 812.6
20	0.051 293	−2.9702	1.866 536	360 629
30	0.033 902	−3.38429	2.189 534	380 658
35	0.028 988	−3.54089	2.311 678	388 232.1
40	0.025 318	−3.67625	2.417 257	394 779.1
50	0.020 203	−3.90194	2.593 296	405 695.2
60	0.016 807	−4.08595	2.736 827	414 595.6
70	0.014 389	−4.24131	2.858 005	422 109.8
75	0.013 423	−4.31078	2.912 196	425 470.1

From the comparison table it is clear that the Gumbel distribution method gives the peak RWH capacity in every return period and the normal distribution method gives the least RWH capacity in each turns of successive return period. In every distribution method, for the initial return period the RWH capacity increases at a faster rate but after some successive years the capacity becomes nearly equal. For the 75-year return period the corresponding values of peak RWH are found to be 382 547.2 cumec, 404 218.7 cumec, 425 470.1 cumec, 465 805.6 cumec with the designed normal distribution, Log Pearson type III, Extreme value type I, and Gumbel method, respectively.

From Figure 13.6, by comparing the discharges we found that the discharge increases in every distribution method with respect to increasing return period. From the Figure 13.5 and Table 13.10, it is found that the Gumbel distribution method gives the peak discharge in every return period. From the comparison table we found that the normal distribution method gives the least discharge in every return period. In every distribution method, for initial return period the discharge increases at faster rate but after some years the discharge becomes nearly equal.

Table 13.10 Comparison of capacity of water harvesting using different techniques.

		Water Harvesting Capacity (Cumec)			
T (Years)	P = 1/T (%)	Normal Distribution Method	Gumbel Method	Extreme value Type I Method	Log Pearson Type III Method
10	10	324 258	345 613.5	325 812.6	329 149.6
20	5	347 201.6	387 622.4	360 629	357 778.3
30	3.33	358 983.5	411 544.1	380 658	372 953.3
35	2.86	362 704.1	420 879.4	388 232.1	377 810.1
40	2.5	366 424.7	428 464.4	394 779.1	382 697.3
50	2	372 005.5	441 883.9	405 695.2	390 084.4
60	1.67	376 966.3	452 386.1	414 595.6	396 706.8
70	1.43	380 686.9	461 721.4	422 109.8	401 707.5
75	1.34	**382 547.2**	**465 805.6**	**425 470.1**	**404 218.7**

Table 13.9 Capacity of water harvesting using Log Pearson type III method.

T (yr)	Cs	K	z (p)	a	b	c	kt	q (cumec)	Q (cumec)
10	−0.29982	−0.04997	1.28	1.248 099	−0.00465	8.75E-05	1.24354	5.517 393	329 149.6
20	−0.29982	−0.04997	1.65	1.563 927	−0.0045	0.000 225	1.559 651	5.553 614	357 778.3
30	−0.29982	−0.04997	1.84	1.720 792	−0.004	0.000 309	1.717 097	5.571 654	372 953.3
35	−0.29982	−0.04997	1.9	1.769 579	−0.00378	0.000 337	1.766 137	5.577 274	377 810.1
40	−0.29982	−0.04997	1.96	1.818 006	−0.00352	0.000 367	1.814 851	5.582 855	382 697.3
50	−0.29982	−0.04997	2.05	1.889 972	−0.00307	0.000 412	1.887 317	5.591 159	390 084.4
60	−0.29982	−0.04997	2.13	1.953 262	−0.00259	0.000 454	1.951 122	5.598 47	396 706.8
70	−0.29982	−0.04997	2.19	2.000 31	−0.00219	0.000 487	1.998 602	5.603 91	401 707.5
75	−0.29982	−0.04997	2.22	2.023 699	−0.001 98	0.000 504	2.022 222	5.606 616	404 218.7

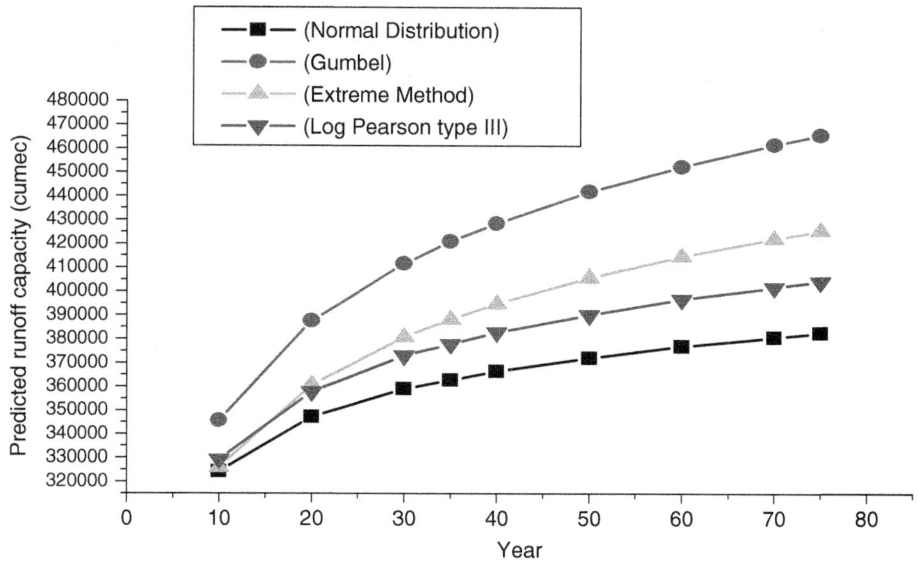

Figure 13.6 Capacity of water harvesting vs return period.

13.6 Conclusions

In this work runoff coefficient is evaluated using the SCS-CN curve for Ramchandrapur. Water harvesting capacity for each successive year is calculated using rational formula. Probability distribution in percentage is evaluated for designing the return period of RWH. Frequency factor is estimated for each method of design. It is found that frequency factor is the key parameter for estimating water harvesting capacity. Among all methods of design for the return period, the Gumbel method suits for peak RWH whereas the Normal distribution method of design for the return period shows poor performance. The impact assessment of RWH may be carried out periodically to find out any impact on a local scale and accordingly the system may be suitably modified to meet the future requirements. The work will help to design RWH structures for small areas, where accessibility of data compilation is difficult. So instead of actual data, the models developed leading to the Gumbel method will be helpful for similar small watersheds without any field data.

References

Agarwal, A. and Narain, S. (1997). *Dying Wisdom: The Rise, Fall and Potential of India's Traditional Water Harvesting System. State of India's Environment, a Citizen Report*, 404. New Delhi, India: Centre for Science and Environment.

Ahmed, W., Gardner, T., and Toze, S. (2011). Microbiological quality of roof-harvested rainwater and health risks: a review. *Journal of Environmental Quality* 40 (1): 1–9.

Angrill, S., Farreny, R., Gasol, C.M. et al. (2012). Environmental analysis of rainwater harvesting infrastructures in diffuse and compact urban models of Mediterranean climate. *The International Journal of Life Cycle Assessment* 17 (1): 25–42.

Athavale, R.N. (2003). *Water Harvesting and Sustainable Supply in India*. Jaipur, India: Rawat Publications.

Basinger, M., Montalto, F., and Lall, U. (2010). A rainwater harvesting system reliability model based on nonparametric stochastic rainfall generator. *Journal of Hydrology* 392 (3–4): 105–118.

Brodie, I.M. (2008). Hydrological analysis of single and dual storage systems for storm water harvesting. *Water Science and Technology: A Journal of the International Association on Water Pollution Research* 58 (5): 1039–1046.

Burns, M.J., Fletcher, T.D., Duncan, H.P. et al. (2015). The performance of rainwater tanks for storm water retention and water supply at the household scale: an empirical study. *Hydrological Processes* 29 (1): 152–160.

Chatfield, R.C., and Coombes, P.J. (2007) Urban rainwater harvesting-Analysis of a decentralized approach for meeting water demand in urban cities. An opportunity for the U.S.? American Rainwater Catchment System Association 2007 Conference, Kilauea Military Camp, Hawaii, USA.

Choudhary, M.K. and Aneja, D.R. (1991). Impact of green revolution on long term sustainability of land and water resources in Haryana. *Indian Journal of Agricultural Economics* 46 (3): 430–431.

City of Bellingham (2012). *Rainwater Harvesting: Guidance toward a Sustainable Water Future*. Washington, USA: City of Bellingham Public Works Department.

City of Tucson. (2009) Development standard no. 10-03.0: Commercial rain water harvesting, City of Tucson, Tucson, Arizona, USA.

De Busk, K.M. (2013) Rainwater harvesting: Integrating water conservation and stormwater management. A dissertation submitted to the Graduate Faculty of North Carolina State University, Biological and Agricultural Engineering, Raleigh, North Carolina, USA.

De Busk, K.M., Hunt, W.F., and Wright, J.D. (2013). Characterization of rain water harvesting utilization in humid regions of the United States. *Journal of the American Water Resources Association* 49 (6): 1398–1411.

Fewkes, A. and Warm, P. (2000). Method of modelling the performance of rainwater collection systems in the United Kingdom. *Building Services Engineering Research and Technology* 21 (4): 257–265.

Gaston, T.L. (2010) Rainwater harvesting in the south western United States: A policy review of the four corner states. 1–26. http://www.stream dynamics.us/resources.

Grady, C., and Younos, T. (2008) Analysis of water and energy conservation of rainwater capture on a single-family home. VWRRC Special Report No. SR39-2008, Virginia Water Resources Research Center, Virginia Polytechnic Institute and State Univ, Blacksburg, Virginia, USA.

Guo, Y. and Baetz, B.W. (2007). Sizing of rainwater storage units for green building applications. *Journal of Hydrologic Engineering* 12 (2): 197–205.

Han, M., and Park, J. (2007) Innovative rainwater harvesting and management in the Republic of Korea. Proc., 13th Int. Conf. On Rainwater Catchment Systems, International Rainwater Catchment Systems Conferences, Sydney, Australia.

Herrmann, T. and Schmida, U. (1999). Rainwater utilisation in Germany: efficiency, dimensioning, hydraulic and environmental aspects. *Urban Water Journal* 1 (4): 307–316.

Horman, M.J., Riley, D.R., Lapinski, A.R. et al. (2006). Delivering green buildings: process improvements for sustainable construction. *Journal of Green Building* 1 (1): 123–140.

Ibraimo, N., and Munguambe, P. (2007) Rainwater harvesting technologies for small scale rainfed agriculture in arid and semi-arid areas. Integrated water resources management for improved rural livelihood, CGIAR program on Water and Food. Waternet.

Jones, M.P. and Hunt, W.F. (2010). Performance of rainwater harvesting systems in the south eastern United States. *Resources, Conservation and Recycling* 54 (10): 623–629.

Jothiprakash, V. and Sathe, M.V. (2009). Evaluation of rain water harvesting methods and structures using analytical hierarchy process for a large-scale industrial area. *Journal of Water Resource and Protection* 1 (6): 427–438.

Kashyap, M., Khalfan, M.M., and Zianul-Abidin, N. (2003). A proposal for achieving sustainability in construction projects through concurrent engineering. In: *Proceeding of COBRA 2003*, 127–138. UK, England: RICS Foudnation, University of Wolverhampton.

Kim, J. and Furumai, H. (2012). Assessment of rainwater availability by building type and water use through GIS-based scenario analysis. *Water Resources Management* 26 (6): 1499–1511.

Kobet, B., Powers, W., Lee, S. et al. (1999). *High Performance Green Buildings*. Harrisburg, PA, USA: Pennsylvania Department of Environmental Protection.

Krishna, H. (2003) An overview of rainwater harvesting systems and guidelines in the United States. Proc., 1st American Rainwater Harvesting Conf., American Rainwater Catchment Systems Association, Austin, Texas, USA.

Ku, K., and Mills, T. (2010) Research needs for building information modeling for construction safety. Proc., 46th ASC Annual Conf., Associated Schools of Construction, Boston, Boston, City in Massachusetts, USA.

Levarios, H.K. (2007). *Designing for Water: Rainwater Harvesting, Storm Water Catchment, and Alternate Water Reuse*. Gabriola Island, BC, Canada: New Society Publishers.

Magent, C.S., Riley, D.R., and Horman, M.J. (2005). High performance building design process model. In: *Proceedings of the Construction Research Congress 2005: Broadening Perspectives*, 424–428. Reston, VA: ASCE., USA.

Mun, J.S. and Han, M.Y. (2012). Design and operational parameters of a rooftop rainwater harvesting system: definition, sensitivity & verification. *Journal of Environmental Management* 93 (1): 147–153.

Nam, C.H. and Tatum, C.B. (2006). Major characteristics of constructed products and resulting limitations of construction technology. *Construction Management and Economics* 6 (2): 133–147.

National Conference of State Legislatures (NCLS). (2012) State rainwater harvesting statues, programs and legislation. http://www.ncsl.org/issues-research/env-res/rainwater-harvesting. aspxi.

Natural Resources Defense Council and Low Impact Development Center (NRDC and LIDC). (2011) Capturing rainwater from rooftops: An efficient water resource management strategy that increases supply and reduces pollution, New York, USA.

Okhravi, S., Eslamian, S., Eslamian, F., and Tarkesh-Esfahani, S. (2014). Indigenous knowledge as a supportive tool for sustainable development and utilisation of rainwater harvesting systems. *Journal of Flood Engineering* 5 (1/2): 39–50.

Oweis, T., Hachum, A., and Kinje, J. (1999). *Water Harvesting and Supplemental Irrigation for Improved Water Use Efficiency in Dry Areas.* Colombo, Sri Lanka: International Water Management Institute (IWMI).

Partzsch, L. (2009). Smart regulation for water innovation – the case of decentralized rainwater technology. *Journal of Cleaner Production* 17 (11): 985–991.

Rahman, A., Hajani, E., and Eslamian, S. (2017). Rainwater harvesting in arid regions of Australia , Chapter 26. In: *Handbook of Drought and Water Scarcity: Environmental Impacts and Analysis of Drought and Water Scarcity*, vol. 2 (eds. S. Eslamian and F. Eslamian), 489–500. USA: Francis and Taylor/CRC Press.

Reed, W.G. and Gordon, E.B. (2000). Integrated design and building process: what research and methodologies are needed? *Building Research and Information* 28 (5/6): 325–337.

Reidy, P.C. (2010) Integrating rainwater harvesting for innovative storm water control, Proc., World Environmental and Water Resources Congress, ASCE, Reston, Virginia. USA.

Texas Rainwater Harvesting Evaluation Committee (TRHEC) (2006). *Rainwater Harvesting Potential and Guidelines for Texas.* Austin, TX, USA: Texas Water Development Board.

USDA (1986) Urban hydrology for small Watersheds, TR-55, United States Department of Agriculture, 210-VI-TR-55, 2nd edn June 1986., USA.

USDA-SCS (1974) Soil survey of Travis County, Texas. College Station, Tex.: Texas Agricultural Experiment Station, and Washington, D.C: USDA Soil Conservation Service, USA.

White, I. (2007) A theoretical framework for understanding factors that contributes to household adoption of rainwater harvesting in Southeast Queensland. Proc., American Rainwater Catchment Systems Association Conf., Univ. of Hawaii, Honolulu, Hawaii, USA.

White, I. (2010). Rainwater harvesting: theorizing and modelling issues that influence household adoption. *Water Science and Technology* 62 (2): 370–377.

Whole Building Design Guide (WBDG). (2012) Architecture. http://http://www.wbdg.Org/design/dd_architecture.phpi.

Zhang, D., Gersberg, R.M., Wilhelm, C., and Voigt, M. (2009). Decentralized water management: rainwater harvesting and grey water reuse in an urban area of Beijing, China. *Urban Water Journal* 6 (5): 375–385.

14

Rainwater Harvesting Impact on Urban Groundwater

A. Jebamalar[1], R. Sudharsanan[1], G. Ravikumar[2], and Saeid Eslamian[3]

[1] *Velammal Engineering College, Chennai, Tamil Nadu, India*
[2] *College of Engineering Guindy, Anna University, Chennai, Tamil Nadu, India*
[3] *Department of Water Engineering, College of Agriculture, Isfahan University of Technology, Isfahan, Iran*

14.1 Introduction

Groundwater is the largest reservoir of fresh water on the planet and is intensively exploited for private, domestic, and industrial uses in many urban centers of the developing world. Exploitation of groundwater resource may lead to several vexing problems like reduced well yields, land subsidence, and intrusion of salty water, especially in coastal areas. In order to overcome these serious environmental implications and to improve the groundwater levels, it is necessary to artificially recharge the depleted aquifers. An underground water banking technique known as aquifer storage and recovery (ASR) (Elise Bekele et al. 2018) has emerged as a means of expanding urban water resources by harvesting waters that would otherwise be foregone.

Groundwater is derived from any surface water source (rain, snow melt, or surface runoff) that infiltrates the land surface and slowly percolates to the water table. This process of adding water to underground storage is called "natural" groundwater recharge. Where the materials in the earths' surface are coarse and the slope of the land is gentle, there is generally more groundwater recharge than in areas where the strata consists of fine-grain material such as shale and clay, or where the slope is steep. The purpose of "artificial recharge" is to increase the rate at which water infiltrates the land surface.

Either absence, unreliable, or too-expensive conventional piped water results in lack of access to clean water for domestic uses for millions of peoples in many parts of the world. In the twenty-first century, overcoming the growing water shortage is one of the biggest challenges. Hence, rainwater harvesting has thus regained its importance as a valuable alternative or supplementary water resource, along with more conventional water supply technologies. Much actual or potential water shortages can be resolved if rainwater harvesting is practiced more widely (Okhravi et al. 2014).

Urbanization increases the area of impervious surfaces, which results in a decrease of infiltration. For a sustainable urban future, society must move toward the goal of efficient and appropriate saving of this surface runoff through roof top water collection, storm runoff collection, and street recharge pits/trenches.

Rainwater harvesting (RWH) means the activity of direct collection of rainwater that can be stored in surface or underground water tanks or recharged into the groundwater (Villarreal and Dixon 2005). RWH systems can be adopted where conventional water supply systems have failed to meet people's needs.

Rain is the first form of water in the hydrological cycle and is the primary source of water for rivers, lakes, and groundwater (secondary sources). The potential of rain to meet water demand is tremendous. Water crises occur only because effective collection and storage of rainwater has been ignored. During the period of rainfall, such unconserved rainwater flows down to other areas without benefiting the place of rainfall. In coastal areas, the rainfall excess will flow to the sea. RWH plays an important role in not only managing the temporary water crises but also in augmenting the groundwater aquifers. Practicing RWH at the community level is very essential for sustainable development.

The concept of RWH lies in tapping the rainwater where it falls. A major portion of rainwater flows as surface runoff in streams and rivers, and finally to the sea. Only 8% of the total rainfall recharges the groundwater aquifers. There are various techniques to conserve rainwater. In rural areas, surface spreading techniques are common since space for such systems is available in abundance and the quantity of recharged water is also large. A few

Handbook of Water Harvesting and Conservation: Basic Concepts and Fundamentals, First Edition. Edited by Saeid Eslamian and Faezeh Eslamian.

techniques to conserve rainwater are contour bund, gully plug, percolation tank, check dams, nalla bunds, dug well recharge, and sub-surface dykes.

In urban areas, rainwater is available from rooftops of buildings, as well as paved and unpaved areas, and can be collected and stored (either recharged to aquifer or stored in underground sumps) that can be utilized gainfully at the time of need. Such structures can be adopted widely by any society for effective conservation. Unless people are involved in conserving rainwater from individual households, it would be very difficult to implement it by any government. RWH system needs to be designed in a way that it does not occupy large spaces for collection and recharge.

A few techniques for rooftop rainwater harvesting in urban areas are recharge pit (with similar dimensions of length, breadth, and depth), recharge trench (similar dimensions of breadth and depth but larger lengths), recharge shaft (smaller diameter and larger depth), and recharge well (larger diameter and larger depth) (Boers and Ben-Asher 1982). Existing dug wells and tube wells can also be used as RWH structures with filtration of rainwater to avoid surface impurities from entering the well.

In India, 18 States have made rooftop RWH systems mandatory for new buildings. The Karnataka state government proposed to give a 5–10% rebate on the water bill for those practicing RWH. Over a period of time, it is expected that this would help to solve 40–45% of the urban water problems. The Delhi government issued directions to install RWH systems in their buildings with an area of 100 m^2 and above. Mumbai City, faced with an acute water shortage, made it mandatory for all buildings with a plot area of 300 m^2 or more to adopt RWH. In Tamil Nadu state, RWH was made mandatory in all buildings irrespective of the size of the rooftop area. Water and sewer connections to all new buildings are provided only after the installation of RWH systems. In Indore City, RWH was made mandatory in all new buildings with an area of 250 m^2 or more. As an incentive for implementing the scheme, a rebate of 6% on property tax has been offered.

Widespread implementation of RWH warrants a huge outlay of funds requiring an assessment of their impacts (Ammar Adham et al. 2016). The impact of RWH could be assessed in terms of improved surface water and groundwater availability. In urban areas, RWH is practiced mostly for recharge purposes through various techniques resulting in improved groundwater potential. Consequently, it converges to estimation of incremental recharge due to implemented RWH structures, which could be estimated as increase in water levels of the wells, as well as increase in net recharge (as the height of water in different strata implies different yields).

Understanding this, the Tamil Nadu government (a state in southern India) vigorously advocated the provision of RWH structures within the city area to recharge the shallow aquifers in the year 2003. People have implemented RWH in their houses. Almost 90% of the houses have implemented only rooftop RWH with source wells through filter. The remaining houses have implemented percolation pits and recharge wells. A study was made to assess the impact of RWH in this urban aquifer by estimating the possible groundwater recharge before and after the implementation of the RWH system.

14.2 State of the Art

Many researchers around the world have attempted to study the groundwater estimation and impact of RWH Systems. Mohammed-Aslam et al. (2010) evaluated groundwater potential of a hard-rock aquifer using remote sensing and geophysics. Israil et al. (2006) carried out groundwater resource evaluation in the Piedmont zone of Himalaya, India, using isotope and GIS techniques.

Glendenning and Vervoort (2011) studied the hydrological impacts of RWH in the Arvari River, Rajasthan, India. This study has examined the catchment-scale RWH impacts using a conceptual water balance model. The simulation results show that RWH has a positive impact on groundwater recharge and the sustainability of irrigated agriculture, but decreases stream flow downstream. Matthew and William (2010) studied the performance of rainwater harvesting systems in the southeastern United States at three rainwater cisterns in North Carolina. A computer model was developed to simulate system performance and simulations were conducted for 2081 rain barrels and larger cisterns. Results of the monitoring study showed that the rainwater harvesting systems were underutilized, which was suspected to result from poor estimation of water usage and public perception of the harvested rainwater. Simulation results showed that a rain barrel was frequently depleted when used to meet household irrigation demands and overflowed during most rainfall events. Yong-chao Zhou et al. (2010) analyzed the rainwater harvesting system for the domestic water supply in Zhoushan, China. They generated a computer model to analyze the performance of the Domestic Rainwater Harvesting System (DRHS) with different ratios of $D/(AR)$ (water demand/average annual collected runoff) and $S/(AR)$ (storage capacity/average annual collected runoff). The performance of the DRHS was analyzed by means of the model simulation.

Pachpute et al. (2009) assessed the sustainability of rainwater harvesting systems in the rural catchment of sub-saharan Africa. A study was undertaken in Makanya catchment of rural Tanzania to assess sustainability of storage type of rainwater harvesting systems including micro dam, dug out pond, sub-surface runoff harvesting tank, and rooftop rainwater harvesting system. They conclude that higher crop production is observed in 12–20 ha area near RWH-type micro dams. Fayez and Al-Shareef (2009) carried out the research to evaluate the potential for potable water savings by using rainwater in residential sectors of the 12 Jordanian governorates, and provided suggestions and recommendations regarding the improvement of both quality and quantity of harvested rainwater. Sturm et al. (2009) investigated rainwater harvesting (RWH) in central northern Namibia, on the basis of hydrological, technical, social, and cultural conditions. Appropriate solutions for RWH are developed, discussed, and evaluated. The calculations reveal that it is economically reasonable to apply decentralized techniques of RWH in terms of the roof catchment systems.

John et al. (2009) studied the effects of rainwater-harvesting-induced artificial recharge on the groundwater of wells in Rajasthan, India. A physical and geochemical investigation utilizing environmental tracers ($\delta^{18}O$ and Cl^-), groundwater level, and groundwater quality measurements and geological surveys was conducted to quantify the proportion of artificially recharged groundwater in wells located near rainwater harvesting structures and to examine potential effects of artificial recharge on the quality of groundwater in these wells. A geochemical mixing model revealed that the proportion of artificial recharge in these wells ranged from 0 to 75%. Groundwater tracer, water table, and geological data provided evidence of complex groundwater flow and were used to explain the spatial distribution of artificial recharge. Furthermore, wells receiving artificial recharge had improved groundwater quality. Statistical analysis revealed a significant difference between the water quality in these wells and wells determined not to receive artificial recharge, for electrical conductivity and SO_4^-. The findings from this study provide quantitative evidence that rainwater harvesting structures in southern Rajasthan influence the groundwater supply and quality of nearby wells by artificially recharging local groundwater.

Deepak et al. (2004) have reviewed the impact assessment of RWH on groundwater quality at Indore and Dewas, India. The impact assessment of rooftop rainwater harvesting on groundwater was carried out with the help working tube wells to improve the quality and quantity of groundwater. The rooftop rainwater was used to put into the ground using a sand filter as pretreatment system. This led to a reduction in the concentration of pollutants in groundwater, which indicated the effectiveness of increased recharge of aquifer by rooftop rainwater. He observes that in certain areas, the amount of total and fecal coliform was observed to be higher in harvested tube well water than normal tube well water. The reason of this increase was poor cleanliness of the rooftop and poor efficiency of the filter for bacterial removal. The author concludes that quality mounting of rainwater harvesting is an essential prerequisite before using it for groundwater recharge.

Sharma and Jain (1997) described the groundwater recharge through rooftop rainwater harvesting in urban habitation. In Nagpur city an experiment was conducted where 80 000 l of water collected from the rooftop of 100 m^2 area was recharged. The rise in water level up to 1 m was recorded in the recharge well and adjusting dug wells. The quality of groundwater was also improved as nitrate concentrations got diluted considerably to desirable limit. They concluded that such a practice, if replicated on a large scale, can bring out sustainable augmentation to groundwater reservoir.

Jaworska-Szulc (2009) carried out the groundwater flow modeling of multi-aquifer systems for regional resources evaluation of the Gdansk hydrogeological system of Poland using Visual MODFLOW software to develop a three-dimensional steady state model on the basis of data from over 1700 boreholes. Chenini and Mammou (2010) carried out the groundwater recharge study in an arid region using GIS techniques and numerical modeling in Maknassy basin, which is located in Central Tunisia. Martínez-Santos et al. (2010) modeled the effects of groundwater based on urban supply in a low permeability aquifer of Madrid in Spain. Senthilkumar and Elango (2011) modeled the impact of a subsurface barrier on groundwater flow in the lower Palar River basin of Southern India. Taheri and Zare (2011) carried out groundwater recharge modeling of Kangavar Basin, a semi-arid region in the western part of Iran using MODFLOW. Groundwater modeling is widely used as a management tool to understand the behavior of aquifer systems under different hydrological stresses, whether induced naturally or by human activities. From the various literatures, it is clearly known that, with the help of models and with various prediction scenarios, management policy can be framed for the future protection/management of aquifer.

14.3 Study Area and Data Collection

Chennai, the capital of Tamil Nadu State and the oldest of the presidential cities in India, is selected as the study area. The index map of study area with the specific region chosen for the study is presented as Figure 14.1.

Chennai is located at 13.04° N and 80.17° E on the southeast coast of India, occupying a total area of 174 sq. km,

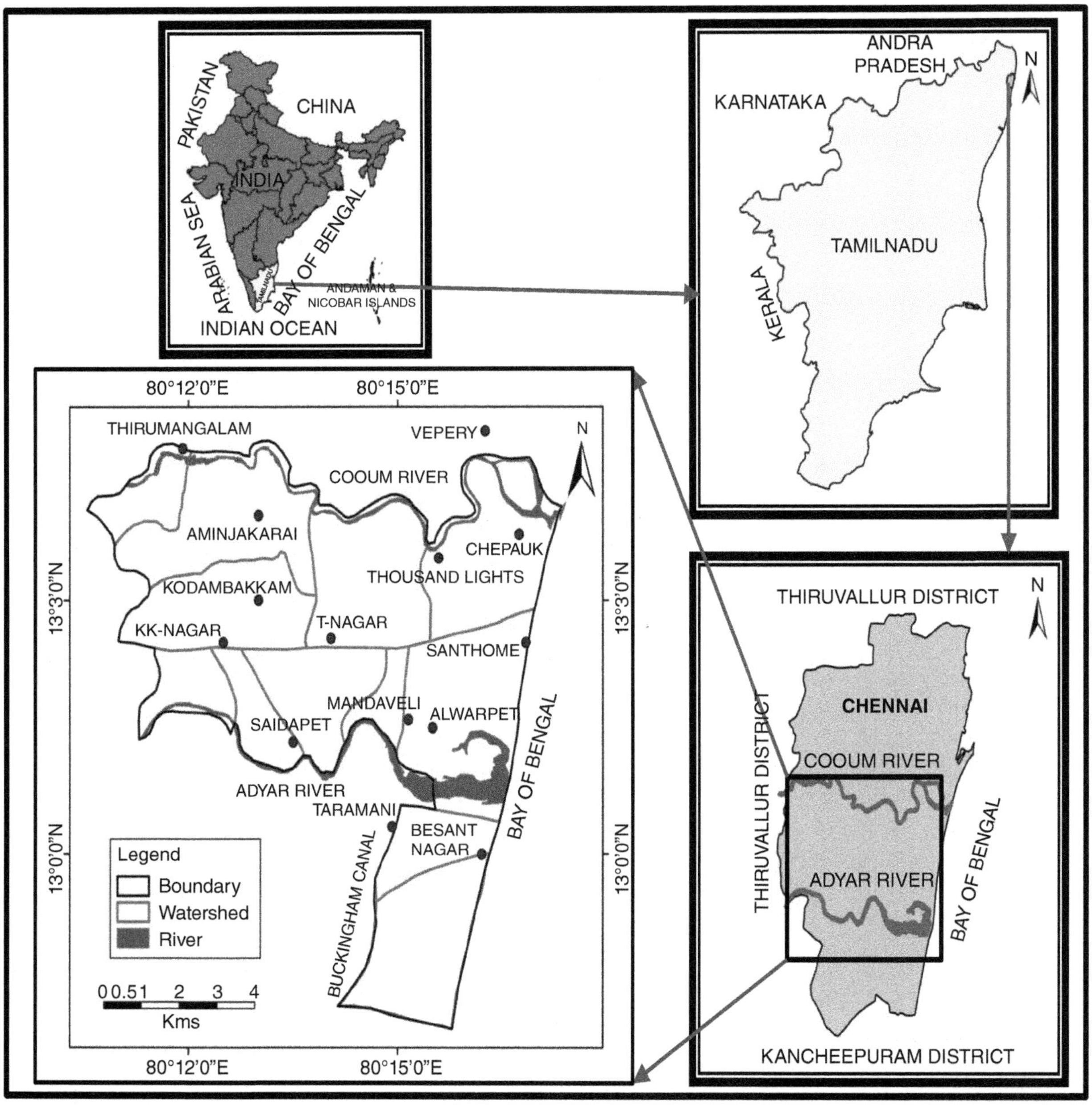

Figure 14.1 Index map of study area.

and is on a flat coastal plain at an average elevation of 6 m. Two rivers meander through Chennai: the Cooum River in the central region and the Adyar River in the southern region. The Buckingham Canal, 4 km inland, travels parallel to the coast, linking the two rivers. Chennai features a tropical wet and dry climate. Chennai lies on the equator and is also coastal, which prevents extreme variation in seasonal temperature. The average annual rainfall is about 1300 mm. The city gets most of its seasonal rainfall from the north–east monsoon winds, from mid-September to mid-December. The geology of Chennai comprises mostly clay, shale, and sandstone. Historically, Chennai has faced problem of water shortages, resulting in over-reliance on annual monsoon rains to replenish reservoirs. The city's groundwater levels were depleted to very low levels in many areas.

Due to the increasing urbanization process, most of the natural features in the city have been modified; mainly the green covers were replaced by impervious concrete buildings, pavements, roads, etc., except the Guindy National Park with an area of 270.57 ha, which is under reserve forest category. This highlights the need to implement RWH in the city. Hence, a scientific development to conserve and protect the resource is stressed upon to regulate the use and abuse of this precious resource. Rooftop RWH in urban areas offers one of the methods/options for conserving the precious groundwater supplementing the drinking water needs particularly.

CMWSSB is the nodal agency for water supply in Chennai city. It has constituted a fully dedicated "RWH cell," with main objectives of creating awareness and to offer technical assistance to the residents to implement suitable cost-effective methods of RWH in their premises. Figures 14.2–14.7 present the RWH structures adopted in Chennai (Sivaraman and Thillai Govindarajan 2002).

Monthly water level data for 19 observation wells were collected from the Central Groundwater Board (CGWB), Government of India, for the periods of pre (1995–2003) and post (2004–2011) RWH implementation periods along with the lithology data. Monthly rainfall data was collected from 1971 to 2011 from India Meteorological Department (IMD).

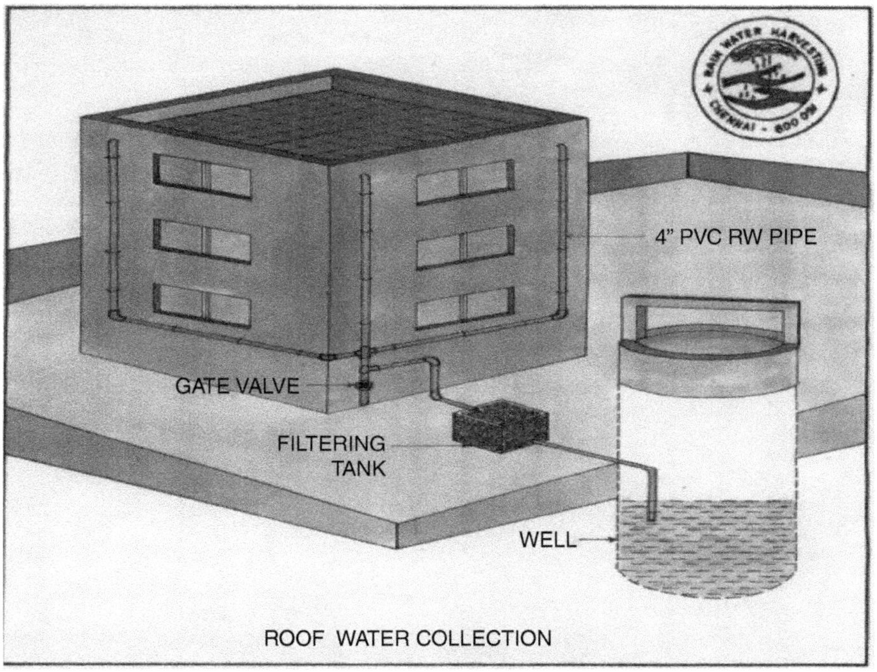

Figure 14.2 Rooftop RWH toward an open well.

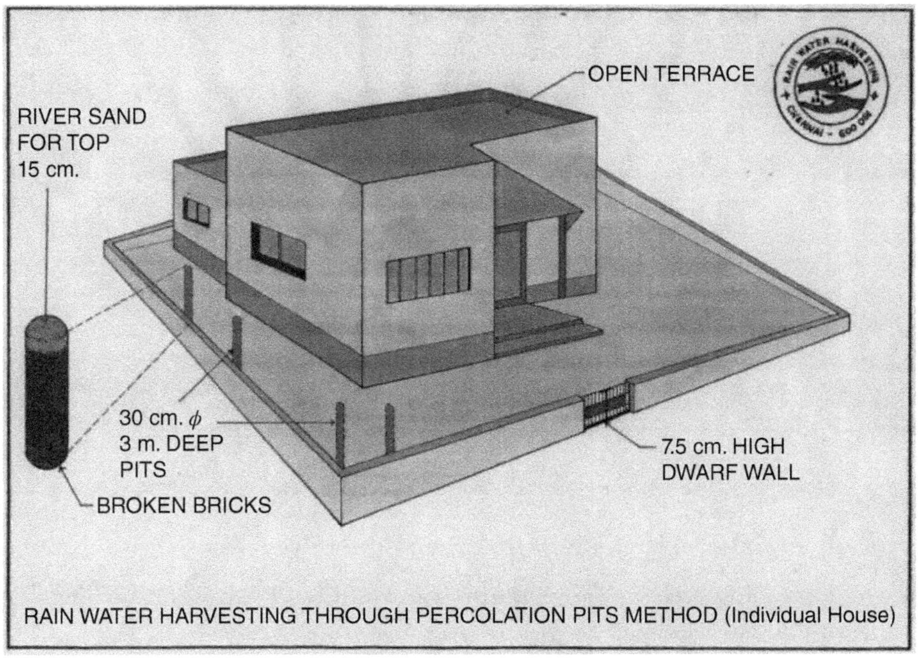

Figure 14.3 Rooftop RWH through percolation pit.

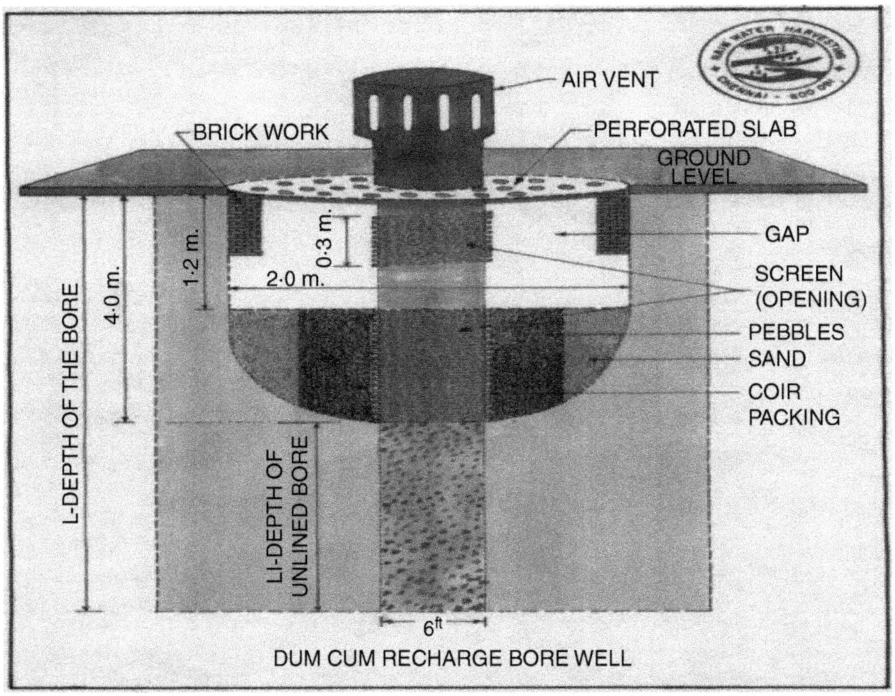

Figure 14.4 RWH through a dug cum recharge borewell.

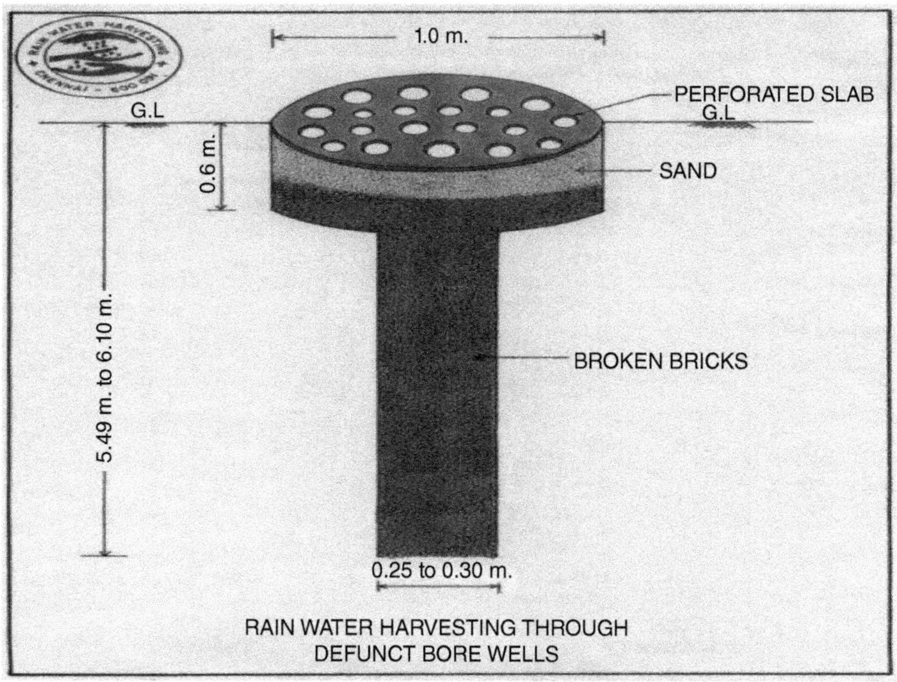

Figure 14.5 RWH through defunct borewells.

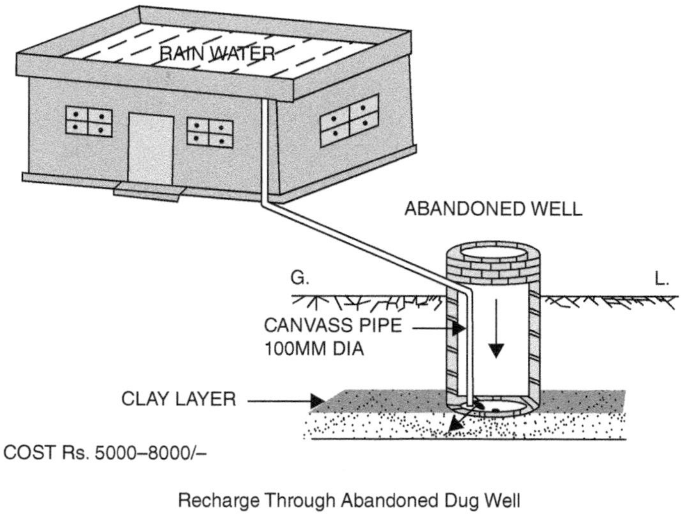

Figure 14.6 RWH through abandoned dug well.

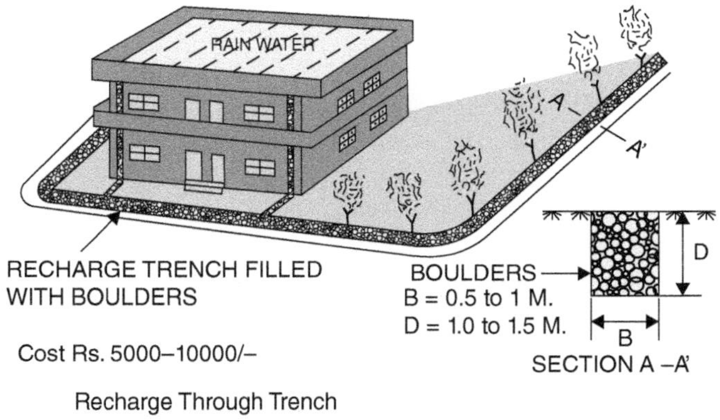

Figure 14.7 RWH through recharge trench.

14.4 Methodology

Preliminary analysis of monthly water level data (through hydrographs and moving average curve) is attempted for any possible change of trend between pre- (1995–2003) and post-RWH (2004–2011) periods. With the application of GIS software, groundwater contours for the study area are drawn to understand the spatial variation within the study area from coastal to inland space.

The impact of RWH represents the improvement of recharge over and above the natural recharge (NR). If the natural recharge is related to input parameters during the pre-RWH period and simulated for the post-RWH period, the natural recharge components could be estimated. Any additional recharge, if any, should have come through artificial recharge due to RWH. This is the basis with which estimating the impact of RWH is analyzed in this study.

Groundwater recharge estimation is done by GEC Norms 1997 (water level fluctuation method). The groundwater balance equation in the non-command area is given by

$$\mathbf{R_G - Dg - B + I_S + I = S(GEC\ Norms\ 1997)} \tag{14.1}$$

where $\mathbf{R_G}$ = Gross recharge due to rainfall and other sources

Dg = Groundwater draft

B = Base flow into the streams from the area

$\mathbf{I_S}$ = Recharge from streams into groundwater body

I = Net groundwater inflow into the area across the boundary

S = Groundwater storage increase

As per the GEC norms, if the area under consideration is a watershed, the net groundwater inflow term, **I** may be taken as zero. Usually, there are difficulties in estimating the base flow and recharge from streams. If the unit of assessment is a watershed in hard rock area, a single-stream gauge monitoring station at the exit of the watershed can provide the required data for the calculation of base flow. As such data is not available in most of the cases, it is recommended that the base flow term and recharge from stream may also be dropped. To signify this, $\mathbf{R_G}$ is replaced with **R** and the equation is rewritten as,

$$\mathbf{R} = \mathbf{S} + \mathbf{D_G} \tag{14.2}$$

Substituting the expression for storage increase S in terms of water level fluctuation and specific yield, the equation becomes,

$$\mathbf{R} = \mathbf{h}^*\mathbf{Sy}^*\mathbf{A} + \mathbf{D_G} \tag{14.3}$$

where **h** is the rise in the groundwater level in the monsoon season, **Sy** is the specific yield, and **A** is the influence area of the well.

The recharge calculated from (14.3) gives the available recharge from rainfall and other sources for the particular monsoon season. For non-command areas, the recharge from other sources may be recharge from tanks and ponds, water conservation structures, and surface irrigation. Influencing area of each well is delineated and calculated based on the Thiessen polygon method with the capabilities of Arc GIS 9.2. Specific yield values are taken from geology map and GEC-97 norms. Using the change in water level, specific yield, and area of influence, groundwater storage increase is estimated. Also, groundwater draft is estimated based on the well density and pumping rate of the study area. Then, possible recharge is calculated by adding groundwater draft with groundwater storage increase.

The recharge values for the entire period (1995–2011) of study are estimated using the above approach. Initially, the relation between rainfall and recharge is determined for pre-RWH implementation period (1995–2003) using Non-linear regression. The recharge estimated for this period represents natural recharge only. Using this relation, the NR component for the post-RWH period is determined for the actual rainfall measurements. The difference of NR (computed) from the recharge (estimated from the water level fluctuation method, explained above) for post-RWH period gives the artificial recharge component due to RWH implementation.

Visual MODFLOW software is used to build a groundwater flow model and to simulate the behavior of flow systems under different stresses such as groundwater recharge rates and pumping based on projected demands. Importance is given to predict the groundwater flow and groundwater head temporally and spatially and to investigate the effect of groundwater abstraction and recharge at a well on the flow regime and predicting the resulting change in groundwater levels with different management scenarios.

Base map of the study area is created in Arc GIS software and the same is imported to the visual MODFLOW. The study area is divided into grids of size 100×100 m. The boundary conditions such as flow boundary, no flow boundary; active and inactive cells are fixed. The elevation data set of the study area is entered. The number of layers of aquifer along with their thickness is also finalized. Observation wells and pumping wells are located along with their thickness of screen and maximum depth of well. Aquifer properties such as hydraulic conductivity, porosity, specific yield, transmissivity, and river stages are entered as input data. Rainfall recharge, pumping rate, and evapotranspiration for the total study area is estimated and given as input.

Model parameters are estimated through a trial and error process until a good match between computed and observed water heads is obtained. Statistical analysis of the calibration results in steady state and transient state are carried out. Then the model is validated by comparing the results of the model with the observed data. The calibrated model (for the pre-RWH period, influenced by natural recharge only) is used to simulate the regional groundwater head and it is compared with the observed head (that is influenced by natural and artificial recharge [due to RWH implementation in the year 2003]). The difference of observed head and simulated head indicates the impact of RWH, if any. The model is also used for the prediction of future hydrogeological conditions of the study aquifer, according to three various scenarios of pumping and recharge.

14.5 Temporal Analysis of Groundwater Level

To understand the temporal variation of groundwater levels and its general trend, groundwater hydrographs of all 19 observation wells are drawn. The moving average graph is plotted for the period 1995–2010 for all the wells and the sample graphs for the wells located in T Nagar and Besant Nagar are shown in Figures 14.8 and 14.9. The curves are drawn to represent 3-year (36-month) moving average values. As the moving average is taken for larger number of months, the curve smoothens to give a better overall picture of the temporal variation. There is also a distinct variation of trend during the pre- and post-RWH

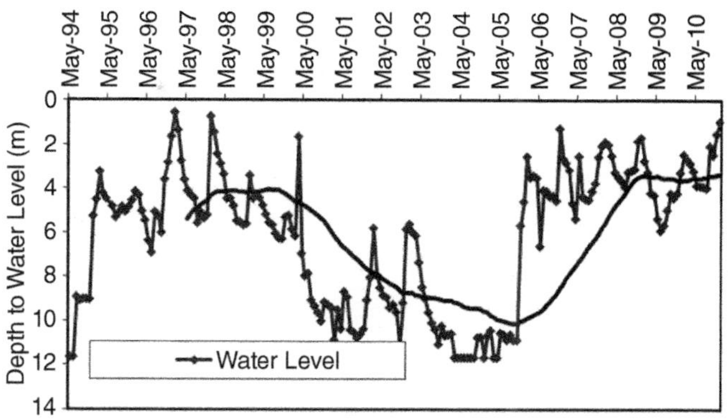

Figure 14.8 Hydrograph of T. Nagar well with moving average curve.

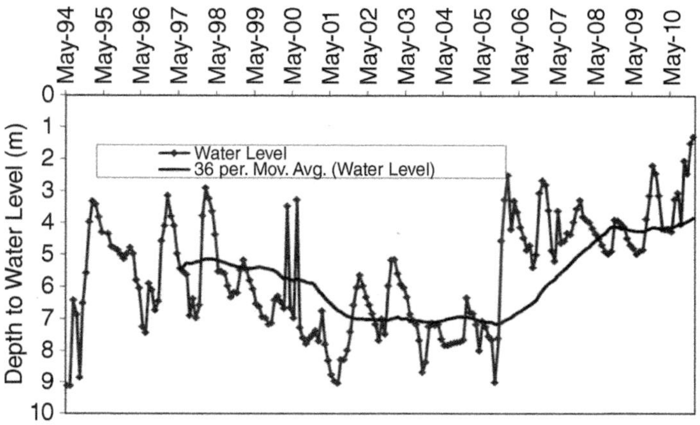

Figure 14.9 Hydrograph of Besant Nagar well with moving average curve.

implementation periods. It shows that the average water level in T. Nagar has gone down by about 6 m during the period of 2000–2005 and then had gone up by 6 m during the period of 2005–2010, whereas in Besant Nagar the average water level has gone down by about 2 m during the period of 2000–2005 and then had gone up by 3 m during the period of 2005–2010.

14.6 Spatial Analysis of Groundwater Table

Spatial variation of groundwater levels (reduced to mean sea level) across the study area is studied for post-monsoon (January) months by drawing water table contours using the Arc GIS 9.2 software for pre-RWH and post-RWH periods. It is observed that the water table has increased from its level in pre-implementation to that of the post-implementation period. The water table contours for the months of January 2001 (pre-RWH) and January 2011 (post-RWH) are presented in Figures 14.10 and 14.11. In the study area, before the implementation of RWH structures, the water table varies from 2 to 11.2 m only after the monsoon (Jan 2001). But, after implementing RWH structures, the water table varies from 3 to 12.7 m during the post-monsoon period (Jan 2011) which shows that the water levels improved to a great extent.

14.7 Impact of RWH on Groundwater Recharge

The water table fluctuation method is employed for computing possible recharge for the monsoon season within a groundwater assessment unit for the years 1995–2011. Recharge calculated during the pre-RWH period (1995–2003) indicates the natural recharge, and recharge calculated during the post-RWH period indicates the combined effect of natural and artificial recharges (due to the RWH implementation). During the water year (June to May), the seven-month period between June and December is taken as monsoon season and the remaining period as the non-monsoon season. The pre-monsoon interval is taken as May of the previous water year and

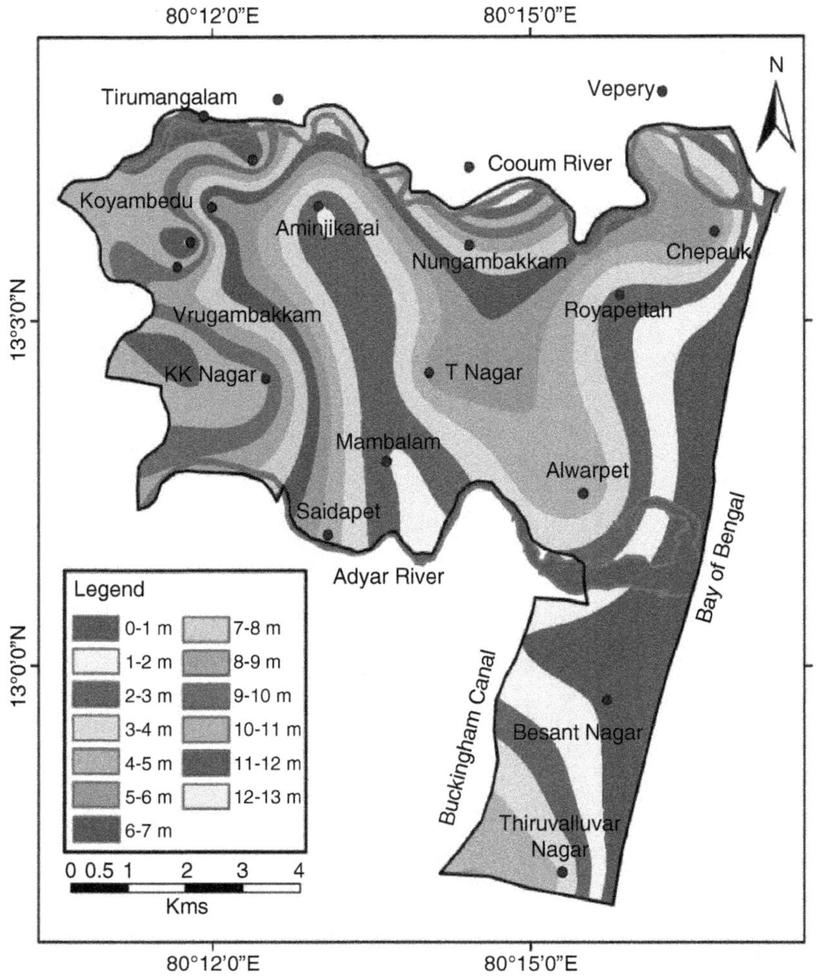

Figure 14.10 Water table contour during Jan 2001.

post-monsoon interval in January. The rise in groundwater level (h) between pre-monsoon (May) and post-monsoon (Jan) levels is estimated from the historic data for each monsoon. Specific yield is considered from the geology map of the study area shown in Figure 14.12 and GEC norms.

Thiessen polygon map is drawn using GIS software for the study area to find the influencing area of each well as presented in Figure 14.13. Using the rise in water level, specific yield, and area of influence, change in storage is estimated for the period 1995–2011. Table 14.1 presents the change in storage calculated for 1995–1996. Groundwater draft was estimated from well density (collected from CGWB) and pumping rate in each area. Recharge during the monsoon period is estimated from change in storage and groundwater draft calculated using Eq. (14.3). The calculated recharge gives the available recharge from rainfall and other sources for the particular monsoon season. The details of recharge estimation are presented in Table 14.2.

A simple analysis of percentage of rainfall becoming groundwater recharge is attempted for the entire study period (1995–2011) as shown in Table 14.2. The average of recharge/rainfall ration during the pre-RWH period was 0.23, whereas the same is 0.31 during the post-RWH period. This signifies the overall impact of the RWH implementation.

The rainfall values are plotted (Figure 14.14) against estimated monsoon recharge using the water level fluctuation method during the pre-RWH period (1995–2003) and a nonlinear regression equation is arrived at with a correlation coefficient 0.963.

$$Y = 0.273\,X^4 - 1.649\,X^3 + 3.503\,X^2 - 2.999\,X + 1.138 \tag{14.4}$$

where Y is the recharge in m; and X is the monsoon rainfall in m.

Using the above nonlinear regression equation (developed for the pre-RWH period), the recharge during the

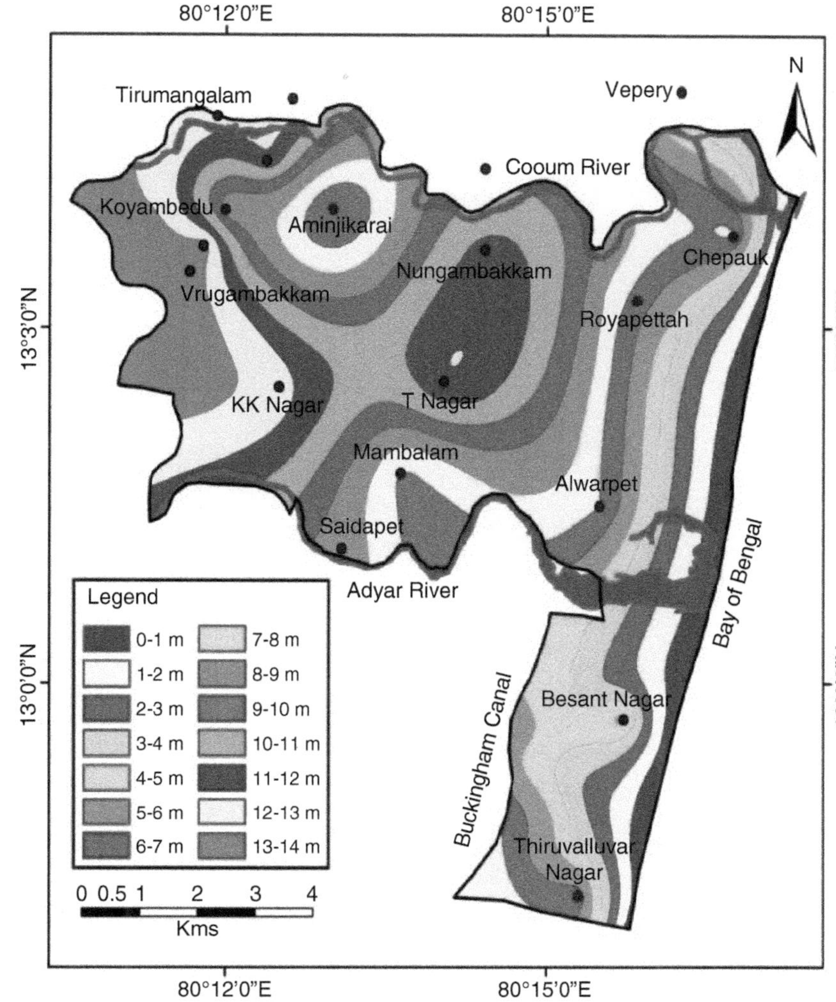

Figure 14.11 Water table contour during Jan 2011.

post-RWH period is calculated for the actual rainfall measurements, which will represent the natural recharge (NR-computed). The difference of NR (computed) from recharge for the post-RWH period (estimated from water level fluctuation method) gives the artificial recharge due to the RWH implementation. Table 14.3 shows the recharge due to RWH during the period 2004–2011. The percentage of artificial recharge ranges from 10% to 30%.

14.8 Model Simulations for Impact of RWH Systems

The model is simulated in transient conditions for a time interval of nine years from January 2004 to May 2012. The calibrated model (the pre-RWH period that is influenced by NR only) is used to simulate the regional groundwater head and it is compared with the observed data (that is influenced by natural and artificial recharge [due to the RWH implementation in the year 2003]) of 15 wells. The difference of observed head and simulated head indicates the impact of RWH, if any.

The results of impact analysis for Jan 2004 are presented in Table 14.4, where the rise of water table is noticed in 10 observation wells and fall in 5 wells. Observation well nos. 1, 3, 7, and 12 are located in suburban watershed, which indicates that the RWH structures provided are not sufficient to improve the water table. But, well no. 14 shows improvement in 2005. Overall, the average rise of water table in Jan 2004 is estimated to be 0.736 m for the monsoonal rainfall of 0.68 m.

Similarly, analysis is carried for all periods and Table 14.5 indicates impact of RWH from 2004 to 2012. A graph shown in Figure 14.15 represents the trend of impact of RWH. In the beginning, the ratio of water table rise to monsoon rainfall is larger but is having a decreasing trend. This is mainly because the implemented RWH structures are losing their efficiency to recharge over a period of time. The top layer of

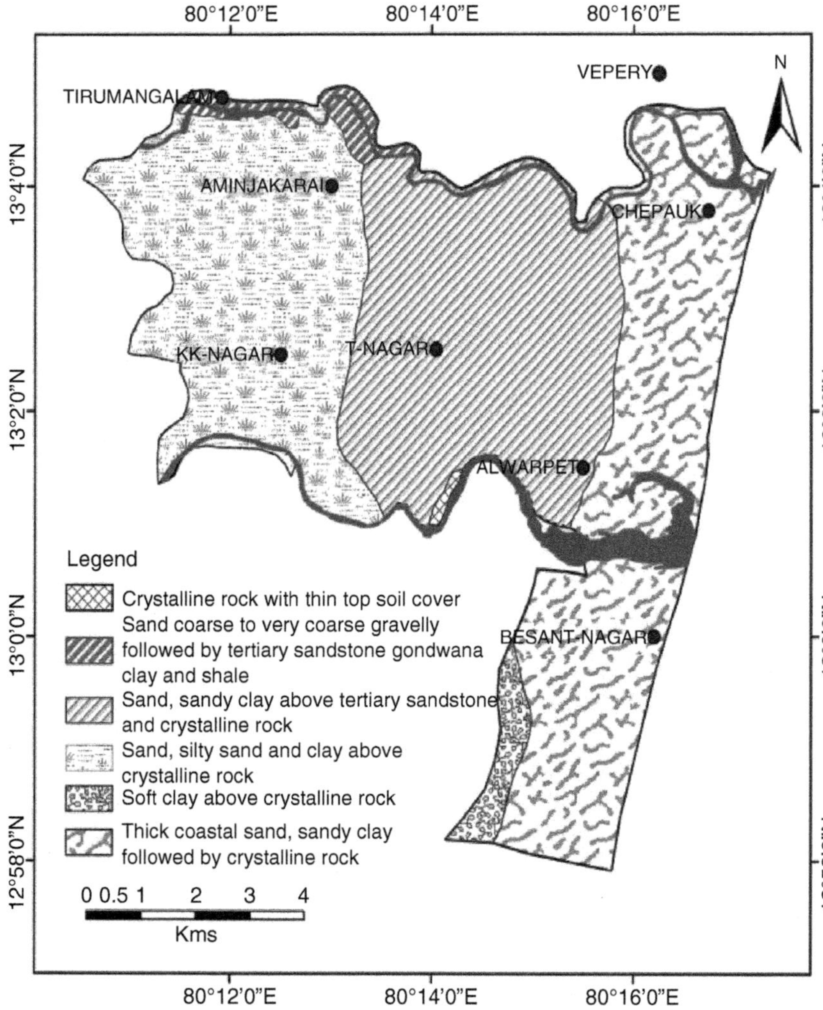

Figure 14.12 Geology map.

RWH structures would have been choked with debris and impurities and are to be removed and replaced at regular intervals. There is a necessity to rejuvenate the RWH structures through regular maintenance for its possible continued impact on the groundwater sustainability.

14.9 Model Predictions for the Future

Prediction scenario on the effect of artificial recharge is carried out after the transient calibration (natural recharge). Improvement in rainfall recharge ratio (after provision of RWH) arrived from regression technique is used for giving the input in Modflow for recharge from 2012 to 2025.

In scenario 1 (S-1), the pumping is given for the future projected population and an average rainfall of 1400 mm is expected to come for recharging of aquifer. But in scenario 2 (S-2), the rainfall expected is less than 20% of average rainfall, which in turn increases the pumping to 20% of scenario 1. In scenario 3 (S-3), the rainfall expected is more than 20% of average rainfall which in turn decreases the pumping to 20% of scenario 1. Then, the model is run to predict the response of aquifer for the above three scenarios. Figures 14.16–14.18 show the mapping of the simulated water table for scenarios 1–3 for the year 2025.

In scenario 1 (average rainfall occurrence), the water table elevation varies from 0 to 12 m, whereas in scenario 2 (rainfall occurrence is less by 20%), it varies from 0 to 10 m, showing a decrease of 2 m. But, in scenario 3 (rainfall occurrence is more by 20%), the water table variation is 0–14 m, indicating a water level rise of 2 m. In these prediction scenarios, additional well drilling and groundwater pumping in the study area have to be monitored and controlled properly for a sustainable future. Efforts could also be in the direction of providing additional RWH structures to improve the water table.

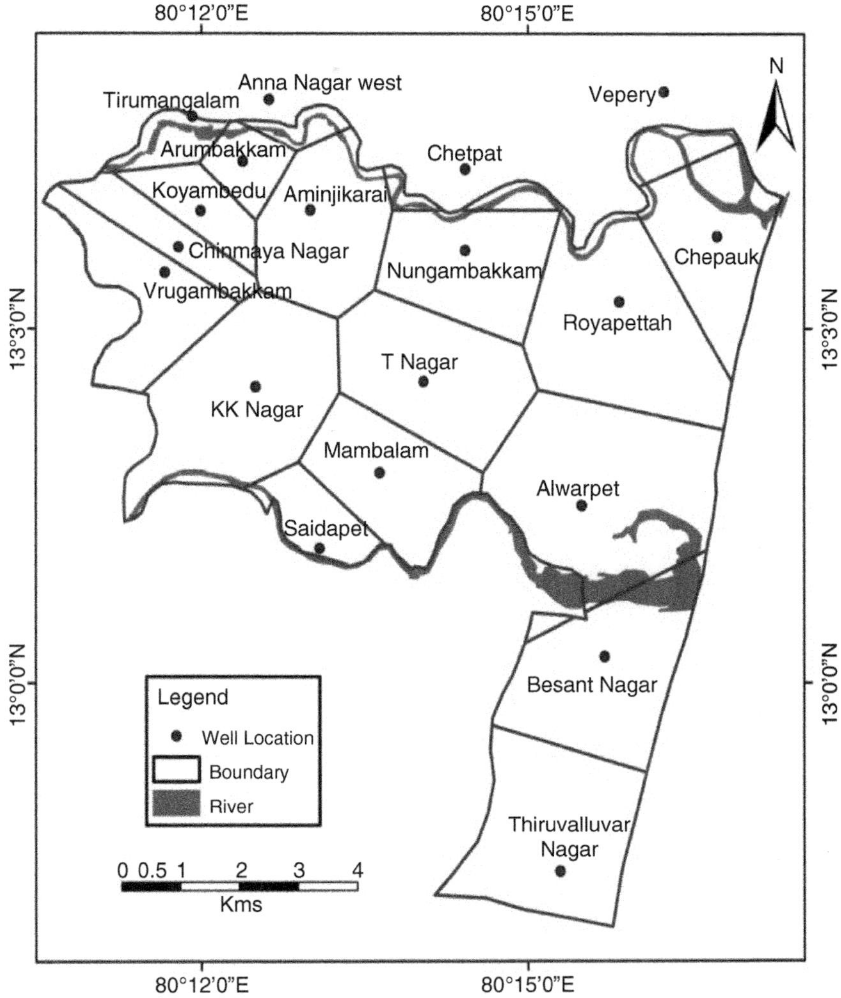

Figure 14.13 Thiessen polygon map.

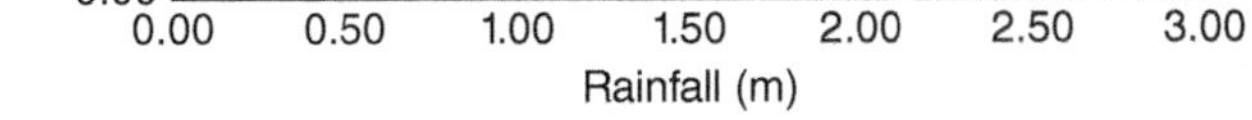

Figure 14.14 Correlation of rainfall and recharge during the pre RWH period.

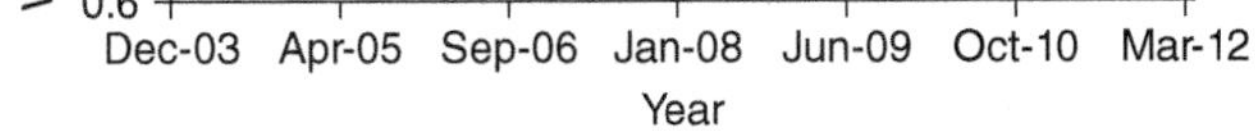

Figure 14.15 Trend of the RWH impact during 2004–2012.

Table 14.1 Change in groundwater storage for the year 1995–1996.

Well no.	Area of influence (km^2)	Water level in May 1995 (m)	Water level in Jan 1996 (m)	Water level Fluctuation (m)	Specific yield	Change in storage (MCM)
1	1.198	4.74	0.99	3.75	0.02	0.067
2	0.664	5.20	4.25	0.95	0.02	0.009
3	1.765	4.95	3.55	1.40	0.02	0.037
4	4.754	6.64	6.01	0.63	0.02	0.045
5	1.319	4.10	1.05	3.05	0.02	0.060
6	2.429	2.60	1.80	0.80	0.02	0.029
7	3.241	2.20	0.90	1.30	0.02	0.063
8	7.523	2.48	1.35	1.13	0.02	0.128
9	5.866	4.62	4.13	0.49	0.03	0.086
10	4.179	3.05	2.25	0.80	0.03	0.100
11	1.729	5.05	4.25	0.80	0.03	0.042
12	4.612	3.00	1.85	1.15	0.03	0.159
13	7.620	7.00	5.05	1.95	0.03	0.446
14	9.666	6.21	2.72	3.49	0.03	1.012
15	4.411	3.20	1.37	1.83	0.10	0.807
16	0.830	6.10	5.70	0.40	0.03	0.010
17	1.134	0.97	0.77	0.20	0.03	0.007
18	7.933	4.33	4.78	−0.46	0.10	−0.361
19	4.566	4.05	4.10	−0.05	0.10	−0.023
	75.440					2.724

Table 14.2 Rainfall recharge ratio during 1995–2011.

Year	Groundwater draft (MCM)	Groundwater storage (MCM)	Recharge (MCM)	Depth of recharge (m)	Rainfall (m)	Rainfall/recharge ratio
1995–1996	18.375	2.724	21.099	0.28	1.07	0.26
1996–1997	18.706	11.093	29.799	0.40	2.42	0.16
1997–1998	19.043	6.295	25.338	0.34	1.98	0.17
1998–1999	19.386	2.043	21.429	0.28	1.03	0.27
1999–2000	19.735	0.921	20.656	0.27	1.11	0.25
2000–2001	20.090	−1.888	18.202	0.24	0.84	0.29
2001–2002	20.513	4.367	24.881	0.33	1.53	0.22
2002–2003	20.883	4.531	25.414	0.34	1.34	0.25
2003–2004	21.259	−1.484	19.774	0.26	0.68	0.38
2004–2005	21.641	3.998	25.639	0.34	0.91	0.37
2005–2006	22.031	13.772	35.803	0.47	1.94	0.24
2006–2007	22.427	3.989	26.416	0.35	1.29	0.27
2007–2008	22.831	2.351	25.182	0.33	1.15	0.29
2008–2009	23.242	2.464	25.706	0.34	1.16	0.29
2009–2010	23.660	7.470	31.131	0.41	1.14	0.36
2010–2011	24.086	4.899	28.986	0.38	1.42	0.27

Table 14.3 Recharge due to RWH during the period 2004–2010.

SL. no	Year	Depth of rainfall (m)	Possible natural recharge from pre RWH equation (m)	Actual recharge observed (m)	Recharge due to RWH alone (m)	% of Recharge due to RWH
1	2004–2005	0.91	0.25	0.34	0.09	25.17
2	2005–2006	1.94	0.33	0.47	0.14	30.35
3	2006–2007	1.29	0.32	0.35	0.03	9.99
4	2007–2008	1.15	0.29	0.33	0.04	12.51
5	2008–2009	1.16	0.29	0.34	0.05	14.16
6	2009–2010	1.14	0.29	0.41	0.12	29.69
7	2010–2011	1.42	0.33	0.38	0.05	13.69

Table 14.4 Impact of RWH in Jan 2004.

SL. no.	Observation well no.	Observed water table (m)	Calibrated water table (m)	Rise/fall (m)
1	5	10.24	8.79	1.45
2	6	12.48	9.67	2.81
3	8	11.03	9.66	1.37
4	9	10.48	8.50	1.98
5	11	7.29	7.27	0.02
6	13	6.69	3.65	3.04
7	15	5.88	2.53	3.35
8	18	3.96	1.52	2.44
9	1	8.43	9.45	−1.03
10	3	6.75	7.24	−0.49
11	4	8.76	8.61	0.15
12	7	8.55	9.97	−1.42
13	10	7.06	6.62	0.44
14	12	5.77	8.50	−2.73
15	14	4.32	4.72	−0.40

Table 14.5 Impact of RWH from 2004 to 2012.

Sl. no.	Month/ Year	Average rise of water table (m)	Monsoon rainfall (m)	Water table/ rainfall ratio
1	Jan-04	0.73	0.68	1.0735
2	Jan-05	0.84	0.91	0.9231
3	Jan-06	1.8	1.94	0.9278
4	Jan-07	1.42	1.29	1.1008
5	Jan-08	1.2	1.15	1.0435
6	Jan-09	1.05	1.16	0.9052
7	Jan-10	0.99	1.14	0.8684
8	Jan-11	1.27	1.42	0.8944
9	Jan-12	1.02	1.19	0.8571

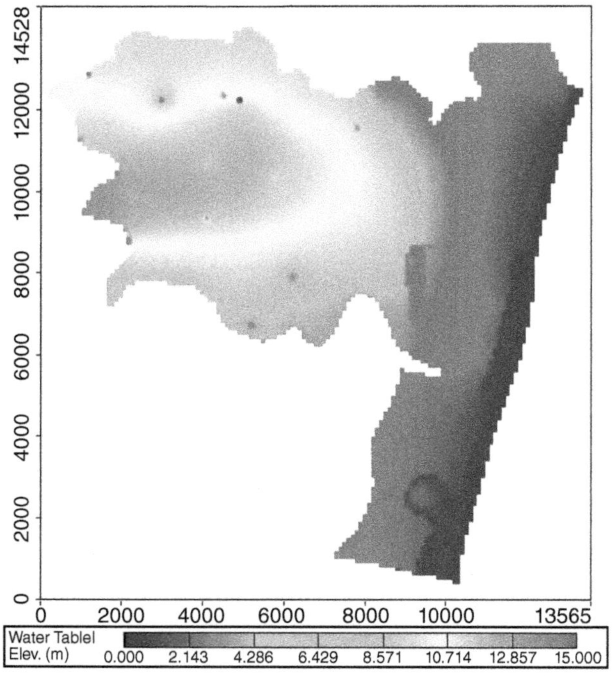

Figure 14.16 Computed water level contours for Jan 2025 S-1.

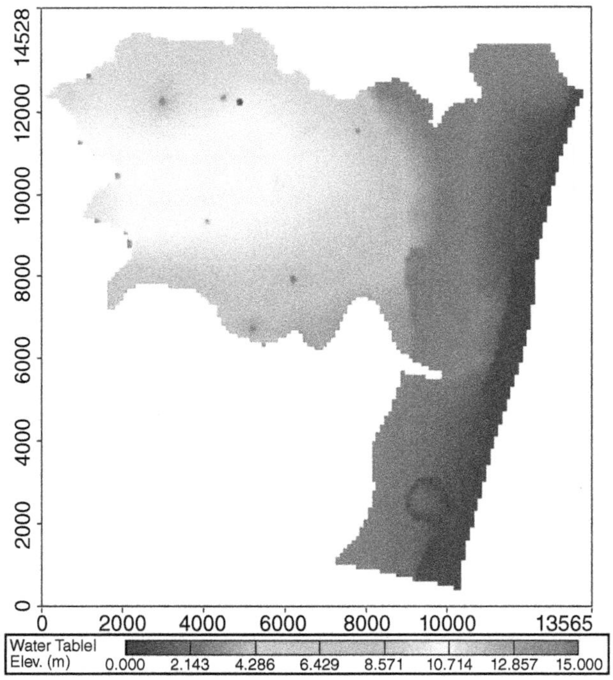

Figure 14.17 Computed water level contours for Jan 2025 S-2.

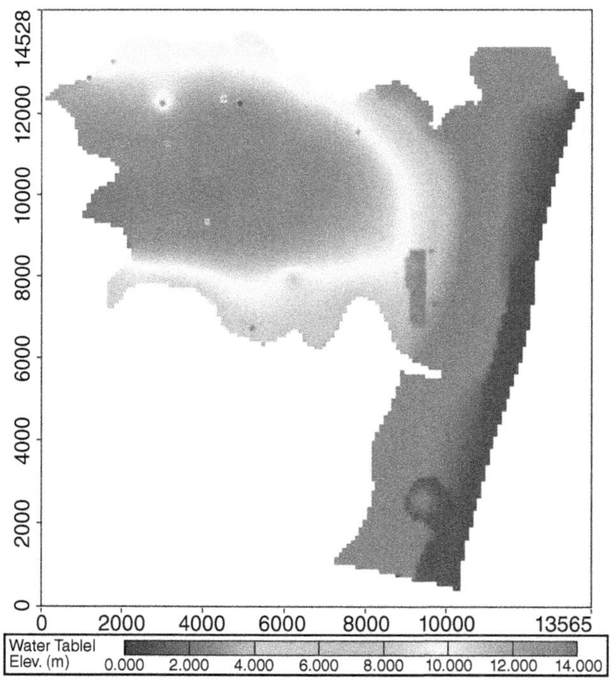

Figure 14.18 Computed water level contours for Jan 2025 (S-3).

14.10 Conclusion

RWH and artificial recharge techniques are useful in water management, and have gained tremendous interest among academicians, institutions, and nonprofessionals in the past few years since they are low-cost solutions to water crisis. This study has examined the impact of the RWH implementation. Groundwater hydrographs show decreasing and increasing trends during the pre- and post-RWH periods, respectively. The spatial variation of water table contours drawn using ArcGIS software shows that many of the urban areas have lower groundwater levels before implementation of RWH. The majority of the study areas have raised groundwater levels during the post-RWH period. Groundwater recharge is assessed based on GEC Norms using the water level fluctuation method. Rainfall recharge ratio is estimated for the pre- and post-RWH periods as 0.23 and 0.31, which signify the RWH impact. Segregation of artificial recharge due to RWH is attempted using nonlinear regression equation. The percentage of artificial recharge ranges from 10% to 30%, which shows the recharge improvement in the post-RWH period. Hence, it is concluded that groundwater recharge is in increasing trend after provision of the RWH structures in the study area.

In reality, it is not possible to see into the sub-surface and observe the geological structure and the groundwater flow processes. It is for this reason that groundwater flow models have been used to investigate the important features of groundwater systems and to predict their behavior under particular conditions. Hence in this study, a groundwater flow model is developed for the urban aquifer wherein the groundwater head over space and time are simulated. The computed groundwater head over space and time matches well with the observed groundwater head. The impact of RWH simulated through modeling shows the improvement of water table. But, the results from the future prediction scenarios indicate that water table will go down if pumping is increased and rainfall is decreased. Hence, there is a necessity to educate the society for proper usage of groundwater and maintaining the RWH structures.

Acknowledgement

The members of Indian National Committee on Groundwater **(INCGW)**, "Ministry of Water Resources, Government of India," are thanked for the financial support of the research project on "Developing a Methodology to Evaluate the Impact of RWH in Urban Areas" and their constructive comments on the research work.

References

Adham, A., Riksen, M., Ouessar, M., and Coen, J.R. (2016). A methodology to assess and evaluate rainwater harvesting techniques in (semi-) arid regions. *Water* 8: 1–23.

Bekele, E., Vanderzalm, J., and Sidhu, J. (2018). Managed aquifer recharge (MAR) in sustainable urban water management. *Water* 10: 1–16.

Boers, T.M. and Ben-Asher, J. (1982). A review of rainwater harvesting. *Agricultural Water Management* 5: 145–158.

Chenini, I. and Mammou, A.B. (2010). Groundwater recharge study in arid region: an approach using GIS techniques and numerical modeling. *Computers & Geosciences* 36: 801–817.

Deepak, K., Ramakand, O., and Srivatsava, R.K. (2004). Impact assessment of rainwater harvesting on groundwater quality at Indore and Dewas, India. *Journal of Indian Water Works Association* XXXVI (2): 123–128.

Fayez, A.A. and Al-Shareef, A.W. (2009). Roof rainwater harvesting systems for household water supply in Jordan. *Desalination* 243: 195–207.

Glendenning, C.J. and Vervoort, R.W. (2011). Hydrological impacts of rain water harvesting (RWH) in a case study catchment: the Arvari River, Rajasthan, India. *Agricultural Water Management* 98: 715.

Israil, M., Al-hadithi, M., and Singhal, D.C. (2006). Application of a resistivity survey and geographical information system. (GIS) analysis for hydrogeological zoning of a piedmont area, Himalayan foothill region, India. *Hydrogeology Journal* 14: 753–759.

Jaworska-Szulc, B. (2009). Groundwater flow modelling of multi-aquifer systems for regional resources evaluation: the Gdansk hydrogeological system, Poland. *Hydrogeology Journal* 17: 1521–1542.

John, M.S., Michael, E.M., and Elizabeth, P.A. (2009). Effects of rain water harvesting induced artificial recharge on the groundwater of wells in Rajasthan, India. *Hydrogeology Journal* 17: 2061–2073.

Martínez-Santos, P., Pedretti, P., Martínez-Alfaro, P.E. et al. (2010). Modelling the effects of groundwater-based Urban supply in low-permeability aquifers: application to the Madrid aquifer, Spain. *Water Resources Management* 24: 4613–4638.

Matthew, P.J. and William, F.H. (2010). Performance of rainwater harvesting systems in the southeastern United States. *Resources, Conservation and Recycling* 54: 623–629.

Mohammed-Aslam, M.A., Kondoh, A., Mohamed, P.R., and Manoharan, A.N. (2010). Evaluating Groundwater Potential of a Hard-rock Aquifer Using Remote Sensing and Geophysics. *Journal of Spatial Hydrology* 10 (2): 29–41.

Okhravi, S., Eslamian, S., Eslamian, F., and Tarkesh Isfahany, S. (2014). Indigenous knowledge as a supportive tool for sustainable development and utilisation of rainwater harvesting systems. *Journal of Flood Engineering* 5 (1–2): 39–50.

Pachpute, J.S., Tumbo, S.D., Sally, H., and Mul, M.L. (2009). Sustainability of rainwater harvesting Systems in Rural Catchment of sub-Saharan Africa. *Water Resources Management* 23: 2815–2839.

Report of Groundwater Estimation Committee. (2009) Groundwater resource Estimation Methodology – 1997, Ministry of Water Resources, Government of India, New Delhi, India.

Senthilkumar, M. and Elango, L. (2011). Modelling the impact of a subsurface barrier on groundwater flow in the lower Palar River basin, southern India. *Hydrogeology Journal* 19: 917–928.

Sharma, S.K. and Jain, S.K. (1997) Groundwater Recharge through Roof Top Rain Water Harvesting in Urban Habitation, International Conference on Management of Drinking Water Resources, CLRI, Chennai, India.

Sivaraman, K.R. and Thillai Govindarajan, S. (2002). *Manual on Rain Water Harvesting*. Chennai, Tamil Nadu: Shruthi Publications., India.

Sturm, M., Zimmermann, M., Schutz, K. et al. (2009). Water for people - rainwater harvesting as an alternative water resource in rural sites in central northern Namibia. *Physics and Chemistry of the Earth, Parts A/B/C* 34 (13–16): 776–785.

Taheri, A. and Zare, M. (2011). Groundwater artificial recharge assessment in Kangavar Basin, a semi-arid region in the Western part of Iran. *African Journal of Agricultural Research* 6 (17): 4370–4384.

Villarreal, E.L. and Dixon, A. (2005). Analysis of a rainwater collection system for domestic water supply in Ringdansen, Norrköping, Sweden. *Building and Environment* 40 (9): 1174–1184.

Zhou, Y.-c., Shao, W.-y., and Zhang, T.-q. (2010). Analysis of a rainwater harvesting system for domestic water supply in Zhoushan, China. *Journal of Zhejiang University-Science A (Applied Physics and Engineering)* 11 (5): 342–348.

15

Effects of Water Harvesting Techniques on Sedimentation

Siavash Fasihi[1] *and Saeid Eslamian*[2]

[1] *Abvarzan Consulting Engineers, Tehran, Iran*
[2] *Department of Water Engineering, College of Agriculture Isfahan University of Technology, Isfahan, Iran*

15.1 Introduction

Water harvesting is defined as the collection of water for its productive use (Critchley et al. 1991). The durability and sustainability of water resources in a catchment, and functionality of storage, however, can be threatened by siltation due to runoff and soil erosion processes, which have on-site and off-site effects (Frankl et al. 2016; Haregeweyn et al. 2008a, 2008b).

For many years, to collect water from rains in certain regions of the world, several different water harvesting techniques (WHTs) were introduced and implemented, such as stone bunds, infiltration or deep trenches, check dams, and percolation ponds (Gebrernichael et al. 2005; Nyssen et al. 2004; Vancampenhout et al. 2006; Walraevens et al. 2015). As these techniques are widely applied in a watershed scale, they can either directly or indirectly affect various aspects of environmental and managerial policies regarding agroeconomy through erosivity control. (Okhravi et al. 2014).

One major influence of WHTs is reducing sedimentation which consequently diminishes the negative impacts on food production, drinking water quality, ecosystem services, mud control, eutrophication, biodiversity, and carbon stock shrinkage (Boardman and Poesen 2007).

Increasing values of surface roughness elements, due to the implementation of WHTs, lead to decreased soil loss by slowing down the flow velocity and trapping eroded soil. Some studies reflect the ability of such techniques to reduce the runoff volume by 40% and sediment yield to an average extent of 60% (Grum et al. 2017).

In some cases, floodwater spreading techniques are referred to as WHTs. These operations are based on expanding decentralized temporary and/or permanent flows over a floodplain in a way to improve soil and water conditions. The capacity of such designs to reduce sediment yield to the downstream watershed and river reaches has both upstream and downstream benefits. They can prevent soil and nutrient loss from upstream fields, thereby restoring soil fertility (Dile et al. 2016). For instance, Gunnell and Krishnamurthy (2003) reported that in dryland peninsular India, farmers transfer fine textured tanked-bed sediment from ex-situ water harvesting systems to the fields of the catchment using bullock carts to balance soil texture and optimize on-site fertility.

While downstream social-ecological systems will benefit from reduced nutrient release, studies showed that nutrients (e.g. nitrogen and phosphorus) are transported to riverine ecosystems embedded through the sediment loads, and these nutrients can degrade water quality downstream (Beusen et al. 2005; Holtan et al. 1988; Ittekkot and Zhang 1989; Lal 2004; Ludwig et al. 1996). Although the global-scale impact of nutrient cycles is still unknown, soil erosion plays an important role in these biogeochemical cycles through lateral fluxes of sediment. These impacts are not only irrevocable but also compared in magnitude as fluxes induced by fertilizers (Regnier et al. 2013).

WHTs like ponds, by trapping sediment loads and nutrients, can improve water quality and avoid downstream ecological problems such as eutrophication of lakes and river reaches. Moreover, they can reduce siltation of lakes and reservoirs downstream. On the other hand, siltation of water harvesting ponds will be a daunting phenomenon (Tamene et al. 2006), and dredging sediment loads from water harvesting ponds to fields will be a challenging task. Releasing sediment-free water to the downstream reaches also has environmental consequences. For instance, sediment-free water released from water harvesting ponds or (floodwater control basins) can erode finer sediments from the receiving channel (Poff et al. 1997). This can result

Handbook of Water Harvesting and Conservation: Basic Concepts and Fundamentals, First Edition. Edited by Saeid Eslamian and Faezeh Eslamian.

in progressing head-ward channel down-cutting and erosion (Chien 1985). Habitat availability for the many aquatic species living in interstitial spaces may also decrease as a result of coarsening of the streambed (Dile et al. 2016).

Thus, to quantify the effectiveness of diverse WHTs some models are developed which pursue the approach of using plot measurements to obtain soil and water conservation measures (Hessel and Tenge 2008). However, these models may basically fall into two categories: empirical and process-based. Undoubtedly the prime example of an empirically based model is USLE (Nearing 2013) and among various process-based models LISEM is notable (Hessel and Tenge 2008).

In order to quantitatively analyze the effects of WHTs on sedimentation, globally or in a catchment scale, one should be able to apply the aforementioned models. So, the following sections of this chapter are devoted to investigating different aspects of such models thoroughly. Firstly, more simple models are introduced and subsequently examples of problems solved by more complex models are brought to attention in more details.

15.1.1 How to Incorporate WHTs in Models

In order to simulate the effect of WHTs on sedimentation and runoff in any of the mentioned models, it is first necessary to collect data on the effect of WHTs on input parameters of the model.

This may induce some changes in model inputs. In catchment scale physically based models, some of these incorporated changes directly affect the value of certain input features. While some of the effects might be reproduced by changing input parameters, they may not directly be affected by the change. If suddenly infiltration ditches are implemented in a catchment model, not all of their effects can be incorporated directly. Their main effect should be that they store water, increase infiltration, and reduce surface flow velocity. Such effects of infiltration ditches might indirectly be simulated by changing certain features like saturated electrical conductivity, random roughness, or Manning's n coefficient, even if the ditch itself does not directly affect those input parameters (Hessel et al. 2003; Hessel and Tenge 2008).

All put together, the input parameter responsible for considering the major effects of WHTs in any model (LISEM [Limburg soil erosion model] or RUSLE [revised universal soil loss equation]) is called practice factor, or P-factor for short. This factor mainly consists of information on the impact of various WHTs on surface flow and soil erosion collected from technical literature and reported as P-factor. To summarize, this factor, which is the ratio of runoff and soil erosion rates in presence of WHTs to the rates in areas without WHTs, is used to quantify the effect of WHTs on runoff and sediment yield for catchments.

15.2 Qualitative Effects and Data Collection

This section is devoted to briefly describing some of the qualitative effects of WHTs, data collection, and field measurement approaches to assess the specific sediment yield of small check dams. Figure 15.1 illustrates examples of WHTs implemented in a catchment scale.

In assessing the effects of WHTs on sedimentation, knowledge of watershed erosion rate and exacerbating factors, along with its recognition and measuring methods, are of much importance. Throughout arid and semi-arid regions, water shortage is the major limiting factor for agricultural development and rangeland improvement. Runoff causes erosion of fertile topsoil, resulting in soil degradation and overexploitation of natural resources for forest and rangeland production (Schiettecatte et al. 2005; Gupta 1995). Therefore, WHTs have long been utilized as an innovative way to reduce soil erosion and sedimentation and to increase soil water storage and soil fertility (Li et al. 2006). WHTs consist of two components: the catchment area, where runoff is collected, and the cultivated area, where the runoff is concentrated (Critchley et al. 1991). Stone trace is an exemplary WHT used in mountainous areas of the Middle East and some other regions in order to avoid soil loss by reducing the amount of runoff and subsequently, soil moisture content and organic matter such as Mg, Ca, and K showed a rising trend (Hammad et al. 2006).

It is notable to mention that reduction in runoff amount does not necessarily lead to sediment yield deduction for each rainfall event (Alseekh and Ghaleb 2009). It might happen due to an unfit WHT selection or simply because of a combination of local temporary hydrological and physiographical responses of the catchment, such as soil moisture remaining from previous rainfalls.

In another case, trapping efficiency of dikes, which leads sediments to accumulate behind them, extended the arable land to be used for cropping. It is notable to mention these lands were uncultivated before implementing the dikes. The sediments that accumulate behind the dikes are used for cropping. Field observations demonstrated that sedimentation due to implementation of WHTs have effects on soil moisture content (Alseekh and Ghaleb 2009). Field measurement showed that the water harvesting structures, of stone terraces and semi-circle, significantly increased the moisture content from that of the control during the first month of the rainy season. The differences between

Figure 15.1 Examples of WHTs implemented in a catchment: (a) percolation pond; (b) deep trench; (c) gabion check dam; (d) concrete check-dam pond (Grum et al. 2017).

the treatments such as stone terraces, soil semi-circle bunds, and soil contour ridges were reported to be small (Alseekh and Ghaleb 2009).

Different WHTs have various effects on soil properties. Studies showed that stone terraces and semi-circle bunds had similar effects on soil texture and mostly increased the coarse sand percentage. Electrical conductivity (EC) measurements were higher in semi-circle bunds than that of stone traces, while the trend was the opposite in the case of soil coarse sand percentage. pH, available nitrogen ($NH4+$ and $NO3-$), and phosphorus were higher than in the control (natural vegetation) (Alseekh and Ghaleb 2009). This latter outcome strengthens the idea that WHTs have positive effects on improving water quality of downstream reaches by controlling nitrogen and phosphorus cycles through trapping sediment yield. Of course, it may take years to reveal the actual effects of WHTs on soil, but chemical properties exhibit small differences between various WHTs (Alseekh and Ghaleb 2009). Some reports indicated that nutrient content of soil experienced an increase (Marx et al. 1999).

15.2.1 Measurements and Data Input

Several parameters are included in estimating soil loss rate using either field investigations or computer-based models such as rainfall, physical and chemical soil properties, moisture content, land cover/land use, and so on. Required data varies in spatial and temporal resolutions for different field approaches or computer-based models. Specific inputs of each model are thoroughly described in relevant sections. Since in some parts of the world, soil conservation and research are notably recent (Haregeweyn et al. 2005), the need for data and scientific approaches has emerged in the present decade. Thus, what follows is a brief description of some of these parameters and a sample methodology of how each of them is measured.

Soil samples are needed to describe physical and chemical properties of soil such as bulk density pH, soil organic carbon, and nutrients. Samples might be collected from different soil depth ranges between 0.2 and 2.0 m depending on the type of parameter under investigation. Number of samples depends greatly on the research budget, time, and the intended accuracy. If there is access to desirable datasets or maps in the case of modeling soil loss in larger areas, it is useful to directly apply them in models. For cases with small areas and/or where data is unavailable, using an ergonomic hand auger is recommended.

Rainfall data for some regions are available online. It is also possible to measure rainfall using a tipping-bucket data-logging RG3-m HOBO rain gauge (precision: 0.2 mm per tip) installed on site. To monitor the soil moisture content the TDR method (time-domain reflectometer) is acceptable. Other sources are remote-sensing products, which might be available in different formats and accuracy levels.

Runoff volume per plot is measured after each rain event or in certain times during the rainy season. A pre-determined portion of liquid is collected from the barrel used to harvest runoff from the experimental plot which then will be used to analyze sediment concentration and probably eroded soil and lost nutrients. By filtering a water sample through appropriate filter paper and drying the residue at room temperature, the sediment concentration of runoff and total amount of eroded soil per plot is calculated.

15.3 Sedimentation in Small Check Dams

Of all WHTs, small dams constructed in catchments are somehow more vulnerable to sedimentation. These structures are mostly earth embankments designed to harvest seasonal runoff. The stored water would then be used at least as a supplementary source for irrigation. These small dams consist of a dam body, an irrigation outlet canal, which is positioned at an elevation above the reservoir to avoid sediments from entering the farm lands, and a spillway to safely pass the floods. Measurement of sediment volume in such structures has some direct approaches which are not highly dependent on factors necessary for soil loss estimate models. Thus, in this section a method that has been developed to assess the specific sediment yield is explained (Verstraeten and Poesen 2001).

In this approach, specific sediment yield is generally related to drainage area and sediment mass which is proposed in Eq. (15.1):

$$SSY = 100 \times \frac{M}{A \times STE \times Y} \tag{15.1}$$

where: SSY = specific sediment yield ($\mathrm{t\,km^{-2}\ y^{-1}}$); M = sediment mass (t); A = drainage area (km2); STE = sediment trap efficiency (%); Y = age of the reservoir (y);

Sediment mass is derived from Eq. (15.2):

$$\mathrm{M = Sv \times dBD} \tag{15.2}$$

where:

Sv = the measured volumetric sediment input in the reservoir ($\mathrm{m^3}$); and

dBD = the area-weighted average dry bulk density of the sediment ($\mathrm{tm^{-3}}$).

Measurement of sediment volume is possible by many methods. One is application of mapping techniques to estimate the deposited extent. The calculation procedure starts with measuring sediment thickness for many numbers of points determined by the size of the reservoir. Next step is to map the deposited sediment boundary using the data generated in the first phase and noting the maximum water level of the dam. Then the boundary of zero sediment thickness is delimited using GIS software which eventually is converted to raster to generate a digital elevation model (DEM). Data including volume of the sediment area is determined by extracting the summary statistics of the raster file. Detailed procedure and relevant modules are fully described in Haregeweyn et al. (2005) and Verstraeten and Poesen (2001). To compare the results with observations and reported data, sediment volume yield should be in mass units. Hence, sediment bulk density is used to make this conversion. The dry bulk density map is determined by the gravimetric method and geographically interpolation of gathered data points. The mass of deposited sediment is determined by multiplying the produced DEM and dBD map. Sediment trap efficiency is regarded as the ratio of retained sediments to that discharged from the reservoir. Particularly in outlying areas, consulting local people and making observations are two helpful options available to estimate sediment trap efficiency.

It is often important for watershed management policy makers to have an estimation of life expectancy of reservoirs. This latter is closely related to effects of the water harvesting structure (dams in this example) on sedimentation. What follows in this section is a hint to calculate life expectancy simply described by (Haregeweyn et al. 2005).

$$\mathrm{LE} = \frac{\mathrm{DSV}}{\mathrm{SR}} \tag{15.3}$$

where: LE = life expectancy of the reservoir (y); DSV = dead storage volume of the reservoir during the design stage ($\mathrm{m^3}$); SR = sediment deposition rate ($\mathrm{m^3\ y^{-1}}$).

$$\mathrm{SR} = \frac{\mathrm{SV}}{\mathrm{T}} \tag{15.4}$$

where: SV is the sediment volume ($m^3\ y^{-1}$); T duration of accumulation (y).

These structures are initially designed to harvest seasonal runoff in mountainous areas. Although a fraction of the reservoir is foreseen for sediment accumulation, the risk of sedimentation still threatens the duration of functionality of these dams.

In the previous sections, qualitative aspects of WHTs' effects on sedimentation, measurements, and data collection, and a relatively basic methodology to calculate specific sediment yield of small dams is presented. In the following parts of this chapter, two of most commonly used computer-based approaches to quantitatively assess the effect of WHTs on sedimentation are presented.

15.4 Revised Universal Soil Loss Equation (RUSLE)

Conservation of soil and water requires both knowledge of the factors affecting these resources, and methods for controlling them (Morgan and Nearing 2011). These methods might as well be used to assess the effects of WHTs quantitatively. For this purpose, the overall approach is to run the model for environmental conditions before and after implementing WHTs. Then by subtracting the amount of calculated sediment yield for these two simulations the effects of WHTs on sedimentation could be measured.

One of the methods that has been widely used over recent decades and demonstrated reasonable estimations for erosion in different land uses is RUSLE (Revised Universal Soil Loss Equation). This model, which is the revised form of USLE, has many versions developed since the late 90s and the program was distributed across the United States in 2004. The most updated version is RUSLE 2. A modification of the RUSLE model was proposed in year 2015 (named RUSLE2015) which improves the quality of estimation by introducing updated, high-resolution, peer-reviewed input layers (Panagos et al. 2015a–e). For detailed information regarding the history of the model, as well as enhancements, capabilities, and tutorials, refer to Handbook of Erosion Modeling by Morgan and Nearing 2011, the RUSLE manual or visit and search the contents of the following websites:

- https://www.ars.usda.gov (ARS Website)
- http://fargo.nserl.purdue.edu (NRCS Website)

RUSLE calculates mean annual soil loss rates by sheet and rill erosion according to the Eq. (15.5) (Panagos et al. 2015a–e):

$$E = R \times K \times C \times LS \times P \tag{15.5}$$

where E: annual average soil loss ($t\ ha^{-1}\ yr.^{-1}$), R: rainfall erosivity factor ($MJ\ mm\ ha^{-1}\ h^{-1}\ yr.^{-1}$), K: Soil erodibility factor ($t\ ha\ h\ ha^{-1}\ MJ^{-1}\ mm^{-1}$), C: cover-management factor

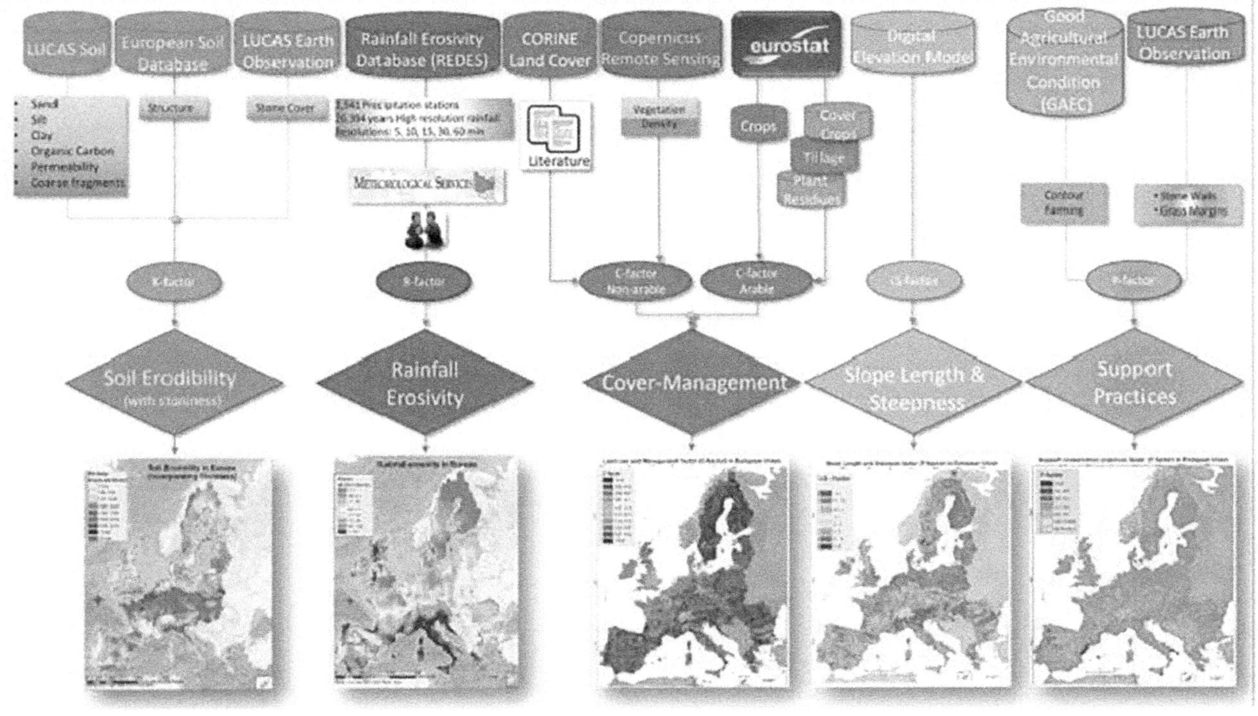

Figure 15.2 Input datasets used for soil loss estimation for Europe in RUSLE2015 (Panagos et al. 2015e).

(dimensionless), LS: slope length and slope steepness factor (dimensionless), and P: support practice factor (dimensionless). Figure 15.2 illustrates the input layers used for estimating erosion rate at the European scale. This figure is only proposed here as an explanation of the sequence of actions and/or possible source of some data.

One important point that should be noted is selecting the appropriate resolution of the output soil loss map. It depends on the resolution of determinant parameter and other maps and data availability of input factors. Also, this resolution is better to fall between coarse values of K-factor, for instance, and high-resolution input layers such as LS-factor. Implementation of WHTs could mostly alter the values of P-factor and C-factor input layers.

What follows is a description of the parameters involved in RUSLE equation leading to supplementary material for these factors and corresponding publications. Eventually some of its abilities and limitations are mentioned.

R-factor: The average annual erosivity value R is an index of erosivity of the climate at a location and the sum of the daily r values. This factor is determined from historical records of individual storms. The erosivity of an individual storm is computed as the product of the storm's total energy, which is closely related to storm amount and the storm's maximum 30-minute intensity. Many examples showed that erosivity values can vary significantly among locations having nearly equal rainfall amounts because of the difference in rainfall intensity among locations (RUSLE User's Reference Guide 2008).

This factor gives the combined effect of the duration, magnitude, and intensity of each rainfall event (Eq (15.6)) (Brown and Foster 1987):

$$R = \frac{1}{n} \sum_{j=1}^{n} \sum_{k=1}^{mj} (EI)_k \tag{15.6}$$

where R = average annual rainfall erosivity, n is the number of years covered by the data records, mj is the number of erosive events of a given year j, and EI_{30} is the rainfall erosivity index of a single event k. The event erosivity EI_{30} is defined as:

$$EI30 = \left(\sum_{r=1}^{k} e_r v_r \right) I_{30} \tag{15.7}$$

where e_r is the unit rainfall energy (MJ ha^{-1} mm^{-1}) and v_r is the rainfall volume (mm) of r_{th} time period of a rainfall event divided in k-parts. I30 is the maximum rainfall intensity during a 30-min period of the rainfall event (mm h^{-1}). The unit rainfall energy (e_r) is calculated for each time interval as follows (Brown and Foster 1987):

$$e_r = 0.29[1 - 0.72 \exp(-0.05 i_r)] \tag{15.8}$$

where i_r is the rainfall intensity during the time interval (mm h^{-1}).

Depending on the spatial scale and data availability of the study area, some considerations and adjustments should be made on the R-factor calculated using the above procedure (Panagos et al. 2015a).

Given the extent of the watershed, several parameters such as low temporal and/or spatial resolution of precipitation data and the huge climatic variability of the study area might affect the value of R.

Since the major force causing soil loss on a global scale is known to be water, it is undoubtedly crucial to be able to measure this type of erosion with extreme caution.

Following the approach (called the empirical approach), equations have been developed to predict R-factor based on rainfall data with lower temporal resolution (de Santos Loureiro and de Azevedo Coutinho 2001; Alcamo et al. 2007; Diodato and Bellocchi 2007; Panagos et al. 2012). Moreover, those empirical equations cannot capture the high rainfall intensities which have significant influence on the average rainfall erosivity. R-factor equations developed for a specific region cannot be applied to the much larger areas, like the whole of Europe, or areas with diverse climatic conditions (Panagos et al. 2015a). Thus, taking the most suitable and pragmatic approach to calculating R-factor needs expert knowledge of regional climate and availability of high-resolution participation data. Further details on assessment procedures for the rainfall erosivity are transparently discussed in (Panagos et al. 2015a). Naipal et al. (2015) stated that the information needed to calculate the R-factor according to the method of Wischmeier and Smith (1978) used in RUSLE is difficult to obtain in large spatial scale in remote areas. In the past, different studies have been done on deriving regression equations for the R-factor (Angulo-Martínez et al. 2009; Meusburger et al. 2012; Goovaerts 1999; Diodato and Bellocchi 2010). Although these studies are however dependent on their local climates, they could be used in similar cases. Many other scientists (Doetterl et al. 2012; Van Oost et al. 2007; Montgomery 2007; Yang et al. 2003) have all used Renard and Freimund (1994), which relates the R-factor to the total annual precipitation. This approach might slightly, or in regions with more complex rainfall patterns, strongly, overestimate the R values. In such cases, users may refer to the latest global rainfall erosivity assessment and its online database (Panagos et al. 2017). Figure 15.3 shows the global erosivity map with 30 arc-second spatial resolution.

K factor: In RUSLE2, the upper case K represents the base soil erodibility as determined using the soil erodibility nomograph proposed by (Wischmeier and Smith 1978) and (Renard et al. 1997) in Eq. (15.9). This nomograph is

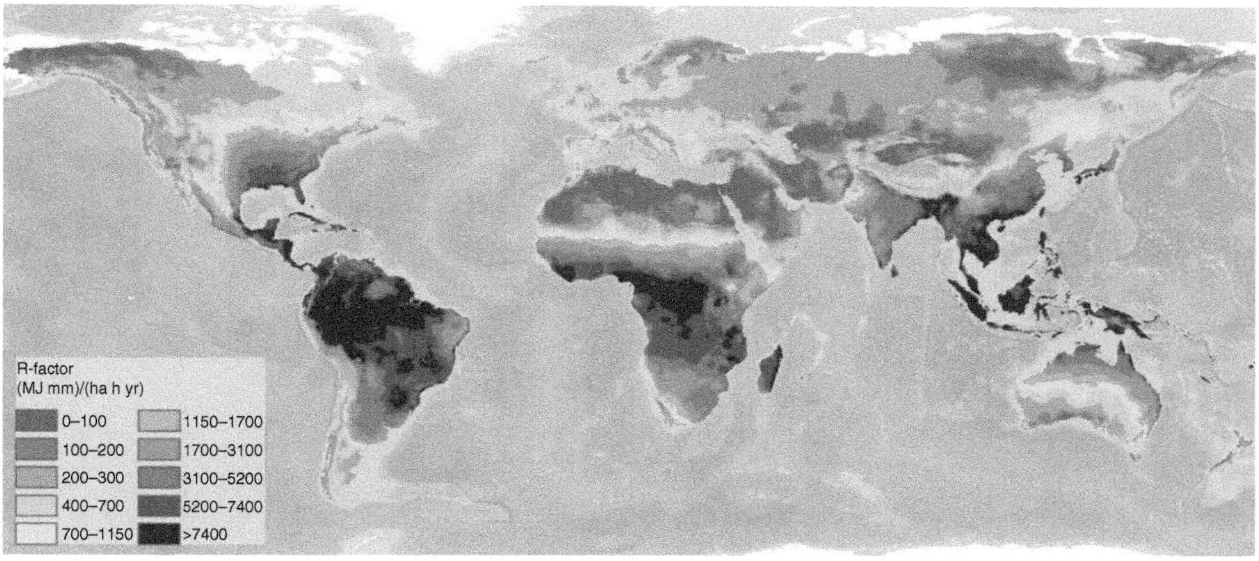

Figure 15.3 Global erosivity map (Panagos et al. 2017).

derived by using K values which were produced from experimentally measured data based on unit plots. The K-factor, which expresses the susceptibility of a soil to erode, is related to soil properties such as organic matter content, soil texture, soil structure, and permeability (Panagos et al. 2014b). In some cases in which the dataset resolution is relatively coarse for most of applications, an interpolation might be necessary. In order to enable a better interpolation of point estimates, Cubist regression-interpolation could be applied (Panagos et al. 2014b).

This factor K, calculated in Eq. (15.9), represents an integrated annual value of the soil profile reaction to the process of soil detachment and transport by raindrops and surface flow (Renard et al. 1997):

$$K = \left[\frac{2.1 \times 10^{-4}M^{1.14}(12 - OM) + 3.25(s - 2) + 2.5(p - 3)}{100}\right] * 0.1317 \quad (15.9)$$

where: M the textural factor with M = (msilt+mvfs)*(100 − mc); mc [%] clay fraction content (<0.002 mm); msilt [%] silt fraction content (0.002–0.05 mm); mvfs [%] very fine sand fraction content (0.05–0.1 mm); OM [%] the organic matter content; s the structure and p is the permeability of the soil. Based on data availability and scale of studies different approaches may be taken to produce K-factor. But at least it should include soil erodibility estimate, to assess the structure and permeability and adjustment of K-factor by inclusion of surface stone cover. Figure 15.4 illustrates the procedure used to generate a K-factor map for Europe (Panagos et al. 2014b).

The **L** and **S** factors jointly represent the effect of slope length, steepness, and shape on sediment production (RUSLE User's Reference Guide 2008). The L represents the effect of slope length for different cover-management conditions. This parameter is an annual value that has been weighted based on the distribution of erosivity during the year (Ars website, last modified 10/7/2016).

RUSLE2 models the total of rill and interrill erosion. Rill erosion increases in a downslope direction because runoff, which is the primary erosive agent for rill erosion, increases in a downslope direction. In contrast, interrill does not vary with location on the slope because it is primarily caused by raindrop impact. Therefore, the slope length factor "l" is greater for those conditions where rill erosion is greater relative to interrill erosion (Ars website, Last modified 10/7/2016).

It should be noted that this model merely predicts rill and interrill erosion. Predicting average annual ephemeral gully erosion is challenging because there is no existing long-term database of ephemeral gully erosion rates comparable to the plot database underlying the USLE, which in turn underlies RUSLE. Ephemeral gully erosion is a process inherently driven by larger-than-average runoff events (Morgan and Nearing 2011). To estimate long-term ephemeral gully erosion, long-term weather records should be applied to an ideal runoff and erosion model in order to create a distribution of erosion events. Eventually, taking the monthly means of this database of ephemeral gully erosion events would help us to have an estimate on long-term ephemeral gully erosion.

As the LS- and C-factors have the greatest impact on modeling soil loss (Risse et al. 1993) improving the understanding and accuracy of calculations are crucial to achieve

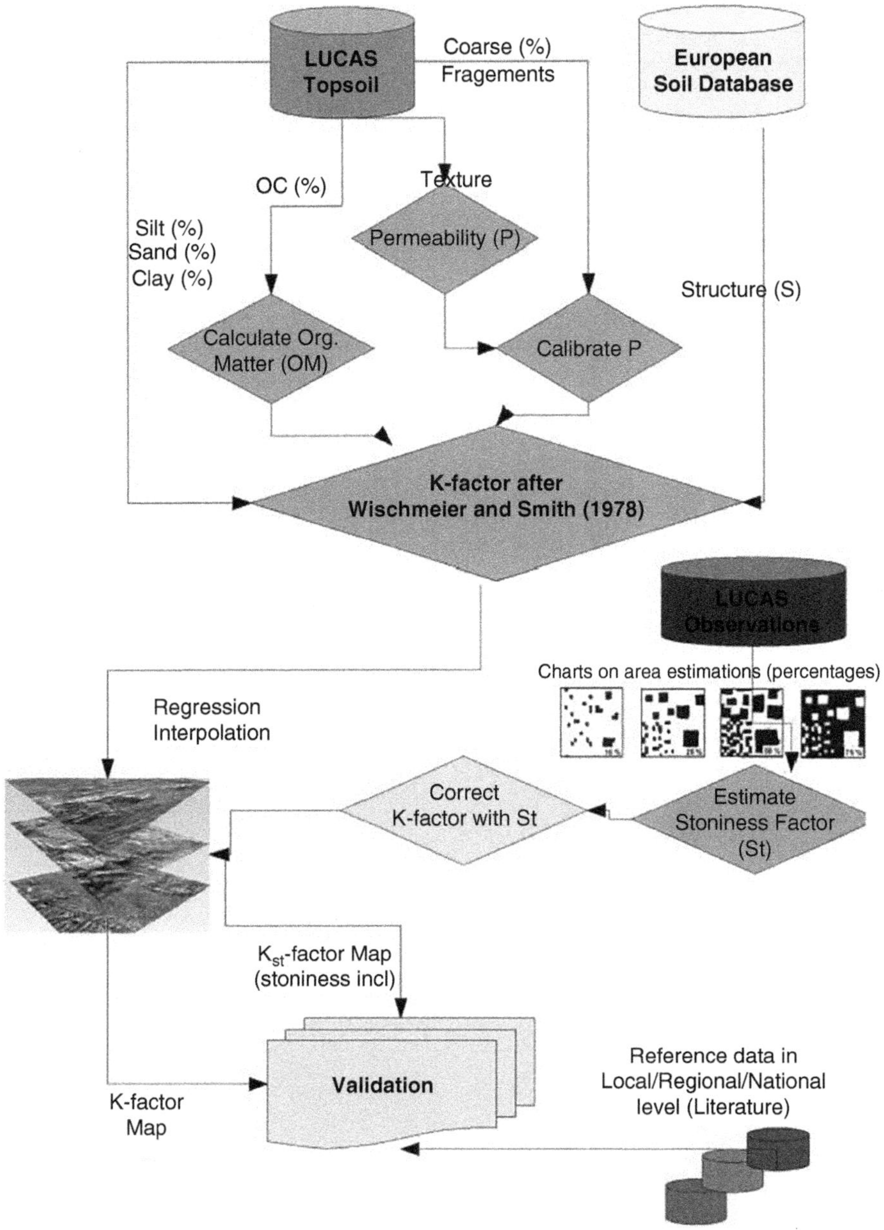

Figure 15.4 Methodology applied to generate K-factor map at European scale (Panagos et al. 2014b).

better results. Although there are several factors influencing the resolution of the soil loss map, the DEM and LS factor share an important contribution in this regard. The DEM quantitatively represents the Earth's surface and as a source, provides us with the information about the terrain including slope, aspect, drainage network, and topographic index (Mukherjee et al. 2013). Most large-scale studies (Van der Knijff et al. 2000; Van Rompaey et al. 2003) have used the algorithm proposed by Desmet 1996 (Desmet and Govers 1996).

There exist various models to be used as tools to produce topographic index maps. One example is SAGA (System for Automated Geoscientific Analyses). This software is a user-friendly platform that provides a comprehensive set of modules for data analysis and numerous applications such as those focusing on DEMs and terrain analysis (Olaya and Conrad 2009; Conrad et al. 2015; Panagos et al. 2015c) have applied it. It is also possible to estimate the LS factor by using a hydrology module implemented in SAGA directly from DEMs, which even carries out calculations much more quickly than the classical ArcGIS toolbox (Schwanghart and Scherler 2014).

Studies found that erosion occurs faster in slopes steeper than 9% (McCool et al. 1987). Although the RUSLE makes no differentiation between rill and interrill erosion in the S factor that computes the effect of slope steepness on soil loss the algorithm adopted in the model considers the slope gradient Θ in degrees (Eqs. (15.10) and Eq (15.11)):

$$S = 10.8 \times \sin\Theta + 0.03 \quad (15.10)$$

where slope gradient <0.09

$$S = 16.8 \times \sin\Theta - 0.5 \quad (15.11)$$

where slope gradient ≥0.09.

There is also another methodology for slopes steeper than 10° developed in China (Liu et al. 2002). The S-factor calculated using this method is lower by around 20% (Panagos et al. 2015c). They stated this difference could be attributed to the fact that 45% of the areas studied in the EU-DEM contained slopes that were greater than 10° and this difference is minimal in arable lands, as these lands are rarely found in steep slopes.

Initially, the L-factor was defined as the ratio of soil lost from a horizontal slope length to the corresponding loss from the slope length of a unit plot (Wischmeier and Smith 1978) L-factor is represented in Eq. (15.12):

$$L = \left(\frac{\lambda}{22.13}\right)^{m} \quad (15.12)$$

where λ is the slope length (in meters) and m is equivalent to 0.5 for slopes steeper than 5%, 0.4 for slopes between 3%–4%, 0.3 for slopes between 1%–3%, and 0.2 for slopes less than 1%. This approach, due to nonuniformity of steepness in an area, extended to a two-dimensional terrain using the concept of the unit-contributing area proposing Eq. (15.13) (Desmet and Govers 1996):

$$L_{i,j} = \frac{(A_{i,j-in} + D^2)^{m+1} - A_{i,j-in}^{m+1}}{D^{m+2} * x_{i,j}^{m} * 22.13^{m}} \quad (15.13)$$

where A *i,−in* is the contributing area at the inlet of grid cell (*i,j*) measured in m2. D is the grid cell size (meters), *Xi,j* = sin *ai,j* + cos *aai,j*, the *aai,j* is the aspect direction of the grid cell (i,j). m is related to the β which is a ratio of the rill to interrill erosion are expressed in (15.15) and (15.15):

$$m = \frac{\beta}{\beta + 1} \quad (15.14)$$

$$\beta = \frac{\frac{\sin\theta}{0.0896}}{[0.56 + 3 * (\sin\theta)^{0.8}]} \quad (15.15)$$

There is also an enhancement in Eq. (15.15) to estimate β which reflects the effects of soil texture, prior land use (soil biomass and soil consolidation), and ground cover and slope steepness (Morgan and Nearing 2011).

The main limitation of this methodology is the existence of landscape features (i.e., roads, paths, fences, contours, stone walls, and grass margins) that may interrupt the water runoff and reduce the slope length, but are not identified in the DEM. Those characteristics may be useful for calibrating LS-factor estimations at national or European scales. However, features which are maintained by farmers, such as grass margins, contours, and stone walls are proposed among the support practices (P-factor) that help reduce soil erosion (Panagos et al. 2015c).

Since large uncertainties exist in estimated rates and the extent of soil erosion by surface runoff on a global scale, there have been studies to improve the global applicability of the RUSLE model by adjustments made to topographical factors (Naipal et al. 2015). Recent studies showed that despite the fractal method (Zhang et al. 1996; Pradhan et al. 2006) being applied on coarse-resolution DEMs to improve the resulting slope, results still show a slight dependence on the original grid resolution (Naipal et al. 2015). It is notable that this approach has had closer estimations for soil erosion than that of the unadjusted RUSLE model.

Of course, due to several factors involved in the soil erosion process, it is strongly recommended to fully study the intended region and gathering as much information as possible as well as calibrating the model before generating final results.

c factor: The c factor accounts for the effects of cover-management. The lowercase c in RUSLE2 refers to the cover-management factor for each day. The uppercase C refers to an average annual C-factor value where the individual daily c factor values have been weighted by the distribution of erosivity during the year (site USDA). The bare plot (no vegetation) with till up and down the slope is taken as a reference condition, with a C-factor value of 1. The soil loss from different land-cover types is compared to the loss from the reference plot and the results are given as a ratio (Panagos et al. 2015b). So the average value of this factor ranges between 0 and 1. Soil loss ratios are computed based on five sub-factors, namely: prior land use, canopy cover, surface cover, surface roughness, and soil moisture (Renard et al. 1997). To quantitatively assess the effects of WHTs on sedimentation by computer-based models, as previously mentioned, one should calculate the difference of sediment yield with and without the presence of WHTs. Therefore, to apply the influence of WHTs on the involved factors, the models should primarily be calibrated. For this purpose, depending on the type of WHT, sub-factors like surface roughness and soil moisture in the case of implementing stone bunds and constructing percolation ponds, respectively, undergo considerable changes. In such cases the calibration might even extend to more than one parameter. P-factor and C are affected the most among all the parameters of the RUSLE model by management practices applied to reduce erosion.

P-factor: According to the manual and online sources of the RUSLE model, the lowercase p refers to a daily value of the support practices factor. The uppercase P is an average annual value determined from the individual daily p values weighted by the erosivity distribution or by taking the ratio of soil loss with the practice to soil loss without the practice (Site USDA). Many parameters are incorporated in P-factor introduced (Renard et al. 1997).

The effect of small impoundments is taken into account in the model by how these practices deposit sediment (ARS Website). Deposition that occurs on concave slopes is taken into account by solving the conservation of mass equation along the flow path. Shape of slope also matters in calculating the erosion. Erosion is greatest for convex slopes that are steep near the end of the slope length where runoff is greatest. Erosion is least for concave slopes where the upper end of the slope is steep and runoff is least (ARS website); such effects may appear as a result of terracing, for instance.

P-values can be derived either from image classifications using remote-sensing data or from previous studies or even from expert knowledge (Panagos et al. 2015d). Of course, image classification requires high-resolution datasets which are not always reachable. In such cases, using various values presented in literature might be a proper primary guess (e.g. Wischmeier and Smith 1978; Renard et al. 1997; Foster et al. 2003). The effectiveness of P-factors derived from plot studies is hardly applicable for large areas. There are also alternative methodologies for estimating the P-factor based on empirical equations (Lufafa et al. 2003; Fu et al. 2005; Terranova et al. 2009); these methods are commonly employed to determine P-factor values using as input the slope gradient %. Such methods are not widely used because the effect of slope is already incorporated in the LS factor (Panagos et al. 2015d).

15.4.1 Abilities and Limitations of RUSLE

The most updated version of RUSLE (Revised-USLE) described in this chapter and in the mentioned references is a computer model to calculate soil loss in large-scale areas based on the Universal Soil Loss Equation (USLE), but goes further beyond the scope of USLE and uses process-based equations derived from fundamental erosion science and professional judgment to enhance its abilities (Morgan and Nearing 2011).

As a major improvement in RUSLE, users are able to define any number of steepness, soil, or management breaks along the slope, and the program will accordingly break the slope into segments to deal nicely with the application of terraces or diversions as a management alternative, and complete the calculations on each combination (Morgan and Nearing 2011).

RUSLE has the ability to reflect the effects of spatial variation of soil erodibility, slope steepness, and cover management along a slope on detachment, transport, and deposition. This approach results in estimates of the long-term average sediment production, erosion rate, transport capacity, deposition, and sediment characteristics along the slope, as well as the sediment amount and characteristics of sediment leaving the slope (Foster et al. 1981; Morgan and Nearing 2011). It approximates backwater effects when it determines the effectiveness of dense narrow vegetative buffers on sediment trapping (USDA-ARS 2008). RUSLE2 also includes the ability to approximate the effect of simple impoundments and channels on sediment delivery and fines enrichment (Morgan and Nearing 2011). The RUSLE benefits handsomely from more than 10 000 plot-years of data and is the most commonly used model worldwide, which makes it simple to find relevant technical guides and experts in case of encountering scientific problems.

Of course, there are many limitations observed by various scientists when applying this model on a global scale. For one, the model was originally developed to be applicable on the agricultural plot scale, which makes it incompatible with the coarse spatial scale of global datasets on soil-erosion influencing factors such as precipitation, elevation, land use, and soil features. Since RUSLE is initially parameterized for US environmental conditions, it is not directly applicable to other regions of the world. This model merely considers sheet and rill erosion (Naipal et al. 2013). Currently RUSLE does not simulate erosion in channels. There are many areas of research to improve some shortcomings of RUSLE but two active fields are residue production in perennial systems and ephemeral gully erosion estimation (Morgan and Nearing 2011). The model predicts average annual soil loss and is not intended to predict soil loss for storms or individual years (Nearing 2013). Although these predictions are described as long term, from the geomorphic perspective they would be referred to as medium term (Desmet and Govers 1996). Some other conflicts are also reported about the units of the equation which are still used in Imperial, but conversion to SI is straightforward.

Recently, major developments were made on RUSLE. One of its most significant applications is producing a global map of erosivity at 30 arc-seconds (~1 km) based on a Gaussian Process Regression (GPR) using the first-ever global rainfall erosivity database for 3625 stations covering 63 countries (Panagos et al. 2017).

All in all, it is the most commonly used erosion model in large areas and has the capacity to have many more developments.

15.5 Limburg Soil Erosion Model (LISEM)

Previous sections of this chapter discussed the introduction, qualitative effects induced by sedimentation in WHTs, a basic method to estimate specific sediment yield in small dams constructed to store runoff and estimating, then estimating average annual soil loss rate in larger areas using the RUSLE model. What follows describes a pragmatic approach to assess the effect of WHTs on event-based hydrological processes and soil erosion.

In certain regions of the world, soil erosion is caused mainly by heavy and infrequent rains and sparse vegetation cover due to absence of soil-preservation services (Medeiros et al. 2010). Thus, an understanding of event-based hydrology and careful runoff measurements for such events which control long-term sedimentation is of much importance (White 2005; Ramsankaran et al. 2013).

To reach to a better understanding of soil erosion specifically under the influence of diverse WTHs, one may use physical event-based hydrological models such as LISEM, which is suitable for the planning and evaluation of measures for mitigating soil erosion (De Roo et al. 1996b). Therefore, this section is devoted to investigating the effect of WHTs on sediment yield resulting from catchment runoff by using LISEM. This model, which is built in a raster-based geographic information system, simulates runoff, erosion, and sediment transport at catchment levels (De Roo et al. 1999a,b). Profiting from several processes, it can be used to predict spatial variability of soil erosion controlled by runoff at catchment scale.

LISEM uses the four point finite-difference solution of kinetic wave and Manning flow resistance equation to simulate channel flows and overland runoff. Sediment transport is calculated, considering the balance between erosion and deposition. Thus, incorporating high-resolution topography maps i crucial in this matter. Erosion estimated by this model might be caused by either runoff flows or splash detachments. It collects random roughness (RR) to measure runoff amount distribution overland (De Roo et al. 1996b). Leaf area index (LAI) and maximum canopy storage are applied to calculate the interception and water storage by plants. Depending on the data availability, infiltration is calculated using various methods such as Green and Ampt (Green and Ampt 1911), Holtan (Beasley and Huggins 1982,) or SWATRE (Belmans et al. 1983).

Needless to say, most of the LISEM input parameters in such modeling are distributed over a relatively large area and extrapolated maps are required to reasonably evaluate the changes in their values. So, increasing the spatial and temporal resolution of the measurement sites leads to more accurate estimates.

If recent soil-texture map is not available, site measurements are required to produce it. These measurements should be distributed well enough to represent the different soil types in the catchment. A DEM is also necessary to generate slope gradients, local drain directions, and catchment outlet locations. Land use/land cover classes might be derived from remote-sensing techniques accompanied by field observations. Since many parameters involved in soil erosion estimates vary by time, it is advisable to produce input maps compatible with the intended period. Several other inputs are necessary to run the simulations; they include initial soil moisture content (SMC), median size of soil texture (d50), saturated hydraulic conductivity (Ksat), soil cohesion, stone fraction, and aggregate stability. These input maps are simply incorporated in the model using available data or in the case of data shortage, should be obtained from site measurements in certain places over the time.

15.5.1 Model Implementation

LISEM, like all other similar models, requires detailed hydrometeorological and spatial input data to eventually estimate soil erosion at a catchment scale. It is notable to mention that the pixel size of maps is determined by various factors, including spatial distribution of available data. Also, a time step for the process of channel flow calculations needs to be considered.

Leaf Area Index (LAI): this parameter is measured over a sample plot during the rainy season for the purpose of land use estimations and time interval of each measurement depends on the plant type and needed accuracy. The leaf area of each plant is firstly averaged for the sample plot plants and then by multiplying this factor by total number of plants, the total leaf area index will be determined. In the case of having trees and/or shrubs in the catchment, larger plots are considered and LAI for them are calculated likewise.

Plant Cover: plant cover is measured using the same plots as for LAI. Plant cover measurements are usually taken using a digital camera and later analyzed by ImageJ (Abràmoff et al. 2004).

Plant height: for any land use, the average value of this factor is derived from representative sample sites using a tape, except for trees whose heights are estimated. It is also notable that data collecting interval for land use related factors depend greatly on the plant type, costs, and required accuracy. One may review literature before commencing sample site construction or data collection.

Initial SMC (soil moisture content): as LISEM uses the Green and Ampt method to model infiltration, initial SMC must be measured for two different soil layers of soil (e.g. 0–40 and 40–70 cm). Each measurement site should be equipped with a time-domain reflectometer and soil moisture sensor of any available and applicable type. Measurements may be taken at many different depths and time intervals, but the average value eventually will be considered in the model. Random Roughness (RR) of each sample site might be measured using a pin meter over at least two rainy seasons. Photographs should be taken using a digital camera and by calculating the SD (standard deviation) of the pin position on the board, the surface roughness is determined.

Ksat which is the hydraulic conductivity of soil is determined using a permeameter.

Soil Cohesion: this parameter is used for classifying soil texture and is measured using a Torvane. Needless to say, this parameter should be measured in both dry and wet seasons.

Stone fraction: this parameter which assumed not to vary with time, estimated based on the areal coverage of stone in a sample site and will be used in land use calculations of the model.

15.5.2 Calibration and Modification of p-Factor

Model calibration is performed by measured suspended sediment concentration (SSC) and water discharge at the catchment outlets. Many statistical methods and hydrologically based criteria are developed to assess the efficiency of a calibration from which are R-squared, RMSE, and Nash-Sutcliffe (NSE).

Ksat and Manning's n values, as the most influential parameters in this model, should be calibrated. The calibrated values may vary significantly with the first implemented ones and/or for outlets of different catchments in similar simulations. Measured and simulated discharges after calibration for most events at the outlets must agree well and a satisfactory hydrological model should have an NSE > 0.5 (Moriasi et al. 2007). It is notable to mention that calibrated Ksat value in particular and SMC might be significantly different from measured values (de Barros et al. 2014; Hessel et al. 2006; Hessel et al. 2003; Grum et al. 2017). This might occur due to difficulties in characterizing them at catchment level and relevant uncertainties (De Roo 1996b).

As mentioned earlier, simulating the effects of WHTs on sedimentation using process-based models can be subdivided in two categories. First are direct effects (affecting model inputs), and second, indirect effects (affecting runoff and erosion). In order to model these effects of WHTs, some data describing the effectiveness of WHTs measures are required.

Although collecting such field data allows us to determine the direct effects of WHTs on LISEM inputs, is not necessarily sufficient for modeling indirect effects since runoff and erosion of such fields are not measured. So, first step is to obtain data on how WHTs influence LISEM inputs and on runoff and erosion. Data gathered from literature and measurements would be expressed as P-factor which is defined as the ratio of runoff and soil erosion rates in areas with WHTs to the rates in areas without WHTs. To reach more realistic outcomes regarding modeling the effect of WHTs on a catchment scale, the value of the above-mentioned P-factor should be calibrated.

Therefore, the procedure should be to first simulate the direct effects changes on the model inputs, and compare the achieved P-factor from simulation results to the P-factor obtained. If this does not result in satisfactory P-factors, calibration is needed on the indirect effects of WHTs.

In this method, LISEM should be calibrated using the data on direct and indirect effects of WHTs for a single pixel and then apply this calibrated model to a catchment scale.

The effects of WHTs measures on LISEM input values are given as multiplication factors. So that a value of 1.2 for plant height (CH) while modeling a catchment implemented with a certain WHT, states that plots having that WHT will have a plant height 1.2 times that of plots without WHT.

The final step of the calibration process on P-factors is to alter the values of some of the LISEM inputs caused by direct effects of WHT measures, so that the resulting P-values approach those estimated P-factors for runoff and soil loss, which is obtained from the literature and or measurements. It is notable to mention that P values increase with storm size, which demonstrates that some techniques might be less effective in case of larger storms happening. Although P-factors hardly depend on slope angle, predicted soil loss indicated an obvious trend with slope angle. So, calibration of WHTs for an on-slope cropland with measured rainfall gives some worthy results to reach a more precise simulation.

Since most WHTs are somehow designed to slow down the flow, Manning's n value will increase in such cases. Ksat would undergo some change as well. Since LISEM is very sensitive to saturated conductivity parameter, in order to correctly predict the amount of runoff, Ksat has to be calibrated too.

Eventually, the calibrated values of LISEM input parameters to form the adapted multiplication factors would be concluded. For further information on the development of this approach, please refer to Hessel and Tenge (2008) and Grum et al. (2017).

15.5.3 Assessing Effects of WHTs on Sedimentation Using LISEM

The evaluation of the effect of WHTs on reduction of runoff and sediment yield is estimated by calculating the differences of the simulations with and without WHTs. Information on these effects reported as P-factors might be gathered from different sources such as literature and technical notes which are used, for instance, in the RUSLE model. Simulation considerations vary depending on the WHTs under investigation and calibration would be terminated when changing the input parameters will not affect the results matched with the reported P-factors for each WHT. For more information regarding the LISEM calibration process, recommended P-factors, and methodology of modeling the effect of WHTs on sedimentation described in this chapter, one may refer to (Hessel and Tenge 2008; Grum et al. 2017).

To assess the effect of WHTs on catchment runoff and subsequently sediment yield, one should obtain the P-factors from the literature and then calibrate the input values of LISEM to replicate those observed. Values of different P-factors for runoff and sediment yield due to various WHTs are summarized in Tables 15.1 and 15.2. The values vary considerably for different techniques which demonstrate their effect on runoff and sedimentation. Ksat and Manning's n need the most to change, which in turn indicates the effect of implementation of WHTs on increasing the resistance and as a result slowing the overland flow velocity.

Lower P-factors for WHTs (e.g. stone bunds+deep trenches) are expected to cause more significant changes to LISEM input parameters such as Ksat and Manning's values. Studies show that increasing rainfall generally reduces the effect of WHTs on runoff and sediment yield reduction (Grum et al. 2017). Trapping eroded soil behind the WHT structures causes improvements in infiltration, which when monitoring the water static level raise, if available, may approve such conclusion. There is a general premise that average sediment yield per unit area is higher for larger catchments. However, there are other factors contributing to the sediment yield (Vanmaercke et al. 2011). For example, the variability of sediment yield in catchments is attributed to gully erosion, surface lithology, land cover, and topography (Tamene et al. 2006). Another study found an inverse correlation between catchment size and sediment yield which highlights the effect of WHTs as sediment sinks (Haregeweyn et al. 2008b). In some cases the combination of different WHTs implemented in a large catchment affects the sediment yield despite the proven

Table 15.1 Estimated P-factors for runoff (Grum et al. 2017).

Water Harvesting Technique	Literature	References
	Runoff	
Stone Bunds	0.2–0.35,0.73	Alseekh and Ghaleb (2009), Taye et al. (2013)
Deep Trenches	0.31	Taye et al. (2013)
Stone Bunds + Deep Trenches	0.16	Taye et al. (2013)
Gabion check dams	0.84–0.91,0,73–0.85	Guyassa et al. (2015), Xu et al. (2013)

Table 15.2 Estimated P-factors for sediment yield (Grum et al. 2017).

Water Harvesting Technique	Literature	References
	Sediment Yield	
Stone Bunds	0.32, 0.36	Gebrernichael et al. (2005), Taye et al. (2013)
Deep Trenches	0.10	Taye et al. (2013)
Stone Bunds + Deep Trenches	0.04	Taye et al. (2013)
Gabion check dams	0.33, 0.37–0.43	Guyassa et al. (2015), Xu et al. (2013)

benefits of individual WHTs (Grum et al. 2016; Mekonnen et al. 2015; Nyssen et al. 2009).

In order to simulate and assess WHT influence on runoff and sedimentation in such a manner to achieve desired outcomes, users should be aware of the capabilities and limitations of the model. LISEM, like any other model, certainly has some inaccuracies in predicting the sediment yield. This may be due to difficulties in modeling sediment transport as it is extremely complicated to gather spatially varied factors controlling erosion in catchment scale (de Barros et al. 2014; Ramsankaran et al. 2013).

15.6 Conclusion

WHTs, by controlling erosivity, influence many aspects of environmental policies. As one major effect of such techniques is reducing sedimentation, the negative impacts on food production, drinking water quality, ecosystem services, mud control, eutrophication, biodiversity, and carbon stock shrinkage will be diminished. Almost every research investigating soil chemical properties states in-situ WHTs significantly reduce nutrient loss associated with soil loss. But various WHTs cause no significant difference in most soil chemical properties in terms of quantity.

Many approaches have been developed to quantitatively assess the effectiveness of diverse WHTs on erosivity reduction. However, these models may basically fall into two categories: empirical and process-based. Despite all limitations and inaccuracies, undoubtedly the prime example of an empirically based model is USLE (Nearing 2013) and among various process-based models LISEM is notable (Hessel and Tenge 2008). Needless to say, the above-mentioned approaches require high spatial and/or temporal resolution data and obviously calibration in order to provide reasonable outcomes. Such models can be used to simulate the hydrological responses as well as sediment yield induced by the implementation of WHTs at a catchment scale.

The described approaches demonstrated that evaluating the effect of WHTs on reduction of runoff and sediment yield is achieved by calculating the differences of the simulations with and without WHTs. The influence of WHTs is considered in the model by the means of using P-factor.

Although the amount of runoff and consequently generated sediment yield is affected if not controlled by the interaction between factors like surface condition, rain amount, and intensity and soil properties, in some cases, WHTs effectively reduced sediment yield and runoff amount up to 60% and 40%, respectively.

Although land management scenarios and WHTs were demonstrated to be useful in decreasing the amount of runoff and sedimentation and in enhancing soil moisture content in arid and semi-arid regions, choosing a specific technique will also be influenced by factors such as climate, land use, and topography and requires in-site prefund studies.

References

Abràmoff, M.D., Magalhães, P.J., and Ram, S.J. (2004). Image processing with ImageJ. *Biophotonics International* 11 (7): 36–42.

Alcamo, J., Flörke, M., and Märker, M. (2007). Future long-term changes in global water resources driven by socio-economic and climatic changes. *Hydrological Sciences Journal* 52 (2): 247–275.

Alseekh, S. and Ghaleb, A. (2009). The Effect of Water Harvesting Techniques on Runoff, Sedimentation, and Soil Properties. *Environmental management* 44: 37–45. https://doi.org/10.1007/s00267-009-9310-z.

Angulo-Martínez, M., López-Vicente, M., Vicente-Serrano, S.M., and Beguería, S. (2009). Mapping rainfall erosivity at a regional scale: a comparison of interpolation methods in the Ebro Basin (NE Spain). *Hydrology and Earth System Sciences* 13 (10): 1907–1920.

de Barros, C.A.P., Minella, J.P.G., Dalbianco, L., and Ramon, R. (2014). Description of hydrological and erosion processes determined by applying the LISEM model in a rural catchment in southern Brazil. *Journal of Soils and Sediments* 14 (7): 1298–1310.

Beasley, D.B., and Huggins, L.F. (1982). ANSWERS User's Manual, West Layette. *Dep. t lf ric, ng., Purdue University, USA*.

Belmans, C., Wesseling, J.G., and Feddes, R.A. (1983). Simulation model of the water balance of a cropped soil: SWATRE. *Journal of Hydrology* 63 (3–4): 271–286.

Beusen, A.H.W., Dekkers, A.L.M., Bouwman, A.F. et al. (2005). Estimation of global river transport of sediments and associated particulate C, N, and P. *Global Biogeochemical Cycles* 19 (4) http://dx.doi.org/10.1029/2005GB0.

Boardman, J. and Poesen, J. (eds.) (2007). *Soil erosion in Europe*, 477–487, Wiley.

Brown, L.C. and Foster, G.R. (1987). Storm erosivity using idealized intensity distributions. *Transactions of the ASAE* 30 (2): 379–386.

Chien, N. (1985). Changes in river regime after the construction of upstream reservoirs. *Earth Surface Processes and Landforms* 10 (2): 143–159.

Conrad, O., Bechtel, B., Bock, M. et al. (2015). System for automated geoscientific analyses (SAGA) v. 2.1. 4. *Geoscientific Model Development* 8 (7): 1991–2007.

Critchley, W., Siegert, K., and Chapman, C. (1991). *A Manual for the Design and Construction of Water Harvesting Schemes for Plant Production*. Rome, Italy: FAO.

De Roo, A.P.J., Wesseling, C.G., and Ritsema, C.J. (1996b). LISEM: a single-event physically based hydrological and soil erosion model for drainage basins. I: theory, input and output. *Hydrological Processes* 10 (8): 1107–1117.

Desmet, P.J.J. and Govers, G. (1996). A GIS procedure for automatically calculating the USLE LS factor on topographically complex landscape units. *Journal of Soil and Water Conservation* 51 (5): 427–433.

Dile, Y.T., Karlberg, L., Daggupati, P. et al. (2016). Assessing the implications of water harvesting intensification on upstream–downstream ecosystem services: a case study in the Lake Tana basin. *Science of the Total Environment* 542: 22–35.

Diodato, N. and Bellocchi, G. (2007). Estimating monthly (R) USLE climate input in a Mediterranean region using limited data. *Journal of Hydrology* 345 (3–4): 224–236.

Diodato, N. and Bellocchi, G. (2010). MedREM, a rainfall erosivity model for the Mediterranean region. *Journal of Hydrology* 387 (1-2): 119–127.

Doetterl, S., Van Oost, K., and Six, J. (2012). Towards constraining the magnitude of global agricultural sediment and soil organic carbon fluxes. *Earth Surface Processes and Landforms* 37 (6): 642–655.

Foster, G.R., McCool, D.K., Renard, K.G., and Moldenhauer, W.C. (1981). Conversion of the universal soil loss equation to SI metric units. *Journal of Soil and Water Conservation* 36 (6): 355–359.

Foster, G.R., Yoder, D.C., Weesies, G.A., et al. (2003). RUSLE 2.0 user's guide. *USDA-Agricultural Research Service, Washington, DC*, USA.

Frankl, A., Deckers, J., Moulaert, L. et al. (2016). Integrated solutions for combating gully erosion in areas prone to soil piping: innovations from the drylands of northern Ethiopia. *Land Degradation and Development* 27 (8): 1797–1804.

Fu, B.J., Zhao, W.W., Chen, L.D. et al. (2005). Assessment of soil erosion at large watershed scale using RUSLE and GIS: a case study in the loess plateau of China. *Land Degradation and Development* 16 (1): 73–85.

Gebrernichael, D., Nyssen, J., Poesen, J. et al. (2005). Effectiveness of stone bunds in controlling soil erosion on cropland in the Tigray highlands, northern Ethiopia. *Soil Use and Management* 21 (3): 287–297.

Goovaerts, P. (1999). Using elevation to aid the geostatistical mapping of rainfall erosivity. *Catena* 34 (3–4): 227–242.

Green, H.W. and Ampt, G.A. (1911). Studies on soil physics. *The Journal of Agricultural Science* 4: 1–24.

Grum, B., Hessel, R., Kessler, A. et al. (2016). A decision support approach for the selection and implementation of water harvesting techniques in arid and semi-arid regions. *Agricultural Water Management* 173: 35–47.

Grum, B., Woldearegay, K., Hessel, R. et al. (2017). Assessing the effect of water harvesting techniques on event-based hydrological responses and sediment yield at a catchment scale in northern Ethiopia using the Limburg soil erosion model (LISEM). *Catena* 159: 20–34.

Gunnell, Y. and Krishnamurthy, A. (2003). Past and present status of runoff harvesting systems in dryland peninsular India: a critical review. *Ambio: A Journal of the Human Environment* 32 (4): 320–325.

Gupta, M.C. (1995). Environmental management and its impact on the operations function. *International Journal of Operations & Production Management.* 15 (8): 34–51. https://doi.org/10.1108/01443579510094071.

Guyassa, E., Frankl, A., Zenebe, A. et al. (2015). Effect of check dams in gullies on runoff response in the headwaters of Tekeze reservoir. In: *Book of Abstracts TropiLakes 2015: Tropical Lakes in a Changing Environment: Water, Land, Biology, Climate and Humans*, 80. Bahir Dar: Bahir Dar University Ethiopia.

Hammad, A.H.A., Børresen, T., and Haugen, L.E. (2006). Effects of rain characteristics and terracing on runoff and erosion under the Mediterranean. *Soil and Tillage Research* 87 (1): 39–47.

Haregeweyn, N., Poesen, J., Nyssen, J. et al. (2005). Specific sediment yield in Tigray-Northern Ethiopia: assessment and semi-quantitative modelling. *Geomorphology* 69 (1-4): 315–331.

Haregeweyn, N., Poesen, J., Deckers, J. et al. (2008a). Sediment-bound nutrient export from micro-dam catchments in northern Ethiopia. *Land Degradation and Development* 19 (2): 136–152.

Haregeweyn, N., Poesen, J., Nyssen, J. et al. (2008b). Sediment yield variability in northern Ethiopia: a quantitative analysis of its controlling factors. *Catena* 75 (1): 65–76.

Hessel, R. and Tenge, A. (2008). A pragmatic approach to modelling soil and water conservation measures with a catchment scale erosion model. *Catena* 74 (2): 119–126.

Hessel, R., Jetten, V., Liu, B. et al. (2003). Calibration of the LISEM model for a small loess plateau catchment. *Catena* 54 (1–2): 235–254.

Hessel, R., Van den Bosch, R., and Vigiak, O. (2006). Evaluation of the LISEM soil erosion model in two catchments in the east African highlands. *Earth Surface Processes and Landforms. The Journal of the British Geomorphological Research Group* 31 (4): 469–486.

Holtan, H., Kamp-Nielsen, L., and Stuanes, A.O. (1988). Phosphorus in soil, water and sediment: an overview. In: *Phosphorus in Freshwater Ecosystems*, 19–34. Dordrecht, Germany: Springer.

Ittekkot, V. and Zhang, S. (1989). Pattern of particulate nitrogen transport in world rivers. *Global Biogeochemical Cycles* 3 (4): 383–391.

Lal, R. (2004). Soil carbon sequestration impacts on global climate change and food security. *Science* 304 (5677): 1623–1627.

Li, X.Y., Gao, S.Y., Xu, H.Y., and Liu, L.Y. (2006). Growth of Caragana korshinskii using runoff-collecting microcatchments under semiarid condition. *Journal of Hydrology* 328 (1-2): 338–346.

Liu, B.Y., Zhang, K.L. and Xie, Y., 2002. An empirical soil loss equation. Proceedings–Process of soil erosion and its environment effect, 2, 21–25.

Ludwig, W., Probst, J.L., and Kempe, S. (1996). Predicting the oceanic input of organic carbon by continental erosion. *Global Biogeochemical Cycles* 10 (1): 23–41.

Lufafa, A., Tenywa, M.M., Isabirye, M. et al. (2003). Prediction of soil erosion in a Lake Victoria basin catchment using a GIS-based universal soil loss model. *Agricultural Systems* 76 (3): 883–894.

Marx, E.S., Hart, J., and Stevens, R.G. (1999). Soil Test Interpretation Guide. *Oregon State University. Journal of Soil Science* 120: 1120–1322.

McCool, D.K., Brown, L.C., Foster, G.R. et al. (1987). Revised slope steepness factor for the universal soil loss equation. *Transactions of ASAE* 30 (5): 1387–1396.

Medeiros, P.H., Güntner, A., Francke, T. et al. (2010). Modelling spatio-temporal patterns of sediment yield and connectivity in a semi-arid catchment with the WASA-SED model. *Hydrological Sciences Journal–Journal des Sciences Hydrologiques* 55 (4): 636–648.

Mekonnen, M., Keesstra, S.D., Stroosnijder, L. et al. (2015). Soil conservation through sediment trapping: a review. *Land Degradation and Development* 26 (6): 544–556.

Meusburger, K., Steel, A., Panagos, P. et al. (2012). Spatial and temporal variability of rainfall erosivity factor for Switzerland. *Hydrology and Earth System Sciences* 16 (1): 167–177.

Montgomery, D.R. (2007). Soil erosion and agricultural sustainability. *Proceedings of the National Academy of Sciences* 104 (33): 13268–13272.

Morgan, R.P.C. and Nearing, M.A. (eds.) (2011). *Handbook of Erosion Modelling*. West Sussex: UK: Wiley-Blackwell.

Moriasi, D.N., Arnold, J.G., Van Liew, M.W. et al. (2007). Model evaluation guidelines for systematic quantification of accuracy in watershed simulations. *Transactions of the ASABE* 50 (3): 885–900.

Mukherjee, S., Mukherjee, S., Garg, R.D. et al. (2013). Evaluation of topographic index in relation to terrain roughness and DEM grid spacing. *Journal of Earth System Science* 122 (3): 869–886.

Naipal, V., Reick, C., and Pongratz, J. (2013). Global Soil Erosion Modelling in the Context of Climate Change. *In EGU General Assembly Conference Abstracts* 15.

Naipal, V., Reick, C.H., Pongratz, J., and Van Oost, K. (2015). Improving the global applicability of the RUSLE model-adjustment of the topographical and rainfall erosivity factors. *Geoscientific Model Development* 8: 2893–2913.

Nearing, M.A. (2013). 22-soil erosion and conservation. In: *Environmental Modelling: Finding Simplicity in Complexity*, 2e (eds. J. Wainwright and M. Mulligan), 365–378. Wiley.

Nyssen, J., Veyret-Picot, M., Poesen, J. et al. (2004). The effectiveness of loose rock check dams for gully control in Tigray, northern Ethiopia. *Soil Use and Management* 20 (1): 55–64.

Nyssen, J., Clymans, W., Poesen, J. et al. (2009). How soil conservation affects the catchment sediment budget–a comprehensive study in the north Ethiopian highlands. *Earth Surface Processes and Landforms* 34 (9): 1216–1233.

Okhravi, S., Eslamian, S., Eslamian, F., and Tarkesh-Esfahani, S. (2014). Indigenous knowledge as a supportive tool for sustainable development and utilisation of rainwater harvesting systems. *Journal of Flood Engineering* 5 (1/2): 39–50.

Olaya, V. and Conrad, O. (2009). Geomorphometry in SAGA. *Developments in Soil Science* 33: 293–308.

Panagos, P., Van Liedekerke, M., Jones, A., and Montanarella, L. (2012). European soil data Centre: response to European policy support and public data requirements. *Land Use Policy* 29 (2): 329–338.

Panagos, P., Meusburger, K., Ballabio, C. et al. (2014b). Soil erodibility in Europe: a high-resolution dataset based on LUCAS. *Science of the Total Environment* 479: 189–200.

Panagos, P., Borrelli, P., Poesen, J. et al. (2015a). The new assessment of soil loss by water erosion in Europe. *Environmental Science and Policy* 54: 438–447.

Panagos, P., Meusburger, K., Ballabio, C. et al. (2015b). Reply to the comment on "rainfall erosivity in Europe" by Auerswald et al. *Science of the Total Environment* 532: 853–857.

Panagos, P., Borrelli, P., Meusburger, K. et al. (2015c). Estimating the soil erosion cover-management factor at the European scale. *Land Use Policy* 48: 38–50.

Panagos, P., Borrelli, P., and Meusburger, K. (2015d). A new European slope length and steepness factor (LS-factor) for modeling soil erosion by water. *Geosciences* 5 (2): 117–126.

Panagos, P., Borrelli, P., Meusburger, K. et al. (2015e). Modelling the effect of support practices (P-factor) on the reduction of soil erosion by water at European scale. *Environmental Science and Policy* 51: 23–34.

Panagos, P., Borrelli, P., Meusburger, K. et al. (2017). Global rainfall erosivity assessment based on high-temporal resolution rainfall records. *Scientific Reports* 7 (1): 4175.

Poff, N.L., Allan, J.D., Bain, M.B. et al. (1997). The natural flow regime: a paradigm for conservation of riverine ecosystems. *BioScience* 47: 769–784.

Pradhan, N.R., Tachikawa, Y., and Takara, K. (2006). A downscaling method of topographic index distribution for matching the scales of model application and parameter identification. *Hydrological Processes: An International Journal* 20 (6): 1385–1405.

Ramsankaran, R.A.A.J., Kothyari, U.C., Ghosh, S.K. et al. (2013). Physically-based distributed soil erosion and sediment yield model (DREAM) for simulating individual storm events. *Hydrological Sciences Journal* 58 (4): 872–891.

Regnier, P., Friedlingstein, P., Ciais, P. et al. (2013). Anthropogenic perturbation of the carbon fluxes from land to ocean. *Nature Geoscience* 6 (8): 597.

Renard, K.G. and Freimund, J.R. (1994). Using monthly precipitation data to estimate the R-factor in the revised USLE. *Journal of Hydrology* 157 (1–4): 287–306.

Renard, K.G., Foster, G.R., Weesies, G.A., et al. (1997). *Predicting soil erosion by water: a guide to conservation planning with the Revised Universal Soil Loss Equation (RUSLE)* (Vol. 703). Washington, DC: United States Department of Agriculture, USA.

Risse, L.M., Nearing, M.A., Laflen, J.M., and Nicks, A.D. (1993). Error assessment in the universal soil loss equation. *Soil Science society of America journal* 57 (3): 825–833.

RUSLE User's Reference Guide (2008). *Revised Universal Soil Loss Equation Version 2 (RUSLE2)*. Washington, D.C, USA: USDA-Agricultural Research Service.

de Santos Loureiro, N. and de Azevedo Coutinho, M. (2001). A new procedure to estimate the RUSLE EI30 index, based on monthly rainfall data and applied to the Algarve region, Portugal. *Journal of Hydrology* 250 (1–4): 12–18.

Schiettecatte, W., Ouessar, M., Gabriels, D. et al. (2005). Impact of water harvesting techniques on soil and water conservation: a case study on a micro catchment in southeastern Tunisia. *Journal of Arid Environments* 61 (2): 297–313.

Schwanghart, W. and Scherler, D. (2014). TopoToolbox 2–MATLAB-based software for topographic analysis and modeling in earth surface sciences. *Earth Surface Dynamics* 2 (1): 1–7.

Tamene, L., Park, S.J., Dikau, R., and Vlek, P.L.G. (2006). Analysis of factors determining sediment yield variability in the highlands of northern Ethiopia. *Geomorphology* 76 (1–2): 76–91.

Taye, G., Poesen, J., Wesemael, B.V. et al. (2013). Effects of land use, slope gradient, and soil and water conservation structures on runoff and soil loss in semi-arid Northern Ethiopia. *Physical Geography* 34 (3): 236–259.

Terranova, O., Antronico, L., Coscarelli, R., and Iaquinta, P. (2009). Soil erosion risk scenarios in the Mediterranean environment using RUSLE and GIS: an application model for Calabria (southern Italy). *Geomorphology* 112 (3–4): 228–245.

USDA-ARS, 2008, Draft science documentation, Revised Universal Soil Loss Equation, version 2. Available at http://www.ars.usda.gov/sp2UserFiles/Place/64080510/RUSLE/RUSLE2_User_Res_guide.pdf

Van der Knijff, J.M., Jones, R.J.A., and Montanarella, L. (2000). Soil erosion risk assessment in Europe, EUR 19044 EN. *Office for Official Publications of the European Communities. Luxembourg, 34.*

Van Oost, K., Quine, T.A., Govers, G. et al. (2007). The impact of agricultural soil erosion on the global carbon cycle. *Science* 318 (5850): 626–629.

Van Rompaey, A.J., Vieillefont, V., Jones, R.J., et al. (2003). Validation of soil erosion estimates at European scale. *European Soil Bureau, Brussels*, Belgium, 25.

Vancampenhout, K., Nyssen, J., Gebremichael, D. et al. (2006). Stone bunds for soil conservation in the northern Ethiopian highlands: impacts on soil fertility and crop yield. *Soil and Tillage Research* 90 (1–2): 1–15.

Vanmaercke, M., Poesen, J., Verstraeten, G. et al. (2011). Sediment yield in Europe: spatial patterns and scale dependency. *Geomorphology* 130 (3–4): 142–161.

Verstraeten, G. and Poesen, J. (2001). Variability of dry sediment bulk density between and within retention ponds and its impact on the calculation of sediment yields. *Earth Surface Processes and Landforms* 26 (4): 375–394.

Walraevens, K., Gebreyohannes Tewolde, T., Amare, K. et al. (2015). Water balance components for sustainability assessment of groundwater-dependent agriculture: example of the Mendae plain (Tigray, Ethiopia). *Land Degradation and Development* 26 (7): 725–736.

White, S. (2005). Sediment yield prediction and modelling. *Hydrological Processes* 19: 3053–3057.

Wischmeier, W.H., and Smith, D.D. (1978). Predicting rainfall erosion losses-a guide to conservation planning. The USDA Agricultural Handbook No. 537, USA.

Xu, Y.D., Fu, B.J., and He, C.S. (2013). Assessing the hydrological effect of the check dams in the Loess Plateau, China, by model simulations. *Hydrology and Earth System Sciences* 17 (6): 2185.

Yang, D., Kanae, S., Oki, T. et al. (2003). Global potential soil erosion with reference to land use and climate changes. *Hydrological Processes* 17 (14): 2913–2928.

Zhang, X.C., Nearing, M.A., Risse, L.M., and McGregor, K.C. (1996). Evaluation of WEPP runoff and soil loss predictions using natural runoff plot data. *Transactions of ASAE* 39 (3): 855–863.

Part E

Hydrometeorological Water Harvesting

16

Principles and Applications of Atmospheric Water Harvesting

Mousa Maleki[1], *Saeid Eslamian*[2], *and Boutaghane Hamouda*[3]

[1] *Department of Civil Engineering, Islamic Azad University, Najafabad Branch, Isfahan, Iran*
[2] *Department of Water Engineering, Isfahan University of Technology, Isfahan, Iran*
[3] *Department of Hydraulics, Universite Badji Mokhtar, Annaba, Algeria*

16.1 Introduction

Water scarcity is the first and most challenging crisis worldwide. With the growing world population, it is imperative to find alternative water resources to meet water demand. Atmospheric water, also known as air humidity, is one of the most accessible resources and therefore could be used as a sustainable resource for water harvesting.

16.1.1 Unconventional Water Resources

Unconventional water resources are a by-product of specific processes or can result from specialized technology to collect (access) water. These resources often need proper pre-use treatment, and when used for irrigation, they require appropriate on-farm management (Qadir et al. 2018). The key examples of unconventional water resources include groundwater confined in deep geological formations; atmospheric moisture harvested through cloud seeding and fog collection; physical transportation of water through icebergs; micro-scale capture of rainwater where it otherwise evaporates; desalinated water; and residual water from urban areas and agriculture (Figure 16.1).

16.2 Atmospheric Water Harvesting Necessity

The world's growing population puts severe demands on freshwater production and supply, and in many areas existing water resources are already overstressed (Kummu et al. 2010). Specifically, existing potable water resources are being depleted in many areas due to climate change (in terms of both precipitation and evapotranspiration), rapid urbanization, exponential population growth, and lack of (or inadequate) treatment of wastewater. By 2025 two-thirds of the worldwide population is expected to live in regions with water scarcity (Macedonio et al. 2012).

Intense water shortages are currently experienced by 470 million people and it is projected that by 2025, the number of people living in water-stressed countries will increase to 3 billion (Obispo 2009). Regarding the estimations, 2.4 billion people in the world lack access to safe drinking water and there are about 1.7 million deaths per year worldwide because of diseases found in poor water quality (Obispo 2009).

Hence, alternative water sources and innovative technologies for drinking water production are sought. Clearly, the availability of an immense amount of ocean water can be utilized (mainly in coastline regions) for seawater desalination. The latter is performed mostly by a reverse osmosis (RO) membrane technology, which has become cost-effective in the last decade and is therefore practiced intensively (Semiat 2008). However, although RO seawater desalination is among the most promising alternative water production technologies, it is not applicable in countries and regions that do not have access to the sea or to underground brackish water. Moreover, transportation of desalinated water from coastline areas to inner continental regions requires large investments in infrastructure, and has high operation energy demands.

In contrast, atmospheric water vapor is a potential source of a plentiful amount of freshwater that is accessible everywhere, yet the process is not currently cost-effective. The atmosphere contains about 13 000 km^3 of freshwater, 98% of which is vapor and only 2% of which in a liquid phase (cloud droplets, fog). This volume can be compared with all the surface and underground freshwater, excluding ice and glaciers (Beysens and Milimouk 2000). As is naturally done by vegetation and animals, liquid water is easier to

Handbook of Water Harvesting and Conservation: Basic Concepts and Fundamentals, First Edition. Edited by Saeid Eslamian and Faezeh Eslamian.

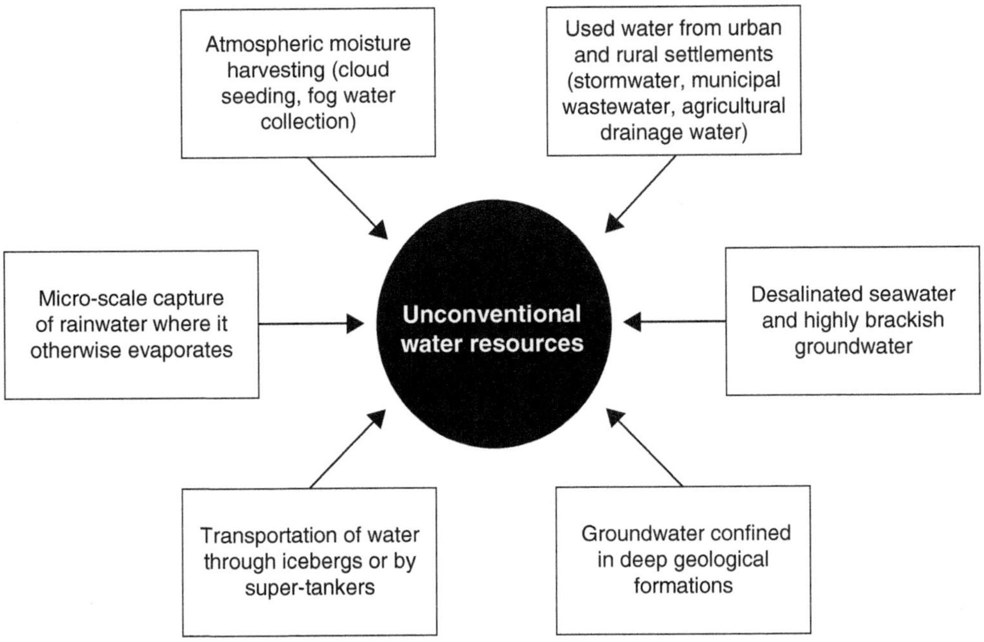

Figure 16.1 Examples of unconventional water resources (Qadir et al. 2018).

Figure 16.2 Large fog collectors (LFCs) made of double layer polyolefin mesh located in Atacama, Chile. Source: Reproduced from: https://www.youtube.com/watch?v=2x2_8RbC2I0; accessed on 22.07.2015. The inset shows the fiber network of a single layer polyolefin mesh. Source: Modified after Klemm et al. 2012; Copyright Royal Swedish Academy of Sciences 2012.

capture (Andrews et al. 2011; Malik et al. 2014; Parker and Lawrence 2001). Man-made fog traps have been tested over the last few decades to some extent (Klemm et al. 2012). However, frequent and predictive fog events occur only in very specific places that enjoy favorable conditions (Domen et al. 2014) (Figure 16.2), thus, on a large scale it is even less accessible than seawater.

16.3 Methods of Atmospheric Water Harvesting

There are some mechanisms for atmospheric water harvesting, including the following methods.

16.3.1 Vapor Condensing

The common way for extracting the atmospheric humidity is by condensing the vapor (i.e. the moist air is cooled to a temperature below its dew point following contact with a cold surface). This process involves a significant latent heat release ($\sim 2500 \frac{kj}{kgw}$) as well as sensible heat interaction between the air and the surface. Hence, the condensation process is limited by the rate of heat loss from the surface, which is necessary to keep its temperature below the dew point. The process of radiative cooling toward the night sky drives natural atmospheric moisture extraction via the formation of dew on surfaces (Muselli et al. 2006; Nikolayev et al. 1996; Sharan 2013). The maximum expected yield of radiative dew harvesting is ~0.8 $L/(m^2\ d)$ (Beysens et al. 2013) but empirical studies of passive dew capturing reveal much lower and varying water yield (Beysens et al. 2005; Guan et al. 2014; Jacobs et al. 2002, 2008; Nilsson 1996).

In general, the process is limited by the rate of radiative heat exchange, the weather, and the surface properties (Beysens 2016). In particular, weather conditions dictate the ratio of latent to sensible heat exchange between the surface and the air. When the dew point temperature, T_d, is much lower than the air temperature, T_a, most of radiative cooling is consumed by a sensible heat exchange and the dew yield may drop to zero. Indeed, if $T_a - T_d > 10K$ it is unlikely to get any significant dew yield (Beysens 2016). Hence, passive dew harvesting can be only a supplementary water source in regions with favorable conditions (Lekouch et al. 2011; Sharan 2002).

16.3.2 Active Cooling of the Ambient Air

The other feasible approach to harvest atmospheric humidity is by active cooling of the ambient air, using an electric compression expansion device (Wahlgren 2001, 2014). Whereas the energy investment liberates the process from the limited passive radiative cooling rate, the efficiency of the process is still highly dependent on the meteorological conditions. Several such atmospheric moisture harvesting (AMH) systems are currently available in the market, mostly for emergency use and/or when relatively small amounts of freshwater are required.

Moreover, the moisture harvesting potential can be used for the selection of optimal time periods for an on-off operation mode, which can increase the overall moisture harvesting efficiency by avoiding operation when the ambient air conditions are highly unfavorable. The suitability of climate conditions for AMH by direct electrical cooling is assessed by a new index, the Moisture Harvesting Index (MHI), which reflects the ratio of the energy invested in the desired water condensation process to the total energy invested in the cooling of the condensable as well as the incondensable gasses in the air bulk. The use of the MHI for estimating water production and its energy requirements will be demonstrated.

16.3.3 Fog Harvesting – Age-Old Practices that Still Work

Historical developments in water treatment systems are significant, from the early days of conventional filtration systems to more modern desalination plants with sophisticated membrane systems. A diversity of water sources is being tapped – mountain streams, springs, lakes, rivers, and where these aren't readily accessible, humidity in the air and even seawater are being harvested. But water collection from the air is not a novel idea. For at least 2000 years it has been in use with air wells in the Middle Eastern deserts and in Europe. Dewponds in the 1400s collected water, and later fog fences.

Fog fences use a technique called fog harvesting or fog collection or even cloud stripping, to collect water from the humidity in the fog. It can be used in coastal areas where inland wind brings in the fog, and high-altitude areas (if water is present in stratocumulus clouds), from 400 to 1200 m (UNEP 1997). How does it work? A mesh material strung tightly on poles is used, which is supported by gutters to collect droplets that are fed into pipes, and then stored in tanks. The size of the mesh can be as small as a meter in length or nearly 100 m long, depending on the lay of the land, space available, and the quantity of water needed.

In eastern Nepal, a fog collection system produces an average 500 l of water daily and about half the quantity in the dry season. A study has shown that in Eritrea (East Africa), 1600 square meters of mesh produced an average of 12 000 l of water a day. Remote areas in Chile, Ecuador, and Peru rely on this technique to draw much-needed water for consumption and irrigation. Other places that can potentially benefit from this technique, according to the International Development Research Centre (1995), include Yemen, Oman, Mexico, Cape Verde, China, the Atlantic coast of southern Africa (Angola, Namibia), South Africa, Sri Lanka, and Eastern Kenya. Researchers are still testing and innovating better quality meshes and configurations that will maximize water production under different conditions.

16.4 Energy Requirements of AMH and Water Production Costs

Due to varying daily and seasonal meteorological conditions, the MHI shows considerable variability. Year-round continuous operation of an AMH device is reasonable only in tropical regions, which experience high relative humidity (RH) and stable temperatures throughout the year. Considerable decrease in energy requirements may be achieved by using time-resolved MHI for informed selection of economically advantageous operation periods, while avoiding operation of the system when the MHI is low (hence, shifting from continuous to non-continuous operation). For example, narrowing the system operation time in non-tropical regions may result in an overall lower energy requirement (per kg of freshwater produced) if the AMH process is switched off when the MHI is low.

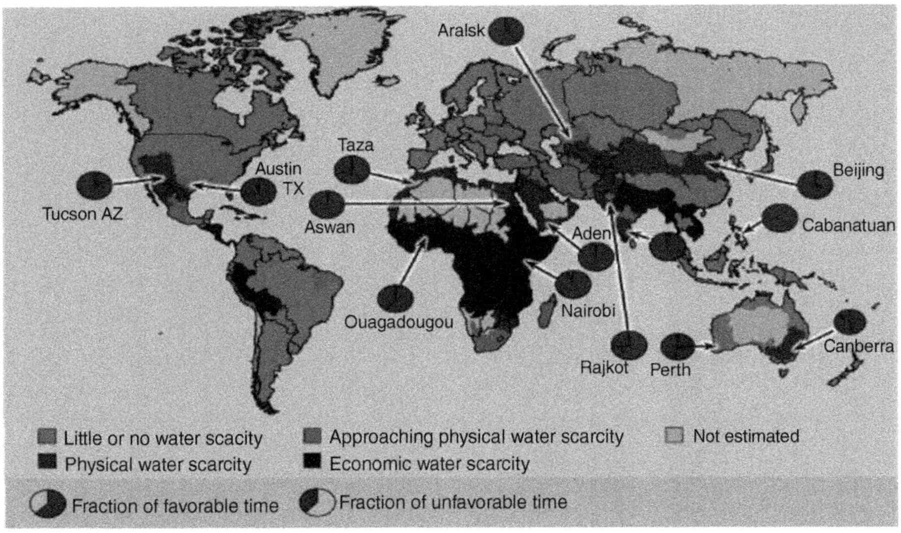

Figure 16.3 Fraction of the time out of the 10 years (2005–2014) meteorological data in which the ambient conditions were estimated to be suitable for AMH (e.g. MHI N 0.3), overlaid on the physical and economical global water scarcity map. Source: Adapted from Molden et al. 2007.

Clearly, the drawback of such an intermittent operation mode may be smaller water production and possibly mechanical problems due to the discontinuous operation cycle. This operation strategy can be implemented based on economic considerations or when apart from freshwater shortage, the energy is also limited. Figure 16.3 presents the hourly MHI variation in different months at four locations, revealing distinct MHI patterns. Aden, Yemen, is characterized by a very large fraction of favorable time for AMH throughout the year, with very small daily and seasonal variation. Perth, Australia, is less suitable for AMH, with a more pronounced daily variation and a small seasonal variation. Ouagadougou, Burkina Faso, is characterized by a very pronounced seasonal variation, with AMH expected to be more suitable in the summer and is clearly unsuitable in the winter. The climate of Sde Boker, Israel, reveals a pattern of both seasonal and daily variation, suggesting that non-continuous operation may result in a significant reduction in energy demands and operation costs.

Daily variation of the MHI is evident in all the seasons, with higher MHI generally characterizing the night, making it the more favorable operation period (note that electricity is also cheaper at night in some locations due to lower load). Details of the energy saving can be approximated by assessing the MHI over the operation hours for different MHI thresholds, in comparison to the overall MHI over the whole period (i.e. continuous operation). Table 16.1 contains estimates of energy saving that may be achieved by MHI-informed operation strategy for different MHI thresholds, and the related decrease in water production relative to continuous operation.

Together, Figure 16.3 and Table 16.1 represent four typical conditions: (i) locations with almost no daily and seasonal variability of the MHI throughout the year (represented by Aden, Yemen), (ii) locations with considerable daily variability but negligible seasonal variability of the MHI (represented by Perth, Australia), (iii) locations with little daily variability but considerable seasonal variability of the MHI (represented by Ouagadougou, Burkina Faso), and (iv) locations where considerable daily and seasonal variability of the MHI is evident (represented by Sde Boker, Israel). These four location types may benefit from distinct AMH operation modes for keeping the AMH process relatively cost-effective. Possible modes of operation include year-round continuous operation, season-specific continuous operation, time-of-day intermittent operation (non-continuous daily operation), and any combination of these options.

16.5 Atmospheric Vapor Harvesting Systems

There are different systems with specific functions to harvest water based on the methods and used materials in the process of atmospheric water generator (AWG) systems.

16.5.1 Water Harvesting from Air with Metal-Organic Frameworks Powered by Natural Sunlight

In this section, water harvesting by vapor adsorption will be demonstrated using a porous metal-organic framework (microcrystalline powder form of MOF-801, [Zr6O4 (OH)4(fumarate)6]) (Furukawa et al. 2014) in ambient air with low RH typical of the levels found in driest regions of the world (down to a RH of 20%). Also, a device based on this metal-organic framework (MOF) is reported that can harvest and deliver water (2.8 l of water per kilogram of MOF per day at 20% RH) under a no concentrated solar

Table 16.1 Estimates of the reduced energy requirements and water production of an AMH process under intermittent mode of operation.

MHI threshold	Percentage of favorable operation time (%)	Average MHI during the operation hours (–)	Estimated specific energy requirement (kWh/L)	Energy saving (%)	Water production reduction
Perth, Australia					
Continuous operation	100	0.29	0.47	–	–
$MHI > 0.1$	83	0.34	0.40	16	1
$MHI > 0.2$	71	0.37	0.36	23	7
$MHI > 0.3$	53	0.42	0.33	31	22
Sde Boker, Israel					
Continuous operation	100	0.32	0.43	–	–
$MHI > 0.1$	69	0.46	0.30	30	1
$MHI > 0.2$	64	0.48	0.28	34	3
$MHI > 0.3$	59	0.5	0.27	37	7
Ouagadougou, Burkina Faso					
Continuous operation	100	0.29	0.47	–	–
$MHI > 0.1$	73	0.39	0.35	26	1
$MHI > 0.2$	4	0.42	0.32	32	5
$MHI > 0.3$	59	0.46	0.30	37	15

The applied assumptions are identical to those in Figure 16.3 (Gido et al. 2016).

flux less than 1 sun (1 kW m^{-2}), requiring no additional power input for producing water at ambient temperature outdoors.

Porous materials, such as zeolites, silica gels, and MOFs can harvest water from air by adsorption over a wide range of humidity values (Canivet et al. 2014; Burtch et al. 2014; Wang et al. 2016). Nevertheless, typical adsorbents (e.g. zeolites and silica gels) suffer from either low uptake of water or requiring high energy consumption to release water. Although MOFs have already been considered in numerous applications – including gas storage, separation, and catalysis (Lee et al. 2009; D'Alessandro et al. 2010; Zhou et al. 2012); heat pumps (Jeremias et al. 2014; de Lange et al. 2015); and dehumidification (Seo et al. 2011), the use of MOFs for water harvesting has only recently been proposed (Furukawa et al. 2014). The flexibility (Eddaoudi et al. 2002; Yaghi et al. 2003; Furukawa et al. 2013) with which MOFs can be made and modified at the molecular level, coupled with their ultrahigh porosity, makes them ideally suited for overcoming the challenges mentioned above.

A critical step is the release of water from the MOF, which can be achieved by applying a low-grade heat-driven (Chu and Majumdar 2012) vapor-desorption process. Solar energy is particularly promising because sunlight is often abundant in arid regions with low RH (>7 kW-hours $m^{-2} day^{-1}$, equivalent to 7 hours of 1 sun per day) where there are limited water resources and where a natural diurnal temperature swing thermally assists the process (adsorption of water during the cooler night and release during the warmer day). This method is much more energy-efficient compared with refrigeration-based dew-harvesting systems because heat is directly used for desorption. The amount of water that can be harvested with MOFs can be much greater than with dew-harvesting systems, which become impractical at RHs less than 50% (25). To use MOFs to harvest water with maximum yield and minimal energy consumption, an isotherm with a steep increase in water uptake within a narrow range of RH is suitable, which enables maximum regeneration with minimal temperature increase. Recent MOFs have exhibited such sorption characteristics (Figure 16.4a).

In particular, MOF-801 is suitable for regions where RH is merely 20% (e.g. North Africa), and UiO-66 (Furukawa et al. 2013) is suitable for regions with ~40% RH (e.g. northern India). Water with MOF-801 and natural sunlight at <1 sun in an environment at regeneration temperatures of ~65 °C is harvested. Once water vapor adsorbed into the MOF, solar energy was used to release the adsorbate. Water was then harvested using a condenser maintained

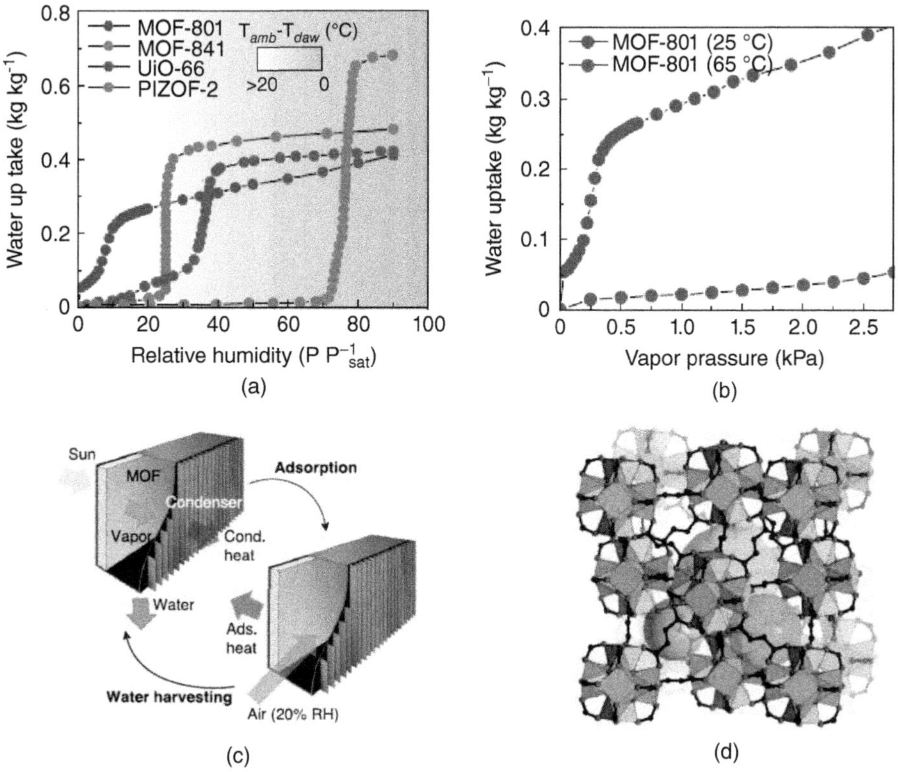

Figure 16.4 (a) Working principle of water harvesting with MOFs. Zr-based MOFs (MOF- 801, MOF-841, UiO-66, and PIZOF-2) water-sorption isotherms at 25 °C, showing a rapid increase in adsorption capacities (in kilograms of water per kilogram of MOF) with a relatively small change in the relative humidity (RH) (P Psat −1, vapor pressure over saturation pressure) (10). (Kim et al. 2017). (b) Working principle of water harvesting with MOFs. Water adsorption isotherms of MOF-801, measured at 25° and 65 °C, illustrating that the temperature swing can harvest greater than 0.25 kg kg^{-1} at >0.6 kPa vapor pressure (20% RH at 25 °C). (Kim et al. 2017). (c) Working principle of water harvesting with MOFs. A MOF water-harvesting system, composed of a MOF layer and a condenser, undergoing solar-assisted water-harvesting and adsorption processes. Within water harvesting (left), the desorbed vapor is condensed at the surrounding temperature and delivered through a passive heat sink, requiring no additional energy input. During water capture, the vapor is adsorbed on the MOF layer, transferring the heat to the ambient surroundings (right). Ads. and cond., adsorption and condensation, respectively. (Kim et al. 2017). (d) Working principle of water harvesting with MOFs. Zr6O4(OH)4(-COO)12 subsidiary building units are linked together with fumarates to form MOF-801. The large yellow, orange, and green spheres are three different pores. Black, C; red, O; blue polyhedra, Zr (Kim et al. 2017).

at temperatures near that of the surrounding environment. For MOF-801, a temperature swing between 25° and 65 °C can harvest more than 0.251 kg^{-1} at >0.6 kPa vapor pressure (20% RH at 25 °C, Figure 16.4b). This water harvesting strategy is completely passive, relying only on the high water uptake capacity, low-grade heat requirement for desorption, and ambient temperatures to condense and collect the water (Figure 16.4c).

For our approach, MOF-801 has several advantages: (i) well-studied water-adsorption behavior on a molecular level, (ii) good performance driven by aggregation of water molecules into clusters within the pores of the MOF, (iii) exceptional stability and recycling, and (iv) constituents that are widely available and low-cost. It is composed of 12-connected Zr-based clusters [Zr6O4(OH)4(-COO)12] joined by fumarate linkers into a three-dimensional, extended porous framework of face-centered cubic topology. The structure of MOF-801 contains three symmetrically independent cavities into which water molecules can be captured and concentrated (Figure 16.4d).

The adsorption-desorption experiments have been carried out for water harvesting with MOF-801 at 20% RH. A powder of MOF-801 was synthesized (as reported in Furukawa et al. 2013) and activated (solvent removal from the pores) by heating at 150 °C under vacuum for 24 hours. Also, the powder was infiltrated into a porous copper foam with a thickness of 0.41 cm and porosity of ~0.95, which was brazed on a copper substrate to create an adsorbent layer (5 by 5 by 0.41 cm) with 1.79 g of activated MOF-801, an average packing porosity of ~0.85 (Figure 16.4a), and enhanced structural rigidity and thermal transport.

This particular geometry with a high ratio of layer area to thickness was selected to reduce parasitic heat loss. Experiments have been performed in a RH-controlled environmental chamber interfaced with a solar simulator. The manufactured MOF-801 layer was placed in the chamber (Figure 16.4a) and evacuated under high vacuum (less than 1 Pa) at 90 °C.

Water vapor was then introduced inside the chamber to maintain a condition equivalent to a partial vapor pressure of 20% RH at 35 °C, matching the steep rise in water uptake for MOF-801 (Figure 16.4a). After getting saturated, the chamber was isolated from the vapor source. A type of solar flux (1 kW m^{-2}, air mass 1.5 spectrum) was introduced to the graphite-coated substrate layer with a solar absorptance of 0.91 to desorb water from the MOF.

16.5.2 Atmospheric Vapor Harvesting Adsorption Materials

Conventional desiccants, such as silica gel, zeolite, and activated alumina typically have wide water vapor sorption windows. However, these desiccants are mainly based on physical adsorption and the energy required to release the adsorbed water is proportional to the water desiccant affinity. In other words, for these desiccants, the more readily the water vapor is adsorbed, the higher the temperature must be to release water. Thus, to efficiently release the captured water, they require high temperatures (>160 ° C), which are typically beyond what simple solar photothermal-based heating devices are capable of offering.(Chua et al. 2002; Dzhigit et al. 1971; Desai et al. 1992; Wang et al. 2009)

In 2017 and 2018, it was demonstrated in lab and field conditions that certain MOFs, due to their outstanding sorption ability with water vapor, delivered potable water from dry air (RH 10–40%) under regular solar irradiation conditions. (Kim et al. 2018, 2017) Anhydrate salt-based water sorbents have also been recently reported for AWG application, which works well under low RH (i.e. 10–35%) and are able to deliver about 20% liquid water of their own weight (Li et al. 2018).

Deliquescent salt has a high affinity with water and is able to sorb water vapor as many as 5–6 times its own weight, which is significantly higher than most porous sorbents (Mauer Taylor 2010; Brien 1948).

As a matter of fact, deliquescent salt is so effective in capturing water vapor that the captured water vapor ultimately dissolves the salt and forms an aqueous solution with a much expanded volume. In a sense, the deliquescent property of the salt would break the physical confinement constraint in terms of the amount of water vapor captured by the conventional porous desiccants where the pore volumes of the solid materials set the upper limits. However, the liquid form poses challenges in handling and engineering design for applying deliquescent salts to water sorption. Recently, Gido et al. (2016) theoretically proposed an atmospheric water harvesting system, which employs lithium chloride solution as water vapor sorbent and can produce water in a continuous process. Based on their simulations, the energy consumption can be reduced up to 65% comparing with conventional condensation-based system.

A proof-of-concept is provided by using calcium chloride ($CaCl_2$), a cheap, stable, eco-friendly, and nontoxic deliquescent salt. The hydration reaction of $CaCl_2$ enables it to capture water at low humidity (i.e. RH 10–25%) while its deliquescence further draws more water vapor into the dissolving salt with the lower range of RH values down to 26% at 25 °C, making it an effective water sorbent over a wide range with a superior capacity.

With a solar photothermal component built in, the deliquescent-salt-hydrogel- photothermal composite material captures 0.74, 1.10, and 1.75 g of water vapor for each gram of the dry composite material under RH of 35%, 60%, and 80%, respectively (mixing ratio 6.5, 11.8 and 16.2 g/kg air), and it releases almost 100% of the captured water under irradiation with regular sunlight intensity. The water capture capacity achieved at 35% humidity is three times higher than the MOF and CuCl2 materials. In outdoor field conditions, (April 2018 at Thuwal, Saudi Arabia), an "easy-to-assemble at household" and all-in-one prototype device was tested with a dry hydrogel disk of 35 g in which 37 g of water vapor sorbed into it from an open air with RH between 60% and 70%. Under natural sunlight, the device quickly delivered ~20 g of fresh water within just 2.5 hours.

Based on these performances and by extrapolation, it is estimated that the material cost of the AWG device to supply a minimum daily water need (3 kg/day) for an adult is only $3.2 USD. This technology provides a promising solution for clean water production in arid and landlocked remote regions. The superior water harvesting capacity within a wide RH range makes the composite hydrogel a versatile AWG for island regions and inland areas with low to high RH as well (Renyuan et al. (2018)).

A group of scientists from University of Texas have improved the process of water collection by creating a material with a hydrophilic lubricant. The key lies in the creation of slippery rough surfaces (SRS) with hydrophilic properties and the ability to guide water droplets in a certain direction. These slippery and coarse areas are much more efficient than the existing technologies, which are based on slippery liquid-infused porous surfaces (SLIPS). Before going any further, though, some of these initials should be clarified. The SLIPS technology makes use of an innovative porous material infused with a lubricant fluid. Thus, it achieves a repellent and self-cleaning surface that repels water, but also dust specks and other pollutants. Inspired by these materials, as well as the Namib beetle and pitcher plants, the SRS technology goes a step further and fosters the accumulation of larger drops, funneling them into reservoirs via lubricated microgrooves.

Therefore, the two main properties of this technological innovation are the ability to attract water droplets, as some desert beetles do, and provide an extremely slippery surface, as the one provided by pitcher plants. To that end, the researchers have leveraged the characteristics of hydroxy functional groups, which are extremely hydrophilic and are the main ingredient composing the lubricant. Additionally, the material where the lubricant is applied sports a set of nanotextures that provide increased mobility for the droplets.

16.5.3 Applications of Superhydrophilic and Superhydrophobic Materials

According to the researchers, the new surface would be able to collect up to an estimated 120 l per square meter daily. Besides the ability to harvest water vapor, the SRS system enjoys a slew of other applications, whether in air conditioning systems, industrial equipment, or as a surface to keep ship hulls clean by preventing the accumulation of rust and debris. Superhydrophilic and superhydrophobic materials are shown in Figure 16.5.

Figure 16.5 Superhydrophilic and Superhydrophobic materials (https://phys.org/news/2018-03-liquid-repellent-surface-maximizes-harvest.html).

On the other hand, there are techniques that employ carbon nanotubes or laser etching to create superhydrophobic surfaces. These achieve the opposite effect, repelling liquids in an extremely efficient manner and even at times causing water droplets to bounce on impact. These new materials can combat rust, ice, and layers over solar panels to prevent water or impurities from accumulating and reducing their efficiency.

16.5.4 Vapor Compression Refrigerating System

This system generates fresh drinking water and also extracts water from humid ambient air by using a cooling condensation process. Cooling and desiccants are the two primary techniques in use. Filtered air is pushed over the coil by a controlled-speed fan.

The gathered water is then passed into a holding tank with a purification and filtration system to keep the water pure. Atmospheric water generating technology offers 99.9% pure drinking water 365 days a year. An AWG is an eco-friendly source of sustainable water. It extracts water from humid ambient air. Air vapor is condensed by cooling the air below its dew point, exposing the air to desiccants, or pressurizing the air. In contrast with a dehumidifier, an AWG is designed to render the water potable. In locations where pure drinking water is difficult or impossible to obtain AWG systems are very useful, as there is almost always a small amount of water in the air.

Cooling and desiccants are the two primary techniques in use. According to the initial unoptimized tests, they generated 18.3 l of water in 24 hours in ambient air conditions with an average relative humidity of 70% RH. During this period, they consumed 2 kW-hours of energy per liter of water generated. A recent research represents that ultraviolet (UV) irradiation is quickly gaining popularity in the consumer market as a safe, effective, and economical approach to disinfection (Wang and Liu 2008).

There are many pathogenic organisms that are more susceptible to UV than they are to chlorine. The need for clean water without any health or environmental concerns can be addressed by AWG technologies. AWGs are already available in locations that have grid power and public waterworks.

16.5.4.1 Water Generation System

AWG systems have the capability of harvesting high-quality fresh water that tastes great from the atmosphere. Yet, the vast majority of the general public has no idea that extracting "water from air" technology exists. Similarly, most people lack even a vague notion of just how much water is readily available in the Earth's atmosphere; in fact, there is over 3100 cubic miles of water, equivalent to more than 3.4 quadrillion gallons at any one time in our atmosphere.

The hydrologic cycle verifies that billions of gallons of water are evaporating into or raining out of the atmosphere every day. This creates a neverending continuous supply of fresh water that is available and accessible almost everywhere in the world. AWGs produce water similar to the way rain is made and are the most environmentally friendly water dispensing systems available on the market.

As the world's freshwater demand increases daily, modern alternative "water production" technologies, particularly those producing water from air, are among the most viable and appealing solutions available today. This modern technology can economically satisfy growing water demand by providing safe drinking water that promotes public health and overall social well-being.

The AWG is an eco-friendly viable and sustainable water source. It is particularly advantageous for agriculture by eliminating the need for sophisticated water purification and water filtration methods. However, because clean drinking water is essential, the water produced by the AWG goes through both UV light and an ozone water purification system, followed by a multi-step carbon water filtration system, to produce 99% pure drinking water.

A prototype of the AWG system in Figure 16.6 depicts a modified dehumidifier, comprised of a used commercial household dehumidifier with its compressor replaced by a 1.1 kW car compressor, and assembled with an electric motor.

Figure 16.7 shows the process of air vapor condensation and water generation in prototype.

16.5.4.2 Operation of Water Generation Systems

Powered by grid electricity, the car compressor liquefies the tetra-fluoroethane refrigerant (R134a) at a pressure of 200–250 per square inch (psi) and raises the coil temperature to about 80–89° Celsius (°C). This hot fluid passes through the filter dryer that removes unwanted sediments and absorbs moisture. It then passes through an expansion valve that automatically regulates the flow of refrigerant with respect to the temperature of the cold suction coil of the compressor. This compressor is run by an AC motor and a car battery to power its magnetic coil. The refrigerant expands at a low pressure of 30–40 psi and at a coil temperature of 10–21 °*C*. At this temperature range, the water vapor from air condenses into liquid to an average of 763 ml per hour at an average 70% RH.

16.5.4.3 Water Treatment System

An off-the-shelf water purification system was used in the system set-up. The initial set-up was a built-in water catchment system which relied on gravity for collection. Water dripped slowly, rendering potable water collection time-inefficient. An automated system of water collection was used where water was pumped into the ceramic filter and exposed to the UV radiation, which accelerated the water collection and treatment process. The collected water is pumped from the storage tank by a submersible pump at a flow rate of up to 83.3 l per minute. This pump is required to provide the minimum pressure of 50 psi necessary to operate the ceramic filter at a production flow

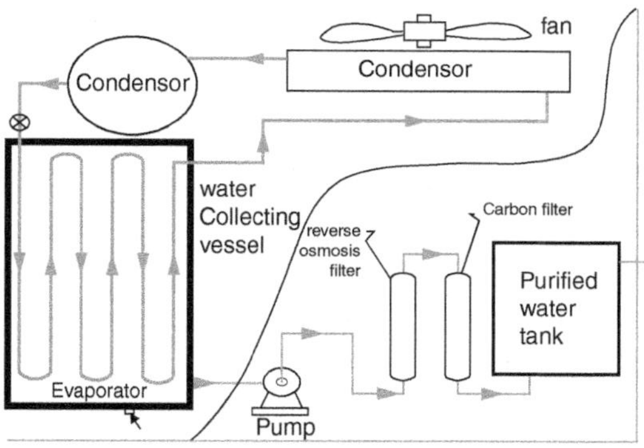

Figure 16.7 Water generation systems (Anbarasu and Pavithra 2011).

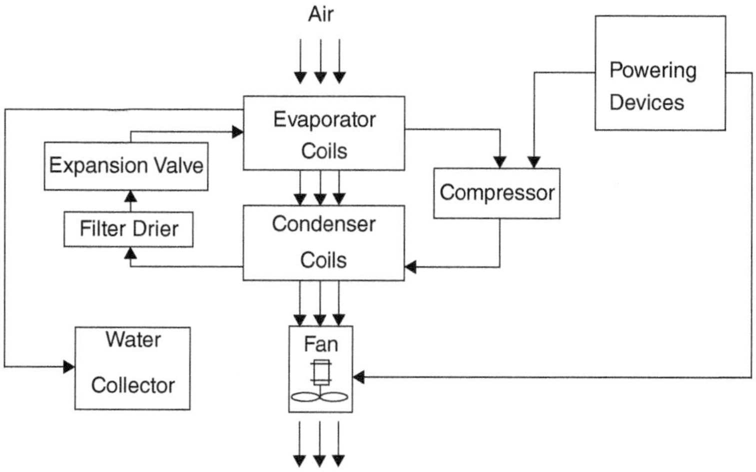

Figure 16.6 Block diagram of prototype Anbarasu and Pavithra 2011

rate of 750 ml per minute, during which sediments are removed. The water then proceeds to the final treatment stage, during which remaining pathogens, bacteria, and viruses are eliminated in the water by exposing it to a UV light with 253.7 nm wavelength.

16.5.4.4 Water Formation in a Humid Atmosphere

The fact that water treeing occurs in humid air is important to current insulation theory because it reveals that water in the liquid state is not required for water treeing to begin. However, if water in the liquid phase can occur as a result of some force in organic materials in humid air, therefore the same explanations which have been proposed for water treeing in the presence of water might be applied to the mechanism in humid air (Harvell et al. 1999).

First, water formation due to ambient temperature change will be considered. After that the feasibility of water formation due to a change of electrical field energy will be considered. The quantitative discussions in this section indicate the two possibilities cannot be accepted. The authors propose a possibility of water formation in the form of induced capillary condensation due to electrical stress.

The absolute water vapor pressure P_1 (Pa) at RH% of relative humidity is calculated. The evaluated numerical values of P_1 are plotted against temperature for the relative humidity of 60%, 80%, and 100% in Figure 16.8. The actual possibility of liquid water formation can be considered from Figure 16.8. For example, in the experiment under the conditions of θ_1=50 °C and RH = 80%, P_1 is equal to 74 mm Hg (9864 Pa).

Thus, a temperature decrease to 45.8 °C corresponds to a saturated vapor pressure of 74 mm Hg, which is required to condense water vapor. Therefore, a temperature change of 4.2 °C is necessary for condensation. Similarly, a temperature decrease of 11.5 °C is necessary for condensation under the experimental conditions of θ_1 = 50 °C and RH = 60%.

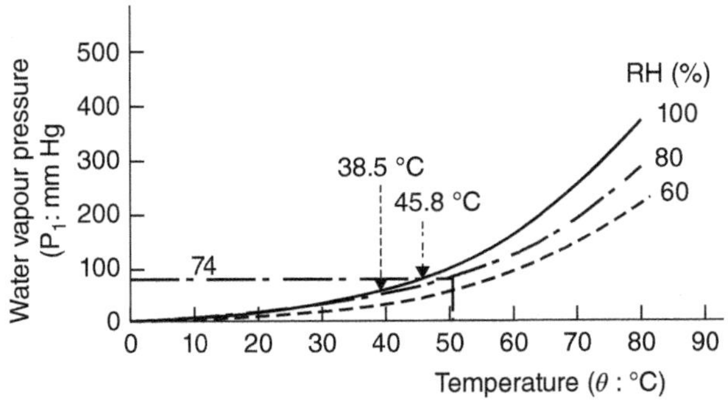

Figure 16.8 Dew point on water vapor pressure curves (Anbarasu and Pavithra 2011).

16.5.4.5 Computations and Estimations

The quantity of water in atmosphere depends on the air temperature and humidity. For relative humidity 100% the maximum partial pressure of water vapor is shown in Table 16.2.

The quantity of water in 1 $m^3$of air may be computed by Eq. (16.1):

$$M_w = 0.00625[p(t_1) - p(t_2)] \tag{16.1}$$

The computation of Eq. (16.1) is presented in Table 16.2. Normal relative humidity of atmosphere air is 0.5–1.

16.5.4.6 Cooling Condensation Process

In a cooling-condensing AWG system, a compressor circulates refrigerant through a condenser and an evaporator coil which cools the air surrounding it, lowering the air's dew point and causing water to condense. Filtered air is pushed over the coil by means of a controlled-speed fan. The captured water is then transferred into a holding tank with purification and filtration system to keep the water pure (Obispo 2009). The water production rate depends on relative humidity and ambient air temperature and size of the compressor. AWGs are nowadays more effective as relative humidity and air temperature are increasing due to climate change.

Table 16.2 Projected volume of water (%) in varied humidity conditions and temperatures (Anbarasu and Pavithra 2011).

Temperature (°C)	Humidity						
	50%	60%	65%	70%	80%	90%	100%
15	5.5	5.7	7.2	8.7	9.5	12.4	14.3
20	7.4	8.4	9.6	11.0	12.4	17.4	23.7
25	9.7	12.4	15.6	18.3	22.0	27.3	33.3
30	12.2	18.0	20.8	25.7	29.2	37.2	47.3
35	15.0	24.4	28.0	32.9	37.2	47.0	57.0

Based on the conventional rule of thumb, cooling-condensation AWG systems do not work efficiently when the air temperature falls below 18.3 °C (65 °F) or the relative humidity drops below 30% (Greenfield 2006). An AWG system's cost-effectiveness depends on the capacity of the machine, local humidity and temperature conditions, and the cost to power the unit.

16.5.4.7 Compressor

On one side of the enclosed casing of the airtight compressor, the various parts of the compressor like cylinder, piston, connecting rod, and the crankshaft are located. If the compressor is a multi-cylinder one, there are more than two cylinders inside the casing. There is an electric winding inside the other side of the casing which the shaft of the motor rotates. The motor can be single speed or multi-speed. In airtight compressors the crankshaft of the reciprocating compressor and the rotating shaft of the motor are common. The motor's rotating shaft extends beyond the motor and forms the crankshaft of the hermetically sealed reciprocating compressor.

In describing how the compressor and motor drive is situated in relation to the gas or vapor being compressed, it can be said that compressors are often open, hermetic, or semi-hermetic. The industrial name of a hermetic is hermetically sealed compressor, while a semi- is commonly called a semi-hermetic compressor. The compressor and motor driving the compressor are integrated into hermetic and most semi-hermetic compressors, and operate within the pressurized gas envelope of the system. This motor is designed to be cooled by the gas or vapor being compressed.

The main difference between the hermetic and semi-hermetic units is that the hermetic uses a one-piece welded steel casing that cannot be opened for repair; if the hermetic unit fails, it is simply replaced by a new unit. In contrast, a semi-hermetic unit uses a large cast metal shell with gasket covers that can be opened to replace motor and pump components.

The major benefit of a hermetic and semi-hermetic is that there is no route for the gas to leak out of the system. Both rely on either natural leather or synthetic rubber seals to retain the internal pressure that exists in open compressors, and these seals require a lubricant such as oil to retain their sealing properties. A benefit of open compressors is that they can be driven by non-electric power sources, such as an internal combustion engine or turbine, whereas open compressors that drive refrigeration systems are not totally maintenance free throughout the life of the system, since some gas leakage will occur over time.

Airtight compressors have inbuilt lubrication systems for the lubrication of the piston and cylinder and crankshaft. The lubricant also acts as the coolant for the piston and cylinder. Moreover, the cool suction refrigerant also provides a cooling effect.

16.5.4.8 Dew Point

When a given parcel of humid air tends be cooled at a constant barometric pressure, this occurrence temperature is called the dew point. The dew point is equivalent to the saturation temperature, and is associated with the relative humidity. The condensed water is called dew, and can often be observed as fine droplets on blades of grass in early morning hours.

The dew point is associated with relative humidity. High percentage of relative humidity indicates that the dew point is closer to the current air temperature. Full relative humidity (100%) indicates the dew point is equal to the current temperature and the air is maximally saturated with water. Whenever the dew point remains constant and temperature increases, the relative humidity will decrease.

16.5.4.9 Relative Humidity

Relative humidity is defined as the ratio of the mass of vapor present in the unit volume of the air at a certain temperature to the maximum mass of the vapor which is capable of being accommodated in the unit volume of air when it is saturated. It was seen that the partial pressure of the water vapor depends on the mass of the water vapor in the air. Hence, relative humidity as the ratio of the partial pressure of water vapor when the air is saturated at the same temperature can also be defined. Relative humidity is a definition used to describe the amount of water vapor that exists in a gaseous mixture of air and water vapor.

The relative humidity (φ) of an air-water mixture is defined as the ratio of the partial pressure of water vapor (e_w) in the mixture to the saturated vapor pressure of water (e^*_w) at a prescribed temperature. Relative humidity is primarily stated as a percentage and is calculated by using the Eq. (16.2):

$$Relative\ humidity = \frac{Actual\ vapor\ density}{Saturation\ vapor\ density} \times 100\% \tag{16.2}$$

16.5.4.10 Comparison Between Various Compression Systems

There are multiple ways for each method of the compression process could be used regarding the climate, accessible facilities, and even energy resource availability.

16.5.4.10.1 Performance Comparison

The coefficient of performances (COPs) for various micro-electronic cooling systems including classical (mechanical compression), with ejector, and with pumpless absorption have been evaluated before (Bolonkin 2004). For this

Table 16.3 COP Comparison (Anbarasu and Pavithra 2011).

Refrigeration type	COP
Mechanical compression	4.24
Pumpless absorption	0.5
Solution pump absorption	0.73

case, the absorption system with solution pump will be compared only to the pumpless (with hydrogen as compensatory gas) absorption system as well as the most efficient mechanical compression refrigeration system. The results are summarized in Table 16.3. This table describes the various refrigeration types and its working comparison.

The largest COP is achieved by the mechanical compression, deemed most efficient of the three refrigeration systems described. The mechanical system does not require additional input energy, as compression occurs through mechanical means in the mechanical compressor ($\sim$ 25 W needed to run the compressor). For the absorption system, an electrical resistance helps boil the ammonia-water mixture, thus separating the ammonia vapors (entering the condenser) from the water, which returns to the absorber to participate in the dissolution (absorption) process.

The electrical resistance amount is equitably high, reaching almost 222 W for almost 100 W of cooling power. The electrical energy is converted into thermal energy further used in the dissolution (associated with the phase change) processing the absorber unit.

The absorption system with solution pump is more useful than the pumpless system due to the reduction of the energy input needed for the boiling in the desorber, due to the preheating of the rich solution during the heat exchange between the rich and weak solution in the heat exchanger, ultimately leading to increased system efficiency.

16.5.4.10.2 Evaporator

Evaporative condensers are usually used in ice plants. They are a combination of air-cooled and water-cooled condensers. In these types of condensers, the hot refrigerant flows through the coils, with water sprayed over these coils. The fan makes air suction from the bottom side of the condenser and discharges it from the top side of the condenser simultaneously. The water spray which is in contact with the condenser coil absorbs the heat from the condenser, cools the refrigerant and condenses it, and finally gets evaporated in the air. Evaporative condensers have the advantages of being water cooled as well as air cooled, hence they occupy less space. One disadvantage is that keeping the evaporative condenser clean and free of scale is extremely difficult and requires significant maintenance.

16.5.4.10.3 Condenser

A condenser is a device or unit used to condense a gaseous substance into a liquid state through cooling. In so doing, the latent heat is released by the substance and transferred to the surrounding environment. Condensers can be made according to numerous designs, and come in many sizes ranging from rather small (handheld) to very large (industrial-scale units used in plant processes). For example, a refrigerator uses a condenser to get rid of heat extracted from the interior of the unit to the outside air.

Air-Cooled Condensers Small facilities such as household refrigerators, deep freezers, water coolers, window air conditioners, split air conditioners, small packaged air conditioners, etc. are equipment that air-cooled condensers are used in. They are also used in plants where the cooling load is small and the total quantity of the refrigerant in the refrigeration cycle is small. Coil condensers is the other name for air-cooled condensers as they are usually made of copper or aluminum coil. Air-cooled condensers engross a relatively bigger space than water-cooled condensers.

Water-Cooled Condensers Water-cooled condensers are capable of being used for large refrigerating plants, big packaged air conditioners, central air-conditioning plants, etc. These condensers are used in plants where cooling loads are overly high and a large quantity of refrigerant flows through the condenser. Three types of water cooled condensers exist: tube-in-tube or double pipe type, shell and coil type, and shell and tube type. The refrigerant flows through one side of the piping while the water flows through the other piping, cooling the refrigerant and condensing it in all these condensers.

16.5.4.10.4 Water Quality Results

Water samples collected from Solar-Powered Atmospheric Water Generation and Purification (SAWGAP) system were submitted for bacteriological analysis at three stages: pre-treatment, post-ceramic filtering, and post-UV irradiation.

16.6 Conclusion

The United Nations Environment Program states that freshwater shortage and stress are increasing in tropical regions as a result of expanding populations, tourism, climate change, and pollution. Approximately 12 900 billion tons of freshwater is held in the Earth's atmosphere and is distributed all over the world with fast replenishment.

Atmospheric water harvesting offers a reliable promising strategy for clean water production in arid regions, landlocked regions, and remote communities. The vapor sorbent is the viral element for atmospheric water harvesting devices based on absorbing-releasing process.

The extraordinary volume of water vapor present in the atmosphere may serve as a potential water resource. The problem then in most cases is not a lack of available water, but rather the inability to obtain it in a cost-effective, reliable manner. Population growth and rising standards of living in many developing countries are increasing demand for clean, safe drinking water.

Atmospheric water harvesters are viable alternatives to existing water supply systems. Atmospheric water harvesting systems can also be considered as supplementary resources and logistical assets for consumers and industries that have limited access to appropriate water. Different types of atmospheric water harvesting methods and systems are introduced and related details are given for each method and the corresponding system.

References

Anbarasu, T. and Pavithra, S. (2011). Vapour compression refrigeration system generating fresh water from humidity in the air. *Langmuir* 27: 3798–3802. http://dx.doi.org/10.1021/la2000014.

Andrews, H.G., Eccles, E.A., Schofield, W.C.E. et al. (2011). Three-dimensional hierarchical structures for fog harvesting. *Langmuir* 27: 3798–3802. http://dx.doi.org/10.1021/la2000014.

Beysens, D. (2016). Estimating dew yield worldwide from a few meteo data. *Atmospheric Research* 167: 146–155. http://dx.doi.org/10.1016/j.atmosres.2015.07.018.

Beysens, D. and Milimouk, I. (2000). The case for alternative fresh water sources. *Pour Les Resour. Altern. En Eau, Sécheresse* 11: 1–17.

Beysens, D., Muselli, M., Nikolayev, V. et al. (2005). Measurement and modelling of dew in island, coastal and alpine areas. *Atmospheric Research* 73: 1–22. http://dx.doi.org/10.1016/j.atmosres.2004.05.003.

Beysens, D., Broggini, F., Milimouk-melnytchouk, I. et al. (2013). New architectural forms to enhance dew collection. *Chemical Engineering Transactions* 34: 79–84. http://dx.doi.org/10.3303/CET1334014.

Bolonkin A.A. (2004) Utilization of Wind Energy at High Altitude, AIAA-2004-5705, AIAA-2004-5756, International Energy Conversion Engineering Conference at Providence., RI, Aug.16–19. 2004. USA. http://arxiv.org.

Brien, F.E.M.O. (1948). The control of humidity by saturated salt solutions. *Journal of Scientific Instruments* 25 (3): 73.

Burtch, N.C., Jasuja, H., and Walton, K.S. (2014). Water stability and adsorption in metal-organic frameworks. *Chemical Reviews* 114 (20): 10575–10612. https://doi.org/10.1021/cr5002589.

Canivet, J., Fateeva, A., Guo, Y. et al. (2014). Water adsorption in MOFs: fundamentals and applications. *Chemical Society Reviews* 43: 5594–5617.

Chu, S. and Majumdar, A. (2012). Opportunities and challenges for a sustainable energy future. *Nature* 488 (7411): 294–303.

Chua, H.T., Ng, K.C., Chakraborty, A. et al. (2002). Adsorption characteristics of silica gel+ water systems. *Journal of Chemical & Engineering Data* 47 (5): 1177–1181.

D'Alessandro, D.M., Smit, B., and Long, J.R. (2010). Carbon dioxide capture: prospects for new materials. *Angewandte Chemie (International Ed. in English)* 49 (35): 6058–6082. https://doi.org/10.1002/anie.201000431.

Desai, R., Hussain, M., and Ruthven, D.M. (1992). Adsorption of water vapour on activated 585 alumina I - equilibrium behaviour. *Canadian Journal of Chemical Engineering* 70 (4): 699–706.

Domen, J.K., Stringfellow, W.T., Camarillo, M.K. et al. (2014). Fog water as an alternative and sustainable water resource. *Clean Technologies and Environmental Policy* 16: 235–249. http://dx.doi.org/10.1007/s10098-013-0645-z.

Dzhigit, O.M., Kiselev, A.V., Mikos, K.N. et al. (1971). Heats of adsorption of water vapour on X-zeolites containing Li+, Na+, K+, Rb+, and Cs+ cations. *Transactions the Faraday Society* 67: 458–467.

Eddaoudi, M., Kim, J., Rosi, N. et al. (2002). Systematic design of pore size and functionality in isoreticular MOFs and their application in methane storage. *Science* 295 (5554): 469–472.

Furukawa, H., Cordova, K.E., O'Keeffe, M. et al. (2013). The chemistry and applications of metal-organic frameworks. *Science* 341 (6149): 1230444. https://doi.org/10.1126/science.1230444.

Furukawa, H., Gándara, F., Zhang, Y.B. et al. (2014). Water adsorption in porous metal–organic frameworks and related materials. *Journal of the American Chemical Society*: 2014136114369–2014136114381.

Gido, B., Friedler, F., and Broday, D.M. (2016). Assessment of atmospheric moisture harvesting by direct cooling. *Atmospheric Research* 182: 156–162.

Greenfield, N. (2006). *Water Extracted from the Air for Disaster Relief*. National Public Radio October 19, USA.

Guan, H., Sebben, M., and Bennett, J. (2014). Radiative- and artificial-cooling enhanced dewcollection in a coastal area of South Australia. *Urban Water Journal* 11: 175–184. http://dx.doi.org/10.1080/1573062X.2013.765494.

Harvell, C.D., Kim, K., Burkholder, J.M. et al. (1999). Emerging marine diseases: climate links and anthropogenic factors. *Science* 285: 1505–1510.

Jacobs, A.F.G., Heusinkveld, B.G., and Berkowicz, S.M. (2002). A simplemodel for potential dewfall in an arid region. *Atmospheric Research* 64: 285–295. http://dx.doi.org/10.1016/S0169-8095(02)00099-6.

Jacobs, F.G., Heusinkveld, B.G., and Berkowicz, S.M. (2008). Passive dewcollection in a grassland area, the Netherlands. *Atmospheric Research* 87: 377–385. http://dx.doi.org/10.1016/j.atmosres.2007.06.007.

Jeremias, F., Fröhlich, D., Janiak, C., and Henninger, S.K. (2014). Water and methanol adsorption on MOFs for cycling heat transformation processes. *New Journal of Chemistry* 38: 1846.

Kim, H., Yang, S., Rao, S.R. et al. (2017). Water harvesting from air with metal-organic frameworks powered by natural sunlight. *Science*: eaam8743.

Kim, H., Rao, S.R., Kapustin, E.A. et al. (2018). Adsorption-based atmospheric water harvesting device for arid climates. *Nature Communications* 9 (1): 1191.

Klemm, O., Schemenauer, R.S., Lummerich, A. et al. (2012). Fog as a fresh-water resource: overview and perspectives. *Ambio* 41: 221–234. http://dx.doi.org/10.1007/s13280-012-0247-8.

Kummu, M., Ward, P.J., de Moel, H., and Varis, O. (2010). Is physical water scarcity a new phenomenon? Global assessment of water shortage over the last two millennia. *Environmental Research Letters* 5: 034006. http://dx.doi.org/10.1088/1748-9326/5/3/034006.

de Lange, M.F., Verouden, K.J., Vlugt, T.J. et al. (2015). Adsorption-driven heat pumps: the potential of metal-organic frameworks. *Chemical Reviews* 115 (22): 12205–12250. https://doi.org/10.1021/acs.chemrev.5b00059. Epub 2015 Oct 23.

Lee, J., Farha, O.K., Roberts, J. et al. (2009). Metal-organic framework materials as catalysts. *Chemical Society Reviews* 38 (5): 1450–1459. https://doi.org/10.1039/b807080f.

Lekouch, I., Muselli, M., Kabbachi, B. et al. (2011). Dew, fog, and rain as supplementary sources of water in South-Western Morocco. *Energy* 36: 2257–2265. http://dx.doi.org/10.1016/j.energy.2010.03.017.

Li, R., Shi, Y., Shi, L. et al. (2018). Harvesting water from air: using 568 anhydrous salt with sunlight. *Environmental Science and Technology* 52 (9): 5398–5406.

Macedonio, F., Drioli, E., Gusev, A.A. et al. (2012). Efficient technologies for worldwide clean water supply. *Chemical Engineering and Processing Process Intensification* 51: 2–17. http://dx.doi.org/10.1016/j.cep.2011.09.011.

Malik, F.T., Clement, R.M., Gethin, D.T. et al. (2014). Nature's moisture harvesters: a comparative review. *Bioinspiration & biomimetics* 9 (3): 031002.

Mauer, L.J. and Taylor, L.S. (2010). Water-Solids Interactions: Deliquescence. *Annu. Rev. Food Sci. Technol* 1 (1): 41–63.

Molden, D., Oweis, T.Y., Pasquale, S., et al. 2007. Pathways for increasing agricultural water productivity (No. 612-2016-40552) IWMI.

Muselli, M., Beysens, D., and Milimouk, I. (2006). A comparative study of two large radiative dew water condensers. *Journal of Arid Environments* 64: 54–76. http://dx.doi.org/10.1016/j.jaridenv.2005.04.007.

Nikolayev, V.S., Beysens, D., Gioda, A. et al. (1996). Water recovery from dew. *Journal of Hydrology* 182: 19–35. http://dx.doi.org/10.1016.0022-1694(95)02939-7.

Nilsson, T. (1996). Initial experiments on dew collection in Sweden and Tanzania. *Solar Energy Materials & Solar Cells* 40: 23–32. http://dx.doi.org/10.1016/0927-0248(95)00076-3.

Obispo, S.L. (2009). System generates drinking water from moisture in the air. *Membrane Technology* 2009 (11): 7.

Parker, A.R. and Lawrence, C.R. (2001). Water capture by a desert beetle. *Nature* 414 (6859): 33.

Qadir, M., Jiménez, G.C., Farnum, R.L. et al. (2018). Fog water collection: challenges beyond technology. *Water* 10 (4): 372. https://doi.org/10.3390/w10040372.

Renyuan, L., Shi, Y., Shi, L. et al. (2018). Harvesting Water from Air: Using 568 Anhydrous Salt with Sunlight. *Environ. Sci. Tech.* 52 (9): 5398–5406.

Semiat, R. (2008). Energy issues in desalination processes. *Environmental Science & Technology* 42: 8193–8201. http://dx.doi.org/10.1021/es801330u.

Seo, Y.K., Yoon, J.W., Lee, J.S. et al. (2011). Energy-efficient dehumidification over hierachically porous metal–organic frameworks as advanced water adsorbents. *Advanced Materials* 24 (6): 806–810.

Sharan, G. (2002). Harvesting dew with radiation cooled condensers to supplement drinking water supply in semi-arid coastal Northwest India. *International Journal for Service Learning in Engineering* 6: 130–150.

Sharan, G. (2013) Harvesting dew water for human use in coastal area of Gujarat (India)., 38, 80–84.

UNEP (1997). *Sourcebook of Alternative Technologies for Freshwater Augmentation in some Countries in Asia*. Washington, DC, USA: UNEP, Unit of Sustainable Development and Environment General Secretariat, Organisation of American States.

Wahlgren, R.V. (2001). Atmosphericwater vapour processor designs for potablewater production: a review. *Water Research* 35: 1–22.

Wahlgren, R.V. (2014). Another water resource for Caribbean countries: water-from-air. In: *Proceeding of the 23rd Caribbean Water and Wastewater Association Annual*

Conference. Atlantis, Paradise Island, Bahamas: http://www.atmoswater.com/cwwa-conference-2014.html.

Wang, S.C. and Liu, Y.H. (2008). High-power-factor electronic ballast with intelligent energy-saving control for ultraviolet drinking-water treatment systems. *IEEE Transactions on Industrial Electronics* 55 (1): 1.

Wang, C., Liu, X., Keser Demir, N. et al. (2016). Applications of water stable metal-organic frameworks. *Chemical Society Reviews* 45 (18): 5107–5134. https://doi.org/10.1039/c6cs00362a.

Wang, Y., Xie, Z., Malhi, S.S. et al. (2009). Effects of rainfall harvesting and mulching technologies on water use efficiency and crop yield in the semi-arid Loess Plateau, China. *Agricultural water management* 96 (3): 374–382.

Yaghi, O.M., O'Keeffe, M., Ockwig, N.W. et al. (2003). Reticular synthesis and the design of new materials. *Nature* 423 (6941): 705–714.

Zhou, H.C., Long, J.R., and Yaghi, O.M. (2012). Introduction to metal-organic frameworks. *Chemical Reviews* 112 (2): 673–674. https://doi.org/10.1021/cr300014x.

17

Dew Harvesting on High Emissive Natural and Artificial Passive Surfaces

Jose Francisco Maestre-Valero[1], Bernardo Martin-Gorriz[1], Victoriano Martínez-Alvarez[1], and Saeid Eslamian[2]

[1] *Escuela Técnica Superior de Ingeniería Agronómica, Universidad Politécnica de Cartagena, 30203 Cartagena, Spain*
[2] *Department of Water Engineering, Isfahan University of Technology, Isfahan, Iran*

17.1 Introduction

Dew is the result of water condensing from the atmospheric humidity on a surface sufficiently radiatively cooled by its own emission of radiation (Beysens et al. 2007). Under natural conditions, this potential water source can be widely used by plants and animals in dry environments and can supply enough moisture to microorganisms for survival (Steinberger et al. 1989; Kidron et al. 2002). This is why its environmental, agricultural, economic, and social role has long attracted researchers and engineers. In fact, collected dew may also serve as a supplementary source of water, which may even be accepted for human consumption, as a result of its chemical qualities (Muselli et al. 2006a; Lekouch et al. 2011; Sharan et al. 2017; Beysens et al. 2017). This is especially significant in regions where the water accessibility and supply becomes difficult (Muselli et al. 2006a; Lekouch et al. 2011; Beysens 2016), such as semi-arid and arid geographical settings and small islands in developing countries (Beysens et al. 2007; Sharan 2007; Malekian Jabali et al. 2017).

There have been a wide range of attempts to collect dew from absorbent material or cloth plates (Kidron 2000), microlysimeters (Jacobs et al. 2002), dew-specific collectors, called passive "radiative dew condensers" (RDCs, Beysens et al. 2005) or small open water surfaces (Maestre-Valero et al. 2015), etc. Tomaszkiewicz et al. (2015) reported a complete review of dew collection experiences in various environments. Among all these methods, RDCs are likely the most suitable techniques to be used at engineering applications, as they allow assessing the performance of different types of foils and supporting structures (shape, tilt, etc.). In addition, such RDCs have been standardized by establishing the methodology and the instrumentation by The International Organization for Dew Utilization (OPUR; http://www.opur.fr). Several field tests have been carried out in this RDC context. For instance, Muselli et al. (2002) tested a 30 m^2 RDC near Ajaccio (Corsica, France), measuring 214 dewy nights over an observation period of 478 days, with an average of 0.120 mm per dewy night and a maximum daily yield of 0.380 mm. Sharan (2005) assessed the suitability of six different materials for dew formation: (i) a plain galvanized iron sheet; (ii) a plain aluminum sheet; (iii) a low-density polyethylene (PE) foil; (iv) a low-density PE foil mixed with TiO_2 and $BaSO_4$; (v) a plain reinforced plastic fiber; and (vi) a plain reinforced corrugated fiber. Results evidenced that PE foils provided the highest dew amounts. Jacobs et al. (2008) compared two types of RDCs, one being a 1 m^2 insulated planar dew condenser set at a 30° angle from horizontal, and the other presenting an inverted-pyramid shape, and they obtained average dew collections of 0.100 and 0.150 mm per dewy night with the planar and the inverted-pyramid shape collectors, respectively. Muselli et al. (2009) again studied the dew yield at the Dalmatian Coast with two 1 m^2 RDCs concluding that it could be worthwhile to rehabilitate the numerous deserted rain collectors (impluviums) existing in the region for the objective of dew harvesting. Maestre-Valero et al. (2011) tested two 1 m^2 RDCs in southeast of Spain, one fitted PE made black and white foils and registered 175 and 163 dewy nights, with a dew production of 17.36 mm and 20.76 mm, respectively, over a year. A few years later, Maestre-Valero et al. (2015) compared the annual performance for dew collection on a black PE foil-lined RDC and an experimental pan. In that study, they collected 17.42 mm and 7.84 mm in the RDC and experimental pan, respectively.

Besides the attempts to collect dew on passive collectors, a wide group of researchers have developed some modeling for its estimation. A valuable review of dew modeling

Handbook of Water Harvesting and Conservation: Basic Concepts and Fundamentals, First Edition. Edited by Saeid Eslamian and Faezeh Eslamian.

methods and their input parameters and references were shown in Tomaszkiewicz et al. (2015). In this sense, Pedro and Gillespie (1982) applied the energy balance to predict wetness duration on single leaves deducing dew duration and amount from a computation of the latent heat flux. Nilsson (1996) calculated dew formation using the energy balance model applied on two RDCs ($1.2 \times 1.2\,m^2$) fitted with pigmented foils in Sweden and Tanzania. Vermeulen et al. (1997) and Moro et al. (2007) used micrometeorological techniques such as the Bowen ratio energy balance or the eddy-covariance techniques. Jacobs et al. (2008) estimated dew collected by two types of RDCs fitted with a white standard foil (WSF)specially designed for dew collection, one being a 1 m^2 insulated planar dew condenser set at a 30° angle from horizontal, and the other presenting an inverted-pyramid shape. Maestre-Valero et al. (2011) found a good correlation between the dew yield on RDCs and the difference between the dew point and the air temperature; i.e. relative humidity (RH), adjusted by the net radiation. Such a good correlation was also observed in other studies such as that of Sharan (2007) carried out in northwest India or that of Muselli et al. (2009) in the Dalmatian Coast. Maestre-Valero et al. (2012) developed and tested a simple night-time RDC energy balance model to estimate surface temperature of the foil and provide a complete description and quantification of dew yield and the energy components. Beysens (2016) provided a simple analytical equation valid for planar dew collectors which only needed cloud coverage, wind velocity, air, and dew point temperature data to be collected, at least once in a day before sunrise.

This chapter presents, on the one hand, the optical basis for the dew formation and a description of three passive dew condensers: a RDC, an experimental pan, and an agricultural pond. Afterwards, the chapter focuses on the presentation of the results of three cases of study for dew collection. Finally, it shows two cases for dew modeling: one based on the correlation with climatic variables and other based on the mass transfer equation.

The chapter's contents will be useful for water managers, planners, and field technicians with paramount interest in predicting the potential dew yield harvested from PE-lined surfaces or open water surfaces as an additional water resource in the water-stressed regions where additional water resources are welcome.

17.2 Passive Surfaces for the Case Studies

17.2.1 Optical Properties

Emitted radiance (W) and emissivity (ε) of a surface play an essential role in dew formation (Nikolayev et al. 1996). The former, W, expresses the amount of energy that is released from a surface at a given surface temperature whereas the latter, ε, indicates the relative ability of a surface to emit energy by radiation. Their accurate determination is indispensable to correctly predict the different energy balance components, including the formation of dew. A spectrophotometer can be used for determining the spectral distribution of the absorptivity ($=\varepsilon$) and transmissivity of the selected surfaces for the mid infrared spectrum (2.5–25 μm).

In this context, for a given wavelength λ, the emitted radiance, W, can be deduced from the Plank's law:

$$W = \frac{C_1}{\lambda^5} \cdot \frac{1}{exp\left(\frac{C_2}{\lambda T_s}\right) - 1} \cdot \varepsilon \tag{17.1}$$

where $C_1 = 3.74{\cdot}10^8$ and $C_2 = 1.44{\cdot}10^4$ are constants, λ is the wavelength, ε is the measured emissivity of each foil configuration in each wavelength interval and T_s (K) is the surface temperature. W and ε values can be then integrated over different wavelength intervals: 2.5–7, 7–14, and 14–25 μm. The range 7–14 μm is considered of special interest as it corresponds to the atmospheric window (Nilsson 1994). The values of the emissivity weighed by the emitted radiance can be determined for all spectrum ranges as:

$$\varepsilon^* = \frac{\sum \varepsilon_i}{\sum W_i} \tag{17.2}$$

where ε_i and W_i are the emissivity and the emitted radiance, respectively, at wavelength λ_i.

In this context, Beysens et al. (2007) determined the infrared ε of corrugated and plane alveolar polycarbonate materials in the visible range (0.2–2.5 μm) with a halon-coated integrating sphere and in the long wave radiation band (7–30 μm) with a Fourier transform type infrared spectrometer "BRUKER Equinox 55." Concerning the PE foils used to fit RDCs, Maestre-Valero et al. (2011) analyzed the emitted radiance and emissivity of two high emissivity PE foils for the entire mid-infrared spectrum (2.5–25 μm) in dry and wet conditions (Figure 17.1). The foils evaluated were the WSF promoted by OPUR and a 0.15 mm thick black low-density PE foil (BF) typically used in agriculture as soil mulching for weed control.

Both had similar values of ε when both foils were compared in the near infrared spectrum range (7–14 μm), but significant differences were observed when comparing them in the entire mid-infrared spectrum (2.5–25 μm), especially in the lower range of the mid infrared spectrum (2.5–7 μm), where the WSF presented lower emissivity and emitted radiance values than the BF. In short, to simulate dew formation, not only are the near infrared spectrum merits analyzed but the entire mid-infrared has to be considered; ε for different materials aimed to collecting dew

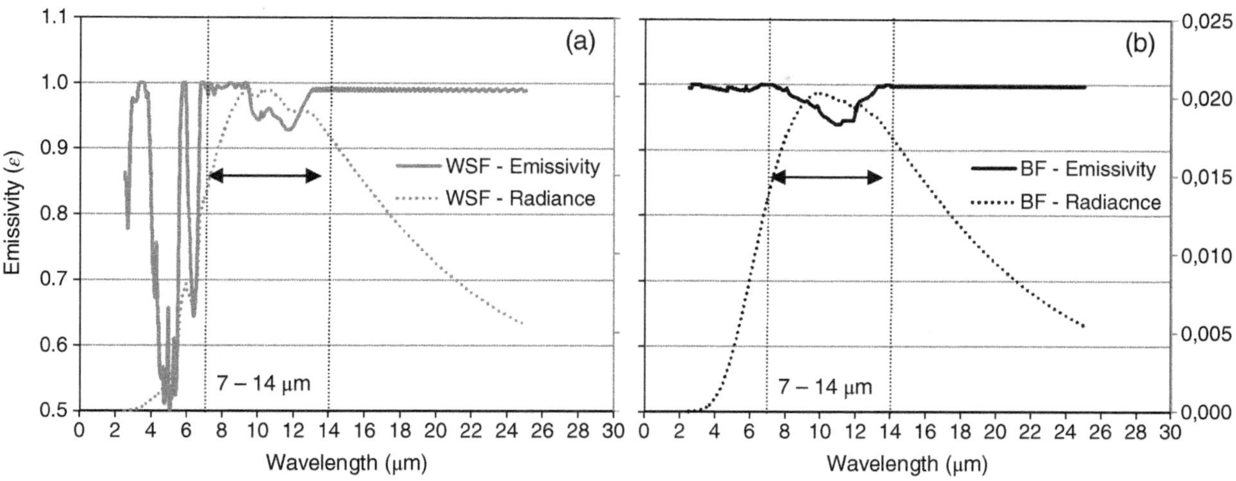

Figure 17.1 Distribution of the foil emissivity (ε) and radiance (W; W/m^2) for (a) the WSF and (b) the BF in the 2.5–25 μm range. Vertical bars delimit the 7–14 μm region (atmospheric window).

as presented in Tomaszkiewicz et al. (2015). In addition, it must be considered that dew formation is mainly driven by the vapor pressure gradient between the surface, at temperature T_s, and the nearby surrounding air temperature. This implies that a precise simulation of T_s, based on the dynamic analysis of the nightly energy balance, is fundamental for the estimation of dew yield (Finch and Gash 2002).

17.2.2 Passive Radiative Condensers and Foils

A RDC consists – following the OPUR international standard regulation – of 1 m^2 insulated flat pans tilted 30° to horizontal to ensure a good compromise between radiative energy loss and water recovery by gravity (Beysens et al. 2003) (Figure 17.2). The water condensing on the surface at night is collected under gravity flow by a gutter and runs to a container where it can be stored and weighed or measured.

Figure 17.2 View of two radiative dew condensers (RDCs). One fitted with the black PE foil (BF) and the other with the white standard foil (WSF) to the 30° tilted flat pans.

In this chapter, field tests were shown on a RDC fitted with the 0.15 mm thick black low-density PE foil and a RDC fitted with the WSF provided by the OPUR. This latter is made of a special white low-density PE foil, with 5% volume of TiO_2 microspheres (diameter 0.19 μm) and 2% volume of $BaSO_4$ microspheres (diameter 0.8 μm) embedded in it. OPUR recommends this material as it provides hydrophilic properties that lowers the nucleation barrier at the onset of the condensation process together with a high emissivity in the near infrared (7–14 μm); two important features that favor dew formation. More information on this material can be found in Nilsson (1994).

17.2.3 Experimental Pan

Field tests on an experimental pan (diameter = 1.21 m and height = 0.10 m) were also conducted to study the dew formation on small water surfaces (Figure 17.3). It was placed on a precision weighing scale (Scaltec SSH 91; precision = 5 g) to determine the water flow variations due to the alternation between evaporation and condensation processes. The pan wall and bottom were insulated with polyurethane to prevent heat inputs and outputs, as well as to avoid condensation on the metallic body of the pan.

17.2.4 Agricultural Pond

Field tests were also conducted on bigger water surfaces. To this aim, an on-farm agricultural pond which covered a surface about 3000 m^2 (50 × 60 m) with a depth of 5.5 m was monitored to measure dew formation (Figure 17.4). Sensors

Figure 17.3 View of the experimental pan placed on a precision weighing scale.

Figure 17.4 Dew monitoring on an on-farm agricultural pond.

were installed into an experimental pan floating into the reservoir to avoid movement by waves.

17.3 Data Collection

17.3.1 Climate Measurements

An automated meteorological station located at the vicinity of the RDCs, the experimental pan, and the agricultural pond provided the meteorological data required for the study (Figure 17.5). The following variables were continuously recorded at 2 m above ground: air temperature (T_a) and RH (Vaisala HMP45C probe), wind speed (U_2) (Vector Instruments A100R anemometer) and downward atmospheric radiation (L_a) (Kipp and Zonen CGR 3 pyrgeometer). Rainfall (P) was measured by means of a tipping bucket gauge (Young 52 203).

Figure 17.5 View of the climatic station.

17.3.2 Dew Measurements

17.3.2.1 RDCs

For each RDC, dew was collected at night (from 20:00 to 8:00). The dew ran along an inclined gutter and passed through a plastic pipe into the container where dew was weighed by means of two high precision balances (COBOS, D-3000-CBJ; precision = 0.1 g). A wiper was used daily at dawn to scrape the extra water that remained on the foils. This quantity was added to the amount recovered in the collecting tanks to give the potential dew recovery. Previous analyses of dew collection on the foils indicated the scraped fraction represented about 15% and 20% of the total yield for the WSF and the BF, respectively, a slightly lower value than the one reported by Muselli et al. (2002). In the following, the analysis results concern the potential dew recovery, which better represents the intensity of the condensation process. No damage due to scraping was noted on the foils during the measurements period. Eventually, dew yield was calculated as the difference between the maximum and the minimum weight of water recorded into the container during the night. Additionally, two infrared radiometers (Campbell Scientific SI – 111) located 30 cm over the foils supplied the foil surface temperature, $T_{s\text{-RDC}}$. Dew point (T_{dew}) was calculated from T_a and RH. The RDCs net radiation ($R_{n\text{-RDC}}$) during the night was calculated as $R_{n\text{-RDC}} = L_a - L_{s\text{-RDC}}$ with $L_{s\text{-RDC}} = \varepsilon^* \sigma T^4{}_{s\text{-RDC}}$, ε^* being the radiance weighed emissivity of each foils.

17.3.2.2 Experimental Pan

For the experimental pan, dew was also quantified from 20:00 to 8:00 according to the differences in weight detected by the precision weighing scale. The pan was refilled at noon every two days to compensate for the water evaporation losses and to alter the overnight water temperature as little as possible. One infrared radiometer (Campbell

Scientific SI – 111) located 20 cm over the water in the pan the water surface temperature, T_{pan}. Then, the pan net radiation ($R_{n\text{-}Pan}$) was calculated as $R_{n\text{-}pan} = L_a - L_{pan}$ with $L_{pan} = \varepsilon\sigma T^4{}_{s\text{-}Pan}$

17.3.2.3 Agricultural Pond

Measurement of dew on water bodies presents some limitations such as: (i) frequent and even continuous water inflows and outflows impeding dew or evaporation measurements by water level variations; (ii) low accuracy of the water pressure sensors to register level variations in the order of tenths of a millimeter when water inflows and outflows are controlled; and (iii) alternation of nightly evaporation and condensation processes that do not allow its separate quantification. Consequently, these drawbacks make dew measurement or modeling extremely complicated. Therefore, in the experiment, dew formation on the agricultural pond was evaluated by analyzing the difference between the dew point (T_{dew}) and the surface temperature (T_{pond}). In fact, if nightly climatic conditions are suitable for dew formation (i.e. high *RH* and low wind) and the difference between the T_{pond} and the T_{dew} becomes negative, dew is very likely to occur. Accordingly, an infrared radiometer (Campbell Scientific SI – 111) located 20 cm over the water was used to determine the T_{pond}, and a wetness sensor (Campbell Scientific 237 - LWS) was installed to check for the presence/absence of dew on the water stored in the pond.

17.3.3 Statistical Analysis

Dew yield data were statistically analyzed by means of the statistical software package Statgraphics Plus (v.5.1), which performs analysis via a variance technique (ANOVA). Moreover, Tukey's range test at a 95% confidence level was calculated for comparison between dew yield data. Data from days corresponding to rainfall events at night were discarded from the data analysis because of the imprecision in measuring dew amount.

All sensors above described were scanned at 10-second intervals and averaged hourly whereas the two precision balances were scanned at hourly intervals. All data were recorded by dataloggers (CR1000 Campbell). The sensors and balances were periodically calibrated.

17.4 Case Studies for Dew Collection

17.4.1 Dew Collection on Passive Radiative Condensers

The experiment was located at the Agricultural Experimental Station of the Technical University of Cartagena, south-eastern Spain (37°41′20″N, 0°57′03″ W). This area is characterized by a Mediterranean semi-arid climate with warm, dry summers and mild winters. Average annual temperature is 17.5 °C, reaching maximum temperatures of 38 °C in summer and minimum temperatures of 0 °C in winter. Annual rainfall averages 320 mm, with high seasonal and interannual variability. Most precipitation occurs during the fall and winter months, but interannual droughts are also common. Average reference evapotranspiration, calculated by the Penman–Monteith method (Allen et al. 1998), is about 1250 mm/year.

Table 17.1 Number of dewy and rainfall nights and total monthly dew yield for the WSF and BF RDCs during the observation period.

Year	Month	Number of days		Total dew yield (mm)	
		Dew on WSF	Dew on BF	WSF	BF
2009	My	22	22	2.47	2.43
	Jn	20	19	1.79	2.29
	Jl	15	15	1.06	1.29
	Ag	10	10	0.51	0.77
	Sp	6	4	0.57	0.69
	Oc	20	19	3.18	3.83
	Nv	17	15	1.84	1.99
	Dc	13	11	1.39	1.71
2010	Ja	13	11	1.12	1.53
	Fb	10	9	0.87	0.99
	Mr	15	15	1.30	1.62
	Ap	14	13	1.26	1.62
Annual		**175**	**163**	**17.36**	**20.76**

From May 2009 to May 2010 dew was collected from two RDCs; one fitted with the WSF and the other with the BF (Figure 17.2). Table 17.1 shows the number of dewy and rainfall nights and total monthly dew yield for the WSF and BF RDCs during the observation period.

During the 1-year experimental period, the number of dewy nights amounted to 175 and 163 with cumulated yields of 17.36 and 20.76 mm for WSF and BF, respectively; the BF was about 16% more efficient in collecting dew than the WSF. Such a better performance could be ascribed to the higher emissivity and emitted radiance of the BF (Figure 17.1). This finding was confirmed with the nightly value of minimum foil temperature, which was on average 0.43 °C lower for BF than for WSF (data not shown).

The maximum dew yield recorded during a dewy night was 0.314 mm in December 2009 for the WSF and 0.316 mm in October 2009 for the BF. These values corresponded to the period from October 2009 to December 2009 when clear sky, low wind speed, and high values of atmospheric

Table 17.2 Dew yield frequency of WSF and BF during the observation period.

Dew yield classes (mm)	WSF	BF
	Number of dewy nights	
0–0.05	42	30
0.05–0.10	43	43
0.10–0.15	32	37
0.15–0.20	30	34
0.20–0.25	14	21
0.25–0.30	1	6
0.30–0.35	1	4
0.35 -	0	0
Total dewy nights	**163**	**175**

humidity were prevailing. Conversely, the lowest dew yield values for both foils were found during the driest months, i.e. July and August 2009 (Table 17.1). On an annual scale, mean values were 0.099 mm/day and 0.127 mm/day for the WSF and the BF, respectively (Table 17.1).

The dew yield frequency by classes of 0.05 mm (Table 17.2) suggested that the higher number of dewy events with low yield (less than 0.05 mm) for the WSF were due to its hydrophilic surface properties. This characteristic allowed the WSF to recover water from small events of dew (less than 0.10 mm) whereas the BF was less effective in this aspect. On the contrary, the BF was more efficient in the upper classes due to its higher emissivity (Figure 17.1). These respective advantages of the WSF and the BF appear to be of the same magnitude in the dew yield range of 0.05–0.10 mm, where dew yield frequency for the two foils was identical. In line with these results, it is worth mentioning that if the experiment had been conducted in another region characterized by smaller dew yield events (less than 0.10 mm), it is very likely that the hydrophilic properties of the WSF would have been allowed to collect more water than the BF.

17.4.2 Dew Collection on the Experimental Pan

This experiment, carried out from January 2011 to January 2012, was located at the same location as the experiment previously carried out on RDCs (see Section 17.4.1). In this case, the dew collected on an experimental pan was compared with that collected on a RDC fitted with a BF.

Table 17.3 presents the monthly distribution of the number of dewy days and the total monthly dew yield for the RDC and the experimental pan during the experiment. The number of dewy days on the RDC was nearly double those of the pan (162 and 90 dewy days respectively) with cumulated yields of 17.42 and 7.84 mm for the RDC and the experimental pan, respectively; the RDC was about 55% more efficient in collecting dew than the experimental pan.

Table 17.3 Number of dewy and rainfall nights and total monthly dew yield for the BF RDC and the experimental pan during the observation period.

Year	Month	Number of days		Total dew yield (mm)	
		Dew on BF	Dew on Class-A	BF	Experimental pan
2011	Ja	12	9	1.95	0.75
	Fb	10	9	1.44	0.82
	Mr	8	4	0.91	0.45
	Ap	12	7	1.28	0.75
	My	16	7	1.89	0.78
	Jn	17	6	0.89	0.27
	Jl	10	4	0.63	0.17
	Ag	17	7	0.90	0.50
	Sp	19	10	2.04	1.13
	Oc	14	11	1.48	0.91
	Nv	10	3	1.49	0.18
	Dc	17	13	2.52	1.14
Annual		**162**	**90**	**17.42**	**7.84**

Total monthly dew yield results clearly differentiated two periods throughout the year (Table 17.3). As expected, one period covered the summer months (from June to August) where the lowest dew yields were observed. The other period, with higher dew yield, covered the rest of the year (from September to May).

The maximum monthly dew yield corresponded to the month of December (2.52 and 1.14 mm for the RDC and the experimental pan, respectively). This was not unexpected since three factors that heavily condition dew formation happened that month: (i) a strong radiative cooling at night due to the prevalence of clear sky conditions; (ii) motion of air masses coming from the Mediterranean Sea (Lekouch et al. 2011) with high water vapor content; and (iii) a great number of rainfall events in November (data not shown) that increased the atmospheric humidity resulting from the high soil evaporation.

Such a significant difference found between the RDC and the experimental pan could be explained by (i) a higher surface temperature of the water stored in the pan that reduced the difference with the dew point temperature throughout the year, and (ii) the possible alternation of condensation and evaporation processes on the pan (Molina-Martínez et al. 2006), which always provides an evaporative water surface. In this sense, Figure 17.6 depicts the evolution of a night of dew formation on the RDC and the pan together with the RH and the surface and the dew temperatures. It should be noted that between 11:00 p.m.

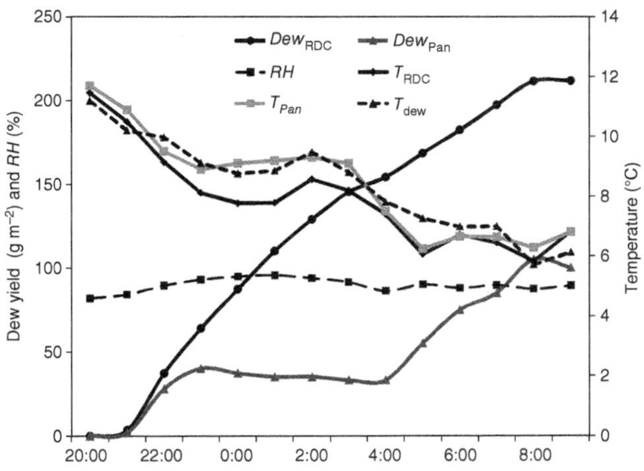

Figure 17.6 Evolution of a dewy night (04–12 to 05–12; 2011) for the RDC (Dew_{RDC}) and the experimental pan (Dew_{Pan}). The figure also shows the evolution of the RDC surface temperature (T_{RDC}), the water surface temperature of the experimental pan (T_{Pan}), the relative humidity (RH) and the dew point temperature (T_{dew}).

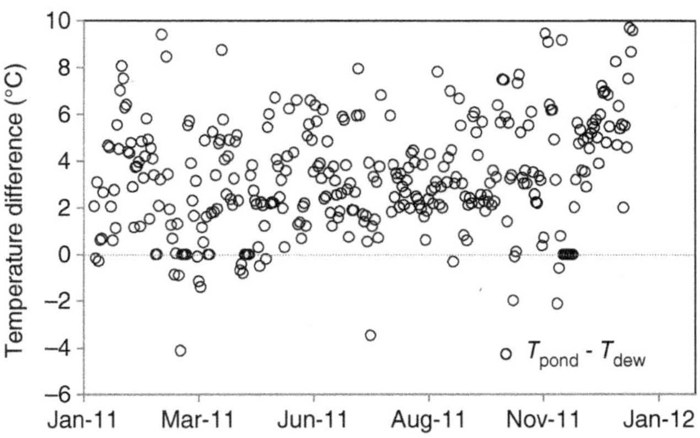

Figure 17.7 Difference between the nightly surface temperature of the water stored in the agricultural pond (T_{Pond}) and the dew point (T_{Dew}). The dashed line indicates the temperature difference threshold at which dew formation can occur.

and 3:00 a.m., $T_{Experimental\,pan}$ is above T_{dew} due to the high water thermal inertia, thus changing the conditions from condensation to evaporation in the pan. However, in the same period for the RDC, T_{RDC} is always below T_{dew} due to the lack of thermal inertia in the film, and hence dew formation is not interrupted.

In this context, the thermal behavior of the water body is a factor that should be borne in mind when analyzing dew formation. In the case of the experimental pan, Molina-Martínez et al. (2006) indicated that the pan thermal stratification is very likely to be avoided by two different factors; during the day, wind speed and turbulence prevailing at the pan surface are the main cause for mixing; whereas during the night, when very low wind conditions frequently occur, the natural convection due to the radiative cooling of the surface appears to be sufficient for homogenizing the temperature. Therefore, the experimental pan is a well-mixed water body and the bulk thermal inertial maintained the surface temperature higher than in the RDC foil up to approximately 3:00 a.m.

17.4.3 Dew Collection on an Agricultural Pond

Figure 17.7 shows the nightly evolution of the difference between the T_{pond} and the T_{dew} of the surrounding air.

Dew on the water surface was anecdotic in the study area as negative values for T_{pond}-T_{dew} only happened 19 days in the one-year experimental period. These results agreed almost perfectly with the dew formation detected by the wetness sensors installed on the water surface. As happened in the experimental pan experiment, the higher thermal inertia of the water than the plastic films as in the RDCs might have prevented the reduction of the surface temperature below the dew point and hence the dew formation.

The thermal behavior of shallow-water ponds and their well-mixed water profile throughout the year has been widely studied (Gallego-Elvira et al. 2010). In spring and summer, part of the incoming radiation is absorbed in the water column, increasing the water temperature (heat storage period), while during the autumn and early winter, as the incoming solar radiation decreases, the previously stored energy is released (heat release period), being utilized primary by the evaporation process (Finch and Gash 2002). Therefore, the pond water surface cooling during night is attenuated by upcoming bulk heat up, avoiding T_{pond} values below T_{dew}.

17.5 Dew Modeling

17.5.1 Correlation with Climatic Variables

Correlation with climatic variables was performed for the dew collected from the RDCs (WSF and BF) in the 2009–2010 experiment and for the RDC (BF) and the experimental pan in 2011 experiments.

The observed night dew yield, Y (mm/night), was first related to the dew point-to-air difference $\Delta T = T_{dew} - T_a$, that is, with the relative humidity, RH. The experimental data were fit to the following linear relationship to get an estimate of Y, Y_{est}, from the knowledge of ΔT (Beysens 2016).

$$Y_{est} = a_1 \cdot (\Delta T - a_2) \tag{17.3}$$

where a_1 (in mm/°C) is the dew yield sensitivity to ΔT and a_2 the threshold value of ΔT below which condensation was not observed.

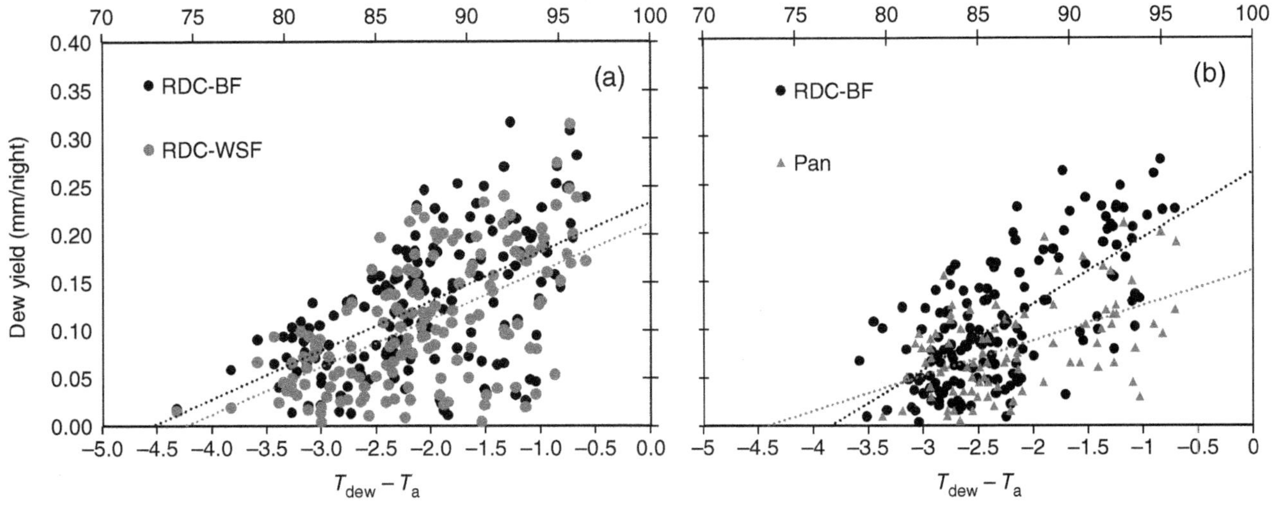

Figure 17.8 Correlation of dew yield Y (mm/night) with $\Delta T = T_{dew} - T_a$ (°C, lower scale) and relative humidity RH (%, upper scale) for (a) the RDC-WSF (gray circles) and the RDC-BF (black circles) and (b) the RDC-BF (black circles) and the experimental pan (gray triangles).

Figure 17.8 shows the correlation of dew yield Y (mm/night) with $\Delta T = T_{dew} - T_a$ (°C, lower scale) and relative humidity RH (%, upper scale) for the RDCs and the experimental pan.

In the case of the comparison between the RDCs, i.e. one fitted with the WSF and the other with the BF, significant differences in the parameter values between the two foils ($a_1 = 0.049 \pm 0.0056$ mm/°C and $a_2 = -4.2$ °C ± 0.26 for the WSF, $a_1 = 0.051 \pm 0.0059$ mm/°C and $a_2 = -4.6$ °C· ± 0.29 for the BF) were not observed (Table 17.4).

In the case of the comparison between the RDC fitted with the BF and the experimental pan, huge differences were found in the dew yield sensitivity to ΔT ($a_1 = 0.068 \pm 0.042$ mm/°C for the RDC and $a_1 = 0.036 \pm 0.019$ mm/°C for the pan; Table 17.4). A maximum cooling for the RDC was found at −3.84 °C whereas such a value was −4.42 °C for the pan.

The dew yield sensitivity for the RDCs was between the values found by Muselli et al. (2006b, 2009), whereas in the case of the pan it was notably lower.

Equation (17.3), applied to the RDCs and the experimental pan, reported low correlation coefficients for both the RDCs (43–47%) and the pan (27%) and hence could not be considered as satisfactory. Additionally, the experimental data presented considerable scatter over the whole range of ΔT, indicating that ΔT alone was a poor descriptor of dew yield (Figure 17.8).

Therefore, to refine the correlation analysis, the residuals of Eq. (17.3) ($r = Y - Y_{est}$) were calculated and related to other climatic variables. In this sense, a strong dependence with the nightly net radiation was revealed only in the case of the RDCs. Subsequently, Eq. (17.3) was multiplied by a function of R_n, $g(R_n)$, to account for this dependence. A decreasing hyperbolic function supplied the best fit (lowest root mean square error between observed and estimated values). Then, the proposed empirical model to predict Y from T_a, T_{dew}, and R_n was:

$$Y_{est} = f(\Delta T) \cdot g(R_n) = (b_1.(\Delta T + b_2)) \cdot \left(1 + \frac{b_3}{R_n}\right) \tag{17.4}$$

Values of the fitted parameters are also presented in Table 17.4.

In the case of the RDCs, the addition of R_n as supplementary predictive variable improved considerably the predictive performance with respect to Eq. (17.3) (Table 17.4).

17.5.2 Mass Transfer Equation

Dew (λZ) in W/m^2 can be also derived from the mass transfer equation.

$$\lambda Z_{est} = h_v \cdot (e_s - e_a) \tag{17.5}$$

where λ (J/kg) is the latent heat of vaporization, e_s (kPa) is the saturated water vapor pressure at the surface temperature, e_a (kPa) the actual vapor pressure of the air, and h_v (W/m^2/kPa) the exchange coefficient for water vapor transfer (W/m^2/kPa) at the air-surface interface.

Assuming analogy between mass and heat transfer, the value of h_v can be considered equal to h_c/γ, where h_c is the exchange coefficient for convective heat transfer (W/m^2/K) and γ is the psychometric constant (kPa/K). The empirical equation proposed by Richards (2009) was used to calculate h_c.

$$h_c = a_1 + a_2 \cdot U_{1m} \cdot \left(\frac{511 + 294}{511 + T_a(K)}\right) \tag{17.6}$$

Table 17.4 Values of fitted statistical parameters characterizing the predictive performance for Eqs. (17.3) and (17.4).

	Experiment May-2009 to May 2010		Experiment Jan-2011 to Jan-2012	
	RDC-WSF	**RDC-BF**	**RDC-BF**	**Experimental pan**
Equation (17.3): $Y_{est} = a_1((T_{dew} - T_a) + a_2)$				
a_1 **(mm/°C)**	0.049 ± 0.005	0.051 ± 0.006	0.068 ± 0.042	0.036 ± 0.019
a_2 **(°C)**	-4.2 ± 0.260	-4.6 ± 0.297	-3.84 ± 0.7	4.42 ± 1.15
R^2	0.33	0.32	0.47	0.27
RMSE (mm/night)	0.043	0.045	0.038	0.034
MBE (mm/night)	0.003	0.003	0.0008	−0.0001
Equation (17.4): $Y_{est} = (b_1(T_{dew} - T_a) + b_2) \cdot (1 + b_3/R_n)$				
b_1 **(mm/°C)**	0.126 ± 0.011	0.129 ± 0.011	0.126 ± 0.011	—
b_2 **(°C)**	3.9 ± 0.128	4.1 ± 0.137	3.73 ± 0.137	—
b_3 **(W/m²)**	19.21 ± 0.94	18.93 ± 0.88	20.8 ± 3.84	—
R^2	0.63	0.65	0.58	—
RMSE (mm/night)	0.035	0.035	0.035	—
MBE (mm/night)	0.002	0.002	−0.0006	—

where U_{1m} (m/s), the wind speed at 1 m height from the surface, was derived from the wind speed measured at 2 m above ground, following Allen et al. (1998).

The empirical coefficients a_1 (W/m^2/K) and a_2 (J/m^3/K) were identified using a preliminary data set collected from March to April 2009. The model was run using different pairs of values of a_1 and a_2, and the pair producing the lowest root mean square error between estimated and observed values of T_s was selected ($a_1 = 7.6 \pm 1.22$ W/m^2/K and $a_2 = 6.6 \pm 1.84$ J/m^3/K). Note that the latter values are somewhat higher than those proposed by Richards ($a_1 = 5.4$ W/m^2/K, $a_2 = 4.1$ J/m^3/K).

To refine the performance of this second method, a simple filtering method was eventually applied to the hourly data and was based on discarding situations where condensation was unlikely to occur, and retaining only the night hours with *RH* above 80%. This filtering was in line with the observation that dew yield is practically non-existent below this threshold value (Sharan 2007; Maestre-Valero et al. 2011). The advantage of applying this filter is that the nightly average values of flux and state variables are calculated only for the hours when condensation occurred, and not for the whole night, thus being more representative of the average intensity of the condensation process.

Once the night-averaged values were filtered in this way, the model ran at a night-time step, with a time span from 20:00 to 8:00. It should be noted that the model provided the night averaged condensation rate, which was applied considering only the hours when $RH > 80\%$ to obtain the total night dew yield (mm/night).

In this study, the model validations for the RDCs (WSF and BF; experiment from May 2009 to May 2010) and the RDC-BF and the experimental pan (experiment from Jan-2011 to Jan-2012) were carried out by comparing the simulated nightly dew yield with the measurements.

In the case of the first experiment (from May 2009 to May 2010) carried out for the RDCs fitted with the RDC-WSF and the RDC-BF, the average latent heat of condensation (λZ) was 6.5 W/m^2 (0.099 mm/night) for the WSF and 9.2 W/m^2 (0.127 mm/night) for BF. In spite of the fact that the lowest radiative losses were observed in October, the highest dew yield value was observed for this month (8.5 W/m^2, 0.159 mm/night, and 11.9 W/m^2, 0.201 mm/night, for WSF and BF respectively). This could be mainly attributed to (i) high atmospheric humidity (monthly average of 92.4%; data not shown) resulting from the high soil evaporation rate after heavy rainfalls (276 mm; see Maestre-Valero et al. 2011) in late September and (ii) low wind speed (monthly average of 0.56 m s^{-1} for this month; data not shown).

A relatively good correlation was found between the predicted and the measured values of dew yield for both RDC-WSF and RDC-BF ($R^2 = 0.66$ and $R^2 = 0.71$, respectively, Figure 17.9a, b), with some scattered points around the 1 : 1 relationship. Both slopes were slightly higher than 1 (slope = 1.02 and 1.09 for WSF and BF respectively).

These results indicated that the model overestimated dew yield for both foils (RMSE = 0.051 mm and 0.053 mm and MBE = +0.009 mm and + 0.016 mm for RDC-WSF and RDC-BF respectively).

In the case of the second experiment (from Jan 2011 to Jan 2012) carried out for the RDC fitted with the BF and the experimental pan, λZ computed by Eq. (17.5) amounted on average to 18.74 W/m^2 (actual dew yield of

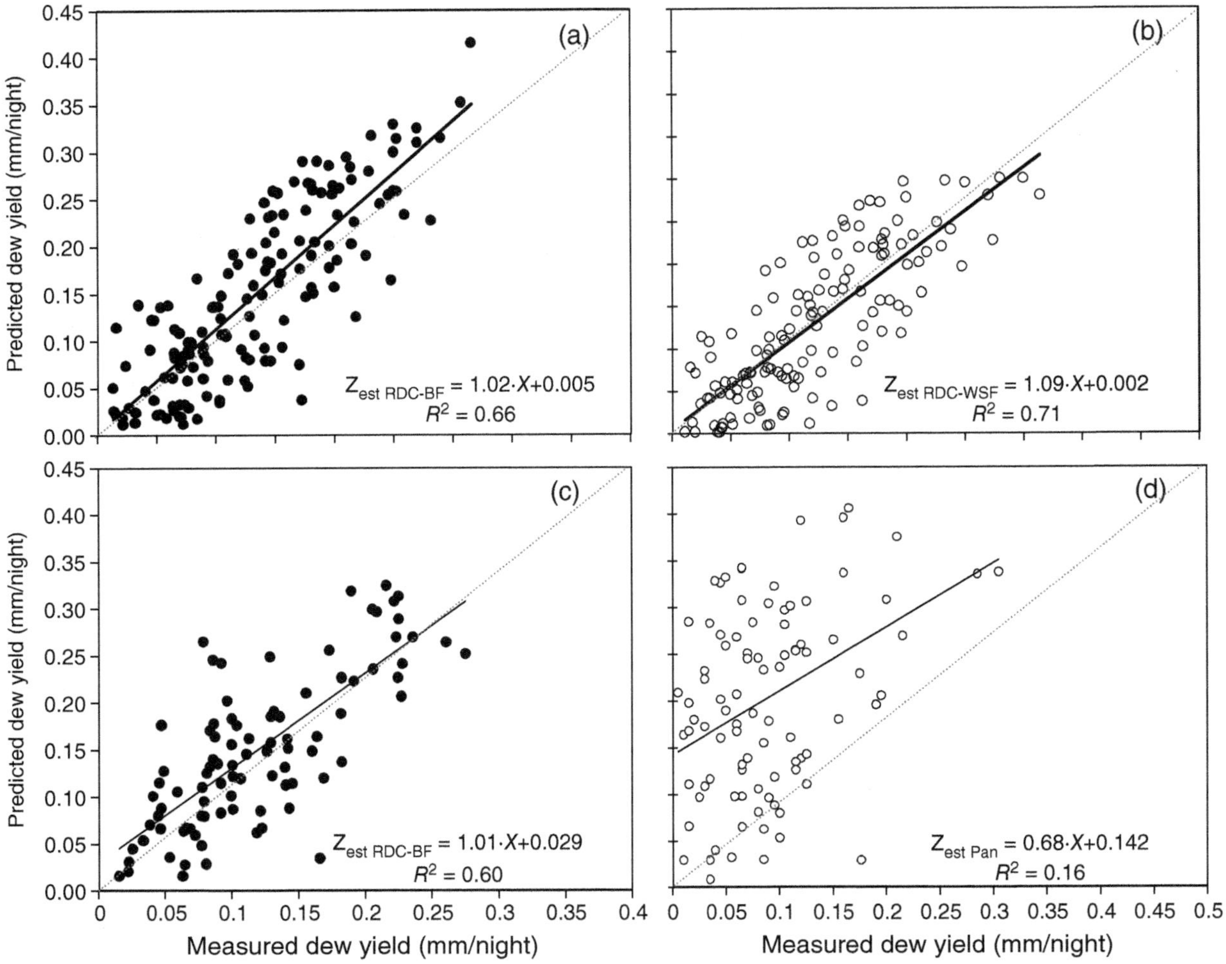

Figure 17.9 Comparison between the nightly predicted and the measured dew yield for (a, b) the RDCs in the experiment carried out from May 2009 to May 2010 and, (c, d) the RDC-BF and the experimental pan in the experiment carried out from Jan 2011 to Jan 2012. The dashed line is the 1 : 1 relationship.

0.108 mm/night) for the RDC and 25.26 W/m^2 (actual dew yield of 0.087 mm/night) for the pan. A relatively good correlation was also found between the predicted and the measured values of dew yield for the RDC (R^2 = 0.60; Figure 17.9c), with some scattered points around the 1 : 1 relationship. In addition, the slope was very close to 1 (slope = 1.01). Statistics manifested that the model slightly underestimated dew yield (RMSE = 0.045 mm and MBE = −0.031 mm).

In contrast, a mediocre correlation was found between the predicted and the measured values of dew yield for the experimental pan (R^2 = 0.16; Figure 17.9d). Consequently, huge errors were detected (RMSE = 0.119 mm and MBE = −0.114 mm) which indicated that the model systematically overestimated dew yield and the mass transfer approach was not able to accurately predict dew on an open-water pan. Such bad performance can also be ascribed to the alternation of condensation and evaporation processes, since when a dew event occurred on the surface water of the pan it did not usually last the whole night and hence the dew collected was immediately evaporated, a circumstance that the model was not able to simulate.

17.6 Conclusion

The experiments presented on passive RDCs have demonstrated the collection of dew as a complementary source of water, mainly in developing countries, rural areas or small islands, where free access to water and energy is expensive.

When comparing two different foils fitted on RDCs, our study, carried out under southeastern Spain semi-arid conditions, has also demonstrated that the potential for dew yield of a low-cost black PE foil (BF) was slightly higher than that of the OPUR-standard foil (WSF), although the BF does not present the hydrophilic properties of the latter. This disadvantage of BF has resulted in less dewy days observed, but was more than compensated on the quantitative aspect, i.e. the amount of annual recollected water, explained by the higher emissivity and radiative cooling power of BF in the lower range (2.5–7 μm) of the mid IR spectrum.

Our results also suggest that the knowledge of the emissivity in the whole IR spectrum is absolutely necessary to correctly assess the performance of the foil. In addition, ensuring a high emissivity over the whole IR spectrum seemed to be more effective for increasing RDC yield than improving surface hydrophilic properties. On a practical point of view, BF could be considered as a suitable material for large-scale RDCs, as in our study, it presented several advantages over the standard reference foil, i.e. higher dew collection performance, longer lifespan, and much lower cost. Dealing with the last two aspects, it must be pointed out that if the WSF were manufactured in large quantities and anti-UV treated, its cost might be reduced and its lifespan extended.

The highest values of daily dew yield have been observed mainly during periods following heavy rainfalls, due to high soil evaporation and high nocturnal atmospheric humidity. Therefore, it is likely that the amount of recollected dew could depend in part on the importance, frequency, and time occurrence of rainfall events that affect the humidity content of the air at the vicinity of the condenser. This has been confirmed by our correlation analysis between nightly yield and atmospheric variables, where the predominant predictive variables were found to be the RH and the net radiation of the foil.

This study has also demonstrated that dew formation is highly conditioned by the type of substrate on which dew occurs and also by the atmospheric conditions in the vicinity of the condensing substrate, especially the humidity content of the air. A clear example of this is the comparison of dew collected between the RDC and the experimental pan or the dew formation at a pond scale. In the first case, the experimental pan collected roughly half of dew that the RDC. In the second case, only anecdotic dew formation events on the pond were observed when the difference between the nightly water surface temperature and the dew point was not positive.

Regarding the modeling approaches, the empirical relationship between yield and the difference between the dew point and the air temperature, corrected by the net radiation, has explained about two-thirds of the total variance in the case of the RDC, and could be used to estimate daily dew yield with reasonable accuracy. Such an equation has not been useful in the case of the experimental pan. Additionally, the implementation of the mass transfer equation is also straightforward and cost effective to be applied, subjected to previous definition of the emissivity (ε) and the empirical coefficients.

The findings would help water managers to predict the potential dew yield that could be harvested and used for an additional water resource in the water stressed regions where additional water resources are welcome, hence avoiding the installation of a passive dew water condenser. It is eventually worth mentioning that increasing the scale of the simulations would probably change the aerodynamic behavior in close proximity of the condenser surface and accordingly, the distribution of surface temperature could not be uniform in large unit as it is in small ones varying the validation of the models. Then, computational fluid dynamics (CFDs) analyses could be performed in future research aimed at studying the possible scale effects.

Acknowledgments

The authors acknowledge the European Community's Seventh Framework Research Programme for the financial support of this study through the grant agreement N° 245159 project SIRRIMED (Sustainable use of IRRIgation water in the MEDiterranean region, FP7-KBBE-2009-1-2-03, www.sirrimed.org), and the Ministry of Science and Innovation and FEDER (Fondo Europeo de Desarrollo Rural) through the research project AGL2010-15001.

References

Allen, R.G., Pereira, L.S., Raes, D., and Smith, M. (1998). *Crop Evapotranspiration. Guidelines for Computing Crop Water Requirements*. Irrigation and Drainage Paper 56. Rome, Italy: FAO, 300 pp.

Beysens, D. (2016). Estimation dew yield from a few meteo data. *Atmospheric research* 167: 146–155.

Beysens, D., Milimouk, I., Nikolayev, V. et al. (2003). Using radiative cooling to condense atmospheric vapor: a study to improve water yield. *Journal of Hydrology* 276: 1–11.

Beysens, D., Muselli, M., Nikolayev, V. et al. (2005). Measurement and modelling of dew in island, coastal and alpine areas. *Atmospheric Research* 73: 1–22.

Beysens, D., Clus, O., Mileta, M. et al. (2007). Collecting dew as a water source on small islands: the dew equipment for water project in Bis˘ evo (Croatia). *Energy* 32: 1032–1037.

Beysens, D., Mongruel, A., and Acker, K. (2017). Urban dew and rain in Paris, France: occurrence and physico-chemical characteristics. *Atmospheric Research* 189: 152–161.

Finch, J.W. and Gash, J. (2002). Application of a simple finite difference model for estimating evaporation from open reservoir. *Journal of hydrology* 255: 253–259.

Gallego-Elvira, B., Baille, A., Martín-Górriz, B., and Martínez-Álvarez, V. (2010). Energy balance and evaporation loss of an agricultural reservoir in a semi-arid climate (South-Eastern Spain). *Hydrological Processes* 24: 758–766.

Jacobs, F.G.A., Heusinkveld, B.G., and Berkowicz, S.M. (2002). A simple model for potential dewfall in an arid region. *Atmospheric Research* 64: 285–295.

Jacobs, F.G.A., Heusinkveld, B.G., and Berkowicz, S.M. (2008). Passive dew collection in a grassland area, The Netherlands. *Atmospheric Research* 87: 377–385.

Kidron, G.J. (2000). Analysis of dew precipitation in three habitats within a small arid drainage basin, Negev highlands, Israel. *Atmospheric Research* 55: 257–270.

Kidron, G.J., Herrnstadt, I., and Barzilay, E. (2002). The role of dew as a moisture source for sand microbiotic crusts in the Negev Desert, Israel. *Journal of Arid Environments* 52: 517–533.

Lekouch, I., Muselli, M., Kabbachi, B. et al. (2011). Dew, fog, and rain as supplementary sources of water in South-Western Morocco. *Energy* 36: 2257–2265.

Maestre-Valero, J.F., Martínez-Alvarez, V., Baille, A. et al. (2011). Comparative analysis of two polyethylene foil materials for dew harvesting in a semi-arid climate. *Journal of Hydrology* 410: 84–89.

Maestre-Valero, J.F., Ragab, R., Martínez-Alvarez, V., and Baille, A. (2012). Estimation of dew yield from radiative condensers by means of an energy balance model. *Journal of Hydrology* 460–461: 103–109.

Maestre-Valero, J.F., Martin-Gorriz, B., and Martínez-Alvarez, V. (2015). Dew condensation on different natural and artificial passive surfaces in a semiarid climate. *Journal of Arid Environments* 116: 63–70.

Malekian Jabali, M., Okhravi, S., Eslamian, S., and Gohari, S. (2017). Water conservation techniques. In: *Handbook of Drought and Water Scarcity, Vol. 2: Environmental Impacts and Analysis of Drought and Water Scarcity* (eds. S. Eslamian and F. Eslamian), 501–520. Boca Raton, USA: Francis and Taylor, CRC Press.

Molina-Martínez, J.M., Martínez-Alvarez, V., González-Real, M.M., and Baille, A. (2006). A simulation model for predicting hourly pan evaporation from meteorological data. *Journal of Hydrology* 318: 250–261.

Moro, M.J., Were, A., Villagarcıa, L. et al. (2007). Dew measurement by Eddy covariance and wetness sensor in a semiarid ecosystem of SE Spain. *Journal of Hydrology* 335: 295–302.

Muselli, M., Beysens, D., Marcillat, J. et al. (2002). Dew water collector for potable water in Ajaccio (Corsica island, France). *Atmospheric Research* 64: 297–312.

Muselli, M., Beysens, D., and Soyeux, E. (2006a). Chemical composition of dew water from passive radiative condenser in Corsica Island (France). *Journal of Environmental Quality.* 35: 1812–1817.

Muselli, M., Beysens, D., and Milimouk, I. (2006b). A comparative study of two large radiative dew water condensers. *Journal of Arid Environments* 64: 54–76.

Muselli, M., Beysens, D., Mileta, M., and Milimouk, I. (2009). Dew and rain water collection in the Dalmatian coast. *Croatia, Atmospheric Research* 92: 455–463.

Nikolayev, V., Beysens, D., Gioda, A. et al. (1996). Water recovery from dew. *Journal of Hydrology* 182: 19–35.

Nilsson, T.M.J. (1994). Optical scattering properties of pigmented foils for radiative cooling and water condensation: theory and experiment. Ph.D. Thesis, Department of Physics, University of Technology, Göteborg, Sweden.

Nilsson, T.M.J. (1996). Initial experiments on dew collection in Sweden and Tanzania. *Solar Energy Materials and Solar Cells* 40: 23–32.

Pedro, M.J. and Gillespie, T.J. (1982). Estimating dew duration. I. Utilizing micrometeorological data. *Agricultural Meteorology* 25: 283–296.

Richards, K. (2009). Adaptation of a leaf wetness model to estimate dewfall amount on a roof surface. *Agricultural and Forest Meteorology.* 149: 1377–1383.

Sharan, G., 2005. Dew yield from passive condensers in a coastal arid Area - Kutch, IIMA Working Papers WP2005-01-05, Indian Institute of Management Ahmedabad, Research a Publication Department, India.

Sharan, G. (2007). Harvesting dew to supplement drinking water supply in arid coastal villages of Gujarat. Indian Institute of Management, India. W.P. No. 08–05.

Sharan, G., Kumar-Roy, A., Royon, L. et al. (2017). Dew plant for bottling water. *Journal of Cleaner Production* 155: 83–92.

Steinberger, Y., Loboda, I., and Garner, W. (1989). The influence of autumn dewfall on spatial and temporal distribution of nematodes in the desert ecosystem. *Journal of Arid Environments* 16: 177–183.

Tomaszkiewicz, M., Najm, M.A., Beysens, D. et al. (2015). Dew as a sustainable non-conventional water resource: a critical review. *Environmental Reviews* https://doi.org/10.1139/er-2015-0035.

Vermeulen, A.T., Wyers, G.P., Romer, F.G. et al. (1997). Fog deposition on a coniferous forest in the Netherlands. *Atmospheric Environment* 31: 375–386.

18

Atmospheric Water Harvesting Using Waste Energy from Landfills and Oilfields

Enakshi Wikramanayake, Onur Ozkan, Aritra Kar, and Vaibhav Bahadur

Walker Department of Mechanical Engineering, The University of Texas at Austin, Austin, TX, USA

18.1 Introduction

Most efforts to source water for industrial or personal consumption gravitate toward surface-based water bodies and groundwater. However, there is significant pressure on such conventional water sources due to rising populations and pollution levels, necessitating a serious look for alternative water sources. Interestingly, the atmosphere is one of the largest freshwater reservoirs on the planet. Depending on the geographical location and proximity to water bodies, 1 km^2 of land holds between 10 000–30 000 m^3 of moisture above it (Bar 2004). While humanity has largely relied on atmospheric precipitation (rain, snow) for water, there have been few attempts at intentionally tapping the atmospheric moisture, other than cloud-seeding. Incidentally, the concept of condensing and harvesting moisture from air is not recent. Dew ponds in the 1400s in Europe were artificial shallow ponds (Algarni et al. 2018; Herschy 2012) that were fed by rainwater and condensation of dew. Fog fences (Algarni et al. 2018), which also date back to prehistoric times, have inspired present-day fog harvesting. This is a passive method to collect water in high-humidity areas by using mesh structures to trap fog droplets (Klemm et al. 2012). Recent studies have looked into optimizing the fiber mesh pattern and chemistry to maximize fog collection (Park et al. 2013). However, fog harvesting techniques work under very limited conditions and require large collection areas. Despite these limitations, small-scale fog harvesting projects have been implemented in many parts of the world, especially in South and Central America, Africa, Europe, and Asia (Klemm et al. 2012).

The water collection rates can be significantly increased by using active water harvesting techniques, which cool air below the dew point to promote condensation. In this work, atmospheric water harvesting (AWH) refers to condensation of water from humid air. The cooling capacity for sustaining condensation can be provided by various kinds of refrigeration systems (Ozkan et al. 2017; Wikramanayake and Bahadur 2016; Wikramanayake et al. 2017). Other technological options including radiative cooling, moisture absorption using a desiccant system, and other moisture absorbing or adsorbing materials have been reported recently (Kim et al. 2018; Wahlgren 2001). Some of the above concepts have been studied for applications in the fields of agriculture and horticulture. As an illustration, studies in early 2000s explored the use of AWH to meet the water requirements of greenhouses (Davies et al. 2004; Wahlgren 2001). More recently, air-cooled, sorbent-based AWH was used with solar thermal energy to generate water at relative humidities as low as 40% (Kim et al. 2018).

Despite the enormous quantity of moisture in air, industrial-scale AWH has never been implemented due to the energy-intensive (2260 $kJ\,l^{-1}$ of water) nature of condensation. Commercial development of AWH systems is limited to electricity powered units that produce hundreds of gallons of water daily, but with high electricity costs, approaching 20 cents gal^{-1} (Wikramanayake and Bahadur 2016).

The economic viability of AWH projects would be significantly enhanced if the energy requirements could be met inexpensively. Excess natural gas, which is currently either vented or flared, can alternately power AWH systems and enable large-scale deployment of this technology. This is the central premise of the present study. This study analyzes utilization of two sources of excess natural gas for AWH: from landfills and from oilfields.

Landfills are responsible for 15% of global anthropogenic methane emissions, and 18% of US greenhouse gas emissions (US Environmental Protection Agency (USEPA) Landfill Methane Outreach Program (LMOP) 2013). Half of the 260 million tons of municipal solid waste (MSW) generated annually in the US is landfilled (U.S. Environmental Protection Agency (USEPA) 2017a). Globally, landfill emissions are expected to continue to increase

Handbook of Water Harvesting and Conservation: Basic Concepts and Fundamentals, First Edition. Edited by Saeid Eslamian and Faezeh Eslamian.

with the rise in population and per capita consumption (Couth and Trois 2011). Importantly, from a technology development standpoint, landfill gas (LFG) emissions continue steadily for many decades (US Environmental Protection Agency (USEPA) Landfill Methane Outreach Program (LMOP) 2013) even after a landfill is closed (after reaching capacity).

LFG is produced by anaerobic decomposition of MSW and contains (US Environmental Protection Agency (USEPA) 2014) 50–55% methane, 45–50% carbon dioxide, and trace amounts of other gases (Brosseau and Heitz 1994). Along with adverse environmental impacts, LFG emissions also constitute a large-scale energy waste. Methane emissions from US landfills equal 14% of US residential natural gas consumption (US Environmental Protection Agency (USEPA) Landfill Methane Outreach Program (LMOP) 2013; USEIA 2017b), and can be valued at US$7.5 billion. There are currently 650+ operational LFG projects in the US (US Environmental Protection Agency (USEPA) Landfill Methane Outreach Program (LMOP) 2013), a majority of which involve electricity generation (US Environmental Protection Agency (USEPA) 2014, 2017a) via gas turbines and engines (Bove and Lunghi 2006; Rajaram et al. 2011). LFG can used in such systems without extensive treatment to remove non-methane components. However, electricity generation projects are not viable everywhere due to inadequate demand and/or access to the grid.

Similarly, there is significant wastage of natural gas in oilfields. Associated natural gas (gas co-produced with oil), is routinely flared due to the lack of gas handling infrastructure and inadequate onsite demand. Investing in gas collection and handling infrastructure (pipelines, compressors, processing facilities) is not economically viable in many oil-producing regions, especially in Shale oilfields. Many Shale fields, like the Bakken Shale in North Dakota and the Eagle Ford Shale in Texas, are predominantly oil plays, where gas has a lower economic value. Furthermore, gas production from hydraulically fractured oil wells declines rapidly (few weeks to months), which makes it challenging to setup long-term infrastructure. All these factors result in unfavorable economics for long-term utilization of excess natural gas, leaving flaring as the only option for operators.

Globally, 4% of natural gas produced is flared (annual flared gas volume: 140 billion m^3). This gas can be valued as much as US$50 billion (The World Bank 2017; USEIA 2017a). Historically, regions (Elvidge et al. 2009; The World Bank 2017) with significant flaring have been Russia, the Middle East, West Africa, and North Africa. The US is the fourth-largest flaring nation, and the recent spike in US flaring can be attributed to an increase in hydraulic fracturing. Texas and North Dakota are responsible for 30% and 40%, respectively, of total US flaring (USEIA 2017a). The Eagle Ford Shale in Texas accounts for more than half of flaring in Texas, despite having just 3% of the state's wells (Horwitt 2014). An average of 340 MCFD (thousand cubic feet per day) of gas is flared per well in the Eagle Ford (Glazer et al. 2014). In the Bakken Shale in North Dakota, one-third of the produced gas is flared, with some producers flaring three-quarters of the produced gas (Horwitt 2014; Trechock 2014). Fifty oilfield sites in the Bakken flare more than 1200 MCFD and more than 275 sites flare in excess of 300 MCFD (Wocken 2014). Flaring is also prevalent in other states including New Mexico, Wyoming, Louisiana, and Oklahoma.

There have been efforts to reduce this energy waste, and find productive uses for the gas from landfills and oilfields. Natural gas can be used to provide onsite power, or as a source for natural gas liquids, or for fertilizer production. However, none of these approaches have been successful in significantly reducing venting or flaring. Alternative utilization concepts are essential to provide economic incentives to reduce this gas wastage. One such use of this excess natural gas can be to power AWH systems to provide high-quality freshwater. While there are many uses for the water generated (including human consumption), this work analyzes the benefits of waste gas-powered AWH to supply water for oilfield operations.

The increase in hydraulic fracturing has significantly increased the water footprint of the oil and gas industry. Hydraulic fracturing of a horizontal well requires six to seven times more water than that used in a conventional vertical well (Scanlon and Reedy 2014). A well can require 1–5 million gallons (Freyman 2014) for fracturing, with the national average being 2.5 million gallons/well. This is equivalent to the water contained in four Olympic-sized swimming pools. In addition to fracturing, water is also needed for operations like drilling and waterflooding. Water challenges are compounded by the fact that a lot of Shale production happens in water-stressed areas. Half of US wells are located in regions facing extreme water stress (Freyman 2014). All wells in the Eagle Ford Shale are located in either medium or extreme water-stress regions. There are similar water issues facing the Barnett Shale (Texas) (Wikramanayake et al. 2017), the Bakken Shale (North Dakota) (Ozkan et al. 2017), and oil production in California (Wikramanayake et al. 2017).

Most water challenges are currently met mainly by trucking water over long distances. The average trucking distance is 50 mi (North Central Texas Council of Governments 2012). Sourcing and transportation costs can approach US$5/barrel of water (Wikramanayake and Bahadur 2016). Furthermore, it takes 450 trucks, on average, to transport the water needed for a single hydraulic fracturing operation. This leads to significant logistical challenges for producers and negative societal consequences in the form of noise, traffic, pollution, and road damage in communities.

This work analyzes the benefits of using these waste natural gas streams to power AWH systems to generate water for oilfield use (Ozkan et al. 2017; Wikramanayake and Bahadur 2016; Wikramanayake et al. 2017). This premise is expanded and analyzed in this work. The benefits of this technology are illustrated via two case studies in Texas. Firstly, LFG-based AWH is analyzed for the Barnett Shale, which is located next to the Dallas-Fort Worth metropolitan area, and can be served by multiple landfills. Next, oilfield gas-based AWH is analyzed for the severely water-stressed Eagle Ford Shale, which has very high flaring activity.

18.2 Refrigeration-Based Atmospheric Water Harvesting Systems

The cooling capacity to enable AWH via condensation of moisture can be provided by waste natural gas-powered refrigeration systems. Vapor compression refrigeration systems rely on a mechanical compressor to drive the cycle. The compressor of a vapor compression-based AWH system can be powered by a gas engine, gas-fired steam turbine, or a gas turbine. Figure 18.1 shows a schematic of an AWH system based on a gas engine-powered vapor compression refrigeration cycle. Waste natural gas from landfills or oil fields is cleaned up in a gas conditioning module and fed to a gas engine which powers the compressor of the refrigeration cycle. The cooling capacity generated by evaporating the refrigerant (R134a) in the evaporator produces chilled water. This chilled water is used to condense moisture from ambient air in a plate and tube heat exchanger. The refrigerant vapor leaving the evaporator is compressed and then condensed in the air-cooled condenser (ACC). The dehumidified and cold air exiting the water condenser can be routed back to the ACC to provide enhanced cooling.

An alternative to vapor compression systems are vapor absorption systems (Srikhirin et al. 2000). Such systems utilize natural gas-based heating to run the cycle. This vapor absorption cycle uses thermal energy or heat as an alternative to a mechanical compressor to create the pressure difference that drives the cycle. Cooling capacity is generated in the evaporator of the cycle via evaporation of a refrigerant, which is absorbed by a secondary liquid in the absorber. The refrigerant-saturated solution is then heated in a vapor generator to release the refrigerant as high-pressure vapor. A natural gas-fired boiler can be used as the vapor generator. This vapor condenses in the ACC, where the heat of condensation is rejected to ambient air. Natural gas-based heating can also be used to run desiccant dehumidification cycles (Nayak et al. 2009).

Various options for the refrigeration cycle are detailed in Table 18.1. These options are compared by evaluating the coefficient of performance (COP) of the system. The COP represents the ratio of the cooling capacity to the total energy input (to power the compressor and auxiliary equipment). Data in the public domain from commercial systems are used for these estimates.

Table 18.1 indicates that a gas engine powered vapor compression cycle will have the highest cooling capacity, and will maximize the yield. In this work, a gas

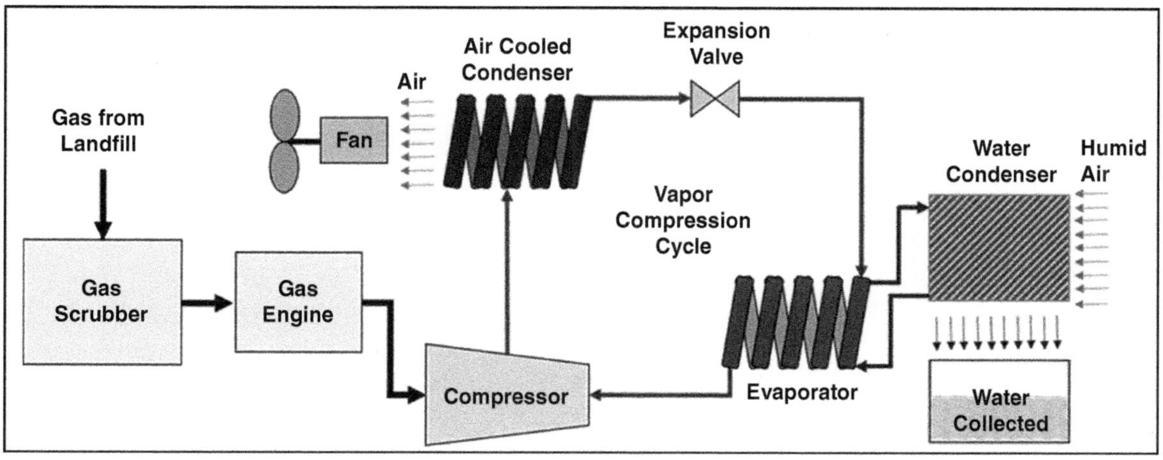

Figure 18.1 Schematic of a vapor compression cycle-powered AWH system that is powered by excess natural gas. A natural gas-fired engine drives the compressor of the refrigeration cycle. Atmospheric moisture condenses on the water condenser (evaporator of the refrigeration cycle). The latent heat of condensation is rejected to the atmosphere by the air cooled condenser (Wikramanayake and Bahadur 2016).

Table 18.1 Various technological pathways to use waste natural gas for atmospheric water harvesting (Wikramanayake and Bahadur 2016).

Technology	Efficiency/COP of individual components	System COP (cooling capacity/energy input)
Gas engine-based vapor compression refrigeration	Gas engine efficiency (GE Power Generation 2017): 38% Vapor compression cycle COP (TRANE 2015): 3.8	1.44
Steam turbine-based vapor compression refrigeration	Steam turbine efficiency (U.S. Environmental Protection Agency (USEPA) Combined Heat and Power Partnership 2007): 30% Vapor compression cycle COP (TRANE 2015): 3.8	1.14
Gas turbine-based vapor compression refrigeration	Gas turbine efficiency (Vicario 2014): 35% Vapor compression cycle COP (TRANE 2015): 3.8	1.33
Gas-fired vapor absorption refrigeration	Vapor absorption cycle COP (Srikhirin et al. 2000): 1	1

engine-based vapor compression cycle was selected for further modeling of AWH systems. The efficiency of the gas engine was selected as 38% (GE Power Generation 2017), and the COP as 3.8. These values are typical of large-scale vapor compression systems (TRANE 2015). It is noted that the COP is not constant. The COP increases at lower temperatures, and will be higher than 3.8 in winter, but lower than 3.8 in summer months. However, routing the cold–dehumidified air from the water condenser to the ACC can keep the temperature of air entering the ACC to less than 20 °C, even in the summer months, thereby preventing the COP from reducing (Ozkan et al. 2017).

18.3 Modeling Waste Natural Gas-Based Atmospheric Water Harvesting

A comprehensive analytical model has been developed by (Ozkan et al. 2017) to estimate the water harvest, which depends on the quantity of excess natural gas, ambient weather, and parameters related to the water harvesting system (efficiency of gas engine, COP of refrigeration cycle, heat transfer coefficients, etc.). The starting point for predicting the water harvest is the estimation of the cooling capacity produced by the refrigeration system. This cooling capacity can be estimated as:

$$q_{cooling} = f\eta_{engine}(HV)(COP) \tag{18.1}$$

where f is the gas flow rate, HV is the heating value of methane, η_{engine} is the efficiency of the gas engine, and COP is the COP of the refrigeration cycle. A constant COP of 3.8 is used in this work, which is typical for large-scale refrigeration systems (Trane 2015). The COP can be higher and lower during the winter and summer months, respectively. During the summer months, cold-dehumidified air from the water condenser can be routed to the ACC to prevent a reduction in the COP. This cold-dehumidified air from the water condenser can also be used to precool the incoming air in another configuration; this will increase the efficiency of the system.

The enthalpy of air leaving the condenser can be calculated from heat transfer calculations based on the mean enthalpy difference and energy balance. However, the humidity of air leaving the condenser should be obtained via estimation of the process line, by finding the ratio of the change in enthalpy to the change in the humidity ratio. The enthalpy of air leaving the condenser can then be calculated as (Ozkan et al. 2017; Threlkeld 1962):

$$h_{air,out} = \frac{h_{C,in}(1 - e^{-(1-c_1)c_2}) + h_{air,in}(1 - c_1)e^{-(1-c_1)c_2}}{1 - c_1 e^{-(1-c_1)c_2}} \tag{18.2}$$

where

$$c_1 = \frac{m_{air} b_C}{\dot{m}_C c_{p,C}} \tag{18.3}$$

$$c_2 = \frac{U_{o,w} A_o}{\dot{m}_a} \tag{18.4}$$

where $\dot{m}_C$ and $c_{p,\,C}$ are the mass flow rate and specific heat of the chilled water, respectively. $h_{C,\,in}$ is the enthalpy of chilled water at its inlet temperature, and $h_{air,\,in}$ and $h_{air,\,out}$ are enthalpies of humid air at its inlet and exit conditions, respectively. A_o is the outer heat transfer area and $U_{o,\,w}$ is the overall heat transfer coefficient based on enthalpy difference. b_C is the slope of the enthalpy curve of humid air at the mean chilled water temperature. The overall heat transfer coefficient based on the enthalpy difference is calculated as (Ozkan et al. 2017; Threlkeld 1962):

$$\frac{1}{U_{o,w}} = \frac{b_C A_o}{h_i A_{T,i}} + \frac{b_{m,F}(1 - \eta_w)}{h_{o,w}\left(\frac{A_{T,o}}{A_F} + \eta_w\right)} + \frac{b_{m,F}}{h_{o,w}} \tag{18.5}$$

where h_i and $h_{o,w}$ are the heat transfer coefficients of the chilled water, and the effective heat transfer coefficient of humid air for wet surface conditions, respectively. $A_{T,o}$ and $A_{t,i}$ are the outer and inner total heat transfer areas, respectively, and A_F is the total fin area. $b_{m,f}$ is the slope of the enthalpy curve of humid air at the mean fin temperature. η_w is the wet fin efficiency, estimated as (Ozkan et al. 2017):

$$\eta_w = \frac{tanh\,(Mr\varphi)}{Mr\varphi} \tag{18.6}$$

where M is modified for wet surface conditions and $r\varphi$ represents the fin length.

The enthalpy of the air leaving the condenser is estimated using Eq. (18.2); however the humidity ratio of the air is needed to estimate the amount of condensed water. The process line for cooling and dehumidification of air can be approximated as (Ozkan et al. 2017; Threlkeld 1962)

$$\frac{\Delta h}{\Delta w} \approx \frac{dh}{dw} = Le\frac{h - h_{m,F}}{\omega - \omega_{m,F}} + (h_{v,a} - 2501Le) \tag{18.7}$$

where $h_{m,F}$ and $\omega_{m,F}$ are the saturated air enthalpy and humidity ratio at the wall temperature, respectively. $h_{v,a}$ is the specific enthalpy of saturated water vapor at the dry bulb temperature, and Le is the Lewis number. Figure 18.2 shows the enthalpy of the air leaving the condenser and the process line associated with the dehumidification process. The slope of the process line is first approximated by using Eq. (18.7) at the inlet conditions. Then, the humidity ratio is decreased by 0.1 g_{water}/kg_{air} and the enthalpy is decreased by 0.1 times the slope of the process line. The slope of the process line is recalculated at each step until the enthalpy becomes equal to the exit enthalpy calculated by Eq. (18.2). Finally, once the relative humidity of the air leaving the condenser is known, the water harvest water can be quantified as:

$$\dot{m}_{harvest} = \dot{m}_{air}(\omega_{in} - \omega_{out}) \tag{18.8}$$

where $\dot{m}_{air}$ is the mass flow rate of dry air, and ω_{in} and ω_{out} are the humidity ratios of the air entering and exiting the heat exchanger, respectively.

The above framework was used to estimate the weather-dependent LFG-based water harvest in the Barnett Shale in Texas. The operating conditions of the refrigeration system were a COP of 3.8, chilled water temperature of 1 °C, chilled water flow rate of 20 l s^{-1}, and a 338 kW compressor. The water harvest is calculated using the hourly weather data for a typical meteorological year (TMY). Analysis for oilfield gas-based water harvest in Eagle Ford Shale was conducted using a first-order thermodynamics model (Wikramanayake and Bahadur 2016), with the same parameters for the refrigeration system being used, as for LFG-based AWH.

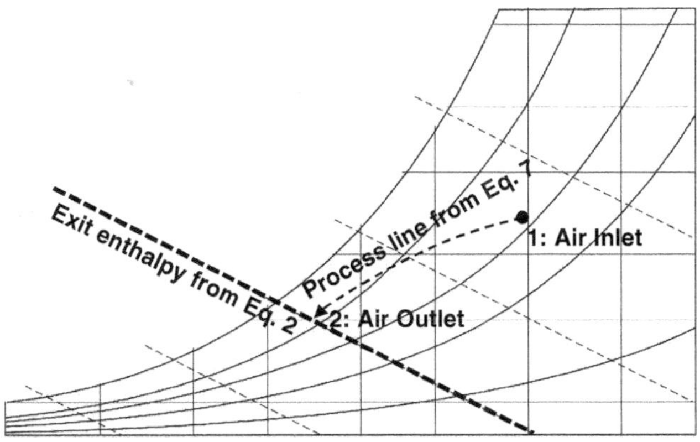

Figure 18.2 Process line associated with dehumidification on a psychrometric chart (Ozkan et al. 2017).

18.4 Landfill Gas-Based Atmospheric Water Harvesting

This section analyzes LFG-based AWH, with a particular focus on the use of the harvested water for oilfield operations. The modeling framework and benefits are illustrated by a case study involving LFG-based AWH to provide water for the Barnett Shale, which is located west of the Dallas-Fort Worth area. Section 18.4.1 details the modeling framework used in this study. Section 18.4.2 quantifies the benefits of this technology, Section 18.4.3 includes a techno-economic analysis of LFG-based AWH, and Section 18.4.4 captures the environmental benefits of this technology.

18.4.1 Modeling LFG-Based AWH in the Barnett Shale

Firstly, a modeling framework was developed to quantify the amount of LFG available for harvesting water. An EPA database (US Environmental Protection Agency (USEPA) 2017c) of more than 2400 US landfills provides details such as location coordinates, landfill capacity, current status (open or closed), and LFG management techniques (generation, capture, and flaring) employed for every landfill. The following relation was used to estimate LFG generation for landfills which did not report a LFG generation rate (US Environmental Protection Agency (USEPA) 2014):

$$Q_{gen} = W \times \frac{432{,}000}{10^6} \tag{18.9}$$

where W is the quantity (tons) of waste in the landfill, and Q_{gen} is the generation rate (standard cubic feet/day).

The generated LFG is either vented (Q_{vent}) or collected (Q_{col}). The collected LFG can be either flared (Q_{flared}), or used (Q_{used}) in an LFG-to-energy project. The database has details on flaring rates and LFG usage for most landfills. For landfills which did not report collection or flaring options, it was assumed that the generated LFG is vented. These considerations are captured as:

$$Q_{gen} = Q_{col} + Q_{vent} \tag{18.10}$$

$$Q_{col} = Q_{used} + Q_{flared} \tag{18.11}$$

Equations (18.9)–(18.11) together predict the LFG available for AWH. It is noted that this work does not exclude the LFG that is currently being utilized, since the objective is to analyze an alternative technology.

Figure 18.3 shows landfills in Texas, along with major Shale fields. Landfills are concentrated around the major population centers of Houston, Dallas-Fort Worth, Austin, and San Antonio. Figure 18.3 shows that there are only three and six landfills within the Eagle Ford Shale and the Permian Basin, respectively, as these oilfields are located in sparsely populated areas. This indicates that LFG-based AWH will not be able to make meaningful contributions to water needs of the Eagle Ford and the Permian Basin. However, there are 30 landfills in the eastern parts of the Barnett Shale, located inside the Shale play or within 25 miles of it, which cater to the Dallas-Fort Worth area. The location of these landfills makes it attractive to transport the harvested water (using LFG-based AWH) to nearby oil fields. This section therefore focuses on LFG-powered AWH in the Barnett Shale.

18.4.2 Benefits of LFG-Based AWH for the Barnett Shale

Before delving into the numbers associated with LFG-based AWH for the Barnett Shale, it is useful to examine them for Texas. LFG generation in Texas landfills ranges from 0.08–30.8 MMSCFD (million standard cubic feet per day), with an average of 3.12 MMSCFD. Cumulative LFG generation from all landfills in Texas is 360 MMSCFD, which translates to 180 MMSCFD of methane emissions (assuming a 50% volume fraction of methane). This volume is comparable to methane currently flared from Shale oil wells in Texas, which is 220 MMSCFD (US Energy

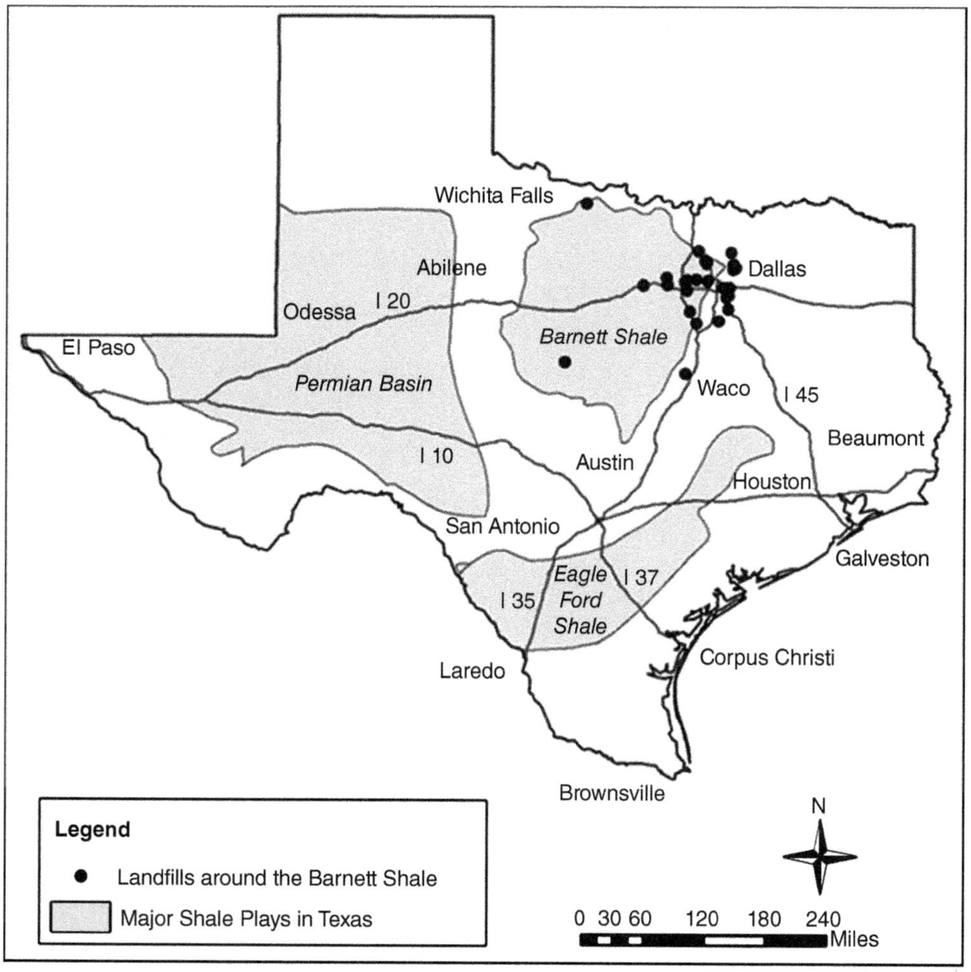

Figure 18.3 Map showing landfills and major oilfields in Texas (Wikramanayake et al. 2017).

Table 18.2 Summary of benefits of LFG-powered AWH for Texas (Wikramanayake et al. 2017).

	Maximum estimated daily water harvest (1000 gal) in Texas
All landfills	22 500
From one landfill (average)	180

Information Administration 2017a). All landfills practice a combination of LFG venting, flaring, and utilization. Based on Eqs. (18.9)–(18.11), 66%, 22%, and 12% of the generated LFG is vented, flared, or used in Texas, respectively. This analysis shows that a majority of LFG (>85%) is currently wasted, which implies the need for new utilization technologies.

Table 18.2 summarizes the water harvest using the LFG generated from landfills in Texas. LFG from all landfills can harvest up to 12 million gallons of water per day (annual average), using the vapor compression-powered AWH system described in Section 18.3 (Ozkan et al. 2017).

The Barnett Shale has 15 000 producing wells (Nicot et al. 2012). On average, a well requires 250 000 and 2.8 million gallons of water for drilling (Freyman 2014) and hydraulic fracturing (Nicot et al. 2012), respectively. In this study, 30 landfills within 25 mi of the Barnett Shale were selected as water sources for Shale-related operations. The total daily LFG generation from these 30 landfills is 108 MMSCFD, with LFG currently being utilized in only 5 landfills. The water harvest from these landfills varies from 26 000–2 704 000 gal d^{-1} with an average of 340 000 gal/d^{-1}.

Figure 18.4a shows the variation in the daily water harvest (monthly average) from the 30 landfills. Significant water can be harvested all year round, with the highest harvests in the summer months (May to September). The peak harvest is 6.7 million gallons/day in August, which can be attributed to higher humidity and temperature relative to other times of the year, as shown in Figure 18.4b. Harvests are lower in winter because of lower temperature and moisture content; the minimum harvest is 250 000 gal d−1. Overall, the average LFG generation from these 30 landfills is 3.6 MMSCFD, which can harvest 45 million gallons of water annually. Figure 18.4 b shows the monthly variation in annual weather in the Barnett Shale. The month averaged temperature and humidity in the Barnett Shale varies from 7 to 30 °C and 56–72%, respectively. The Barnett has relatively steady humidity throughout the year and such humid conditions result in high AWH rates, especially during summer.

The gas utilization fraction (GUF) (Ozkan et al. 2017) represents the fraction of the available gas that is utilized to harvest water. GUF is lower in winter because lower ambient temperatures result in less heat transfer, and therefore lower power and gas requirements. In this analysis, the water condenser was sized to ensure 100% gas utilization in summer when the GUF is highest. Figure 18.4c shows that the GUF is close to 1 during the summer months, but decreases in winter. The GUF is zero whenever the dry bulb temperature is lower than the chilled water temperature. The annual-averaged GUF for these landfills in Texas is 65%. Another parameter of interest is the moisture condensation fraction (MCF), which is the percentage of moisture in the incoming airstream that is condensed and represents the efficiency of water harvest. Figure 18.4c shows that up to 50% of the incoming moisture in air can be condensed in the summer months in Texas; however the condensation fraction decreases at lower temperatures.

Overall, 1.3 billion gallons of water can be harvested annually from 30 landfills near the Barnett Shale. This is equivalent to 34% of the annual water consumption of the Barnett Shale (Freyman 2014). This water is sufficient to hydraulically fracture 480 wells or drill 4495 new wells. It can eliminate 243 000 trucking roundtrips, for a truck capacity (Wikramanayake and Bahadur 2016) of 5550 gal. There are several other benefits to nearby communities such as reduced road damage, accidents and noise pollution. The harvested water could still need to be trucked into oil fields, but this will be relatively short distance trucking due to the proximity of the selected landfills to oilfields.

18.4.3 Techno-Economic Analysis of LFG-Powered AWH

The economic viability of LFG-based AWH is ultimately the basis for any related project. This section details a techno-economic assessment of LFG-powered AWH for the Barnett Shale. Similar analyses (Leme et al. 2014; Shin et al. 2005) have been conducted on alternative applications for LFG. While there are several ways of evaluating project economics, this study evaluates the Pay Back Period (PBP) and Net Present Value (NPV) of a LFG-powered AWH project. The PBP is the time required to recover the investment, while the NPV quantifies the time-adjusted returns from the project.

In this study, the techno-economic analysis was carried out in the context of a landfill installing water harvesting infrastructure and using it to supply water to nearby oilfields. Calculations were done for a landfill with a capacity that is the average of the 30 landfills near the Barnett Shale.

The capital expenditure for such a project includes the gas engine, vapor compression system, water condenser, and a water storage system. Annual maintenance costs were included, and a 5% depreciation of capital equipment was factored in. The cost of transporting water to oilfields

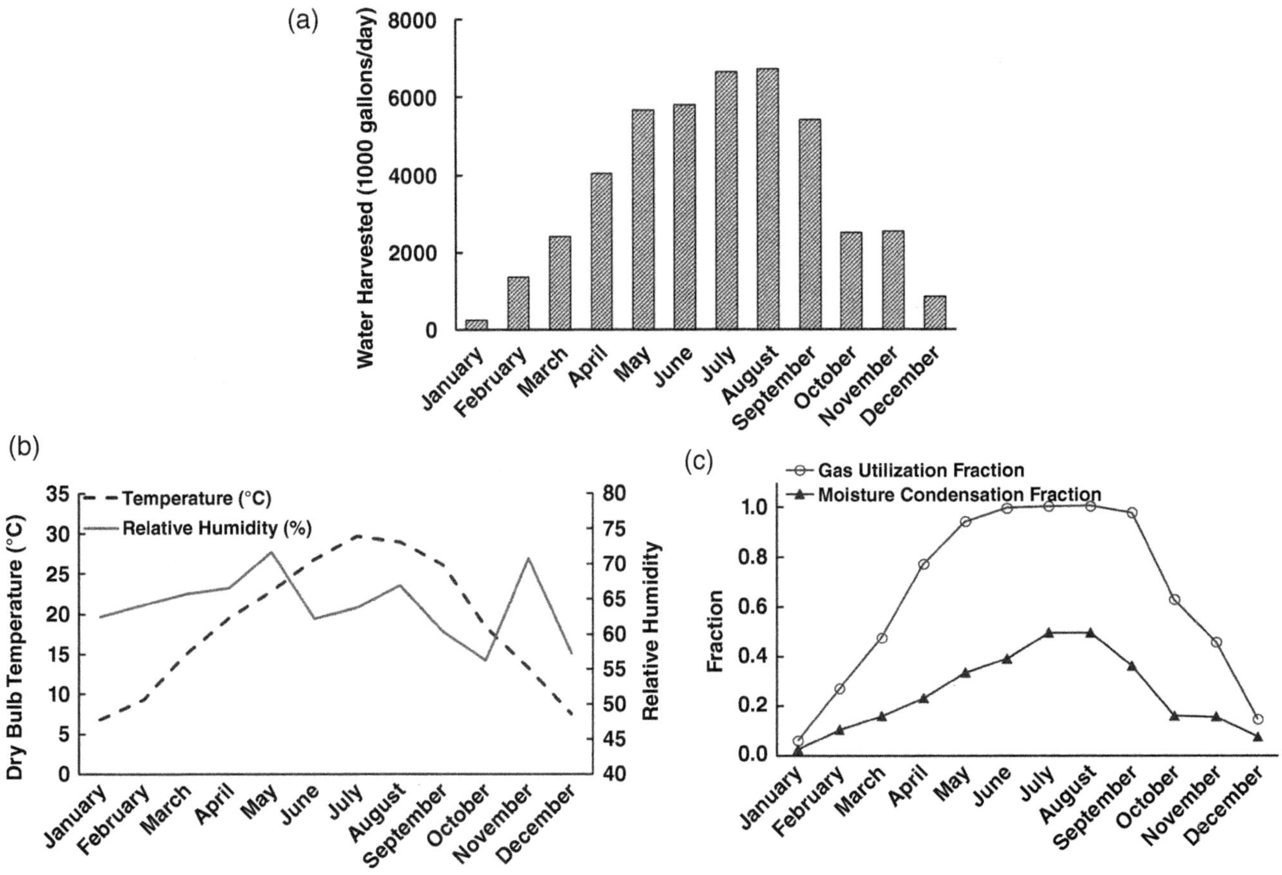

Figure 18.4 (a) Variation in daily water harvest using the LFG generated by 30 landfills in the Barnett Shale area (Wikramanayake et al. 2017). (b) Annual variation of dry bulb temperature and relative humidity in the Barnett Shale (Wilcox 2012). (c) Gas utilization fraction (GUF) and moisture condensation fraction (MCF) associated with LFG-powered AWH in the Barnett Shale (Ozkan et al. 2017).

is borne by the landfill operator and is included in the analysis. It is assumed that the generated water will be stored in tanks (Plastic-Mart 2017). The storage capacity was estimated to be 30 times the maximum daily water harvest (it is assumed that the harvested water will be utilized within a month). Furthermore, it is assumed that the water is trucked to oilfields as pipelines will not be feasible, due to the short-term nature of Shale production. However, the trucking distances will be less than present-day values, owing to the proximity of landfills to oilfield sites. Also, analysis shows that the value proposition of this technology does not significantly depend on the trucking distance.

Table 18.3 details the costs associated with transporting water, which includes the truck leasing costs, fuel costs, and labor expenses. The revenue for the landfill operator depends on the price of harvested freshwater. Since prices fluctuate, this analysis estimates the benefits assuming various price levels of water. The annual water harvest was estimated using the model described in Section 18.3. Table 18.3 lists the key parameters and cost inputs for this techno-economic analysis. A discount rate of 10% was used in this analysis; this is consistent with established practices on valuations of energy and water-related projects (Oxera 2011). The numbers in Table 18.3 suggest that this technology is CAPEX-intensive, with water trucking costs being much lower than equipment costs.

The time horizon for this analysis was 30 years, which is consistent with the lifetime of infrastructure components, and is also consistent with 30+ years of steady emissions from landfills (US Environmental Protection Agency (USEPA) 2014). The NPV of such an LFG-powered AWH project can be estimated as:

$$NPV = \sum_{j=0}^{n} \frac{I_{t=j}}{(1+r)^j} - \left[C_0 + \sum_{j=1}^{n} \frac{M_{t=j} + D_{t=j} + T_{t=j}}{(1+r)^j} \right] \tag{18.12}$$

where I is the income (from water), C_0 is the capital expenditure, M is the maintenance cost, D is depreciation, T is the cost of trucking water, and n is the number of years. The PBP is obtained for the time that the $NPV = 0$. A larger NPV indicates a more valuable long-term investment. A shorter PBP indicates faster return on investment and reduced risk.

Table 18.3 Key parameters and inputs in the techno-economic assessment of LFG-powered AWH for the Barnett Shale.

	Barnett (TX)	Units
LFG flowrate	3.61	MMSCFD
Water harvest (annual average)	44.9	Million gallons/year
Discount rate (r) (Oxera 2011)	10	%
Costs associated with gas engine operation		
CAPEX (US Environmental Protection Agency (USEPA) Combined Heat and Power Partnership 2007)	550	\$/kW
Maintenance	10	% of CAPEX
Depreciation	5	% of CAPEX
Costs associated with vapor compression refrigeration system		
CAPEX (Goetzler et al. 2016)	483	\$/kW
Maintenance (Al-Ugla et al. 2016)	4	% of CAPEX
Depreciation	5	% of CAPEX
Costs associated with water condenser (fin and tube heat exchanger)		
CAPEX (Loh et al. 2002)	20	\$/ft^2
Total surface area	27 375	ft^2
Cost of water storage		
CAPEX (Plastic-Mart 2017)	0.5	\$/gallon
Total storage capacity	6720	Thousand gallons
Break up of CAPEX requirements		
Gas engine	19	%
Refrigeration system	65	%
Water condenser	2	%
Water storage	14	%
Cost of transporting water by trucking		
Truck lease (American Transportation Research Institute 2015)	0.21	\$/mile
Truck maintenance (American Transportation Research Institute 2015)	0.16	\$/mile
Gasoline price (U.S. Energy Information Administration 2017b)	2.3	\$/gallon
Truck driver wages (Bureau of Labor Statistics 2015)	19.8	\$/hour
Truck capacity (Glazer et al. 2014)	5550	gallons
Trucking distance (North Central Texas Council of Governments 2012)	50	miles
Mileage (U.S. Energy Information Administration (EIA) 2016b)	6.4	miles/gallon
Average truck speed	50	miles/hour

These numbers correspond to one landfill (average capacity of all landfills in the region) (Wikramanayake et al. 2017).

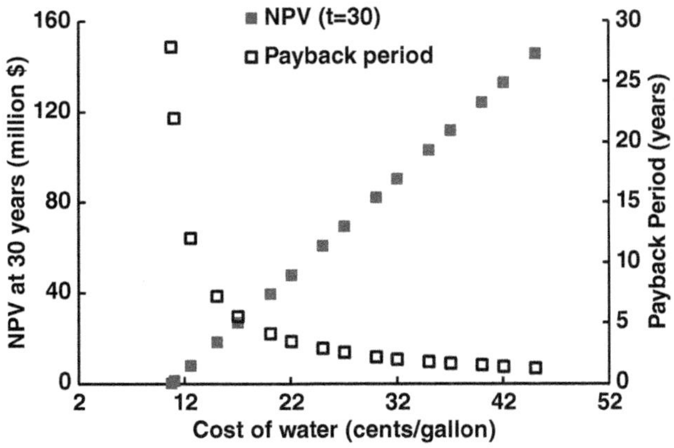

Figure 18.5 Net present value at 30 years and payback period for LFG-powered AWH projects which provide water to oilfields in the Barnett Shale (Wikramanayake et al. 2017).

Typical payback periods for LFG based-electricity generation projects range from three to six years (Rajaram et al. 2011).

Figure 18.5 shows the estimated NPV and PBP for LFG-powered AWH versus the price of water. The 30-year NPV is positive for water prices exceeding 11 cents/gal in the Barnett. This corresponds to the higher end of the price range that oil producers currently pay for water. The PBP for these projects is price dependent, with a six-year PBP requiring prices of 23 cents gal^{-1} for this case study in the Barnett. It was also seen that the NPV and PBP values were not very sensitive to the trucking distance. For example, doubling the trucking distance (from landfill to oilfields) will increase the PBP for a Barnett Shale AWH project from 72 months to 77 months (when the price of water is 19 cents gal^{-1}).

This analysis does not consider other factors which could increase the value proposition of this technology. Currently, landfills do not pay any carbon tax, and there are no financial repercussions on flaring and venting. Any regulatory push toward taxing emissions would enhance the economic viability of this technology. There are other intangible benefits of this technology including reduced pressure on surface and groundwater resources, reductions in air, light, and noise pollution (from flares), and reduced truck traffic. Such benefits cannot be easily monetized, nevertheless they do contribute toward decision making.

18.4.4 Environmental Benefits of LFG-Powered AWH

Venting natural gas is very detrimental to the environment due to the high global warming potential of methane. Most landfills vent natural gas as it is the least costly option of handling excess gas. Flaring natural gas converts methane into carbon dioxide, which has a much lower global warming potential. Although using LFG for AWH will generate the same volume of CO_2 as flaring, the key benefit is the water being harvested. In 2016, the methane emitted from landfills generated 107.7 MMT CO_2e (carbon dioxide equivalent) (Inventory of U.S. Greenhouse Gas Emissions and Sinks: 1990–2016 2018). If all this methane was converted to CO_2 via combustion in a gas engine for AWH, the CO_2e would reduce by 95%.

Table 18.4 details the carbon footprint associated with various options for handling LFG from the 30 landfills in the Barnett Shale. The emission analysis considers a scenario in which the water generated via AWH replaces the water currently trucked in (and does not supply additional water). Annual CH_4, CO_2, and CO_2e emissions are estimated from reported LFG emissions, and the estimated trucking trips. The cases analyzed in Table 18.4 include complete venting of LFG, complete flaring of LFG, the present state of LFG management, and complete utilization of LFG for AWH. The first three cases assume that water for Barnett Shale is trucked in, with an average trip being 50 miles (North Central Texas Council of Governments 2012).

Emissions are estimated using EPA provided guidelines (US Environmental Protection Agency (USEPA) 2017b). LFG-powered AWH will lead to CO_2 emissions from methane combustion. Equation (18.13) estimates the CO_2e emissions resulting from the venting of LFG. This accounts for the fact that CH_4 has a 25 times higher global warming potential than CO_2 (Howarth et al. 2011). LFG flowrate is converted to the quantity of CO2e generated from venting as:

$$CO_2e = \left[CO_2 \times \left(\frac{1CO_2e}{1CO_2}\right)\right] + \left[CH_4 \times \left(\frac{25CO_2e}{1CO_2}\right)\right] \tag{18.13}$$

Equation (18.14) estimates the CO_2e emissions associated with flaring, power generation, and AWH (US Energy Information Administration (EIA) 2016a), since all these processes involve the combustion of CH_4 to produce CO_2.

$$\begin{aligned} CO_2e\left(\frac{metric\ ton}{day}\right) &= \left[CO_2\frac{scf}{day} \times \frac{1\ metric\ ton CO_2e}{19{,}300\ scf}\right] \\ &+ \left[CH_4\frac{scf}{day} \times 1{,}000\frac{Btu}{scf} \times (117 \times 10^6)\frac{lb\ CO_2}{Btu}\right. \\ &\left.\times (4.5 \times 10^{-4})\frac{metric\ ton\ CO_2e}{lb\ CO_2}\right] \end{aligned} \tag{18.14}$$

An average truck transports 5550 gal of water (Glazer et al. 2014; Wikramanayake and Bahadur 2016) and travels 50 mi (North Central Texas Council of Governments 2012).

Table 18.4 Annual carbon emissions from various LFG management techniques in the Barnett Shale (TX) (quantities in million metric tonnes) (Wikramanayake et al. 2017).

	CH_4 released ($MM\,t\,yr^{-1}$)	CO_2 released ($MM\,t\,yr^{-1}$)	Trucking emissions (CO_2) ($MM\,t\,yr^{-1}$)	CO_2e released ($MM\,t\,yr^{-1}$)
	TX	TX	TX	TX
Complete venting of LFG	0.38	1.03	0.14	10.64
Complete flaring of LFG	0.00	2.08	0.14	2.21
Present state of LFG management (venting, flaring, and electricity)	0.25	1.38	0.14	7.80
LFG-powered AWH	0.00	2.08	0.12	2.20

This information can be used to estimate the total annual vehicle miles traveled (VMT). Eq. (18.15) estimates the CO_2e emissions, based on the VMT miles as (North Central Texas Council of Governments 2012):

$$CO_2e\left(\frac{metric\ ton}{year}\right) = VMT\frac{miles}{year} \div 6.3\frac{miles}{gallon} \times 22.4\frac{lb\ CO_2}{gallon\ of\ diesel} \times (4.5\times 10^{-4})\frac{metric\ ton\ CO_2e}{lb\ CO_2} \quad (18.15)$$

Table 18.4 shows that *LFG-powered AWH can reduce CO_2e emissions by 72% in the Barnett Shale* when compared to the current state of LFG management (venting, flaring, and partial utilization). These benefits are primarily due to elimination of methane venting (and replacement with CO_2 which is a less potent greenhouse gas). The emissions from trucking are insignificant when compared to other activities. However, any reduction in trucking is beneficial from a traffic, logistics and noise pollution standpoint.

18.5 Oilfield Gas-Based Atmospheric Water Harvesting

This section analyzes oilfield gas-based AWH, with the objective of using the harvested water for oilfield operations. Estimates of oilfield-gas based AWH are obtained using a thermodynamics-based model (Wikramanayake and Bahadur 2016). The benefits of this technology are illustrated by a case study of the Eagle Ford Shale in south Texas.

The Eagle Ford Shale accounts for half of the flaring in Texas despite having only 3% of the state's wells. Recent estimates show that on average 340 MCFD of gas is flared per well from newly completed wells in the Eagle Ford. A thermodynamic model developed by (Wikramanayake and Bahadur 2016) predicts that a maximum of 43 000 gal $(d\ well)^{-1}$ can be harvested in the Eagle Ford Shale. Most production sites have multiple wells that are flared, which will increase the total water output.

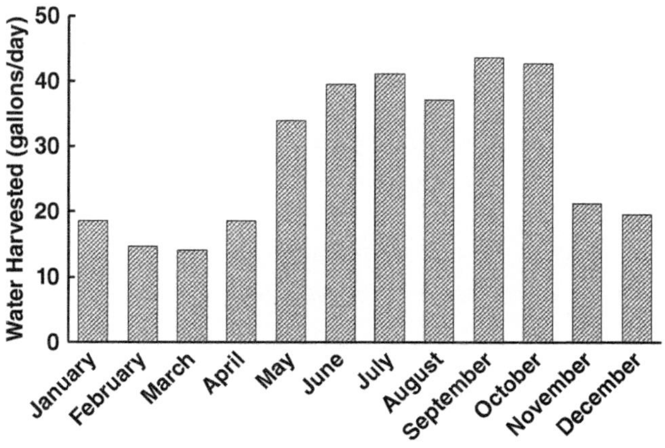

Figure 18.6 Annual variation of daily water harvest using the gas currently flared per well in the Eagle Ford Shale (Wikramanayake and Bahadur 2016).

Figure 18.6 shows monthly variations in the daily water production from the gas currently flared per well in the Eagle Ford Shale. Significant quantities of water can be harvested all year round, with the harvest peaking in summer. Like the Barnett Shale, the Eagle Ford also has ideal conditions for AWH during the summer months (high temperature and humidity).

Overall, 2.9 billion gallons of water can be harvested annually in the Eagle Ford using the gas currently being flared. This is equivalent to 36% of the annual water consumption in the Eagle Ford (Freyman 2014). This quantity of water can be used to hydraulically fracture 680 wells. The water required for one hydraulic fracturing operation can be obtained using the excess gas produced in a nearby well over a period of 100 days. The benefits of AWH are amplified if the water is used for drilling operations, since drilling is 10 times less water intensive than hydraulic fracturing, and requires an average of 250 000 gal well^{-1} (Nicot et al. 2012). The water harvested annually in the Eagle Ford can be used to drill 11 500 wells. The water required for drilling one well can be obtained using the excess gas produced in a nearby well over a period of six days. All these timeframes are consistent with the fact (Glazer et al. 2014) that newly completed wells have high gas production rates in the first few weeks and months. All these observations indicate that AWH can meet a large percentage of the water requirements for Shale production. Additional benefits of AWH to local communities include reduced truck traffic, accidents, and elimination of light and environmental pollution. Based on the water harvest values, it is estimated that oilfield gas-based AWH in the Eagle Ford Shale can eliminate 520 000 trucking roundtrips annually.

18.6 Sensitivity of the Water Harvest to Various Parameters

This section details a sensitivity analysis to quantify the dependence of the water harvest on environmental and system parameters. The effectiveness of the water harvest is quantified by a dimensionless figure of merit which is the volume of water harvested per unit volume of LFG utilized.

Figure 18.7a clearly shows that the water harvest is very sensitive to the temperature and humidity. It also shows the minimum temperature below which AWH will not work due to insufficient heat transfer (as a result of the low dry bulb temperature). This minimum temperature depends on the humidity level and will decrease with increasing relative humidity. For example, AWH will not work below 20 °C if the relative humidity is only 40%. However, AWH can work at temperatures approaching 10 °C if the relative humidity is 85%.

Figure 18.7b shows the sensitivity of the water harvest to system parameters like the COP, heat exchanger area, heat transfer coefficient, and maximum air speed. The baseline for Figure 18.7b is the annual water harvest in the Barnett Shale. It is seen that the harvest is not very sensitive to variations in the heat exchanger area, heat transfer coefficient, and maximum air speed, unless the deviations are more than 25% from the baseline values. However, the water harvest rate is a strong function of the refrigeration cycle COP. To further elaborate, Figure 18.7c shows the effect of doubling and halving the baseline COP on water harvests throughout the year. An increase in

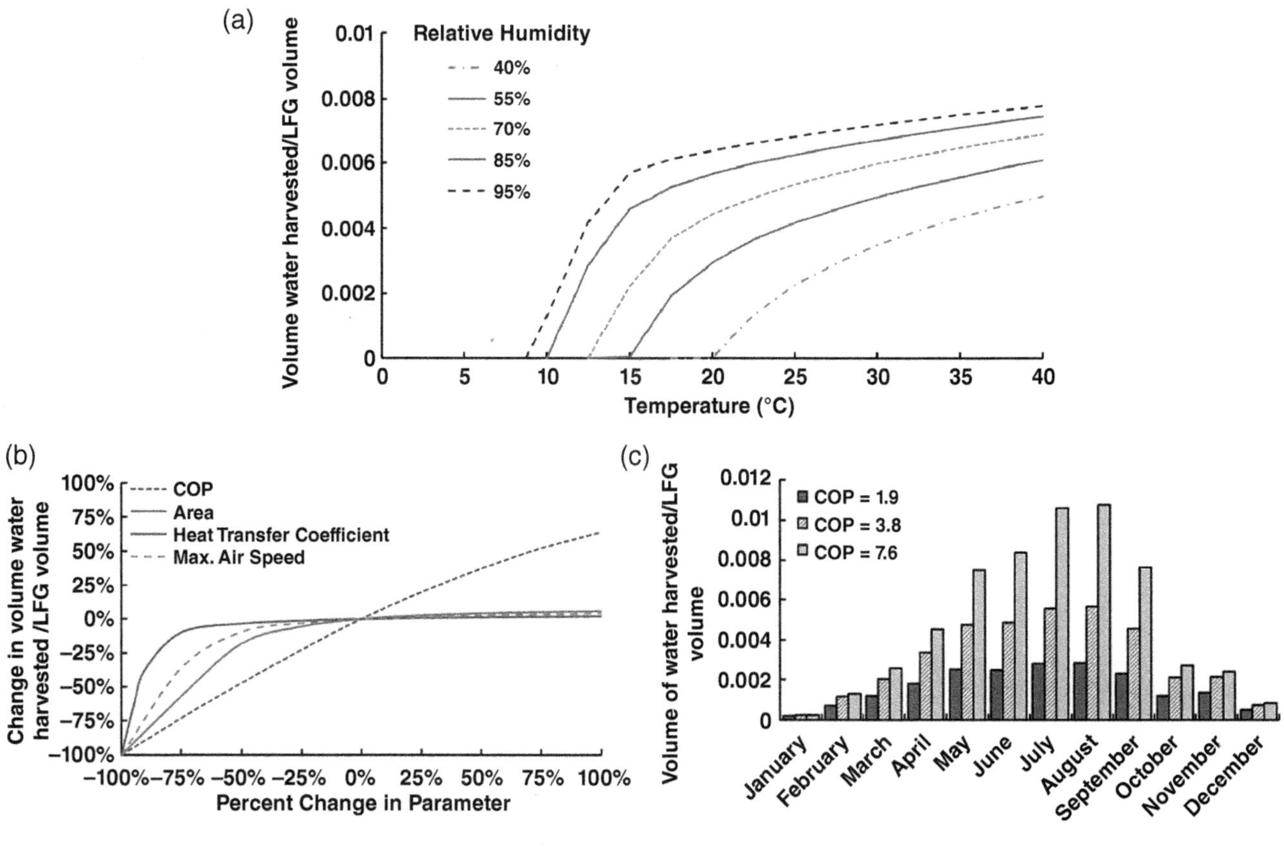

Figure 18.7 (a) Sensitivity of the water harvest to environmental conditions (temperature, humidity). (b) Sensitivity of the water harvest to key parameters of the AWH system. (c) Influence of the COP on water harvest across an entire year (Wikramanayake et al. 2017).

the COP proportionally increases the water harvest in the summer months. However, in the winter, the effect of an increase in the COP is limited, since the low heat transfer cannot be improved by increasing the air speed beyond acceptable limits.

Overall, this analysis indicates that the water harvest strongly depends on the ambient weather and the COP (or cooling capacity) of the refrigeration system. The benefits of this technology will be maximized for regions near the equator due to the convergence of multiple factors which strengthen the value proposition of LFG-powered AWH. These include extensive landfilling activity (due to large populations), high year-round humidity levels, and strong freshwater demand.

18.7 Comparison of AWH to Other Techniques for Producing Water

This section compares AWH with other techniques for producing water by using the energy of waste natural gas. While the moisture in air is a convenient source of water, there are significant quantities of brackish water available worldwide. Hence, it is important to understand the advantages and disadvantages of water treatment methods, and compare it with water conservation techniques, such as rainwater harvesting (Okhravi et al. 2014; Okhravi et al. 2015). Waste natural gas can provide the energy required to treat brackish water into freshwater. The top two technologies to produce freshwater from brackish water are reverse osmosis (RO) and thermal desalination. In RO systems, a majority of the energy requirements are consumed by the pump. An excess natural gas-fired gas turbine can provide the mechanical power required to drive this pump. In thermal desalination technologies, such as multistage flash distillation (MSF) and mechanical vapor recompression (MVR), the energy from excess gas can be converted to heat to desalinate brackish water.

Table 18.5 compares the water produced from excess natural gas using three techniques: AWH, thermal desalination, and RO. It is seen that RO yields the greatest amount of freshwater, and the yields can be 100X higher than AWH. These numbers are based on first order approximations (Glazer et al. 2017) of the water harvest using these techniques and clearly highlight the energy-intensive nature of AWH. However, although the other two techniques have higher water yields, they are limited by other considerations. Firstly, both RO and thermal desalination rely on the availability of brackish water, which is not always possible. Secondly, the water harvest from both these techniques strongly depends on the quality of the brackish water. RO-based techniques become ineffective at TDS (total dissolved solids) levels >50 000. In many Shale fields the TDS levels of brackish water are much higher, (as high as 250 000), and RO-based water treatment becomes unrealistic. AWH, on the other hand, does not rely on the availability of any surface water. Finally, the selection of any technique for producing freshwater is ultimately based on techno-economic analysis. While AWH is very energy intensive compared to RO and thermal desalination, it is possible that the economic analysis could be more favorable after capital and maintenance costs of various technologies are compared.

Table 18.5 Estimates of water production by AWH, thermal desalination, and reverse osmosis (Glazer et al. 2017; Ozkan et al. 2017).

	Amount of water produced per m^3 of gas (gal)
Atmospheric water harvesting	0.5–2.5
Thermal desalination	30–100
Reverse osmosis	150–400

18.8 Perspectives on Atmospheric Water Harvesting

This section discusses various aspects and considerations of AWH that will determine large-scale adaptation of the technology discussed in this chapter. The key to increasing the attractiveness of waste gas-based AWH systems is to reduce the capital cost, which will result in more favorable techno-economic analysis. The capital expenditures for AWH systems can be reduced by making the ACC and the water condenser more compact. In particular, the ACC is responsible for ultimately rejecting the heat released during condensation. Low heat transfer coefficients on the air-side drive up the size and capital costs; this heat exchanger is typically the most voluminous component of the refrigeration system. Similarly, reducing the size of the water condenser will also benefit the overall system. Currently, water condenses as a film which acts as a thermal barrier and impedes heat transfer. The use of hydrophobic coatings to facilitate drop-wise condensation can enable an order of magnitude enhancement in heat transfer which will reduce the water condenser area. It is also important to design the water condenser to enable rapid drainage of the condensed water.

There are some important distinctions between the AWH systems that run on oilfield gas versus the ones that run on LFG. The rate of gas production from oilfields is highly variable, and declines rapidly. Most gas production occurs in the first few weeks and months after a hydraulically fractured well is completed. The short-term nature of gas production makes it economically unviable to deploy any permanent AWH infrastructure. Instead, the availability of mobile and modular AWH units, which can be daisy-chained, is critical to technology adoption. Ideally, the AWH equipment should be compact enough such that it can be accommodated on a large trailer. On the other hand, gas emissions from landfills are steady for many decades, which makes it attractive to deploy permanent AWH infrastructure.

Ultimately, the deployment of AWH systems at any site will be contingent on favorable techno-economic analysis, which involves details of capital expenditures, equipment depreciation rates, labor costs, compliance and permitting costs, and projections of the benefits. Since the compact and mobile AWH units required to enable oilfield gas-based AWH do not exist, a techno-economic analysis was not conducted. For LFG-based AWH, where stationary units can be used and size is not a constraint, a detailed techno-economic analysis was conducted.

This study focused on the benefits of waste gas-based AWH for unconventional oil production. While the case studies involved Shale fields in Texas, these technologies have applicability in other parts of the US. In the Bakken Shale (North Dakota), one-third of the gas produced is currently flared. Oilfield gas-based AWH in the summer months can make meaningful contributions to the water requirements of the Bakken Shale (Wikramanayake and Bahadur 2016). Similarly, there are oilfields that are located close to large population centers and can use the LFG generated from nearby landfills to harvest water. One specific region is Kern County in California (Wikramanayake et al. 2017), which is California's top oil-producing county. Other oilfields in the US can also benefit, but to a lesser extent, as they are located in sparsely populated regions.

18.9 Conclusions

While there is increasing focus on reducing venting and flaring, and the presence of LFG-to-energy projects is encouraging, only 12% of the LFG is currently used in Texas. This suggests the need to develop new technologies to incentivize excess gas utilization and reduce venting and flaring. The technology proposed in this study impacts the water sector, benefits the environment, and fulfills a critical need of the oil-gas industry. LFG-powered AWH can meet 34% of the water requirements in the Barnett Shale. Similarly, oilfield gas-powered AWH can meet 36% of the water requirements of the Eagle Ford Shale. This technology can reduce emissions by over 60% in these two regions. Depending on the price of freshwater, LFG-powered AWH projects can be implemented in the Barnett Shale with payback periods of less than 10 years.

On a technical front, results indicate that 1 cubic meter of natural gas can condense 2.3 gal of water (on an annual average basis) in Texas. The weather-dependent condensation rates range from 3 to 10 gal $(d\,m^2)^{-1}$ of condenser area. Interestingly, the results indicate that only 20–50% moisture in the airstream is being condensed in Texas. This suggests opportunities to enhance the harvest via design optimization of the water condenser.

While LFG-based AWH was focused on the Barnett Shale, which can be catered to by 30 landfills, there are other US oilfields that can benefit even with a sparser distribution of landfills. Landfills exist everywhere on the planet and this work can be extended to other regions in the US and worldwide. As populations increase, and extensive landfilling takes place, the resulting LFG can provide a sizeable, steady, and long-term energy source for AWH.

Overall, this waste-to-value technology can address significant issues related to water, energy and environmental protection. This technology can benefit regions which have an abundance of gas and face significant water access issues, but lie in hot-humid weather zones (within the Tropic of Cancer and the Tropic of Capricorn). There are several regions that fall in this category, including the southern US, Central and South America, the Middle East, South East Asia, and East/North Africa. All these regions can benefit significantly from the adaptation of waste natural gas-based AWH.

Acknowledgements

The authors acknowledge a research grant from the Vice President for Research's office at The University of Texas at Austin for partial support of this work.

References

Algarni, S., Saleel, C.A., and Mujeebu, M.A. (2018). Air-conditioning condensate recovery and applications—current developments and challenges ahead. *Sustainable Cities and Society* 37: 263–274.

Al-Ugla, A.A., El-Shaarawi, M.A.I., Said, S.A.M., and Al-Qutub, A.M. (2016). Techno-economic analysis of solar-assisted air-conditioning systems for commercial buildings in Saudi Arabia. *Renewable and Sustainable Energy Reviews* 54: 1301–1310.

American Transportation Research Institute. (2015). *Analysis of operational costs of trucking*.

Bar, E. (2004). Extraction of water from air - an alternative solution for water supply. *Desalination* 165: 335.

Bove, R. and Lunghi, P. (2006). Electric power generation from landfill gas using traditional and innovative technologies. *Energy Conversion and Management* 47 (11–12): 1391–1401.

Brosseau, J. and Heitz, M. (1994). Trace gas compound emissions from municipal landfill sanitary sites. *Atmospheric Environment* 28 (2): 285–293.

Bureau of Labor Statistics. (2015). Heavy and Tractor-Trailer Truck Drivers.

Couth, R. and Trois, C. (2011). Waste management activities and carbon emissions in Africa. *Waste Management* 31 (1): 131–137.

Davies, P., Turner, K., and Paton, C. (2004). Potential of the seawater greenhouse in middle eastern climates. International Engineering Conference, Mutah University 2004, pp. 523–540, Jordan.

Elvidge, C.D., Ziskin, D., Baugh, K.E. et al. (2009). A fifteen year record of global natural gas flaring derived from satellite data. *Energies* 2 (3): 595–622.

Freyman, M. (2014). Hydraulic Fracturing and Water Stress : Water Demand by the Numbers, (February), 85. Lender & Operator Guide to Water Sourcing February, A CERES Report.

GE Power Generation. (2017). Jenbacher Type 2 Gas Engines | GE Power.

Glazer, Y.R., Kjellsson, J.B., Sanders, K.T., and Webber, M.E. (2014). Potential for using energy from flared gas for on-Site Hydraulic Fracturing Wastewater Treatment in Texas. *Environmental Science and Technology Letters* 1: 300–304.

Glazer, Y.R., Davidson, F.T., Lee, J.J., and Webber, M.E. (2017). An inventory and engineering assessment of flared gas and liquid waste streams from hydraulic fracturing in the USA. *Current Sustainable/Renewable Energy Reports* 4 (4): 219–231.

Goetzler, W., Guernsey, M., Young, J., et al. (2016). *The Future of Air Conditioning for Buildings*.

Herschy, R.W. (2012). Dew Ponds. In: *Encyclopedia of Lakes and Reservoirs*, 1e (eds. L. Bengtsson, R.W. Herschy and R.W. Fairbridge), 217–219. Dordrecht, Germany: Springer.

Horwitt, D. (2014). *Up in Flames*. Earthworks.

Howarth, R.W., Santoro, R., and Ingraffea, A. (2011). Methane and the greenhouse-gas footprint of natural gas from shale formations a letter. *Climatic Change* 106 (106).

Inventory of U.S. Greenhouse Gas Emissions and Sinks (2018). Report of GGES (1990–2016), USA.

Kim, H., Rao, S.R., Kapustin, E.A. et al. (2018). Adsorption-based atmospheric water harvesting device for arid climates. *Nature Communications* 9 (1): 1–8.

Klemm, O., Schemenauer, R.S., Lummerich, A. et al. (2012). Fog as a fresh-water resource: overview and perspectives. *Ambio* 41 (3): 221–234.

Leme, M.M.V., Rocha, M.H., Lora, E.E.S. et al. (2014). Techno-economic analysis and environmental impact assessment of energy recovery from municipal solid waste (MSW) in Brazil. *Resources, Conservation and Recycling* 87: 8–20.

Loh, H.P., Lyons, J., and White, C.W. (2002). *Process Equipment Cost Estimation Final Report United States: doi:10.2172/797810.*

Nayak, S.M., Hwang, Y., and Radermacher, R. (2009). Performance characterization of gas engine generator integrated with a liquid desiccant dehumidification system. *Applied Thermal Engineering* 29 (2–3): 479–490.

Nicot, J.-P., Reedy, R.C., Costley, R.a., and Huang, Y. (2012). Oil and Gas Water Use in Texas: Update to the 2011 Mining Water Use Report (September), USA, 1–117.

North Central Texas Council of Governments. (2012). *Development of Oil and Gas Mobile Source Inventory in the Barnett Shale in the 12-County Dallas-Fort Worth Area,* USA.

Okhravi, S., Eslamian, S., Eslamian, F., and Tarkesh-Esfahani, S. (2014). Indigenous knowledge as a supportive tool for sustainable development and utilisation of rainwater harvesting systems. *Journal of Flood Engineering* 5 (1/2): 39–50.

Okhravi, S., Eslamian, S., and Adamowski, J. (2015). Water reuse in rainwater harvesting. In: *Urban Water Reuse Handbook*, Ch. 59 (ed. S. Eslamian), 779–796. Taylor and Francis, CRC Group.

Oxera. (2011). Discount rates for low-carbon and renewable generation technologies, Prepared for the Committee on Climate Change. Oxera Consulting Ltd, London, UK.

Ozkan, O., Wikramanayake, E.D.E.D., and Bahadur, V. (2017). Modeling humid air condensation in waste natural gas-powered atmospheric water harvesting systems. *Applied Thermal Engineering* 118: 224–232.

Park, K.-C., Chhatre, S.S., Srinivasan, S. et al. (2013). Optimal Design of Permeable Fiber Network Structures for fog harvesting. *Langmuir* 29 (43): 13269–13277.

Plastic-Mart. (2017). Vertical Liquid Storage Tanks https://www.plastic-mart.com/category/109/plastic-storage-tanks.

Rajaram, R., Faisal Zia, S., and Emran Khan, M. (2011). *From Landfill Gas to Energy*. USA: CRC Press.

Scanlon, B.R., Reedy, R.C., and J.-P. Nicot (2014). Comparison of Water Use for Hydraulic Fracturing for Unconventional Oil and Gas versus Conventional OilEnviron. Sci. Technol., 48, 20, 12386–12393.

Shin, H.-C., Park, J.-W., Kim, H.-S., and Shin, E.-S. (2005). Environmental and economic assessment of landfill gas electricity generation in Korea using LEAP model. *Energy Policy* 33 (10): 1261–1270.

Srikhirin, P., Aphornratana, S., and Chungpaibulpatana, S. (2000). A review of absorption refrigeration technologies. *Renewable and Sustainable Energy Reviews* 5 (4): 343–372.

The World Bank. (2017). Global Gas Flaring Reduction Partnership (GGFR).

Threlkeld, J. (1962). *Thermal Environmental Engineering*. Prentice Hall.

Trane (2015). Air-Cooled Series R. Chillets Model RTAC.

Trechock, M. (2014). *The Flaring Boom. Western Organization of Resource Councils*. USA.

US Energy Information Administration. (2017a). Natural Gas Gross Withdrawals and Production, USA.

US Energy Information Administration. (2017b). Texas Gasoline and Diesel Retail Prices, USA.

US Energy Information Administration (EIA). (2016a). ENVIRONMENT: Carbon Dioxide Emissions Coefficients, USA.

US Energy Information Administration (EIA). (2016b). Heavy-Duty Truck Fuel Economy, USA.

US Environmental Protection Agency (USEPA) (2014). *Landfill gas energy basics*. In: *LFG Energy Project Development Handbook*. USA: US Environmental Protection Agency.

US Environmental Protection Agency (USEPA). (2017a). Advancing Sustainable Materials Management: Facts and Figures, USA.

US Environmental Protection Agency (USEPA). (2017b). Greenhouse Gases Equivalencies Calculator - Calculations and References, USA.

US Environmental Protection Agency (USEPA). (2017c). Landfill Gas Energy Project Data and Landfill Technical Data. Retrieved March 21, USA.

US Environmental Protection Agency (USEPA) Combined Heat and Power Partnership. (2007). *Biomass Combined Heat and Power Catalog of Technologies*, USA.

US Environmental Protection Agency (USEPA) Landfill Methane Outreach Program (LMOP). (2013). *An Overview of Landfill Gas Energy in the United States*, USA.

USEIA. (2017a). Natural Gas Vented and Flared USA.

USEIA. (2017b). U.S. Natural Gas Consumption by End Use, USA.

Vicario, D. (2014). *High Efficiency 370kW Microturbine with Integral Heat Recovery. Capstone Turbine Corporation/Oak Ridge National Laboratory/NASA, USA*.

Wahlgren, R.V. (2001). Atmospheric water vapour processors designs for potable water production: a review. *Water Research* 35 (1): 1–22.

Wikramanayake, E.D. and Bahadur, V. (2016). Flared natural gas-based onsite atmospheric water harvesting (AWH) for oilfield operations. *Environmental Research Letters* 11 (3): 034024.

Wikramanayake, E.D., Ozkan, O., and Bahadur, V. (2017). Landfill gas-powered atmospheric water harvesting (AWH) for oilfield operations in the United States. *Energy* 138: 647–658.

Wilcox S. (2012). National Solar Radiation Data Base: 1991–2010 Update National Renewable Energy Laboratory, US Department of Energy., USA.

Wocken, C. (2014). Assessment of Remote Capture Technologies to Improve Gas Utilization. In *2014 Williston Basin Petroleum Conferenc*. USA.

Part F

Environmental Aspects of Water Harvesting

19

Treatment Techniques in Water Harvesting

Brandon Reyneke[1], Monique Waso[1], Thando Ndlovu[1], Tanya Clements[1], Sehaam Khan[2], and Wesaal Khan[1]

[1] *Department of Microbiology, Faculty of Science, Stellenbosch University, Stellenbosch, South Africa*
[2] *Faculty of Health Sciences, University of Johannesburg, Doornfontein, South Africa*

19.1 Introduction

Water scarcity is a growing concern worldwide and is directly impacted by the contamination and misappropriation of available water sources, a lack of adequate water supply infrastructure, an increase in population size, and climate change (Zavala et al. 2018). Globally, numerous countries are utilizing alternative water sources to compensate for the depletion of renewable freshwater, and roof-harvested rainwater (RHRW) has subsequently been implemented as a source of potable and non-potable water (Dobrowsky et al. 2014; Campisano et al. 2017).

Rainwater harvesting refers to the collection and storage of rainwater by roof-catchment systems, land surfaces, or other artificial catchments for domestic and agricultural uses (Li et al. 2010; Campisano et al. 2017). Despite the general perception that rainwater can be utilized for potable purposes, numerous studies have demonstrated that harvested rainwater is contaminated and does not comply with drinking water standards (World Health Organization [WHO] 2011; Dobrowsky et al. 2015a). Atmospheric deposition and the materials used to construct the roof-catchment system may contribute to the chemical contamination of RHRW (De Kwaadsteniet et al. 2013). However, to date the chemical contaminants present in rainwater have not been associated with human disease, while the outbreak of disease has been linked to the utilization of rainwater contaminated with pathogenic microorganisms (Simmons et al. 2008; Franklin et al. 2009). Microbial contaminants may enter the rainwater as it traverses polluted air (Kaushik et al. 2012). Microorganisms present in organic matter and in the feces of animals and birds localized on the roof-catchment and in the gutter system can consequently enter the storage tank during a rain event (Ahmed et al. 2008; Simmons et al. 2008; Rahman and Eslamian 2015). The microbial quality of rainwater is thus routinely monitored by testing for indicator bacteria, such as fecal coliforms, *Escherichia coli* (*E. coli*), and *Enterococcus* spp. (Ahmed et al. 2010). Bacterial and protozoan pathogens and opportunistic pathogens, such as *Pseudomonas* spp., *Legionella* spp., *Klebsiella* spp., *Campylobacter* spp., *Cryptosporidium* spp., and *Giardia* spp. have also been detected in untreated harvested rainwater (Ahmed et al. 2008; Kaushik and Balasubramanian 2012; Dobrowsky et al. 2014).

The presence of both indicator and pathogenic microorganisms in harvested rainwater has resulted in numerous research groups investigating various disinfection systems. These technologies have been implemented in order to reduce microbial contamination in water sources (Figure 19.1) and include treatment that focuses on the prevention of contaminant entry (gutter screens/first-flush diverters), treatment within the storage tank (sedimentation/ultraviolet [UV] treatment), and post-collection treatment (chemical [chlorination] and physical treatments [filtration, solar disinfection, UV treatment, thermal disinfection, solar pasteurization]) (Nolde 2007; Li et al. 2010; Gikas and Tsihrintzis 2012; Moreira Neto et al. 2012; Dobrowsky et al. 2015a, b; Strauss et al. 2016; Shaheed et al. 2017). This chapter will focus on the current primary chemical and physical treatment methods utilized for harvested rainwater and will also review the potential of biological methods as treatment.

Handbook of Water Harvesting and Conservation: Basic Concepts and Fundamentals, First Edition. Edited by Saeid Eslamian and Faezeh Eslamian.

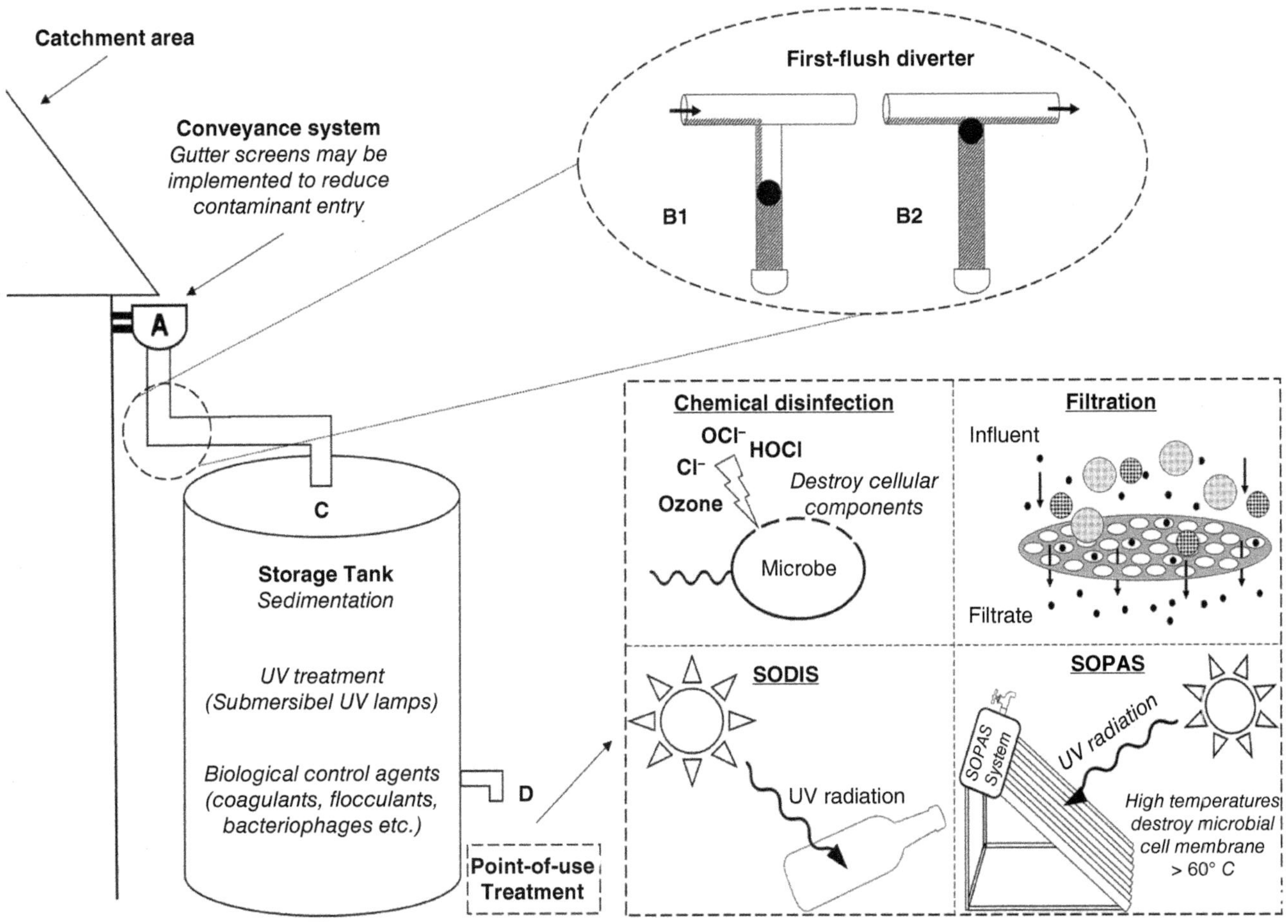

Figure 19.1 Strategies that may be employed to improve the quality of harvested rainwater. (a) Gutter screens may prevent large organic debris from entering the conveyance system, (b) First-flush diverters can be installed as part of the conveyance system and collect the initial roof runoff water considered to contain the most contaminants, (c) Rainwater may be treated inside the storage tank, (d) Various possible point-of-use treatment methods that could be applied after the rainwater has been collected from the storage tank.

19.2 Pretreatment of Harvested Rainwater: Prevention of Debris Entry and Sedimentation

The quality of harvested rainwater may be significantly improved if contaminants present on the roof-catchment area (i.e. decaying organic matter, dust, leaves, bird/rodent feces, etc.) are prevented from entering the storage tank (Ahmed et al. 2010; Okhravi et al. 2015). Various rainwater pretreatment systems, including the implementation of gutter screens and first-flush diverters, have thus been investigated (Mendez et al. 2011; Gikas and Tsihrintzis 2012). While gutter screens mainly prevent large organic debris from entering the conveyance system, first-flush diverter systems redirect the initial roof runoff water, which is considered to have the highest concentration of pollutants (Sánchez et al. 2015). In principle, the first rainwater runoff is collected in the flush pipe (collection system of first-flush diverter) and as the flush pipe fills with contaminated water, a valve (normally a ball valve) automatically blocks access to the flush pipe (Figure 19.1b1). The remaining water flowing through the conveyance system is then diverted into the rainwater tank (Figure 19.1b2).

Sedimentation may occur inside the rainwater harvesting tank, thereby decreasing water turbidity and potentially also improving the microbial and chemical quality of the harvested rainwater (Spinks et al. 2005; Despins et al. 2009). This may be attributed to physicochemical interactions facilitating the aggregation of dissolved organic matter (DOM) with bacteria, viruses, nutrients, metals, and chemicals, etc., which subsequently increases particle size and the efficiency of sedimentation. While assessing the efficiency of a multi-component treatment system (first-flush diverter, sedimentation tank, biological treatment and, UV disinfection), Nolde (2007) noted that there was a "slight decrease" in *E. coli* concentrations as a result of sedimentation. Similarly, Spinks et al. (2005) reported that heterotrophic bacteria were 50–100 times more concentrated in the tank sediment as compared to the tank water, while concentrations of lead were also significantly higher in the tank sediment.

Due to the simplicity of the sedimentation technique, the use of coagulants and flocculants have been assessed to promote the association of microbial pathogens with larger particles and thereby increase the efficiency of sedimentation (Sobsey 2002). The use of chemical coagulants, such as ferric sulfate, have been combined with chlorine to produce commercially available disinfectants (e.g. P&G Purifier of Water™ and WaterMarker™ "Chlor-Floc"), with a combined coagulation/chlorination treatment method resulting in baseline log reductions of 7.0, 2.0–4.5, and 3.0 for the removal of bacteria, viruses, and protozoa, respectively (Sobsey et al. 2008).

The implementation of these pretreatment systems thus generally improves the quality of harvested rainwater (Lee et al. 2010). However, Gikas and Tsihrintzis (2012) reported that the use of first-flush diverters and other pretreatment techniques primarily improve the physicochemical quality of rainwater (e.g. turbidity) and not the microbial quality. Additional treatment may thus be required to reduce particularly the microbial load in harvested rainwater and produce water of a potable standard.

19.3 Chemical Disinfection

Chemical disinfection is widely used during the final stages of water treatment, before drinking water is distributed to consumers, and has been employed in the treatment of harvested rainwater to improve its microbiological quality (Sobsey 2002; Moreira Neto et al. 2012). For the disinfection of drinking water, chlorine and chlorine-based compounds, as well as non-chlorine-based chemicals, such as ozone, are utilized. Chemicals used for the disinfection of drinking water may not only eradicate microorganisms but may also oxidize taste and odor compounds and reduce color through the oxidation of highly colored organic compounds, and thus improve the overall quality of water (Leopold and Freese 2009).

19.3.1 Chlorination

In conventional treatment processes chlorine is added to drinking water as elemental chlorine (chlorine gas), chlorine dioxide, monochloramine/chloramine, sodium hypochlorite (NaOCI) solution (bleach), or dry calcium hypochlorite (Seidel et al. 2011). Chlorine gas reacts with water to form hypochlorite (OCl^-) and hypochlorous acid (HOCl), which are extremely reactive with the cellular components of microorganisms (Figure 19.1). Hypochlorous acid has been shown to affect several cellular processes within microbial cells via oxidation, hydrolysis, and deamination reactions, which result in microbial cell death (LeChevallier and Au 2004). In addition, chlorine has been reported to display destructive effects on the sulfhydryl groups of proteins and convert several α-amino acids into a mixture of corresponding nitriles and aldehydes by oxidation (Venkobachar et al. 1977). A study conducted by Venkobachar et al. (1977) showed that chlorine induced changes in the permeability of the microbial cell membrane, which resulted in the leakage of macromolecules. Moreover, the specific reactions that cause lesions in the microbial cell membrane are dependent on the concentration of the chlorine used and the pH of the treated water (Fukuzaki 2006). Chlorine dioxide has also been reported to exhibit strong oxidizing properties that aids in the inactivation of microorganisms through the direct oxidation of certain amino acids mainly found in the structural regions of metabolic enzymes or membrane proteins (Gates 1998).

Chlorination thus aids in the reduction of odors and the removal of algae, molds, and slime bacteria, which generally thrive in water supply reservoirs (material surfaces in water mains and storage containers). Additionally, chlorine-based compounds facilitate the removal of certain chemical compounds that hinder disinfection and aid in the removal of iron and manganese from untreated water. Notably, only chlorine-based compounds provide residual disinfectant levels that inhibits the regrowth of microorganisms in water distribution systems (Leopold and Freese 2009). For the treatment of harvested rainwater, chlorine-based compounds should preferably be utilized after the rainwater has been extracted from the harvesting tank, as chlorine has the potential to react with organic matter that could settle at the base of the tank and form undesirable by-products (De Kwaadsteniet et al. 2013). These by-products include trihalomethanes, chloroform, chlorite, and chlorate ions, amongst others, which result in adverse health effects in humans when they react with organic matter originating from the natural environment. This implies that drinking water treatment systems must incorporate measures for the control of the formation of these undesirable chemical compounds (Amy et al. 2000). For disinfection using chlorine-based compounds to be deemed effective, 0.4–0.5 $mg\,l^{-1}$ of free chlorine should be detected for at least 15 minutes following the addition of the chemicals to the water (De Kwaadsteniet et al. 2013).

Sodium hypochlorite, which is an active ingredient in commercial laundry bleach solutions, has also been shown to be a safe, effective, and inexpensive chemical disinfectant for point-of-use water treatment (Fukuzaki

2006). Fagerli et al. (2017) compared the effect of a commercial chlorination product (brand name Air RahMat) to traditional boiling practices for the treatment of stored drinking water obtained from various sources (borehole, rainwater, protected, and unprotected wells) in Tangerang, Indonesia. The application of Air RahMat improved the microbial quality of all the stored water types and the authors noted that there were lower incidences of diarrhea reported in households that treated their water with Air RahMat compared to those who merely boiled the water as a treatment method. Molla et al. (2009) then investigated the use of sodium dichloroisocyanurate (NaDCC) tablets as an alternative to NaOCI for the treatment of water (obtained from a community tap or standpipe) in the Lalbagh thana area of Dhaka in Bangladesh. Results from the study showed that 84% (n = 50) of the samples collected from households using NaDCC tablets to treat their water, had lower levels of fecal coliforms with a maximum of 23 colony forming units (CFUs)/100 ml detected. In comparison 1000–2400 CFU/100 ml fecal coliforms were enumerated in the pre-intervention (untreated) source water.

Chemical disinfection can also successfully be employed in combination with a physical method for the removal of total suspended solids (TSS) and microorganisms, such as protozoa. Moreira Neto et al. (2012) investigated the efficiency of slow-sand filtration followed by chlorination for the treatment of stored rainwater. The authors showed that the quality of the treated rainwater improved in terms of pH, turbidity, total hardness, TSS, and the chemical oxygen demand (COD). The authors also indicated that there was a 3.0 and 4.0 log reduction in the *E. coli* and total coliform counts, respectively in the treated rainwater. In contrast, a study conducted by Norton and LeChevallier (2000) showed that certain Gram-positive bacteria (spore-forming), such as *Bacillus* or *Clostridium*, and acid-fast bacteria, such as *Mycobacterium* and *Norcadia,* were resistant to chlorine disinfection due to the presence of extracellular capsules. Overall however, chlorination used for drinking water treatment can easily be upscaled, provide cost-effective and reliable water disinfection for large cities, mid-sized communities, and remote villages alike and thus effectively assists in supplying safe potable water to various communities.

19.3.2 Non-Chlorine Disinfectants

Non-chlorine-based chemicals that have been proposed for water disinfection include halogens, such as iodine and bromine, as well as a variety of metals. Bromine and iodine exhibit a strong oxidant potential, which makes them suitable disinfectants. However, bromine has been reported to readily form undesirable by-products, such as trihalomethanes, in the presence of naturally occurring organic matter. Non-chlorine-based disinfectants have however, been shown to function more optimally when used in combination with chlorine or other compounds such as certain metal ions. For example, a combination of copper (Cu) and silver (Ag) ions successfully inactivated bacteria and viruses in water, with the required inactivation contact time ranging from a few hours up to a few days, depending on the quantity of organic matter in the water source (Thurman et al. 1989). A study conducted by Thurman et al. (1989) reported a 5.0 log reduction in *E. coli* in tap water within 120 seconds after a combined dosage of 0.1 mg l chlorine, 38 μg l Ag ions, and 380 μg l of Cu ions. The majority of these chemicals (Ag and Cu ions) are, however, not applied in routine water treatment as they are costly and readily produce undesirable by-products which may pose significant health risks to the consumer.

Additional examples of non-chlorine disinfectants that could be utilized in water treatment include peracetic acid (PAA) and hydrogen peroxide. PAA has been reported to display powerful oxidation properties due to the production of free hydroxyl radicals (OH^-), which damages the microbial macromolecules including amino acids, carbohydrates, nucleic acids, and lipids. Additionally, when PAA enters the microbial cell, it has been shown to oxidize the essential enzymes, therefore leading to the impairing of vital biochemical pathways and active transport membranes (Kitis 2004). Koivunen and Heinonen-Tanski (2005) conducted a pilot study to determine the efficiency of disinfection by hydrogen peroxide, NaOCI and PAA, using a synthetic wastewater medium spiked with enteric pathogens (*E. coli*, *Enterococcus faecalis* [*E. faecalis*], *Salmonella enteritidis* [*S. enteritidis*], and coliphage MS2). The authors showed that dosages of 3 and 7–15 mg l of PAA resulted in 2–3 and 1–1.5 log reductions in enteric bacteria and coliphage MS2, respectively. Higher concentrations of NaOCI and hydrogen peroxide were required to obtain similar enteric bacterial log reductions. Therefore, PAA was shown to be a preferred alternative to the chlorine-based compounds and hydrogen peroxide for the inactivation of microorganisms and it was suggested that PAA could be used as a point-of-use disinfectant for the treatment of harvested rainwater.

Ozone has been utilized for the disinfection of drinking water in European countries since the early twentieth century and is currently applied as a common disinfectant worldwide (Gray 2014). The principle of ozonation is reliant on the OH^- radicals generated after their disintegration in water which facilitates the oxidization of pollutants (Hoigné 1988; Gardoni et al. 2012). The process of ozonation thus incorporates molecular ozone and OH^-

radicals for the destruction of pollutants, such as microorganisms, in water (Langlais et al. 1991; Gardoni et al. 2012). Moreover, ozone is known to target unsaturated bonds forming aldehydes, ketones, or carbonyl compounds (Langlais et al. 1991), which implies that microorganisms are inactivated by targeting the functional proteins in the cytoplasmic membrane of bacteria, the protein structure of virus capsids, or the nucleic acids of microorganisms. A concentration between 2 and 4 mg l of residual ozone has been reported as the most effective for the disinfection of water (Leopold and Freese 2009).

El Araby et al. (2009) investigated the use of ozone for the removal of iron and manganese from contaminated groundwater intended for potable purposes. The results of the study showed that 3 mg l of ozone improved the removal of ferrous (Fe^{2+}) and manganese (Mn^{2+}) ions by more than 96% and 83%, respectively. The disadvantage of using ozone for disinfection is that it has a short half-life (1–20 minutes in drinking water) (Hoigné 1988) and it is therefore recommended that it must be used in combination with other chemical disinfectants, such as chlorine (Leopold and Freese 2009). The combination of chlorination and ozone treatment may reduce the potential of undesirable by-product formation and is effective against protozoa, including *Giardia* cysts and *Cryptosporidium* oocysts (Leopold and Freese 2009).

The efficiency of chemical disinfection methods for the treatment of water may further be improved by incorporating physical treatment methods as part of a combination treatment approach. For example, filtration methods may be used to reduce the concentration of organic matter within water samples before chemical disinfection is applied.

19.4 Physical Disinfection

The physical disinfection of water sources not only refers to the physical removal of microbial contaminants by filtration systems but also to the application of physical stressors (e.g. heat, UV radiation) that may be detrimental to microorganisms. Examples of physical disinfection methods that have been used for the treatment of harvested rainwater include filtration, solar disinfection (SODIS), UV treatment, boiling, and solar pasteurization (SOPAS).

19.4.1 Filtration Techniques

Various types of filtration systems including microfiltration, ceramic filtration, and gravity membrane filtration have been assessed for the treatment of rainwater where the microbial removal efficiency is based on size exclusion (viruses 20–80 nm; bacteria 0.5–2 μm; protozoa 4–20 μm) (Sobsey et al. 2008; Li et al. 2010). Ceramic filters are generally constructed from clay and sawdust, with the pores of the filter formed during the combustion process when the sawdust and other organic molecules are burnt off. These filters are then able to remove microbial contaminants through size exclusion and/or sorption (Kallman et al. 2011). In comparison, filters with a known pore size are used during microfiltration, where the small pore size (0.1–10 μm) of the filter also allows for the removal of contaminants (Baker 2012). In order to decrease treatment time, these microfiltration systems are generally used in combination with a pump, whereas gravity flow is used to pass contaminated water through a filter when applying gravity-driven membrane filtration (Kus et al. 2013; Ding et al. 2017). Certain filtration systems are then viewed as biological treatment processes (e.g. slow-sand or bio-sand filtration) where the increased surface area of the filter bed enables biofilm formation and the subsequent removal of microorganisms (Section 19.5.1). The availability and affordability of filtration materials will thus predominantly determine which systems are used in developing countries.

While assessing the efficiency of microfiltration systems (combined polyvinyl alcohol nanofiber and activated carbon filtration system) to treat harvested rainwater, Dobrowsky et al. (2015a) reported log reductions of 1.0, 2.40, and 2.30 for the removal of *E. coli*, total coliforms, and heterotrophic bacteria, respectively. In comparison, Ding et al. (2017) reported a 1.30 log reduction in heterotrophic bacteria using gravity-driven membrane filtration, while Sobsey et al. (2008) reported baseline log reduction values of 2.0, 0.5, and 4.0, for the removal of bacteria, viruses, and protozoa, respectively, using ceramic filtration. Although the use of biological filtration systems (such as slow-sand filtration) have been shown to be effective in reducing microbial contamination in harvested rainwater (Section 23.5.1), the obtained log reduction values are generally lower than those obtained by physical filtration systems (Sobsey et al. 2008; Islam et al. 2010; Dobrowsky et al. 2015a). Additionally, the efficiency of slow-sand filtration systems is dependent on filter-bed depth as Islam et al. (2010), reported log reduction values of 0.15, 0.26, and 0.40 for the removal of total coliforms when using filter-beds with a depth of 30, 45, and 60 cm, respectively.

In order to increase the treatment efficiency of filtration systems, the addition of antimicrobial compounds (e.g. Cu or Ag nanoparticles, colloidal Ag) to filters have also been investigated (Kallman et al. 2011). These antimicrobial compounds increase the efficiency of the filtration system by directly inactivating susceptible microorganisms and thereby decrease biofouling of the filters. Contact between the metals and bacterial cell results in the disruption of the bacterial cell envelope, whereafter Cu and Ag ions cause the generation of reactive oxygen species (ROS), which leads to irreversible damage to cellular components (Mathews et al. 2013). Kallman et al. (2011) used ceramic filters coated with Ag nanoparticles to improve the microbiological quality of water at the point-of-use in rural areas of Guatemala. *E. coli* and total coliforms were reduced by 1.10 and 0.87 log, respectively. While assessing the efficiency of cost-effective filter materials (anionic and cationic resins, fiber glass, sand, and zeolite) coated with Ag nanoparticles (0.1 mM $AgNO_3$), Mpenyana-Monyatsi et al. (2012) identified $AgNO_3$ coated cationic resin as a potential alternative cost-effective filter, as a 100% bacterial removal efficiency was obtained.

Although the major drawbacks associated with using filtration systems include prolonged treatment time and the continuous replacement of system components when the filters become saturated (Sobsey 2002; De Kwaadsteniet et al. 2013), the biggest advantage of using these systems is that they may also reduce chemical contamination in harvested rainwater (Kim et al. 2005; Nolde 2007). While investigating the efficiency of slow-sand filtration, Ahammed and Meera (2006) reported that lead and zinc concentrations in RHRW were reduced by 1.0 and 1.56 log, respectively, following filtration treatment, with the rainwater turbidity also reduced between 0.53 and 1.21 log. Moreover, various filtration systems may be combined or be incorporated as part of a multi-treatment system when treating harvested rainwater, thereby increasing treatment efficiency (Areerachakul et al. 2009; Moreira Neto et al. 2012). For example, Areerachakul et al. (2009) combined granular activated carbon (GAC) filtration with microfiltration to reduce heterotrophic bacteria to below the detection limit in filtered rainwater.

19.4.2 SODIS/UV Treatment

In developing countries, rainwater treatment systems that are easy to use, require minimal maintenance, and are constructed from readily available cost-effective materials are required to ensure sustained user compliance (Sobsey et al. 2008; McGuigan et al. 2012; De Kwaadsteniet et al. 2013). One such treatment method is SODIS, which utilizes the synergistic effects of light (UV-A [315–400 nm] and UV-B [280–315 nm]) and solar-mild heat to inactivate microbial contaminants (McGuigan et al. 2012). In its simplest form, polyethylene-terephthalate (PET) bottles are filled with contaminated water and are exposed to sunlight for 6–48 hours, with treatment time being dependent on sunlight intensity and the sensitivity of the contaminating microorganisms to UV radiation and solar-mild heat. UV radiation inactivates microorganisms by damaging deoxyribonucleic acid (DNA) and the cell membrane as a result of the formation of ROS (McGuigan et al. 2012). Additionally, the water temperature will increase as water molecules absorb the UV radiation, with the increase in temperature also contributing to the disinfection process, effectively leading to cell membrane damage (McGuigan et al. 2012). Numerous research groups have subsequently reported on the efficiency of SODIS to reduce bacterial, viral, and protozoan contaminants in rainwater, with baseline log reduction values of 3.0, 2.0, and 1.0, reported, respectively (Sobsey et al. 2008; McGuigan et al. 2012).

However, certain microorganisms may initiate photoreactivation mechanisms (repair of DNA following UV damage) or utilize the formation of survival structures (e.g. Gram-positive endospore formers or protozoan cysts) and display slower SODIS inactivation rates (Sinton et al. 2002; Boyle et al. 2008). As a result, it is recommended that SODIS treated water be used within 24 hours to avoid post-treatment regrowth (McGuigan et al. 2012).

Due to the simplicity of the SODIS treatment mechanism, various research groups have reported on the use of SODIS-enhancement-technologies to further improve the efficiency of the SODIS treatment technique and thereby negate the influence of poor weather conditions (decreased UV radiation) or increased water turbidity (may shield microorganisms from UV radiation) (McGuigan et al. 2012). These SODIS-enhancement-technologies include the use of flow reactors and larger reactor tubes (increases the treatment volume) (Ubomba-Jaswa et al. 2010), solar mirrors (concentrates UV) (Amin and Han 2009), the addition of heterogeneous photocatalysts (e.g. iron oxide [Fe_2O_3], titanium dioxide [TiO_2], and zinc oxide [ZnO]), which increase the production of ROS (Herrera Melian et al. 2000; Ibáñez et al. 2003), and chemical additives (e.g. sodium percarbonate, riboflavin) which have antimicrobial properties (Heaselgrave and Kilvington 2010; Fisher et al. 2012) (Figure 19.2).

The small volume (2 l) of water that can be treated with conventional SODIS is one of the major drawbacks associated with the technique and larger reactor tubes have subsequently been investigated, with non-colored glass (e.g. borosilicate glass) being preferred as a result of its higher UV transmittance capabilities (McGuigan et al. 2012). Amin and Han (2009) then reported log reductions of 0.51–2.62 and 0.46–1.51 for the removal of *E. coli* and total coliforms, respectively, when investigating the efficiency of SODIS using a solar collector (solar mirrors). In comparison, Strauss et al. (2016) demonstrated that SODIS using a solar cooker effectively reduced *E. coli* by >2 logs and heterotrophic bacteria by >6 logs. However, intact *Legionella* spp. (>1 log reduction) and *Pseudomonas* spp. (<1 log reduction) were still detected using the

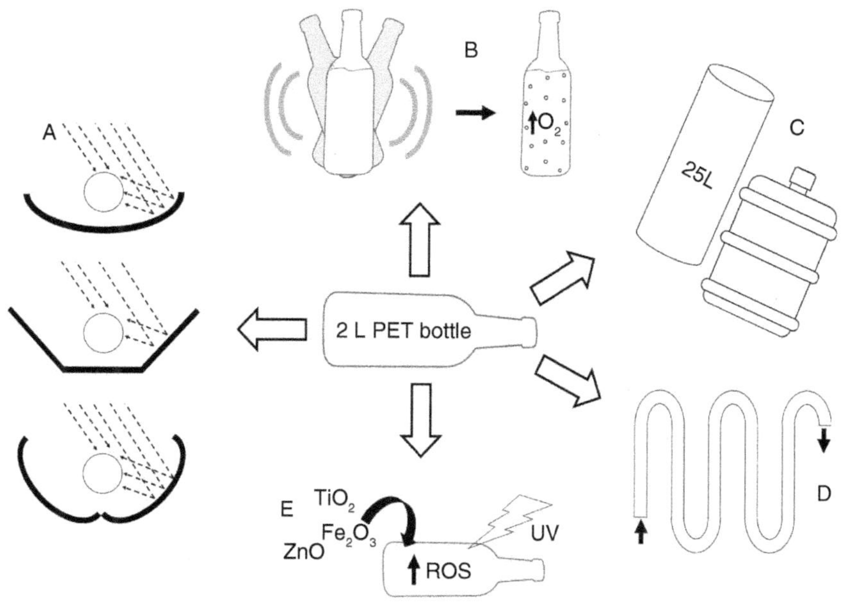

Figure 19.2 SODIS enhancement technologies. (a) Reflective surfaces may be used to concentrate UV radiation on the reactor tube. (b) Aeration of the water will increase the concentration of dissolved oxygen, which may lead to the generation of ROS. (c) Large reactor tubes and bottles and (d) flow reactors may be used to increase treatment volume. (e) Heterogeneous photocatalysts may be used to increase the generation of ROS.

ethidium monoazide bromide quantitative polymerase chain reaction (EMA-qPCR) technique following SODIS treatment for eight hours.

Photocatalysts induce a series of electron transfer reactions when irradiated, leading to the formation of various ROS. Numerous research groups have subsequently reported on the efficiency of photocatalysts, such as TiO_2, to inactivate total coliforms, spore-forming bacteria, fungi, and viruses (Herrera Melian et al. 2000; Ibáñez et al. 2003; McGuigan et al. 2012). The addition of chemical additives (e.g. sodium percarbonate, riboflavin) have also been investigated as an enhancement to SODIS, with Fisher et al. (2012) reporting on the accelerated inactivation of *E. coli*, *Enterococcus* spp., and MS2 coliphages, when combinations of sodium percarbonate, citric acid, and Cu plus ascorbate were added to SODIS systems.

As UV radiation has proven to be effective at reducing microbial contamination in water, the use of artificial UV radiation (generated using UV lamps) has also been investigated, as the optimal lethal UV dose may be more easily obtained, thereby negating the variability associated with using natural sunlight (Kim et al. 2005; Nolde 2007; Jordan et al. 2008). Moreover, the use of artificial UV treatment may readily be combined with other treatment mechanisms in a rainwater harvesting system, as submersible UV lamps may be installed inside the rainwater harvesting tank. While investigating substrate filtration combined with UV treatment, Nolde (2007) reported that *E. coli* were reduced by 2.16 log when the combination treatment was used, while substrate filtration only resulted in a 0.90 log reduction. Similarly, Jordan et al. (2008) assessed filtration combined with UV disinfection (22 W UV lamp) for the treatment of harvested rainwater, with results indicating that *E. coli*, total coliforms, heterotrophic bacteria, and MS2 coliphages, were reduced by 6.0, 2.0, 1.82, and 5.0 log, respectively.

19.4.3 Thermal Disinfection

The thermal disinfection or pasteurization of contaminated water has long been recognized as a safe water treatment method, with the removal of pathogens being independent of turbidity, pH, and additional parameters that may influence other proposed water treatment systems (Burch and Thomas 1998; Helmreich and Horn 2009). Numerous studies have subsequently reported on the inactivation of opportunistic pathogens commonly associated with rainwater, including *Enterococcus* spp., *Escherichia* spp., *Aeromonas* spp., *Pseudomonas* spp., and *Klebsiella* spp. at temperatures between 55 and 65 °C (Spinks et al. 2006; Despins et al. 2009), with the time required to pasteurize water decreasing with increasing treatment temperatures (Feachem et al. 1983). For example, treatment time decreases approximately 10-fold for every 10 °C temperature increase above 50 °C (Feachem et al. 1983; Burch and Thomas 1998).

Traditionally, the process of boiling was used in rural communities to treat contaminated water. However, the high costs associated with the utilization of firewood or other fuels (e.g. paraffin) and the potential detrimental effect on the environment (deforestation) are the major drawbacks associated with using this water treatment method (Sobsey 2002; Islam and Johnston 2006). In order to overcome these drawbacks, it was proposed that affordable and sustainable sources of fuel be identified within a region for households to use, or that other innovative ways of utilizing thermal disinfection be identified.

The Chulli Water-Treatment System, as described by Islam and Johnston (2006) is one such innovative method, where a hollow aluminum coil was built inside traditionally used clay ovens (Chullis) in Bangladesh. During conventional cooking processes, contaminated water is passed through the coil, which is heated by the increased temperatures within the clay ovens (70 °C effluent water). Field trials subsequently indicated that the treatment system was able to inactivate thermotolerant coliforms and produce 90 l of treated water per day without any additional time or fuel requirements by the households.

Solar pasteurization systems then utilize solar energy to treat rainwater at temperatures >60 °C (Abraham et al. 2015). Various types of SOPAS systems have been designed, which enable either the direct (e.g. combined solar collector storage tank system) or indirect (e.g. secondary heat transfer) heating of water (Nieuwoudt and Mathews 2005). While investigating the efficiency of SOPAS systems to treat RHRW, Dobrowsky et al. (2015b) and Reyneke et al. (2018) reported that pasteurization temperatures above 71 and 66 °C, respectively, were sufficient to reduce indicator organism numbers (e.g. *E. coli*, total coliforms, heterotrophic bacteria) to below the detection limit (<1 CFU/100 ml). However, using the EMA-qPCR technique, Reyneke et al. (2016) and Strauss et al. (2016) showed that intact *Legionella* spp. (1.4×10^4 gene copies/mL DNA) and *Pseudomonas* spp. (7.31×10^4 gene copies/mL DNA) were still detected at temperatures greater than 90 °C after SOPAS treatment. The survival of these opportunistic pathogens was attributed to potential biofilm formation (Murga et al. 2001) within the SOPAS systems and the ability of *Legionella* and *Pseudomonas* spp. to live as intracellular parasites of protozoa (Thomas et al. 2010).

As chemical contaminants in rainwater also pose a potential human health risk, it is important to note that SOPAS does not improve the chemical quality of the treated water (Islam and Johnston 2006). Additionally, it has been shown that metal components used within SOPAS systems may leach into the treated water as a result of the increased temperatures (Dobrowsky et al. 2015b; Reyneke et al. 2016). It is therefore advised to combine SOPAS with a filtration system capable of removing metal contaminants in water, which may then also serve as an additional barrier for microbial contaminants that could have survived SOPAS treatment (e.g. protozoan cysts, bacterial endospore formers).

19.5 Biological Treatment

Biological water treatment refers to the removal of contaminants from water sources by mainly employing microorganisms or compounds naturally produced by microorganisms. This is accomplished by the biodegradation of substances (such as micro-pollutants, organic matter, or ammonium) present in contaminated water sources, by microorganisms (Rittmann et al. 1989). Although a number of biological treatments have been utilized for water purification, few studies have focused on the application of these treatment methods for harvested rainwater.

19.5.1 Slow-Sand and Granular Activated Carbon Filters

The best-studied methods employed for the treatment of harvested rainwater are biological-physical combination treatment systems, such as slow-sand filtration (also referred to as bio-sand filters). In these systems, the sand acts as an initial barrier which traps particles present in the water. A biofilm or "schmutzdecke" is then allowed to form on the top layer of the sand, which aids in the removal of microbial and chemical contaminants from the water source (WHO 2004; Helmreich and Horn 2009). Indigenous microorganisms are most commonly utilized to form the biofilms on slow-sand filters as this results in a mixed microbial community (which may consist of various bacteria, yeast, fungi, and protozoa), which is able to reduce microbiological contamination in water, through strategies such as predation (Clark et al. 2012). In addition, particles or pollutants are trapped in the biofilm matrix and organic pollutants may be metabolized by the microorganisms in the biofilm community (WHO 2004). Established and well-maintained slow-sand filters may reduce microorganism levels in the treated water by 90–99% and depending on the degree of contamination, the system may remain functional for a few weeks to a couple of months (WHO 2004; Helmreich and Horn 2009). However, once the accumulation of debris and biological growth on the top layer of the sand filter increases, the head loss (resistance of flow) and the filter itself no longer operates efficiently. Subsequently, the top layer of the filter needs to be removed and the biofilm should be re-established. Despite the maintenance required for slow-sand filtration systems, this treatment technique is widely used and still regarded as an easy and cost-effective rainwater treatment method (De Kwaadsteniet et al. 2013). Shaheed et al. (2017)

investigated the use of an activated carbon slow-sand combination filter system to treat harvested rainwater and lake water. The combination filter efficiently reduced the COD by 85–100% in the lake water samples, while the COD for the harvested rainwater samples was below the limit of detection before and after filtration. The TSS for the harvested rainwater and lake water samples was also significantly reduced (90–100% reduction) (Shaheed et al. 2017). In addition, the water samples were screened for *E. coli* using culture-based methods and it was observed that the concentration of *E. coli* in the harvested rainwater and lake water was significantly reduced after filtration, with the removal efficiency of the filter ranging from 92% to 100% for both water sample types (Shaheed et al. 2017).

Similarly, GAC filters in part rely on the formation of a biofilm in the filter matrix, which reduces dissolved organic carbon (DOC) and may reduce microbial concentrations in treated water. GAC filters consist of raw organic material or coal, which contains high concentrations of carbon. Heat is then used to activate (increase) the surface area of the carbon. The filter effectively removes dissolved chemicals from the water by adsorption (Minnesota Department of Health 2013). In addition, the formation of biofilms on the organic matter in the filters may further increase the adsorption capacity of the filters (Servais et al. 1994). Kus et al. (2013) investigated the effect of GAC filters for the treatment of harvested rainwater prior to the application of membrane filtration. The GAC filters significantly reduced the turbidity of the harvested rainwater by 78% to below 1 Nephelometric Turbidity Unit (NTU), which is the recommended limit for water turbidity as outlined in the Australian Drinking Water Guidelines (ADWG) (National Health and Medical Research Council [NHMRC] and National Resource Management Ministerial Council [NRMMC] 2004; Kus et al. 2013). In addition, the GAC filters reduced the DOC in the rainwater by up to 99% over the course of the study. This finding is of great importance as high concentrations of DOC may cause biofouling of membrane filters. It was concluded that GAC filters should be used in combination with membrane filtration systems to treat harvested rainwater (Kus et al. 2013).

19.5.2 Coagulation and Bioflocculants

Alternative biological treatment methods which have been investigated for water purification and which could be applied to treat harvested rainwater, include the use of bioflocculants and coagulants, bacteriophages, and proteins produced by bacteriophages. Coagulation is a common method utilized to reduce particle number and organic matter during water treatment, with aluminum and iron salts being the most widely used coagulants (Aljuboori et al. 2013). However, due to the potential human health risks associated with many coagulants, such as aluminum- (aluminum sulfate and aluminum chloride) and iron- (ferric sulfate and ferric chloride) based coagulants currently used in the water industry, research has shifted to the investigation of bioflocculants as natural and biodegradable alternatives (Deng et al. 2002; Aljuboori et al. 2013). Bioflocculants are metabolic products produced by various microorganisms and may consist of glycoproteins, proteins, and polysaccharides (Ma et al. 2008; Aljuboori et al. 2013). These compounds have been investigated for the treatment of wastewater to reduce turbidity, remove humic acids, and to separate oil in oil–water emulsions (Deng et al. 2002; Ma et al. 2008). Ma et al. (2008) investigated the ability of a bioflocculant produced by *Bacillus* sp. F6 to reduce turbidity in kaolin clay solutions and raw water supplies. The bioflocculant reduced turbidity by an average of 85% in kaolin clay solutions (concentrations ranging from 6 to 20 mg l). In addition, the turbidity in raw water was reduced by 94.2% when the bioflocculant was used in combination with a commercial coagulant, iron sulfate hydrate [$Fe_2(SO_4)_3$].

Bioflocculants have also been shown to precipitate microorganisms from culture media and water sources (Nie et al. 2011; Zhao et al. 2013). For example, Nie et al. (2011) observed that a bioflocculant produced by *Klebsiella pneumoniae* (*K. pneumoniae*) strain NY1 not only reduced the TSS in raw wastewater by 72%, but precipitated 54% of cyanobacteria in culture media. In addition, Zhao et al. (2013) spiked 1 l of water with 1×10^6 *Acanthamoeba* cysts and treated these spiked water samples with a bioflocculant produced by *K. pneumoniae*. It was found that the bioflocculant precipitated up to 84% of the *Acanthamoeba* cysts after one hour of treatment (Zhao et al. 2013). Bioflocculants can thus be utilized as environmentally friendly alternative water treatment methods, which could be employed to treat wastewater, produce drinking water, and could potentially be used for on-site rainwater treatment.

Similar to the principle of bioflocculation, *Moringa oleifera* (*M. oleifera*) seeds contain low molecular weight cationic proteins, which exhibit coagulation characteristics (Keogh et al. 2017). When added to turbid water with dissolved and suspended particles, the proteins contained in the *Moringa* seeds precipitate the particles and subsequently decreases the water turbidity (Keogh et al. 2017). In addition, the use of *Moringa* seeds has minimal effects on the pH and the conductivity of treated water (Keogh et al. 2017). It has also been shown that extracts of *Moringa* seeds exhibit antimicrobial activity against a number of pathogenic microorganisms. Keogh et al. (2017) evaluated the efficiency of *M. oleifera* seeds as a pretreatment for SODIS. Results indicated that powdered *Moringa* seeds were more effective in reducing water turbidity in comparison to extracts produced from *Moringa* seeds. In addition, a 24-hour *Moringa* pretreatment resulted in a 2.1 log reduction in viable *E. coli* before SODIS treatment, which subsequently resulted in a 6 log reduction after a six-hour SODIS treatment (Keogh et al. 2017). This treatment strategy could thus easily be applied to disinfect harvested rainwater as SODIS is recognized as one of the primary methods to treat rainwater in developing countries.

19.5.3 Bacteriophages and Bacteriophage Proteins

Bacteriophages are viruses that infect prokaryotes and may be applied in bioremediation, specifically for the removal of bacteria from water (Whitey et al. 2005). They exert their antimicrobial effect by infecting bacteria and lysing the cells (Wu et al. 2017). In water research, bacteriophages have predominantly been used as indicators of water quality or for the monitoring of water treatment system efficiencies and have only recently been proposed as biological agents for water treatment (Wu et al. 2017). In particular, bacteriophages may be used to prevent membrane fouling during water and wastewater treatment by infecting and lysing bacteria that may form biofilms on membrane filters. Alternatively, they can produce enzymes which prevent the production of or degrade the polymeric matrix (Wu et al. 2017). Goldman et al. (2009) spiked water with *Pseudomonas aeruginosa* (*P. aeruginosa*), *Acinetobacter johnsonii* (*A. johnsonnii*), and *Bacillus subtilis* (*B. subtilis*), filtered the water through a bench-scale ultrafiltration system, and studied the effect of bacteriophages on membrane fouling in the system. In comparison to the water samples which were not treated with bacteriophages, membrane fouling was reduced by 40 to 60% in the bacteriophage-treated samples. Zhang et al. (2013) investigated whether bacteriophages could selectively remove *P. aeruginosa* from two biological filter systems (GAC and anthracite filters). The bacteriophages reduced the *P. aeruginosa* concentrations (monitored with the use of quantitative polymerase chain reaction [qPCR]) by 70% and 56% in the biofilm formed in the GAC and anthracite filter matrices, respectively, while no *Pseudomonas* could be detected in the effluent collected from the filters. It was concluded that the bacteriophages were able to reduce specific bacteria on biological filters with no adverse effect on the beneficial bacterial community established, subsequently improving effluent quality.

Bacteria are, however, able to rapidly develop resistance to bacteriophages and therefore bacteriophage-based enzymatic (protein) antibiotics have been proposed as alternatives in water treatment. These protein antibiotics are produced by bacteriophages and may rapidly lyse cells or interrupt the formation of the cell wall in susceptible bacteria (Wu et al. 2017). Well-known protein antibiotics include lysins (peptidoglycan hydrolases), which break down the bacterial cell wall by targeting the bonds in the peptidoglycan layer, and holins (hydrophobic proteins), which form pores in the cytoplasmic membrane (Wu et al. 2017). Theoretically, these protein antibiotics could be incorporated into membrane filters in water treatment systems as these compounds are able to reduce dense bacterial communities at low doses (Wu et al. 2017). These protein antibiotic compounds have also been shown to be effective in dispersing and removing biofilms and may therefore be utilized as anti-biofouling agents in water treatment systems (Wu et al. 2017). Additionally, bacteriophages could be applied as a pretreatment for SODIS or SOPAS for rainwater treatment, in order to reduce the initial bacterial load.

It should however be noted that many of these biological treatment methods have not been applied to specifically treat harvested rainwater. Future studies could thus focus on investigating the efficacy of these treatment strategies in reducing contaminants in stored rainwater.

19.6 Conclusion

Research has indicated that harvested rainwater is not pure and may contain organic and inorganic contaminants as well as pathogenic and opportunistic pathogenic microorganisms. Pretreatment or contamination prevention strategies (such as implementing gutter screens or first-flush diverters) may be utilized to reduce or divert the initial contamination load. They are constructed from readily available materials and are considered robust, easy to install and maintain, and are inexpensive add-ons to rainwater harvesting systems (Helmreich and Horn 2009). However, while gutter screens and first-flush diverter systems may improve the physicochemical quality of RHRW, post-harvesting treatment would be required to produce water of a potable standard.

Chlorine and chlorine-based compounds offer a cost-effective and efficient chemical-based method to disinfect rainwater, however caution should be exercised as the addition of chlorine to rainwater high in organic matter may lead to the formation of hazardous by-products, which may pose a health risk to the consumer. Physical treatment systems such as membrane filters, SODIS, and SOPAS are also considered effective treatment strategies, however viable microorganisms have been detected in harvested rainwater post-treatment. Quantitative microbial risk assessment (QMRA) may consequently be employed to estimate the human health risks associated with utilizing water sources (including rainwater) where pathogens or opportunistic pathogens have been detected. This is achieved by considering the intended use of the water source, the ingestion or inhalation volumes, and the concentration of the pathogenic or opportunistic pathogenic microorganisms present in the treated water. Finally, biological treatment methods, such as slow-sand filters, have effectively been employed to treat harvested rainwater, and while the maintenance required to ensure the efficiency of these filters may be time-consuming, these systems can be constructed from low-cost, readily available materials, rendering it an attractive treatment strategy. While extensive optimization and analysis will be required, alternative biological treatments, such as the application of bacteriophages or bioflocculants, should also be investigated to effectively reduce microbial contamination in harvested rainwater.

In conclusion, the provision of a continuous and safe water source is essential to human health and well-being. While harvested rainwater could serve as a natural resource to supplement rapidly depleting fresh- and surface water resources, targeted research should be conducted on treatment strategies, which increase the disinfection efficiency and produce rainwater that could be utilized for potable and domestic purposes. Where required, these treatment systems could combine the physical, chemical, and biological treatment strategies outlined in this chapter.

References

Abraham, J.P., Plourde, B.D., and Minkowycz, W.J. (2015). Continuous flow solar thermal pasteurization of drinking water: methods, devices, microbiology, and analysis. *Renewable Energy* 81: 795–803.

Ahammed, M.M. and Meera, V. (2006). Iron hydroxide-coated sand filter for household drinking water from roof-harvested rainwater. *Journal of Water Supply: Research and Technology--AQUA* 55 (7–8): 493–498.

Ahmed, W., Huygens, F., Goonetilleke, A., and Gardner, T. (2008). Real-time PCR detection of pathogenic microorganisms in roof-harvested rainwater in Southeast Queensland, Australia. *Applied and Environmental Microbiology* 74 (17): 5490–5496.

Ahmed, W., Goonetilleke, A., and Gardner, T. (2010). Implications of faecal indicator bacteria for the microbiological assessment of roof-harvested rainwater quality in Southeast Queensland, Australia. *Canadian Journal of Microbiology* 56 (6): 471–479.

Aljuboori, A.H.R., Idris, A., Abdullah, N., and Mohamad, R. (2013). Production and characterization of a bioflocculant produced by *Aspergillus flavus*. *Bioresource Technology* 127: 489–493.

Amin, M.T. and Han, M.Y. (2009). Roof-harvested rainwater for potable purposes: application of solar disinfection (SODIS) and limitations. *Water Science and Technology* 60 (2): 419–431.

Amy, G., Bull, R., Craun, G.F., Pegram, R.A., Siddiqui, M., and World Health Organization. (2000) Disinfectants and disinfectant by-products. WHO, Geneva, Switzerland, ISBN 92 4 157216 7.

Areerachakul, N., Kitiphatmontree, M., Kandasamy, J. et al. (2009). Submerged membrane system with biofilter as a treatment to rainwater. *Water, Air, and Soil Pollution* 9 (5-6): 431–438.

Baker, R. (2012). Microfiltration. In: *Membrane Technology and Applications*, 3e, 303. San Francisco: Wiley.

Boyle, M.A., Sichel, C., Fernández-Ibánez, P. et al. (2008). Identifying the bactericidal limits of solar disinfection (SODIS) of water under real sunlight conditions. *Applied and Environmental Microbiology* 74 (10): 2997–3001.

Burch, J.D. and Thomas, K.E. (1998). Water disinfection for developing countries and potential for solar thermal pasteurization. *Solar Energy* 64 (1): 87–97.

Campisano, A., Butler, D., Ward, S. et al. (2017). Urban rainwater harvesting systems: research, implementation and future perspectives. *Water Research* 115: 195–209.

Clark, P.A., Pinedo, C.A., Fadus, M., and Capuzzi, S. (2012). Slow-sand water filter: design, implementation, accessibility and sustainability in developing countries. *Medical Science Monitor* 18 (7): RA105–RA117.

De Kwaadsteniet, M., Dobrowsky, P.H., Van Deventer, A. et al. (2013). Domestic rainwater harvesting: microbial and chemical water quality and point-of-use treatment systems. *Water, Air, and Soil Pollution* 224 (7): 1629.

Deng, S.B., Bai, R.B., Hu, X.M., and Luo, Q. (2002). Characteristics of a bioflocculant produced by *Bacillus mucilaginosus* and its use in starch wastewater treatment. *Applied Microbiology and Biotechnology* 60: 588–593.

Despins, C., Farahbakhsh, K., and Leidl, C. (2009). Assessment of rainwater quality from rainwater harvesting systems in Ontario, Canada. *Journal of Water Supply: Research and Technology--AQUA* 58 (2): 117–134.

Ding, A., Wang, J., Lin, D. et al. (2017). A low pressure gravity-driven membrane filtration (GDM) system for rainwater recycling: flux stabilization and removal performance. *Chemosphere* 172: 21–28.

Dobrowsky, P.H., De Kwaadsteniet, M., Cloete, T.E., and Khan, W. (2014). Distribution of indigenous bacterial pathogens and potential pathogens associated with roof-harvested rainwater. *Applied and Environmental Microbiology* 80 (7): 2307–2316.

Dobrowsky, P.H., Lombard, M., Cloete, W.J. et al. (2015a). Efficiency of microfiltration systems for the removal of bacterial and viral contaminants from surface and rainwater. *Water, Air, and Soil Pollution* 226 (3): 33.

Dobrowsky, P.H., Carstens, M., De Villiers, J. et al. (2015b). Efficiency of a closed-coupled solar pasteurization system in treating roof harvested rainwater. *Science of the Total Environment* 536: 206–214.

El Araby, R., Hawash, S., and El Diwani, G. (2009). Treatment of iron and manganese in simulated groundwater via ozone technology. *Desalination* 249 (3): 1345–1349.

Fagerli, K., Trivedi, K.K., Sodha, S.V. et al. (2017). Comparison of boiling and chlorination on the quality of stored drinking water and childhood diarrhoea in Indonesian households. *Epidemiology and Infection* 145 (15): 3294–3302.

Feachem, R.E., Bradley, D.J., Garelick, H., and Mara, D.D. (1983). *Sanitation and Disease: Health Aspects of Excreta and Wastewater Management*. New York: Wiley.

Fisher, M.B., Iriarte, M., and Nelson, K.L. (2012). Solar water disinfection (SODIS) of *Escherichia coli*, *Enterococcus* spp., and MS2 coliphage: effects of additives and alternative container materials. *Water Research* 46 (6): 1745–1754.

Franklin, L.J., Fielding, J.E., Gregory, J. et al. (2009). An outbreak of *Salmonella Typhimurium* at a school camp linked to contamination of rainwater tanks. *Epidemiology and Infection* 137 (3): 434–440.

Fukuzaki, S. (2006). Mechanisms of actions of sodium hypochlorite in cleaning and disinfection processes. *Biocontrol Science* 11 (4): 147–157.

Gardoni, D., Vailati, A., and Canziani, R. (2012). Decay of ozone in water: a review, ozone: science and engineering. *The Journal of the International Ozone Association* 34 (4): 233–242.

Gates, D. (1998). *The Chlorine Dioxide Handbook: Water Disinfection Series*. Denver: American Water Works Association USA.

Gikas, G.D. and Tsihrintzis, V.A. (2012). Assessment of water quality of first-flush roof runoff and harvested rainwater. *Journal of Hydrology*: 466, 115–467, 126.

Goldman, G., Starosvetsky, J., and Armon, R. (2009). Inhibition of biofilm formation on UF membrane by use of specific bacteriophages. *Journal of Membrane Science* 342 (12): 145–152.

Gray, N.F. (2014). Ozone disinfection. In: *Microbiology of Waterborne Diseases*, 2e), Ch. 33 (eds. S.L. By Percival, M.V. Yates, D.W. Williams, et al.), 599–615. London: Elsevier Academic Press.

Heaselgrave, W. and Kilvington, S. (2010). Antimicrobial activity of simulated solar disinfection against bacterial, fungal, and protozoan pathogens and its enhancement by riboflavin. *Applied and Environmental Microbiology* 76 (17): 6010–6012.

Helmreich, B. and Horn, H. (2009). Opportunities in rainwater harvesting. *Desalination* 248 (1): 118–124.

Herrera Melian, J.A., Dona Rodriguez, J.M., Viera Suarez, A. et al. (2000). The photocatalytic disinfection of urban waste waters. *Chemosphere* 41 (3): 323–327.

Hoigné, J. (1988). The chemistry of ozone in water. In: *Process Technologies for Water Treatment*, Ch. 6 (ed. S. Stucki), 121–141. New York: Plenum Press.

Ibáñez, A.J., Litter, I.M., and Pizarro, A.R. (2003). Photocatalytic bactericidal effect of TiO2 on *Enterobacter cloacae* comparative study with other gram negative bacteria. *Journal of Photochemistry and Photobiology A: Chemistry* 157 (1): 81–85.

Islam, M.F. and Johnston, R.B. (2006). Household pasteurization of drinking-water: the Chulli water-treatment system. *Journal of Health, Population and Nutrition* 24 (3): 356–362.

Islam, M.M., Chou, F.N.F., Kabir, M.R., and Liaw, C.H. (2010). Rainwater: a potential alternative source for scarce safe drinking and arsenic contaminated water in Bangladesh. *Water Resources Management* 24 (14): 3987–4008.

Jordan, F.L., Seaman, R., Riley, J.J., and Yoklic, R.M. (2008). Effective removal of microbial contamination from harvested rainwater using a simple point of use filtration and UV-disinfection device. *Urban Water Journal* 5 (3): 209–218.

Kallman, E.N., Oyanedel-Craver, V.A., and Smith, J.A. (2011). Ceramic filters impregnated with silver nanoparticles for point-of-use water treatment in rural Guatemala. *Journal of Environmental Engineering* 137 (6): 407–415.

Kaushik, R. and Balasubramanian, R. (2012). Assessment of bacterial pathogens in fresh rainwater and airborne particulate matter using real-time PCR. *Atmospheric Environment* 46: 131–139.

Kaushik, R., Balasubramanian, R., and Armah, A. (2012). Influence of air quality on the composition of microbial pathogens in fresh rainwater. *Applied and Environmental Microbiology* 78 (8): 2813–2818.

Keogh, M.B., Elmusharaf, K., Borde, P., and McGuigan, K.G. (2017). Evaluation of the natural coagulant *Moringa oleifera* as a pretreatment for SODIS in contaminated turbid water. *Solar Energy* 158: 448–454.

Kim, R., Lee, S., and Kim, J. (2005). Application of a metal membrane for rainwater utilization: filtration characteristics and membrane fouling. *Desalination* 177 (1-3): 121–132.

Kitis, M. (2004). Disinfection of wastewater with peracetic acid: a review. *Environment International* 30 (1): 47–55.

Koivunen, J. and Heinonen-Tanski, H. (2005). Inactivation of enteric microorganisms with chemical disinfectants, UV irradiation and combined chemical/UV treatments. *Water Research* 39 (8): 1519–1526.

Kus, B., Handasamy, J., Vigneswaran, S. et al. (2013). Gravity driven membrane filtration system to improve the water quality in rainwater tanks. *Water Science and Technology: Water Supply* 13 (2): 479–485.

Langlais, B., Reckhow, D.A., and Brink, D.R. (1991). *Ozone in Water Treatment, Applications and Engineering*. Chelsea, MI: Lewis Publishers, Inc.

LeChevallier, M.W. and Au, K.K. (2004). *Water Treatment and Pathogen Control: Process Efficiency in Achieving Safe Drinking Water*. Geneva: World Health Organization and IWA.

Lee, J.Y., Yang, J.S., Han, M., and Choi, J. (2010). Comparison of the microbiological and chemical characterization of harvested rainwater and reservoir water as alternative water resources. *Science of the Total Environment* 408 (4): 896–905.

Leopold, P., and Freese, S. (2009) *A simple guide to the chemistry, selection and use of chemicals for water and wastewater treatment*. WRC Report No. TT405\09. Water Research Commission, Pretoria, South Africa.

Li, Z., Boyle, F., and Reynolds, A. (2010). Rainwater harvesting and greywater treatment systems for domestic application in Ireland. *Desalination* 260 (1): 1–8.

Ma, F., Zheng, L.N., and Chi, Y. (2008). Applications of biological flocculants (BFs) for coagulation treatment in water purification: turbidity elimination. *Chemical and Biochemical Engineering Quarterly* 22 (3): 321–326.

Mathews, S., Hans, M., Mücklich, F., and Solioz, M. (2013). Contact killing of bacteria on copper is suppressed if bacterial-metal contact is prevented and is induced on iron by copper ions. *Applied and Environmental Microbiology* 79 (8): 2605–2611.

McGuigan, K.G., Conroy, R.M., Mosler, H. et al. (2012). Solar water disinfection (SODIS): a review from bench-top to roof-top. *Journal of Hazardous Materials*: **235**, 29–236, 46.

Mendez, C.B., Klenzendorf, J.B., Afshar, B.R. et al. (2011). The effect of roofing material on the quality of harvested rainwater. *Water Research* 45 (5): 2049–2059.

Minnesota Department of Health. (2013) Water treatment using carbon filters (GAC). Minnesota Department of Health, St. Paul, Minnesota, United States of America. Available: http://www.health.state.mn.us/divs/eh/hazardous/topics/gac.html [Accessed: 12 February 2018].

Molla, N.A., Mollah, K.A., Hossain, A. et al. (2009). Drinking water treatment at home: a trial in Dhaka, Bangladesh. *Bangladesh Journal of Scientific and Industrial Research* 44 (2): 233–240.

Moreira Neto, R.F., Calijuri, M.L., de Castro Carvalho, I., and da Fonseca Santiago, A. (2012). Rainwater treatment in airports using slow sand filtration followed by chlorination: efficiency and costs. *Resources, Conservation and Recycling* 65: 124–129.

Mpenyana-Monyatsi, L., Mthombeni, N.H., Onyango, M.S., and Momba, M.N.B. (2012). Cost-effective filter materials coated with silver nanoparticles for the removal of pathogenic bacteria in groundwater. *International Journal of Environmental Research and Public Health* 9 (1): 244–271.

Murga, R., Forster, T.S., Brown, E. et al. (2001). Role of biofilms in the survival of *Legionella pneumophila* in a model potable-water system. *Microbiology* 147 (11): 3121–3126.

National Health and Medical Research Council (NHMRC) Australia, National Resource Management Ministerial Council (NRMMC) Australia. (2004) Australian drinking water guidelines (ADWG) Paper 6 National Water Quality Management Strategy. National Health and Medical Research Council, National Resource Management Ministerial Council Commonwealth of Australia, Canberra Australia.

Nie, M., Yin, X., Wang, Y. et al. (2011). Production of a novel bioflocculant MNXY1 by *Klebsiella pneumoniae* strain NY1 and application in precipitation of cyanobacteria and municipal wastewater treatment. *Journal of Applied Microbiology* 111: 547–558.

Nieuwoudt, M.N. and Mathews, E.H. (2005). A mobile solar water heater for rural housing in southern Africa. *Building and Environment* 40 (9): 1217–1234.

Nolde, E. (2007). Possibilities of rainwater utilisation in densely populated areas including precipitation runoffs from traffic surfaces. *Desalination* 215 (1-3): 1–11.

Norton, C.D. and LeChevallier, M.W. (2000). A pilot study of bacteriological population changes through potable water treatment and distribution. *Applied and Environmental Microbiology* 66 (1): 268–276.

Okhravi, S., Eslamian, S., and Adamowski, J. (2015). Water reuse in rainwater harvesting. In: *Urban Water Reuse Handbook*, Ch. 59 (ed. S. Eslamian), 779–796. Boca Raton: CRC Press, Taylor and Francis Group USA.

Rahman, A. and Eslamian, S. (2015). Rainwater tanks as a means of water reuse and conservation in urban areas. In: *Urban Water Reuse Handbook*, Ch. 60 (ed. S. Eslamian), 797–808. Boca Raton: CRC Press, Taylor and Francis Group USA.

Reyneke, B., Dobrowsky, P.H., Ndlovu, T. et al. (2016). EMA-qPCR to monitor the efficiency of a closed-coupled solar pasteurization system in reducing *Legionella* contamination of roof-harvested rainwater. *Science of the Total Environment* 553: 662–670.

Reyneke, B., Cloete, T.E., Khan, S., and Khan, W. (2018). Rainwater harvesting solar pasteurization treatment systems for the provision of an alternative water source in peri-urban informal settlements. *Environmental Science: Water Research & Technology* 4: 291–302.

Rittmann, B.E., Huck, P.M., and Bouwer, E.J. (1989). Biological treatment of public water supplies. *Critical Reviews in Environmental Control* 19 (2): 119–184.

Sánchez, A.S., Cohim, E., and Kalid, R.A. (2015). A review on physicochemical and microbiological contamination of roof-harvested rainwater in urban areas. *Sustainability Water Quality and Ecology* 6: 119–137.

Seidel, C., Ghosh, A., and Morley, K. (2011). Selecting disinfectants in a security-conscious environment: six utility case studies. *American Water Works Association Journal* 103 (3): 30–33.

Servais, P., Cauchi, B., and Billen, G. (1994). Experimental study and modelling bacterial activity in biological activated carbon filters. *Water Science and Technology* 14 (2): 223–231.

Shaheed, R., Mohtar, W.H.M.W., and El-Shafie, A. (2017). Ensuring water security by utilizing roof-harvested rainwater and lake water treated with a low-cost integrated adsorption-filtration system. *Water Science and Engineering* 10 (2): 115–124.

Simmons, G., Jusry, S., Thornley, C. et al. (2008). A Legionnaires' disease outbreak: a water blaster and roof-collected rainwater systems. *Water Research* 42 (6): 1449–1458.

Sinton, L.W., Hall, C.H., Lynch, P.A., and Davies-Colley, R.J. (2002). Sunlight inactivation of fecal indicator bacteria and bacteriophages from waste stabilization pond effluent in fresh and saline waters. *Applied and Environmental Microbiology* 68 (3): 1122–1131.

Sobsey, M.D. (2002) Managing Water in the Home: Accelerated Health Gains from Improved Water Supply. Water, Sanitation and Health: Department of Protection of the Human Environment, World Health Organization, Geneva, Swizerland.

Sobsey, M.D., Stauber, C.E., Casanova, L.M. et al. (2008). Point of use household drinking water filtration: a practical, effective solution for providing sustained access to safe drinking water in the developing world. *Environmental Science and Technology* 42 (12): 4261–4267.

Spinks, A., Berghout, B., Dunstan R., et al. (2005) Tank sludge as a sink for bacterial and heavy metal contaminants and its capacity for settlement, re-suspension and flocculation enhancement. In: *Proceedings of the 12th International Rainwater Catchment Systems Association Conference*, New Delhi, India, November 2005.

Spinks, A.T., Dunstan, R.H., Harrison, T. et al. (2006). Thermal inactivation of water-borne pathogenic and indicator bacteria at sub-boiling temperatures. *Water Research* 40 (6): 1326–1332.

Strauss, A., Dobrowsky, P.H., Ndlovu, T. et al. (2016). Comparative analysis of solar pasteurization versus solar disinfection for the treatment of harvested rainwater. *BMC Microbiology* 16 (1): 289–305.

Thomas, V., McDonnell, G., Denyer, S.P., and Maillard, Y. (2010). Free-living amoeba and their intracellular pathogenic microorganisms: risks for water quality. *FEMS Microbiology Reviews* 34 (3): 231–259.

Thurman, R.B., Gerba, C.P., and Bitton, G. (1989). The molecular mechanisms of copper and silver ion disinfection of bacteria and viruses. *Critical Reviews in Environmental Science and Technology* 18 (4): 295–315.

Ubomba-Jaswa, E., Fernández-Ibáñez, P., Navntoft, C. et al. (2010). Investigating the microbial inactivation efficiency of a 25 L batch solar disinfection (SODIS) reactor enhanced with a compound parabolic collector (CPC) for household use. *Journal of Chemical Technology and Biotechnology* 85 (8): 1028–1037.

Venkobachar, C., Iyengar, L., and Rao, A.P. (1977). Mechanism of disinfection: effect of chlorine on cell membrane functions. *Water Research* 11 (8): 727–729.

Whitey, S., Cartmell, E., Avery, L.M., and Stephenson, T. (2005). Bacteriophages - potential for application in wastewater treatment processes. *Science of the Total Environment* 339 (1-3): 1–18.

World Health Organization (WHO). (2004) Water treatment and pathogen control: Process efficiency in achieving safe drinking water. International Water Association Publications, London, United Kingdom, 1-40. ISBN 1 84339 069 8.

World Health Organization (WHO) (2011). *Guidelines for Drinking-Water Quality*. Rev., 4e World Health Organization. Geneva, Switzerland: WHO Press. ISBN: 978-92-4-154815-1.

Wu, B., Rong, W., and Fane, A.G. (2017). The roles of bacteriophages in membrane-based water and wastewater treatment processes: a review. *Water Research* 110: 120–132.

Zavala, M.Á.L., Prieto, M.J.C., and Rojas, C.A.R. (2018). Rainwater harvesting as an alternative for water supply in regions with high water stress. *Water Science and Technology: Water Supply* 18 (6): 1946–1955.

Zhang, Y., Hunt, H.K., and Hu, Z. (2013). Application of bacteriophages to selectively remove *Pseudomonas aeruginosa* in water and wastewater filtration systems. *Water Research* 47 (13): 4507–4518.

Zhao, H., Liu, H., and Zhou, J. (2013). Characterization of a bioflocculant MBF-5 by *Klebsiella pneumoniae* and its application in *Acanthamoeba* cysts removal. *Bioresource Technology* 137: 226–232.

20

Water Recycling from Palm Oil Mill Effluent

Hossein Farraji[1,2], Irvan Dahlan[3], and Saeid Eslamian[4]

[1] School of Physical and Chemical Department, University of Canterbury, Canterbury, New Zealand
[2] Institute of Scientific Research Aminazmay Shargh (ISRAS), Mashhad, Iran
[3] School of Chemical Engineering, Universiti Sains Malaysia, Nibong Tebal, Pulau Pinang, Malaysia
[4] Department of Water Engineering, Isfahan University of Technology, Isfahan, Iran

20.1 Introduction

Nowadays, oil palm production in Malaysia has reached 20 000 000 tons (MPOB 2016). Production and by-products from the activities of 426 mills have a positive effect on the Malaysian economy. However, the other side of this industry is overpopulation (Farraji et al. 2016) and environmental impacts of the palm oil industry, which means 60 000 000 tons of effluent are a major concern for Malaysia as the leader of scientific research in green improvement.

The conventional treatment method for palm oil mill effluent (POME) is a ponding system, which is an easy-to-operate but time-consuming process in several ponds or a tank digestion system, with a high cost of operation using a large land area and having an environmental impact (Mansor et al. 2017). This method is highly dependent on land application and rivers as the final destination of effluent discharge. A zero-discharge policy or water recycling from POME could be an effective and high-performance method for sustainable development and removal of the palm oil industry from rivers as the final discharge media. The ponding system has been applied in 85% of the palm oil industry and is the most common commercial treatment method (Poh and Chong 2009), while the remaining 15% are operated using a tank digestion system. Finding effective treatment method for POME will be more effective when Malaysian authorities aim to strengthen standard limits to BOD_5 equals to $20\,mg\,l^{-1}$ (Liew et al. 2015).

Raw POME is an acidic media with a temperature of 80–90 °C (Chin et al. 2013), which causes low microbial population count. A long period of time is required for the available microbial population to adapt and start the digestion process (Farraji et al. 2019). However, low biodegradability of lignin compounds (Oswal et al. 2002) makes the digestion process longer, as well as decreases the pollutant removal efficiency and methane production. Further studies should be conducted to develop innovative methods to reduce the cost of treatment and increase treated water quality (Mansor et al. 2017) especially by the phytoremediation method in constructed wetlands for polishing pond effluent (Sa'at et al. 2019).

20.2 Problem Statement

Approximately 50% of POME pollutants are low or non-biodegradable organic matters, which should be separated from fine, easily degradable organic compounds such as fatty acids by a suitable coagulation mechanism (Ahmad et al. 2005a). Augmentation of municipal wastewater is an advanced treatment technique for enhancing the efficiency of pollutant removal in wastewaters and often have been used in landfill leachate (Mojiri et al. 2014; Aziz et al. 2011a). Activated sludge is recognized as one of the most effective aerobic treatments, however it is the least applied by palm oil mills because of its higher operation cost (Chan et al. 2010). On the other hand, high concentration of ammonia nitrogen is another cause of low efficiency of sequencing batch reactor (SBR) treatment, and high concentration of ammonia nitrogen is also very toxic for microorganism populations and could gradually be protected by the sludge acclimatization process (Aziz et al. 2011a). POME has nontoxic properties (Rupani et al. 2010) and contains very low concentration or absence of hazardous heavy metals such as Pb and Cu (Agustin et al. 2008), Cd, and Cr (Ubani et al. 2017). In POME the main pollutants are BOD_5, chemical oxygen demand (COD), total suspended solids (TSS), and color (Liew et al. 2015).

Handbook of Water Harvesting and Conservation: Basic Concepts and Fundamentals, First Edition. Edited by Saeid Eslamian and Faezeh Eslamian.

Table 20.1 World's main palm oil production countries (USDA 2016).

Production[a)]	2012/13	2013/14	2014/15	2015/16	2016/17
Indonesia	28 500	30 500	33 000	33 000	35 000
Malaysia	19 321	20 161	19 879	18 750	21 000
Thailand	2135	2000	2068	2100	2300
Colombia	974	1041	1110	1174	1175
Nigeria	970	970	970	970	970
Other	4478	4670	4614	4809	4945
Total	56 378	59 342	61 641	60 803	65 390

a) All production in thousand metric tons.

20.3 Palm Oil Production

Palm oil is a major global source of renewable and sustainable raw material for food, biofuel, and oleochemical industries (Basiron 2007). Oil palm is a highly productive crop with an output-to-input energy ratio of 9 : 1 compared with 3 : 1 for other oil seed crops, such as soybean or rapeseed (Basiron 2007). The world's total palm oil production (Table 20.1) reaches 65 390 000 tons (USDA 2016), and oil production per hectare of land per year in palm has several times the efficiency in comparison with other vegetable oil seeds (Corley and Tinker 2008).

Malaysia is the second-largest oil palm producer in the world, producing 40% of the global crude palm oil (CPO) (Johari et al. 2015). The production of Malaysian CPO reached 19 216 459 tons in 2013; 19 667 016 tons in 2014; and 19 960 703 tons in 2015 (MPOB 2016).

20.4 POME as an Agro-Industry Wastewater

POME is one of the greatest pollutants in colloidal agro-industry wastewater. The three sources for POME in oil palm mills are sterilizer condensate, hydrocyclone wastewater, and separator sludge (Borja et al. 1996). Approximately three tons of POME for each ton of palm oil are produced in oil palm mills (Wu et al. 2010). Other types of agro-industry wastewater characteristics are presented to clarify the pollutant range of POME, as listed in Table 20.2.

20.5 Characteristics of POME

In POME, 95% is water and the remaining 5% are pollutants. The combined suspended (2%) and dissolved solids (2%) compose 80% of POME pollutant content; the oil content of POME reaches (1–2%) (Igwe and Onyegbado 2007). The centrifugal fraction of POME is displayed in Table 20.3.

POME in raw form consists of large amounts of lignin and several other large molecular weight structural solids (Oswal et al. 2002). Most of the particle sizes in POME are in the range of 0–100 μm (Figures 20.1 and 20.2). The rod-like colloidal particles that include suspended solids can be removed only with 112 gravity power centrifuge (Table 20.3). A correlation occurs between suspended solids, turbidity, and color (Wah et al. 2002); consequently, the color in POME is apparently (Tan et al. 2017) caused by suspended solids. Sia et al. (2017) reported the same removal trend and model for color, TSS, and COD. Wah

Table 20.2 Different agro-industry wastewater characteristics.

Parameter	POME	Sweet potato	Coffee	Olive oil	Potato	Sugar beet
COD	95 465–112 023	73 260–79 540	22 000	15 528–17 343	6769 ± 970	6621 ± 113
BOD_5	62 500–69 215	7160–8954	12 000	—	—	—
TS	68 854–75 327	15 800–34 800	2050	—	756 ± 101	6062 ± 35
TSS	44 680–47 140	1620–2180	700	11 469–11 700	2187 ± 1046	665 ± 21
TDS	—	13 620–33 260	1350	—	—	—
TP	—	—	—	—	36 ± 28	2.7
pH	4.24–4.66	4.7–6.5	4.5	5.19	4.9 ± 0.8	6.82
Temperature	—	26–31	25	42–45	35 ± 15.5	—
Location	Thailand	Thailand	India	Greece	Zimbabwe	Turkey
Reference	(Choorit and Wisarnwan 2007)	(Tantipaibulvut et al. 2015)	(Devi et al. 2008)	(Paraskeva and Diamadopoulos 2006)	(Manhokwe et al. 2015)	(Alkaya and Demirer 2011)

All units are mg/l except temperature (°C) and pH.

Table 20.3 The centrifugal fraction of POME (Ho and Tan 1983).

Pictorial representation of fractions	Centrifuge force	Dimension	Weight percentage, w/w (Wet)	Characteristics
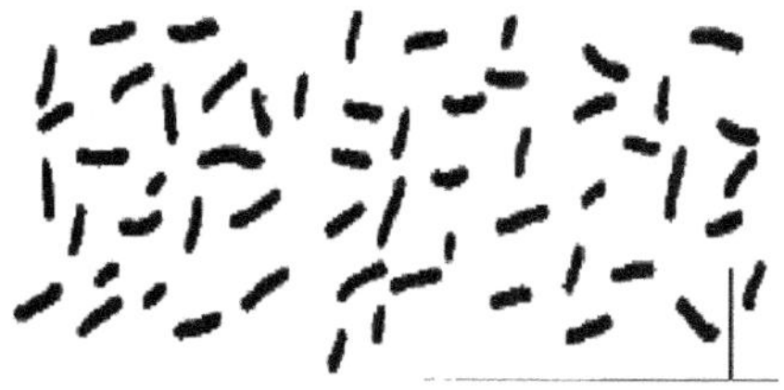	112 g	Length: 2.32 ± 0.52 μm Width: 0.59 ± 0.03 μm	0.12–0.15	Rod-like particles of colloidal dimension
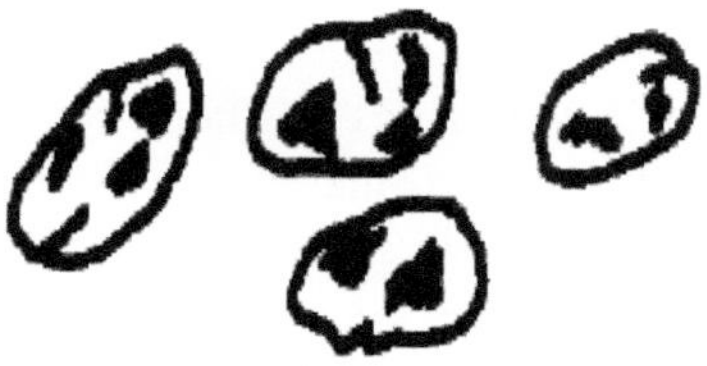	40 g	Length: 10–40 μm Breadth:50–130 μm	1.5–2.4	Plant cell debris: ruptured cell-walls and cell debris with entrapped oil drops
	28 g	Length:50–150 μm Breadth:50–130 μm		
	10 g	Length:50–130 μm Diameter:2–3 μm		Crystal-like particles. Light polarizable
	Normal gravity	Not determined	0.04–0.07	Fiber and sand

et al. (2002) presented the COD, ammonia nitrogen, and total nitrogen related to fine soluble compounds. Said et al. (2015) compared the pretreatment processes containing decantation, adsorption, and ultrafiltration prior to the nano-filtration membrane, thereby proving that only ultrafiltration and adsorption can reduce all collected pollutants to more than 80%. The research of Ahmad et al. (2008) indicated that particle sizes of POME are in the range of 0 to 400 μm (Figure 20.1).

Khanam et al. (2016) conducted the particle size analyzing (PSA) system research, which improved the previous data (Ahmad et al. 2008) on POME particle size. A particle size larger than 100 μm is hardly observed using PSA. Figure 20.2 illustrates two major areas of POME particle size distribution. These small particles, which are mostly about 0.1 μm, had not been reported by Ahmad et al. (2008).

Lignin and its related degradation products are chemically stable; thus, these substances are resistant to biological degradation and become intractable to removal by cost-effective methods before any treatment is applied (Mohan and Karthikeyan 1997). Based on the literature, POME cannot settle without the aid of coagulant (Ho and Tan 1989). Alrawi et al. (2015) conducted a particle size

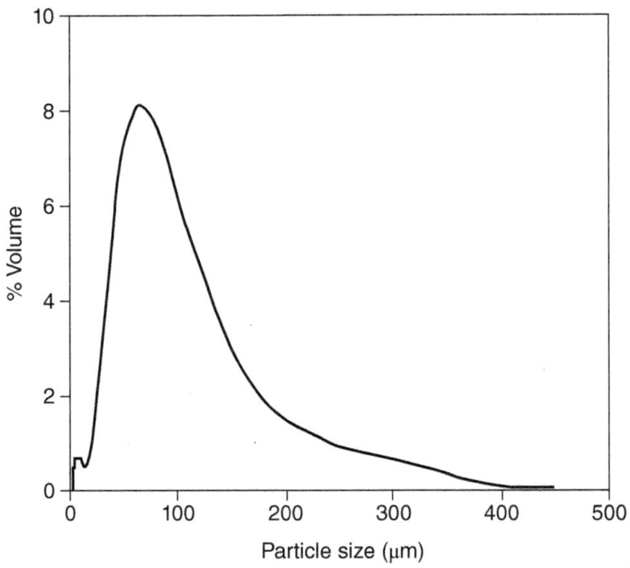

Figure 20.1 Particle size distribution data for raw POME (Ahmad et al. 2008).

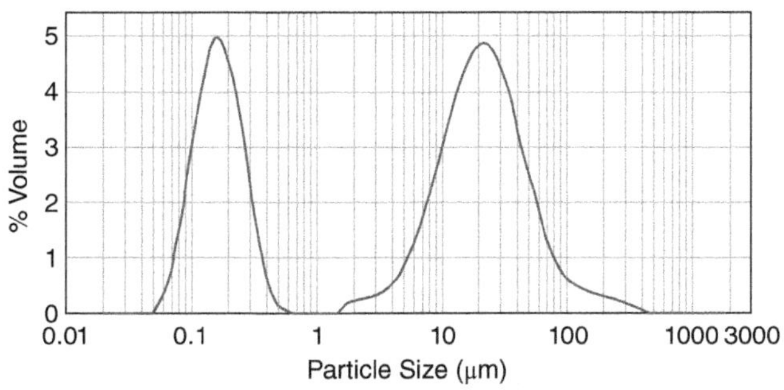

Figure 20.2 PSA system of POME (Khanam et al. 2016).

distribution study on non-oily POME particles (lignin, cellulose, and plant parts) with a size range of 38–100 μm. This study indicates that non-rising particles indicate a settling behavior because of their higher density versus oily particles that have a lower density than water and normally rise (floating) on the surface of POME. The result of a predictive study for hard-degradable particles of POME indicates that the large size of these particles causes increased velocity and settling time. The high BOD_5 and colloidal nature of suspended solids cause difficulty in presenting effective degradation or treatment by conventional methods. Consequently, the ponding method is not well adapted as an effective treatment method in many oil palm mills (Wu et al. 2010). Suspended solids are the origin of pollutants in POME; POME contains lignin compounds (difficult to biodegrade) and suspended solids (biodegradable). Thus, the removal of the two parts of pollutants as the main objectives of POME decontamination will be the basis of the treatment method for POME degradation and coagulation.

20.5.1 Total Suspended Solids

The concentration of TSS that contains suspended solids and volatile suspended solids (VSS) will change because of the oil palm extraction process and is dependent on the quality of numerous processes of final production preparation. Suspended solids are naturally lignin-based and non-biodegradable; VSS are mostly volatile fatty acids and short carbon chain, thereby making them prone to degradation by microorganisms. High suspended solids or humus (Mara 2013) in wastewater could be considered one of the most critical characteristics that can be affected by the treatment method and trend of decontamination. The highest TSS collected for raw POME is 47 140 mg l^{-1} (Choorit and Wisarnwan 2007), and the discharge regulation for TSS is 400 mg l^{-1} (MPOB 2016). In medium-strength wastewater, approximately 40% of filterable solids and approximately 75% of suspended solids are organic materials (Metcalf and Eddy et al. 2014). Nevertheless, this high-strength ratio of POME (Najafpour et al. 2005) and agricultural colloidal suspension (Wu et al. 2009) is approximately 50% filterable solids and almost 100% organic materials. Carbohydrates, greases, and oil are the major organic compounds present in wastewater. These compounds are recognized as the biodegradable parts of suspended solids in POME. Furthermore, Drinan and Spellman (2012) reported that these compounds are lignin-based suspended solids in POME. Lignin and its related degradation products are chemically stable; thus, these substances are resistant to biological degradation and become intractable to removal by cost-effective methods before any treatment is applied (Mohan and Karthikeyan 1997).

20.5.1.1 Volatile Suspended Solids

VSS, which is a biomass concentration (Zinatizadeh et al. 2006), can be considered a critical element in determining the operational behavior and biological concentration throughout the system, especially in the effluent. VSS is a part of the final production of oil palm mills that could not be extracted and could be decreased by enhancing the quality of oil extraction. In POME, the concentration of the VSS is approximately half of TSS (Choorit and Wisarnwan 2007). Volatile fatty acid decreased significantly throughout the anaerobic bacterial biogas production from POME, and the biodegradation of the volatile chain of fatty acid by microorganisms has been cited (Borja et al. 1996). Volatile fatty acid is a major VSS compound in POME and is considered a substrate for polyhydroxyalkanoate production in anaerobic microbial digestion (Gumel et al. 2012).

20.5.2 Biological Oxygen Demand

BOD_5 is a type of oxygen (dissolved oxygen) required for degrading organic matter by microorganisms at a certain time (frequently five days) and specific temperature (20 °C) in a dark place (Jouanneau et al. 2014). BOD_5 was selected as an important characteristic of pollutants for river pollution in the United Kingdom in 1908 and was adopted by the American Public Health Association (APHA) in 1936 (APHA 2005). BOD_5 is the amount of oxygen divided by the total volume of the treatment system that is processed in the respiration period of microorganisms. Aeration equipment will be sized in wastewater treatment on the basis of the value of BOD_5. Repeatability and inhabitation by wastewater toxic compounds and ions are the major drawbacks of BOD_5 measurement (Dubber and Gray 2010). This characteristic of POME depends on discharge standards (100 mg l^{-1}) and has a strength limitation of 20 mg l^{-1} in certain areas of Malaysia (Sabah and Sarawak). Ahmad et al. (2005b) considered this characteristic as a bioCOD. POME with $BOD_5 < 5000$ mg l^{-1} can be used for land application (Corley and Tinker 2008). Nontoxic and organic characteristic of POME is the main reason for degradation of POME pollutants in soil even with high concentration of BOD5 < 5000 mg l^{-1}.

20.5.3 Chemical Oxygen Demand

COD represents the ultimate oxygen demand for oxidizing organic carbon without any dependence on the degradation capability of microorganisms. It is also defined as a required oxygen for decomposing organic and inorganic decontaminations through chemical pathways (Darajeh et al. 2014), and high COD is an indicator for circumstances that include biologically resistant organic compounds. This wastewater measurement indicator data can be quickly collectible within 24 hours, and significant results can be expected without influence from any toxic compounds or ions. Furthermore, reliability and extremely high concentration of COD can be considered merits of this method. COD is recognized as one of the acceptable discharge standards for industrial effluents (Kim et al. 2007). Based on the MPOB (2016) the Malaysian discharge standard for COD of POME, which is 400 mg l^{-1}, has not been updated after 1984. The removal of COD is a combination of physical reduction (settling) and biological decontamination (biodegradation). Less than half of the organic matter in POME is degradable (Chou et al. 2016). COD as the representative of the organic matter should be considered "food" for cellular activity (Alrawi et al. 2011), and microbial growth is decreased because of decreasing COD (Gerardi 2002).

20.5.4 Color

The presence of colors is esthetically unpleasant and is associated with effective treatment method. Color compounds in aquatic media can penetrate sunlight and inhibit self-purification in aquatic biota. Self-purification is the most important part of the food chain. Chelate metal ions can react with color compounds and induce effects on aquatic biota (Mohan and Karthikeyan 1997). Inhibition of anaerobic digestion (Zouari and Ellouz 1996), harder degradation of chlorinated lignin compounds (Annibale et al. 2004), and consideration as a specific character in oily wastewater (Paraskeva and Diamadopoulos 2006) are other aspects of color in wastewater treatment engineering. The existence of carotene (α and β) in POME produces a specific red/brownish dark color (Ahmad and Chan 2008). The concentration of carotenoids widely differs from 500 to 4000 ppm in palm oil depending on the origin of the oil palm plant species. The characteristics of carotene are recognized as a compound that is fat-soluble and/or less soluble in aquatic media. The color of POME is well-known as a hardly biological degradable compound because of the structure and properties of POME. The esthetic problematic aspect of the residual color in POME will be the first visual specification in water bodies in addition to the discharged effluent (Yuniarto et al. 2013). TSS and dissolved components are the root sources of color in POME. However, the removal trend for color compared with other organic contaminants located at the least capacity (COD > BOD_5 > SS > lignin and tannin > color) in the biological treatment using biosorbent (Mohammed and Chong 2014). The anaerobic treatment method, which has been recognized as an effective treatment method for POME, is incapable of fully decolorizing POME (Zhang et al. 2008). Only a chain of combined treatments containing membrane technology can be considered a true decolonization treatment method (Ahmad et al. 2003). Meanwhile, aerobic bacterial communities are highly capable for effective decolonization processes through wastewater treatment (Mohana et al. 2007). However, the cost of electrical energy for oxygen supply is a drawback of the aerobic treatment for the decolonization process (Limkhuansuwan and Chaiprasert 2010). Malaysian palm oil discharge limits include nutrients such as total nitrogen and ammonia (Table 20.3) which were not discussed here; meanwhile, color is not included in discharge standard and discussed as one of the main characteristics of POME. High concentration of selected factor (BOD [biological oxygen demand], COD, TSS, and color) and problematic process of decontamination caused them to be considered as the major pollutants of POME (Liew et al. 2015).

20.5.5 Biodegradability of POME

The biodegradability of wastewater highly depends on the availability of oxygen for microorganism activities, and the ratio of BOD_5 to COD in wastewaters is called the biodegradability index (Metcalf and Eddy et al. 2014). An old landfill is source of stabilized leachate with low BOD/COD ratio as well as low COD (mg/l) as well-known low biodegradable-strength wastewater (Rivas et al. 2004). In other words, the low biodegradability ratio is a suitable indicator to show that municipal landfill leachate belongs to stabilized landfill (Kurniawan et al. 2006). For treatment of such landfill leachate, two or three physical, chemical, or combined methods should be conducted (Li et al. 2010). Meanwhile, POME is cited as either a biodegradable wastewater (Rupani et al. 2010) or easily biodegradable (Sompong et al. 2012). The mean collected BOD_5/COD ratio from 15 studies conducted on Malaysian POME is 0.46 (Farraji et al. 2015). Based on Metcalf and Eddy et al. (2014), if the biodegradability index was ≥ 0.5, then the target wastewater can be an eligible source for biological treatment.

Most of the previous studies on the aerobic SBR treatment of POME (Chan et al. 2010, Chan et al. 2011, Chou et al. 2016, and Zahrim et al. 2009) did not clearly discuss BOD_5/COD ratio as a biodegradability index. In the research of Chin et al. (1987) and Chan et al. (2011), BOD_5/COD were 0.45 and 0.1, respectively. Clearly, the SBR system operates for a wide range of biodegradability. However, this effective factor has never been collected before and after treatment. Thus, the BOD_5/COD of POME before treatment and after decontamination process should be collected and reported as the biodegradability index. Consequently, the hardness of the biological treatment process can be predicted by collecting the biodegradability of raw POME, and conducting the treatment process and method can be discussed for decontaminated effluent.

20.6 POME Treatment Methods

Discharging POME to water resources causes temporary and permanent environmental footprints. From 1978, when the first discharge standards were published for the oil palm industry, until the present, these standards have been continuously improved to reach eco-friendly industry conditions and the highest level of environmental protection (Wu et al. 2010). By contrast, the increased global demand for palm oil consumption results in a greater focus on environmental pollution introduced by additional oil palm plantations (Mansor et al. 2017). The other influence factor for additional oil palm plantations is the high and extensive productivity of the oil palm as a hardwood oil tree (Basiron 2007). Since 2006, the BOD_5 concentration level for discharging POME has decreased to 20 $mg\,l^{-1}$ (Ng and Cheng 2017) in Sabah and Sarawak.

20.6.1 Commercial Treatment Method

The most common treatment method for POME in 426 oil palm mills is the ponding system, which contains several specific ponds. The cooling pond is used for decreasing temperature of raw POME from 80 to 90 °C to about 30 °C. The acidification pond is used for increasing the pH of raw POME from 3–4 to 6–7. Alternative aerobic and anaerobic ponds are used for POME digestion in several processes and all aforementioned treatment systems are conducted in about 60 days. About 85% of current oil palm mills of Malaysia are using the ponding treatment method, which is a very easy-maintenance treatment method. However, it requires a large land area, depends highly on rivers as discharge destinations, as well as has a high environmental footprint containing greenhouse gases emissions, bad odor production, and water resource pollution. The other main emerging treatment method for POME is an anaerobic digestion system with methane gas production, which is mostly the same as the ponding system but with the addition of a covered section for methane gas collection added through the anaerobic ponds. This system faces a high maintenance fee and high cost of construction, as well as requiring a large land area. The process of treatment takes up to 18 months. The current approach of anaerobic digestion of POME has been well reviewed by (Poh and Chong 2009). Unfortunately, both current conventional treatment methods for POME can't meet the environmental discharge standards of Malaysia. Malaysian discharge standards and characteristics of POME are presented in Table 20.3. This table illustrates current discharge standards of Malaysia (DOE 1999) and the average results of numbers of research studies.

The final efficiency of both current commercial treatment methods, anaerobic digestion and ponding systems, is unable to pass the Malaysian discharge standards. A rapid treatment method with high efficiency that is cost-effective and is applicable for 60 000 000 tons of POME as a commercial treatment system is highly necessary for the oil palm industry (Said et al. 2015). Current conventional treatment methods of POME are inefficient and causing a large environmental footprint (Iskandar et al. 2017).

20.6.2 Non-Commercial Treatment Method

Several lab-scale treatment methods have been formulated for POME treatment. These methods include an open digestion tank for methane collection and decreasing greenhouse gas emissions through anaerobic digestion (Yacob et al. 2005; Loh et al. 2017; Beaudry et al. 2018); electrocoagulation, which is an effective and slightly expensive post-treatment for color removal (Nasrullah et al. 2017);

Fenton oxidation, which was developed as a quality treatment method for COD and color, even in end-pipe POME (Aris et al. 2008); and phytoremediation as a plant-based post-treatment that is often used for anaerobically treated and diluted POME (Darajeh et al. 2014).

Another treatment method of POME is membrane technology. In the membrane treatment system, several processes of pre-treatment contain anaerobic, aerobic, adsorbent augmentation, and even reverse osmosis techniques that could be used as an additional process for enhancing the quality of decontamination and increasing the lifetime of the required membranes (Zhang et al. 2008). Membrane technology is recently conducted on a pilot scale (Tabassum et al. 2015).

20.7 Water Recycling by Membrane Technique

The membrane technique has been widely used for collecting high-quality treated water and wastewaters. The high cost of the treatment membrane and maintenance fee, short life of membrane filters, and specific characterization of each membrane for particular type of pollutants, as well as the wide range of pollutant sizes, cause too many drawbacks for application of this treatment method. The zero-discharge policy for POME treatment has been intensively studied (Ahmad et al. 2003; Ahmad et al. 2006; Loh et al. 2013; Madaki and Seng 2013; Othman et al. 2014; Tabassum et al. 2015; Wang et al. 2015). Extremely high concentration of pollutants, wide range of degradability in POME pollutants, absence of microbial communities, and the oily colloidal characteristic of POME are some of the reasons for the difficulty of presenting unique treatment methods for water recycling. The application of membrane system for POME is a post-treatment process and it should be located at the end of a chain process of numerous treatment methods. In other word, water recycling in POME contains several systems which have been conducted; even long-term (about two years) pilot studies (Tabassum et al. 2015; Wang et al. 2015) contain:

(i) Pre-treatment of POME
(ii) Biological treatment (anaerobic, aerobic)
(iii) Biogas power generation process
(iv) Sludge disposal
(v) Water reclamation system

High-cost equipment and a high-energy consumer system in a long-term treatment requires more thinking on the myth or reality (Tsai et al. 2017) of water recycling from POME. On the other hand, an "all in one" refinery system, which provides zero discharge and valuable byproducts, and environmental protection via biogas production, indicates that water recycling from POME is a valuable emerging policy (Loh et al. 2017). Figure 20.3 illustrates the process of POME treatment in a pilot study with water recycling target. In fact, water recycling in POME is a complicated multi-process technique which needs further improvement in order to decrease the cost of treatment, as well as shortening the required time of recycling.

It seems a short-term and cost-effective treatment method could be considered as a short cut for aforementioned merits and drawbacks. Enhancing the biodegradability of POME by mixing municipal wastewater as microbial source through the aerobic SBR system (Farraji et al. 2015) showed high efficiency in BOD and COD removal and POME was converted to a good biodegradable effluent. The application of phytoremediation as a green post-treatment system for anaerobically treated POME (Darajeh et al. 2014) provides a wide range of decontamination by capable plant species, concerning positive influenced augmentation for achieving high-performance decontamination (Farraji et al. 2017a, 2017b). Combining two simple treatment processes for enhancing the efficiency of decontamination such as electrocoagulation (Agustin et al. 2008) of anaerobically

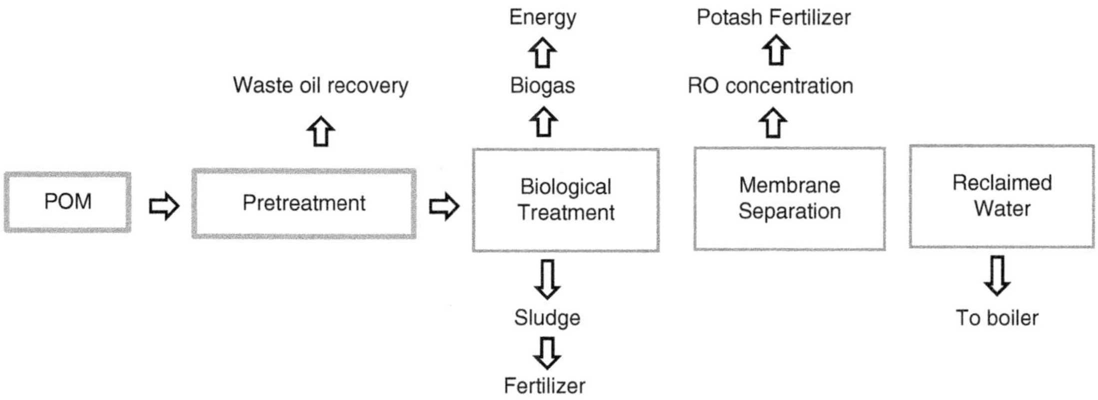

Figure 20.3 Zero-discharge treatment method flowchart. Source: Improved from Tabassum et al. (2015).

treated POME through the biogas production treatment method (Qamaruz-Zaman et al. 2016) may present capable pre-treated POME for final post-treatment process in order to recycle water.

The quality of recycled water is a major influencing factor for management of the treatment process. Recycled water could be used for several objects based on the standard qualities, including agricultural irrigation, animal drinking water, industrial applications, and recycled crystal-clear water for human drinking. Aerobic SBR, as the most influenced short-term treatment system for POME which has provided acceptable discharge limits of Malaysian standards (Chan et al. 2010; Chan et al. 2011), will be assessed in Section 20.8.

20.7.1 Benefits and Drawbacks of Membrane Treatment Method for POME

The membrane treatment method for POME couldn't be considered as a capable method for water recycling of water from POME, since the membrane bioreactor could not remove color properly (Ahmad et al. 2008; Facta et al. 2014; Sulaiman and Ling 2004). Because of the importance of color in recycled water, replacement methods for color removal from POME should be found. There are some promising color removal methods in POME treatment, such as banana peel (Mohammed and Chong 2014), and granular activated carbon (Alkhatib et al. 2015). The adsorbent techniques are suitable in the short term but after a while release the adsorbed color because of short life span and so require continued replacement (Facta et al. 2014). In other research, *Phanerochaete chrysosporium* mycelium was used for decoloring anaerobically treated POME through a shaking system. There was 83.4% color removal after 3 days of shaking in 130 rpm at room temperature (Rakamthong and Prasertsan 2011). Growing cultures of *Curvularia clavata* showed 80% color removal after 5 days of treatment (Neoh et al. 2014), which needs to be improved to be considered a cost-effective or valuable system for commercialization in 60 million tons of Malaysian POME.

20.8 Application of the SBR in POME Treatment

The historical background of aerobic treatment indicates that the post-treatment of the anaerobically treated POME by aerobic technologies should be applied to reduce POME specification within the discharge standards (Ho and Tan 1983). Table 20.4 The effectiveness of the current aerobic SBR treatment method is summarized in Table 20.5.

Table 20.4 Characteristics of raw POME, Malaysian discharge standards.

Factors	Raw POME (mg l)$^{-3}$	Standard (mg l)$^{-1}$
pH	4.7	5–9
BOD_5	25 000	100[a)]*
COD	50 000	—
TSS	18 000	400
Oil	4000	50
NH_3-N	35	150
TN	750	200

a) In some states of Malaysia (Peninsular) standard limit is 20 mg l^{-1}.
Source: (Bello and Raman 2017; DOE 1999).

The research by Chin et al. (1987) on the aerobic SBR treatment system illustrates POME application. These researchers used a POME treated by lime and alum as coagulator and additional treatment by polymers for decreasing pollutant rates as a base of study. The aerobic SBR system treatment is then used as a target decontamination method (Chin et al. 1987). The research of Fun et al. (2007) and Vijayaraghavan et al. (2007) presented low efficiency for an extended period, an extended treatment time requirement, and high energy consumption.

The research of Zahrim et al. (2009) indicated that the environmental adaptation of microorganisms supplied from activated sludge is a time-consuming process through the aerobic SBR of anaerobically treated POME. Their research was conducted in 60 days, and removal efficiency had no improvement after the 41st day.

Cheng et al. (2010) reached acceptable discharge standards; an extended treatment period and high energy consumption remained the major drawbacks of the aerobic SBR system for POME treatment. Other researchers could achieve high efficiency with high temperature and extensive setting time to enhance the degradation rate in SBR (Chan et al. 2011); the research of Chou et al. (2016) replaced the acclimatization process by applying the inoculums that were under the aerobic digestion system for six months. Hydraulic retention time was 20–60 days, and the optimum removal efficiency was achieved in the thermophilic condition.

All conducted research was time-consuming. The starting step depends on the adaptation of microorganisms throughout the biological digestion (Chan et al. 2010); the extended time of aeration (Fun et al. 2007), necessary thermophilic augmentation (Chan et al. 2011), and high cost of activated sludge application process (Fun et al. 2007; Cheng et al. 2010) are the main drawbacks of this

Table 20.5 Effectiveness of the aerobic SBR treatment for POME.

Parameters	(Fun et al. 2007)	(Chin et al. 1987)	(Zahrim et al. 2009)	(Chan et al. 2010)	(Chan et al. 2011)	(Chou et al. 2016)
System	SBR	SBR	SBR	SBR	SBR	SBR
Augmented with	Activated sludge	Activated sludge	Activated sludge	Activated sludge	Activated sludge	Activated Sludge[a)]
Source of POME	Anaerobic digested	Adsorbent Pre-treated	Anaerobic digested	Anaerobic digested	Anaerobic digested	Anaerobic digested
Influent COD ($mg\,l^{-1}$)	—	1550	1141	13 650	13 532	10 030
Influent BOD_5 ($mg\,l^{-1}$)	—	700	—	1355	1355	—
Biodegradability[b]	—	0.45	—	0.1	0.1	—
COD removal (%)	82	31–50	70	95–96	63–86	75–93
BOD_5 removal (%)	—	50–70	—	97–98	65–87	—
TSS removal (%)	62	—	—	98–99	79.2–89.1	81–95
Color removal (%)	—	50	41	—	—	—
T (°C)	—	42		28	50	30
pH	—	—	7.7–8.3	7.4	7.4	7.7
Treatment time	14 d	25 d	60 d	30 d	>11 d	7.2–18

a) Equal inoculum used as activated sludge.
b) Calculated by the author.

method. A gap exists in presenting a rapid, cheap, effective, and easy aerobic SBR treatment method for POME.

20.8.1 Factors Affecting the SBR System

SBR, as a biological system, is highly influenced by the presence and concentration value of toxic compounds, such as ammonia nitrogen, and hazardous metallic elements (Aziz et al. 2011b). The biodegradability of wastewater, which is highly dependent on BOD_5/COD ratio, is an important factor that affects the efficiency of the SBR system. A positive correlation exists between biodegradability and SBR performance (Metcalf and Eddy et al. 2014). A high level of DO causes growth and development of protozoan predators, which consume nitrifying bacteria. These bacteria are one of the major groups of microorganisms which form biofilm. Consequently, a high concentration of dissolved oxygen as the aeration result causes minimized biomass growth, limited sludge damping, and increased treatment cost (Sperling 2007). Janczukowicz et al. (2001) indicated that sludge volume index and sedimentation time are negatively correlated. Some operational factors that influenced SBR process and performance are (Aziz et al. 2013):

- Characteristics of wastewater (WW)
- Ratio of treating WW to working volume of the SBR
- Contact time, aeration ratio, cycle time, hydraulic retention time, organic loading rate
- Concentration of dissolved oxygen, mixed liquor VSS
- Presence and dosage of adsorbent

20.8.2 Microbial Augmentation for POME

Preliminary microbiological treatment is a fundamental requirement for enhancing the efficiency of the adsorption treatment process in wastewater containing residual oil (Ahmad et al. 2005c). The augmentation substrate, such as rumen fluid (Alrawi et al. 2011) and condensation water (Borja et al. 1995), was utilized for enhancing biogas production and pollutant removal application throughout the anaerobic digestion of POME. The microbial growth rate started from the 21st day to the 60th day, and this growth decreased with the decrease in COD. Favorable microbial growth was reported as the reason for high efficiency of POME pollutant removal in the anaerobic–aerobic bioreactor (Chan et al. 2012). Granular sludge exhibits a positive effect on enhancing the microbial aerobic biodegradation of COD, and improvement of settling properties was reported by Abdullah et al. (2011). Better efficiency of mesophilic (28 °C) aerobic SBR treatment than the thermophilic (55 °C) condition was reported by Chan et al. (2010) for activated sludge-augmented anaerobically treated POME. The low number of filamentous microorganisms should be presented in aerobically treated POME because the lack of foaming and sludge bulking are the drawbacks in the system (Vijayaraghavan et al. 2007).

20.9 Discussions

A reasonable water recycling project from POME should be arranged based on engineering policy, concentrating on the following points:

- The main pollutants of POME (BOD, COD, TSS, and color) have organic characteristics;
- Very acidic and high temperature of end-pipe effluent as compulsory pre-treatment process;
- Wide range of biodegradability in suspended solids as root source of POME pollutants;
- Extremely high concentration of pollutants even in comparing with other agricultural effluents;
- Absence of microbial communities as basic requirement for biological treatment system;
- Availability of required metallic elements for microorganism growing such as Ca, Mg, and Fe as well as absence of hazardous heavy metals.

By considering the aforementioned specifications of POME, a couple of basic microbiological pre-treatment systems such as aerobic SBR augmented with raw municipal wastewater as microbial sources (Farraji et al. 2017a, 2017b) and a suitable polishing post-treatment method such as phytoremediation, constructed wetland (Darajeh et al. 2014; Farraji 2017; Sa'at and Zaman 2017; Sa'at et al. 2017), or electrocoagulation (Agustin et al. 2008) can provide capable recycled water for industrial application in oil palm mills. Meanwhile, the cooling process and acid pond process should be passed before all water recycling arranged systems. Water recycling can provide required water for oil palm mills and directly influence dependence on rivers as discharge destination. Environmental protection (direct and/or indirect effects) will be the most important merit of water recycling in the oil palm industry.

20.10 Conclusion

Geographic information system assessment for relocation of oil palm mills based on independence from rivers and application of recycled wastewater is warranted. Greenhouse gas emission through water recycling from POME in comparison with the current ponding treatment system may provide additional attraction for a zero-discharge policy. Furthermore, squishing current vast treatment areas (ponding systems) into reasonable compact and industrialized water recycling based treatment procesess, high value byproducts such as biogas and settlements of the flocculation-coagulants stage could be considered cost-effective aspects of water recycling.

References

Abdullah, N., Ujang, Z., and Yahya, A. (2011). Aerobic granular sludge formation for high strength agro-based wastewater treatment. *Bioresource technology* 102 (12): 6778–6781.

Agustin, M.B., Sengpracha, W.P., and Phutdhawong, W. (2008). Electrocoagulation of palm oil mill efluent. *International Journal of Environmental Research and Public Health* 5 (3): 177–180.

Ahmad, A.L. and Chan, C.Y. (2008). Recovery of oil and carotenes from palm oil mill effluent (POME). *Chemical Engineering Journal* 141 (1): 383–386.

Ahmad, A.L., Suzylawati, I., and Subhash, B. (2003). Water recycling from palm oil mill effluent (POME) using membrane technology. *Desalination* 157 (1): 87–95.

Ahmad, A.L., Ismail, S., and Bhatia, S. (2005a). Membrane treatment for palm oil mill effluent: effect of transmembrane pressure and crossflow velocity. *Desalination* 179 (1): 245–255.

Ahmad, A.L., Ismail, S., and Bhatia, S. (2005b). Optimization of coagulation–flocculation process for palm oil mill effluent using response surface methodology. *Environmental Science and Technology* 39 (8): 2828–2834.

Ahmad, A.L., Sumathi, S., and Hameed, B.H. (2005c). Adsorption of residue oil from palm oil mill effluent using powder and flake chitosan: equilibrium and kinetic studies. *Water Research* 39 (12): 2483–2494.

Ahmad, A.L., Chong, M.F., Bhatia, S., and Ismail, S. (2006). Drinking water reclamation from palm oil mill effluent (POME) using membrane technology. *Desalination* 191 (1–3): 35.

Ahmad, A.L., Chong, M.F., and Bhatia, S. (2008). Population balance model (PBM) for flocculation process: simulation and experimental studies of palm oil mill effluent (POME) pretreatment. *Chemical Engineering Journal* 140 (1): 86–100.

Alkaya, E. and Demirer, G.N. (2011). Anaerobic acidification of sugar-beet processing wastes: effect of operational parameters. *Biomass and Bioenergy* 35 (1): 32–39.

Alkhatib, M.F., Mamun, A.A., and Akbar, I. (2015). Application of response surface methodology (RSM) for optimization of color removal from POME by granular activated carbon. *International journal of environmental science and technology* 12 (4): 1295–1302.

Alrawi, R.A., Ahmad, A., Ismail, N., and Kadir, M.O.A. (2011). Anaerobic co-digestion of palm oil mill effluent with rumen fluid as a co-substrate. *Desalination* 269 (1–3): 50–57.

Alrawi, R.A., Ahmad, A., Ismail, N., and Omar, A.K. (2015). Predicted free settling velocity of spherical non-oily palm oil mill effluent particles at different particle sizes. *Journal of Solid Waste Technology and Management* (4): 41, 395–405.

Annibale, A., Ricci, M., Quaratino, D. et al. (2004). Panus tigrinus efficiently removes phenols, color and organic load from olive-mill wastewater. *Research in Microbiology* 155 (7): 596–603.

APHA (2005). *Standard Methods for the Examination of Water and Wastewater*. USA: American Public Health Association.

Aris, A., Ooi, B.S., Kon, S.K., and Ujang, Z. (2008). Tertiary treatment of palm oil mill effluent using Fenton oxidation. *Malaysian Journal of Civil Engineering* 20 (1): 12–25.

Aziz, S.Q., Aziz, H.A., Yusoff, S.M., and Bashir, M.J.K. (2011a). Landfill leachate treatment using powdered activated carbon augmented sequencing batch reactor (SBR) process: optimization by response surface methodology. *Journal of Hazardous Materials* 189 (1): 404–413.

Aziz, S.Q., Aziz, H.A., and Yusoff, M.S. (2011b). Powdered activated carbon augmented double react-settle sequencing batch reactor process for treatment of landfill leachate. *Desalination* 277: 313–320.

Aziz, S.Q., Aziz, H.A., Mojiri, A. et al. (2013). Landfill leachate treatment using sequencing batch reactor (SBR) process: limitation of operational parameters and performance. *International Journal of Scientific Research in Knowledge* 1 (3): 34–43.

Basiron, Y. (2007). Palm oil production through sustainable plantations. *European Journal of Lipid Science and Technology* 109 (4): 289–295.

Beaudry, G., Macklin, C., Roknich, E. et al. (2018). Greenhouse gas assessment of palm oil mill biorefinery in Thailand from a life cycle perspective. *Biomass Conversion and Biorefinery* 8 (1): 43–58.

Bello, M.M. and Raman, A.A.A. (2017). Trend and current practices of palm oil mill effluent polishing: application of advanced oxidation processes and their future perspectives. *Journal of Environmental Management* 198: 170–182.

Borja, R., Banks, C.J., Martin, A., and Khalfaoui, B. (1995). Anaerobic digestion of palm oil mill effluent and condensation water waste: an overall kinetic model for methane production and substrate utilization. *Bioprocess Engineering* 13 (2): 87–95.

Borja, R., Banks, C.J., and Sánchez, E. (1996). Anaerobic treatment of palm oil mill effluent in a two-stage up-flow anaerobic sludge blanket (UASB) system. *Journal of Biotechnology* 45 (2): 125–135.

Chan, Y.J., Chong, M.F., and Law, C.L. (2010). Biological treatment of anaerobically digested palm oil mill effluent (POME) using a lab-scale sequencing batch reactor (SBR). *Journal of Environmental Management* 91 (8): 1738–1746.

Chan, Y.J., Chong, M.F., and Law, C.L. (2011). Optimization on thermophilic aerobic treatment of anaerobically digested palm oil mill effluent (POME). *Biochemical Engineering Journal* 55 (3): 193–198.

Chan, Y.J., Chong, M.F., and Law, C.L. (2012). An integrated anaerobic–aerobic bioreactor (IAAB) for the treatment of palm oil mill effluent (POME): start-up and steady state performance. *Process Biochemistry* 47 (3): 485–495.

Cheng, J., Zhu, X., Ni, J., and Borthwick, A. (2010). Palm oil mill effluent treatment using a two-stage microbial fuel cells system integrated with immobilized biological aerated filters. *Bioresource Technology* 101 (8): 2729–2734.

Chin, K.K., Ng, W.J., and Ma, A.N. (1987). Palm oil refinery effluent treatment by sequencing batch reactors. *Biological Wastes* 20 (2): 101–109.

Chin, M.J., Poh, P.E., Tey, B.T. et al. (2013). Biogas from palm oil mill effluent (POME): opportunities and challenges from Malaysia's perspective. *Renewable and Sustainable Energy Reviews* 26: 717–726.

Choorit, W. and Wisarnwan, P. (2007). Effect of temperature on the anaerobic digestion of palm oil mill effluent. *Electronic Journal of Biotechnology* 10 (3): 376–385.

Chou, K.W., Tan, S.W., Morad, N. et al. (2016). Aerobic post-treatment of different anaerobically digested palm oil mill effluent (POME). *International Journal of Environmental Science and Development* 7 (7): 511.

Corley, R.H.V. and Tinker, P.B.H. (2008). *The Oil Palm*. Wiley.

Darajeh, N., Idris, A., Truong, P. et al. (2014). Phytoremediation potential of vetiver system technology for improving the quality of palm oil mill effluent. *Advances in Materials Science and Engineering*: 1–10.

Devi, R., Singh, V., and Kumar, A. (2008). COD and BOD reduction from coffee processing wastewater using Avacado peel carbon. *Bioresource technology* 99 (6): 1853–1860.

DOE (1999) Industrial processes and the environment (Handbook No. 3). Crude Palm Oil Industry. Department of Environment, Ministry of Science, Technology and the Environment, Malaysia.

Drinan, J.E. and Spellman, F. (2012). *Water and wastewater treatment: A guide for the nonengineering professional*. CRC Press.

Dubber, D. and Gray, N.F. (2010). Replacement of chemical oxygen demand (COD) with total organic carbon (TOC) for monitoring wastewater treatment performance to minimize disposal of toxic analytical waste. *Journal of*

Environmental Science and Health Part A 45 (12): 1595–1600.

Facta, M., Salam, Z., and Buntat, Z. (2014). Design of Single Switch Resonant Converter with parallel load resonant for colour removal in palm oil mill effluent. *Proceeding of the Electrical Engineering Computer Science and Informatics* 1 (1): 337–340.

Farraji, H. (2017). Phytoremediation of suspended solids and turbidity of palm oil mill effluent (POME) by Ipomea aquatica. *Galeri Waris. Kejuruter.* 1 (1): 36–40.

Farraji, H., Zaman, N.Q., Abdul Aziz, H. et al. (2015). Enhancing BOD/COD ratio of POME treatment in SBR system. In: *Applied Mechanics and Materials*, vol. 802 (eds. H.A. Aziz, M.O. Hamzah, F. Ahmad, et al.), 437–442. Trans Tech Publications.

Farraji, H., Zaman, N.Q., Vakili, M., and Faraji, H. (2016). Overpopulation and sustainable waste management. *International Journal of Sustainable Economies Management (IJSEM)* 5 (3): 13–36.

Farraji, H., Zaman, N.Q., Aziz, H.A., and Sa'at, S.K.M. (2017a). Palm oil mill effluent and municipal wastewater co-treatment by zeolite augmented sequencing batch reactors: turbidity removal. In: *AIP Conference Proceedings*, vol. 1892, No. 1, 040013. Kuala Lumpur.: AIP Publishing.

Farraji, H., Zaman, N.Q., Aziz, H.A., and Sa'at, S.K.M. (2017b). Palm oil mill effluent treatment: influence of zeolite, municipal wastewater and combined aerobic SBR system. In: *AIP Conference Proceedings*, vol. 1892 , No. 1, 040014. Kuala Lumpur: AIP Publishing.

Farraji, H., Mojiri, A., and Yusoff, M.S. (2019). Smart microbial sources Management for Treatment: palm oil mill effluent and landfill leachate. In: *Handbook of Research on Implementation and Deployment of IoT Projects in Smart Cities* (eds. K. Saravanan and J. Golden), 350–369. IGI Global.

Fun, C.W., Haq, M.R.U., and Kutty, S.R.M. (2007). Treatment of palm oil mill effluent using biological sequencing batch reactor system. In: *River Basin Management IV, WIT Transactions on Ecology and the Environment*, vol. 104 (eds. C.A. Brebbia and K.L. Katsifarakis), 511–518. Southampton, UK: WIT Press.

Gerardi, M.H. (2002). *Settleability Problems and Loss of Solids in the Activated Sludge Process.* Wiley.

Gumel, A.M., Annuar, M.S.M., and Heidelberg, T. (2012). Biosynthesis and characterization of polyhydroxyalkanoates copolymers produced by pseudomonas putida Bet001 isolated from palm oil mill effluent. *PLoS One* 7 (9): e45214.

Ho, C.C. and Tan, Y.K. (1983). Centrifugal fractionation studies on the particulates of palm oil mill effluent. *Water Research* 17 (6): 613–618.

Ho, C.C. and Tan, Y.K. (1989). Comparison of chemical flocculation and dissolved air flotation of anaerobically treated palm oil mill effluent. *Water Research* 23 (4): 395–400.

Igwe, J.C. and Onyegbado, C.C. (2007). A review of palm oil mill effluent (POME) water treatment. *Global Journal of Environmental Research* 1 (2): 54–62.

Iskandar, M.J., Baharum, A., Anuar, F.H., and Othaman, R. (2017). Palm oil industry in South East Asia and the effluent treatment technology—a review. *Environmental Technology & Innovation. Volume* 9: 169–185.

Janczukowicz, W., Szewczyk, M., Krzemieniewski, M., and Pesta, J. (2001). Settling properties of activated sludge from a sequencing batch reactor (SBR). *Polish Journal of Environmental Studies* 10 (1): 15–20.

Johari, A., Nyakuma, B.B., Nor, S.H.M. et al. (2015). The challenges and prospects of palm oil based biodiesel in Malaysia. *Energy* 81: 255–261.

Jouanneau, S., Recoules, L., Durand, M.J. et al. (2014). Methods for assessing biochemical oxygen demand (BOD): a review. *Water Research* 49: 62–82.

Khanam, T., Ata, S.W., Nur, W., and Rashedi, A. (2016). Particle size measurement in waste water influent and effluent using particle size analyzer and quantitative image analysis technique. In: *Advanced Materials Research*, vol. 1133 (ed. K. Noorsal), 571–575. Trans Tech Publications.

Kim, H., Lim, H., and Colosimo, M.F. (2007). Determination of chemical oxygen demand (COD) using ultrasound digestion and oxidation-reduction potential-based titration. *Journal of Environmental Science and Health Part A* 42 (11): 1665–1670.

Kurniawan, T.A., Lo, W.H., and Chan, G.Y. (2006). Degradation of recalcitrant compounds from stabilized landfill leachate using a combination of ozone-GAC adsorption treatment. *Journal of Hazardous Materials* 137 (1): 443–455.

Li, W., Hua, T., Zhou, Q. et al. (2010). Treatment of stabilized landfill leachate by the combined process of coagulation/flocculation and powder activated carbon adsorption. *Desalination* 264 (1–2): 56–62.

Liew, W.L., Kassim, M.A., Muda, K. et al. (2015). Conventional methods and emerging wastewater polishing technologies for palm oil mill effluent treatment: a review. *Journal of Environmental Management* 149: 222–235.

Limkhuansuwan, V. and Chaiprasert, P. (2010). Decolorization of molasses Melanoidins and palm oil mill effluent phenolic compounds by fermentative lactic acid bacteria. *Journal of Environmental Sciences* 22 (8): 1209–1217.

Loh, S.K., Lai, M.E., Ngatiman, M. et al. (2013). Zero discharge treatment technology of palm oil mill effluent. *Journal of Oil Palm Research* 25 (3): 273–281.

Loh, S.K., Nasrin, A.B., Mohamad Azri, S. et al. (2017). Biogas capture–a means of reducing greenhouse gas emissions from palm oil mill effluent. *Oil Palm Bulletin* 75: 27–36.

Madaki, Y.S. and Seng, L. (2013). Pollution control: how feasible is zero discharge concepts in Malaysia palm oil mills. *American Journal of Engineering Research* 2 (10): 239–252.

Manhokwe, S., Parawira, W., and Zvidzai, C. (2015). Aerobic mesophilic treatment of potato industry wastewater. *International Journal of Water Resources and Environmental Engineering* 7 (7): 92–100.

Mansor, U.Q., Yahya, A., Shafie, N.F. et al. (2017). An overview of anaerobic treatment processes performance treating palm oil mill effluent (POME)-past, present and future. *Advanced Science Letters* 23 (5): 4179–4183.

Mara, D. (2013) *Domestic Wastewater Treatment in Developing Countries*, Routledge.

Metcalf and Eddy, Inc., Burton, F.L., Stensel, H.D., and Tchobanoglous, G. (2014). *Wastewater Engineering: Treatment and Reuse*. McGraw Hill.

Mohammed, R.R. and Chong, M.F. (2014). Treatment and decolorization of biologically treated palm oil mill effluent (POME) using banana peel as novel biosorbent. *Journal of Environmental Management* 132: 237–249.

Mohan, S.V. and Karthikeyan, J. (1997). Removal of lignin and tannin colour from aqueous solution by adsorption onto activated charcoal. *Environmental Pollution* 97 (1): 183–187.

Mohana, S., Desai, C., and Madamwar, D. (2007). Biodegradation and decolourization of anaerobically treated distillery spent wash by a novel bacterial consortium. *Bioresource Technology* 98 (2): 333–339.

Mojiri, A., Aziz, H.A., Zaman, N.Q. et al. (2014). Powdered ZELIAC augmented sequencing batch reactors (SBR) process for co-treatment of landfill leachate and domestic wastewater. *Journal of Environmental Management* 139: 1–14.

MPOB. 2016. Sumarry of the Malaysian Oil Palm Industries (2015) Retrieved from. http://bepi.mpob.gov.my/index .php/summary/729-summary-2015.html (January 1, 2016).

Najafpour, G., Yieng, H.A., Younesi, H., and Zinatizadeh, A. (2005). Effect of organic loading on performance of rotating biological contactors using palm oil mill effluents. *Process Biochemistry* 40 (8): 2879–2884.

Nasrullah, M., Singh, L., Mohamad, Z. et al. (2017). Treatment of palm oil mill effluent by electrocoagulation with presence of hydrogen peroxide as oxidizing agent and polialuminum chloride as coagulant-aid. *Water Resources and Industry* 17: 7–10.

Neoh, C.H., Lam, C.Y., Lim, C.K. et al. (2014). Decolorization of palm oil mill effluent using growing cultures of Curvularia clavata. *Environmental Science and Pollution Research* 21 (6): 4397–4408.

Ng, K.H. and Cheng, C.K. (2017). Photocatalytic degradation of palm oil mill effluent over ultraviolet-responsive Titania: successive assessments of significance factors and process optimization. *Journal of Cleaner Production* 142: 2073–2083.

Oswal, N., Sarma, P.M., Zinjarde, S.S., and Pant, A. (2002). Palm oil mill effluent treatment by a tropical marine yeast. *Bioresource Technology* 85 (1): 35–37.

Othman, M.R., Hassan, M.A., Shirai, Y. et al. (2014). Treatment of effluents from palm oil mill process to achieve river water quality for reuse as recycled water in a zero emission system. *Journal of Cleaner Production* 67: 58–61.

Paraskeva, P. and Diamadopoulos, E. (2006). Technologies for Olive Mill Wastewater (OMW) treatment: a. *Review. Journal of Chemical Technology and Biotechnology* 81 (9): 1475–1485.

Poh, P.E. and Chong, M.F. (2009). Development of anaerobic digestion methods for palm oil mill effluent (POME) treatment. *Bioresource Technology* 100 (1): 1–9.

Qamaruz-Zaman, N., Yaacof, N., and Faraji, H. (2016). Anaerobic digestion of palm oil mill residues for energy generation. In: *Recycling of Solid Waste for Biofuels and Bio-Chemicals* (eds. P. Karthikeyan, K. Heimann and S. Muthu), 187–210. Singapore: Springer.

Rakamthong, C., and Prasertsan, P. (2011) Decolorization and phenol removal of anaerobic palm oil mill effluent by phanerochaete chrysosporium ATCC 24725. In *TIChE international conference*. Hatyai, Songkhla Thailand.

Rivas, F.J., Beltrán, F., Carvalho, F. et al. (2004). Stabilized leachates: sequential coagulation–flocculation+ chemical oxidation process. *Journal of Hazardous Materials* 116 (1–2): 95–102.

Rupani, P.F., Singh, R.P., Ibrahim, M.H., and Esa, N. (2010). Review of current palm oil mill effluent (POME) treatment methods: vermicomposting as a sustainable practice. *World Applied Sciences Journal* 11 (1): 70–81.

Sa'at, S.K.M. and Zaman, N.Q. (2017, 2017). Phytoremediation potential of palm oil mill effluent by constructed wetland treatment. *Engineering Heritage Journal/Galeri Warisan Kejuruteraan* 1 (1): 49–54.

Sa'at, S.K.M., Zaman, N.Q., Yusoff, S.M., and Ismail, H.A. (2017). Investigation of the potential of *Cyperus alternifolius* in the phytoremediation of palm oil mill effluent. In: *AIP Conference Proceedings*, vol. 1892, No. 1 (eds. U. Wagenknecht, P. Potschke, S. Wiessner and M. Gehde), 040009. Kuala Lumpur, Malaysia: AIP Publishing.

Sa'at, S.K.M., Zaman, N.Q., Yusoff, S.M., and Farraji, H. (2019). The selection of palm oil mill effluent for phytoremediation treatment system using *Cyperus alternifolius. Int. J. Environmental Engineering* 10 (2): 106–117.

Said, M., Mohammad, A.W., Nor, M.T.M. et al. (2015). Investigation of three pre-treatment methods prior to nanofiltration membrane for palm oil mill effluent treatment. *Sains Malaysiana* 44 (3): 421–427.

Sia, Y.Y., Tan, I.A.W., and Abdullah, M.O. (2017). Adsorption of colour, TSS and COD from palm oil mill effluent (POME) using acid-washed coconut shell activated carbon: kinetic and mechanism studies. In: *MATEC Web of Conferences*, vol. 87, 03010. EDP Sciences.

Sompong, O., Boe, K., and Angelidaki, I. (2012). Thermophilic anaerobic co-digestion of oil palm empty fruit bunches with palm oil mill effluent for efficient biogas production. *Applied Energy* 93: 648–654.

Sperling, M.V. (2007). *Activated Sludge and Aerobic Biofilm Reactors*, 322. IWA Publishing.

Sulaiman, N. and Ling, C.K. (2004). Membrane ultrafiltration of treated palm oil mill effluent (POME). *J Teknol* 41: 113–120.

Tabassum, S., Zhang, Y., and Zhang, Z. (2015). An integrated method for palm oil mill effluent (POME) treatment for achieving zero liquid discharge–a pilot study. *Journal of Cleaner Production* 95: 148–155.

Tan, Y.H., Goh, P.S., Ismail, A.F. et al. (2017). Decolourization of aerobically treated palm oil mill effluent (AT-POME) using polyvinylidene fluoride (PVDF) ultrafiltration membrane incorporated with coupled zinc-iron oxide nanoparticles. *Chemical Engineering Journal* 308: 359–369.

Tantipaibulvut, S., Pinisakul, A., Rattanachaisit, P. et al. (2015). Ethanol production from desizing wastewater using co-culture of Bacillus subtilis and Saccharomyces cerevisiae. *Energy Procedia* 79: 1001–1007.

Tsai, J.H., Macedonio, F., Drioli, E. et al. (2017). Membrane-based zero liquid discharge: myth or reality? *Journal of the Taiwan Institute of Chemical Engineers.* 80: 192–202.

Ubani, S.C., Chukwudi, O., Okpashi, V.E. et al. (2017). Palm oil mill effluent effect on soil fertility: a longitudinal assessment of Zea mays plant. *EQA-International. Journal of Environmental Quality* 23 (1): 43–53.

USDA (2016) Oil Seed World Markets and trade, Retrieved from. *United States Department of Agriculture*. http://apps.fas.usda.gov/psdonline/circulars/oilseeds.pdf. (June 1, 2016)

Vijayaraghavan, K., Ahmad, D., and Aziz, M.E.A. (2007). Aerobic treatment of palm oil mill effluent. *Journal of Environmental Management* 82 (1): 24–31.

Wah, W.P., Sulaiman, N.M., Nachiappan, M., and Varadaraj, B. (2002). Pre-treatment and membrane ultrafiltration using treated palm oil mill effluent (POME). *Songklanakarin Journal Science Technology* 24: 891–898.

Wang, J., Mahmood, Q., Qiu, J.P. et al. (2015). Zero discharge performance of an industrial pilot-scale plant treating palm oil mill effluent. *BioMed Research International* 2015: 1–9.

Wu, T.Y., Mohammad, A.W., Jahim, J.M., and Anuar, N. (2009). A holistic approach to managing palm oil mill effluent (POME): biotechnological advances in the sustainable reuse of POME. *Biotechnology Advances* 27 (1): 40–52.

Wu, T.Y., Mohammad, A.W., Jahim, J.M., and Anuar, N. (2010). Pollution control technologies for the treatment of palm oil mill effluent (POME) through end-of-pipe processes. *Journal of Environmental Management* 91 (7): 1467–1490.

Yacob, S., Hassan, M.A., Shirai, Y. et al. (2005). Baseline study of methane emission from open digesting tanks of palm oil mill effluent treatment. *Chemosphere* 59 (11): 1575–1581.

Yuniarto, A., Noor, Z.Z., Ujang, Z. et al. (2013). Bio-fouling reducers for improving the performance of an aerobic submerged membrane bioreactor treating palm oil mill effluent. *Desalination* 316: 146–153.

Zahrim, A.Y., Rachel, F.M., Menaka, S. et al. (2009). Decolourisation of anaerobic palm oil mill effluent via activated sludge-granular activated carbon. *World Appl. Science Journal* 5 (Special Issue for Environment): 126–129.

Zhang, Y., Li, Y.A.N., Xiangli, Q.I.A.O. et al. (2008). Integration of biological method and membrane technology in treating palm oil mill effluent. *Journal of Environmental Sciences* 20 (5): 558–564.

Zinatizadeh, A.A.L., Mohamed, A.R., Abdullah, A.Z. et al. (2006). Process modeling and analysis of palm oil mill effluent treatment in an up-flow anaerobic sludge fixed film bioreactor using response surface methodology (RSM). *Water Research* 40 (17): 3193–3208.

Zouari, N. and Ellouz, R. (1996). Toxic effect of coloured olive compounds on the anaerobic digestion of olive oil mill effluent in UASB-like reactors. *Journal of Chemical Technology and Biotechnology* 66 (4): 414–420.

Part G

Green Water Harvesting

21

Vegetation Advantages for Water and Soil Conservation

Hadis Salehi Gahrizsangi[1], Saeid Eslamian[2], Nicolas R. Dalezios[3], Anna Blanta[3], and Mohadaseh Madadi[4]

[1] *School of Food, Plant and Animal Industries, Melbourne Polytechnic, Melbourne, Australia*
[2] *Department of Water Engineering, College of Agriculture, Isfahan University of Technology, Isfahan, Iran*
[3] *Department of Civil Engineering, University of Thessaly, Volos, Greece*
[4] *Department of Architecture, Islamic Azad University, Shahinshahr Branch, Isfahan, Iran*

21.1 Introduction

Humanity has an enormous advantage on Earth: the ability to grow food. Every other lifeform spends most of their day gathering and hunting for nutrition. Today, billions of people are sustained by what farmers grow. When it comes to survival, growing food is clearly an asset. However, this upper hand is fragile (Monsanto).

Soil and water conservation constitute a worldwide strategy for sustainable natural resource management. "Soil and water conservation are those activities at the local level, which maintain or enhance the productive capacity of the land including soil, water and vegetation in areas prone to degradation through: prevention or reduction of soil erosion, compaction, salinity; conservation or drainage of water; and maintenance or improvement of soil fertility" (Department of Earth Sciences, Free University of Berlin). For all farmers and people engaged in conservation, reducing soil erosion has long been a priority. On the other hand, there is significant research effort to evaluate different methods for soil conservation and deduction of soil loss. Traditionally, soil erosion is described as a natural incident and usually occurs at a very low level. However, it can become a serious issue when human activities or extreme weather events change the ecological balance (Department for Environment, Food and Rural Affairs [DEFRA] 2005). It has also been proven that soil erosion can remove the soil much faster than soil-forming procedures can replace it (European Soil Data Centre 2009). It has been stated that soil erosion is a two-stage procedure consisting of detachment of single soil particles from the soil mass followed by their transport via running water or wind (Morgan 2005). When adequate energy is not available anymore to transport the particles, deposition (third phase) occurs. Furthermore, in specific areas, environmental conditions control the type and rate of erosion. These conditions consist of climate, topography, soil, and land cover as the primary components (Toy et al. 2002).

Worldwide, several basic methods have been used to control soil erosion. Maintaining ground cover, reducing of soil disturbance, and application of barriers are examples of these methods. It has been indicated that crop and vegetation management are among the most useful methods, since they are highly efficient and comparatively less expensive. Grass is the most common plant chosen for this purpose. Significant attention has been given to *Vetiveria zizanioides* L (known as Vetiver grass), which is presented in detail in this chapter. The chapter is organized as follows: First, background information is presented on theoretical research aspects of erosion. The next section describes the vegetation advantages for water and soil conservation in artificial plots. This is followed by the presentation of a case study that has been conducted on Vetiver grass for soil and water conservation in artificial plots in Malaysia.

21.2 Background

A brief review of selected theoretical research is presented in this chapter.

21.2.1 Soil Erosion Concepts

Soil erosion has constituted an environmental issue worldwide for millennia. Especially for countries surrounding the Mediterranean basin, this concern has been more serious. The major motivation for scientific research on soil erosion and conservation, particularly in the United

Handbook of Water Harvesting and Conservation: Basic Concepts and Fundamentals, First Edition. Edited by Saeid Eslamian and Faezeh Eslamian.

States, came from Hugh Hammond Bennett, who led the soil conservation movement in the 1920s and 1930s. In Western Europe, there was a growing realization from the 1970s that soil erosion could have a major effect on soils, even in lowland areas (Morgan 2005).

Soil erosion is a form of soil degradation and occurs where mankind's direct or indirect activities cause soil to become less healthy. Generally, background erosion removes soil at approximately the same rate as soil is formed. Nevertheless, accelerated soil erosion is a recent problem due to humans' unwise actions, such as overgrazing or inappropriate cultivation methods. In accelerated erosion, soil loss happens much faster than soil formation (Wall 2013). Generally, mass movement, water erosion, and wind erosion are the three types of soil erosion. Indeed, soil erosion is a natural procedure affecting various landforms (Ritter and Eng 2012). In agricultural terms, soil erosion is described as tearing away of a field's topsoil by the natural physical forces of water (Figure 21.1) and wind (Figure 21.2) or through forces related to agricultural activities, such as tillage.

An estimate of the ability of soil to resist erosion is called soil erodibility, which depends on the physical characteristics of each soil. Soil structure, percentage of organic matter, and soil permeability contribute to erodibility. However, texture is the main characteristic affecting the ability of soil to resist erosion. Generally, good infiltration rate, high levels of organic matter, and a proper soil structure provide great resistance to erosion. It is expressed that soils with sand, sandy loam, and loam texture tend to be less erodible compared to soils with silt, very fine sand, and certain clay textures (Ritter and Eng 2012). It has also been stated that soil erosion normally occurs in three phases (O'geen and Schwankl 2006). Detachment of soil particles and transport of those particles are the first two phases. Soil erosion usually ends with the third phase, which is the deposition of soil particles in a new location. Bare soils with less cover of living or dead plant biomass are at high risk for soil erosion.

Figure 21.1 Water erosion. Source: Pixabay 2012.

Figure 21.2 Wind erosion. Source: Wikimedia Commons 2009.

21.2.2 Water-Induced Erosion

Water-induced erosion is a natural phenomenon and its main motivators are severe rainfall, topography, low soil organic matter content, percentage, and type of vegetation cover. However, human activities like inappropriate cultivation techniques and cropping methods, changes in hydrological conditions, deforestation, and land marginalization can affect the ability of soil to resist erosion (Sustainable Agriculture and Soil Conservation [SOCO] 2009).

Three main types of water-induced soil erosion have been reported so far. These types are described as sheet, rill, and gully erosion. Natural processes, such as heavy rain, contribute toward soil erosion, however, it has been shown that human activities like inappropriate irrigation is expediting this procedure. There is a strong evidence that the soil of the past millennium is being washed away by unconscious human activities. It is believed that this type of erosion threatens human capacity to provide food and fiber for the increasing world population (O'geen and Schwankl 2006). This ability is intimately related to economic vitality, environmental quality, and human health concerns. Every year billions of tons of fertile soil are wasted worldwide from agricultural systems. Soil's productivity is degraded by erosion in several ways. Soil erosion increases runoff, but reduces the efficiency of plant nutrient use, plants' rooting depth, and soil infiltration rate, as well as soil's water-holding capacity and permeability. It also damages seedlings. Deposition of sediments by erosive water can bury seedlings and cause the formation of surface crusts that reduce seedling emergence, finally resulting in reduced crop yield. Eventually, a combination of soil degradation and poor plant growth can cause an even greater erosion. These effects happen at or close to the erosion site. Transport of sediments, nutrients, and agricultural chemicals are the off-site impacts and might be even more costly than on-site impacts (O'geen and Schwankl 2006).

There are on-site and off-site damages due to water-induced erosion (SOCO 2009). Some of the on-site damages are loss of organic matter, soil structure degradation, soil surface compaction, and reduction of water infiltration, as well as supply to the water table. In addition, soil loss at the surface, nutrient removal, increase of the coarse fraction of soils, rill and gully formation, plant uprooting, and decrease in soil productivity are all reported as on-site damages. On the other hand, off-site damages are noticeable as well. These off-site damages include water pollution, water eutrophication, floods, burial of infrastructure, obstruction of drainage networks, changes in shape of watercourses, and finally, silting up of waterways and ports (SOCO 2009).

It is very important to be able to distinguish evidence of soil erosion in the field, and visual indicators of soil erosion are revealed by the United States Department of Agriculture (USDA)–Natural Resources Conservation Service (NRCS) (O'geen and Schwankl 2006). Bare soil, plants or rocks on pedestals, exposed roots, small benches of soil behind obstacles, and surface soil crusts are some of these indicators. Another sign of soil erosion could be deposition of soil, where there is a change in field's slope. More indicators are reported as: decreased thickness of topsoil, exposed subsoil at the soil surface, and visible rills or gullies, as well as silt-clouded water or sediment deposits in surface water bodies and irrigation canals. The volume of soil lost is much more than it appears; for example, 1 mm of soil loss over 1 ha is equivalent to 11–16 ton. The false opinion that the eroded soil stays within the field happens when deposition of coarse sand is observed at the foot of a field. The small material, along with nutrients and pesticides, travels much further. Where erosion stops depends on the farm and landscape features, like roads and tracks, and natural slopes.

21.2.3 Water-Induced Erosion in the Slope and Agricultural Farms

DEFRA (2005) stated that wherever soils, especially those with a high sand or silt content, are exposed to heavy or protracted rainfall, the risk of erosion by water becomes higher. In localized areas, apparent signs of soil loss from fields, such as deposition on roads or in watercourses can be seen when erosion occurs. Problems with erosion have become more obvious in recent years due to changes in cropping patterns, field sizes, soil-management practices, and livestock enterprises. Discolored runoff conveying soluble nutrients, pesticides, and lately applied animal manures can enter watercourses and cause pollution even though no sediment is carried off farmland. In addition, soil erosion can be a noticeably destroy the ecology of watercourses, ponds, lakes, etc. Research results have indicated that the majority of the soil splashed due to heavy rain is not instantly lost from the field (Al Kaisi 2008). Most of the splashed soil particles do not leave the field; they block surface pores that decrease water infiltration and expedite water runoff and soil erosion.

In a case study carried out in the Northern Ethiopian Highland, assessment of erosion and soil erosion processes were conducted in two catchment areas. Direct linkage between natural processes and human intervention was concluded from the field study and the mapping and modeling of soil erosion in these two catchment areas. Because of the high pressure on landscape stability, processes caused by human intervention to livestock farming intensify natural erosion processes. Cultivation, livestock farming, density of settlements, and the migration of people and livestock are the man-made factors that cause and intensify soil erosion processes. It was also revealed in this research that soil erosion processes are altogether very similar. Because of the compaction, the rate of the soil infiltration of water decreases and surface runoff increases. Disturbed soil properties also increase erodibility of the soil. Hence, it can be concluded that several erosion forms and badlands develop leading to clearance, overgrazing, and migration of people and livestock, as well as the poor maintenance of existing soil conservation measures. The alarming impact of channeled runoff along footpaths and cattle treads shows the development of almost all gully-system (Thiemann et al. 2005).

It is noticeable that significant soil erosion-related research has been conducted worldwide through framework simulation and soil erosion plots. This simulation helps researchers to set up a desirable environment. Following is the summary of two case studies, which are conducted by simulation of soil erosion plots. Nowocień et al. (2004) simulated a framework in order to estimate outflow and sediment uptake. Soil susceptibility to surface washout based on simulated rain conditions was examined on a field consisting of 10 experimental plots at terrain slope 10%. The experimental area included 20 bottomless chests (0.3, 1.0, 2.0 m) (Figure 21.3), which were placed in trenches organized in two rows, separated by a 3 m wide technological path (Figure 21.4). Intervals between

Figure 21.3 Experimental micro-plots. Source: Nowocień et al. 2004.

individual plots in each row were 1 m from each other. Surface steepness of all plots was 10%. The hutches have been filled up with soil material taken from upper horizons of 10 chosen soil types, which were then held in black fallow (with no plant cover) during the entire experimental period. Two probes (Time Domain Reflectometer [TDR]) were installed on each of the experimental plots to measure humidity. Separated soil matter was piped away by 5% steep runnel through release pipe to the measurement collector. The soil loss in experimental plots was filled up with fresh soil of the same origin. Each time the surface of the micro-plot was arranged to achieve fallow-like soil surface structure (Nowocień et al. 2004).

Moreover, in another case study, the efficacy of using dairy manure compost as an erosion control method is evaluated by plot simulation. In this study, eight (0.9 m by 1.8 m) plots were installed on a custom-built steel bed (9.1 m × 1.8 m × 228.6 mm deep) divided with metal margins (Figure 21.5). At the downslope end of each plot, a tray and downspout were built to carry runoff to a sampling container (Mukhtar et al. 2008).

It is good to remember that the key factors of controlling hillslope erosion are: the use of land based on its capability, protecting the soil surface with different types of cover, and controlling runoff before it develops into an erosive force (Queensland Department of Science, Information Technology, Innovation and the Arts [DSITIA] 2012).

Figure 21.4 Experimental area. Source: Nowocień et al. 2004.

Figure 21.5 Simulated plots designed for the evaluation of soil erosion control by using dairy manure compost. Source: Mukhtar et al. 2008.

21.2.4 Soil and Water Conservation by Crop Management

Soil is the earth surficial substantial that is considered a medium for plant growth. Objectives of land management must be cleared before questions, such as "where to invest erosion control methods" and "how to measure return-on-investment" can be answered (Bui et al. 2011). In a case study, it was stated that surface erosion can reduce crop yield, pasture growth, and tussock biomass. However, appropriate soil conservation management can reduce most of these production losses. According to this study, it is significant to ensure that a healthy, vegetative ground cover exists over the soil for as long as possible. Thus, methods that enhance the soil fertility, soil organic matter, and soil flora and fauna, as well as application of the best plant species for the site, should be approved (Cairns et al. 2001).

It was shown that ground cover has an important function in controlling soil erosion. DSITIA (2012) described ground cover as: vegetation, living and dead, biological crusts, and stone that is in contact with the soil surface. Ground cover plays a significant role in reducing the loss of topsoil by storm rainfall and surface runoff through protecting the surface. According to both time and space, the amount and type of protective groundcover will change, because of the physical features of the soils, the climate and land use, and the type of land management practices being used by the landowner. Hence, the level of ground cover prior to storm events may vary from high to bare ground. It has been assessed in this research that high levels of surface and ground cover can reduced the amount of soil erosion from hill slopes during storm rainfall. It was also reported that sufficient surface cover is a significant factor in controlling erosion, since it can slow runoff by reducing the erosive effect of raindrops falling onto bare soils. Once soil cover levels reach more than 30–40%, a remarkable reduction in the risk of erosion is expected to occur. Nevertheless, higher levels of cover should be aimed for in order to achieve noticeable results. Cover levels of 100% are accessible for plenty of grazing and cropping systems DSITIA 2012.

Grismer and Davis (2012) described the very interesting concept of the effectiveness of cover crop in reducing soil erosion. Specifically, the effect of plant canopy cover on reducing runoff and erosion is mostly related to improving the litter cover, soil macro-porosity, and soil structure, rather than the direct protection against rainfall. Osmond et al. (2012) believed that knowledge of land use, management, and conservation practices are extremely necessary to understanding the effectiveness of conservation programs. Based on this study, it is shown that farmers more often perform conservation methods to control pollutants they can see. For instance, farmers can see soil loss, and try to control soil erosion either by conservation tillage or by terraces and grassed waterways. There is a strong belief that coverage of grass can reduce surface runoff and soil erosion more effectively than other vegetation types. It is observed in a case study that high coverage of grass and deciduous trees could supplement each other to attain a good restoring effect, which would not only decrease surface runoff and soil erosion, but also increase the establishment of fertile islands and enhance the stability of subsurface soils (Fusun et al. 2013).

According to Al Kaisi (2008), critical components for reducing raindrop impact on soil particles are tillage and cropping management systems, because of the accessibility of crop residue to protect the soil surface. High amounts of crop residue, such as corn or fall cover crop, can perform as an effective conservation system and provide abundant residue cover to protect the soil surface from spring rains. It is also stated by these studies that cover crops, permanent vegetation, strip cropping, and planting on the contour can reduce the speed of water runoff and slow soil erosion in steep slope areas. "The faster the crop is growing, the sooner a crop canopy will develop, a partial crop canopy is better than none at all" (Al Kaisi 2008). Furthermore, in this study, it is stated that for minimizing soil erosion, it is advisable to evaluate the residue cover, the uniformity of residue distribution, and residue effectiveness. It is also recommended to evaluate plant populations, the damage that the field experienced, and the alternatives for replanting.

Hacisalihoglu (2007) carried out a study in a steep hill slope of the village of Mertesdorf/Ruwertal in Germany and mentioned that improper land use is one of the main reasons for soil erosion and land degradation. In this study, Allgemeine Bodenabtragsgleichung (ABAG) have been applied to specify and compare the soil erosion between the different land use types, like vine growing, forestlands, grasslands, shrubs, and new forestations. It is concluded that the soil erosion varies in a high ratio according to the land use types. It has also been concluded that soil erosion in the vine growing areas were the highest (6.47 tonnes per hectare per year). In addition, grazed grasslands with 1.19 tonnes per hectare per year had the second highest soil erosion and finally the lowest erosion have been determined in the forestlands (0.66 tonnes per hectare per year). When the effects of plant coverage existence on the soil surface were examined, it was concluded that the soil erosion rate in areas with plant coverage is six times lower than in areas with no plant coverage. Hacisalihoglu (2007) revealed that the forestlands have one of the lowest amounts of soil erosion among other lands. In forestlands, trees can prevent direct falling of the rain on the soil surface by leaves, branches, and stems. Soil erosion amount can increase in high slope lands when protective plant coverage is removed from the soil surface or agricultural field works are undertaken without protective precautions. It is also stated (Hacisalihoglu 2007) that land use type when dealing with soil erosion is the most important environmental problem. In another case study, the effects of ground cover on soil loss were analyzed during extreme flood conditions in Queensland, Australia. This study examined the effects of landscape and ground cover on soil loss under the extreme pressures of the floods experienced in Queensland during the period of December 2010 to January 2011. Prior to the floods, the level of ground cover, as well as existing land use and land management, affect the amount of damage experienced during the floods. Areas with low levels of surface ground cover were reported to have higher levels of soil loss and damage to paddocks and stream channels by water flow (Queensland Department of Science, Information Technology, Innovation and the Arts [DSITIA] 2012).

21.2.5 Conservation by Vetiver Grass

As mentioned earlier, grass is one of the best methods for soil and water conservation purposes. The website of USDA revealed useful information of the required characteristics of these grasses NRCS 2010, such as the following: (i) should be perennial, quick to establish, and be able to survive both drought and flood; (ii) should have deep-rooted system and a uniform density of top growth; (iii) should either be sterile or propagate very slowly, so that they do not become weeds when they are planted next to other crops.

Numerous types of erosion control methods have been recommended to develop the land in ways that are suitable for its geography. The application of Vetiver grass is one of these methods and it is gradually becoming popular in several countries. The Vetiver System (VS) was applied for the first time by the World Bank for soil and water conservation in India in the mid-1980s and it is based on the utilization of some phenomenal features of Vetiver grass (*Vetiveria zizanioides* L). These features consist of a massive and deep root system and tolerance to extreme climatic variations like prolonged drought, flood, submergence, fire, frost, and heat waves. It is also tolerant to an extensive range of soil acidity, alkalinity, salinity, sodicity agrochemicals, and high levels of heavy metals in the soil (Truong 2000).

Acording to Yong (2000), Vetiver grass' natural habits, physiology, architecture, self-rising ability, underground networking, and many other characteristics make this grass very unique and effective for soil conservation purposes. Joy (2009) has mentioned that there are alternative names for Vetiver grass, such as kus-kus, Vetiver (both English), mulimuli (Fijian), chiendent odorant (French), garara (Hindi), patchuli-falso (Portuguese), and pacholi (Spanish). It is also expressed by Truong and Loch (2004) that Vetiver is a noninvasive grass and it does not have runners and rhizomes. Vetiver is being widely applied for steep slope stabilization in many countries, such as countries in the Caribbean, Fiji, India, Africa, Malaysia, and Thailand. It is noticeable that other related grasses, like lemongrass and citronella grass, have also been used as vegetative erosion hedges, however they have been less successful than Vetiver grass. Truong and Loch (2004) have also revealed the sketch of a Vetiver plant in their search (Figure 21.6) in which: (i) silt-loaded runoff is slowed down by the plant; (ii) silt drops out of the water behind the plant; (iii) sediment free water continues down the slope; and (iv) the dense root system holds the soil together to a depth up to 3 m.

Nowadays, Vetiver grass is widely applied for soil conservation on steep lands worldwide, including Africa, Asia, central and south America, and southern Europe (Figure 21.6). Moreover, it is extensively used in Australia to stabilize steep batters of road and railway in north, central, and southeast Queensland. On the other hand, VS has been used in China for erosion and sediment control on more than 150 000 km of railway, highway, and road batters (Truong and Loch 2004).

According to Brokish (2011), reducing soil erosion has long been a priority for Hawaii's farmers and people involved in conservation. It is explained in this study that effective approaches to prevent soil erosion normally involve land grading to make various clay berms in order to reduce or direct runoff. However, these methods generally are very expensive and reduce space available for farming in Hawaii's many small farms. Specifically, Vetiver grass

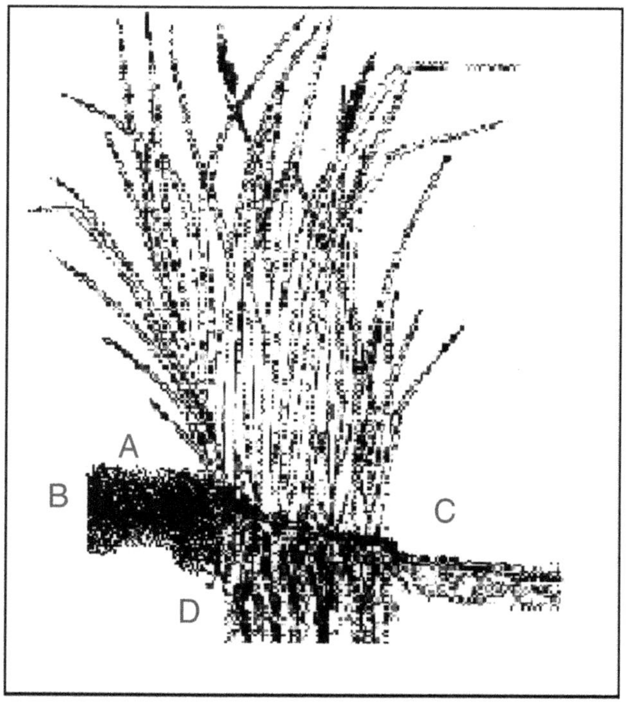

Figure 21.6 Sketch of a Vetiver plant. Source: Truong and Loch 2004.

Figure 21.7 Vetiver hedges enhanced productive area on this Hawaii Kai Farm. Source: Brokish 2011.

is named as a "promising alternative" to conventionally constructed berms, as well as a significant low-cost erosion and sediment control technology (Figure 21.7).

Truong (2000) conducted a research to study the competition of soil moisture between crops and Vetiver hedges. This study clearly indicated that there was no competition between Vetiver and both shallow-rooted crops (bean) and deep-rooted crops (sorghum) under irrigation. It is also expressed that when Vetiver is planted in rows it can form a thick hedge with its stiff stems. These hedgerows can stand up to water flow at least 0.6 m deep by forming a living barrier that slows and spreads runoff water (Truong 2000).

Besides, application of Vetiver for soil erosion prevention in Cassava fields was also conducted in Thailand. It is indicated in this research that Cassava is a plant with high rates of soil erosion, particularly if it is grown in sloping sandy soils. The results of 10 years of application of this project have had a major impact on farmers' knowledge of the importance of reducing soil erosion. According to this study, after examining different options to reduce soil losses by erosion, the planting of Vetiver grass hedgerows was selected as the most appropriate and efficient erosion control method. As a result, implementation of a project that has its targets to conserve the soil, water, and the environment, must include the people of the entire community, or at least, it must start with some parts of the community that participate in the plan. Moreover, the seriousness of the problems that need to be solved must be clear for the villagers. Encouraging the villagers to share their opinions and make joint decisions was another point that should not be ignored (Vongkasem et al. 2003).

21.3 Vegetation Advantage for Soil and Water Conservation in Artificial Plots

In the Malaysian peninsula, soil erosion is a common natural incident, because of the specific topography, soils, and corresponding vegetation that predominate and the extensive rainfall that the country experiences. Nevertheless, due to prompt land use developments, accelerated soil erosion is becoming a serious issue. The focus of this study was to demonstrate that Vetiver can be used for soil conservation and water retention, especially for agricultural farms on the slope, without worrying about the negative effects of Vetiver on the adjacent crops. The main concentration in this research was on water erosion.

21.3.1 Soil Erosion in Malaysia

Peninsular Malaysia has a moderate erosion hazard as judged by the concept of drainage density, such as the length of streams per unit. It is also shown that extensive areas of Malaysia land are available for agriculture, both in Peninsular Malaysia, Sabah, and Sarawak (Wong and Chen 2002). In fact, in recent years, uncontrolled clearance of forestlands and replacing them with agricultural farms, absence of over crop during heavy rain, as well as concentrating on man-made soil erosion control methods rather than natural methods, such as cover crops, were some of the most common factors that increased soil erosion in Malaysia.

On the other hand, there is a strong belief that most of the studies related to soil erosion, which have previously been virtually ignored, must now be highlighted to avoid uncontrolled replacement of forest by cultivated crops. It is described that Malaysia suffered serious sedimentation problems early in the 1990s from indiscriminate logging and tin mining, as well as bad cultural practices. Sediment suppressed the area of Kuala Kubu under several feet of silt. Erosion hazards are usually high if tree and ground cover plantings are not completed before the monsoon season, or on small plantations, where soil protection either is not prescribed or is ignored. In recent years, Malaysia has been trying in different ways to control soil erosion in the country. However, this burgeoning issue is not resolved yet and every year a large amount of fertile soil, which can be used for effective agricultural purposes, gets eroded resulting in a massive quantity of sediments. In response to this problem, the current study has proposed to investigate an efficient way of soil conservation in Malaysia. The aim of this research is to carry out a participatory investigation into various options for soil conservation and water retention which are easy to maintain, affordable for farmers, as well as offering high levels of efficiency in the results.

This study also gives benefactors of the research outcomes the assurance that this method is not going to affect the adjacent crops and can harness this method at ease with peace of mind.

Gregersen et al. (2003) conducted a study to look at the linkage between land use and soil erosion in Tikolod, Sabah, Malaysia. In this area, rice and ginger are mainly grown for home consumption. Hence, this study was conducted for rice and ginger that were planted on steep lands and it revealed that the risk of erosion was high in ginger and hill rice fields mostly because of the steep slopes, especially when no conservation measures were used. It also showed that farmers did not relate soil erosion to land management, but to rainfall. Erosion risk expressed as annual soil loss rates for hill rice and ginger and mainly determined by high LS (L: slope length factor, S: slope steepness) varied from 68.6 t $(\text{ha yr}^{-1})^{-1}$. to 669.5 t $(\text{ha yr}^{-1})^{-1}$. As a result, and based on the land use in the study area, it was concluded that the tendency in the Tikolod watershed to cultivate plots near the river may be the reason for higher turbidity in these areas as compared to the other watersheds.

Another study was performed to predict the soil erosion risk by the Universal Soil Loss Equation/Geographical Information System (USLE/GIS) method for planning conservation measures in the site. The USLE/GIS method was applied to predict potential soil erosion in the Tasik Chini catchment. It is proven that soil erosion within catchment varied both spatially and temporally. Generally, soil erosion increases with yearly rainfall, slope, and land use with open canopies. Spatial distributions of different erosion-prone areas were identified by using the USLE/GIS methodology in order to measure erosion control in the affected areas. The outcomes showed that soil erosion in the Tasik Chini catchment was higher than the previous classification in areas with severe soil loss. It is remarkable that the high soil erodibility potential and the lack of conservation practices on open surfaces was the reason of high soil erosion in this catchment. On the other hand, the greatest threat to the environment was human activities. This study concluded that an effective way to control the negative effects to the environment due to human interruption is by controlling human activities (Sujaul Islam et al. 2010).

According to Soo (2011), sedimentation is one of the main problems in Malaysia. In addition, one of the worst sedimentation problems in Malaysia is recorded in the Cameron Highlands. It is mentioned in this study that due to extensive deforestation and indiscriminate land clearing, soil erosion over the land surface of Cameron Highlands can lead to sedimentation of rivers and of the ringlet reservoir. This study conducted research to determine the mean annual soil loss rate by applying the Revised Universal Soil Loss Equation (RUSLE) for the upper Catchment of Cameron Highlands. In this study, data, such as rainfall pattern, soil type, topography, cover management, and support practice, were used for soil modeling. This experiment covered few sub-catchments in the Cameron Highlands. It was concluded that sediments were separated and conveyed from the upper catchments and were finally deposited in one of reservoirs. The amount of sediment was gradually expected to increase as agriculture activities and deforestation continues to take place. Therefore, the life expectancy decreased enormously due to the increasing sediment yield with time.

As mentioned earlier, application of simulated farm plots is quite common among researchers. For instance, Othman et al. (2011) designed a wash trap to estimate surface wash and runoff using close system erosion in Tekala Forest Reserve, Hulu Langat, Selangor, Malaysia. The traps were made from sheets of zinc tin consisting of four parts: a collection tank (100 cm × 40 cm × 50 cm), a lip (100 cm × 25 cm), a cover (100 cm × 60 cm), and a divisor (Figure 21.8). The purpose of using the cover was to prevent direct rainfall from entering, as well as collecting after evaporating. It is mentioned in this study that the divisor was installed at the back of the highest place to transfer the overflow to a lower collection bin. Six erosion plots were installed, along with two slope profiles in the Tekala river catchment. Three plots for each profile were installed at three different slope units.

Outcomes of this study showed that surface wash has a close connection with surface runoff. Moreover, it is stated that the amount of rainfall can be considered as the best parameter linked to both surface wash and surface runoff. It is also mentioned that the rates of surface wash

Figure 21.8 Installed wash trap. Source: Othman et al. 2011.

and surface runoff are not significantly associated with physical properties of soil, like bulk density, organic matter content, and soil texture. Lastly, it was concluded that the relationships between physical features of soil and surface wash and runoff are very complex and require inclusive samples and more detailed research (Othman et al. 2011).

21.3.2 Soil and Water Conservation in Malaysia

Moradi et al. (2012) stated in their study that in Malaysia, four soil conservation methods are often suggested for non-terraced oil palm plantations. These methods are: oil palm Empty Fruit Bunches (EFB), Ecomatand, and pruned oil palm fronds. These three oil palm remainders are used as organic mulching materials. The fourth method is silt pits (SIL), which are soil trenches to gather nutrients from runoff water and then redistribute them back into the soil. In order to determine their relative effects on increasing soil chemical properties, such as pH and cation exchange capacity, as well as oil palm nutrition levels like N and P, a three-year-long field experiment was conducted. In addition, biomass decomposition rate, as well as nutrients release rate in the field by the three mulching materials, was evaluated. Outcomes showed that EFB mulching was remarkably better than the other three methods with regard to improving almost all the measured soil and plant parameters. It is reported that EFB was the most efficient option due to the combined effects of higher amounts of dry matter added and the higher nutrient concentrations in the EFB compared to the other mulching materials. It was also concluded that SIL is not as effective as EFB, since SIL could only trap and return nutrients back into the soil, while EFB could do both, trapping nutrients and releasing additional nutrients into the soil (Moradi et al. 2012).

In another study, Ahmad (2001) stated that cover crops, usually legume plants such as *Calopogoniumcaeruleum*, *Puerariajavanica*, and *Mucunabracteata*, are often planted in most large plantations and smallholdings of oil palm and rubber in Malaysia. This method provides an additional complement of nutrient N to the soil. Instant planting of legumes allows support to cover the shredded stem of oil palm and prevent the spread of the Oryctes beetle. Mulching is a common practice particularly in oil palm plantations and vegetable crops (Ahmad 2001). In oil palm, dead leaves are aligned along the rows of planted palms. The use of EFB as mulch is very common. Previously, most of the EFB used to be burned to generate ash as a substitute for potash. The ideal rate of EFB as mulching is 25 ton ha^{-1} for newly transplanted seedlings.

21.3.3 Case Study: Application of Vetiver Grass for Soil and Water Conservation in Artificial Plots

The main contribution of this research was to investigate the efficiency of Vetiver grass for soil and water conservation. In order to ensure the benefits and advantages of Vetiver grass for the aforementioned purposes, the following steps were carried out. Firstly, the role of Vetiver grass and its effectiveness for soil and water conservation in previous research were studied. In order to understand the entire cycle of Vetiver management techniques for soil and water conservation, it is vital to undergo this process as a stepping stone to build a solid base and rationale for the research. Secondly, application of Vetiver management techniques for soil and water conservation were adjusted specifically for the Malaysian climatic conditions and environment. In order to understand the contribution of Vetiver grass use for soil and water conservation, it is important to account for its application and effectiveness under various climatic conditions and environment. This allowed the selection of techniques that had been used in countries with similar climatic conditions to Malaysia. In the next step, preparation of the research environments was conducted to support the execution of the experimentation. This model environment consisted of four separate plots arranged on a slope of 10%, with rain shelter installed over the entire experiment to protect the plots from heavy rain and have control of the irrigation as well. Several observations of the research outcomes were conducted, specifically for measuring soil texture, soil loss, and runoff water, as well as soil moisture retention. At this stage, the implementation of the experiment was finalized and analysis and observations were performed. Further, findings were consolidated and outcomes were generated to support the use of Vetiver grass as a measure of soil conservation that is cheap, reproduceable, and sustainable. Lastly, in order to finalize and confirm Vetiver grass applications to meet the objectives of this research, it was important to choose the methodologies and the techniques carefully to allow for a cheap, replaceable, and sustainable method for soil and water conservation in Malaysia.

Running this experiment in the field was difficult since it was very hard to have control of the rain and amount of water that the soil samples receive. Hence, simulation of farm plots was carried out in the lab, although there was no control on the temperature and sunlight. In fact, simulated farm plots were installed in the open area with a rain shelter on top to protect the plots from heavy rain. However, temperature and sunlight were not manipulated in the study area. Simulated farm plots, rain shelter, and plot stand (slope simulator) were built in the lab according to the required measurement for this specific study. This

Figure 21.9 Study area after installation of rain shelter and simulated farm plots.

(a) (b) (c)

Figure 21.10 Sweet corn samples after (a) one week, (b) two weeks, and (c) one month.

equipment was installed in an open area near the lab and growing of Vetiver, as well as sweet corn samples, started after that, and were respectively occupied with sweet corn only (P_1), Vetiver only (P_2), sweet corn and Vetiver together (P_3), and soil only (P_4) (Figures 21.9 and 21.10).

Since soil texture is one of main requirements for growing sweet corn, and has an important role in erosion, sieving method analysis and hydrometer tests were carried out to conclude what soil texture was used in this study

Figure 21.11 Soil weight evaluation in the lab.

(Figure 21.11). Results from sieving and hydrometer methods showed that the soil sample used in this experiment consisted of 67% sand, 20% silt, and 13% clay. By applying these results in the soil texture triangle, the texture of soil sample is confirmed to be sandy loam.

Westerfield (2012) stated that sweet corn grows best in loamy, well-drained soils. In addition, it is shown that sandy loam texture is one of the main requirements for sweet corn to thrive (Wells 2001). Hence, it was tried in this experiment to deliver the basic requirements of sweet corn in order to attain the maximum growth level and conduct the research with matured sweet corn samples to reach maximum accuracy in the results.

On the contrary, it has been proven that Vetiver can grow in any type of soil and its growth is not affected by soil texture. A research was conducted to study the effectiveness of Vetiver grass for removing heavy metals in contaminated soils. In this study, two different soil textures were applied with four various soil types (sandy and clay soil with

Figure 21.12 Irrigation by hand sprinkler, runoff and sediment collection.

Figure 21.13 ECH2O sensors remain in the plots for one week.

abundant and poor organic matter). The results of this research showed that Vetiver grass can grow in all four soil types (Minh and Khoa 2009). Therefore, the sandy loam soil texture used in this research was the most suitable for growing of sweet corn and Vetiver samples.

Almost three months of sampling time was allowed to provide enough time for the grass to mature and become suitable for soil conservation. On the fourth month of the plants age, sediment and runoff started being collected. Wasted water and soil were under lab processing for drying and weight comparison. This data collection has provided enough information for comparison of sediment and runoff, which were conveyed by each plot to the sediment collector devices. Comparing the amount of collected sediment and runoff in each plot helps to have a better understanding of the role of Vetiver and sweet corn in reducing runoff and sediment (Figure 21.12). The weight of collected sediment and runoff were examined in the lab and further analysis was conducted via randomized complete block design (RCBD).

Furthermore, soil moisture retention by Vetiver and sweet corn were monitored using ECH2O probes. This probe has a 5 cm sensor that is designed to install inside the soil and scan soil moisture, soil temperature, and soil

Figure 21.14 Summary of the methods.

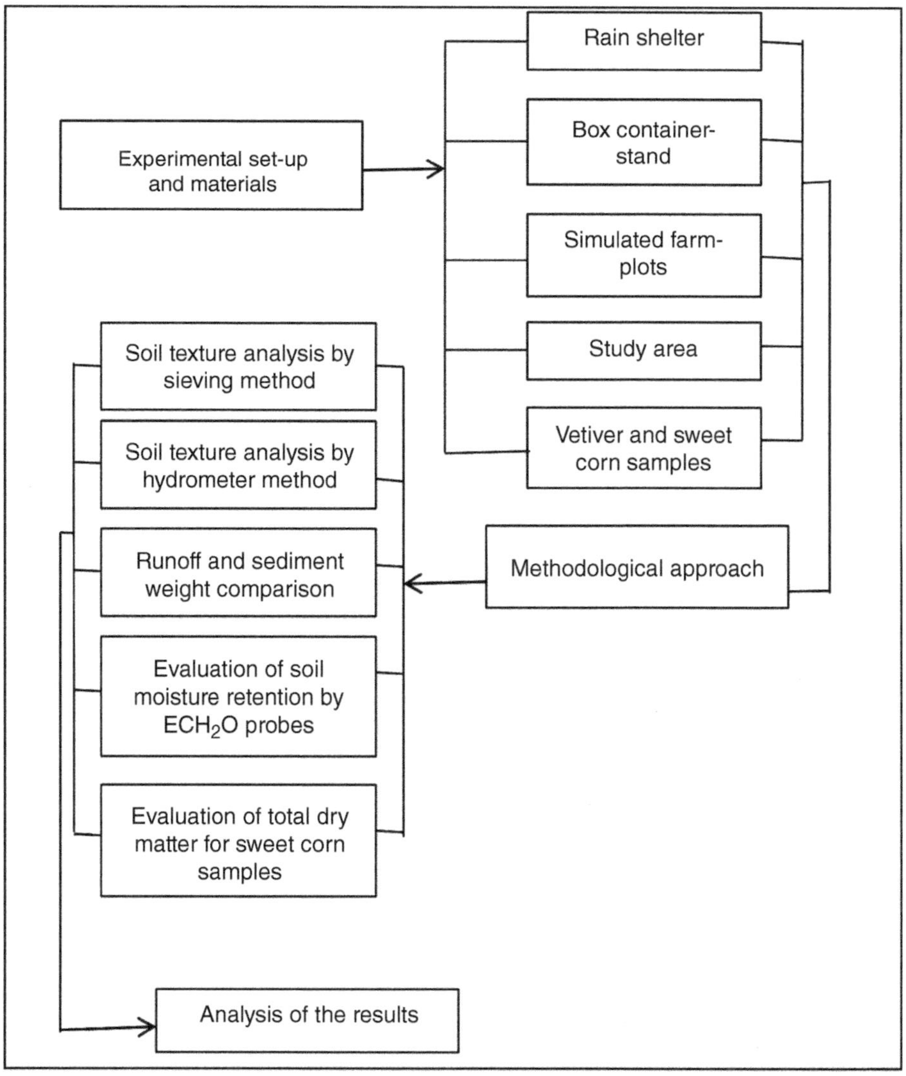

electrical conductivity. ECH20 Utility software comes with these probes and allows users to download the raw data from the data logger to the computer. A data logger is a device that records all data during the experiment, and it is connected to the sensors through a cable. Raw data collected from this experiment was used for further analysis to compare soil moisture retention in the simulated plots. Instantly after irrigation, one probe was installed in each plot and they remained inside the soil for seven days to record the amount of soil moisture every two hours. All the plots were observed with no more irrigation for the remaining six days. It is noticeable that this experiment has been carried out for the first replicate only (Figure 21.13). Figure 21.14 presents a summary of the methods that were used in this research.

Based on this experimental study, the main findings of this study can be summarized as follows: Firstly, the volume of runoff and sediment were significantly reduced in plot number two and three, which were occupied with Vetiver grass only and a combination of Vetiver grass and sweet corn, respectively. In addition, the weight of the collected sediment and runoff from the three replicates showed that the amount of sediment and runoff in P_4 (plot 4) was the highest amount among the rest of experimental plots. On the other hand, P_2 has the lowest volume of sediment and runoff. Secondly, RCBD analysis revealed that there are meaningful differences among the treatments of runoff and sediment weight comparison. Hence, Vetiver grass could effectively control the amount of sediment and runoff in these two plots. Thirdly, the highest recorded soil moisture value in the first day of the experiment belonged to the plot with Vetiver grass. Eventually, it was also concluded that the highest recorded soil moisture value in the last day of the experiment belonged to plot number two. It is noticeable that even though the amount of irrigation water for each plot was the same, the amount of recorded soil moisture differed for each plot at the beginning of the experiment. In fact, less absorption of water, more runoff and sediment, as well as less amount of soil moisture, belonged to the plots without Vetiver (P1 and P4). On the other hand, runoff and sediment were controlled in other plots (P2 and P3). There is no doubt that the presence of Vetiver grass could effectively conserve the soil and water, hence increase the soil moisture retention in these plots.

21.4 Conclusions

Conservation planning is becoming necessary day by day, mainly due to the current weather challenges that the world is experiencing and the prediction uncertainty, as well as the uncertainty in the severity and frequency of rain events. Soil erosion has always been associated with the lack of vegetation cover or residue to protect the soil surface from the rain intensity. Thus, in order to sustain soil quality and maintain high agricultural productivity, conservation practices need to be part of the portfolio of agriculture systems, not only for environmental concerns, but for economic viability. Economic viability comes through the increase in the resilience of agricultural production systems.

The reality is that conservation practices, if treated in a systemic approach, and not as single separated practices, can provide high economic and environmental rewards. Moreover, conservation practices must be an integral and essential component of nutrient reduction and developing sediment and nutrient loading plans should be used as an effective solution in protecting soils and water quality (Al-Kaisi 2012).

According to Agriculture Victoria (2017), land degradation, such as wind and water erosion, salinity, and acidification can significantly reduce farm's productive potential and impact the health of neighboring rivers, wetlands, and native vegetation. Loss of surface soil from erosion has a major impact on productivity through the depletion of soil organic matter and associated soil nutrients, since it destroys the surface structure of the soil resulting in surface sealing and compaction.

As mentioned earlier, water-induced erosion has been an environmental concern worldwide for millennia. The Vetiver grass case study was carried out to provide an efficient way for soil and water conservation that is easy to maintain, affordable for farmers, as well offering high levels of efficiency in the results. The study was set out to explore the concept of soil and water conservation in artificial plots by the application of Vetiver grass. In this study, four simulated farm plots were installed on 10% slope with different treatments consisting of Vetiver grass and sweet corn. Runoff and sediment, as well as soil moisture retention, were evaluated for different treatments to explore the role of Vetiver grass in soil and water conservation in comparison to sweet corn. It was identified that Vetiver grass can be used as an effective bioengineering tool for soil conservation purposes and to enhance the soil moisture retention, particularly in the agricultural farms that have been placed on slope. It was also shown that Vetiver grass is one of the most recommended plants for soil and water conservation purposes.

Having a clear understanding of soil erosion, its procedure, and effects, can assist researchers in discovering new ways to prevent undesirable effects of erosion on environment. In other words, designing various conservation methods to save the natural resources should be considered as one of the most required pieces of knowledge at present. The agricultural community must come out with a new strategy to protect the environment; otherwise, sooner or later all the natural resources will be destroyed by unwise human activities. The bottom line is that, as Theodore Roosevelt once said: "The nation that destroys its soil destroys itself."

References

Agriculture Victoria (2017) *Soil conservation.* [Online] Available at: http://agriculture.vic.gov.au/agriculture/farm-management/land-health/soil-conservation [Accessed 15 March 2019].

Ahmad, F. (2001) Sustainable Agriculture System in Malaysia. Bangkok: Integrated Plant Nutrition System (IPNS), Development in Rural Poverty Alleviation, Thailand.

Al Kaisi, M., (2008) *Cropnews: Iowa State University.* [Online] Available at: http://www.extension.iastate.edu/CropNews/2008/0604MAlKaisiMHelmers.htm [Accessed 11 July 2014].

Al-Kaisi, M. (2012) *Conservation Practices Save Soil and Sustain Agriculture.* [Online] Available at: https://crops.extension.iastate.edu/cropnews/2012/05/conservation-practices-save-soil-and-sustain-agriculture [Accessed 15 March 2019].

Brokish, J. (2011) *Articles: University of Hawaii.* [Online] Available at: http://www.ctahr.hawaii.edu/sustainag/news/articles/V6-Brokish-vetiver.pdf [Accessed 17 January 2014].

Bui, E.N., Hancock, G.J., and Wilkinson, S.N. (2011). Tolerable Hillslope soil Erosion rates in Australia: linking science and policy. *Agriculture, Ecosystems and Environment* 144 (1): 136–149.

Cairns, I., Handyside, B., Harris, M., et al. (2001) *Soil Conservation Technical Handbook.* The Ministry for the Environment. Wellington, New Zealand, *soilerosion-lowlandmanual: defra.* [Online] Available at: www.defra.gov.uk [Accessed 14 July 2014].

Department of Earth Sciences, Free University of Berlin, n.d. *Geolearning - e - learning in the environment and geosciences.* [Online] Available at: https://www.geo.fu-berlin.de/en/v/geolearning/watershed_management/soil_and_water_conservation/introduction_swc/index.html [Accessed 6 March 2019].

Department for Environment, Food and Rural Affairs (DEFRA). (2005). soilerosion- lowlandmanual: defra. Retrieved July 17, 2014, from DEFRA website: www.defra.gov.uk

European Soil Data Centre (2009) *Fact sheet no. 5: Conservation agriculture.* [Online] Available at: http://esdac.jrc.ec.europa.eu/projects/SOCO/FactSheets/ENFactSheet‐05.pdf [Accessed 13 March 2019].

Fusun, S., Jinniu, W., Tao, L. et al. (2013). Effects of different types of vegetation recovery on runoff and soil Erosion on a Wenchuan earthquake-triggered landslide, China. *Journal of Soil and Water Conservation* 68 (2): 138–145.

Gregersen, B., Aalbæk, J., Lauridsen, P.E. et al. (2003). *Land Use and Soil Erosion in Tikolod, Sabah, Malaysia, Tikolod (Sabah).* ASEAN Review of Biodiversity and Environmental Conservation (ARBEC).

Grismer, M.E. and Davis, U. (2012). Standards vary in studies using rainfall simulators to evaluate Erosion. *California Agriculture* 66 (3): 102–107.

Hacisalihoglu, S. (2007). Determination of soil erosion in a steep hill slope with different land-use types: a case study in Mertesdorf (Ruwertal/Germany). *Journal of Environmental Biology* 28 (2): 433–438.

Joy, R.J. (2009). *Sunshine VetiverGrass.* USA: Hoolehua: USDA NRCS Plant Materials Center.

Minh, V. and Khoa, L. (2009). Phytoremediation of cadmium and Lead contaminated soil types by vetiver grass. *VNU Journal of Science, Earth Science* 25: 98–103.

Monsanto, n.d. *Soil Conservation.* [Online] Available at: https://monsanto.com/company/sustainability/soil-conservation [Accessed 6 March 2019].

Moradi, A. et al. (2012). Evaluation of four soil conservation practices in a non-terraced oil palm plantation. *Agronomy Journal* 104 (6): 1727–1740.

Morgan, R. (2005). *Soil Erosion and Conservation.* Malden: Blackwell.

Mukhtar, S., McFarland, M.L., and Wagner, C.A. (2008). Use of dairy manure compost as Erosion control material under vegetated and non vegetated conditions. *American Society of Agricultural and Biological Engineers* 51 (3): 927–936.

Natural Resources Conservation Service (NRCS) (2010) *Cover Crop.* [Online] Available at: http://www.nrcs.usda.gov [Accessed 6 February 2015].

Nowocień, E., Podolski, B., and Wawer, R. (2004). Estimating outflow and sediment uptake chosen polish soil kinds in simulated conditions. *Electronic Journal of Polish Agricultural Universities* 7.

O'geen, T.A. and Schwankl, L.J. (2006) *ANR Publication 8196: Univerity of California.* [Online] Available at: https://anrcatalog.ucanr.edu/pdf/8196.pdf [Accessed 13 March 2019].

Osmond, D. et al. (2012). Improving conservation practices programming to protect water quality in agricultural watersheds: lessons learned from the National Institute of Food and Agriculture-conservation effects assessment project. *Journal of Soil and Water Conservation* 67 (5): 122A–127A.

Othman, J., Sharifah Mastura, S.A., and Toum, S. (2011). The Tekala Forest reserve: a study on surface wash and runoff using close system erosion plots. *Malaysian Journal of Society and Space* 7 (1): 1–13.

Pixabay (2012) *Coastline water shore erosion* [Online] Available at: https://pixabay.com/photos/coastline-water-shore-erosion-54971 [Accessed 13 March 2019].

Queensland Department of Science, Information Technology, Innovation and the Arts (DSITIA), (2012) *Ground Cover Under Pressure,* s.l.: Australian Government Caring for our Country.

Ritter, J. and Eng, P. (2012) *Fcatsheet: omafra.* [Online] Available at: www.omafra.gov.on.ca [Accessed 12 May 2014].

Soo, H.T. (2011) *Soil Erosion Modeling using RUSLE and GIS on Cameron Highlands, Malaysia for Hydropower Development, Master Thesis,* University of Iceland and University of Akureyri, Iceland.

Sujaul Islam, M., Muhammad Barzani, G., Sahibin Abd, R., and Mohd Ekhwan, T. (2010). Soil loss assessment in the Tasik Chini catchment, Pahang, Malaysia. *Bulletin. Geological Society of Malaysia* 56: 1–7.

Sustainable Agriculture and Soil Conservation (SOCO) (2009) *Sustainable agriculture and soil conservation introduction, Fact sheet No. 1.* [Online] Available at: https://esdac.jrc.ec.europa.eu/projects/SOCO/FactSheets/EN%20Fact%20Sheet.pdf [Accessed 13 March 2019].

Thiemann, S., Schütt, B., and Förch, G. (2005). Assessment of Erosion and soil Erosion processes – a case study from the northern Ethiopian Highland. *FWU Journal of Social Science* 3: 173–185.

Toy, T., Foster, G., and Renard, K. (2002). *Soil Erosion: Processes, Prediction, Measurement, and Control.* New York: Wiley.

Truong, P. (2000) *The Vetiver Grass System: Potential Application for Soil and Water Conservation in Northern California.* Brisbane, STIFF Grass Technology Seminar, Australia.

Truong, P. and Loch, R. (2004) *Vetiver System for Erosion and Sediment Control,* Brisbane, Conserving Soil and Water for Society: Sharing Solutions, Australia.

Vongkasem, W., Klakhaeng, K., Watananonta, W. and Howeler, R.H. (2003) *Articulos: International Center for Tropical Agriculture.* [Online] Available at: http://ciat-library.ciat.cgiar.org/Articulos_Ciat/THE_USE_OF_VETIVER_FOR_SOIL.p df [Accessed 4 January 2014].

Wall, G. (2013) *Soil Erosion: Omafra.* [Online] Available at: www.omafra.gov.on.ca [Accessed 25 Jun 2014].

Wells, O.S. (2001) *Resources: University of New Hampshire.* [Online] Available at: https://extension.unh.edu/resources/files/Resource000619_Rep641.pdf [Accessed 17 January 2014].

Westerfield, R.R. (2012) Extension: UGA. [Online] Available at: Univeristy of Georgia Website [Accessed 25 January 2015].

Wikimedia Commons (2009) *Wind erosion.* [Online] Available at: https://commons.wikimedia.org/wiki/File:Wind_Erosion,_Wadi_Hitan.jpg [Accessed 13 March 2019].

Wong, C.C. and Chen, P.C. (2002) [Online] Available at: http://www.fao.org/ag/AGP/AGPC/doc/counprof/malaysia.htm [Accessed 12 January 2014].

Yong, L.H. (2000). *Evaluation of Bioengineering Soil Erosion Control Techniques in Standard USLE Plots.* Malaysia: Universiti Putra Malaysia.

22

Water Harvesting in Forests: An Important Step in Water-Food-Energy Nexus

Rina Kumari[1] *and Saeid Eslamian*[2]

[1] *School of Environment and Sustainable Development, Central University of Gujarat, Gandhinagar, India*
[2] *Department of Water Engineering, College of Agriculture, Isfahan University of Technology, Isfahan, Iran*

22.1 Introduction

Former United Nations Secretary-General Kofi Annan:

> Lack of access to water for meeting basic needs such as health, hygiene, and food security undermines development and inflicts enormous hardship on more than a billion members of the human family. And its quality reveals everything, right or wrong, that is done, in safeguarding the global environment

All major human civilization started on the bank of river, and for livelihood, many people were dependent on forests. Water is an essential commodity for all living beings for various purposes such as agriculture, industrial activity, to sustain life, and for maintenance of the range of ecosystem services needed to support and sustain economic and social activities. The ecosystem services are the benefits which humans receive from nature and natural processes (MEA 2003). Among the various ecosystem services being mapped, those related to water are of prime importance. Characterizing, assessing, and valuing ecosystem services can support forest management and livelihood of local community.

Forest–water relationships are complex and highly contextual, and sustainable development of any area depends on judicious use of these two resources. Over the last few decades, change in land use/land cover and climatic conditions has greatly affected these two systems. Forested ecosystems play an important role in global and local water budgets, besides playing a major role in determining the quality and quantity of water generated. Forests return approximately 40% of total annual precipitation back to the atmosphere in the form of evapotranspiration, which in turn governs the climate of the area. Forests and trees must be recognized as primary players in the regulation of water, energy, and carbon cycles. Forested catchments provide a high proportion of water for the domestic, agricultural, industrial, and ecological needs of both upstream and downstream areas. This is very important for ensuring food security and better livelihood opportunities for rural communities. Forest-driven water and energy cycles are poorly understood at regional, national, and global decision-making levels. Although forests play a very important role in hydrology, a conceptual framework regarding water harvesting and its impact is missing in regional, national, and international policies. A large knowledge gap exists in this regard, and a proper conceptual framework still needs to be developed as far as water harvesting and its impact is concerned. Integrative studies, including all related dimensions, needs to be discussed in the international arena. Also, temporal and spatial assessment using Geographic Information Systems (GIS) and remote sensing will give a true insight on the overall picture. In the present study, a comprehensive framework to improve the assessment efficacy of hydrological services provided by forests has been discussed.

22.2 Global Water Scarcity

Water is an important natural resource for the survival of all living beings and is core to sustainable development of the ecosystem. Food security is also dependent on water security. It is estimated that by 2025, 1.8 billion people will be living in regions with absolute water scarcity and that two-thirds of the world's population will experience water-stress conditions. At the global scale, agriculture is the main water consumer, using 75% of abstractions, far outweighing the amounts used for industrial (20%) or domestic purposes (5%). However, significant regional disparity also exists due to local climatic and

Handbook of Water Harvesting and Conservation: Basic Concepts and Fundamentals, First Edition. Edited by Saeid Eslamian and Faezeh Eslamian.

physiographic features (Shiklomanov 2000), imprinting an often-complex picture of change when trying to assess the strong impact that global climate change has on freshwater supplies.

Combined effects of economic growth and rapidly growing global human population has led to increased demand for freshwater, which in turn has resulted in the over-exploitation of water withdrawal from aquifers as well as from surface freshwater bodies. Human activities have caused global water scarcity, particularly in arid and semi-arid regions. Determining the extent and severity of impacts of human activities and climate change on water resources across the globe, particularly in ungauged basins is a major issue.

The relationship between forests and water is of crucial importance for survival, especially for rural livelihood, and their interactions have been studied and documented since ancient times. About one-third of the earth's land surface is covered by forests. As the single-largest ecosystem type, forests significantly affect the global hydrological cycle and provide a myriad of ecosystem services to humans (Sedell et al. 2000; Chang et al. 2013). Hydrological services or the water-related services provided by forest ecosystems are very crucial for human well-being (MEA 2003). From arid deserts to the humid tropical rainforests, the flow of water through the ecosystem shapes the characteristic fauna and flora as well as the soil systems.

Food security is dependent on water security, which in turn depends on forests. Forests produce approximately 75% of the world's accessible fresh water, which is used for domestic, agricultural, industrial, and ecological needs as well as other purposes. It is also estimated that a significant proportion of the drinking water required for the world's biggest cities comes from forested catchments and watersheds (Tidwell 2016).

Forests act as natural water filters. Forests (soil and vegetation) promote infiltration, increasing soil moisture content and groundwater recharge, contributing to the gradual release of water (Bruijnzeel 2004). Due to interaction with tree canopies and the root system, surface runoff is reduced, which in turn maintains soil stability and improves water quality in terms of sediment loading (Ilstedt et al. 2007; Lele 2009). Furthermore, it is also evident forests contribute in regulating disasters like floods and landslides (Bredemeier 2011; Calder and Aylward 2006). On the other hand, forests may reduce the annual water yield through increased loss by evapotranspiration, consequently limiting the amount of water available in the system (Bosch and Hewlett 1982; Bredemeier 2011; Bruijnzeel 2004; van Dijk and Keenan 2007). The density of forests also influences the hydrological services, which are site specific and vary as a function of climate and biophysical conditions (Calder 2002). A conceptual framework for the hydrological services provided by forests is given in Figure 22.1.

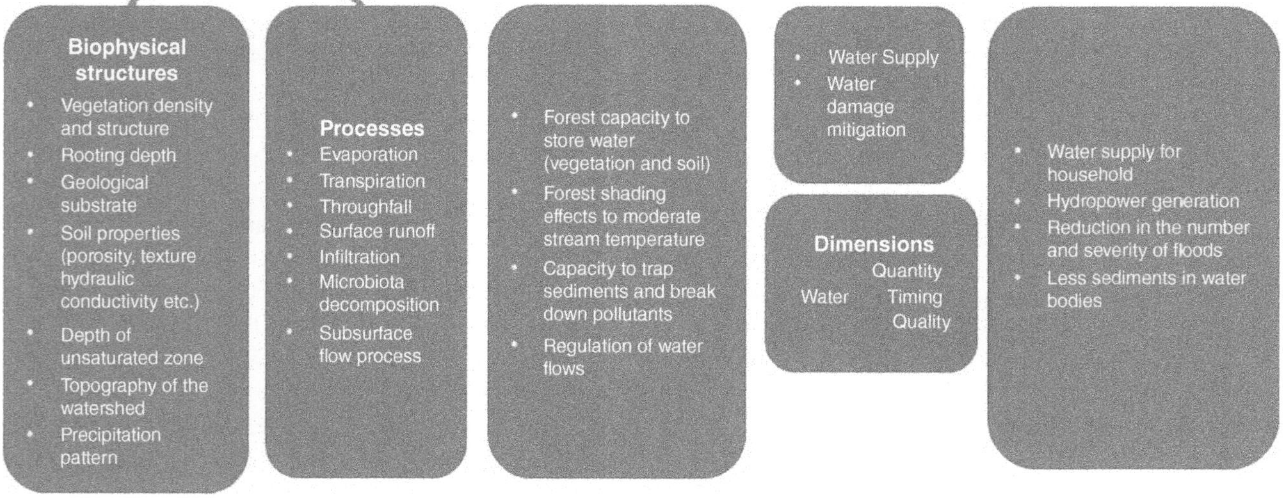

Figure 22.1 Conceptual framework for hydrological services provided by forests. Source: Adapted and modified from Carvalho-Santos et al. 2014.

22.3 Change in Land Use-Land Cover and its Impact on Forest and Water Resources

Forests play a major role in global and local water budgets by returning approximately 40% of total annual precipitation back to the atmosphere in the form of evapotranspiration (ET). In the catchment area it supplies a large proportion of water for domestic, agricultural, industrial, and ecological needs in both upstream and downstream areas. From the past few decades, change in Land Use-Land Cover (LULC) has resulted in rapid rates of deforestation, which in turn has also affected hydrology. This has particularly altered groundwater and baseflow locally and precipitation regionally, which ultimately affects the food security and total ecological balance of the ecosystem. So, the key challenges facing land, forest and water management experts is to maximize the wide range of multi-sectoral forest benefits without them being detrimental to water resources and normal ecosystem function.

Deforestation and anthropogenic land-use transformations have important implications for climate, ecosystems, the sustainability of livelihoods, and the survival of species, raising concerns about long-term damage to natural earth system functions (Steffen et al. 2015a,b). Mean warming due to land cover change may explain as much as 18% of current global warming trends (Alkama and Cescatti 2016). Deforestation exerts an influence on warming at the local scale and alters rainfall and water availability, not to mention the emission of greenhouse gases.

Also, changing climate is affecting the global hydrological cycle, so the relationship between forests and water is of critical importance. The interaction between forests and water and their impacts on ecosystem must be addressed at different scales.

22.4 Forest Hydrology

The interaction of forest and hydrology is studied under the umbrella of forest hydrology, which basically describes the structure and function of watersheds and their influence on water movement and storage. This encompasses conceptual theory, observations, methodologies, and processes, which finally results in the formation of a feasible policy. The scale varies from a smaller scale, like a watershed, to a global scale, like an ocean. It involves the wider disciplines of science such as meteorology, geology, hydrology, forestry, soil science, and plant physiology. This field may be interest for scientists, economists, or for politicians of land-use management.

Today, forest hydrological science has expanded from understanding the meteorological and hydrological influences of forests in small watersheds, to quantifying the eco-hydrological impacts of global changes (Amatya et al. 2011; Vose et al. 2011).

22.4.1 Hydrologic Processes in Forest

In hydrology, the land unit is the *watershed*, which also may be referred to as a *basin* or *catchment*. A watershed is defined as an area of land in which all the incoming precipitation drains to the same place – toward the same body of water or the same topographic low area (e.g., a sinkhole) – due to its topographical features.

The study of water in forests is termed *forest hydrology* and includes the distribution, storage, movement, and quality of water; hydrologic processes within forested areas; and the delivery of water from forested areas (NRC 2008). Forest hydrology is viewed as one of the foundational sciences in integrated watershed management (IWM) (Black 1996; Brooks et al. 2012).

Water balance in a forest watershed within a geographic area can be expressed by the following equation:

$$\Delta S = P - \mathrm{ET} - Q$$

where ΔS, P, ET, and Q represent the change in soil water storage, precipitation, evapotranspiration (the combination of evaporation and transpiration), stream flow (surface and subsurface flow), and groundwater recharge generated within the watershed boundary, respectively. All these four variables can change dramatically in a short period of time. However, in the long-term, the change in storage (ΔS) can be minor and are often assumed to be negligible. Climate change can alter the water balance and thus ΔS may not be zero before the ecosystem reaches a new steady state. Major hydrologic processes in the forest watershed are given in Figure 22.2.

Hydrologic processes can be divided into vertical water movement (including precipitation, snow accumulation and melt, infiltration, and deep seepage) and lateral flow. Once water has entered the soil, it can move laterally either as shallow lateral flow through the forest duff layer or through surface soil horizons and macropores. It may also move laterally as groundwater. A diagrammatic representation of this is given in Figure 22.3.

Groundwater impacts may be limited at the hillslope scale, but they become important in larger watersheds. A diagrammatic representation of hydrological processes in a forest hillslope is given in Figure 22.4. Infiltration rates depend on soil properties, such as the soil surface cover and soil water content. If the soil is not saturated, then the infiltration rate is a function of water content and

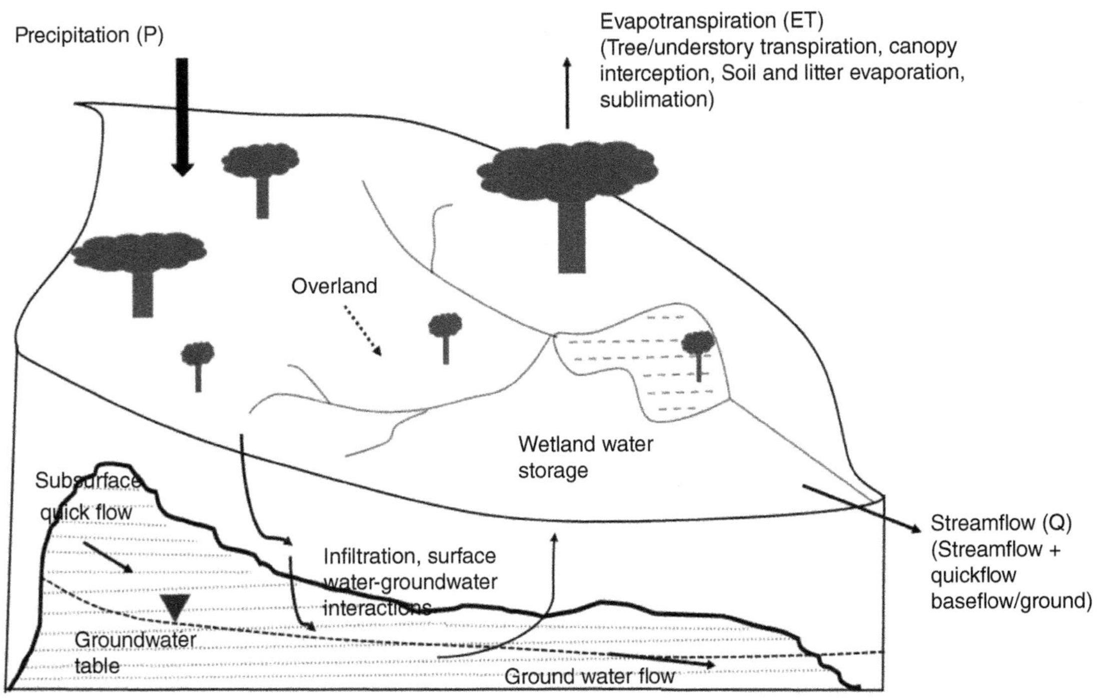

Figure 22.2 Schematic of major hydrologic processes in forest watershed. Source: Adapted and modified from Sun et al. 2016.

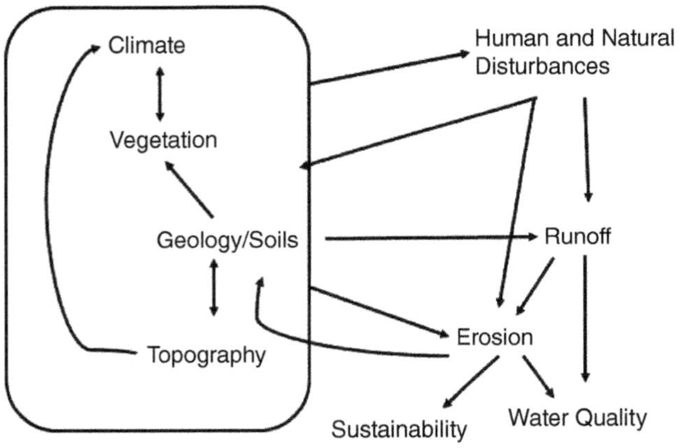

Figure 22.3 Diagram showing complex interactions that drive forest watershed processes.

soil properties, such as noncapillary porosity, texture, bulk density, water repellency, and surface cover. If the soil is saturated, then the infiltration rate will be limited by the saturated hydraulic conductivity of the soil (Luce 1995).

Vegetation plays a significant role in forest hydrology (Hubbart 2007) through evapotranspiration. Evapotranspiration varies with age and condition of vegetation, with very young and mature forests tending to have lower transpiration rates, whereas healthy growing forests have the greatest rates.

22.4.2 Effects of Forest Structure on Hydrological Processes

The first and most important step that forests act on in the water cycle is rainfall partitioning (Livesley et al. 2014). However, it is difficult to predict the interactions between forest structure and rainwater pathways (Herwitz 1985; Zimmermann et al. 2007). In summer during rainfall events, raindrops are intercepted by the canopy. Other raindrops fall through the canopy directly to the ground and this is called throughfall (Crockford and Richardson 2000; Brauman et al. 2010). Figure 22.5 describes the input and output schematic pathways in forests.

22.4.2.1 Stemflow

Stemflow is an important resource of ground water; stemflow also contributes to soil chemistry (Johnson and Lehmann 2006). Horizontal precipitation is an important hydrological input (Ingraham and Matthews 1988).

Horizontal precipitation plays a crucial role in mountain forests (Ingraham and Matthews 1988) as 0–17% of precipitation input is from horizontal precipitation (Dawson 1998). Forests also rerouted the pathway of horizontal precipitation by canopy interception (Ingwersen 1985). The process was influenced by forest structure, humidity, wind speed, temperature, etc. Compared to vertical precipitation, precipitation in the form of dew and fog

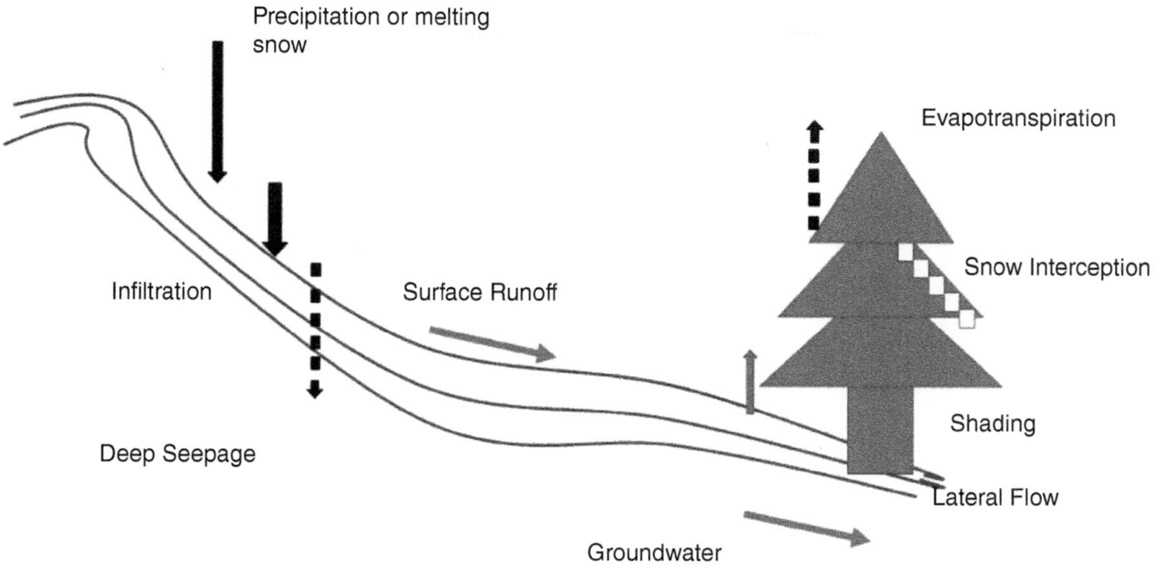

Figure 22.4 Dominant hydrologic processes in a forest hillslope.

Figure 22.5 Rainwater input and output schematic pathways in forest.

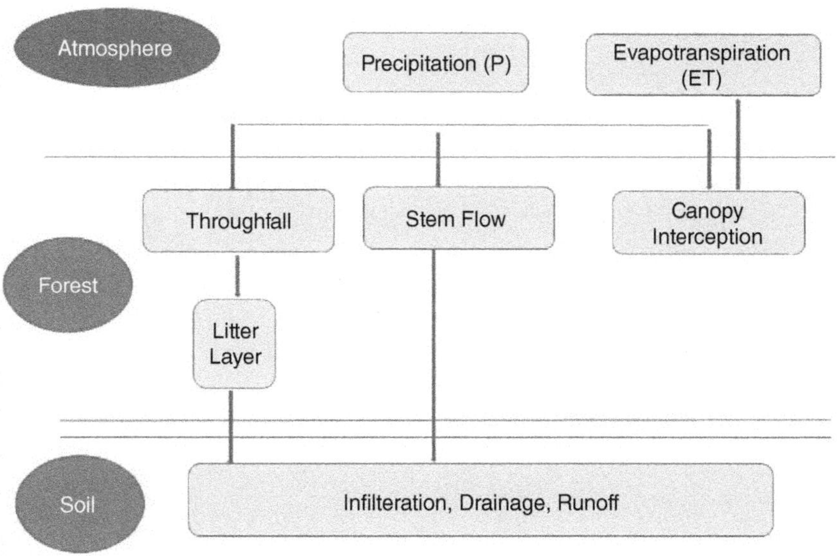

were less than 0.5 mm each night, much less than vertical precipitation.

22.4.2.2 Litterfall

Litter refers to all dead and decaying forestry materials above the soil surface (Helvey and Patric 1965). It performs important ecological functions such as to conserve soil and nutrition, intercept precipitation, and provide habitat for organisms (Sangha et al. 2006). Compared to canopy interception, litter interception accounted for only a small fraction of gross precipitation. However, litter interception played a significant role in the water cycle process (Bulcock and Jewitt 2012). Litter also helps in maintaining soil moisture, reduction in evaporation, and amplitude of soil temperature by insulating the surface (Sangha et al. 2006). Furthermore, the function of soil conservation should be kept note of, particularly in the condition where canopy extension leads to increased raindrop sizes, and consequently accelerated soil erosion (Putuhena and Cordery 2000). Thus, the presence of litter will effectively prevent soil particle detachment by absorbing the energy of raindrops (Bulcock and Jewitt 2012).

22.4.3 Preconditions for Rainwater Infiltration

The following parameters help with infiltration of rainwater. (The interactions are shown graphically in Figure 22.6.)

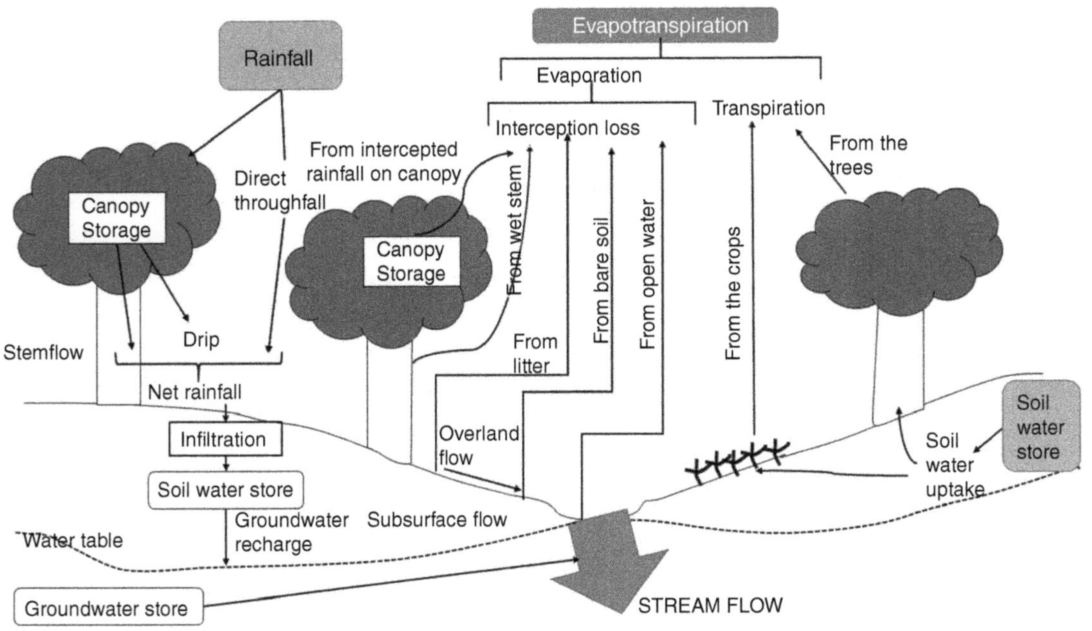

Figure 22.6 The hydrological cycles and their interaction with various components.

22.4.3.1 Vegetative Cover

Rainwater is absorbed by plant roots and returned to the atmosphere through plant respiration. The soil-vegetation complex functions to a certain degree to an extent that it acts as a filtration system that reduces clogging of the surface pores of the soil.

22.4.3.2 Soil Type

Effective porosity and permeability are soil parameters that influence the infiltration process. Bedrock should not be within less than 1.2 m of the infiltration surface.

22.4.4 Groundwater Conditions

Distance from groundwater and variations in groundwater levels are some of the data needed for planning for infiltration facilities. It is recommended that the distance from groundwater is at least 1 m.

22.4.5 Dimensions of Hydrological Services Governed by Forest

Various hydrological services provided by forests are given here.

22.4.5.1 Water Quantity and Forests

Forests regulate the water cycle by maintaining soil health and moisture, supporting soil infiltration, and groundwater recharge. Forests also evaporate intercepted precipitation and transpired water from soil and groundwater storages. Forest diversity and type greatly influences the hydrological pathways in it. Whether it will be a net source of water or utilize water for its own functions depends on various factors.

22.4.5.2 Water Quality and Forests

Forests cover reduces sedimentation rate; thus, it helps to improve water quality as water travels through its course. They may also help to reduce pollutants entering water courses through reducing erosion and slowing down the rate of surface water runoff, improving natural infiltration into soils and rocks. However, improvement in water quality in due course will depend on the amount of forest as well as the place where forest cover is in the landscape.

22.4.5.3 Evapotranspiration, Precipitation, and Water Loss

Accounting for the water budget, evapotranspiration (ET) is generally seen as water loss from the system. ET contributes to atmospheric humidity, cloud cover, and precipitation downwind. This helps to operate at many scales: global, regional, local, and within catchments and forests; all biophysical processes that contribute to local and larger-scale cooling and water availability. This issue is receiving bigger attention since scientific evidence has suggested large-scale deforestation in the tropics is linked to increases in regional drought frequency and severity (Elision et al. 2017).

22.4.5.4 Erosion/Sediment Control and Forests

Within forests, the complex system of ground cover with shallow roots and organic litter, combined with deep tree roots, contribute to forests providing effective erosion and sediment control. This function influences water quality and supports soil protection. Forests may complement gray infrastructure, reducing costs associated with dredging dams or higher levels of water treatment.

Everyone agrees that well-managed, forested watersheds produce less sediment that can affect storage capacity of reservoirs, water quality, irrigation systems, and hydroelectric dams. However, the frequently asked question is the extent to which forest cover can reduce sediment when extreme events are taking place.

22.4.5.5 Forests and Flood Control, Drought, and Fire Risks

Forests serve as buffers against natural disasters, such as floods, landslides, storm surges, etc. For example, one prevailing generalization is that forests reduce flooding. This is true in some contexts, but there are limitations dependent on spatial and temporal scales, including the size, duration, and intensity of a precipitation event, as well as the geographic area. Generally, forests noticeably reduce the magnitude of small floods. Forests contribute to mitigating climate change by acting as carbon sinks and buffering the effects of extreme events associated with climate change, such as floods and droughts. However, extreme events are also reducing the ability of forests to be effective in this role. Droughts, flooding, and forest fires are increasing in occurrence and are affecting hydrology, such as increasing erosion and degradation of soils, and resulting in altered landscapes, sedimentation of water bodies, etc. (Sedaei et al. 2017).

However, it is widely accepted that forests contribute in regulating river flows through controlling peaks and volumes. Due to their high infiltration capacity, forests can store higher rainfall quantities, reduce runoff rates, and therefore minimize, to some extent, floods. The question is to what extent the forest cover can contribute to the protection from floods. At a larger scale, it is believed that forests have no substantial impact on reducing flood damages. Studies have shown that beyond certain spatial scales, forests/land use don't have effects on river flows.

22.4.5.6 Forests and Groundwater

Inversely correlated to surface flow and erosion, the relationship between forests and groundwater is related to soil health and quality. Forests and their soils slow down water movement, allowing for soil infiltration and improved soil water moisture condition and groundwater recharge. Organic matter from forest vegetation, above and below ground, plays an important role. However, this net recharge depends on many factors, including forest type (e.g. natural versus planted), species composition, tree density, and other land uses.

22.4.5.7 Forests and Their Effect on Rainfall

Research on how afforestation/deforestation affects rainfall remains inconclusive. Simulation models predict that massive deforestation will decrease rainfall in some areas and increase it in others (Kaimowitz 2002). On this subject, Calder (1999) outlined that deforestation has little effect on regional precipitation and Lee (1980) noted that the removal of all forest cover would only reduce global precipitation by at most 1–2%.

22.4.5.8 Forests and Riparian Management

Vegetated riparian buffer zones play an important role in the transport of nutrients from farms into surface water bodies. Riparian zones also serve many functions: reducing floods, erosion, and sedimentation of water; regulating water temperature, as well as nutrients and sediments; providing biodiversity corridors that connect upstream and downstream ecosystems, etc. Riparian zones are also important for providing water access to inland areas and users. Forestry tends to emphasize the protection and conservation of natural vegetation in riparian zones to maintain natural water supply. Water management, on the other hand, tends to emphasize water-user access, including agricultural, domestic, and industrial users, and water-related risks, such as flooding, in order to maintain water supply for all those reliant on the water source. Integrative approaches that allow for both utilitarian and conservation purposes are needed that are in turn supported in legislation (Swanson et al. 2017).

Three pathways can be distinguished by which incident precipitation arrives at the forest floor: (i) a small fraction of the water reaches the forest floor without touching the leaves or stems – this is known as direct throughfall; (ii) another, also small, fraction of the water flows down the tree trunks as stemflow (SF), some of which will evaporated back to the atmosphere; and (iii) the rest of the water hits the canopy and continues to fall as crown drip or is also evaporated during and shortly after the rainstorm.

22.5 Rainwater Harvesting in Forests

22.5.1 Definition and Typology of Rainwater Harvesting Systems

Empirically and theoretically, it is well-established scientific fact that forests use more water than lower vegetation and annual crops in rainfed agriculture. Consequently, there is strong empirical evidence that cutting forests results in increased stream flows. Typically, when forest cover is regenerated, more rainfall tends to (once again) be partitioned through soil infiltration and to green water (used for food and fiber production), reducing its availability as blue water (available for human consumption) downstream.

Older forests may work as "sponges" to better retain its role in recharge groundwater and to maintain dry season streamflow. Due to large amounts of litterfall and soil protection, forests maintain a high soil infiltration capacity. Increasing surface runoff after deforestation increases surface runoff and possible soil deterioration, leading to more "blue water," but water that is often polluted by soil erosion. The higher surface runoff during rainfall events at a deforested location means that less water is contributing to long-term groundwater recharge onsite. Depending on the location, shallow groundwater is often linked to lower-lying stream flows, regulating the river base flow during dry seasons.

Decreasing shallow groundwater recharge through deforestation may thus deplete surface water sources in times of high demand. The reduction of stream flow after deforestation has often been observed by rural people, but only a few studies have reported the expected long-term decline in dry season flows (Bruijnzeel 1989).

There are various technologies to harvest, store, and provide water to meet demand by humans and/or human activities. These technologies can be divided into two main types depending on source of water collected; namely, the in situ and the ex situ types of rainwater harvesting, respectively. In situ rainwater harvesting technologies are soil management strategies that enhance rainfall infiltration and reduce surface runoff, whereas the ex situ systems are those systems which have rainwater harvesting capture areas external to the point of water storage. The rainwater capture area varies from being a natural soil surface with a limited infiltration capacity, to an artificial surface with no infiltration capacity. The rainwater harvesting structure of the forest ecosystem is given in Figure 22.7.

Dug wells recharged by in situ water harvesting

Check dam

Figure 22.7 Rainwater harvesting structures in forests.

Rainwater harvesting in forests can be instrumental to decentralized water supplies and local food security. Its management and development refer to the conservation, regeneration, and the judicious use of the water by nearby human habitats within a shared ecosystem (geological, hydrological-aquatic, and ecological) located within a common drainage system.

The water can be stored in storages of different construction and dimensions; for example, large reservoirs with large catchments and small tanks and ponds with small catchments, or use of natural or artificial ground water recharge to store water in the soil.

22.6 Deforestation and its Impact

The role of rain forests in the water cycle is to add moisture to the atmosphere through the process of transpiration (in which plants release water from their leaves during photosynthesis). This moisture contributes to the formation of clouds, which release the water back into the rainforest. When forests are cut down, less moisture goes into the atmosphere and rainfall declines, sometimes leading to drought. These have been made worse by deforestation. Impact of individual tree in rainwater infiltration is given in Figure 22.8.

Forest loss and degradation reduce ET, with important implications for rainfall thousands of kilometers downwind (Andrich and Imberger 2013). Changes in Earth's surface albedo, temperature, ET, and surface roughness also alter moisture and heat fluxes between terrestrial surfaces and the atmosphere. These observations have led climate modelers to predict that large-scale deforestation will reduce rainfall in some regions by as much as 30% (Lawrence and Vandecar 2015). Deforestation also impacts local thermodynamics, resulting in a decrease in heat released to the atmosphere. This impacts atmospheric circulation and its associated rainfall (Werth and Avissar 2005a,b). Deforestation weakens the local hydrological cycle, and a new pattern of heat release occurs due to the changed land cover (Werth and Avissar 2005a,b). Deforestation, which results in warming and altered rainfall patterns due to climate change, can lead to feedback effects on remaining vegetation, reduced biomass accumulation, drought, die-off, and fires (Brienen et al. 2015; Duffy et al. 2015).

Reductions in precipitation have implications for regional economies and livelihoods. Further expansion of agriculture in the Amazon could lead to reductions in total agricultural output due to deforestation-driven declines in precipitation (Lawrence and Vandecar 2015; Oliveira et al. 2013). Large-scale deforestation of the Amazon may further reduce hydropower generation through declining precipitation and river discharge (Stickler et al. 2013). LULC change and its impact on water parameter on a basin size is given in Table 22.1.

In the contiguous US, forested watersheds generate about 50% of runoff and management decisions include both afforestation and deforestation to create desired water yields (Ellison et al. 2012).

Water stored in the soil profile is a central component of the forest hydrologic cycle and globally accounts for over

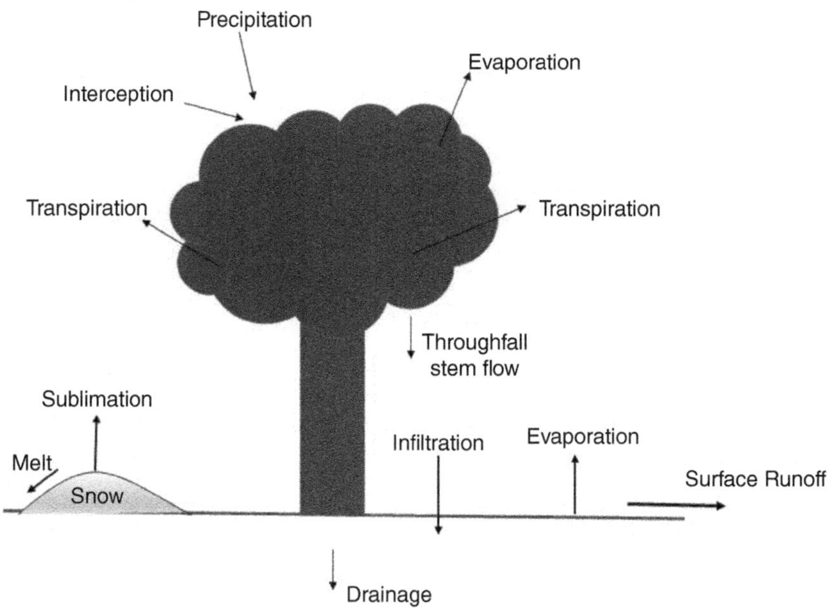

Figure 22.8 Impact of individual tree in rainwater infiltration.

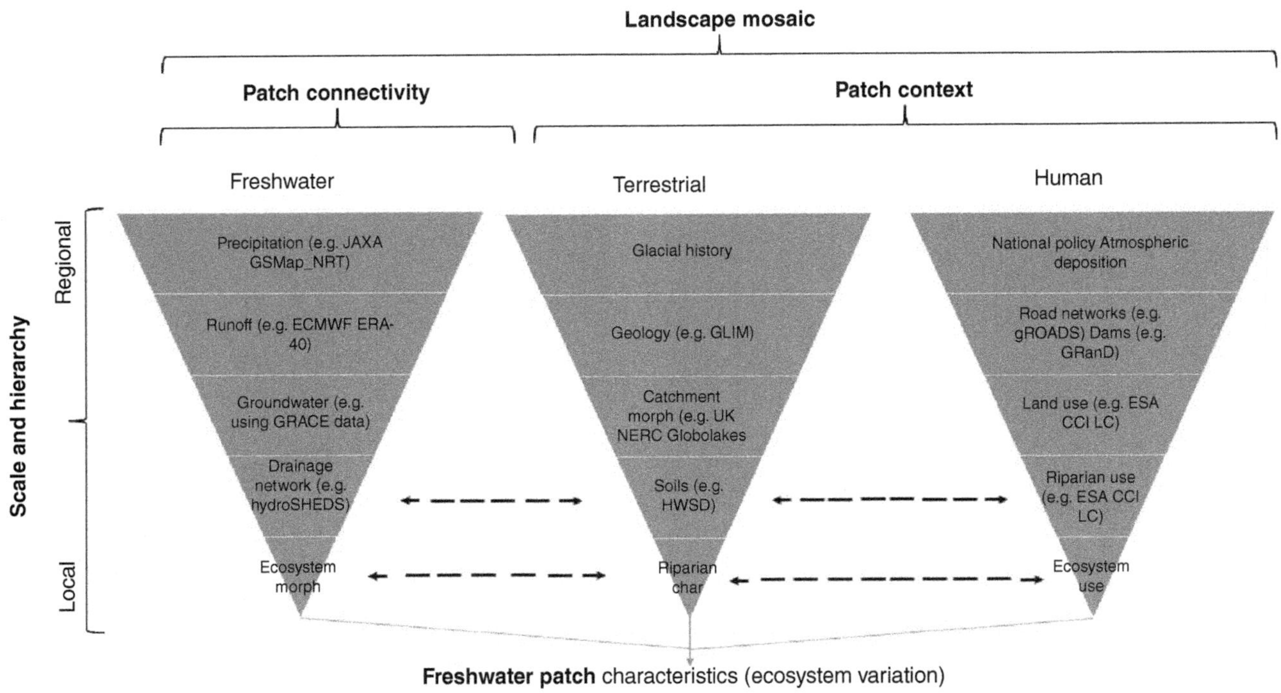

Figure 22.9 Holistic approach to manage climate-water-forest nexus for sustainable development.

Table 22.1 Impact of change in land use on water parameter by basin size.

Parameter		Basin size (km^2)					
	0.1	1	10	100	1000	10 000	100 000
Average flow	×	×	×	×	0	0	0
Peak flow	×	×	×	×	0	0	0
Base flow	×	×	×	×	0	0	0
Groundwater recharge	×	×	×	×	0	0	0
Sediment load	×	×	×	×	0	0	0
Nutrients	×	×	×	×	×	0	0
Organic matter	×	×	×	×	0	0	0
Pathogens	×	×	×	0	0	0	0
Salinity	×	×	×	×	×	×	×
Pesticides	×	×	×	×	×	×	×
Heavy metals	×	×	×	×	×	×	×
Thermal regime	×	×	0	0	0	0	0

$120 \times 103\ km^3$ of water per year (Trenberth et al. 2007). However, current knowledge of how soil water storage dynamics change over time and influence forested processes is limited, particularly due to the lack of long-term in situ measurements of soil moisture. By intercepting precipitation, evaporating moisture from vegetative surfaces, transpiring soil moisture, capturing fog water, and maintaining soil infiltration, forests influence the amount of water available from groundwater, surface watercourses, and waterbodies.

22.7 Forest Management and Watershed Development

Sustainable forest management and watershed development are closely linked. Upstream management of forests strongly affects downstream use of water resources. These upstream–downstream linkages and related interactions are well recognized within a watershed management perspective. Effective watershed management that includes all interactions and implications related to the upstream–downstream relationship is required for forest and water sustainable management. In addition to the technical aspects, it takes into account the ecological, social, and economic dimensions. The role of forests in the conservation and sustainable use of water resources is still an issue of debate. Effective watershed management provides a framework where the role of forests and forested watersheds in the protection of uplands and in the sustainable use and conservation of lowlands resources are better understood. This perspective, which recognizes the opportunities and limitations of forest management on freshwater, translates this recognition into more effective planning, implementation, and monitoring of forest, water resources, and agricultural development programs.

Forest hydrology has developed to a more comprehensive concept known as watershed management. Watershed management implies land resource management within social and economic contexts. With time, the scope of watershed management evolved from the initial concept concentrating on water resource management to a participatory integrated approach aiming to maintain productivity and improve conditions for people upland.

Forest watersheds play an important role in providing domestic, agricultural, and commercial water, and therefore the conservation of forests is key to sustaining the availability and quality of water. The water conservation, soil protection, and flood/drought mitigation provided by forests are paramount to human welfare and environmental sustainability. Forests capture and store water by increasing absorption, infiltration, abating runoff velocity, and reducing water erosion. Filtration of water pollutants and flow regulation are other important hydrological services provided by forest.

Rainwater harvesting in the context of a watershed means collecting runoff from within a watershed area, storing it, and employing it for different purposes. Runoff collection is generally distinguished as in situ management, when the water is collected within the area of harvesting, and ex situ when it is diverted outside of the harvesting area. The storage is of crucial importance: for in situ rainwater harvesting the soil acts as the storage, whereas for ex situ rainwater harvesting the reservoir can be natural or artificial, where natural generally means groundwater recharge and artificial means surface/subsurface tanks and small dams. The differentiation between the two is often minor, as water collection structures are generally placed in a systematic relation with each other; hence, the runoff from certain structures may be a source of recharge for others.

It is empirically and theoretically well-established paradigm that forests use more water than lower vegetation and annual crops in rainfed agriculture. Consequently, empirical evidence is strong that cutting forests results in increased stream flows (Bosch and Hewlett 1982). Typically, when forest cover is regenerated, more rainfall tends to (once again) be partitioned through soil infiltration and to green water (used for food and fiber production), reducing its availability as blue water (available for human consumption) downstream (Farley et al. 2005).

Forests have been shown to maintain a high soil infiltration capacity by superior litterfall and soil protection. Increasing surface runoff after deforestation increases surface runoff and possible soil deterioration, leading to more "blue water," but water that is often polluted by soil erosion. The higher surface runoff during rainfall events at a deforested location means that less water is contributed to long-term groundwater recharge on site during the wet season. Depending on the location, shallow groundwater is often linked to lower-lying stream flows, regulating the river base flow during dry seasons. Decreasing shallow groundwater recharge through deforestation may thus deplete surface water sources in times of high demand.

Over the years, watershed management has been a key concept in the rural development processes in arid and semi-arid areas, in particular in rainfed ecosystems – combining projects for ecological sustainability with those for socioeconomic development. Theoretically, it attempts to integrate sectors such as water management, agriculture, forestry, wasteland development, off-farm livelihood development, etc., and to establish a foundation for rural development. The approach aims to be flexible enough to be adapted to varying sociological, hydrological, and ecological conditions (Joy et al. 2006). Apart from the purely environmental concerns, i.e. restoring ecosystem functions, the watershed framework often focuses on livelihood improvements, poverty alleviation, and a general increase in human well-being.

22.8 Knowledge Gaps

Despite its various important roles, the interplay of forest, water, and climate change requires a holistic approach to study and there is a great challenge to both scientist and policy maker in making forest-smanagement strategies. Several authors have reported hydrological responses to forest change in small watersheds ($<1000\,km^2$), mostly on the impact of forest change on annual runoff in small watersheds (Bosch and Hewlett 1982; Sahin and Hall 1996; Bruijnzeel 2004).

There are different schools of thought; some studies suggest deforestation can increase annual runoff while some studies suggest afforestation affects stream flow in the opposite way. However, these responses can be inconsistent, suggesting the response intensity of annual runoff to forest cover change can be variable among watersheds, especially for watersheds with afforestation or reforestation (Stednick 2008; Lacombe et al. 2016). However, the relationship between forest change and water yield has been less investigated in large watersheds (>1000 km^2).

The interaction between forests and water takes place on multiple scales and its impact varies in different regions and in different forest types, with diverse species under different management regimes. Studies from small watersheds suggested that forest change can generate pronounced effects on annual runoff and small- to medium-sized floods – in the interactions between forest and many other hydrological variables, still a consistent conclusion is lacking. A large debate is going on the effects of forests on precipitation, flood regulation, and mitigation.

22.9 Forests and Water in International Agreements

In recent years, change in climatic conditions has affected both the forest system as well as various parts of hydrological processes, which in turn is affecting food security in many parts of the world. The recurrence of extreme weather events, climate change, and the need for adaptation strategies is gaining attention worldwide. Many national and international organizations are bringing attention to water security and its impact on ecosystem processes. All are considering urgent policy and management measures, pointing to the interrelationship between forests and water. There is a growing number of water-related initiatives worldwide, such as the International Network of Basin Organizations (INBO, http://www.inbo-news.org) or the World Water Council (www.worldwatercouncil.org). These are progressively taking into account the role of trees, forests, riparian ecosystems, and their management in achieving targets of freshwater quality, quantity, timing, and hazard prevention.

There are various international agreements and initiatives, such as UNECE Convention on the Protection and Use of Transboundary Watercourses and International Lakes, REDD+ (Reducing Emissions from Deforestation and forest Degradation), and LULUCF (Land Use, Land-use Change and Forestry); the Ramsar Convention is addressing the linkages between water, wetlands, and forests and their importance.

In many regional policy goals and initiatives, the significance of the interactions between forests and water is also embodied, such as the Warsaw Resolution 2 "Forests and Water" of Forests Europe (formerly the Ministerial Conference on the Protection of Forests in Europe). The issue plays an important part in ensuring environmental sustainability, one of the eight UN Millennium Development Goals.

In an International Expert Meeting on Forests and Water in Shiga, Japan, held in November 2002 in the framework of the Third World Water Forum in Kyoto, Japan, a major step toward improved understanding and effective implementation of policies, planning, and management initiatives worldwide related to forests and water has been adopted. Convened jointly by the FAO, the International Tropical Timber Organization (ITTO), the United Nations Educational, Scientific and Cultural Organization (UNESCO), and the Forestry Agency of Japan, the expert meeting focused on new challenges and perspectives concerning forest and water interactions, such as the need for better understanding of the hydrological and environmental services provided by forest ecosystems, more effective management tools integrating forest and water resources, and clearer national strategies and policies to guide stakeholders in the field.

The meeting at Shiga considered four broad areas:

- integrated, participatory and cross-sectoral approaches to planning and management;
- understanding of biophysical processes;
- economics of watershed services;
- effective collaborative arrangements and partnerships among stakeholders.

22.10 Role of Geospatial Technologies

Remote sensing has the potential to provide an important source of information for quantifying and mapping dynamic terrestrial and aquatic ecosystem services and hydrological processes. In large catchment areas of most of important river and lake systems in the world (particularly those with large populations dependent upon them and their attendant ecosystem services), the synoptic, wide area coverage and frequent observations provided by satellite-based remote sensing are an important source of information, however there is a variability at the national and regional level regarding in situ data collection methodologies and standards.

Geospatial technology applications in forest hydrological processes management include: (i) forest cover mapping and change analysis; (ii) forest soil water/moisture estimation and forested wetlands analysis; (iii) forest vegetation and biomass mapping; (iv) forest ET estimation; (v) forest hydrology attributed geohazard analysis, such as forest fires, landslides, and flooding; and (vi) forest stream water quality management.

As a result, remote sensing has become an important geospatial technology for deriving information about lakes and their catchments, which alongside advances in situ sensor systems to capture spatial information in addition to measurements across time and analysis with GIS, provides an important step forward in our ability to model lake and catchment status and behavior.

Through both passive (e.g. multispectral and hyperspectral) and active (LiDAR and RADAR) remote sensing systems, it is possible to retrieve the state of important lake and catchment variables that may vary across space at particular points in time. Unmanned aerial vehicles (UAV) and unmanned aircraft systems (UAS) are making forest hydrology management more efficient through the acquisition of centimeter-scale spatial resolution images with user-specified band widths – thus helping in SSFMDS by mapping soil moisture, plant stomatal conductance, canopy temperature, and leaf area index (LAI) to measure forest evapotranspiration (ET) and by monitoring forest fires (Grenzdörffer et al. 2008).

RS imagery provides information on drought, vegetation vigor, flood damage, forest fires, deforestation, and other natural disasters that are directly or indirectly influenced by forest-hydrology. D'urso and Minacapilli (2006) used a semi-empirical approach for forest surface soil water content estimation using radar data.

Potential RS systems, such as color infrared (CIR) aerial photography, most other multispectral scanners (MSS) (Landsat, QuickBird), and hyperspectral systems (AVIRIS, HyMap, CASI), bathymetric LiDAR, MISR, Hyperion, TOPEX/Poseidon, MERIS, AVHRR, and CERES, are being used by scientists to remotely estimate the hydrological flux on the earth's surface, including forest land cover.

Recent development in remote sensing (RS), with high-resolution images and GIS with their combination with land use models integrating the forest cover and various hydrological processes, helps to evaluate the hydrological responses of a watershed. Various numerical models such as MIKE SHE (Water balance modeling), SWAT (Water balance and non-point source pollution), HSPF (Water balance and non-point source pollution), VIC (Water balance) have been used worldwide. All these models use various thematic process such as LULC map of the area, infiltration, soil moisture storage, runoff generation processes, flow routing, and soil evaporation which give the water balance and river discharge simulations results of that area.

GIS provides the tools to accurately map this information globally and locally, including development of automated geospatial models for precise and proficient forest hydrology management decision support (FHMDS). Improved computational capabilities, in conjunction with digital elevation models (DEMs), digital data on soil type and land use, and GIS tools, are offering new possibilities for hydrological research, helping us to better understand the fundamental physical processes underlying the hydrological cycle and develop various numerical models representing those processes.

The Soil Conservation Service curve number (SCS-CN) method can be calculated worldwide to assess the effects of land cover change or forest cover change on surface runoff. The runoff coefficient can be defined as either the ratio of total runoff depth to total rainfall depth or the ratio of the peak rate of runoff to rainfall intensity for the time of concentration. The runoff CN is widely used in hydrology to predict direct runoff or infiltration from excess rainfall. The rainfall-runoff relations within a watershed are driven primarily by the interplay of such factors as climate, land cover, and soil. Physically based distributed hydrological models have become a more feasible approach to flood prediction and rainfall-runoff computation in recent years.

The Copernicus satellites build on the success and capabilities of other currently operational (e.g. Landsat series) and non-operational (e.g. Envisat MERIS and AATSR) instruments and include the Sentinel-2 Multispectral Imager (MSI) (launched June 2015), Sentinel-3 Ocean and Land Color Instrument (OLCI), and Sentinel-3 Sea and Land Surface Temperature Radiometer (SLSTR), the first of which launched in late 2015. Sentinel-2 MSI has 13 wavebands in the visible and NIR at 10–60 m spatial resolution and a 5-day revisit time. Sentinel-3 OLCI at a 300 m spatial resolution will employ 21 wavebands in the visible and NIR. Sentinel-3 SLSTR will employ 9 wavebands, with a nominal 500 m spatial resolution in the visible and NIR and 1 km at the TIR. In addition, Sentinel-3 will have a revisit time of one to two days.

22.11 Managing the Climate-Water-Forest Nexus for Sustainable Development

Sustainable development is dependent on the forest-water nexus. To achieve SDG 15, water and food security should be at the heart of forest management and the restoration of multi-functional landscapes. Similarly, water management should incorporate appropriate forest management as a natural infrastructure solution to achieve SDG 6. Simply recognizing the forest-water nexus is not enough. We must improve our ability to design, implement, and learn from landscape approaches that both rely on the relationships between forests and water and impact them.

Due to its importance to total landscape, forest management strategies should lead to preservation of hydrological flows, mitigation of extreme hydrological events, retention of soils and sediments, support productivity, and biodiversity, as well as maintenance and purification of water supply. Effective conservation of water resources, two major barriers exists: lack of a well-articulated conceptual framework and lack of practical strategies for implementing such a framework. The framework should consist of a well-defined set of principles based on hydrological theory, which could form the basis of ecosystem management to ensure sustainability of water and related resources in forested landscapes, since hydrological processes simultaneously drive geomorphic, biogeochemical, and ecological processes in forest ecosystems. Hydrological principles will enable the forest hydrological community (including industry, governments, academia, and citizens) to develop sustainable management policies and practices that lead to safe and secure water resources.

The experts shared their knowledge and tried to explain why the climate-water-forest nexus is so important in terms of environmental conservation and well-being of the society. What are the new findings around the interrelations of these three topics, which are the misconceptions of the current models, and what can be done in order to head toward a more sustainable development?

With 2018 celebrating the themes "Forests and Sustainable Cities" and "Nature for Water," it is high time we strategize an inter-sectoral approach toward advocacy and collaborative engagement, championing the forest-water nexus.

A holistic approach is required to understand the whole environment, taking into account that each case offers unique opportunities and challenges (Figure 22.9).

22.12 Case Studies

22.12.1 Combating Water Scarcity in Latin America

Latin America is one of the richest water regions worldwide but has suffered from several water crises in the past, and climate change is expected to further exacerbate the problem. The region is also facing critical rates of deforestation and forest degradation. Water scarcity is often assumed to be closely linked with poverty, bad sanitation infrastructure, and inequality, as well as with extensive droughts – observed in countries such as Argentina, Guatemala, and Mexico – and other weather-related disasters. Regional and global efforts in counteracting water scarcity as well as forest degradation and poverty need to address the relationships between water and forests to develop sustainable and appropriate measures. In order to safeguard water resources and access in the long run, we need to highlight and optimize the various environmental services provided by watershed landscapes and ecosystems. This must be done through adequate landscape and watershed management. For example, water scarcity as a consequence of decreases in precipitation is worsened by the reduced infiltration capacity of degraded sloping environments. Forests and trees can help to keep erosion rates low and to improve the soil's infiltration capacity.

22.12.2 Amazon River

Forests and trees play an important role in providing diverse environmental, social, economic, and cultural benefits to people in both rural communities and urban areas. The interactions between forests and water influence the provisioning and filtering of water, regulation of floods, the conservation of soils, and climate regulation, among others.

22.12.3 Case Study of Southeast Asia

The Mekong River Commission (MRC) in Southeast Asia is one of the largest-scale and most complex examples of integrated transboundary forest and water management programs. It deals with 795 000 km^2 in 6 riparian countries and over 60 million people. Ninety percent of the area's population lives in rural areas where they supplement food crops with fish from forests and wetlands, including large areas of flooded forests. One of the three main goals of the strategic plan for 2006–2010 was the implementation of an integrated approach to watershed management in which forest conservation plays a pivotal role in relation to biodiversity; water quality, availability, timing, use, and monitoring; and individual and institutional capacity building. MRC is part of the aforementioned INBO, which brings together watershed management authorities worldwide. Forested watersheds are exceptionally stable hydrological systems shown in Figure 22.10.

22.13 Conclusions

Forests play a vital role in regulating the water cycle and providing clean water. The interaction between forests and water at multiple scales has to be given higher priority when thinking about ensuring water quantity and quality. The understanding of interactions between forests and water is mainly drawn from small watersheds, which impedes the design of sustainable natural resource management strategies on large spatial scales. The study of large watersheds is urgently required. There is a dynamic interplay between forests, water, and climate change.

Figure 22.10 Forested watersheds are exceptionally stable hydrological systems.

Forests can be managed to mitigate the negative effects of climate change on water resources and ecosystems and consequently to ensure a safe and clean water supply and to reduce the risk of floods and droughts. It is crucial to maximize forest ecosystem services without threatening water resources, especially with regard to climate change mitigation measures. Rainwater harvesting can be a vital intervention in the rehabilitation of ecosystem services for enhancing human well-being in the context of watershed management. Its appropriate application can influence changes in the well-being of both human-oriented and ecosystem services. The changes are triggered through synergies across sectors; for instance, through interactions between agricultural practices, rainwater recharge, soil conservation, and food security needs. However, it is important to recognize that the approach of harvesting rainwater in watershed management, through major and minor schemes, has its own limitations, both in terms of appropriateness of the precise interventions, their techno-economic feasibility, and their practical method of implementation. Therefore, close monitoring of the impacts is required in environmental, economic, social, and technical terms during all the phases of the implementation is required.

References

Alkama, R. and Cescatti, A. (2016). Biophysical climate impacts of recent changes in global forest cover. *Science* 35 (6273): 600–604.

Amatya, D.M., Douglas-Mankin, K.R., Williams, T.M. et al. (2011). Advances in forest hydrology: challenges and opportunities. *Transactions of the ASAE* 54: 2049–2056.

Andrich, M.A. and Imberger, J. (2013). The effect of land clearing on rainfall and fresh water resources in Western Australia: a multi-functional sustainability analysis. *The International Journal of Sustainable Development and World Ecology* https://doi.org/10.1080/13504509.2013.850752.

Black, P.E. (1996). *Watershed Hydrology*. CRC Press.

Bosch, J.M. and Hewlett, J.D. (1982). A review of catchment experiments to determine the effect of vegetation changes on water yield and evapotranspiration. *Journal of Hydrology* 55 (1–4): 3–23.

Brauman, K.A., Freyberg, D.L., and Daily, G.C. (2010). Forest structure influences on rainfall partitioning and cloud interception: a comparison of native forest sites in Kona, Hawai'i. *Agricultural and Forest Meteorology* 150 (2): 265–275.

Bredemeier, M. (2011). *Forest Management and the Water Cycle: Introduction to the Challenge. Forest Management and the Water Cycle: An Ecosystem-Based Approach*, ix–xv. Heidelberg: Springer. Germany.

Brienen, R.J., Phillips, O.L., Feldpausch, T.R. et al. (2015). Long-term decline of the Amazon carbon sink. *Nature* 519 (7543): 344.

Brooks, K.N., Ffolliott, P.F., and Magner, J.A. (2012). *Hydrology and the Management of Watersheds*. Wiley.

Bruijnzeel, L.A. (1989). Forestation and dry season flow in the tropics: a closer look. *Journal of Tropical Forest Science*: 229–243.

Bruijnzeel, L.A. (2004). Hydrological functions of tropical forests: not seeing the soil for the trees? *Agriculture, Ecosystems and Environment* 104 (1): 185–228.

Bulcock, H.H. and Jewitt, G.P.W. (2012). Field data collection and analysis of canopy and litter interception in commercial forest plantations in the Kwa Zulu-Natal midlands, South Africa. *Hydrology and Earth System Sciences* 16 (10): 3717–3728.

Calder, I.R. (1999). *The Blue Revolution: Land Use and Integrated Water Resources Management*. Earthscan.

Calder, I.R. (2002). Forests and hydrological services: reconciling public and science perceptions. *Land Use and Water Resources Research* 2 (2): 1–12.

Calder, I.R. and Aylward, B. (2006). Forest and floods: moving to an evidence-based approach to watershed and integrated flood management. *Water International* 31 (1): 87–99.

Carvalho-Santos, C., Honrado, J.P., and Hein, L. (2014). Hydrological services and the role of forests: conceptualization and indicator-based analysis with an illustration at a regional scale. *Ecological complexity* 20: 69–80.

Chang, J., Wu, X., Wang, Y. et al. (2013). Does growing vegetables in plastic greenhouses enhance regional ecosystem services beyond the food supply? *Frontiers in Ecology and the Environment* 11 (1): 43–49.

Crockford, R.H. and Richardson, D.P. (2000). Partitioning of rainfall into throughfall, stemflow and interception: effect of forest type, ground cover and climate. *Hydrological Processes* 14 (16–17): 2903–2920.

Dawson, T.E. (1998). Fog in the California redwood forest: ecosystem inputs and use by plants. *Oecologia* 117 (4): 476–485.

Duffy, P.B., Brando, P., Asner, G.P., and Field, C.B. (2015). Projections of future meteorological drought and wet periods in the Amazon. *Proceedings of the National Academy of Sciences* 112 (43): 13172–13177.

D'urso, G. and Minacapilli, M. (2006). A semi-empirical approach for surface soil water content estimation from radar data without a-priori information on surface roughness. *Journal of Hydrology* 321 (1–4): 297–310.

Ellison, D., Futter, M.N., and Bishop, K. (2012). On the forest cover–water yield debate: from demand-to supply-side thinking. *Global Change Biology* 18 (3): 806–820.

Farley, K.A., Jobbágy, E.G., and Jackson, R.B. (2005). Effects of afforestation on water yield: a global synthesis with implications for policy. *Global Change Biology* 11 (10): 1565–1576.

Grenzdörffer, G.J., Engel, A., and Teichert, B. (2008). The photogrammetric potential of low-cost UAVs in forestry and agriculture. *The International Archives of the Photogrammetry, Remote Sensing and Spatial Information Sciences* 31 (B3): 1207–1214.

Helvey, J. and Patric, J.H. (1965). Canopy and litter interception of rainfall by hardwoods of eastern United States. *Water Resources Research* 1 (2): 193–206.

Herwitz, S.R. (1985). Interception storage capacities of tropical rainforest canopy trees. *Journal of Hydrology* 77 (1–4): 237–252.

Hubbart, J.A. (2007) Measuring and modeling hydrologic responses to timber harvest in a continental/maritime mountainous environment. PhD dissertation, University of Idaho, Moscow, USA. 170 p.

Ilstedt, U., Malmer, A., Verbeeten, E., and Murdiyarso, D. (2007). The effect of afforestation on water infiltration in the tropics: a systematic review and meta-analysis. *Forest Ecology and Management* 251 (1–2): 45–51.

Ingraham, N.L. and Matthews, R.A. (1988). Fog drip as a source of groundwater recharge in northern Kenya. *Water Resources Research* 24 (8): 1406–1410.

Ingwersen, J.B. (1985). Fog drip, water yield, and timber harvesting in the bull run municipal watershed, Oregon 1. *JAWRA Journal of the American Water Resources Association* 21 (3): 469–473.

Johnson, M.S. and Lehmann, J. (2006). Double-funneling of trees: stemflow and root-induced preferential flow. *Ecoscience* 13 (3): 324–333.

Joy, K. J., Shah, A., Paranjape S., et al. (2006) Reorienting the Watershed Development Programme in India, Occasional Paper, Forum for Watershed Research and Policy Dialogue. SOPPECOM0, Pune. India.

Kaimowitz, D. (2002). Amazon deforestation revisited in. *Latin American Research Review* 37 (2): 221–235.

Lacombe, G., Ribolzi, O., Rouw, A.D. et al. (2016). Contradictory hydrological impacts of afforestation in the humid tropics evidenced by long-term field monitoring and simulation modelling. *Hydrology and Earth System Sciences* 20 (7): 2691–2704.

Lawrence, D. and Vandecar, K. (2015). Effects of tropical deforestation on climate and agriculture. *Nature Climate Change* 5 (1): 27.

Lee, R. (1980). *Forest Hydrology*. New York: Columbia University Press.

Lele, S. (2009). Watershed services of tropical forests: from hydrology to economic valuation to integrated analysis. *Current Opinion in Environmental Sustainability* 1 (2): 148–155.

Livesley, S.J., Baudinette, B., and Glover, D. (2014). Rainfall interception and stem flow by eucalypt street trees–the impacts of canopy density and bark type. *Urban Forestry and Urban Greening* 13 (1): 192–197.

Luce, C.H. (1995). Forests and wetlands. In: *Environmental Hydrology* (eds. A.D. Ward and W.J. Elliot), 253–284. Boca Raton, FL, USA: CRC Lewis Publishers.

MEA (2003). *Millennium Ecosystem Assessment: A Framework for Assessment*. Washington, DC, USA: Island Press.

NRC (2008). *Hydrologic Effects of a Changing Forested Landscape*. Washington, D.C., USA: National Research Council.

Oliveira, L.J., Costa, M.H., Soares-Filho, B.S., and Coe, M.T. (2013). Large-scale expansion of agriculture in Amazonia may be a no-win scenario. *Environmental Research Letters* 8 (2): 024021.

Putuhena, W.M. and Cordery, I. (2000). Some hydrological effects of changing forest cover from eucalypts to Pinus radiata. *Agricultural and Forest Meteorology* 100 (1): 59–72.

Sahin, V. and Hall, M.J. (1996). The effects of afforestation and deforestation on water yields. *Journal of Hydrology* 178 (1–4): 293–309.

Sangha, K.K., Jalota, R.K., and Midmore, D.J. (2006). Litter production, decomposition and nutrient release in cleared and uncleared pasture systems of Central Queensland, Australia. *Journal of Tropical Ecology* 22 (2): 177–189.

Sedaei, L., Sedaei, N., Cox, J.P. et al. (2017). Forest fire mitigation under water shortage, Ch. 26. In: *Handbook of Drought and Water Scarcity Vol. 3: Management of Drought and Water Scarcity* (eds. S. Eslamian and F. Eslamian), 527–550. USA: Francis and Taylor/CRC Press.

Shiklomanov, I.A. (2000). Appraisal and assessment of world water resources. *Water international* 25 (1): 11–32.

Steffen, W., Richardson, K., Rockström, J. et al. (2015a). Planetary boundaries: guiding human development on a changing planet. *Science* 347: 1259855.

Steffen, W., Richardson, K., Rockström, J. et al. (2015b). Planetary boundaries: guiding human development on a changing planet. *Science* 347 (6223): 1259855.

Stickler, C.M., Coe, M.T., Costa, M.H. et al. (2013). Dependence of hydropower energy generation on forests in the Amazon Basin at local and regional scales. *Proceedings of the National Academy of Sciences* 110 (23): 9601–9606.

Sun, Y.H., Meng, L., Tian, L., Li, G.L., and Sun, O.J. (2016). Assessing current stocks and future sequestration potential of forest biomass carbon in Daqing Mountain nature Reserve of Inner Mongolia, China. J Forestry Res 27: 931–938.

Swanson, S., Kozlowski, D., Hall, R. et al. (2017). Riparian proper functioning condition assessment to improve watershed management for water quality. *J Soil Water Conserv.* 72 (2): 168–182. https://doi.org/10.2489/jswc.72.2.168.

Tidwel, T.L. (2016, 2016). Nexus between food, energy, water, and forest ecosystems in the USA. *J Environ Stud Sci* 6: 214–224. https://doi.org/10.1007/s13412-016-0367-8.

Trenberth, K.E., Smith, L., Qian, T. et al. (2007). Estimates of the global water budget and its annual cycle using observational and model data. *Journal of Hydrometeorology* 8 (4): 758–769.

Van Dijk, A.I.J.M. and Keenan, R.J. (2007). Planted forests and water in perspective. *Forest Ecology and Management* 251: 1–9.

Vose, J.M., Sun, G., Ford, C.R. et al. (2011). Forest ecohydrological research in the 21st century: what are the critical needs? *Ecohydrology* 4: 146–158.

Werth, D. and Avissar, R. (2005a). The local and global effects of southeast Asian deforestation. *Geophysical Research Letters* 32 (20): L20702. https://doi.org/10.1029/2005GL022970.

Werth, D. and Avissar, R. (2005b). The local and global effects of African deforestation. *Geophysical Research Letters* 32 (12): L12704. https://doi.org/10.10292005GL022969.

Zimmermann, A., Wilcke, W., and Elsenbeer, H. (2007). Spatial and temporal patterns of throughfall quantity and quality in a tropical montane forest in Ecuador. *Journal of Hydrology* 343 (1–2): 80–96.

23

Rainwater and Green Roofs

Sara Nazif[1], *Seyed Ghasem Razavi*[1], *Pouria Soleimani*[1], *and Saeid Eslamian*[2]

[1] *School of Civil Engineering, College of Engineering, University of Tehran, Tehran, Iran*
[2] *Department of Water Engineering, College of Agriculture, Isfahan University of Technology, Isfahan, Iran*

23.1 Introduction

In urban areas with a considerable amount of impervious areas, green roofs can be effectively used for natural runoff management in urban areas in addition to other runoff management practices. In other words, the green roofs help in enhancing or restoring the natural hydrologic cycle in urban area (Dayani et al. 2017). A small amount of water is used by vegetation on the green roof; however, the growth media will store a significant amount of water. The stored water in the saturated soil will be slowly released; therefore, it helps reduce the risk of a flash flood. Furthermore, they can be effective in improving the quality of urban runoff.

Green roofs, or living roofs, have been used for a long time, but recently, with the advancement of technology and improvement of their installation and construction, people are particularly attracted to these roofs. These roofs are more likely to be built on low-rise buildings because of security and safety issues (Herrera-Gomez et al. 2017). By understanding the benefits of green roofs in recent decades, they are widely spreading among countries around the world such as Singapore, Canada, the US, Japan, and Hong Kong. Many cities around the world have decided to use green roofs and have considered some supports for this purpose. For example, in Munich, Germany, all buildings with a roof area of more than 100 m^2 must be equipped with a green roof and building owners who use green roofs receive €5000 as a prize; in Vienna, Austria, government pays €8–25 per square meter as support for buildings with green roofs (Brudermann and Sangkakool 2017).

Beside the effects of green roofs on urban runoff volume, green roofs and vegetables on them can be effective in many ways for the Earth. They can mitigate the temperature of the Earth by reducing the amount of carbon and other destructive matters. Green roofs make shadows by their leaves on the building and so there is less heat on the building equipped with a green roof; the temperature inside the building can be conserved due to the leaves' insulating effect on the roof (Erdemir and Ayata 2017; Tang and anf Qu 2016; Rowe 2018). They can also reduce the air pollution by absorbing carbon and other harmful matter via their leaves and converting them to organic compounds (Johannessen et al. 2017; Rowe 2018).

In this chapter, after an introduction about green roof types and characteristics (components and performance), their effects on urban runoff quantity and quality are investigated. Then the other functions of green roofs, such as temperature reduction, reducing building energy demand, and reducing air pollution, are discussed. The given information in this chapter is provided based on a comprehensive review of the related studies, papers, and standards to provide a basis for decision making about green roof development in different areas around the world. Finally, a summary and conclusion are given.

23.2 Green Roof Components

Unlike conventional roofs, green roofs are made of several parts of which their characteristics depend on the regional climate and the usage of green roofs. Mostly, a green roof is made of the following components, as shown in Figure 23.1 (Vijayaraghavan 2016; Pérez et al. 2012):

- Vegetation: The outer layer of green roofs consists of vegetation;
- Growth substrate: It has the responsibility to feed the plants and retain water;
- Filter layer: This layer prevents soil particles from going to the drainage layer while it lets water flow, and is commonly made of geotextiles;

Handbook of Water Harvesting and Conservation: Basic Concepts and Fundamentals, First Edition. Edited by Saeid Eslamian and Faezeh Eslamian.

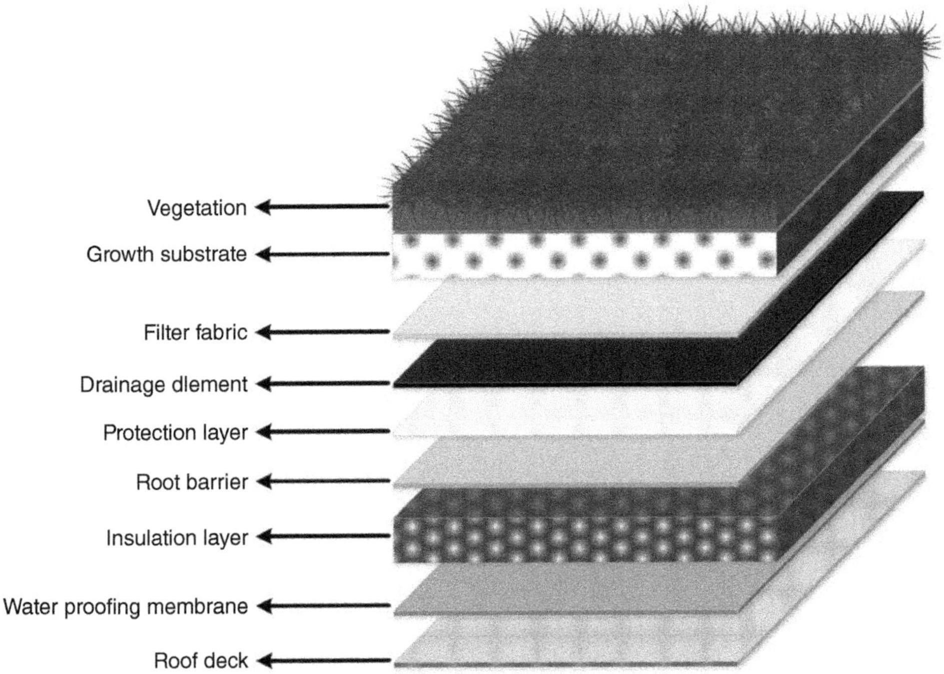

Figure 23.1 Components of common types of green roofs. Source: Adapted from Besir and Cuce, 2018.

- Drainage layer: Its responsibility is to balance air and water in the green roof in order to improve the growth of plants;
- Insulation layer: This layer is used as a thermal insulation layer;
- Root barrier: It must prevent the root of plants from penetrating to roof;
- Waterproof layer: As leakage is possible for green roofs, an insulation layer is essential. This layer prevents damage to the roof because of leakage;
- Protection layer: This layer protects the lower layer of green roof structure.
- It should be mentioned that in different regions, some components of the green roofs such as insulation layer and waterproof layer, might change depending on the climatic condition.

23.2.1 Vegetation

Vegetation, as the outermost part of the green roof, has a vital role on the success of green roof. It is important to know that green roofs are not an appropriate place for growing vegetation due to several reasons, such as minimum amount of nutrients and water limitations (Pérez et al. 2012; Vijayaraghavan 2016). The plants which can be used in green roofs, are mainly defined by the climate, temperature, and type of the green roof (Table 23.1). An appropriate vegetation type should have the following specifications:

Resist dry conditions;
Survive with the minimum amount of nutrients;
Cover properly the considered part of roof for green roof development;

Table 23.1 Different types of green roofs and plants used around the world.

Author/year	Type of green roof	City	Vegetation type
Smith et al. 2011	Extensive	Chicago, USA	Grass
Sun et al. 2016	Extensive	Beijing, China	Sedum
Ng et al. 2012	Intensive	Hong Kong, China	Grass and tree
Ouldboukhitine et al. 2014	Extensive	Rochelle, France	Sedum, grass, herbs
Alcazar et al. 2016	Extensive	Madrid, Spain	Sedum with LAD 0.5 and Lucerne with LAD 1.5.
Nagase et al. 2017	Semi-intensive	Rotherham, Northern England, UK	Forbs, grasses, and bulbs

LAD, Leaf Area Density.

Needs the minimum amount of maintenance;
Has short and soft roots.

As it is not possible for a plant to have all the above characteristics, it has been found that the best option for green roofs is vegetation from succulent types such as sedum. Sedums can resist dry conditions and survive without water for days and even weeks because of their crassulacean acid metabolism (CAM). Because of the fact that sedums are not native to many parts of the world, researchers are trying to find different types of vegetation that are native and have the appropriate factors for green roofs. The other important factor that can improve the effectiveness of the green roof is to use a different type of plant in a roof, but it should not be exaggerated (Vijayaraghavan 2016).

Weed growing can mainly increase the maintenance needs and cost in all types of green roofs because they interfere with the other plants' growth, and therefore it is essential to remove them (Vijayaraghavan 2016). In order to prevent weed growth in green roofs Nagase et al. (2013) suggested three ways as follows:

- Use tall plants or a cover in order to prevent light to reach the substrate layer;
- Expand the diversity of plants;
- Removing parent plants before seeds are physiologically capable of germination.

Green roofs, even those with low grass, flowers, shrubs, or food gardens, attract insects and animals. This can be beneficial as it results in biodiversity in cities, however integrated pest management should be considered to help building residents stay ahead of pest activity.

23.2.2 Growth Substrate

Growth substrate plays a crucial role on different behaviors of green roofs such as thermal behavior, plants growing, water storage, and sound insulation. Furthermore, the growth substrate should be resistant to the different weather conditions that happen during the year. Recently, a commercial growth substrate that is ready to use has been used in green roof development. However, it should be considered that these substrates are appropriate for specific types of climate and are not useful in all climates. In some cases, garden soil is used as a growth substrate, but this action can decrease the water storage capacity of the green roof and increases the probability of weed growth. Furthermore, due to the heavy weight of garden soil, the chance of green roof failure is increased. As growth substrate adds an external load to the building, it is very important to reduce the weight of the substrate as much as possible.

In general, the substrate is made of organic and inorganic materials. The suggested material for growth substrate of green roofs is low-density inorganic material such as perlite. These materials decrease the weight of the substrate and make it possible to use thicker growth substrate. Therefore, a wider variety of plants can be used on the green roof. The main weakness of using this material is that it is not stable enough in severe climate conditions such as raining and massive storms. This material floats in the water and after a while they move to the top part of the green roof. Then, after drying, the wind can easily move them. This procedure also causes air pollution.

Another characteristic of growth substrate is sorption capacity, which is absolutely effective on the runoff quality produced from the green roof. Green roof runoff is phosphorus runoff due to the use of fertilizer. To address the phosphorus runoff, the growth substrate must be made of materials that can absorb phosphorus ions. But as green roofs are made of inorganic material, their sorption ability is weak. To overcome this issue, it is essential to use other materials such as mulch, natural clays, or some biomaterials that have potential sorption ability. Biomaterials can also provide nutrients for vegetation. By all accounts, selecting material is really important and special attention should be given to the absorption mechanism in order to control the runoff. The other two features of the growth substrate are water holding capacity (WHC) and root aeration, which not only are essential for the growth of plants under tough situations throughout the year but also can prevent roof leakage (Vijayaraghavan 2016; Jennett and Zheng 2018).

In general, the characteristics of an ideal growing medium are as follows (Nagase and Dunnett 2011):

- High stability under different conditions;
- Supportive of different plants;
- Availability;
- Cost-effectiveness;
- Minimum organic contents;
- High water holding capacity;
- Light weight;
- High hydraulic conductivity;
- High capacity in sorption and less leaching;
- Good root aeration mechanism and flow properties;
- Water quality amendment.

23.2.3 Filter Layer

The filter layer is placed between the growth substrate and drainage layer in green roofs. The filter layer prevents entrance of the soil particles of the growth substrate into the drainage layer, which would damage the drainage ability of the drainage layer. Also, the filter layer must allow water to flow at the desired velocity. The layer is mostly made of geotextile materials to have enough bearing capacity for the weight of the upper layers. For green roofs which consist of plants with short and soft roots, which mostly happen for extensive green roofs, the filter layer can be a root-barrier at the same time (Vijayaraghavan 2016).

23.2.4 Drainage Layer

The aim of using a drainage layer is to balance air flow and water in the structure of green roof. The other function of drainage layer is to retain and drain water at the time of necessity. As a balance of air and water is needed for proper growth of plants, this function ensures the good growth of plants in the green roof. Drainage layers are mostly made of polypropylene or polyester, but rubber crumbs can also be used with the same features, such as water retention and thermal behavior. They are divided in two types of modular panels and granular materials. The difference between these two types is that modular can retain more water compared to granular. Costs, building structure, and vegetation types are the factors that specify the type of drainage layer. The only limitation for granular type is that they cannot be used on steep roofs (Pérez et al. 2012; Shafique et al. 2018).

23.2.5 Root Barrier

The goal of the root barrier is to prevent root from penetrating to the structure. Root barrier mostly is made of concrete (Shafique et al. 2018; Vijayaraghavan 2016; Pérez et al. 2012). In green roofs where the plants have big and thick roots, the root barrier is necessary while in other cases root barrier is not essential.

23.2.6 Waterproof Layer

The likelihood of leakage always threatens the green roof and roof structure, and can damage them. As it is hard to fix the leakage when it happens, it is essential to implement a waterproof protection layer as a prevention layer (Shafique et al. 2018; Vijayaraghavan 2016; Pérez et al. 2012).

23.2.7 Insulation Layer

In some cases, thermal insulation is also considered in green roofs. This layer is typically made of a layer of extruded polystyrene. This layer can be installed below the roof deck, however it is preferred to be positioned above the waterproof layer (known as an inverted green roof), because it protects the membrane from condensation and physical damage.

23.2.8 Protection Layer

To protect the waterproof membrane from damage, protection boards are used. The common materials used for this purpose are water-permeable, hard wearing, and dense synthetic fibers: polyester and polypropylene. The protection boards may provide some noise-absorbing capability and also help in more water retention on the roof (from $3\,l\,m^{-2}$ to $12\,l/m^{-2}$) when the slope is below 15°.

23.3 Green Roof Types

Green roofs are divided in four different categories (Morakinyo et al. 2017):

- Intensive;
- Semi-intensive;
- Semi-extensive;
- Extensive.

Intensive green roofs are made of a thick substrate layer so they can hold a variety of different plants such as shrubs and even small trees. The substrate depth of intensive green roofs reaches to 200 mm and the weight of them are around $300\,kg\,m^{-2}$. It is important to know that costs of fertilizing, irrigation, and maintenance of them are very high. The finished depth of intensive green roofs can vary between 200 and 350 mm, or even can be more than 350 mm depending on the vegetation type (Downton, 2013). Extensive green roofs are made of a thinner substrate layer, so the range of their plants is more limited. In extensive green roofs, the substrate depth is about 150 mm and the weight varies from 60 to $150\,kg\,m^{-2}$. Some plants like grass or sedums and herbs can be used on these kinds of green roofs. Because of low costs of maintenance and their low weight, they are more popular than other types (Vijayaraghavan 2016; Erdemir and Ayata 2017). The finished depth of this kind of green roofs vary between 50 and 200 mm (Downton, 2013). Semi-intensive green roofs, like semi-extensive ones, are something between intensive and extensive green roofs. These types of green roofs gather some beneficial properties of intensive and extensive green roofs. The maintenance costs in semi-intensive green roofs are more than extensive type (Mahdiyar et al. 2018; Nagase et al. 2017), while their growing media is thicker than extensive green roofs.

Although all types of green roofs mentioned above can be used in any buildings, they are not cost effective.

Intensive green roofs are mostly considered in areas with public access but extensive green roofs are used in private buildings (Vijayaraghavan 2016). To choose the right type of green roof two important factors, in order of importance, are: (i) structural capability of the building; and (ii) expected function of the green roof (discussed in the next sections) should be considered. It is essential to keep the weight of the green roof under the limit assigned by various standards; the maximum load for a green roof is 1.46 kN/m^2 according to the European standard (Cascone et al. 2018). Also, the green roof user expectation is important, too. In Table 23.1, a summary of different types of green roofs that have been used in different regions is given.

23.4 Green Roof Irrigation

The irrigation water demand of green roofs, due to water scarcity, can't be always provided from urban water systems, and some cities have set some laws against using urban water for irrigation. Therefore, it is essential to have some alternative water sources for green roof irrigation that can supply the demand of irrigation and can also be sustainable. Harvesting rainwater and using it again for the irrigation of green roofs is one way that can be considered for this purpose. In order to collect the water from rain it is possible to use tanks with a special filter to prevent harmful substances in the water. Another option is to use gray water (wastewater collected from kitchen sinks, bathroom sinks, and showers).

Improving the irrigation system is also a fundamental status that needs special attention. The most important factor that can affect the irrigation of the green roof is the type of the plants which are selected. Plants with high resistance against dry condition such as succulent and sedums can decrease the demand of the water. The result of a study shows that such plants need to be irrigated at least every two or three weeks to survive (Nagase and Dunnett 2010; Schweitzer and Erell 2014). Another way to decrease the water for irrigation is to use some kind of special material in green roof layers such as sheets made of foam or substrates made of perlite to increase the WHC of green roof.

Furthermore, irrigation quantity can be easily controlled by monitoring factors like humidity of the weather, moisture of the substrate, and hours of exposure to the sun. Monitoring these factors can help to prevent extra usage of water for irrigation. It is important to note that irrigation is essential during the time that the moisture of the soil is below a specific level, and usually in places where precipitation is effective there is no need to irrigate unless there is a dry situation where watering is a must.

23.5 Green Roof Standards

With the expansion of the use of green roofs around the world, different countries have developed their own regulations and standards for these systems. One of the pioneering countries in the field of developing green roof standards is the US. Germany and Great Britain have also provided standards for green roof development and implementation. All the standards mainly focus on the following issues in developing green roofs (Dvorak 2011):

- System design: The most important point in green roof function is coordination of the system. System design is the knowledge of how each single part of a green roof should appropriately work together and improve the function of the green roof. So in system design, several issues including wind load, fire resistance, filtration, and slope of the roof should be considered (Weiler and Scholz-Barth 2009).
- Vegetation: Plant selection has a significant role on the success of green roofs. Vegetation standards consist of vegetation selection and vegetation implementation (Vijayaraghavan 2016).
- Structural issues: Structural considerations play a crucial role in green roof construction and are mostly related to different types of loads that roofs should tolerate. Structural standards consider issues of load designing and material attributes.
- Maintenance: Maintenance mainly consists of irrigation, fertilizing, and weed protection. Therefore, the issues considered in maintenance-related standards include irrigation and weed protection in the green roof (Lee and Jim 2018).
- Growing media: As a growing layer for plants, it is important to adhere to all the necessities requires in an appropriate substrate and it is essential to pay attention to the items such as material of the layer, composition of the material, absorbency management, and depth of the layer.

As an example, in Table 23.2 a comparison between German and US green roof-related standards on some of the most important characteristics of green roofs is given. In northern and central Europe the FLL guideline is used for green roof development (Van Mechelen et al., 2015). The green roof standards in Germany and the US, are well developed and consider different aspects of these systems' performance, therefore they are commonly considered in other countries. However, it worth mentioning that the green roof standards in different countries vary depending on the regional conditions and the functions that are expected from green roofs.

Table 23.2 A comparison of green roof standards in the US and Germany.

Item	Germany	USA standard
Vegetation coverage	Based on DIN, after 12–15 months from plantation, 80% of the green roof must be covered with the planted vegetation.	Based on WBDG standard, after 12 months from plantation, 60% of the green roof area must be covered with the planted vegetation.
Slope	FLL suggests that slope ratio be more than 2%. For slope ratio less than 2% special maintenance should be considered.	ANSI suggests maximum ratio in general for slope is 2 : 12. ASTM suggests maximum ratio for slope is 15%.
Growth substrate depth	FLL suggests the following depths for different types of green roofs: Intensive: from 15 to 50 cm. Extensive: from 2 to 20 cm.	ANSI suggests less than 15.2 cm depth for extensive green roofs while more than 15.2 cm for intensive green roofs. ASTM suggests around 2.5 cm depth for extensive green roofs and 25 cm depth for intensive green roofs.
Vegetation type	FLL suggests lawn, shrubs, coppices for intensive green roofs and moss, sedum, herbaceous, grass for extensive green roofs.	ASTM suggests small tree and shrubs for intensive green roofs and herb, grass, moss, and some type of succulent for extensive green roofs.
Wind	–	According to ANSI the maximum speed of the wind where green roofs are implemented must be less than 140 mph.
Irrigation	–	According to ASTM two types of irrigation can be used: Passive: storing water in the drainage layer and pump it to media layer Active: pumping water to vegetation
Permeability of growing media	Intensive: hold more than 45% of water inside and water flow should be 0.3 mm min^{-1}. Extensive: hold more than 35% of water inside and water flow should be 0.6 mm min^{-1}.	Intensive: – Extensive: hold 35% of water and 15% of air inside them and water flow should be 0.05 in min^{-1}.

23.6 Green Roofs for Rainwater Collection and Storage

Urbanization is rapidly growing all over the world. As the buildings have risen up over the years, they have affected the hydrological cycle and runoff rates in cities. More urbanization causes less pervious media and it means more runoff volume. To overcome this, there are several approaches which can help in managing urban runoff; one of the sustainable approaches is the green roof. Green roofs are one of the greatest ways to use roofs for runoff reduction. One of the capabilities of the green roof is the ability to collect and store water (Okhravi et al. 2015). The general mechanism of green roofs for runoff mitigation is their porous media of soil, which retain water in their content. However, there are other mechanisms in the green roof that help in runoff volume reduction which are listed as follows (Speak et al. 2013):

- As the water is absorbed in different parts of green roof, it postpones the starting time of runoff;
- As some water is retained in green roof temporary or permanently green roof has effect on the runoff quantity;
- Runoff will be produced in a longer time due to the slow release of water in green roof.

The studies also show that by using green roofs instead of normal roofs in cities even on a small scale means the annual runoff volume is decreased (Johannessen et al. 2017; Mentens et al. 2006).

23.6.1 Hydrologic Modeling of Green Roof Performance

The annual water balance of green roofs can be developed using mass balance as illustrated in Figure 23.2. The main component inflows of the system would be the precipitation and irrigation and the outflows of the system are evapotranspiration and drained water. The difference of inflows and outflows will show the changes in retained water volume in green roof.

However more detailed models are developed for analysis of hydrologic performance of green roofs. The considered

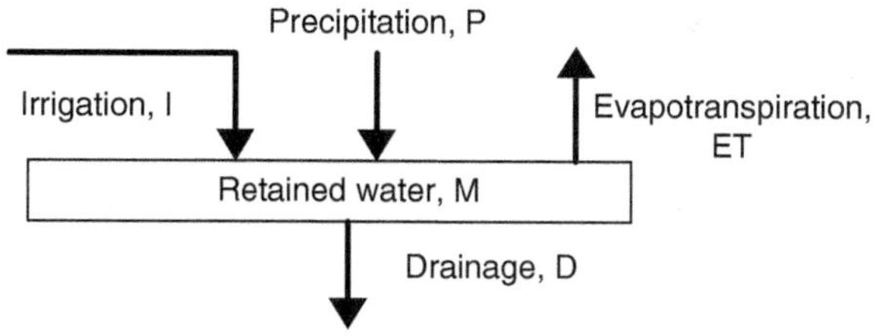

Figure 23.2 Water mass balance components in a green roof.

modeling approaches vary, from models for single event runoff to conceptual models applied to simulate distributed urban runoff. Less attention is paid to developing models that can continuously simulate both single events with fine time resolution and the annual water balance of single green roofs. The single event hydrological response of green roofs is highly affected by initial moisture conditions, which are dependent on evapotranspiration rates and are estimated from a continuous simulation, but the long-term performance is influenced by single event runoff and evapotranspiration rates. Therefore, it would be beneficial to develop a model that can continuously simulate both single events and long-term response. Therefore, the deterministic lumped rainfall–runoff conceptual model developed by Locatelli et al. (2014) is described here. They have proposed a simple and computationally effective hydrological model for green roofs that is capable of simulating the response to individual rainfall events and long term runoff.

The proposed model divides the green roof into three different water storages that are connected to each other as shown in Figure 23.3. A model based on the mass balance equation is implemented for each storage part. Some of the model parameters such as R_{max}, S_{max}, K_c (they are described in the next paragraphs) are estimated from physical data.

The mass balance for each of the storages is described as:

$$\frac{ds}{dt} = q_{in} - q_{out} \tag{23.1}$$

where s is storage depth, q_{in} and q_{out} are the inflow and outflow intensities normalized by the green roof area, and t is time. Retained water by the vegetation layer of the green roof is called surface storage. S_{max} corresponds to the maximum surface storage. The capacity of the surface storage is continuously changed because of evaporation. When the maximum surface storage S_{max} is exceeded, the effective precipitation, P_{eff}, is driven as infiltration into the subsurface storage.

The volume of retained water in the green roof substrate and in the drainage layer (and eventually other built-in layers) is referred to as subsurface storage, R. It should be mentioned that the drainage layer is commonly constructed to ensure that the roof has sufficient underdrain and there is no water retention capacity. In that case, the drainage layer is not considered in the subsurface storage. R_{max} represents the maximum capacity of both the substrate and drainage layer (if it has water storage capacity); this can be estimated as the difference between the water content at field capacity θ_{FC} (after drainage) and the permanently retained water content θ_{WP} (comparable to permanent wilting point). $R = 0$ represents the permanently retained moisture in the green roof (it does not correspond to oven dry conditions) and may vary in response to atmospheric variables. The water content in the subsurface storage is continuously reduced by evapotranspiration. From the subsurface storage Q_{ss} is diverted into the detention storage.

The maximum volume of detention storage is D_{max}. The detention storage corresponds to the excess water that cannot be held by the green roof and therefore will be drained through the drainage layer. The maximum capacity of the detention storage D_{max} is equal to the difference between the saturated water content and the water content at field capacity. When the maximum capacity D_{max} is exceeded, overland flow runoff occurs. The runoff from the detention storage can be described by the non-linear reservoir method

Figure 23.3 . Conceptual mass balance model of a green roof. Source: Adapted from Locatelli et al., 2014.

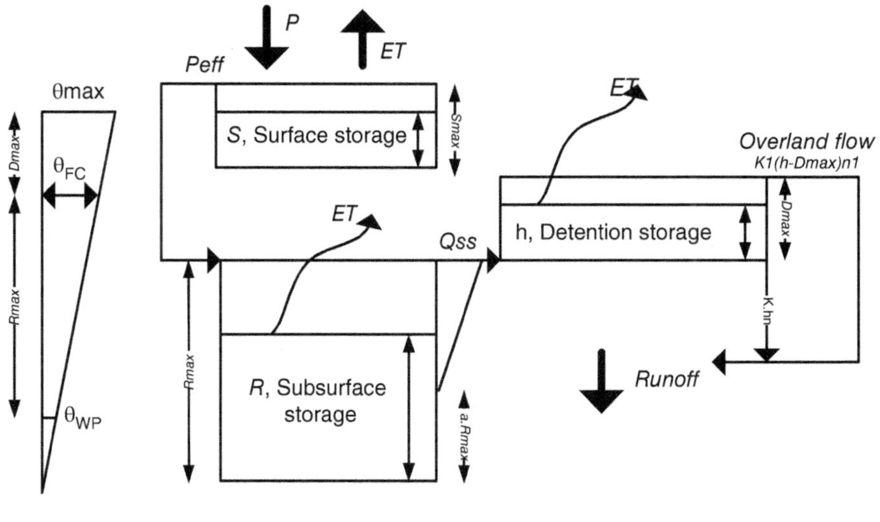

proposed by Zimmer and Geiger (1997) as follows:

$$Runoff = \begin{cases} k \cdot h^n \; h \leq D_{\max} \\ k \cdot D_{\max}^n + k_1(h - D_{\max})^{n1} \; h > D_{\max} \end{cases} \tag{23.2}$$

where n, k are the routing parameters for subsurface runoff through the drainage layer and n_1 and k_1 are the routing parameters for saturated overland flow.

Due to the high conductivity of the subsurface materials, all the rain is assumed to infiltrate and percolate vertically to the drainage layer. However, the water can accumulate in the system up to a certain volume before overland flow observation. Evaporation and transpiration continuously affect the capacity of all the storage. Evaporation firstly happens from the surface storage, if the surface storage falls below zero then water is withdrawn from the detention storage and when the detention storage is empty evapotranspiration takes place in the subsurface storage.

The actual ET rates can be calculated based on the FAO Penman- Monteith equation (Allen et al. 1998) as follows:

$$ET = ET_0 \cdot K_c.F \tag{23.3}$$

where K_c is the crop coefficient, F is the reduction factor due to water deficit and ET_0 is the reference evapotranspiration. The input parameters for ET_0 are the observed wind speed, atmospheric pressure, relative humidity, global radiation, and air temperature. The crop coefficient K_c depends on the type of vegetation. The reduction factor F is equal to 1 for both the surface and detention storages. For the subsurface storage F is equal to when water content R is more than wR_{max} and then decreases linearly to zero. The wR_{max} shows water content at the stomatal closure point.

The flow recharging the detention storage depends on the filling ratio of the subsurface storage.

$$Q_{ss} = \begin{cases} P_{eff} \dfrac{\dfrac{R}{R_{\max}} - a}{1-a} \; \dfrac{R}{R_{\max}} > a \\ 0 \; \dfrac{R}{R_{\max}} \leq a \end{cases} \tag{23.4}$$

When water content is more than aR_{max}, water is drained from the subsurface storage ($0 < a < 1$). When the subsurface storage is full then $Q_{ss} = P_{eff}$. The parameter a is a threshold value that changes the hydrograph at the beginning of the rain event by allowing runoff from the green roof even if the subsurface storage still has some capacity.

23.6.2 Green Roof Rainwater Retention Potential

Deutsch et al. (2008) says, "If 20% of all existing green roof-ready buildings had green roofs, the roofs would remove the same amount of pollutants as 17 000 trees, and reduce runoff from the roof surfaces by 68%." Based on the EPA research (OConnor et al. 2014), the rainfall retention of extensive green roofs varies between 12% and 86%. This large range of green roof performance in rainfall retention shows the wide variety of green roof configurations and the multitude of parameters that impact their WHC. Factors like green roof age, soil moisture, season, material used in layers, and regional climate affect the retention potential of green roofs. Based on the results of an experiment at the University of Nottingham, it is found that water retention potential of green roofs is highly dependent on rainfall depth. Based on the results of this study, the planted roof retained 100% of a 3 mm rainfall, 80% of a 3–23 mm rainfall, and for one period where 41 mm of rain fell over 47 hours, 73% of the rain was retained (Green roofs, 2018). The climate factors that affect the water retention potential of green roofs are as follows:

- Length of proceeding dry period;
- Season/Climate (air temperature, wind conditions, humidity);
- Characteristics of the rain event (intensity and duration).

One of the most important ones is soil moisture, which directly affects the water retention ability of the green roof. Normally, green roofs' water storage performance is better when rain occurs after a dry period or rain does not last for a long time period. In long rain events, the soil will be saturated and won't have appropriate retention ability. That is the reason why the irrigation practices of green roofs have always been of great importance and regional climate condition should be considered in green roof irrigation planning.

Part of the water that is stored in different parts of the plant and layers is evacuated through evapotranspiration. Evapotranspiration potential depends on variety of factors, of which the most important ones are the amount of water available, air temperature, air humidity, and wind. Also, the plant's type, which is considered through product coefficients (Kc), is effective in evapotranspiration. In dry seasons, due to the higher rates of evapotranspiration, green roofs have more potential to retain water, which leads to a reduction in runoff production.

Aging of the green roof due to some phenomena like compaction of the soil in the green roof or root penetration, means the water retention potential of the green roof changes and it is important to monitor these items (Speak et al. 2013; Mentens et al. 2006).

23.6.3 Green Roof Characteristics and Rainwater Retention Potential

Many researchers have investigated the effect of characteristics of green roofs on rainwater retention (Table 23.3).

Table 23.3 List of some studies about the role of green roofs on runoff volume reduction.

Reference	Dimensions (m)	Slope depth	Average retained rainfall (%)		Length of study period	Location
Monterusso et al. (2004)	2.4 × 2.4	2% (2 or 6 and 10 cm)	49		4 rainfall events	East Lansing, Mich.
Bengtsson et al. (2005)	5m^2	2.6% – 3 cm	46		17 months	Malmo, Sweden
VanWoert et al. (2005)	2.44 × 2.44	Study1: 2% – 2.5 cm Study 2: 2% – 4 cm	60.6 87		14 months	East Lansing, MI, USA.
Carter and Rasmussen (2007)	5.2 × 8.2	<2% – 7.62 cm	78		13 months	Athens, Georgia
Berghage et al. (2009)	1.8 × 2.4	1 : 12 and 9–10 cm depth	Over 50%	Winter <20 Summer 95	January 2005– November 2005	Rock Springs, PA, USA.
Stovin (2010)	3 × 1	1.5^0–8 cm	34		Spring 2006	Sheffield, UK
DeNardo et al. (2005)	1.8 × 2.4	8.9 cm	45		April 2002–February 2003 (11 months)	Pennsylvania, USA.

VanWoert et al. (2005) investigated the relationship between characteristics of green roofs and rainwater retention capacity. They used three platforms with equal dimensions for this purpose. The considered platforms were green roof with vegetation, green roof without vegetation, and conventional roof with 2-cm depth gravel ballast. In 430 days of simulation, they documented meteorological parameters and water drainage from all platforms (to calculate runoff) every 5 minutes in 24 hours. They also considered the effect of slope and depth of media. As statistics show, vegetated green roof performance for runoff reduction is better than others. On the other hand, when considering slope and depth of media in this study, the system with 2% of slope and 4 cm of depth performed better than other combinations. In another study, in summer, with the goal of assessing green roof performance in runoff mitigation, a roof by 1.5% slope was considered. They demonstrated that after heavy rains, the soil or media will saturate and their efficiency in runoff storage will decrease; for example, runoff retention was 33% at a high rate of precipitation and 45% at a lower rate (Stovin 2010).

By reviewing studies evaluating green roofs' effect on runoff mitigation, it was found that the main effective factors on runoff reduction in green roofs are as follows (Berndtsson, 2010):

- Number of layers and type of materials used in green roof
- Thickness of soil;
- Type of soil;
- Vegetation cover;
- Vegetation type;
- Geometry of roof, especially slope;
- Position of roof (e.g. shadowed or not, which direction faced);
- Green roof age.

23.7 Green Roof Effect on Runoff Quality

Green roofs' effect on the runoff water quality due to their material. Vegetation and substrate can absorb a range of pollutants including nitrogen, phosphorus, and heavy metals such as cadmium, copper, lead, and zinc. This ensures that rainwater is filtered for pollutants before entering the stormwater system. However, fertilizing can affect the performance of green roofs, which should be taken into account.

Table 23.4 lists the studies which have been conducted in field of quality of produced runoff from green roofs. Some studies show the amount of sulfate, calcium and magnesium salts in the runoff produced from green roofs is always more than the normal roofs. Otherwise, the amount of phosphorus and nitrogen is higher in green roofs in heavy rain (Teemusk and Mander 2007). In another study, besides investigating the quantity of produced runoff from green roofs, the quality of runoff was considered. The considered water quality measures were pH, EC, color, turbidity, and nitrate. A summary of results shows that

Table 23.4 Effectiveness of green roofs on runoff quality based on literature.

Reference	Scope of research	Assessed parameters	Results of some parameters	Place
Bliss et al. (2009)	Green roof effect on runoff quantity and quality	Phosphorus, chemical oxygen demand (COD), sulfate, nitrogen, turbidity, pH, and heavy metals	Turbidity and sulfate concentrations reduced and majority of metal concentrations were in order of 0.1 $mg\,l^{-1}$	Pennsylvania, USA
Alsup et al. (2010)	The exchangeability of metals from selected green roof growth substrates	Cd, Cr, Fe, Cu, Mn, Ni, Pb, Zn	Cr, Cu, Fe, Ni, or Zn were not found in substrates.	Illinois, USA
Carpenter and Kaluvakolanu (2011)	Effect of roof surface type on storm runoff	Total solids, Total phosphate (TP), Total Nitrate (TN)	TP and TN concentration in produced runoff from green roofs were lower than conventional roofs	Southfield, Michigan, USA
Berndtsson et al. (2006)	The influence of extensive vegetated roofs on runoff quality	Cd, Cr, Cu, Fe, K, Mn, Pb, Zn, NO3–N, NH4–N, TN, PO4–P, and TP.	Decrease in nitrogen concentration and increase in potassium phosphorus, Cr and Cu concentrations don't change, Cd concentrations are less than 1 $\mu g\,l^{-1}$ and Iron content in green roofs is lower than conventional roofs	Sweden
Hathaway et al. (2008)	Investigation of green roof impact on runoff quality	TN and TP	TN: 2.7 mg/l higher than rainfall, and TN: 1 mg/l higher than rainfall	North Carolina, USA
Barr et al. (2017)	Impact of green roofs on the runoff quality in comparison with other vegetated sites	NO2-N, NOX, PO4-P, TKN-N, TKP-P, NO3-N, TN	N and P concentrations in produced runoff from green roof were often higher than conventional roof	Villanova, Pennsylvania, USA
Matlock and Rowe (2017)	Green roof substrate effect on runoff quality	Nitrate and phosphate concentrations	Nitrate concentration ranges from 311 to 645 ppm and phosphorus concentration ranges from 6.5 to 8.2 ppm (more than other studies)	Michigan, USA
Monteiro et al. (2017)	Effect of growing substrates composition on runoff quality	Turbidity, pH, conductivity, NH4, NO3, PO4, and COD	No change happened in water quality.	Porto, Portugal

produced runoff from green roofs have a higher range of pH in acid rain conditions, yellow color, higher EC, and also equal nutrients or more in case of comparison with bare roofs (Berghage et al., 2009).

Monteiro et al. (2017) showed that what makes a difference in water quality of runoff in green roofs is their growing media. So, it can be concluded that using green roofs with any growing media is not sufficient to reduce water contaminants available in runoff. The best vegetation and growing media should be determined based on the regional climate. It should also be mentioned that age of the roof is very important in the produced runoff quality. Investigations have shown that nutrients will decrease in the case of aging roofs (Matlock and Rowe 2017).

Factors which affect the quality of green roof runoff can be summarized as follows (Berndtsson, 2010):

- Type of material used in green roof (composition of soil, material of drainage, and/or underlying hard roof material, rain pipe material);
- Soil thickness;
- Type of drainage;
- Maintenance/chemicals used;
- Type of vegetation, season (biomass using nutrients);
- Dynamics of precipitation;
- Wind direction;
- Pollution sources;
- Physico-chemical properties of pollutants.

23.8 Other Functions of Green Roofs

Beside runoff management, other functions can be also considered for green roofs, of which the most important ones are as follows (Silva et al. 2016; Solcerova et al. 2017; and Johannessen et al. 2017):

- Improving energy efficiency in buildings;
- Air pollution reduction;
- Improve human feelings;
- Affecting the urban heat island;
- Noise pollution reduction.

These issues are discussed briefly in the next subsections.

23.8.1 Improving Energy Usage Efficiency

It has been reported that in developed countries about 20–40% of total energy is consumed by buildings, and it is interesting that it is even higher than industry and transportation in the US and EU (Pérez-Lombard et al. 2008; Palut and Canziani 2007; Ma et al. 2011). So, to achieve sustainability in cities special attention should be given to energy consumption in buildings. One of the sustainable solutions to reach great energy efficiency in buildings is the green roof. Green roofs can be effective in energy usage reduction through (Castleton et al. 2010):

- Heat transfer reduction and solar reflectivity;
- Thermal mass addition to help internal temperature stay constant;
- Local temperature effect on air conditioning.

Table 23.5 summarizes some studies about the effect of green roofs on building energy demand. Green roofs have positive effects on the energy savings of the building. For green roofs, plant types, seasonal performance, materials, and type of building are important factors that play significant roles for energy-saving purposes in buildings (Niachou et al. 2001). Most heat attraction is from the roof, so by using green roofs, especially with their soil and water content, there will be a good insulation (Wong et al. 2003a,2003b). By absorbing internal heat and operating as an effective shield to prevent external heat from coming inside during summer (or spring), green roofs decrease interior temperature (up to 20 °C) and peak cooling load (reduce air conditioning energy between 25% and 80%) (Saadatian et al. 2013; Wong et al. 2003aa,b).

A green roof helps the temperature control through evapotranspiration and shading. As the substrate layer of intensive green roofs is thicker than semi-intensive and extensive types of green roofs, the potential of evapotranspiration is higher. The green roof during winter increases heat absorption, while in summer decreases its absorption (Ayata et al. 2017). A comparison between white and black roofs shows that during winter, black roofs act better than white roofs due to its lesser albedo. The comparison between black and green roofs shows that green roofs (intensive and semi-intensive) can act better due to their evapotranspiration and better insulation. The result also shows that by adding an insulation layer the energy needs of different types of roofs decreased, except intensive ones, due to high evapotranspiration ability that has some contradiction with appropriate insulation layer (Silva et al. 2016).

23.8.2 Air Pollution Reduction

Green roofs can affect air pollution. Plants can absorb CO_2 and other destructive particles of the air. Since the ability

Table 23.5 A review on effect of green roofs on building energy demand.

Reference	type	Indoor air temperature (mean)(°C)	Cooling demand (kWhm^{-2} year^{-1})	Heating demand (Kwhm^{-2} year^{-1})	Total energy demand (Kwhm^{-2} year^{-1})	Place
Jaffal et al. (2012)	Green roof	26.2	0.1	36.1	36.2	La Rochelle, France
	Conventional roof	28.3	2.5	36	38.5	
Niachou et al. (2001)	Green roof	–	22–45% (annual energy saving)	45–46% (annual energy saving)	31–44% (total annual energy saving)	Loutraki, Athens, Greece
Jaffal et al. (2012)	Green roof	33.9	12.5	15.2	27.7	Athens, Greece
	Conventional roof	31.3	26.4	14.7	40.5	
Jaffal et al. (2012)	Green roof	24.2	0	120.3	120.3	Stockholm, Sweden
	Conventional roof	25.6	0	131	131	

For non-insulated buildings.

Table 23.6 A summary on literature findings about green roof effect on air pollution reduction.

Reference	Scope of research	Assessed parameters	Summary of results	Place
Currie and Bass (2008)	Estimation of air pollution mitigation with green plants and green roofs	O3, SO2, NO2, CO, and PM10 (particulate matter)	Removal of 3.14 mg O3, 1.6 mg NO2, 2.17 mg PM10, and 0.61 mg SO2	Toronto, Canada
Getter et al. (2009)	Quantifying the carbon storage potential of extensive green roofs	Carbon storage of green roofs	Average of 168 gC m^{-2}·removal	Michigan, USA
Yang et al. (2008)	Quantifying air pollution removal by green roofs	O3, SO2, NO2, PM10	52% O3, 27% NO2, 14% PM10, 7% SO2 removal (of the total)	Illinois, USA
Li et al. (2010)	Evaluation of green roof effect on the ambient CO2 concentration	CO2 concentration, plant's CO2 absorption velocity and emission rate, simulation of the CO2 concentration distribution around a green roof	2% reduction of CO2 concentration in nearby, in a sunny day	Hong Kong, China
Speak et al. (2012)	Estimation of urban particulate pollution reduction by green roof vegetation	PM10	2.3% remediation of 9.18 tons of input PM10	Manchester, UK
Pugh et al. (2012)	Effectiveness of green infrastructure for improvement of air quality in urban street canyons	NO2 and PM	40% of NO2 and 60% of PM10 concentration reduction	London, UK

Under a maximum sedum green roof installation scenario for 325 ha.

of green roofs to reduce contamination is dependent on the plantation, plant selection and depth of substrate layer are important. A thicker substrate layer can absorb more CO_2. A study has shown that the tobacco plant reduces NO_2 by up to 30 times as much as the succulents. Furthermore, studies show areas that buildings equipped with green roofs have a better air flow, which allows dust to diffuse (Rowe 2018).

Table 23.6 shows a summary on findings of previous studies on air pollution reduction by green roofs. Based on the given information in Table 23.6, it is concluded that by selection of appropriate plants for green roofs, they can be in air pollution reduction.

23.8.3 Human Feelings

Huang et al. (2018) have shown that green spaces have a beneficial effect on human feelings and are effective in reducing stress. As a result of hard work in cities, people become stressed. Because accessibility to green places in cities in order to relieve the quotidian stress is hard for people, a green roof can be an alternative environment to increase happiness and relieve stress.

23.8.4 Green Roof Effect on Urban Heat Island

The temperature of cities is higher than rural areas due to storage of solar radiation by building materials, lack of green spaces, and inappropriate airflow. This phenomenon is called urban heat island (UHI) (Herrera-Gomez et al. 2017; Vijayaraghavan 2016; Santamouris 2014). Recently mitigation technologies have developed in order to control the temperature of cities. Most of these technologies focus on increasing the albedo in urban areas and increasing the green places in order to overcome excess heat in cities. In cities that are suffering from UHI, roofs can provide a great opportunity to use mitigation technologies. Green roofs help to decrease the air temperature in two different ways (Solcerova et al. 2017; Collins et al. 2017):

- By reducing the thermal load of the buildings, and acting as an insulation layer and reducing energy consumption;
- By cooling the temperature of the outside of buildings.

Table 23.7 A summary on available literature on the effect of green roofs on temperature.

Reference	Type of green roof	City	Reduction of temperature
Smith et al. (2011)	Extensive	Chicago, USA	2–3 °C
Savio et al. (2006)	Extensive	New York, USA	0.3–0.86 °C
Chen et al. (2009)	Extensive	Tokyo, Japan	almost 0 °C
Ng et al. (2012)	Intensive	Hong Kong, China	0.6 °C

Table 23.7 shows results of some studies about green roof effect on temperature. Growth substrate, vegetation, and maintenance are three main factors that define the cooling ability of green roofs (Lee and Jim 2018).

Studies show that green roofs as an insulator layer (not as a cooler instrument) help to reduce the interior temperature. They do not have the cooling effect 24 hours a day, even in some hours in the morning, because the temperature above the green roof is higher than normal roofs due to the lower albedo. The low albedo of the green roof causes higher attraction of sun radiation, which makes the temperature rises. They mostly have effect on temperature at night. Important factors which affect the cooling ability of green roof are moisture of the soil and cloudy sky. When the soil is dry or is in high moisture level, the cooling ability is weak. An extensive green roof covered by sedum showed that due to the low albedo they are weak in cooling the air, so using sedums is not logical for this reason (Erdemir and Ayata 2017; Solcerova et al. 2017).

Some studies on the thermal behavior of green roofs have emphasized the freezing and melting of the volumetric water content (VWC), which is inside the soil in winter season. The results of studies show that due to the freezing temperature during the day the water inside the soil at specific and constant temperatures starts changing phase. At the time the frost penetrates into soil during the 24 hours in winter and the water starts freezing, the K-value of the whole components of the green roof start to decrease, this means that the green roof works as an insulation layer and transfers less heat in comparison to the normal roof. The relation between K-value and VWC in temperatures above 0 °C can be explained by the bridge water effect and for the temperature below 0 °C it can be explained by particle discontinuity. Also, the vegetation acts the same as the soil and the K-value decreases by reduction of temperature. This phenomenon causes the temperature and the heat flux of the outside of the green roof to stay constant. It has been found that the average temperature at night for the green roof was less than in comparison to the conventional roof and during nighttime in winter when the demand for heat is at the highest point, green roofs lose heat less than conventional roofs (Arkar et al. 2015; Collins et al. 2017).

23.8.5 Interior Noise Pollution Reduction

There are many studies that have investigated the role of green roofs on noise pollution reduction. Connelly and Hodgson (2013) concluded that vegetated roofs increase the transmission loss of roof systems. By evaluating substrate depths, water content, and plant species it has been proven that transmission loss of vegetated roofs is more than non-vegetated roofs (up to 10 dB in low and 20 dB in the mid-frequency range). Van Renterghem (2018) has introduced the green roof as a practical solution for roof absorption and mitigating diffracting sound waves. Their experiment showed a 3 dB noise reduction in urban road traffic noise by the green roof. The green roof area, depth and type of the substrates, position of the green roof system, and also vegetation are important factors that influence noise reduction performance of green roofs.

23.9 Cost and Benefit Analysis of Green Roofs

The use of green roofs in recent years in cities and countries around the world has increased because of their benefits, such as mitigating air pollution in urban areas, improving water quality of runoff and better control on urban runoff, reducing building energy consumption, and also noise reduction (Bianchini and Hewage 2012). Many researchers have surveyed the benefits and costs of green roofs and at last compared it with conventional roofs (Table 23.8). Almost all studies report their results based on NPV (net present value) of both green and conventional roofs. It has known that less NPV value reflects lower cost.

Based on Table 23.8, it is understood that although installation cost for green roofs are higher than conventional roofs, after some years (dependent to the location), the NPV for green roofs become less than the NPV for conventional roofs. As cost and benefits are totally different from city to city and country to country (because of different discount rates, inflation, labor, green roof efficiency, cost of materials, and energy consumption/savings), the NPV for each case study is different from another one. However

Table 23.8 The costs and benefits of green roofs based on the literature.

Reference	Scope of research	Type of roof	Installation costs ($/m^2)	Benefits related to stormwater ($/m2 year)	Benefits related to energy consumption	Air pollution mitigation benefits	Net present value analysis	Location
Sproul et al. (2014)	Economic comparison of white, green, and black flat roofs in USA	Green	172	0	Energy costs of green roofs are \$11 m^{-2} less than white roofs	Green roofs efficiency in global cooling is less than white roof (CO2 offsets)	Net saving of green roofs is less than \$71 m^{-2} and white roofs are more than \$25 m^{-2}	22 case studies
		Conventional roofs in white color	22	0.9 (relative to green)				
		Conventional roofs in black color	22	0.9 (relative to green)				
Clark et al. (2008)	Economic analysis of environmental benefits of green roofs	Green	approximately 248	\$0 per year	Green roofs save \$1670 more than conventional roofs	From \$890 to \$3390	For 40 years, NPV is 25.2% less than conventional roof	Michigan, USA
		conventional	167	\$520 per year				
Carter and Keeler (2008)	Life-cycle cost–benefit analysis of extensive vegetated roof systems	Green	158.82	\$780.82	0.37 \$ m^{-2} saving	0.11 \$ m^{-2} saving	\$15 547 094 for green roof and \$21 552 206 for conventional roof	Athens, Atlanta, GA
		conventional	83.78					

some studies have shown that green roof benefits can't compensate for its installation cost during its life cycle (Lee, 2004; Sproul et al. 2014). The list below will show the parameters which affect on the NPV of the green roof (Bianchini and Hewage 2012):

- Initial installation cost: As mentioned above, initial costs of green roofs are higher than conventional roofs.
- Property value increment: Usually a place which is naturalized will increase its value and more people are likely to buy, so using green roofs, could achieve a higher property value.
- Tax reduction: many cities around the world have tax incentive policies.
- Stormwater volume reduction and increment in its quality: green roofs are effective in the reduction of runoff volume and actually, by considering some issues, could have a positive effect on water quality of runoff.
- Better energy performance: green roofs will decrease the cooling and heating demands;
- More lifespan than conventional roofs: many researchers believe that green roof lifespans are more than conventional roofs; for instance Saiz et al. (2006) showed that the green roof lifespan is about 50 years.
- Operation and maintenance cost.
- Air pollution during construction of green roofs.
- Air quality improvement and carbon mitigation.
- Creation of habitat.
- Aesthetics.
- Urban heat island effect reduction.

23.10 Conclusion

In this chapter, after giving an introduction on green roofs and their components, the potential of water retention in green roofs has been discussed. Green roofs are composed of different layers, of which the most important ones that affect their functionality are growth media and vegetation. The composition, thickness, and other characteristics of this medium highly affects the water retention potential of the green roof. Furthermore, the quality of released runoff from green roofs is different from bare roofs.

A review of the literature shows that the quantity and quality of produced runoff from green roofs are highly dependent on green roof composition, type of vegetation, and growth media, as well as regional climate and other physical conditions. The reported values for runoff volume reductions by green roofs are highly different, even though all of the studies show considerable reduction in runoff volume, especially in normal rainfall events. It is important to formulate the relation between runoff volume and green roof characteristics in order to find the best composition of these roofs for this purpose. The review of the literature shows that green roofs may also have inverse effects on runoff quality. This is because of the fertilizers that are used in green roof maintenance, which could increase the nutrient concentration in produced runoff. This is an important issue that should be taken into account in design of these systems, especially regarding where the produced runoff is drained.

The other functions of green roofs are helping in more efficient energy use, air pollution reduction, reducing urban heat island impacts, and better human feeling, as well as noise pollution reduction. The effectiveness of green roofs in each of these areas is highly dependent on its composition and characteristics. The cost–benefit analysis of green roof implementation through different studies show that by just considering the tangible costs from the developer point of view, the green roofs may not be economic choice in comparison with conventional roofs in lots of regions where governmental supports are not available. However, the reasoning that has emerged from the usage of these systems is that the intangible benefits of these systems, as well as the benefits that they can provide for the entire city such as runoff reduction, improvement of air quality, and reduction of the air temperature, are worth it.

References

Alcazar, S.S., Olivieri, F., and Neila, J. (2016). Green roofs: Experimental and analytical study of its potential for urban microclimate regulation in Mediterranean–continental climates. *Urban Climate* 17: 304–317.

Allen, R.G., Pereira, L.S., Raes, D., and Smith, M. (1998). Crop evapotranspiration-Guidelines for computing crop water requirements. *FAO Irrigation and drainage paper, No. 56*. Rome, Italy.

Alsup, S., Ebbs, S., and Retzlaff, W. (2010). The exchangeability and leachability of metals from select green roof growth substrates. *Urban Ecosystem* 13 (1): 91–111.

Arkar, C., Domjan, S., Majkovič, D. et al. (2015). Lightweight Green Roofs' Thermal Response under Freezing Conditions. *Energy Procedia* 78: 1189–1194.

Ayata, T., Erdemir, D., and Ozkan, O.T. (2017). An investigation for predicting the effect of green roof utilization on temperature decreasing over the roof surface with Gene Expression Programming. *Energy and Buildings* 139: 254–262.

Barr, C.M., Gallagher, P.M., Wadzuk, B.M., and Welker, A.L. (2017). Water Quality Impacts of Green Roofs Compared with Other Vegetated Sites. *Journal of Sustainable Water in the Built Environment* 3 (3): 04017007.

Bengtsson, L., Grahn, L., and Olsson, J. (2005). Hydrological function of a thin extensive green roof in southern Sweden. *Hydrology Research* 36 (3): 259–268.

Berghage, R., Beattie, D., Jarrett, A. et al. (2009). *Green Roofs for Stormwater Runoff Control.* Washington, DC: U.S. Environmental Protection Agency EPA/600/R-09/026, 2009.

Berndtsson, J.C., Emilsson, T., and Bengtsson, L. (2006). The influence of extensive vegetated roofs on runoff water quality. *Science of the Total Environment* 355 (1–3): 48–63.

Berndtsson, J.C. (2010). Green roof performance towards management of runoff water quantity and quality: A review. *Ecological engineering* 36 (4): 351–360.

Besir, A.B. and Cuce, E. (2018). Green roofs and facades: A comprehensive review. *Renewable and Sustainable Energy Reviews* 82: 915–939.

Bianchini, F. and Hewage, K. (2012). Probabilistic social cost-benefit analysis for green roofs: A lifecycle approach. *Building and Environment* 58: 152–162.

Bliss, D.J., Neufeld, R.D., and Ries, R.J. (2009). Storm Water Runoff Mitigation Using a Green Roof. *Environmental Engineering Science* 26 (2): 407–418.

Brudermann, T. and Sangkakool, T. (2017). Green roofs in temperate climate cities in Europe – An analysis of key decision factors. *Urban Forestry & Urban Greening* 21: 224–234.

Carpenter, D.D. and Kaluvakolanu, P. (2011). Effect of Roof Surface Type on Storm-Water Runoff from Full-Scale Roofs in a Temperate Climate. *Journal of Irrigation and Drainage Engineering* 137 (3): 161–169.

Carter, T. and Keeler, A. (2008). Life-cycle cost–benefit analysis of extensive vegetated roof systems. *Journal of Environmental Management* 87 (3): 350–363.

Carter, T.L. and Rasmussen, T.C. (2007). Hydrologic Behavior of Vegetated Roofs1. *JAWRA Journal of the American Water Resources Association* 42 (5): 1261–1274.

Cascone, S., Catania, F., Gagliano, A., and Sciuto, G. (2018). A comprehensive study on green roof performance for retrofitting existing buildings. *Building and Environment* 136: 227–239.

Castleton, H.F., Stovin, V., Beck, S.B.M., and Davison, J.B. (2010). Green roofs; building energy savings and the potential for retrofit. *Energy and Buildings* 42 (10): 1582–1591.

Chen, H., Ooka, R., Huang, H., and Tsuchiya, T. (2009). Study on mitigation measures for outdoor thermal environment on present urban blocks in Tokyo using coupled simulation. *Building and Environment* 44 (11): 2290–2299.

Clark, C., Adriaens, P., and Talbot, F.B. (2008). Green Roof Valuation: A Probabilistic Economic Analysis of Environmental Benefits. *Environmental Science and Technology* 42 (6): 2155–2161.

Collins, S., Kuoppamäki, K., Kotze, D.J., and Lü, X. (2017). Thermal behavior of green roofs under Nordic winter conditions. *Building and Environment* 122: 206–214.

Connelly, M. and Hodgson, M. (2013). Experimental investigation of the sound transmission of vegetated roofs. *Applied Acoustics* 74 (10): 1136–1143. 3.

Currie, B.A. and Bass, B. (2008). Estimates of air pollution mitigation with green plants and green roofs using the UFORE model. *Urban Ecosystem* 11 (4): 409–422. https://doi.org/10.1007/s11252-008-0054-y.

Dayani, S., Sabzalian, M.R., Hadipour, M., and Eslamian, S. (2017). Water Scarcity and Sustainable Urban Green Landscape, Ch. 30 in. In: *Handbook of Drought and Water Scarcity, Vol. 2: Environmental Impacts and Analysis of Drought and Water Scarcity* (eds. S. Eslamian and F. Eslamian), 557–604. USA: Francis and Taylor/CRC Press.

DeNardo, J.C., Jarrett, A.R., Manbeck, H.B. et al. (2005). Stormwater Mitigation And Surface Temperature Reduction By Green Roofs. *Transactions of ASAE* 48 (4): 1491–1496.

Deutsch, B., Whitlow, H., Sullivan, M., Savineau, A. (2008). Re-greening Washington DC : a green roof vision based on environmental benefits for air quality and stormwater management, Conference, ETDEWEB, 379–384, Canada.

Downton, P. 2013. Green roofs and walls. Your Home Manual. Retrieved on 5 May 2017 from: http://www.yourhome.gov.au/materials/green-roofs-and-walls.

Dvorak, B. (2011). Comparative Analysis of Green Roof Guidelines and Standards in Europe and North America. *Journal of Green Building* 6 (2): 170–191.

Erdemir, D. and Ayata, T. (2017). Prediction of temperature decreasing on a green roof by using artificial neural network. *Applied Thermal Engineering* 112: 1317–1325.

Getter, K.L., Rowe, D.B., Robertson, G.P. et al. (2009). Carbon Sequestration Potential of Extensive Green Roofs. *Environmental Science and Technology* 43 (19): 7564–7570.

Hathaway, A.M., Hunt, W.F., and Jennings, G.D. (2008). A Field Study of Green Roof Hydrologic and Water Quality Performance. *Transactions of the ASABE* 51 (1): 37–44.

Herrera-Gomez, S.S., Quevedo-Nolasco, A., and Pérez-Urrestarazu, L. (2017). The role of green roofs in climate change mitigation. A case study in Seville (Spain). *Building and Environment* 123: 575–584.

Huang, Y.-Y., Chen, C.-T., and Liu, W.-T. (2018). Thermal performance of extensive green roofs in a subtropical metropolitan area. *Energy and Buildings* 159: 39–53.

Jaffal, I., Ouldboukhitine, S.-E., and Belarbi, R. (2012). A comprehensive study of the impact of green roofs on building energy performance. *Renewable Energy* 43: 157–164.

Jennett, T.S. and Zheng, Y. (2018). Component characterization and predictive modeling for green roof substrates optimized to adsorb P and improve runoff quality: A review. *Environmental Pollution* 237: 988–999.

Johannessen, B.G., Hanslin, H.M., and Muthanna, T.M. (2017). Green roof performance potential in cold and wet regions. *Ecological Engineering* 106: 436–447.

Lee, A. (2004). Life cycle cost analysis-green roofs form an investment. In: *Proceedings of the second North American green roof conference: greening rooftops for sustainable communities*, 2–4. Portland, OR: The Cardinal Group, Toronto. Canada.

Lee, L.S.H. and Jim, C.Y. (2018). Thermal-cooling performance of subtropical green roof with deep substrate and woodland vegetation. *Ecological Engineering* 119: 8–18.

Li, J., Wai, O.W.H., Li, Y.S. et al. (2010). Effect of green roof on ambient CO2 concentration. *Building and Environment* 45 (12): 2644–2651.

Locatelli, L., Mark, O., Mikkelsen, P.S. et al. (2014). Modelling of green roof hydrological performance for urban drainage applications. *Journal of hydrology* 519: 3237–3248.

Ma, L., Liu, P., Fu, F. et al. (2011). *Integrated Energy Strategy for the Sustainable Development of China*. Elsevier.

Mahdiyar, A., Tabatabaee, S., Abdullah, A., and Marto, A. (2018). Identifying and assessing the critical criteria affecting decision-making for green roof type selection. *Sustainable Cities and Society* 39: 772–783. 7.

Matlock, J.M. and Rowe, D.B. (2017). Does Compost Selection Impact Green Roof Substrate Performance? Measuring Physical Properties, Plant Development, and Runoff Water Quality. *Compost Science and Utilization* 25 (4): 231–241.

Mentens, J., Raes, D., and Hermy, M. (2006). Green roofs as a tool for solving the rainwater runoff problem in the urbanized 21st century? *Landscape and Urban Planning* 77 (3): 217–226.

Monteiro, C.M., Calheiros, C.S.C., Palha, P., and Castro, P.M.L. (2017). Growing substrates for aromatic plant species in green roofs and water runoff quality: pilot experiments in a Mediterranean climate. *Water Science and Technology* 76 (5): 1081–1089.

Monterusso, M.A., Rowe, D.B., Rugh, C.L., and Russell, D.K. (2004). Runoff water quantity and quality from green roof systems. *Acta Horticulturae* (639): 369–376.

Morakinyo, T.E., Dahanayake, K.W.D.K.C., Ng, E., and Chow, C.L. (2017). Temperature and cooling demand reduction by green-roof types in different climates and urban densities: A co-simulation parametric study. *Energy and Buildings* 145: 226–237.

Nagase, A. and Dunnett, N. (2010). Drought tolerance in different vegetation types for extensive green roofs: Effects of watering and diversity. *Landscape and Urban Planning* 97 (4): 318–327.

Nagase, A. and Dunnett, N. (2011). The relationship between percentage of organic matter in substrate and plant growth in extensive green roofs. *Landscape and Urban Planning* 103 (2): 230–236.

Nagase, A., Dunnett, N., and Choi, M.-S. (2013). Investigation of weed phenology in an establishing semi-extensive green roof. *Ecological Engineering* 58: 156–164.

Nagase, A., Dunnett, N., and Choi, M.-S. (2017). Investigation of plant growth and flower performance on a semi-extensive green roof. *Urban Forestry & Urban Greening* 23: 61–73.

Ng, E., Chen, L., Wang, Y., and Yuan, C. (2012). A study on the cooling effects of greening in a high-density city: An experience from Hong Kong. *Building and Environment* 47: 256–271.

Niachou, A., Papakonstantinou, K., Santamouris, M. et al. (2001). Analysis of the green roof thermal properties and investigation of its energy performance. *Energy and Buildings* 33 (7): 719–729.

OConnor, T., Culligan, P., Carson, T. et al. (2014). *Evaluation of Green Roof Water Quantity and Quality Performance in an Urban Climate*. Washington, DC, EPA/600/R-14/180: US EPA Office of Research and Development.

Okhravi, S., Eslamian, S., and Adamowski, J. (2015). *Water Reuse in Rainwater Harvesting, Urban Water Reuse Handbook, Ch. 59* (ed. S. Eslamian), 779–796. Taylor and FrancisCRC Group.

Ouldboukhitine, S.-E., Belarbi, R., and Sailor, D.J. (2014). Experimental and numerical investigation of urban street canyons to evaluate the impact of green roof inside and outside buildings. *Applied Energy* 114: 273–282.

Palut, M., and Canziani, O. (2007). Contribution of working group II to the fourth assessment report of the intergovernmental panel on climate change. IPCC.

Pérez, G., Vila, A., Rincón, L. et al. (2012). Use of rubber crumbs as drainage layer in green roofs as potential energy improvement material. *Applied Energy* 97: 347–354.

Pérez-Lombard, L., Ortiz, J., and Pout, C. (2008). A review on buildings energy consumption information. *Energy and Buildings* 40 (3): 394–398.

Pugh, T.A.M., MacKenzie, A.R., Whyatt, J.D., and Hewitt, C.N. (2012). Effectiveness of Green Infrastructure for Improvement of Air Quality in Urban Street Canyons. *Environmental Science and Technology* 46 (14): 7692–7699.

Rowe, B. (2018). Green Roofs for Pollutants' Reduction. In: *Nature Based Strategies for Urban and Building Sustainability*, 141–148. Butterworth-Heinemann.

Saadatian, O., Sopian, K., Salleh, E. et al. (2013). A review of energy aspects of green roofs. *Renewable and Sustainable Energy Reviews* 23: 155–168.

Saiz, S., Kennedy, C., Bass, B., and Pressnail, K. (2006). Comparative Life Cycle Assessment of Standard and Green Roofs *Environmental Science & Technology* 2006, 40, (13): 4312–4316.

Santamouris, M. (2014). Cooling the cities – A review of reflective and green roof mitigation technologies to fight heat island and improve comfort in urban environments. *Solar Energy* 103: 682–703.

Savio, P., Rosenzweig, C., Solecki, W.D. et al. (2006). *Mitigating New York city's heat island with urban forestry, living roofs, and light surfaces, New York city regional heat island initiative.* Albany: New York State Energy Research and Development Authority Retrieved from https://www.coolrooftoolkit.org/wp-content/uploads/2012/04/Mitigating-New-York-Citys-Heat-Island-with-Urban-Forestry-Living-Roofs-and-Light-Surfaces.pdf.

Schweitzer, O. and Erell, E. (2014). Evaluation of the energy performance and irrigation requirements of extensive green roofs in a water-scarce Mediterranean climate. *Energy and Buildings* 68: 25–32.

Shafique, M., Kim, R., and Rafiq, M. (2018). Green roof benefits, opportunities and challenges – A review. *Renewable and Sustainable Energy Reviews* 90: 757–773.

Silva, C.M., Gomes, M.G., and Silva, M. (2016). Green roofs energy performance in Mediterranean climate. *Energy and Buildings* 116: 318–325.

Smith, K.R., Roebber, P.J., Smith, K.R., and Roebber, P.J. (2011). Green Roof Mitigation Potential for a Proxy Future Climate Scenario in Chicago, Illinois. *Journal of Applied Meteorology and Climatology* 50 (3): 507–522. https://doi.org/10.1175/2010JAMC2337.1.

Solcerova, A., van de Ven, F., Wang, M. et al. (2017). Do green roofs cool the air? *Building and Environment* 111: 249–255.

Speak, A.F., Rothwell, J.J., Lindley, S.J., and Smith, C.L. (2012). Urban particulate pollution reduction by four species of green roof vegetation in a UK city. *Atmospheric Environment* 61: 283–293.

Speak, A.F., Rothwell, J.J., Lindley, S.J., and Smith, C.L. (2013). Rainwater runoff retention on an aged intensive green roof. *Science of the Total Environment* 461–462: 28–38.

Sproul, J., Wan, M.P., Mandel, B.H., and Rosenfeld, A.H. (2014). Economic comparison of white, green, and black flat roofs in the United States. *Energy and Buildings* 71: 20–27.

Stovin, V. (2010). The potential of green roofs to manage Urban Stormwater. *Water Environment Journal* 24 (3): 192–199.

Sun, T., Grimmond, C.S.B., and Ni, G.-H. (2016). How do green roofs mitigate urban thermal stress under heat waves? *Journal of Geophysical Research-Atmospheres* 121 (10): 5320–5335.

Tang, X. and anf Qu, M. (2016). Phase change and thermal performance analysis for green roofs in cold climates. *Energy and Buildings* 121: 165–175.

Teemusk, A. and Mander, Ü. (2007). Rainwater runoff quantity and quality performance from a greenroof: The effects of short-term events. *Ecological Engineering* 30 (3): 271–277.

Van Mechelen, C., Dutoit, T., and Hermy, M. (2015). Adapting green roof irrigation practices for a sustainable future: A review. *Sustainable Cities and Society* 19: 74–90.

Van Renterghem, T. (2018). Improving the noise reduction by green roofs due to solar panels and substrate shaping. *Building Acoustics* 25 (3): 219–232.

VanWoert, N.D., Rowe, D.B., Andresen, J.A. et al. (2005). Green Roof Stormwater Retention. *Journal of Environmental Quality* 34 (3): 1036.

Vijayaraghavan, K. (2016). Green roofs: A critical review on the role of components, benefits, limitations and trends. *Renewable and Sustainable Energy Reviews* 57: 740–752.

Weiler, S.K. and Scholz-Barth, K. (2009). *Green Roof Systems : A Guide to the Planning, Design, and Construction of Landscapes Over Structure.* Wiley Retrieved from https://books.google.com/books?hl=en&lr=&id=_V_tXyuyZm0C&oi=fnd&pg=PA1&dq=+Green+Roof+Systems:+A+Guide+to+the+Planning,+Design,+and+Construction+of+Landscapes+over+Structure&ots=SkZfU2xjm1&sig=bRR0AphYpWl1R3xskUtnExwyPWQ23v=onepage&q=Green Roof Systems%3A A Guide to the Planning%2C Design%2C and Construction of Landscapes over Structure&f=false.

Wong, N., Cheong, D.K., Yan, H. et al. (2003a). The effects of rooftop garden on energy consumption of a commercial building in Singapore. *Energy and Buildings* 35 (4): 353–364.

Wong, N.H., Chen, Y., Ong, C.L., and Sia, A. (2003b). Investigation of thermal benefits of rooftop garden in the tropical environment. *Building and Environment* 38 (2): 261–270.

Yang, J., Yu, Q., and Gong, P. (2008). Quantifying air pollution removal by green roofs in Chicago. *Atmospheric Environment* 42 (31): 7266–7273. 3.

Zimmer, U. and Geiger, W.F. (1997). Model for the design of multilayered infiltration systems. *Water Science and Technology* 36 (8–9): 301–306.

24

Green Landscaping and Plant Production with Water Harvesting Solutions

Saeid Eslamian[1], Saeideh Parvizi[1], and Sayed Salman Ghaziaskar[2]

[1] *Department of Water Engineering, College of Agriculture, Isfahan University of Technology, Isfahan, Iran*
[2] *Department of Geography, Islamic Azad University, Najafabad Branch, Iran*

24.1 Introduction

In modern times, humanity confronts the major challenge of water scarcity, mainly due to the rapidly growing world population and global climatic changes (Schewe et al. 2014). This is particularly true for urban environments, which are among the most vulnerable systems. These bear great environmental pressures, associated with large ecological footprints, and depend to a great extent on water from distant sources, which are transported by means of large infrastructures. Approximately 53% of the world's population is concentrated in cities and more than 75% of the population is in cities of North America, Europe, and Oceania (Ghimire et al. 2014). Climate change, together with cities' growing population, further increases the stress on water resource availability in urban areas (Ghimire et al. 2014).

Figure 24.1 A sample of green landscaping and plant production with water Harvesting solutions. Source: Reference: Rainwater harvesting for beneficial use, Green Seattle permits, Seattle Department of Construction and Inspections, Tip 520.

The major hydrological concerns in urban environments are the reduction of infiltration and groundwater recharge due to the existence of large impervious areas, and changes in the pattern of surface runoff and river flow (Niemczynowicz 1999). These changes impose high peak flows and large runoff volumes that increase the risk of flood events and accelerate the transport of pollutants and sediment from the urban areas. In order to cope with these challenges and provide a sustainable urban future, there is a need to move toward the goal of efficient and appropriate water use and management, mainly in arid and semi-arid climates (Abdulla and Al-Shareef 2009; Dayani et al. 2017).

Water harvesting is an ancient practice that has been used, mainly in dry environments, to increase efficiency of water collection. This is achieved by directing water from a large natural watershed or man-made collection surface into a small basin where the water can be stored in underground reservoirs or can be used directly for irrigation or domestic uses. Water harvesting can significantly increase plant production in the arid and semi-arid tropics (Vohland and Barry 2009).

More recently, water harvesting has been neglected, particularly in developed countries, due to the technological achievements in the fields of water production and transport. Nevertheless, in recent years, water harvesting in modern-urban environments has become a necessary practice (Figure 24.1) (Nachshon et al. 2016). The urban regions are being paved and built up, resulting in reduction of groundwater recharge areas. Consequently, large amounts of water that rain over cities is withdrawn from recharge as it is directed into the municipal drainage system. Moreover, in extreme rain events the drainage systems may be overflooded which may lead to ecologic and economic hazards (Pandey 2003).

This chapter reviews the history of rainwater harvesting (RWH) and discusses its impacts in modern urban environments on the hydrological system. Two types of RWH

Handbook of Water Harvesting and Conservation: Basic Concepts and Fundamentals, First Edition. Edited by Saeid Eslamian and Faezeh Eslamian.

methods are discussed. They are the methods used in the past and the methods used recently.

24.2 Water Harvesting

Water harvesting is the capture and storage of water for beneficial use. It can be accomplished anywhere water supply is available for collection and a water source is desired or required. Water harvesting is an ancient practice that has been used, mainly in dry environments, to increase efficiency of water collection and use by directing water from a large natural watershed or man-made collection surface into a small basin where the water can be stored in underground reservoirs or be used directly for irrigation or domestic uses. In modern times water harvesting has been neglected, particularly in developed countries, due to the technological achievements in the fields of water production and transport. Nevertheless, over recent years, water harvesting in modern urban environments has become a necessary practice.

To understand the process fully, it is important to understand water harvesting terms, which include rainwater harvesting, gray water harvesting, reclaimed water, and potable water that are briefly explained below (SDCI 2009):

- *Rainwater harvesting* is the capture and storage of rainwater; it is considered the cleanest form of harvested water.
- *Gray water harvesting* is the capture and storage of water that has already been used for non-sewage purposes from baths and showers to washing machines, sinks, and vehicle washing runoff. Reuse of gray water triggers more code requirements and design regulations than the use of rain water.
- *Reclaimed water* is wastewater treated to levels that allow it to be used for non-drinking water purposes. Reuse of reclaimed water triggers more code requirements and design regulations than reuse of rainwater.
- *Potable water* is clean water satisfactory for drinking, culinary, and domestic purposes, and meets the drinking water standards established by the Washington State Department of Health.

24.3 Rainwater Harvesting

RWH, which is collection, storage, delivery, and use of rainwater for various purposes (Stec and Kordana 2015), has great potential as a sustainable way to cope with the hydrological challenges imposed by the urban environment (Angrill et al. 2012; Stec and Kordana 2015).

In other words, RWH is a common and old practice in which rainwater is being collected and stored in order to be used for domestic and small-scale agricultural uses. While RWH has been used in rural and urban places for centuries, in modern cities its use is more limited. Nevertheless, in recent years there has been a growing trend to use RWH in modern and urban environments as part of the solution to the growing challenges associated with the supply of good quality water to the world's population which is concentrated in cities (Buhaug and Urdal 2013). Moreover, most urban environments have a deleterious impact on the hydrological cycle and sustainable management of the urban hydrological system is needed (Buhaug and Urdal 2013). Since urban development is not likely to be halted for water considerations, it is crucial to guide urban planners on how to manage urban development with minimal damage to groundwater resources (Carmon et al. 1997).

24.3.1 Rainwater Harvesting in the Past

Historically, RWH was used mainly in arid environments. Le Houérou and Lundholm (1976) estimated that 3–5% of the arid zones worldwide could be cultivated by proper use of RWH, which is on an equal footing with traditional irrigation techniques, in these regions, from rivers, springs, or lakes (Bruins et al. 1986). The most ancient archeological evidences of primitive RWH systems are estimated to be 9000 years old and found in the city of Beidha, Jordan (Bruins et al. 1986). Similar evidence exists in the Negev Desert, Israel, where today annual precipitation is in the order of 100 mm. The archeological evidence from the Negev, dated to 2000 BCE, points to the existence of agriculture practices that required the use of flood water irrigation (Evenari et al. 1958) which is the most basic application of RWH (Abdulla and Al-Shareef 2009; Bruins et al. 1986). In flood water irrigation, surface runoff, following rain events is being routed toward topographical depressions where it is being used for irrigation as presented in Figure 3.2a (Young et al. 2002). In the Negev desert, full exploiting of surface runoff from small watersheds for water storage, which included complex channel systems to deliver rain and surface runoff water toward underground cisterns (Figure 3.2b) dates as far back as 850–600 BC (Evenari et al. 1958). Kedar (1967) and Evenari et al. (1982) estimated that during the Nabatean-Byzantine period (100–700 AD) RWH supported the cultivation of 4000 ha, out of an area of 200 000 ha, in the Negev desert, and wheat, barley, grapes, olives, dates, and other crops were successfully grown (Bruins et al. 1986). In Yemen, a complex RWH system dates back to 750 BC and some RWH systems are still being used today (Brunner and Haefner 1986). Other ancient RWH systems with various degrees of complexity were found in Egypt,

Algeria, Tunisia, Sahel and West Africa, India, China, Turkmenistan region, North and South America, and others. Nevertheless, even though RWH does not require high technological skills, its contribution for agriculture and domestic water use is significant, still, in many regions worldwide.

24.3.2 Modern Rainwater Harvesting

In modern times, RWH consists of rainwater collection from large surfaces, mainly rooftops (Liaw and Tsai 2004), and storage of the water in under- or aboveground reservoirs (Figure 24.2). Based on water quality, which is mainly affected by the quality and state of the water collection surfaces and the delivery systems, the water can be used for drinking, domestic uses, and irrigation (Jones and Hunt 2010; Sturm et al. 2009). RWH is a renewable source of clean water that is ideal for domestic and small-scale agricultural uses and the greater attraction of a RWH system is in its low cost, accessibility, and simple maintenance at the household level (Abdulla and Al-Shareef 2009). RWH can promote significant water saving in residences in different countries. For example: In Germany, Herrmann and Schmida (2000) showed that potential savings of potable water in a house might vary from 30% to 60%, depending on the demand and size of the rain water collection area. In Newcastle, Australia, it was concluded that RWH would save 60% of potable water whereas Brazil showed potential water savings from using water harvesting in the range of 34–92%, with an average of 69% Ghisi (2006). Abdulla and Al-Shareef (2009) reported potential water saving of up to 20% of drinking water by applying RWH in urban environments in Jordan. While traditional RWH systems consist of storing of the harvested rainwater in storage containers at the site and self-use of the water by the property owners, recently, few works have discussed the option of allowing the harvested rainwater to recharge the local aquifers through infiltration wells, usually after overfilling the storage tanks (Dillon 2005; Mun and Han 2012). Stec and Kordana (2015) described a state-of-the-art RWH system at a four-story building with 24 apartments, where rainwater is being collected from the rooftop and stored in storage tanks. The water is being used for various non-potable uses and any excess water which exceeds the storing capacity of the tanks is routed to infiltration wells located at the site. In addition, harvested water can be directly routed to gardens and consumed by vegetation and/or infiltrated to groundwater (Gao et al. 2016). Construction of infiltration wells for the harvested rainwater has to account for the hydrological and climatological conditions at the site, as well as the surface area of the collection surface. In order to ensure efficient infiltration of harvested rainwater into the aquifer with no overflood of the infiltration well system, it is required to construct the infiltration well sufficiently deep with a filter length long enough to enable sufficiently high water flow from the well into the ground, while considering the hydraulic properties of the medium at the specific site of infiltration. Hereon the discussion will focus on RWH systems that consist of infiltration of the harvested water to groundwater and it will be termed "rainwater harvesting and recharge system" (RWHR).

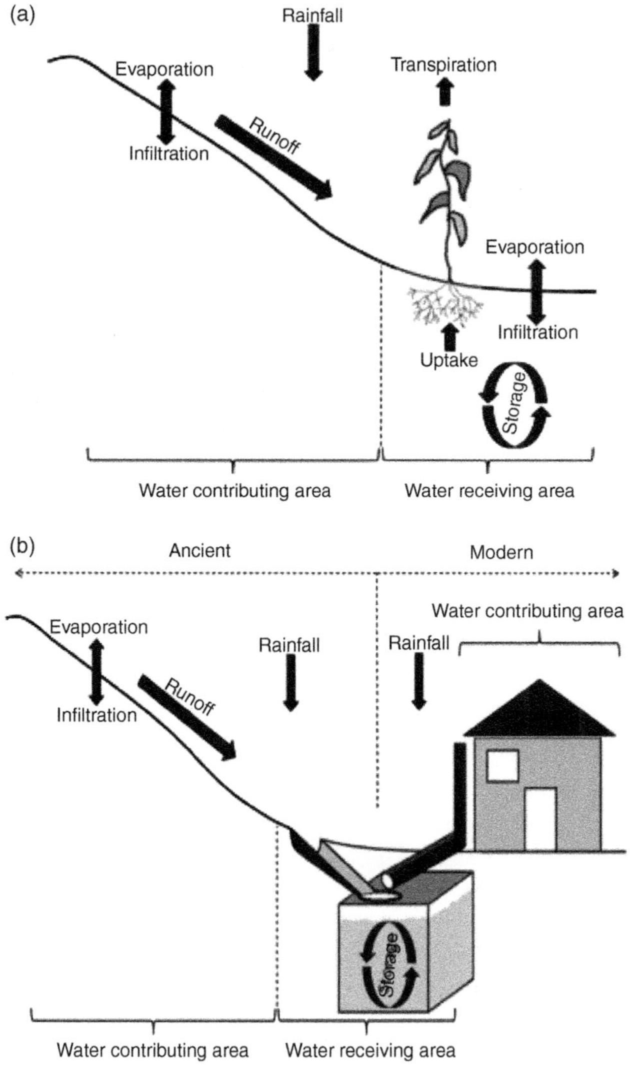

Figure 24.2 (a) primitive RWH consists of runoff collection and routing toward lower areas where the water is being used directly for irrigation: and (b) more complex RWH systems where rain and surface water are routed by systems of channels (in ancient times) or pipes and tubes (in modern times) toward reservoirs, where the water is being used for domestic and agricultural purposes, based on water quality and needs (Nachshon et al. 2016).

24.4 The Goals and Benefits of Rainwater Harvesting

RWH provides a host of design benefits that range from reducing owner utility rates and improving landscape health, to reducing combined sewer overflows into Seattle's water supply and reducing demand on the city's potable water systems (SDCI 2009):

- *Financial benefits* Use of collected rainwater helps residents save on potable water use fees and on sanitary sewer use fees.
- *Plant health* Landscape plants flourish with irrigation from collected rainwater. Rainwater does not contain chlorine (an important additive that keeps potable water safe for drinking), benefiting many ecologically sensitive plants.
- *Water quality* Removing rainwater from the combined sewer systems and redirecting it to cisterns has the potential to lower peak flows and reduce the amount of pollutants that find their way into our natural water bodies.
- *Water supply* Combine a rapidly increasing population with lower annual rainfall and higher temperatures and the demand for potable water increases dramatically, putting greater pressure on municipal water supplies. RWH reduces this demand.
- *Green building credits* Many green building systems offer credits for rainwater harvest systems.

24.5 Impact of RWHR on Infiltration and Surface Runoff Processes

As mentioned, implementation of RWHR systems has two major benefits with respect to the urban hydrological system: (i) it increases groundwater recharge; and (ii) decreases surface runoff volumes.

24.5.1 Groundwater Recharge

To estimate the volume of rainwater that infiltrates into the vadose zone and groundwater it is common to use infiltration coefficients that are dependent on physical properties at the point of interest and include: soil hydraulic properties, topography, vegetation, and land use (Pauleit and Duhme 2000). For any given area, either urban, natural, or cultivated, the effective infiltration coefficient $I_{c(eff)}$ can be estimated by calculating the weighted arithmetic mean of the different infiltration coefficients of the various land cover components in the area ($I_{c(i)}$) and their corresponding surface area ($A_{(i)}$).

$$I_{c(eff)} = \frac{\sum(A_{(i)} \cdot I_{c(i)})}{\sum A_{(i)}} \tag{24.1}$$

Actual volume of infiltrated water into groundwater, I (m^3), is defined by:

$$I = R.A.I_{c(eff)} \tag{24.2}$$

where R is annual precipitation (m) and A (m^2) is the surface area through which infiltration is taking place. Assuming that in the urban environment, parks, asphalt areas, and buildings are the three major land cover components, by knowing their I_c values it is possible to compute $I_{c(eff)}$ for any given area with any composition of the three components.

24.5.2 Surface Runoff Estimation

The combined impact of global climate change and the impervious nature of modern cities results in an increased frequency of extreme events of flooding in urban environments due to failure of drainage systems at extreme rain events (Schreider et al. 2000). The flooding events cause large damage to buildings, other public and private infrastructure, and risk people's lives. Reduction of surface runoff is essential to reduce the risk of flooding and cut down the cost of drainage systems in urban environments. RWHR may play a key role in this aspect as it increases the volume of water penetrating the subsurface instead of flowing as surface runoff. A common model to estimate surface runoff was developed by the United States Soil Conservation Service (SCS) and it has been widely applied to estimate storm runoff depth for every patch within a watershed based on runoff curve numbers (CNs) (Weng 2001). The SCS equation for storm runoff depth is expressed as:

$$\begin{aligned} Q &= \frac{(P - 0.2S)^2}{(P + 0.8S)} \quad (for\, P > 0.2S) \\ Q &= 0 \quad (for\, P > 0.2S) \end{aligned} \tag{24.3}$$

Where Q is the daily storm runoff (mm), P is the daily storm precipitation (mm), and S is the potential maximum storage of the region (mm). S is defined by CN, which is the runoff curve number of hydrologic soil group and land cover combination that describes how much of the precipitation is routed to the surface runoff (Weng 2001) by the following correlation:

$$S = \frac{25400}{CN} - 254 \tag{24.4}$$

24.6 Climate Change and RWH

With respect to groundwater recharge and surface runoff, global climate change has two major impacts: (i) change of precipitation patterns; and (ii) alteration of soil properties

(Pandey et al. 2003). Recent years' evidence, as well as climatic models, indicate that climate variability is expected to increase, with a high chance of extreme rainfall events, droughts, heat waves, floods, cyclones, and hurricanes (Nachshon et al. 2014). These extreme events are likely to have a much greater impact on human society than gradual changes in climate (Gill et al. 2007). Moreover, increase of global temperatures results in changes to soil properties over a range of time scales, such that the soils of the future may not have the same infiltration properties as existing soils. Holman (2006) pointed out that in natural and cultivated lands, future warmer climates will lead to an increase in the length of the growing season, so that the soils will return to field capacity later in the autumn and start drying out sooner in the spring. This will lead to a reduction in the length of the groundwater recharge period and the overall effect would be a reduction in recharge. Moreover, an increase in annual temperatures and changes in precipitation patterns are expected to lead to peat soil degradation (Holman 2006), reduction of microbial activity and organic matter in soils, increase of salt concentrations in soils (Nachshon et al. 2014), and over longer time scales clay content in the soil may decrease. These processes will reduce soil aggregate stability (Lavee et al. 1998; Reid and Goss 1981), water holding capacity, permeability, and will increase the probability of crust formation. These processes will lead to a dramatic decrease in infiltration rates and increase in surface runoff, even if rainfall intensity will not increase. Even though it is impossible to estimate the exact impact of the global climate change on I_c and CN, it is clearly seen that I_c is expected to decrease and CN is expected to increase for many parts of the world, including the populated urban environments. Consequently, implementation of RWHR practices is crucial to ensure a sustainable management of the changing hydrological system. Proper use of the RWHR will enable to maintain relatively high groundwater recharge, while reducing the risk of flood events in urban areas, with reasonable costs.

24.7 Landscape Functions and RWH

According to the Millennium Ecosystem Assessment (2005), landscape functions comprise ecosystem functions, and the provision of goods and services. Ecosystem functions such as photosynthesis, litter turnover, or water cycling are natural processes inherent to ecosystems. Ecosystem goods such as crop biomass or clean water can be derived from ecosystems. Ecosystem services denote processes such as purifying water or maintaining soil fertility and biodiversity. In dry-land ecosystems, water is the major limiting factor in agricultural production systems and the performance of landscape functions relies heavily on the availability of water.

For example, semi-arid areas such as Africa are modified arid and semi-arid landscapes fulfill a range of functions, ranging from ecological processes such as water and nutrient cycling, through biomass production and biodiversity conservation to socioeconomic services such as providing the basis for sustainable rural livelihoods. Landscape function processes differ in their importance with regard to their spatial scale, e.g. biomass production at the plot scale up to carbon sequestration at the global scale (Hein et al. 2006). The introduction of the RWH structures modifies landscape functions at different spatial scales. In the following sections of this paper, the effects of the RWH practices will be reviewed and evaluated with respect to the appropriate landscape scale. The major issues are presented according to the processes affected by RWH, indicators as well as the spatial scales are considered, and conclusions are made about the impact of the RWH practices on landscape functions.

24.8 Hydrological Functions and RWH

24.8.1 Infiltration

The RWH structures modify water flows in the landscape mainly by enhancing water infiltration at plot scale (Wakindiki and Ben-Hur 2004). The velocity of runoff is reduced, and the water is collected behind the structures. Soil moisture increased significantly below semi-circular bunds in Burkina Faso and in run-on basins in South Africa (van Rensburg et al. 2004). Furthermore, the greater number of soil macro pores enhances infiltration due to increased biological activity at vegetation strips, as shown for Australian dry lands (Ludwig et al. 2005). Infiltration at the RWH structures might require enhanced runoff up slope (Rahman et al. 2017). Soil physical characteristics are important but biological crusts might also be crucial in arid ecosystems to extend the runoff area in order that sufficient run-on collects at the RWH structures (Eldridge et al. 2002). A negative effect might occur under poor drainage conditions that might lead to waterlogging.

24.8.2 Groundwater Recharge

Very few direct field measurements of groundwater recharge are available. When local people were asked directly about the impact of RWH on groundwater recharge, they stated that the RWH practices enhanced water availability (e.g. Mutekwa and Kusangaya 2006).

24.8.3 Water Competition

The RWH practices not only trap surface water but also reduce runoff. Aquatic and wetland ecosystems downstream of the RWH structures might not receive enough surface runoff to maintain and sustain their functionality. Surface runoff might not always reach the next stream (Woyessa et al. 2005). Very little is known about the effects of in situ RWH practices on the hydrology, or on downstream freshwater ecology and biodiversity, although early studies, including hydrological modeling, indicate a positive impact on groundwater. To take account of the competing demands for water, adaptive research increasingly integrates agronomic, legal, socioeconomic, and ecological aspects at the watershed scale, for example in southern Africa (Rockstrom 2004; Usman and Reason 2004), as well as for compensating mechanisms as with the Green Water Credits Project (Dent and Kauffman 2007).

24.9 Soil Fertility and Biomass Production

24.9.1 Soil Fertility

Depending on slope, precipitation, and farming practices, as well as on the quality of design and maintenance, soil erosion is reduced under RWH (Schiettecatte et al. 2005). The RWH structures act as sediment traps, and therefore can enhance nutrient availability at the structures, as was shown for traditional Zai in Niger (Fatondji 2002). Although another study on soil under half-moons in Burkina Faso did not show significant differences in P, K, Ca, or N between the control and the various treatments consisting of stone rows, grass strips, and manure. The most significant effect of the RWH practices on biomass and grain yield was obtained with mulching. Irrigation has often been emphasized as a regional planning measure because it has appeared to be the best way of making farmers less dependent on erratic rainfall. However, this policy had limited success because of the build-up of salt in irrigated soils, a problem which does not usually occur in runoff farming because of the better quality of runoff water from very small catchments (FAO 2003). However, the quality of irrigation water varies greatly between different sources, locations, and seasons.

24.9.2 Crop Yields and Biomass Production

RWH has been shown to enhance the production of biomass for Zai in Burkina Faso (Kabore and Reij 2004). In the Central Plateau of Burkina Faso, application of Zai contributed to the restoration of large degraded encrusted sites and reduced the impact of dry spells (Rockstrom 2004). Generally, biomass production is enhanced by the RWH structures. Studies have mainly been conducted on cereal crops such as sorghum (Kabore and Reij 2004) and maize (Kayombo et al. 2004; Pretorius et al. 2005). Biomass production generally starts with low precipitation, with more biomass produced per mm precipitation. The RWH structures do not lead to increases in crop yields under all conditions. In a one-field experiment with Fanya juu in Kenya, yields were stabilized but increased only at locations with lower precipitation. As far as we know, there have so far been no studies considering total biomass changes, including non-crop plants and pastures, due to the RWH practices in Africa.

24.9.3 Biodiversity Conservation

Landscapes are a repository of genetic diversity for current ecosystem processes and for future demands on ecosystem processes. The impact of the RWH practices on three aspects of biodiversity, namely, floral diversity, structural heterogeneity, and animal diversity are now considered in the following sections.

24.9.3.1 Changes in Floral Diversity

Though mainly adopted for crop and fodder purposes, the RWH practices are known, in principle, to enhance floral diversity, by the fact that plants start to grow in places where there were bare degraded soils. It has been anecdotally reported that native trees and bushes increase in abundance as the RWH practices supported the replenishment of aquifers (Kongo and Jewitt 2006). But an ecological conflict might arise because RWH allows cropping in areas that were formerly exclusively used by nomads/herders and wild animals. Transformation of land is often carried out without the necessary knowledge of the RWH practices and might negatively impact soils and the productivity of the land. In many arid or semi-arid areas of Africa, a change from pastures to cropland is not sustainable in the long term, especially where the rural population is heavily dependent on livestock production systems (FAO 2006).

24.9.3.2 Changes in Structural Heterogeneity/Patchiness

RWH practices enhance biomass production per unit area, and therefore modify the spatial structure of the ecosystems in various ways. The RWH practices can completely change the character of an arid landscape as reported in the "Savannization" project in the northern Negev, where trees have been planted in a largely treeless semi-arid landscape (Warren et al. 1995). The relationship between the RWH practices and landscape heterogeneity refers to the number of plant patches, the spatial and temporal structure of the patches, as well as their composition. The impact of the RWH practices on the number of plant patches per unit area has so far scarcely been investigated. In Burkina Faso, the number of plant patches certainly increased, as plants were grown on totally degraded soil following the introduction of RWH. The height of plants and, therefore, the three-dimensional size of plant patches, are also enhanced. Plants grow to larger sizes, as data for crop plants, fodder grasses, and trees indicate (cf. crop yields). In dry land areas the distribution of plants is more strongly clumped in drier environments because water and nutrients are concentrated on micro-sites (Ludwig and Tongway 1995). The RWH structures improve water availability locally by creating artificial patches, i.e. at crop plant sites. Therefore, as in any agro-ecosystem, the spatial distribution of crops is strongly man-made, as farmers decide the spacing between plants. In general, crops are planted more regularly than natural vegetation occurs, which means that a random dispersal of patches is displaced by more equally dispersed patches, which might impact dispersal success of different plants and animals (Wiens 1976). Most RWH structures are installed for crop plants, and much less frequently for rangeland improvement or ecosystem conservation. Consequently, the special structure of indigenous grasses and herbs are replaced by crops such as maize and millet. Further, weeding is carried out around these crops. Thus, managed plant patches differ from natural ones in their composition, not only due to enhanced water availability induced by RWH but also to agronomic measures. In some cases, multiple crop species are planted within a patch created by the RWH structures. In Kenya, the edge of the RWH structure is planted with grasses surrounding the (cash) crop. In the Zai structures of West Africa, in spiny bushes are planted to protect the crop against livestock.

24.9.3.3 Changes in Animal Diversity

Animal diversity should be enhanced as more biomass becomes available for food and shelter, as shown for trees in Rajasthan, India (Pandey 2001). For most taxa occurring in dry lands, a positive relation exists between biomass and species diversity and abundance. More biomass also allows more complex trophic chains. Several field studies, including some for Africa, show that the numbers of animal species and individuals in an area increase due to more and larger vegetation patches. Other studies report the opposite situation, that decreases in animal species and abundance are due to missing plant patches (Hoffman 2000; Nangula and Oba 2004). In the context of the RWH structures an increased species biodiversity is anecdotally reported for increased rodent numbers in the Fanya juu terracing systems. This is a matter that farmers and nature conservationists might consider differently. If the RWH structures are not suitably designed, standing water may lead to increases in vector-borne parasitic diseases such as malaria or bilharzias through insect vectors (Bangoura 2002). Changes in the composition of animal communities may occur as natural vegetation is partly replaced by crop plants. Generally, natural vegetation hosts more animal species than (monospecific) crops, so this probably is true for transformed rangelands, too. Even more, it should be considered that if combined with only few crops the RWH practices might attract pests.

24.9.4 Sustainable Livelihoods

24.9.4.1 Food Security

Increased and sustained crop yields are vital for food security in rural communities. Several studies have shown that in many cases, crop yields are higher when the RWH practices are applied (Ellis-Jones and Tengberg 2000). Other studies showed that the RWH structures such as tera mainly serve to reduce crop failure during dry spells and droughts and thus help to enhance food security. Valuable mechanisms in this context are a prolonged vegetation period caused by early planting as well as an improved ability of crops to survive dry spells (Falkenmark et al. 2001). Increased efficiency of rainwater use can be reached by supplementary irrigation (Rockstrom 2004; Rockstrom et al. 2002). Over and above a site-specific amount of precipitation, the RWH structures do not necessarily lead to increased crop yields and improved food security because the RWH structures tend to occupy precious cropping areas that are subsequently lost for cropping. Weeding might be complicated and unsatisfactory, which might even permit rodents to become established within the fields (Herweg and Ludi 1999).

24.9.4.2 Conflicts Concerning Water Resources

Enhanced infiltration favors groundwater recharge and refilling of wells. Nevertheless, due to reduced downstream runoff, conflicts might arise between neighbors competing for available water resources. In densely populated areas of Kenya, such communal conflicts have been minimized through the adoption of a sophisticated technique of collecting street runoff and distributing the rainwater (Mutunga and Critchley 2001). Furthermore, social conflicts might arise, mainly concerning land and water rights when RWH systems facilitate cropping in areas formerly used exclusively by nomads (Tesfay and Tafere 2004).

24.9.4.3 Income/Social Balance

The adoption of RWH proved to have a positive effect on incomes, measured in return to labor (Cofie et al.

2004). However, RWH adoption rates are still low. Farmers hesitate to invest time and money in setting up the RWH structures, as they often have no security of land ownership and/or limited access to local markets where they could sell surpluses of food crops or cash crops (Drechsel et al. 2005). An increase in market access, measured as travel time to the capital, was identified as a driving force in improving agricultural production and productivity in densely populated areas in Kenya, and this could improve investments in RWH practices such as terracing. The Machakos and Kitui regions of Kenya are nowadays frequently cited as areas where successful RWH management practices have been put into practice. Case studies show that under specific conditions, application of the RWH practices is not profitable as yields are not high enough to justify the investments in labor and materials, although in the long term, gross margins are generally higher with RWH (van Rensburg et al. 2004).

24.10 Discussions

Contemporary society faces the necessity to reduce its consumptions: energy consumption is increasingly followed by water consumption. The sometimes-indiscriminate use of the water resources must face the problems of shortage and droughts occurring in several countries. Less than 1% of the water on the planet is drinkable. The water of atmospheric derivation, which corresponds to rainwater, water from the melting of snow, and from the fog, has always been considered an important source of supply, although this type of water needs to undergo certain treatments in order to meet the legal requirements of potability under the chemical, microbiological, and sometimes precise bacteriological profiles. The main issues emerging from several world reports on the planet's health and water consumption concern the qualitative and quantitative conditions of the resource: limited availability, uneven distribution on the territory, and increasing pollution, which are being aggravated each year by the growth of the world population, the indiscriminate rise of consumption, and by the lowering of water aquifers due to excessive withdrawals and the poor management of the territory.

In the other words, awareness and efficiency in the management of the water resources are addressed in different ways in countries of the world, depending on the presence and availability of the resource in each country. Countries with a good presence of underground, surface, or atmosphere water have always shown a minor sensitivity in the use of the resource compared to countries that have long dealt with serious shortages or total lack of water for their population. This has led to a different development over time of water harvesting and water collection systems that would provide for their needs.

Also, RWH practices are mostly simple and do not leave much space for technical improvement. However, major challenges lie in improving nutrient management, through mulching, and in stronger mechanization, with animal tracking (Raffaella 2011). Further, best practices and traditional methods may be combined. Another challenge remains to assess the potential and impact of the RWH practices with regard to future climate and other global and regional changes. Thornton et al. (2002) combined population scenarios and climate until 2050 to outline cropping boundaries together with the expected range of livestock and other production systems. Similar studies for the potentials of RWH have been fragmentary, covering, e.g. the potential of upscaling Zai (Freeman 2009), and studies on larger spatial scales are only now emerging (Senay and Verdin 2004). More in-depth research is also required at local scales. The socioeconomic and political conditions are qualitatively known (Critchley et al. 1992; Oweis et al. 1999) but not applied quantitatively to understand and support the individual decisions of farmers. Crop and risk assessment models and approaches developed so far for the RWH practices (Cohen et al. 1995; Young et al. 2002) rely mainly on ecological and technical information. Economic, social, political, or cultural factors are neglected; the spatial analyses of biophysical factors are not combined with socioeconomic analyses, e.g. for access to markets, available labor, and microcredit systems. The scanty literature on the impact of the RWH practices on biodiversity conservation shows that the effects of RWH on landscape functions are still poorly understood. Research aimed at improving agronomic practices does not necessarily focus on biodiversity-conserving landscape functional aspects such as animal biodiversity, or on the spatial landscape structure. This goes beyond issues of food security and poverty alleviation. Despite some attention gained, e.g. through The Economics of Ecosystems and Biodiversity (TEEB) process (Sukhdev et al. 2008), biodiversity conservation is not deeply embedded in agricultural research. Studies on landscape ecology and conservation biology consider humans less as acting and shaping entities, rather than as a disturbance against which the natural ecosystem is defined and defended. More integrated adaptive research to combine societal and ecological demands is needed to minimize trade-offs and escape the trap of combined poverty and land degradation.

In this manner, RWH practices have an overall positive effect on landscape functions. Hydrologic improvements concern the recharge of aquifers and increase in soil water. But competition may arise in communities between different water users, between different ecosystems (including natural and agroecosystems), and between ecosystems and humans (Falkenmark and Rockstrom 2004). The data basis needs to be improved, and research efforts should aim at reducing unproductive evaporation. Different scales of spatial diversity (heterogeneity) have to be considered. At the scale of the individual plants, spatial diversity is enhanced as larger plants provide more three-dimensional structures. At the field scale, spatial diversity is enhanced as more biomass is present, but species diversity could be reduced as plants might belong to only one or a few crop species, to the exclusion of the natural plant community. The same is true for the landscape scale. RWH systems aim to minimize seasonal variation in water availability such as droughts and dry spells (Rockstrom et al. 2002). Consequently, temporal diversity is modified as natural patch dynamics are replaced by an anthropogenic patch system. Fostering and improving these technologies in a sustainable way, while taking biodiversity aspects into account, offers a means of minimizing the risk of drought, crop failure, and ecological refugees. The social and economic sustainability of the RWH practices depend largely on the extent of involvement by farmers and the general communities. This might be the weakest link in the chain of sustainability issues. The more local communities are involved in planning, the higher the possibility that the RWH structures will be maintained and benefits are shared (Bangoura 2002). Possible "off-site" effects are competition for water between natural and agroecosystems as well as between different users, i.e. between upstream and downstream users. A particularly important example is the competition and conflicts that RWH practices could potentially cause between pastoralists and sedentary farmers. RWH is related to largescale changes of land use patterns, turning rangeland and natural vegetation into cropland. This might strengthen conflicts such as in Darfur (Nyong 2005).

24.11 Conclusions

This chapter presented the concepts of RWH systems and their high potential to support a sustainable water management system in different areas. Implementing RWH and RWHR systems in already built neighborhoods is likely not to be practical in most cases. Nevertheless, cities are persistently growing, and new areas are rapidly being converted from natural or agricultural areas into urban environments. Therefore, it is essential to implement RWH methods in these areas to reduce the impact of urbanization on the local, as well as regional, water cycle. The positive impacts of RWH on reducing surface runoff and the risk of floods in urban environments was presented, alongside with its impact on groundwater recharge. Two methods of RWH were discussed: (i) RWH where the harvested water is being stored in container reservoirs; and (ii) the RWHR system where the harvested water is being used to recharge the local aquifer. It is suggested here, that for most cases in modern cities, where a local aquifer is found below the city, it is more efficient to use the RWHR approach as it save the need of complicated pumping systems, save space needed for the water storing tanks, and fairly and equally distribute the rain water resource between the entire city populations. In addition to RWH of buildings and rooftops, it is important to construct harvesting systems in larger scales to collect surface water from public areas and other impervious areas such as roads and pavements. This water could be also directed into groundwater or be used for municipal purposes such as parks irrigation.

The RWH practices improve hydrological indicators such as infiltration and groundwater recharge. Soil nutrients are enriched and biomass production increases, with subsequent higher yields. Higher biomass supports a higher number of plants and animals, although native species might be replaced by crops as the landscape might change as a whole. This might strengthen conflicts between a nomadic and a sedentary population. Farmers applying in situ the RWH practices profit from higher food security and higher income. However, some aspects are only poorly covered within the scientific literature. More integrative research concepts are needed.

References

Abdulla, F.A. and Al-Shareef, A.W. (2009). Roof rain water harvesting systems for household water supply in Jordan. *Desalination* 243 (1–3): 195–207.

Angrill, S., Farreny, R., Gasol, C.M. et al. (2012). Environmental analysis of rain water harvesting infrastructures in diffuse and compact urban models of Mediterranean climate. *The International Journal of Life Cycle Assessment* 17 (1): 25–42.

Bangoura, S. (2002). Water harvesting techniques in West and Central Africa. In: *Proceedings of the International Seminar on Combating Desertification: Freshwater Resources and the Rehabilitation of Degraded Areas in the Drylands, Samantha Wauchope, Fushia Publishing Paris, N'Djamena, Chad, October 30 to November 4, 2000* (eds. A. Dupuy, C. Lee and T. Schaaf), 20–26. UNESCO.

Bruins, H.J., Evenari, M., and Nessler, U. (1986). Rain water-harvesting agriculture for food production in arid zones: the challenge of the African famine. *Applied Geography* 6 (1): 13–32.

Brunner, U. and Haefner, H. (1986). The successful floodwater farming system of the Sabeans, Yemen Arab Republic. *Applied Geography* 6 (1): 77–86.

Buhaug, H. and Urdal, H. (2013). An urbanization bomb? Population growth and social disorder in cities. *Global Environmental Change* 23 (1): 1–10.

Carmon, N., Shamir, U., and Meiron-Pistiner, S. (1997). Water-sensitive urban planning: protecting groundwater. *Journal of Environmental Planning and Management* 40 (4): 413–434.

Cofie, O., Barry, B., and Bossio, D. (2004) Human resources as a driver of Bright Spots: thecase of rain water harvesting in West Africa. In: Paper presented at the NEPAD/IGAD Regional Conference “Agricultural Successes in the Greater Horn of Africa”. Nairobi, November 22–25, 2004 (online). Kenya.

Cohen, I.S., Lopez, V., Slack, D.C., and Yanez, C.H. (1995). Assessing risk for water harvesting systems in arid environments. *Journal of Soil and Water Conservation* 50: 446–449.

Critchley, W., Gowing, J., and Mollee, E. (1992). *Water Harvesting in Sub-Saharan Africa*. Taylor and Francis Group. London and New York: Routledge.

Dayani, S., Sabzalian, M.R., Hadipour, M., and Eslamian, S. (2017). Water scarcity and sustainable urban green landscape, Chapter 30. In: *Handbook of Drought and Water Scarcity: Environmental Impacts and Analysis of Drought and Water Scarcity*, vol. 2 (eds. S. Eslamian and F. Eslamian), 557–604. USA: Francis and Taylor/CRC Press.

Dent, D.L. and Kauffman, J. (2007). The spark has jumped the gap – green water credits – proof of concept. In: *Green Water Credits Report*, vol. 7, 1–64. Wageningen, the Netherlands: ISRIC – World Soil Information Center.

Dillon, P. (2005). Future management of aquifer recharge. *Hydrogeology Journal* 13: 313–316.

Drechsel, P., Olaleye, A., Adeoti, A. et al. (2005). *Adoption Driver and Constraints of Resource Conservation Technologies in SubSaharan Africa*, 1–21. International Water Management Institute. Sri Lanka.

Eldridge, D.J., Zaady, E., and Shachak, M. (2002). Microphytic crusts, shrub patches and water harvesting in the Negev Desert: the Shikim system. *Landscape Ecology* 17: 587–597.

Ellis-Jones, J. and Tengberg, A. (2000). The impact of indigenous soil and water conservation practices on soil productivity: examples from Kenya, Tanzania and Uganda. *Land Degradation & Development* 11: 19–36.

Evenari, M., Aharoni, Y., Shanan, L., and Tadmor, N.H. (1958). The ancient desert agriculture of the Negev: III. Early beginnings. *Israel Exploration Journal* 8 (4): 231–268.

Evenari, M., Shanan, L., Tadmor, N., and Shkolnik, A. (1982). *The Negev: the challenge of a desert*. Cambridge, MA: Harvard University Press. USA.

Falkenmark, M. and Rockstrom, J. (2004). *Balancing Water for Humans and Nature. The New Approach in Ecohydrology*. London, UK: Earthscan.

Falkenmark, M., Fox, P., Persson, G., and Rockstro, M.J. (2001) Water Harvesting for Upgrading of Rainfed Agriculture. Problem Analysis and Research Needs. SIWI Report 11. Stockholm Environmental Institute, Sweden.

FAO (2003) A perspective on water control in southern Africa. Support to regional investment initiatives. FAO, Land and Water Discussion Papers 1, pp. 1–36.

FAO (2006). *Livestock's Long Shadow—Environmental Issues and Options*, 1–319. Rome, Italy: FAO.

Fatondji, D. (2002). *Organic fertilizer decomposition, nutrient release and nutrient uptake by millet crop in a traditional land rehabilitation techniques (Zai), in the Sahel*, ZEF Ecology and Development Series No. 1. Bonn: Cuvillier Verlag., Germany.

Freeman, P. (2009). Mapping the range of proven soil management practices in Africa. *Agriculture, Ecosystems and Environment* 131: 119–127.

Gao, Y., Fan, R., Li, H. et al. (2016). Thermal performance improvement of a horizontal ground-coupled heat exchanger by rain water harvest. *Energy and Buildings* 110: 302–313.

Ghimire, S.R., Johnston, J.M., Ingwersen, W.W., and Hawkins, T.R. (2014). Life cycle assessment of domestic and agricultural rain water harvesting systems. *Environmental Science and Technology* 48 (7): 4069–4077.

Ghisi, E. (2006). Potential for potable water savings by using rainwater in the residential sector of Brazil. *Building and Environment* 41 (11): 1544–1550.

Gill, S.E., Handley, J.F., Ennos, A.R., and Pauleit, S. (2007). Adapting cities for climate change: the role of the green infrastructure. *Built Environment* 33: 115–133.

Hein, L., van Koppen, K., de Groot, R.S., and van Ierland, E.C. (2006). Spatial scales, stakeholders and the valuation of ecosystem services. *Ecological Economics* 57: 209–228.

Herrmann, T. and Schmida, U. (2000). Rain water utilisation in Germany: efficiency dimensioning, hydraulic and environmental aspects. *Urban Water Journal* 1 (4): 307–316.

Herweg, K. and Ludi, E. (1999). The performance of selected soil and water conservation measures – case studies from Ethiopia and Eritrea. *Catena* 36: 99–114.

Hoffman, B.D. (2000). Changes in ant species composition and community organisation along grazing gradients in semi-arid rangelands of the Northern Territory. *Rangeland Journal* 22: 171–189.

Holman, I.P. (2006). Climate change impacts on groundwater recharge-uncertainty, shortcomings and the way forward. *Hydrogeology Journal* 14 (5): 637–647.

Jones, M.P. and Hunt, W.F. (2010). Performance of rain water harvesting systems in the southeastern United States. *Resources, Conservation and Recycling* 54 (10): 623–629.

Kabore, D. and Reij, C. (2004). *The Emergence and Spreading of an Improved Traditional Soil and Water Conservation Practice in Burkina Faso*. Washington, USA: International Food Policy Research Institute.

Kayombo, B., Hatibu, N., and Mahoo, H.F. (2004) Effect of micro-catchment in rain water harvesting on yield of maize in a semi-arid area. In: 13th International Soil Conservation Organisation Conference. Conserving Soil and Water for Society, Sharing Solutions, Brisbane, Soil and Water for Society: Sharing Solutions, Australia.

Kedar, Y. (1967). *The Ancient Agriculture in the Negev Mountains*. Jerusalem, Israel: Bialik Institute.

Kongo, V.M. and Jewitt, G.P.W. (2006). Preliminary investigation of catchment hydrology in response to agricultural water use innovations: a case study of the Potshini catchment—South Africa. *Physics and Chemistry of the Earth* 31: 976–987.

Lavee, H., Imeson, A.C., and Sarah, P. (1998). The impact of climate change on geo-morphology and desertification along a mediterranean arid transect. *Land Degradation and Development* 9: 407–422.

Le Houérou, H.N. and Lundholm, B. (1976). Complementary activities for the improvement of the economy and the environment in marginal drylands. *Ecological Bulletins* 24: 217–229.

Liaw, C. and Tsai, Y.-L. (2004). Optimum storage volume of rooftop rain water harvesting systems for domestic use. *JAWRA Journal of the American Water Resources Association* 40 (4): 901–912.

Ludwig, J.A. and Tongway, D.J. (1995). Spatial-organization of landscapes and its function in semiarid woodlands, Australia. *Landscape Ecology* 10: 51–63.

Ludwig, J.A., Wilcox, B.P., Breshears, D.D. et al. (2005). Vegetation patches and runoff-erosion as interacting ecohydrological processes in semiarid landscapes. *Ecology* 86: 288–297.

Millennium Ecosystem Assessment (2005). *Ecosystems and Human Well-Being: Synthesis*, 1–137. Washington, USA, DC: Island Press.

Mun, J.S. and Han, M.Y. (2012). Design and operational parameters of a rooftop rain water harvesting system: definition, sensitivity and verification. *Journal of Environmental Management* 93 (1): 147–153.

Mutekwa, V. and Kusangaya, S. (2006). Contribution of rain water harvesting technologies to rural livelihoods in Zimbabwe: the case of Ngundu ward in Chivi District. *Water SA* 32 (3): 437–444.

Mutunga, K., and Critchley, W. (2001) Farmers' initiatives in land husbandry. UNDP— Office to combat desertification and drought (UNSO/ESDG/BDP) and Sida's Regional Land Management Unit.

Nachshon, U., Ireson, A., Van Der Kamp, G. et al. (2014). Impacts of climate variability on wetland salinization in the North American prairies. *Hydrology and Earth System Sciences* 18 (1): 1251–1263.

Nachshon, U., Netzer, L., and Yakov, L. (2016). Land cover properties and rain water harvesting in urban environments. *Sustainable Cities and Society* 27: 398–406.

Nangula, S. and Oba, G. (2004). Effects of artificial water points on the Oshana ecosystem in Namibia. *Environmental Conservation* 31: 47–54.

Niemczynowicz, J. (1999). Urban hydrology and water management -present and future challenges. *Urban Water Journal* 1 (1): 1–14.

Nyong, A. (2005) Nigerian Scholar links drought, climate change to conflicts in Africa. Press Release of "The United States Diplomatic Mission to Nigeria."

Oweis, T., Hachum, A., and Kijne, J. (1999). *Water Harvesting and Supplemental Irrigation for Improved Water Use Efficiency in Dry Areas*, 1–51. Colombo, Sri Lanka: International Water Management Institute.

Pandey, D.N. (2001). A bountiful harvest of rain water. *Science* 293: 1763–11763.

Pandey, D.N. (2003). Cultural resources for conservation science. *Conservation Biology* 17: 633–635.

Pandey, D.N., Gupta, A.K., and Anderson, D.M. (2003). Rain water harvesting as an adaptation to climate change. *Current Science* 85: 46–59.

Pauleit, S. and Duhme, F. (2000). Assessing the environmental performance of land cover types for urban planning. *Landscape and Urban Planning* 52 (1): 1–20.

Pretorius, E., Woyessa, Y.E., Slabbert, S.W., and Tetsoane, S. (2005). Impact of rain water harvesting on catchment hydrology: a case study of the Modder River basin, South Africa. *Journal of Ecology and Environment* 80: 1–59.

Raffaella, R. (2011). Water harvesting and water collection systems in Mediterranean area. The case of Malta. *Procedia Engineering* 21: 81–88.

Rahman, A., Hajani, E., and Eslamian, S. (2017). Rainwater harvesting in arid regions of Australia. Chapter 26. In: *Handbook of Drought and Water Scarcity: Environmental Impacts and Analysis of Drought and Water Scarcity*, vol. 2 (eds. S. Eslamian and F. Eslamian), 489–500. USA: Francis and Taylor/CRC Press.

Reid, J.B. and Goss, M.J. (1981). Effect of living roots of different plant species on the aggregate stability of two arable soils. *Journal of Soil Science* 32: 521–541.

van Rensburg, L.D., Botha, J.J., and Anderson, J.J. (2004) The nature and function of the infield rain water harvesting system to improve agronomic sustainability. ICIDFAO International Workshop.

Rockstrom, J. (2004). Making the best of climatic variability: options for upgrading rainfed farming in water scarce regions. *Water Science and Technology* 49: 151–156.

Rockstrom, J., Barron, J., and Fox, P. (2002). Rain water management for increased productivity among small-holder farmers in drought prone environments. *Physics and Chemistry of the Earth* 27: 949–959.

Schewe, J., Heinke, J., Gerten, D. et al. (2014). Multimodel assessment of water scarcity under climate change. *Proceedings of the National Academy of Sciences of the United States of America* 111 (9): 3245–3250.

Schiettecatte, W., Ouessar, M., Gabriels, D. et al. (2005). Impact of water harvesting techniques on soil and water conservation: a case study on a micro catchment in southeastern Tunisia. *Journal of Arid Environments* 61: 297–313.

Schreider, S.Y., Smith, D.I., and Jakeman, A.J. (2000). Climate change impacts on urban flooding. *Climatic Change* 47: 91–115.

Seattle Department of Construction and Inseptions (SDCI), Tip 520, (2009) Rain water harvesting for beneficial use. Pp., 1–11.

Senay, G.B. and Verdin, J.P. (2004). Developing index maps of water-harvest potential in Africa. *Applied Engineering in Agriculture* 20: 789–799.

Stec, A. and Kordana, S. (2015). Analysis of profitability of rain water harvesting, gray water recycling and drain water heat recovery systems. *Resources, Conservation and Recycling* 105: 84–94.

Sturm, M., Zimmermann, M., Schütz, K. et al. (2009). Rain water harvesting as an alternative water resource in rural sites in central northern Namibia. *Physics and Chemistry of the Earth, Parts A/B/C* 34 (13–16): 776–785.

Sukhdev, P., Schauer, M., Neuville, A. et al. (2008). *The Economics of Ecosystems and Biodiversity (TEEB) an Interim Report*, 1–64. European Communities.

Tesfay Y., and Tafere K. (2004) Indigenous Rangeland Resources and Conflict Management by the North Afar Pastoral Groups in Ethiopia. A Pastoral Forum Organized by the Drylands Coordination Group (DCG) in Ethiopia, June 27–28, 2003, Mekelle, Ethiopia. DCG Report No. 31, pp. 1–94.

Thornton, P.K., Kruska, R.L., Henninger, N. et al. (2002). *National Programme to Combat Desertification.* New Delhi, India: Ministry of Environment and Forests, Government of India.

Usman, M.T. and Reason, C.J.C. (2004). Dry spell frequencies and their variability over southern Africa. *Climate Research* 26: 199–211.

Vohland, K. and Barry, B. (2009). A review of in situ rain water harvesting (RWH) practices modifying landscape functions in African drylands. *Agriculture, Ecosystems and Environment* 131: 119–127.

Wakindiki, I.I.C. and Ben-Hur, M. (2004). Indigenous soil and water conservation techniques: effects on runoff, erosion, and crop yields under semi-arid conditions. *Australian Journal of Soil Research* 40: 367–379.

Warren, A., Montalvo, J., Casado, M.A. and Pienda, F.D. (1995) Comments on conservation reclamation and grazing in the Northen Negev: contradictory or complementary concepts, Food and Agriculture Organization of the United Nations, Paper 38a, Italy.

Weng, Q. (2001). Modelling urban growth effects on surface run-off with the integration of remote sensing and GIS. *Environmental Management* 28 (28): 737–748.

Wiens, J.A. (1976). Population responses to patchy environments. *Annual Review of Ecology and Systematics* 7: 81–120.

Woyessa, Y.E., Pretorius, E., van Heerden, P.S. et al. (2005). Implications of rain water harvesting in a river basin management: evidence from the Modder Basin, South Africa. *Journal of Ecology and Environment* 83: 257–266.

Young, M.D.B., Growing, J.W., Wyseure, G.C.L., and Hatibu, N. (2002). Parched-thirst: development and validation of a process-based model of rain water harvesting. *Agricultural Water Management* 55: 121–140.

Part H

Reliable Rainwater Harvesting and Storage Systems

25

Comparing Rainwater Storage Options

Sara Nazif[1], Hamed Tavakolifar[1], Hossein Abbasizadeh[2], and Saeid Eslamian[3]

[1] School of Civil Engineering, College of Engineering, University of Tehran, Tehran, Iran
[2] Department of Water Engineering, Science and Research Branch, Islamic Azad University, Tehran, Iran
[3] Department of Water Engineering, College of Agriculture, Isfahan University of Technology, Isfahan, Iran

25.1 Introduction

Water conservation usually conveys negative feelings to people, as it is thought to use less water through rationing. However, water conservation is not just about cutting water consumption, but it is about proper management of water supply and implementation of water-saving technologies, such as rainwater harvesting facilities (Jabali et al. 2017). Rainwater harvesting is considered as collecting surface run-off from a structure or other impervious surface and storing it for future uses like irrigation and toilet flushing (Rahman et al. 2017). Also, the stored water can be utilized for other different objectives namely gardening, domestic, aquifer recharge, and others. There are many well-known names widely used to describe collection of rainwater such as rainwater collection, rainwater harvesting, and rainwater catchment. Various expressions are used in different countries; for instance, roof water collection or rooftop water collection are common names in some countries (Innovative Water Solutions LLC 2018). Traditionally, rainwater harvesting refers to gathering rain from a rooftop. The rainwater will collect in gutters in which the water flows through downspouts and then is stored in a container. Rainwater collection systems can be simple and undemanding, like collecting rain in a rain barrel, or can be intricate and complex such as harvesting rainwater into large tanks to supply the entire domestic demand.

The notion of rainwater harvesting generally brings thoughts of an old farm cistern or impressions of developing countries. However, the fact is that rainwater harvesting is turning into a reliable alternative to conventional water supply systems in urban and non-urban areas. This method could help supply household demand and provide businesses with water even in developed countries. Today, in many developed countries such as Germany, Australia, and the United States, rainwater harvesting is widely used. The Australian government funded the National Rainwater and Greywater Initiative and a plan entitled *Water for the Future* to help people use water more efficiently and provide them with rainwater harvesting systems. Also, rainwater harvesting systems have become more popular in the US after the need for having more environmentally friendly buildings increased. As a more practical example in Germany, in the period of 1989–1999, German business Mall-Beton GmbH, has installed over 100 000 rainwater storage tanks (Herrmann and Schimida 2000). Water from these tanks is used in applications such as schools and car washing businesses. Types and characteristics of rainwater storage systems are determined regarding to the selected rainwater harvesting method and the usages of stored water. In other words, in comparison with rainwater storage systems, it is necessary to consider the characteristics of rainwater harvesting methods as well as the purpose of water storage. Therefore, in this chapter, first a brief history of rainwater harvesting practices is given. Then main rainwater harvesting and storage options are reviewed and the most important parameters and considerations about them are introduced. Finally, a comparison between different types of rainwater harvesting and storage options, and advantages and disadvantages of each type is presented and discussed. Finally, a summary and conclusion are given.

25.2 History of Rainwater Harvesting

The first use of rainwater harvesting goes back to around the third century BCE to the society of farmers in Baluchistan, which is now located in Iran. Pakistan, Afghanistan, Kutch, and India employed rainwater harvesting to irrigate their agricultural crops. In old Tamil Nadu in India

Handbook of Water Harvesting and Conservation: Basic Concepts and Fundamentals, First Edition. Edited by Saeid Eslamian and Faezeh Eslamian.

rainwater was collected by Chola kings. Rainwater from the Brihadeeswarar temple, which is located in Balaganpathy Nagar, Thanjavur, India, was collected in Shivaganga reservoir. During the later Chola era, the Vīrānam tank was constructed (1011–1037 CE) in the Cuddalore district of Tamil Nadu state to store water for drinking and irrigation purposes. Vīrānam is a 16-km-long reservoir with a storage capacity of 41.5 MCM (million cubic meters) (Morey et al. 2016).

Although there is little information about ancient Venice, for centuries, the town of Venice was said to be dependent on rainwater harvesting. Ancient Venice city dwellers founded a system of rainwater collection which was based on artificial insulated collection wells. Venice's underground water table was not easily accessible. Up until a century ago, fresh water was harvested from the springs on the mainland and carried in barrels to the lagoon by boat.

The traditional rainwater collection system consisted of a wellhead and an underground tank filled with clean sand; a watertight layer of clay all around that provided protection against saltwater penetration. The rainwater infiltrated into the ground by means of collectors which are located around the well in slightly lower levels than the rainfall collection surface; water filtered through the sand flows down to the waterproof clay bottom of the tank. The tank, which was waterproofed by a layer of clay spread over its entire length, filled up from bottom level with the harvested rainwater, which had been purified as it drained through the sand. This method supplemented the traditional rainwater collection system used in the Campi and Campielli. The water was carried with casks. Later, as Venice claimed sovereignty of territories over the mainland, residents of Venice started to import water from local rivers by boat, but the wells were maintained in use and were especially important during the wars when adversaries blocked access to the mainland water (Venice Backstage 2018).

Therefore, it can be concluded that using rainwater harvesting and storage systems for supplying water for different purposes is not a new approach, and it has been applied and available whenever it is required. Furthermore, a variety of storage options have been used for this purpose, mainly depending on regional characteristics.

25.3 Benefits of Rainwater Storage

Rainwater harvesting is a simple and primary way of collecting and storing water from natural precipitation. This method can be used in both wet and dry regions even though its type and purposes would be different. In regions with enough precipitation, rainwater harvesting is used as an alternative to available water resources. But the situation in dry areas is different. In dry areas, besides suffering from limited rainfall, the majority of rainwater – about 90% – is lost directly or indirectly as "natural leakage," and is unavailable for agriculture or domestic use. In these areas, water harvesting is a practical and economic way to recover a large part of this lost water. During water crises like periodic droughts, the water which is stored by this method could be the most easily adaptable alternative method of alleviating water scarcity. This system is also an appropriate technique for regular water uses in normal circumstances. It is an eco-friendly technique which contains effective rainwater collecting and efficient storage of water that provides remarkable help for local people (The Constructor Civil Engineering Home 2017). The main advantages of rainwater harvesting are as follows:

(i) It can reduce the load on the public water supply system, which is the main source of water in cities. The collected water can be easily used for supplying part of non-potable water demand and therefore, the withdrawal from public water supply resources is decreased.

(ii) It can be utilized at the time of an emergency. When water supply runs out or is contaminated, the stored rainwater could be considered in an emergency contingency plan to be put into place as another water source (Pennington 2014).

(iii) It is an economical approach due to its low installation and maintenance cost, and it can mitigate expenses that a household has to pay for water bills; for example, rainwater barrels usually need no maintenance and any repairs and maintenance can be done by the homeowner without any significant cost (EPA 2013).

(iv) It is commonly used to raise soil moisture levels to develop soil vegetation, especially in arid and semi-arid climates where the rainfall frequency is limited and rainfall is usually heavy. These rainfalls produce huge amounts of surface runoff and soil cannot absorb the rain during the short time. To use these rainfalls efficiently, is important to provide soil moisture retention or conservation by encouraging rainfall to infiltrate and store (Anschutz et al. 2003).

(v) It can be used to recharge groundwater during and after rainfalls by using artificial recharge instruments like wells, recharge basins, and direct injection wells, where the area has high groundwater infiltration rate and good geographical features. This water supplies adequate water with good quality for urban and non-urban areas (ICIMOD 2009).

(vi) It cuts power consumption for water supply and distribution. Rainwater harvesting systems can be used as the decentralized system in the urban water supply system which could be the most energy-efficient method (Kenway et al. 2008).

25.4 Main Rainwater Storage Options

The main factors of a rainwater harvesting project that have to be considered are where and how to store the harvested water. There are some key points that should be kept in mind before making a decision about a storage system, which are as follows:

(i) Volume of water which is intended to be stored and the main purpose of its usage (How much water is wanted to be stored and for what purpose? Domestic, irrigation, livestock, industries, or aquifer recharge?)
(ii) Location of storage reservoir or tank related to land (Where will the tanks stand? Above, on, or under the ground?)
(iii) Material of the storage system (What are the tanks built from?)
(iv) Methods of operation and maintenance of storage system (How will the tanks connect to capture system? How will the whole system work? Manual or automated?)
(v) Available budget (Is the capital and operation cost of the system affordable?)
(vi) Water quality consideration according to the harvested water usage (For what purpose will the stored water be used and what are the qualities necessary for that purpose?)

Harvested rainwater can be used for several purposes, from gardening and irrigation to domestic supply and groundwater recharge. Refreshing the pot plants just requires a small tank. But irrigating a lawn or washing needs more storage capacity. Annual precipitation pattern (amount, duration, and seasonality) is another consideration in storage system volume and type. In a wet climate area where there are many rainy days, tanks can refill regularly, so there is no need for huge storage capacity. But in arid or semi-arid climates with few rainy days, large tanks may be better for rainwater storage. Although having an estimation about storage capacity beforehand is important, the final storage capacity of system can be scalable. Adding more tanks to the system in the future is an option to make a rainwater harvesting system affordable.

Bearing capacity of the underlying soil, simplicity of piping and proximity of the captured water to the consumption spots, public appearance of the system, and available space are the main parameters which determine the location and appropriate design of the storage tanks. Storage tanks can be made from different materials such as high density plastic, fiberglass, wood, metal, concrete, and more. Plastic tanks are popular and affordable. Having proper UV protection, availability of material, being opaque to reduce bacterial growth, and workability (easy to fix or replace fitments such as taps and pipes) are the most important considerations for choosing storage material.

Essentially, the different rainwater harvesting methods affect the available storage options. In general, there are three ways of harvesting rainwater:

(i) Surface runoff harvesting
(ii) Rooftop rainwater harvesting
(iii) Rainwater harvesting in situ

In the following sections, various methods of rainwater harvesting are described briefly (The Constructor Civil Engineering Home 2017).

25.4.1 Surface Runoff Harvesting

In urban areas, rainwater turns into surface runoff and flows over the catchment's surface. This runoff can be collected and used for recharging aquifers, as well as domestic and agricultural uses, by choosing appropriate methods (Morey et al. 2016). Surface runoff harvesting can be divided into two main sectors: (i) surface runoff harvesting using surface and underground structures; and (ii) surface runoff harvesting using paved and unpaved roads.

25.4.1.1 Surface Runoff Harvesting Using Surface and Underground Structures

Surface runoff can be appropriately gathered in artificial reservoirs, or intercepting and impounding the runoff using small dams. There are numerous methods and extensive literature on impoundments and dam design, and this technology has been widely applied throughout the world. Local impoundments are artificial storage lakes excavated into the ground to reserve surface runoff for use in the future. Dams are designed and constructed to escalate the storage capacity of rivers or streams, intercepting runoff and maintaining water in their reservoirs for later use. The main difference is that dams are built where stream flow is constantly exists, while local impoundments are essentially for collecting and storing seasonal streams and local rainfall runoff. In contrast to dam reservoir these impoundments can dry out during dry periods but dams' reservoirs usually do not dry out (Salzedo et al. 1997).

Artificial impoundments, such as the one shown in Figure 25.1, are often excavated lower than the ground surface in a naturally impervious soil, or the soil is treated to become impervious. The impoundment could be built in various shapes such as rectangular, square, circular,

Figure 25.1 An artificial impoundment located in the central part of Iran.

or quasi-circular, depending on the required depth and capacity of the pool. The side slopes of the pool could range from 2 : 3 to 1 : 2 (vertical: horizontal) based on the type and soil repose angle. The excavation process of impoundments could be done either mechanically or manually. The volume of conventional impoundments ranges from 500 m^3 to 1 million m^3 according to the accessibility of runoff and the demand for water. If the water is needed to be used for domestic purposes, a filtration plant or chlorination unit should be provided at the system outlet (Salzedo et al. 1997).

25.4.1.2 Surface Runoff Harvesting Using Paved and Unpaved Roads

Paved and unpaved roads tend to shed water to their outside edges because the roads have a camber that helps water drain off them. The runoff can be collected in road edges by drainage channels or underground networks. There are numerous methods which have been used for this purpose. The common components of these systems are a collection area, drainage system, storage system, and distribution system. Most channels in these systems are trapezoidal. They are generally either parallel or perpendicular to the roads. The roadside ditches collect surface water temporarily and the kinetic energy is gradually dissipated using stones or other structures designed to reduce the flow velocity at which the runoff flows into gutters from the road surface, and the gutters transfer the runoff to storage systems. Storage systems are usually built perpendicular to the drainage channels, and take the form of other conduits or underground galleries. A stone masonry wall is constructed at the inlet of the drainage network. This wall is solid and is perforated to let the water enter the network while keeping large particulates, animals, or debris out of the network (Pedrani et al. 1997). The schematic of drainage and storage system for this kind of surface runoff harvesting is shown in Figure 25.2.

25.4.2 Rooftop Rainwater Harvesting

Rooftop rainwater harvesting is a system of collecting rainwater where it falls. In this method, the effective impervious area (EIA) is the building's roof, which directs the runoff to the drainage channels. This roof is considered as the catchment and the rainwater is harvested from house or building's roof. The collected water can be stored in a tank or can be directed to an artificial groundwater recharge system. This method is economically efficient and effective and it can help to increase the local region's groundwater level if it is applied appropriately (The Constructor Civil Engineering Home 2017).

25.4.2.1 Components of Rooftop Rainwater Harvesting

The typical components of a rooftop rainwater harvesting system are shown in Figure 25.3. The system primarily consists of the following subcomponents.

25.4.2.1.1 Catchments

The area that receives precipitation and drains it to the storage system, EIA, is considered as the catchment of the rainwater harvesting system. The collection surface in most cases is the house or building's roof. The catchment may also be house's terrace, courtyard, or paved ground. The catchment's area and the material used in constructing the rooftop influence the rainwater harvesting efficiency and the stored water quality (Torres et al. 1997). The terrace

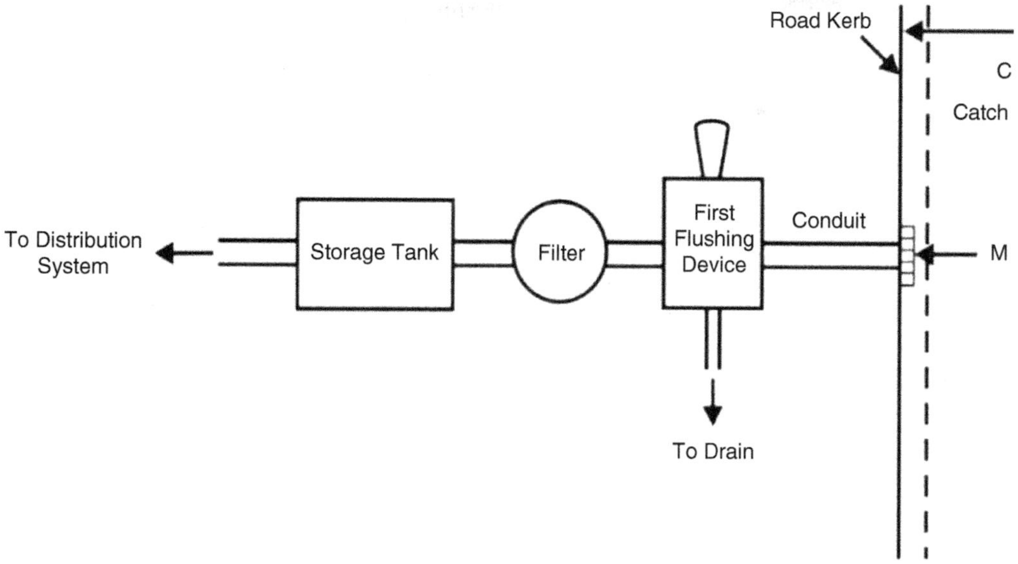

Figure 25.2 Surface runoff harvesting system for an urban paved road area.

Figure 25.3 Components of a rainwater harvesting system.

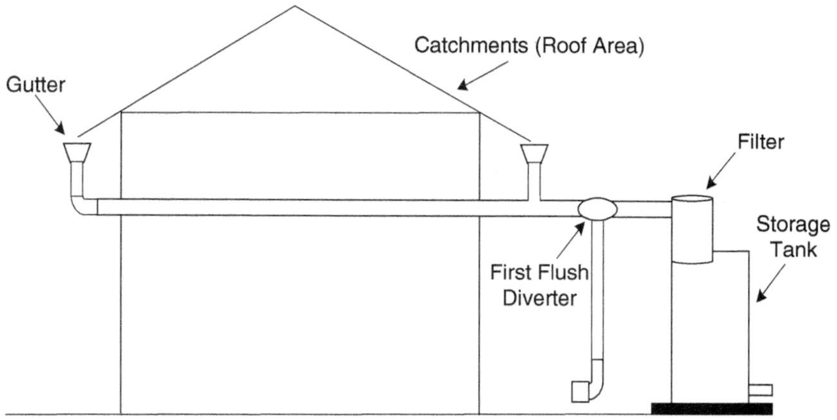

could be either a flat or sloping roof. Therefore, the catchment is the area which is actually involved in collecting rainwater to the harvesting system (The Constructor Civil Engineering Home 2017).

25.4.2.1.2 Transportation

Collected rainwater from the rooftop should be channeled into the bottom storage tank through water pipes or a harvesting system. Chemically inert materials like wood, plastic, aluminum, or fiberglass should be used in constructing drainpipes and roof surfaces to lessen harmful effects on water quality (Torres et al. 1997). Water pipes used in this system should be resistant to UV, such as ISI (Indian Standards Institute) HDPE (high-density polyethylene) or PVC pipes, according to the required capacity. The collected water from steep roofs is channeled through gutters and down-take pipe. At terraces, the inlet of the each drain pipe should have wire mesh to screen out floating material (The Constructor Civil Engineering Home 2017).

25.4.2.1.3 First Flush

First flush is a device used to flush off the water from the first rainfall. The first rainfall should be flushed off to keep storable or rechargeable water away from the probable contaminants which exist in the atmosphere and the catchment surface. It also helps to clean sediment and silt and other materials deposited on rooftop during dry seasons. Therefore, a system for separating first rainfall should be provided at the outlet of each drainpipe (The Constructor Civil Engineering Home 2017).

25.4.2.1.4 Filter

There is always some doubt about the existence of contaminants in harvested rooftop rainwater. This possibility can come true if no appropriate filtering mechanism is applied. Make sure that underground sewer drains are not punctured and there is no leakage occurring in close proximity.

Filters are used to remove turbidity, color, and microorganisms from water. Specially after first flushing of rainfall, rainwater should channel through filters. The gravel, sand, and "netlon" mesh filters are designed and usually located on top of the storage tank. These filters play an essential role in keeping the rainwater clean in the storage tank. They eliminate silt, dust, leaves, and other organic matter and prevent them from entering the storage tanks (The Constructor Civil Engineering Home 2017).

The filter media have to be washed after every rainfall event. If the filters are clogged it may prevent rainwater from entering the storage tank easily and in this case the filters will overflow. The sand or gravel media should be taken out and cleaned before they are placed in the filter (The Constructor Civil Engineering Home 2017). A typical scheme of a filter is shown in Figure 25.4.

There are different types of filters in practice, but the basic function is to purify water. Some of these filters are described below (The Constructor Civil Engineering Home 2017):

(i) Sand gravel filter: These are commonly used filters, constructed by brick masonry and filleted by pebbles, gravel, and sand. Each layer should be separated by wire mesh. This is the most common type of filter used in rainwater harvesting systems. Biswas and Mandal (2014) used sand filters for rainwater harvesting systems in Khulna, Bangladesh. The results show that the stored water meets the regional and WHO (World Health Organization) standards for drinking water.

(ii) Charcoal filter: Charcoal filter can be made in-situ or in a drum. Pebbles, gravel, sand, and charcoal should fill the drum or chamber. Each layer should be separated by wire mesh. A thin layer of charcoal is used to absorb odor whenever it is needed.

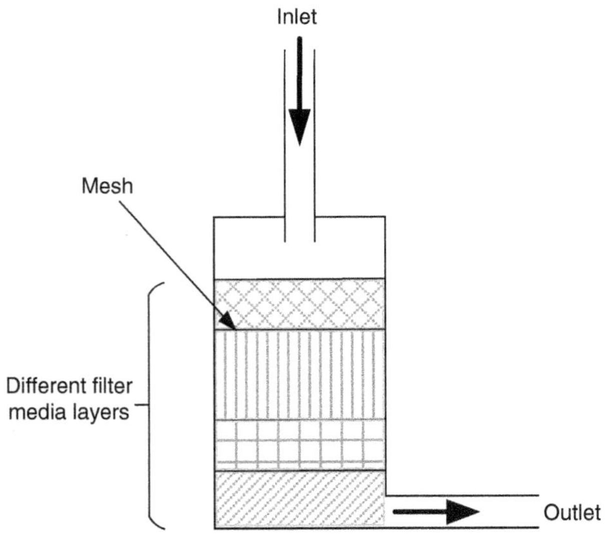

Figure 25.4 A typical scheme of a filter used in rainwater harvesting system.

(iii) PVC Pipe filter: This filter can be made by PVC pipe of 1–1.20 m length; the diameter of the pipe depends on the area of roof. A 15 cm diameter pipe is enough for a 140 m^2 roof and 20 cm diameter pipe is used for a roof more than 140 m^2. The pipe is divided into three compartments by wire mesh. Each component should be alternately filled with gravel and sand. A layer of charcoal could also be inserted between two layers. Both ends of the filter should have required size to connect inlet and outlet. This filter could be placed horizontally or vertically in the system.

(iv) Sponge Filter: It is a simple filter made from PVC drum having a layer of sponge in the middle of drum. It is the easiest and the cheapest form of filter, suitable for residential units. This filter is the easiest and cheapest for residential units, which is used in rural areas.

It should be kept in mind that a combination of above filters as well as advanced types of filters such as membrane filters can be used in rainwater harvesting systems. As an example, Lumbera et al. (2013) proposed a filtering system that consists of three varying layers – a layer of sand, activated carbon, and sponge – to improve the pH, turbidity, and remove impurities from stored rainwater.

25.4.2.1.5 Storage Tank

In the rooftop rainwater harvesting method, rainwater harvested from the rooftop of a house or building is channeled to a storage tank or cistern, which should also be built of an inert material. Storage tanks can be built as part of the house or building, or can be constructed separately as a unit situated away from the building (Torres et al. 1997). The storage tank has to be designed based on the water demand, considering the rainfall pattern and distribution over the year and catchment availability. It is recommended that each tank should have an excess water overflow system. Excess water has to be directed to the recharge system. Water from the storage tank can also be utilized for secondary aims such as washing and gardening. This is the most economical way of rainwater harvesting.

However, there are different regulations and codes for using rainwater harvesting systems in different countries that could limit the storage system capacity and type. In different countries, depending on water right and water law, governments may not allow the use of cisterns over a specific size. This roots in differences in stormwater runoff right in different countries and who owns the rainfall runoff. For example, in San Francisco, US, the barrel capacity is limited to 5000 gal and for bigger storage system a permit from the Department of Building Inspection (DBI) is needed (San Francisco Water, Power, Sewer 2016). In Australia, there are many regulations and codes for rainwater harvesting tanks. These regulations

Figure 25.5 A storage tank on a platform in Iran.

consider many aspects from identifying potential hazards and health risks to construction materials, size, and installation (Commonwealth of Australia 2010).

The main advantage of harvesting and utilizing the rainwater during the rainy season is not only to store water from typical sources, but also to save energy, which a household must expend for transportation and distribution of water at the doorstep. This also protects groundwater, if it is being used to meet the demand during rainfall (The Constructor Civil Engineering Home 2017). A conventional figure of storage tank is depicted in Figure 25.5.

Storage tanks, generally, are the most expensive part of the rainwater harvesting system and are available in a wide range of sizes and types. Different types of storage tanks and cisterns have been used over the centuries around the world in different geographical regions: earthenware cisterns in India, large pottery containers in Africa, aboveground vinyl-lined swimming pools in Hawaii, concrete or brick cisterns in the central United States, and, in Texas and Colorado, galvanized steel tanks and site-built stone-and-mortar cisterns.

The main factors to choose the type of storage tanks are the budget and the tank's location. When it comes to choosing the size of storage tank or cistern, there are many variables that play essential roles, including rainwater supply or local precipitation, water demand, predicted length of dry spells without rain, catchment surface area, esthetics, personal preference, and the available budget.

Tanks can be located either above or below ground. To determine the location of the tank, some factors should be considered such as soil condition, variety of outside temperature, and implementation cost. Some tanks are appropriate for aboveground placement i.e. vinyl-lined swimming pools; other types can be applied both above and below ground, i.e. polyethylene. Some types of tanks are built to be buried, such as polyethylene tanks.

As a result, knowing accurate information about the available alternatives is crucial to choosing the best type of tank efficiently and economically, since it should prove to be something to be used for a long time. Table 25.1

Table 25.1 The summary of storage tank materials' characteristics.

Material	Expected life	General availability	Transportability	Maintenance requirement	Build your own	Cost (Pushard 2014)	Environmental impacts based on LCA (life cycle assessment) (For a 1000 l storage tank)
Fiberglass	4	4	4	1	0	High	Low
Above-ground polyethylene	4	5	5	1	0	Low	Low (Shah et al. 2016)
Below ground polyethylene	5	4	5	1	0	High, due to the cost of excavation and heavily reinforced tank	Low (Shah et al. 2016)
Cement	4	3	0	4	3	High	High (Shah et al. 2016)
Ferro-cement	4	3	0	4	4	Low	High (Shah et al. 2016)
Plastered tires	5	2	0	4	4	Very low	Moderate
Stone	5	2	0	3	3	High due to labor cost	Low
Wood	3	2	4	3	2	High	Low
Above-ground metal	5	4	4	4	0	Relatively high	Moderate (Shah et al. 2016)
Below-ground metal	5	4	3	4	0	High	Moderate (Shah et al. 2016)

Source: Adapted from Pushard (2014).

provides a comparison between different types of storage tanks based on seven factors of expected life, availability, transportability, maintenance requirement, possibility of building them by house owners, cost, and environmental impacts. In this table number 5 shows high desirability while number 0 shows complete undesirability. Based on the table, only wood tanks could provide a minimum of desire in all factors. Among fixed storage tanks, those made of plastered tires have the most desirability in all factors.

Rainwater tank capacity is determined based on the average rainfall in the study area and the catchment area (EIA); for example, in Wellington, New Zealand, during a summer with average rainfall, a 750 l rainwater tank can be refilled by rain collected through house roof guttering. This volume of water could supply the water demand of a family of three (with 20 l per day water demand) for around 80 days. However, during a dry summer, the same tank could only last 13 days for a family of three before it runs dry. To maximize the system efficiency, it is necessary to connect the roof guttering of the house to the tank to get the maximum amount of rainfall (Wellington Water 2017).

The maximum amount of rainwater which could be harvested from rooftops is determined as follows:

$$V = A \times R \times C \tag{25.1}$$

Where V is the volume of harvestable water, A is catchment area (rooftop area), and C is the runoff coefficient. Table 25.2 shows the basis used for determining the rainwater tanks' size based on average rainfall and number of household members in Wellington, New Zealand.

25.4.2.2 The Usage of Harvested Water

Another issue that should be considered in this type of rainwater harvesting is the usage of stored water which is different among countries depending on their regulations. In Brazil, water collected through rainwater harvesting projects is used for agriculture, water for livestock, and improving sanitary standards (ASA 2015). In Germany, to cut down on potable water consumption, rainwater harvesting system was introduced to public and private buildings and industrial premises (Cleanawater 2015). They replace potable water with rainwater for appropriate uses like toilets, schools' facilities, and car washing businesses (Cleanawater 2015).

25.4.3 Rainwater Harvesting In Situ

In arid and semi-arid regions, where there is mostly lack of precipitation during the dry season, it is essential to store water as much as possible from rainwater during the wet season for use of agricultural and domestic water supply later in the future. Water storage in situ is one of the common solutions to harvest rainwater. The principle of in situ rainwater harvesting is to collect and conduct runoff from an area that is not cultivated to the area where crops are grown. Topographically low areas are ideal sites for in situ harvesting of rainfall. This technique has been tested in the arid and semi-arid regions such as northeastern Brazil, Argentina, and Paraguay, mainly for irrigation purposes.

In situ rainfall harvesting systems consist of three components: a collection area, a transmission system, and a storage area. In this application, collection and storage area is based on the region's landscape. Topographic depressions represent ideal collection and storage areas. In many conditions, the presence of clay soil, which reduces infiltration, makes the area impermeable. Making furrows and strips against the runoff flow direction is another technique that is used widely in the world for in situ rainwater harvesting. After the site is chosen and prepared, little maintenance is required for operation. Maintenance primarily consists of keeping the collection area free of debris and unwanted vegetation (Anjos et al. 1997).

25.4.3.1 Use of Topographic Depressions as Rainfall Harvesting Areas

In areas where there is impermeable soil with enough depth (for example, clay soils at least 3 m deep), rainwater can be collected and compounded for storage and future usage. In Paraguay, this kind of storage is known as Tajamare. The distribution conduits and pipes are used

Table 25.2 Estimation of the size of rainwater storage tank considering average rainfall in Wellington, New Zealand.

Tank size (l)	The number of days that a rainwater storage tank could supply water considering a water demand of 20 l/person/day in a summer with average rainfall		
	Household occupants		
	2	3	4
500	80 d	40 d	30 d
750	unlikely to run out of water	80 d	55 d
1000	unlikely to run out of water	Unlikely to run out of water	60 d
2000	Unlikely to run out of water		

Source: Adapted from Wellington Water (2017).

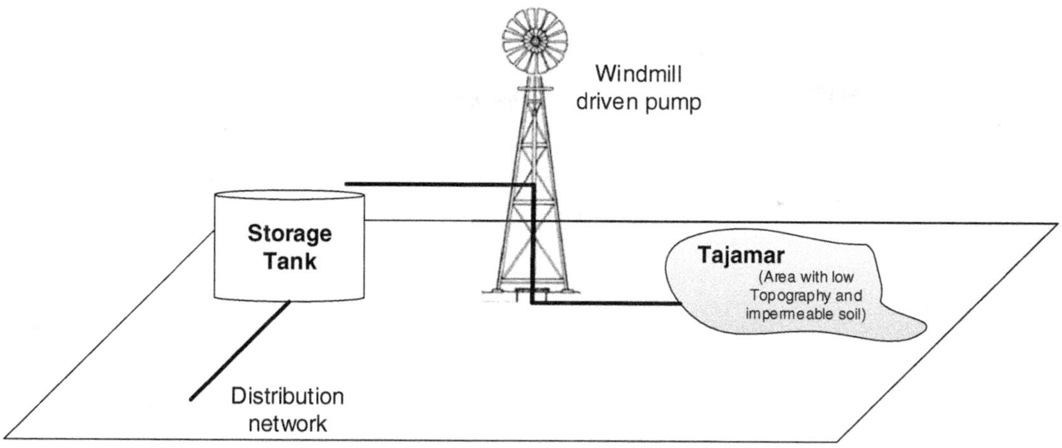

Figure 25.6 Typical scheme of a low topography rainfall harvesting system (Tajamare).

to convey water from the storage area to the points of use. The collection and storage areas must be protected against contamination caused by animals. Clay tanks are usually used for storage of in situ harvested rainwater. In some cases, it is necessary to pump water from the in situ rainfall collection area to the storage tanks, usually by a windmill, as shown in Figure 25.6.

25.4.3.2 Use of Furrows as Rainwater Storage Areas

Furrows and strips may be used for in situ storing of harvested rainwater. They are built prior to or after planting to store water for future use by the plants. There are different patterns of topographic depressions to store rainfall, but generally in this method there are flattened trenches between the rows of crops to store water. In order to retain water for longer periods and avoid excessive surface runoff and erosion, mud dams or barriers may be made every 2–3 m along the row. Uncultivated areas may be left between rows, spaced at 1 m apart, to assist in capturing rainwater falling on the land surface between furrows.

25.5 Comparing Rainwater Storage Options

In this section the pros and cons of each method of rainwater storage are described and the popularity and practicality of different methods are discussed. Furthermore, the measures that should be taken into account to develop these methods are given.

(i) **Surface runoff harvesting using surface and underground structures**: These methods are applicable in regions where the time and spatial distribution of rainfall are highly variable and storage is required to meet specific demands, such as water supply for irrigation and hydroelectric power generation. Their suitability depends on favorable topography, geology, and economic conditions (Salzedo et al. 1997).

(ii) **Surface runoff harvesting using paved and unpaved roads**: The technique is suitable for use in arid and semi-arid rural areas where runoff from paved and unpaved roads can be collected and stored.

(iii) **Rooftop rainwater harvesting**: This technology is suitable for use in all areas, especially urban areas with high population density, as a means of augmenting the amount of water available. It is most useful in arid and semi-arid areas where other sources of water are scarce.

(iv) **Rainwater harvesting in situ**: This technology is applicable to low-lying areas in arid or semi-arid climates (Anjos et al. 1997).

In Tables 25.3 and 25.4, different rainwater storage options are compared according to their advantages, disadvantages, and popularity regarding the findings of the previous studies in related fields.

Based on the above tables and information, Table 25.5 depicts the main features of each rainwater storage option that can be considered in selection of the most appropriate method for a region. There is almost no limitation in usage of any rainwater storage option except the limitations forced by regional regulations. Surface and underground structures are suitable for large amounts of rainwater harvesting, but the capital cost of them is generally more than the other options. In addition, undesirable environmental impacts of structures, which developed in this approach, are more serious. Harvesting by road surfaces have very low cost because the capital cost of the road construction

Table 25.3 Comparison of advantages and disadvantages of different rainwater storage options.

Method		Advantages	Disadvantages
Surface runoff harvesting	Surface and underground structures	• Providing water for different water usages (agricultural, domestic, industrial, recreational, hydroelectric power generation, non-consumptive uses ...) especially in arid and semi-arid regions. • Enhancement of the flora and fauna. • Decreasing water pollution by dilution of contaminants. • Increasing minimum flows and water levels. • Suitable for multiple-use water projects.	• Requiring lands with the proper topography (generally valuable agricultural land). • Requiring impermeable soil for seepage control • Water losses due to evaporation and infiltration especially in arid and semi-arid areas. • High construction costs. • Failure (brake) risk. • Impoundments can flood adjacent lands during wet periods. • Negative environmental impacts.
	Paved and unpaved roads	• Enhancement of the flora and fauna. • Employment for cultivation in arid and semi-arid regions. • Low operating cost. • Subsumed capital cost in the cost of road constructing. • Simple operation and maintenance. • Erosion reduction and sediment control.	• Requiring supplemental irrigation during dry periods. • Requiring protection of the planting areas from animals • Appropriate soil conditions requirements. • Contamination (litter, debris, and chemical pollutants from vehicles).
Rooftop rainwater harvesting		• Providing water at the need point. • Providing water source for emergencies. • Simple construction for locals. • Flexible technology. • Improving the engineering of building foundations as in the case of masonry cisterns. • Better quality in comparison to available groundwater or surface. • Low operation costs. • Construction, operation, and' maintenance are not labor-intensive.	• Depending on frequency and amount of rainfall (Unreliable in dry weather or prolonged drought) • Low storage capacities • Deterioration of bed-bearing capacity due to leakage from storages. • Risk of drowning in case of inappropriate access protection. • Risk of water contamination resulting from animal wastes and vegetable matter. • Health risks due to lack of water purification for domestic use. • Possibility of growth of insect pests. • Increasing construction costs (up to 30–40% additional cost) and needing governmental subside. • Possible reduction in public utilities revenues.
Rainwater harvesting in situ		• Minimal labor. • Flexibility of construction. • Better utilization of rainwater for irrigation purposes. • Compatible with agricultural best management practices • Flexibility in soil utilization. • Can be used as a method of groundwater recharging.	• Cannot be implemented in the slope of greater than 5%. • Difficult to implement in rocky soils. • Stones and trees should be cleared before implementation. • The additional costs for farmers. • Impermeable soils and low topographic relief are required. • Evaporation is a limitation.

is not including runoff harvesting system. The main limitation of this option is availability of them just beside roadways, which may result in considerable costs for transfer of collected water to the usage point. The best and most affordable method for small-scale water harvesting especially for household usage is rooftop rainwater harvesting. This method is applicable in both rural and urban area. In situ water harvesting is actually a wise usage of a land plowing scheme. It can increase the productivity of dry farming.

Table 25.4 Comparison of popularity of different rainwater storage options and suggestions for improving.

Method		Popularity	Suggestions for improving the technology
Surface runoff harvesting	Surface and underground structures	• Widely accepted as a water supply augmentation method for developed and developing countries. • Have used in small-scale (e.g. farm dams) and large-scale projects (Salzedo et al. 1997).	• The design of storage systems (especially small-scale ones) must be improved to be more efficient in retaining water, preventing failures, and reducing evaporative losses. • Improvements in operation can be very beneficial. Further reduction of evaporation and infiltration is needed (Salzedo et al. 1997).
	Paved and unpaved roads	• This technology is easily accepted by public works departments and communities in arid and semi-arid areas because of simplicity and low-cost (Pedrani et al. 1997).	• Combining with in situ techniques is needed to improve the system efficiency
Rooftop rainwater harvesting		• Projects which have been predominantly run by local people have had a much higher rate of success than those operated by people foreign to an area. • Attitudes toward the use of rainwater for domestic consumption differ (household purposes, drinking and cooking, washing and gardening, or flushing toilets) and are related to the level of education of the users as well as to their traditional preferences (Torres et al. 1997).	• Development of first-flush bypass devices. • Greater involvement of the public health department in the monitoring of water quality. • Monitoring the quality of construction. • Provision of assistance from governmental sources for appropriate-sized cisterns. • Promotion of rainwater harvesting with emphasis on the savings on water bills. • Provision of assistance to the public in construction and standardized plumbing and monitoring code. • Development of new materials to lower the cost of storage. • Preparation of guidance materials.
Rainwater harvesting in situ		• Communities in many arid and semi-arid regions can easily increase their irrigation levels and increase their production using this technique (Anjos et al. 1997).	• The equipment of construction of the furrows must be improved. • Relatively inexpensive plows and tractors can reduce the cost of implementation. • New methods of soil conservation should be explored (Anjos et al. 1997).

Table 25.5 Comparing different methods of rainwater storage.

Method		Usages	Expense	Public/private structure	Required space
Surface runoff harvesting	Surface and underground structures	Domestic, industry, irrigation, groundwater recharge, and hydroelectric power generation	High	Public	Vast land with appropriate topography
	Paved and unpaved roads	Mostly irrigation and groundwater recharge	Low	Public	Available beside all roadways and pavements
Rooftop rainwater harvesting		Domestic, irrigation, and groundwater recharge	Low (depends on the size of cistern)	Private	Small space
Rainwater harvesting in situ		Mostly irrigation and groundwater recharge	Low	Public	Vast land with the slope less than 5% and low topography

25.6 Conclusion

Rainwater harvesting is a simple, eco-friendly, and effective way of collecting water from natural precipitation, which has been applied for many years in different countries. It is also a reliable and adaptable alternative to other water supply systems in the critical periods. Today there are many developed countries, like Germany and Australia, and developing countries, like Bangladesh and Thailand, that are using this method widely. Its usage is not limited to a specific climate and there are successful experiences of rainwater harvesting implementations in wet and arid regions even though their specifications are different.

This method stores rainwater and this amount of water can be used for different purposes like domestic use, irrigation, gardening, aquifer recharge, hydroelectric power generation, and industrial uses. So the type and features of the system are chosen according to the system's application and it could be simple like a rain cistern or a complex system like a rainwater harvesting system that supplies all demands of a domestic, industrial, or agricultural unit. There are many advantages in using this method due to its low cost for installation and application. Therefore, it is worth using this system in all countries to alleviate water scarcity, especially for those living in arid and semi-arid climates. However, some countries have considered some limitations on the size of private harvesting systems and its end use which should be taken into account when developing these systems. There are different components in these systems which commonly include collection, transition, storage, and treatment systems. In selection of system components, types, and characteristics, different aspects of technical, economic, social, and environmental issues should be considered.

Rainwater harvesting could be divided into three main methods. The first method is "surface runoff harvesting." In this method, rainwater is harvested using two different ways: (i) surface runoff harvesting using surface and underground structures, and (ii) surface runoff harvesting using paved and unpaved roads. These methods collect runoff from the surface of catchments which is produced by rainfall and flows into impoundments through channels. The second main method of rainwater collecting is "rooftop rainwater harvesting" and the last one is "rainwater harvesting in situ." These two methods collect rainfall directly in cisterns, impoundments, and furrows in situ. Each method has its own applications, benefits, and drawbacks. To reach the maximum efficiency of each method, first, the main objective of rainwater harvesting and stored water usage must be determined. Furthermore, economic and environmental aspects should be considered.

In conclusion, it can be said that rainwater harvesting can be used as a sustainable and reliable choice for water supply for different purposes in both developing and developed countries. The low cost of collection and treatment has increased this method's popularity and considering the vast variety of its methods and types, it can be well used in different regions with different characteristics, needs, and limitations.

References

Anjos, J., Porto, E., Brito, L. et al. (1997). *Rainwater harvesting in situ*. In: *Source Book of Alternative Technologies for Freshwater Augmentation in Latin America and the Caribbean Unit of Sustainable Development and Environment General Secretariat* (eds. J.A. Thornton and B. Robinson), 70. Washington, D.C, USA: Organization of American States.

Anschutz, J., Kome, A., Nederlof, M. et al. (2003). *Water Harvesting and Soil Moisture Retention*, vol. 9. Wageningen, the Netherlands: Stoas Digigrafi.

Ações- P1MC Programa Um Milhão de Cisternas, Articulação Semiarido Brasileiro, 2015. Available on: www.asabrasil.org.br/acoes/p1mc#categoria_img [18th October 2015].

Biswas, B.K. and Mandal, B.H. (2014). Construction and evaluation of rainwater harvesting system for domestic use in a remote and rural area of Khulna, Bangladesh. *International Scholarly Research Notices* 2014 (6): 1–6.

Cleanawater. (2015) The Need to Tackle Rainwater Scarcity around the World. Available on: https://cleanawater.com.au/information-centre/rainwater-harvesting-solutions-which-countries-lead-the-way. [4 December 2015].

Guidance on Use of Rainwater Tanks. (2010) Commonwealth of Australia. Available on: www.health.gov.au/internet/main/publishing.nsf/Content/0D71DB86E9DA7CF1CA257BF0001CBF2F/$File/enhealth-raintank.pdf.

Herrmann, T. and Schimida, U. (2000). Rainwater Utilisation in Germany: Efficiency, Dimensioning, Hydraulic and Environmental Aspects. *Urban Water* 1: 307–316.

Innovative Water Solutions LLC. (2018) What Is Rainwater Harvesting?Available on: http://www.watercache.com/education/rainwater-harvesting-101.

International Center for Integrated Mountain Development (ICIMOD). (2009) Water Storage; A Strategy for Climate

Change Adaptation in the Himalayas. Sustainable Mountain Development. No. 56, Winter 2009.

Jabali, M.M., Okhravi, S., Eslamian, S., and Gohari, S. (2017). Water conservation techniques, , Chapter 27. In: *Handbook of Drought and Water Scarcity: Environmental Impacts and Analysis of Drought and Water Scarcity*, vol. 2 (eds. S. Eslamian and F. Eslamian), 501–520. USA: Taylor and Francis/CRC Press.

Kenway, S.J., Priestley, A., Cook, S. et al. (2008). *Energy Use in Provision and Consumption of Urban Water in Australia and New Zealand*, 54. Water Services Association of Australia.

Lumbera, M.P., Macuha, R.N., Castro, P.P.M. (2013) *Integrating Storage Sizing and Water Treatment for Rainwater Harvesting in the Philippines*. SB13 Manila-Sustainable Buildings, Infrastructures and Communities in Emerging Economies Conference, Rotterdam, Philippines.

Morey, A., Bhushan, D., Haste, V., and Bhupesh, W. (2016). Rainwater harvesting system. *International Research Journal of Engineering and Technology (IRJET)* 3 (4): 2158–2162.

Pedrani, A., Porto, E., Brito, L. et al. (1997). Runoff collection from paved and unpaved roads. In: *Source Book of Alternative Technologies for Freshwater Augmentation in Latin America and the Caribbean Unit of Sustainable Development and Environment General Secretariat* (eds. J.A. Thornton and B. Robinson), 89. Washington, D.C, USA: Organization of American States.

Pennington, T. (2014). *The Prepper's Blueprint: The Step-by-Step Guide to Help You through Any Disaster*. USA: Create Space Independent Publishing Platform.

Pushard, D. (2014) Comparing Rainwater Storage Options. Available on: http://www.harvesth2o.com/rainwaterstorage.shtml.

Rahman, A., Hajani, E., and Eslamian, S. (2017). *Rainwater harvesting in arid regions of Australia*, , Chapter 26. In: *Handbook of Drought and Water Scarcity: Environmental Impacts and Analysis of Drought and Water Scarcity*, vol. 2 (eds. S. Eslamian and F. Eslamian), 489–500. USA: Taylor and Francis/CRC Press.

Salzedo, C., Fermin, C., Porto, E. et al. (1997). Runoff collection using surface and underground structures. In: *Source Book of Alternative Technologies for Freshwater Augmentation in Latin America and the Caribbean Unit of Sustainable Development and Environment General Secretariat*. Washington, D.C, USA: Organization of American States.

San Francisco Water, Power, Sewer. (2016) Installing a Rain Barrel for Non-spray Irrigation. Service of the San Francisco Public Utilities Commission. Available on: http://sfwater.org/modules/showdocument.aspx?documentid=8605.//./

Shah, N., Varandani, N.S., and Panchani, M. (2016). Life cycle assessment of household water tanks – a study of LLDPE, mild steel and RCC tanks. *Journal of Environmental Protection* 7: 760–769.

The Constructor Civil Engineering Home. (2017) Methods of Rainwater Harvesting -Components, Transport and Storage, https://theconstructor.org/water-resources/methods-of-rainwater-harvesting/5420

Torres, E., Brito, L., Marco, J. et al. (1997). *Rainwater harvesting from rooftop catchments*. In: *Source Book of Alternative Technologies for Freshwater Augmentation in Latin America and the Caribbean Unit of Sustainable Development and Environment General Secretariat* (eds. J.A. Thornton and B. Robinson), 56. Washington, D.C, USA: Organization of American States.

United States Environmental Protection Agency (EPA). (2013) *Rainwater Harvesting, Conservation, Credit, Codes,* and Cost, Literature Review and Case Studies, USA.

Venice Backstage. (2018) Available on: http://www.venicebackstage.org/en/159/i-pozzi-veneziani.

Wellington Water. (2017) Available on:https://wellingtonwater.co.nz/your-water/drinking-water/water-conservation/rainwater-tanks.

26

Rainwater Harvesting Storage-Yield-Reliability Relationships

John Ndiritu

University of the Witwatersrand, South Africa

26.1 Introduction

Rainwater harvesting (RWH) from roofs or other impervious surfaces for on-site use is a common source of water supply in many regions of the world. However, decisions on storage sizing and assessments of the yield and reliability of supply in many regions are not based on the comprehensive hydrologic analysis that is typically applied to larger centralized water supply systems (McMahon and Adeloye 2005; Basson et al. 1994). Design and operational decisions of centralized water resource systems are usually based on in-depth studies conducted by the responsible authorities or consultants, but RWH is often installed by individuals or entities that do not have the skills to conduct detailed hydrological analysis or the resources to involve specialists. Consequently, simplified methods based on monthly average rainfalls are sometimes used or recommended (Handia et al. 2003; Abdulla and Al-Shareef 2009; Denison et al. 2011) and RWH users are at times uncertain about the capabilities of their systems (Thomas et al. 2014). The optimization of system sizing is also considered a requirement for effective RWH development (Duncker and Matsebe 2015). Although RWH systems are often regarded as backup supply in many countries, they also serve as primary water sources in several countries of the world (Campisano et al. 2017). If RWH serves as a primary water source, it is essential that comprehensive storage-yield-reliability assessments be carried out to inform design and operational decision-making. This is, however, also advisable even when RWH serves as a backup. Storage-yield-reliability analysis informs us about the hydrologic and economic feasibility of RWH, as system costs and benefits are highly dependent on the size of storage, yield, and reliability.

26.2 The Rainwater Harvesting Storage-Yield-Reliability Problem

The typical RWH system consists of a rainfall catchment, a conveyance, a storage tank, a conduit to deliver the water to supply, and an overflow for spillage as shown on Figure 26.1. It is common for conveyance to be gravity-driven to a single storage, although systems with pump-driven conveyance to single or multiple storages exist. Delivery could also be gravity or pump-driven. The quantity and variability of water conveyed to the storage depends on the area of the catchment, the rainfall intensity, and the efficiency of rain collection and conveyance. The supply available from storage depends on the quantity and variability of the incoming (conveyed) water, the size of the storage, and how this storage is operated in supplying the demand. The level of supply is typically quantified by yield and its reliability. In practice, the catchment area for RWH is usually fixed or confined to a few known values, and the typical hydrologic analysis problem involves relating storage, yield, and reliability. Once obtained, this relationship is used to inform decision making relating to the feasibility of RWH, the expected supply level and the appropriate storage sizes. The RWH hydrologic analysis could, however, also be used to determine the catchment area and to obtain an appropriate system operation policy.

The RWH system design and operational decisions are for use in the future for which the rainfall and water demand are not known. If there is a reliable and adequately long historic rainfall time series at the RWH site or close to it, this is often assumed to be adequately representative of the future. In a rainfall event, a certain proportion of the rainfall is retained on the catchment and additional losses occur due to leakages and overflow in conveyance and in first-flush

Handbook of Water Harvesting and Conservation: Basic Concepts and Fundamentals, First Edition. Edited by Saeid Eslamian and Faezeh Eslamian.

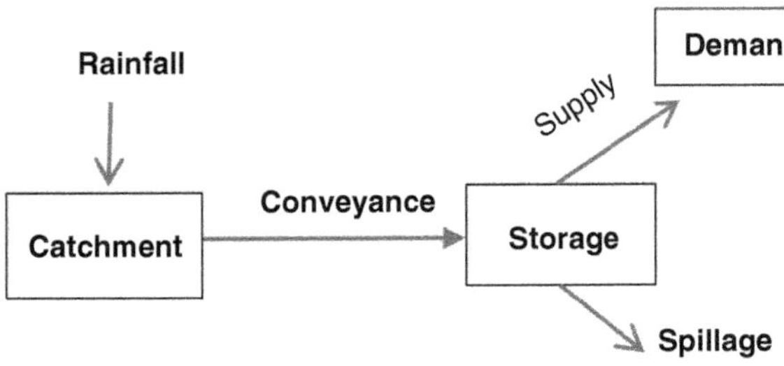

Figure 26.1 Typical components of a rainwater harvesting system.

devices if these are installed. The storage could be confined to the commercially available tank sizes or could be cast in-situ. In both options, availability of space could constrain the installable storage capacity.

The RWH water supply could be used for a variety of purposes, including domestic use, irrigation, air conditioning, toilet flushing, and cleaning. Precise determination of demand is generally not achievable (Campisano et al. 2017) and overestimating the demand sometimes leads to oversizing of storage capacity (Ward et al. 2012). Guidelines for demand estimation usually specify inexact ranges of possible demands (CSIR Building and Construction Technology 2003; Hofkes 1983) and demands obtained from metered water consumption (van Zyl et al. 2017) and field measurements (Saunders 2012) are highly variable. Comprehensive analysis for demand estimation is therefore required for storage-yield-reliability analysis. The RWH supply could be for complementing other sources, a backup, or as the main water source. Reliability of supply is set on the basis of the expected severity of the impacts of unavailability or shortage of water, and high reliabilities are required when RWH is the main water source or a vital backup. For situations where no reliable or adequately long rainfall data exists close to the RWH site, interpolation methods could be applied to obtain an artificial rainfall time series at the site (Taylor et al. 2012; Hengl et al. 2007; Tan and Xu 2014). Geraldi and Ghisi (2017) found 10 and 30 years of simulation to match, while Mitchell (2007) also found 10 year-long simulations to compare reasonably well with those of 50 years. Liaw and Tsai (2004) found a length of at least 50 years to be required for accurate analysis, while McMahon and Adeloye (2005) recommended streamflow lengths of at least 25 years for reservoir storage-yield analysis of typical streams. There are therefore large variations in what is considered adequately long, but the selected length should ideally include the observed regional variability of rainfall (dry, normal, and wet multi-annual periods).

Rainfall records obtained by instrumentation do not include the inter-centennial variability that has been inferred from paleo-reconstructed data (Tozer et al. 2016; Shi et al. 2016), and long paleo-reconstructed data could therefore be applied in RWH analysis to incorporate inter-centennial variability. However, as Prairie et al. (2008) noted, paleo-reconstructed data may be considerably uncertain and could be used to provide a time series of the climatic state (wet or dry) rather than the actual hydro-climatic data. Prairie et al. (2008) provided an approach of using such a time series with instrumental measurements to generate stochastic hydroclimatic data.

26.3 Modeling Storage-Yield-Reliability Relationships

26.3.1 Modeling Approaches and Methods

Modeling RWH storage-yield-reliability relationships is fundamentally similar to the modeling of these relationships for water supply reservoirs (McMahon and Adeloye 2005), and the commonly used RWH storage-yield analysis methods are adaptations of reservoir storage-yield analysis methods. Continuous simulation of reservoir mass balance, usually termed the behavior analysis method (McMahon and Adeloye 2005), provides an easily understandable and versatile method that simulates the operation of the actual system closely. The literature indicates that this approach is arguably the most widely used method for RWH storage-yield analysis (Campisano et al. 2017; Ward et al. 2012; Ward 2010; Youn et al. 2012; Khastagir and Jayasuriya 2010; Taffere et al. 2016). McMahon and Mein (1978) recommended the modified Gould's probability matrix method and the Behavior Analysis approach (with the effect of the assumed initial storage recognized) for water supply reservoir systems planning. The modified Gould's probability matrix method is based on Moran's theory of storage (Moran 1959) and applies behavior analysis within years while modeling the variation of yearly storage state as a Markov chain. Subsequently, McMahon and Adeloye (2005) recommended the sequent peak algorithm with the reliability norm over the behavior analysis method. The sequent peak algorithm applies mass balance over the periods of deficit and obtains the capacity as the largest cumulative deficit out of these periods. By controlling the severity of shortages to supply and the proportion of time over which they occur, the reliability norm is implemented (Adeloye et al. 2001). The preference of the sequent peak algorithm with the reliability norm was due to the perceived inability of the behavior analysis to determine storage capacity for specified levels of volumetric and time-based reliability. McMahon and Adeloye (2005) stated that the behavior analysis method has no ability to specify the magnitude of shortfalls and can therefore not obtain the volumetric reliability for a specified time-based reliability. Adeloye et al. (2001) also perceived this as one of the reasons for the observed bias ([mis]-behavior) in storage estimates by the behavior analysis method for over-year storage systems at high drafts (Pretto et al. 1997). The sequent peak algorithm with the reliability norm solved this problem by applying a method of controlling shortages during the years when the full demand could not be met. Reducing supply during low storage states (hedging) is however easily incorporated into the behavior analysis method (Tu et al. 2008; Wang and Liu 2013; Ndiritu et al. 2017b; Ndiritu and Sinha 2009). The use of behavior analysis and evolutionary optimization is also able to optimize operating rules to maximize yield whilst meeting multiple reliability constraints and levels of shortage (Ndiritu 2005). Furthermore, critical periods are easily identifiable in behavior analysis and the reliability norm implemented on the sequent peak algorithm can easy be incorporated into behavior analysis. The sequent peak method has been applied for the RWH storage-yield-analysis (Lee et al. 2000) and so has the well-known mass curve method (Handia et al. 2003; Schiller and Latham 1992). The mass curve method has been used to size reservoirs in several regions of the world (McMahon and Adeloye 2005) and has been widely attributed to Rippl (1883) in several textbooks (Loucks and van Beek 2017; Chow et al. 1980; Shaw et al. 2011; Linsley and Franzini 1979; Raudkivi 1979). However, as Klemes (1979) highlighted, Rippl's method (Rippl 1883) is not the classical mass curve method but a sequent trough algorithm that is equivalent to the sequent peak algorithm.

Storage-yield-reliability analysis methods based on Moran's theory of storage (Moran 1959) have been developed and Gould's probability matrix method (Gould 1961) was found to match the behavior analysis method using Australian streamflow data (McMahon and Mein 1978). Gould's method has been studied for reservoir storage-yield analysis (Srikanthan and McMahon 1985a, b; Ndiritu 1992; Otieno and Ndiritu 1997) but only one application for RWH has been found in the literature (Liaw and Tsai 2004). This method applies continuous simulation within years and assumes that the variation of storage across years is a Markov chain. It is independent of an assumed initial storage state and its main limitation is the assumption of independence of inflows and storage behavior across years. It could therefore be particularly suitable for within-year RWH storage-yield-reliability analysis or for over-year analysis where serial correlation of annual rainfall is negligible.

For RWH storage-yield-reliability analysis, a daily time-step has been found to be adequate unless very small storages are used (Fewkes and Butler 2000; Mitchell 2007; Campisano and Modica 2014). Although the monthly time-step has also been used (Taffere et al. 2016), and recommended for large storages (Fewkes and Butler 2000), it has been found to be too coarse when compared with daily time step analysis (Ndiritu et al. 2011). The standard operating rule which supplies all water demanded if available in storage is assumed for most RWH storage-yield-reliability analysis although more complex rules that incorporate hedging (reduction of supply during low storage states) could be applied. Generalized RWH storage-yield-reliability relationships have been developed in several regions of the world (Fewkes 2000; Gathenya et al. 2010; Hanson and Vogel 2014; Liaw and Chiang 2014; Ndiritu et al. 2017a, 2018) as statistical models, nomographs, or charts that relate rainfall characteristics, storage size, yield, and reliability. The generalized relationships enable preliminary assessment of the RWH potential and estimation of component sizes using much less time and computation than detailed analysis.

Storage-yield reliability analyses are used to support decision-making for systems that are intended for installation or operation in the future. The rainfall that will occur during the operational life of the RWH system is unknown at this stage, although it is certain that the available historic record will not be replicated. Therefore, although most analyses use the single historic rainfall time series for RWH hydrologic analysis, ensembles of stochastic rainfall have also been used (Lopes et al. 2017; Basinger et al. 2010; Cowden et al. 2008). Using multiple stochastic sequences of plausible future rainfalls helps to inform us better about the likely performance of the system during the operational life and could be particularly useful for over-year storage systems that may face water shortages over the few multi-annual drought periods in a single historic record. Figure 26.2 illustrates over-year and within-year storage behavior and it is seen that for over-year storage systems, substantial quantities of water are carried across years and the system typically fails during the major multi-annual droughts. For within-year storage systems, the storage typically dries out every year and water is supplied for only part of the year.

The determination of reliability of the RWH systems could be undertaken at different levels of rigor as illustrated in Figure 26.3. In Figure 26.3a, a stochastic rainfall generator is used to generate ensembles of stochastic rainfall and each of these is used as the input for storage-yield-analysis. Each analysis obtains an output (e.g. yield or storage) and frequency analysis of the population of these outputs using plotting position (Weibull 1951; Cunnane 1978) or statistical distribution fitting (Gumbel 1941) provides the relationship between reliability (probability of exceedance) and output values. In Figure 26.3b, the analysis uses the

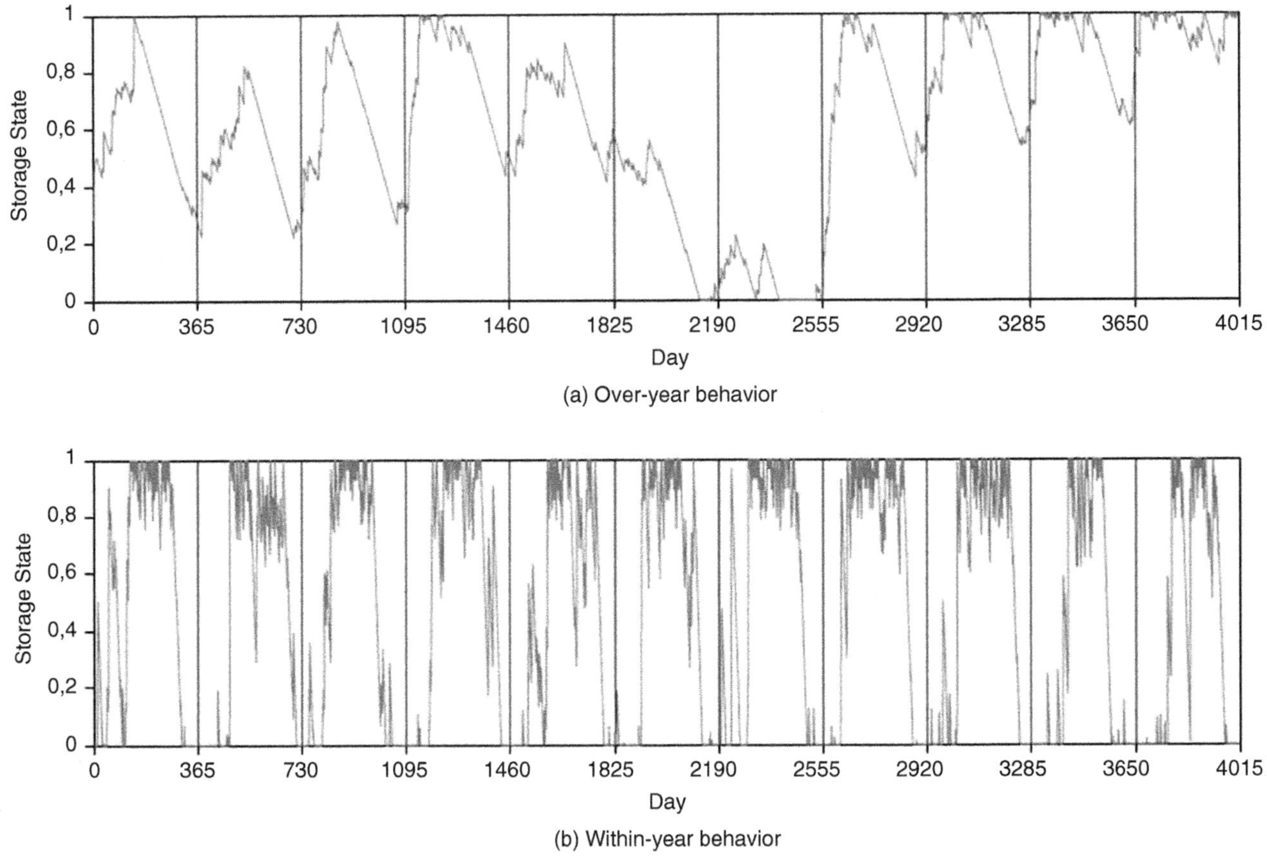

Figure 26.2 Illustration of over-year and within-year storage behavior. (a) Over-year behavior. (b) Within-year behavior.

historic rainfall sequence and obtains a time series of outputs (e.g. yield) for each year of simulation. Frequency analysis of these outputs obtains the relationship between reliability and the outputs. For this analysis, a single overall measure of performance for the complete record (e.g. expected yield) could also be obtained. This approach is suited for within-year storage systems and it has been applied in some RWH studies and applications (Campisano and Modica 2014; Fonseca et al. 2017; Su et al. 2009; Ndiritu et al. 2014). For over-year storage systems where substantial carryover of water across years occurs, the approach illustrated in Figure 26.3b would not be realistic. Figure 26.3c shows the common approach in which a single output is obtained from storage-yield-analysis (Geraldi and Ghisi 2017; Mitchell 2007; Taffere et al. 2016; Fewkes 2000; Seo et al. 2012; Sample et al. 2013). This output is generally not associated with an exceedance probability, although it could be considered an indicator of the median of the within-year yields.

Continuous simulation via behavior analysis is a commonly applied and versatile method for RWH storage-yield analysis and is presented in more detail in Section 26.3.2. Rippl's method (Rippl 1883) is considered the first comprehensive storage-yield analysis method (Klemes 1987; McMahon et al. 2007) and was initially demonstrated using a 33 year-long rainfall series. Rippl's method is practically identical to the sequent peak algorithm which is also widely used for storage-yield analysis (McMahon et al. 2007). The sequent peak algorithm and Rippl's method are therefore described together in Section 26.3.3. In Section 26.3.4, the basic approach used to develop generalized relationships of storage-yield-reliability relationships for RWH systems is outlined. This is then followed by a description of some of the generalized relationships that have been formulated in various regions of the world. The generation of daily stochastic rainfall data is well developed (Srikanthan and McMahon 2001) and there are several stochastic weather generators available for this (Vu et al. 2018). For these reasons, and also because stochastic rainfall is not routinely used for RWH hydrologic analysis, this chapter does not review stochastic rainfall generation. Srikanthan and McMahon (2001) provide an in-depth review of these methods and the reader can also refer to more recent studies on stochastic rainfall generation (Basinger et al. 2010; Cowden et al. 2008; Vu et al. 2018; Chowdhury et al. 2017; Apipattanavis et al. 2007). The assessment of the effect of

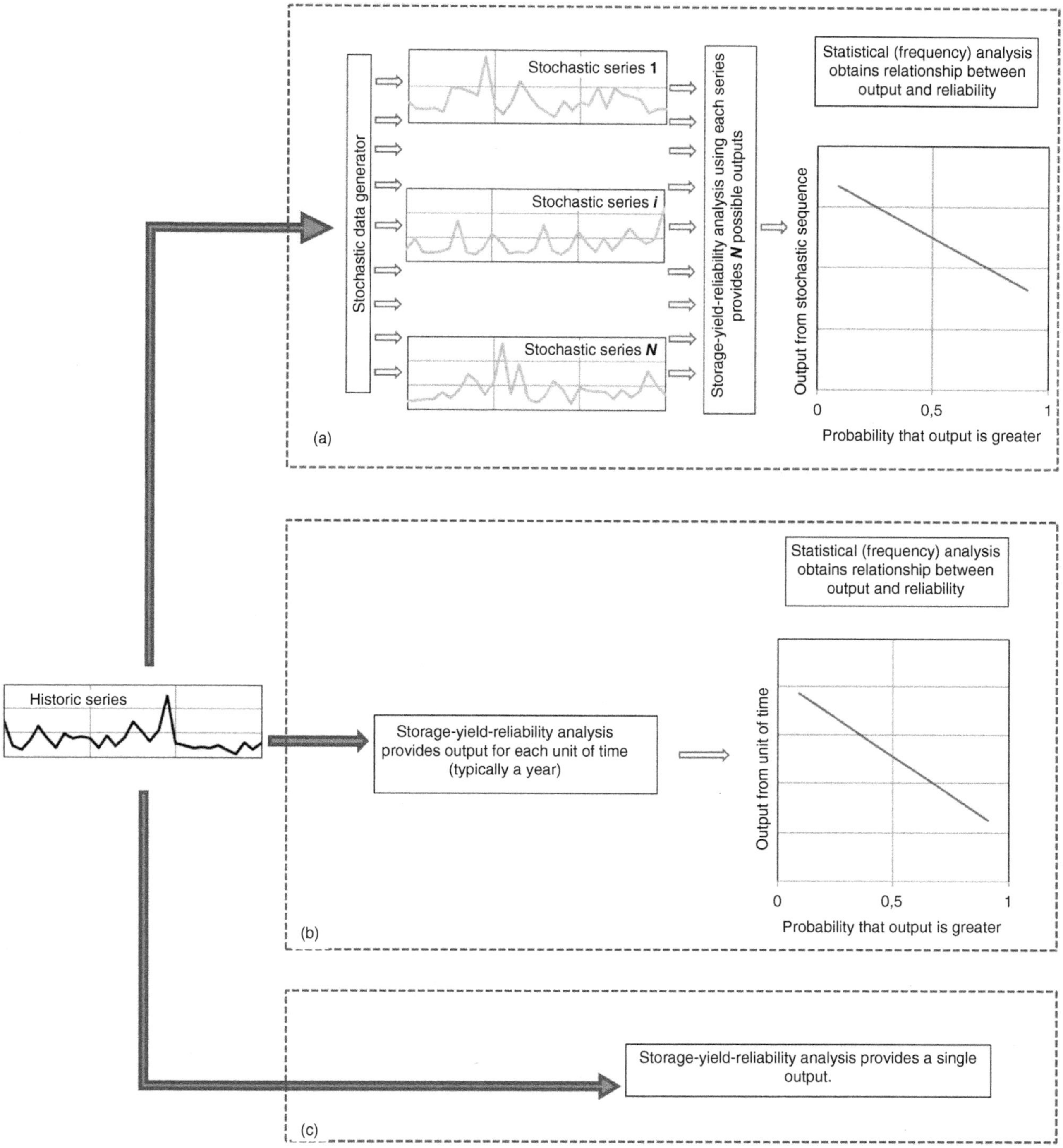

Figure 26.3 Approaches for storage yield reliability analysis.

climate change on RWH is important and has been studied in several regions (Kisakye et al. 2018; Haque et al. 2016; Schulze 2012). Regardless of climate change-related variation in rainfall and water demand, storage-yield-reliability analysis methods essentially remain unchanged and a review and incorporation of the effects of climate change on RWH is therefore not included in this chapter.

26.3.2 Behavior Analysis (Continuous Simulation) Method

The behavior analysis, Rippl's method, and the sequent peak algorithm are described for a typical RWH system with components and fluxes as shown on Figure 26.4. On this Figure, $\boldsymbol{R_t}$ is the rainfall depth in period $\boldsymbol{t}$, $\boldsymbol{A}$ is the effective roof area, $\boldsymbol{\eta}$ is the water collection efficiency, $\boldsymbol{Q_t}$ is

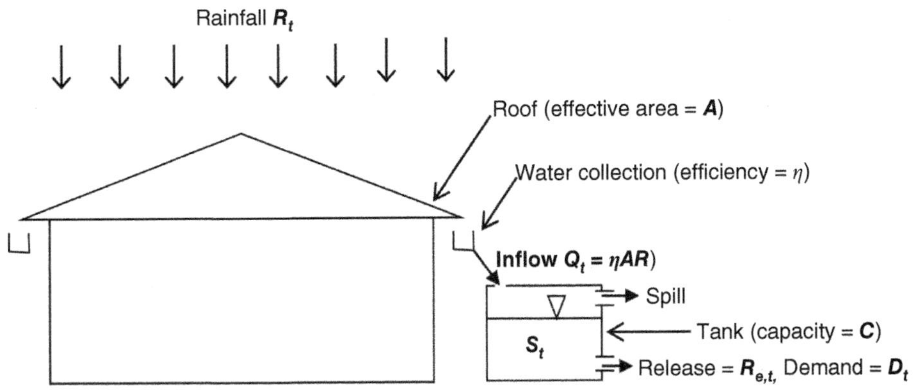

Figure 26.4 Rainwater harvesting components and fluxes for storage-yield-reliability analysis.

the inflow into the tank in period t and is equal to ηAR_t, S_t is the volume of water in storage at the start of period t, C is the live storage capacity of the tank, $R_{e,t}$ is the volume of water released to meet the demand in period t, and D_t is the demand in period t. Behavior analysis by continuous daily time step simulation applies mass balance assuming a storage capacity and constraining the volume in storage between the full and the empty state.

The mass balance starts assuming an initial storage state and the inflow sequence is sometimes concatenated (used in two cycles joining the end with the start of the series) with itself to minimize the effect of assuming an initial storage state (Pretto et al. 1997). Alternatively, iteration using repeated simulation runs could be used to ensure that the initial storage state equals the final one. The mass balance could assume that during each day, all the conveyed rainwater flows into storage before any supply from storage happens. If this inflow would lead to spillage, the spillage is assumed to also occur before any supply from the storage and this algorithm is therefore termed as yield after spillage (YAS) Conversely, if it assumed that the entire demand is supplied from storage before any inflow into storage, any spillage would happen after supply and this procedure is termed as yield before spillage (YBS). Mass balance assuming YAS leads to more conservative estimates of yield, as it results in larger spillage volumes. Mitchell (2007) found daily time step YAS analysis to match six minute time-step simulation of the RWH systems better than YBS and Fewkes and Butler (2000) also found daily YAS to match hourly YAS simulation better than daily YBS simulation. Latham (1983) formulated a more generalized mass balance in which the inflow into storage and the supply from storage could happen simultaneously for parts of the day. Eqs. (26.1) and (26.2) describe this generalized mass balance algorithm in which θ takes a value of 0 for YAS and 1 for the YBS approach.

$$R_{e,t} = minimum \begin{Bmatrix} D_t \\ S_t + \theta Q_t \end{Bmatrix} \tag{26.1}$$

$$S_{t+1} = min \begin{Bmatrix} C - (1-\theta)R_{e,t} \\ S_t + Q_t - R_{e,t} \end{Bmatrix} \tag{26.2}$$

where $R_{e,t}$ is the volume of water released to meet the demand in period t, D_t is the demand in period t, Q_t is the inflow into storage in period t, S_t is the volume of water in storage at the start of period t, and C is the live storage capacity of the tank.

The release $R_{e,t}$ is the yield at each time step and the ratio of the total release to the total demand for the complete simulation is the volumetric reliability (Eq. (26.3)). Volumetric reliability, widely applied for RWH storage-yield analysis (Geraldi and Ghisi 2017; Mitchell 2007; Taffere et al. 2016; Fewkes 2000; Campisano and Modica 2014; Seo et al. 2012; Sample et al. 2013) has also been termed as water supply or saving efficiency (Fonseca et al. 2017; Palla et al. 2011; Palla et al. 2012) and as "1-deficit rate" (Youn et al. 2012; Lopes et al. 2017; Su et al. 2009). The proportion of time that the demand is fully met during simulation is termed as time-based reliability has also been used as a measure of RWH performance (Cowden et al. 2008; Haque et al. 2016).

$$v_r = \frac{\sum_{t=1}^{N} R_{e,t}}{\sum_{t=1}^{N} D_t} \tag{26.3}$$

Where v_r is the volumetric reliability, $R_{e,t}$ is the volume of water released to meet the demand in period t, D_t is the demand in period t, and N is the number of periods of analysis.

While this measure informs the expected performance over the life span of the system, for within-year storage systems, it does not indicate how this performance varies within years. This variation may be critical in regions with high interannual rainfall variability (Ndiritu et al. 2017a) and frequency analysis of within-year volumetric reliabilities can be applied to provide storage-yield-reliability relationships. In deriving these relationships, volumetric reliability is considered as the yield and the exceedance probabilities from frequency analysis as reliabilities (Fonseca et al. 2017; Su et al. 2009). For over-year storage systems, applying multiple stochastic sequences provides ensembles of volumetric reliabilities and frequency analysis provides the reliabilities required for storage-yield-reliability relationships. This could also be carried out for within-year storage systems although there would still be the need to assess the impacts of interannual rainfall variability.

26.3.3 Sequent Peak Algorithm and Rippl's Method

Rippl's method and the sequent peak algorithm are illustrated graphically on Figure 26.5a,b. The function of storage is to hold water during periods of surplus (when demand is lower than inflow) and to supply this water during periods of deficit. In Figure 26.5a, the deficit periods are $\boldsymbol{t_2 - t_3}$ and $\boldsymbol{t_5 - t_6}$ and it is in these periods that storage is required to meet the deficits. The volume of water that needs to be available in storage for each deficit period therefore equals the magnitude of the cumulative deficit. The largest value of storage from all the deficit periods ($S_{t_2-t_3}$ on Figure 26.5a) is then the required storage for the system. Both sequent peak and Rippl's method assume that the reservoir will be full at the start of this deficit period ($\boldsymbol{t_2}$) and that the

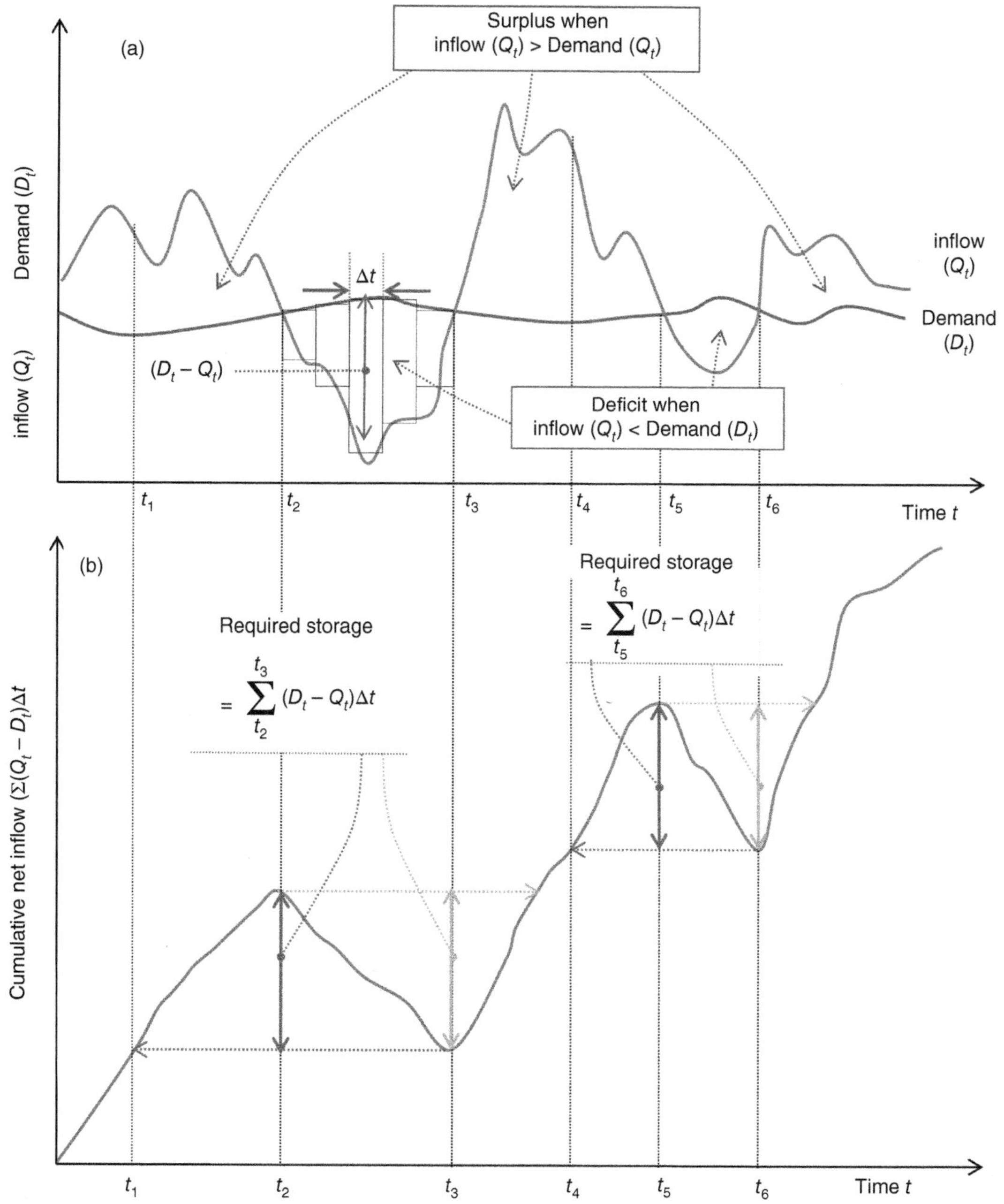

Figure 26.5 Illustration of the sequent peak algorithm and Rippl's method.

reservoir will only empty at the end of the critical period (t_3). Rippl's method further requires that prior to period t_2, the cumulative inflows should be sufficient to fill up the reservoir and to therefore adequately supply the deficit in period $t_2 - t_3$. Figure 26.5b shows how a time series of cumulative net inflows can be used to obtain the required storage. The sequent (succeeding) peak needs to be higher than the previous one and the trough is located at the lowest point between two peaks. The reduction in cumulative net inflows during the critical period ($t_2 - t_3$) is the required capacity. The sequent peak algorithm obtains this capacity by projecting a horizontal line (dotted green line) from the peak (t_2) into the future until it crosses the cumulative net inflow curve. The sequent peak method also requires the final deficit in water to equal the initially assumed deficit. If the two are different, the initial deficit is replaced by the final one and an additional sequent peak analysis run is made. If the initial and final deficits do not match, then the demand imposed cannot be met by the inflows. Rippl's method projects a horizontal line (dotted red line) from the trough (t_3) into the past until it crosses the cumulative net inflow curve (t_1). If the horizontal projection from the trough (t_3) fails to cross the cumulative net inflow curve, this could mean that (i) the demand imposed is higher than the inflows can supply or, (ii) a substantial deficit period occurs close to the start of the inflow time series. If early location of the deficit period is the reason, a sufficiently long representative inflow data (e.g. from the end of the inflow series) could be placed at the start of the inflow time series to enable the projection from the trough to cross the cumulative inflow curve (Rippl 1883). The sequent peak method therefore requires the storage to fill up after the critical period, while Rippl's method requires the reservoir to have been filled up before the critical period started. As Figure 26.5b shows, the two methods obtain identical storage capacities unless large deficits cause (i) a deep trough close to the beginning or (ii) a final cumulative inflow lower than the last peak. If the recommended practice of concatenating the inflow with itself (McMahon and Adeloye 2005) is included, the two methods obtain identical storage capacities. The sequent peak algorithm is usually implemented numerically by locating the deficit periods and computing the deficits using Eq. (26.4). Rippl's method was developed in the pre-computer era and was therefore graphical, but it is more convenient to implement it numerically.

$$S_{t_2-t_3} = \sum_{t_2}^{t_3} (D_t - Q_t)\Delta t \tag{26.4}$$

where $S_{t_2-t_3}$ is the required storage for period $t_2 - t_3$, D_t is the demand in period t, Q_t is the inflow in period t and Δt is the time step.

The description of the two methods to this point has obtained a single storage capacity using a single historic sequence and a statistical assessment of reliability cannot therefore be made. However, for within-year systems where the storage fills up and empties in most of the years, the analysis obtains the storage capacities for meeting deficits in each year as illustrated in Figure 26.6. Frequency analysis of these storages using an empirical plotting position formula (Weibull 1951; Cunnane 1978) or a probability distribution (Gumbel 1941) could determine a relationship between storage and probability of exceedance. For the 10 within-year storages shown on Figure 26.6, the Weibull plotting position formula (Eq. (26.5)) would obtain probabilities of exceedance ranging from 91% (for the highest storage of 81 m^3) to 9% (for the lowest storage of 34 m^3). These probabilities could practically be interpreted as the reliabilities at which the demand would be met if the associated storage capacity is installed. The assessment of within-year reliabilities could be enhanced by stochastically extending the inflow sequence or by the use of multiple stochastically generated sequences.

$$p = \frac{m}{n+1} \tag{26.5}$$

where p is the exceedance probability (reliability) of the storage ranked m with the ranking in ascending order (Figure 26.5), and n is the total number of storages.

For over-year storage where only few critical periods occur, Rippl's method or the basic sequent peak algorithm cannot obtain a reliability measure if a single historic record is used.

By using multiple stochastic sequences, an ensemble of storages can be obtained, and frequency analysis of these storages would provide the exceedance probabilities required to obtain storage-yield-reliability relationships. A further development of the basic sequent peak, termed as sequent peak algorithm with the reliability norm (Adeloye et al. 2001), could be used to obtain a specified time-based and volumetric reliability for a single historic sequence. In this approach, the critical periods are identified and the ratio of the duration of the non-critical periods to the simulation period is specified as the time-based reliability. The supplies over the critical periods are then reduced to levels of shortage that result in the required volumetric reliability. Since this process may cause shifts in the critical period, iterations are applied until convergence to the specified time-based and volumetric reliability is achieved. By applying this method with multiple stochastic sequences, storage-yield-reliability relationships can be obtained as for the basic sequent peak method. For this, the set volumetric reliability, time-based reliability, and demand would together be considered as what constitutes yield. The sequent peak algorithm could also be modified

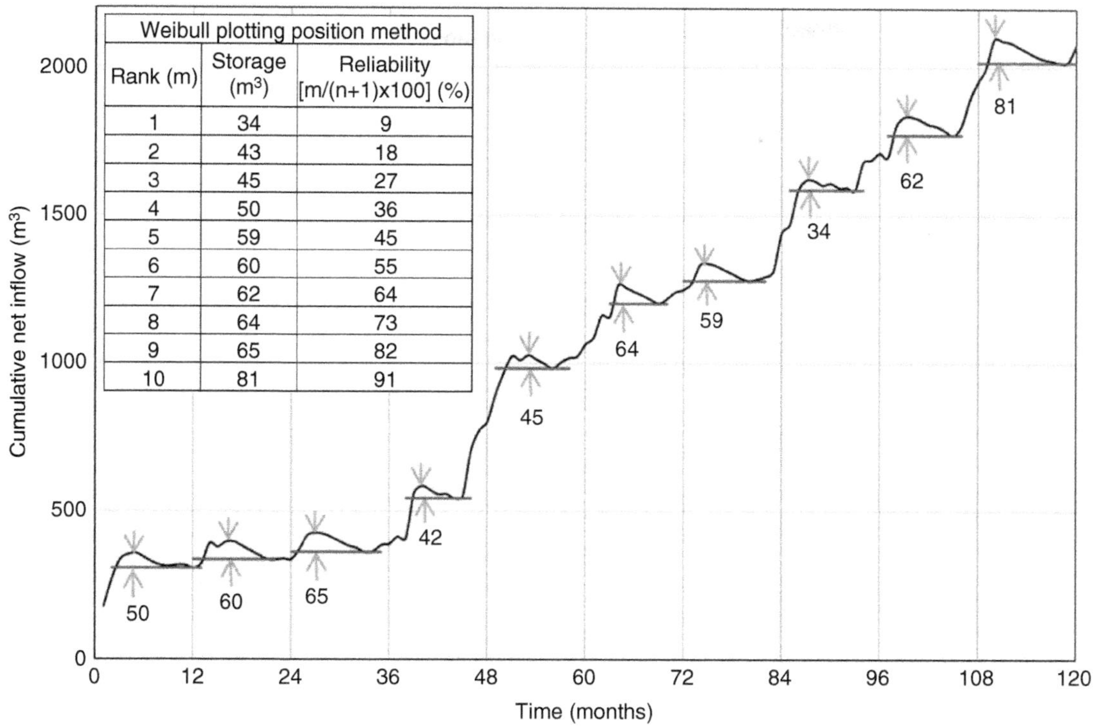

Figure 26.6 Rippl's method and sequent peak algorithm for within-year analysis.

to allow periods of non-supply and the proportion of the periods with full supply to the total period of analysis could then be defined as the time-based or volumetric reliability (Lee et al. 2000).

26.3.4 Generalized Storage-Yield-Reliability Relationships

Generalized storage-yield-reliability relationships are used for preliminary sizing and assessment of RWH system potential. They could also serve other purposes such as analysis of the regional potential of RWH but are not alternatives to detailed hydrologic analysis (Hanson and Vogel 2014). They mostly take the form of non-linear models of non-dimensionless relationships of storage, yield, reliability, and one or more rainfall statistic (mean or standard deviation), although they are sometimes presented as nomographs (Gathenya et al. 2010) or as charts (Ndiritu et al. 2018). The data on yields, reliabilities and storage required for generalizing are usually obtained by continuous simulation (Fewkes 2000). The relationships are fitted statistically and verification tests are usually included as part of model development (Hanson and Vogel 2014; Liaw and Chiang 2014; Ndiritu et al. 2017a). Because of their statistical basis, their application is usually confined to the regions from which the rainfall data was sourced, but could be applicable to other regions that have similar rainfall characteristics. Following are some generalized storage-yield-reliability relationships that have been recently developed in different regions of the world.

Gathenya et al. (2010) applied monthly time step behavior analyses to obtain nomographs relating tank size, roof area and water demand for a time-based reliability of 67%. The analysis applied 20 years of monthly rainfall data and nomographs were developed for 50 towns in Kenya.

Liaw and Chiang (2014) delineated northern Taiwan into four rainfall regions and obtained dimensionless relationships of the form of Eq. (26.6) for specific volumetric reliabilities for each region. The analysis used data from 58 rainfall stations that had records of at least 50 years. From the study, graphs of dimensionless storage-yield relationships were developed for volumetric reliabilities ranging from 50% to 95%.

$$d_r = a s_r^{\,b} \qquad d_r = \frac{D_d}{R} \qquad s_r = \frac{C_d}{R} \qquad S_d = \frac{C}{A} \tag{26.6}$$

Where $\boldsymbol{d_r}$ is the demand fraction, $\boldsymbol{s_r}$ is the supply ratio, $\boldsymbol{D_d}$ is the annual demand as depth, $\boldsymbol{R}$ is the mean annual rainfall, $\boldsymbol{C_d}$ is the storage capacity as a depth, $\boldsymbol{C}$ is the storage capacity, $\boldsymbol{A}$ is the effective roof area, and $\boldsymbol{a}$ and $\boldsymbol{b}$ are statistically fitted parameters.

Hanson and Vogel (2014) developed generalized storage-reliability-yield relationships for the USA using rainfall data from 231 stations that had a median record length of 59 years. The USA was delineated into three regions and relationships of the form given in Eq. (26.7) were formulated for volumetric reliabilities of 80%, 90%, 95%, and 98%.

$$S_r = \begin{cases} e^a e^{b(1-\ln(d_r))^d} \sigma_w{}^{c_1} \left(\dfrac{1+\rho}{1-\rho}\right)^{c_2} & \text{East region} \\ e^{b(1-\ln(d_r))^d} \sigma_w{}^{c_1} C_v{}^{c_2} & \text{Midwest region} \\ e^a e^{b(1-\ln(d_r))^d} \mu_w{}^{c_1} \left(\dfrac{1+\rho}{1-\rho}\right)^{c_2} & \text{West region} \end{cases} \tag{26.7}$$

Where $\boldsymbol{S_r}$ is the storage ratio in units of $m^3/100\ m^2$ of roof area, $\boldsymbol{d_r}$ is the demand fraction, $\boldsymbol{\sigma_w}$ and $\boldsymbol{\mu_w}$ is the standard deviation and mean of the daily wet-day rainfall series, $\boldsymbol{\rho}$ is the lag1 annual serial correlation, $\boldsymbol{C_v}$ is the coefficient of variation of daily rainfall depth, and $\boldsymbol{a}$, $\boldsymbol{b}$, $\boldsymbol{d}$, $\boldsymbol{c_1}$ and $\boldsymbol{c_2}$ are statistically fitted parameters.

Ndiritu et al. (2017a) used continuous simulation of the within-year RWH systems that could be implemented in 19 shopping centres located in 4 regions of South Africa to develop generalized storage-yield-reliability relationships. Rainfall data of an average length of 117 years from 14 rainfall stations were used. Reliability was obtained from frequency analysis of within-year yields and was defined as the probability that the specified supply levels (proportion of days with full supply) would be met in any year. Three-dimensional pareto optimal fronts at which yield and reliability are maximized while storage is minimized were identified for each system and used to formulate the generalized relationships (Eqs. (26.8)–(26.12)). Using the same approach, Ndiritu et al. (2018) developed generalized storage-yield-reliability charts (Figure 26.7) for within-year RWH analysis for Johannesburg, South Africa. The analysis used data from eight rainfall stations that averaged 123 years in length. Ninety percent of the years in this data set had continuous data.

$$\begin{aligned} &S_{p-r} + (r - r_t)S_{LR} \\ &r \geq r_t \\ &0.85 \leq r, r_t \leq 0.99 \end{aligned} \tag{26.8}$$

$$\begin{aligned} &S_{p-r} = a\left(\frac{1}{d_r}\right)^b \\ &a = 1.1428(1-r)^{0.1514} \\ &b = 1.2416(1-r)^{-0.037} \\ &0.85 \leq r \leq 0.99 \end{aligned} \tag{26.9}$$

$$\begin{aligned} &S_{LR} = eS_{p-r}{}^f \\ &e = 0.6629(1-r)^{-0.184} \\ &f = -1.7615r + 2.3725 \\ &0.85 \leq r \leq 0.99 \end{aligned} \tag{26.10}$$

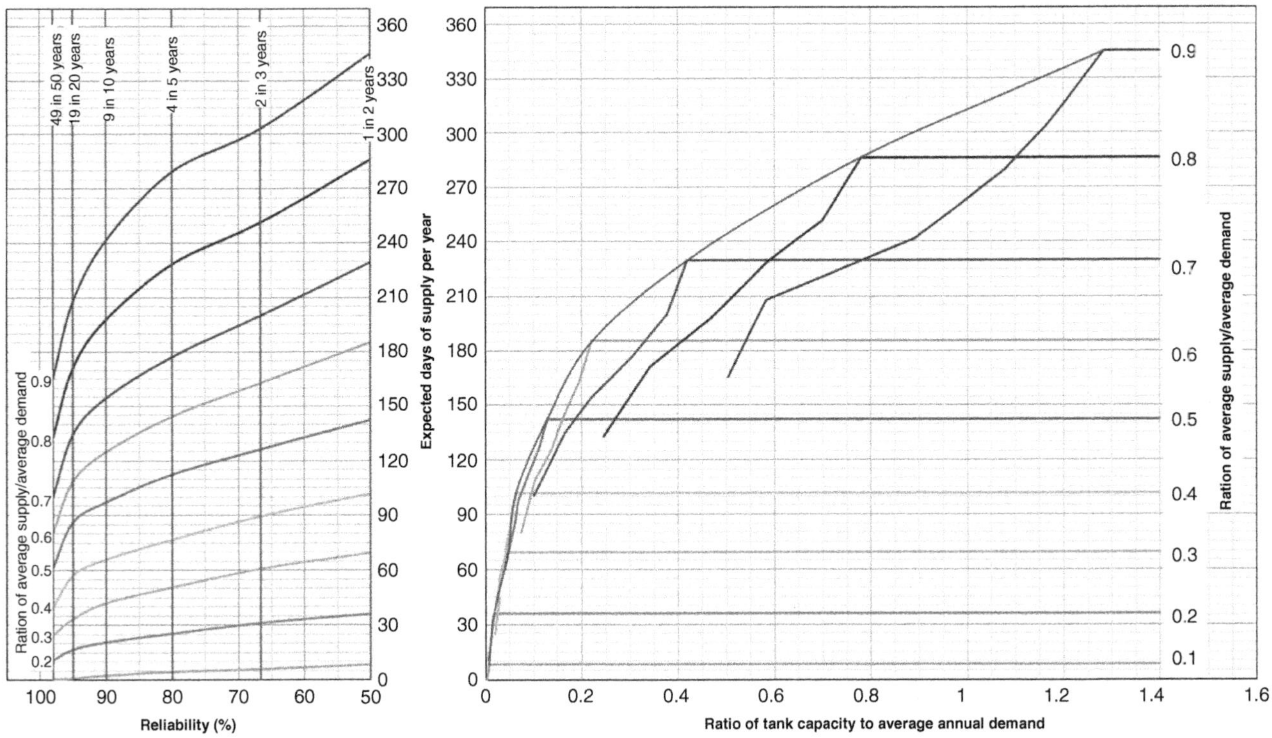

Figure 26.7 Generalized within-year RWH design and assessment charts for Johannesburg (Ndiritu et al. 2018).

$$R_{TD-r} = cS_{p-r}{}^{d}$$
$$c = 1.4365(1-r)^{0.5703}$$
$$d = 2.0065(1-r)^{0.2131} \quad (26.11)$$
$$0.85 \le r \le 0.99$$

$$R_{TD-r} = \frac{C_r}{D_a} \quad (26.12)$$

where $S_{p\text{-}rt}$ is the proportion of the year fully supplied at reliability r_t, $S_{P\text{-}r}$ is the proportion of the year fully supplied at reliability r, r is the reliability at the Pareto front, S_{LR} is the slope of the yield–reliability plot for the storage capacity that is optimal (located at the Pareto front) for reliability r, d_r is the demand fraction, $R_{TD\text{-}r}$ is the ratio of storage capacity that locates at the pareto front at reliability r (C_r) to the volume of annual demand (D_a), a, b, c, d, e, and f are functions of reliability r.

26.4 Key Considerations

26.4.1 How Is the Adequacy of the Rainfall Time Series Assessed?

The rainfall time series needs to include periods that reflect the variability of climate in the area or the region and typically include normal, dry, and wet periods. Short records are not expected to include these as inter-decadal climatic variability is predominant in many regions of the world. Liaw and Tsai (2004) found 50 years as the required length to capture long-term variability while Geraldi and Ghisi (2017) found 10 year-long simulations to closely compare with 30 year-long ones. McMahon and Adeloye (2005) recommended a minimum streamflow length of 25 years for reservoir storage-yield analysis and this may be a reasonable guide for RWH as well. The possibility of extending rainfall data to create as long and reliable a time series needs to be always considered as long-term variability of rainfall is usually high and rainfall is a primary input of RWH storage-yield-reliability analysis.

26.4.2 What Modeling Methods are Best Suited for Use?

Daily time-step continuous simulation via behavior analysis is the preferred method of RWH storage-yield-reliability analysis and it has been widely applied for this. In order to prevent bias because of assuming an initial storage state, repeated simulations to ensure that the initial and the final storage state are equal could be incorporated into behavior analysis. Alternatively, concatenation of the inflow record with a warm-up period could be included in the simulation. The sequent peak algorithm or the sequent peak with the reliability norm recommended by McMahon and Adeloye (2005) could also be used for RWH system modeling but these are not as versatile as behavior analysis. The sequent peak algorithm is indeed a special case of the behavior analysis method for which the storage capacity empties just once in the simulation period, as the following example demonstrates.

Suppose the storage capacity for a constant demand of 300 m^3/year needs to be determined using the annual inflow data given in Table 26.1. For behavior analysis, the capacity required so that the storage empties only once is determined and a single iteration is carried out to set the starting storage state to equal the final storage state. A single iteration is also carried out for the sequent peak algorithm to set the initial cumulative deficit to equal the final cumulative deficit. The two methods obtain an identical storage capacity of 916.5 m^3 and plots of cumulative deficit (from the sequent peak algorithm) and volume in storage (from behavior analysis) also perfectly mirror each other about the 50% storage state, as shown in Figure 26.8. The two analyses are therefore identical.

Since behavior analysis can easily incorporate many forms of operating rules of varied complexity (Tu et al. 2008; Wang and Liu 2013; Ndiritu 2005; Ndiritu and Sinha 2009), the sequent peak algorithm with the reliability norm (Adeloye et al. 2001) can also be perceived as a special case of behavior analysis that applies an operation rule that restricts the demand to a set level during critical periods.

Gould's probability matrix method (Gould 1961) or its modified form (Srikanthan and McMahon 1985a) have not been widely applied for reservoir or RWH storage-yield-reliability analysis but have been found to match the behavior analysis method if annual serial correlation of the input data is not significant (Srikanthan and

Table 26.1 Annual inflow data for storage capacity determination.

Year	Annual inflow (m^3)	Year	Annual inflow (m^3)	Year	Annual inflow (m^3)
1	735	12	99	23	185
2	230	13	221	24	216
3	143	14	117	25	198
4	351	15	206	26	251
5	1271	16	365	27	160
6	819	17	272	28	83
7	385	18	159	29	307
8	972	19	549	30	179
9	758	20	1280	31	311
10	855	21	279	32	259
11	758	22	257	33	511

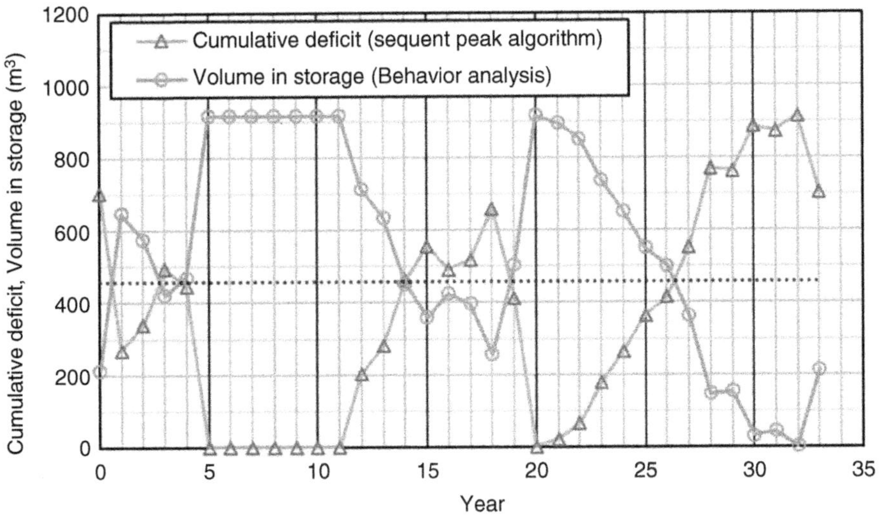

Figure 26.8 Mirror imaging of cumulative deficit from sequent peak algorithm and volume in storage from behavior analysis.

McMahon 1985b). Therefore, Gould's probability matrix method probably holds much untapped potential for RWH storage-yield-reliability analysis.

26.4.3 When Is It Essential to Apply Statistically-Based Reliability? How Is this Done?

It is advisable to always assess reliability statistically and this is essential if the RWH system needs to provide supply at high levels of reliability. In order to select an appropriate approach of assessing reliability, it is necessary to find out if the RWH system will predominantly operate as a within-year storage or as an over-year storage system. For over-year storage systems (Figure 26.2a), use of a single historic sequence typically results in a low number of critical periods (during the multi-annual drought occurrences) and a realistic statistical assessment of how the system would perform during critical periods is generally not possible. If such systems are required to operate at high levels of reliability (certainty of water supply), the use of multiple stochastically generated sequences and frequency analysis of the resulting ensembles of performance (Figure 26.3a) is recommended. For systems where high levels of certainty of supply are not required, the historic sequence may suffice although using multiple stochastic sequences would improve confidence in the analysis. For within-year storage systems (Figure 26.2b), a single performance measure from the simulation (Figure 26.3c) is an indicator of the expected average performance and a reasonable estimate of the median of the within-year performances. The median has an approximate exceedance probability of one in two years and this average performance would therefore be achieved in about half of the operational life of the RWH system.

For regions with high interannual rainfall variability, the system's performance may be much lower than the median for extended and intermittent periods. For such situations, statistically based reliability of system performance therefore needs to be obtained from frequency analysis of within-year performance measures (Figure 26.3b). It is important to recognize that the widely-applied volumetric reliability value (also termed as water supply or saving efficiency or as "1-deficit rate") is not a statistical measure of reliability but a measure of yield.

26.4.4 When Do Generalized Storage-Yield-Reliability Relationships Need to Be Used?

Generalized storage-yield-reliability relationships can be used for preliminary design, feasibility analysis, and regional assessment of the RWH potential. However, they do not obviate the need for detailed hydrological analysis if the data required for such analysis is obtainable. For situations where rainfall data is unavailable or is highly inadequate at the site of interest, generalized regional relationships that are based on much longer rainfall data may be the realistic choice for RWH storage-yield-reliability analysis.

26.5 Conclusions

Storage-yield-reliability analysis of RWH systems is an essential aspect of RWH-related decision-making, as RWH system size and performance have significant implications on the feasibility, benefits, and costs of RWH. Section 26.2 of this chapter describes the main features of the RWH storage-yield-reliability problem. In Section 26.3, a description of the general approach for solving the problem and the modeling methods that are applicable for this are presented. The key considerations for enabling effective storage-yield-reliability analysis are discussed in Section 26.4 and highlight the need to: obtain and apply as long and reliable dataset of rainfall as possible; apply a realistic modeling approach such as continuous daily simulation, and; incorporate a statistics-based assessment of reliability.

References

Abdulla, F.A. and Al-Shareef, A.W. (2009). Roof rainwater harvesting systems for household water supply in Jordan. *Desalination* 243: 195–207. https://doi.org/10.1016/j.desal.2008.05.013.

Adeloye, A.J., Montaseri, M., and Garmann, C. (2001). Curing the misbehavior of reservoir capacity statistics the modified sequent peak algorithm. *Water Resources Research* 37 (1): 73–82.

Apipattanavis, S., Podesta, G., Rajagopalan, B., and Katz, R.W. (2007). A semiparametric multivariate and multisite weather generator. *Water Resources Research* 43. https://doi.org/W11401 doi: 10.1029/2006WR005714.

Basinger, M., Montalto, F., and Lall, U. (2010). A rainwater harvesting system reliability model based on nonparametric stochastic rainfall generator. *Journal of Hydrology* 392 (3–4): 105–118. https://doi.org/10.1016/j.jhydrol.2010.07.039.

Basson, M.S., Allen, R.B., Pegram, G.G.S., and Van Rooyen, J. (1994). *Probabilistic Management of Water Resource and Hydropower Systems.* Littleton, CO, USA: Water Resources Publications.

Campisano, A. and Modica, C. (2014). Selecting time scale resolution to evaluate water saving and retention potential of rainwater harvesting tanks. *Procedia Engineering* 70: 218–227. https://doi.org/10.1016/j.proeng.2014.02.025.

Campisano, A., Butler, D., Ward, S. et al. (2017). Urban rainwater harvesting systems: research, implementation and future perspectives. *Water Research* 115: 195–209. https://doi.org/10.1016/j.watres.2017.02.056.

Chow, V.T., Maidment, D.R., and Mays, L. (1980). *Applied Hydrology*. USA: McGraw Hill.

Chowdhury, A.F.M.K., Lockart, N., Willgoose, G. et al. (2017). Development and evaluation of a stochastic daily rainfall model with long-term variability. *Hydrology and Earth System Sciences* 21: 6541–6558.

Cowden, J.R., Watkins, D.W., and Mihelcic, J.R. (2008). Stochastic rainfall modeling in West Africa: parsimonious approaches for domestic rainwater harvesting assessment. *Journal of Hydrology* 361 (1–2): 64–77. https://doi.org/10.1016/j.jhydrol.2008.07.025.

CSIR Building and Construction Technology (2003). *Guidelines for Human Settlement Planning and Design*, vol. 2. Pretoria, South Africa: CSIR, Building and Construction Technology.

Cunnane, C. (1978). Unbiased plotting positions – a review. *Journal of Hydrology* 37: 205–222.

Denison, J., Botha, M., and Kruger, E. (2011) *Water Harvesting and Conservation, Volume 2 Part 1 : Technical Manual and Farmer Handouts*. Report No. TT 493/11, Water Research Commission, South Africa.

Duncker, L., and Matsebe, G. (2015) *Research Framework for Supporting Domestic Rainwater Harvesting in South Africa*. Report No. TT 625/14. Water Research Commission, South Africa.

Fewkes, A. (2000). Modelling the performance of rainwater collection systems: towards a generalised approach. *Urban Water Journal* 1: 323–333. https://doi.org/10.1016/S1462-0758(00)00026-1.

Fewkes, A. and Butler, D. (2000). Simulating the performance of rainwater collection and reuse systems using behavioural models. *Building Services Engineering Research and Technology* 21 (2): 99–106.

Fonseca, C.R., Hidalgo, V., Díaz-Delgado, C. et al. (2017). Design of optimal tank size for rainwater harvesting systems through use of a web application and geo-referenced rainfall patterns. *Journal of Cleaner Production* 145: 323–335. https://doi.org/10.1016/j.jclepro.2017.01.057.

Gathenya, J.M., Kinyari, P.K., and Home, P. (2010). Domestic roof rainwater harvesting tank sizing calculator and nomograph. *JAGST* 12 (2): 115–125.

Geraldi, M.S. and Ghisi, E. (2017). Influence of the length of rainfall time series on rainwater harvesting systems: a case study in Berlin. *Resources, Conservation and Recycling* 125: 169–180. https://doi.org/10.1016/j.resconrec.2017.06.011.

Gould, B.W. (1961). Statistical methods for estimating the design capacity of dams. *Journal of the Institution of Engineers, Australia* 33 (12): 405–416.

Gumbel, E.J. (1941). The return period of flood flows. *The Annals of Mathematical Statistics* 12 (2): 163–190.

Handia, L., Madalitso, J., and Mwiindwa, C. (2003). Potential of rainwater harvesting in urban Zambia. *Physics and Chemistry of the Earth* 28: 893–896. https://doi.org/10.1016/j.pce.2003.08.016.

Hanson, L.S. and Vogel, R.M. (2014). Generalized storage–reliability–yield relationships for rainwater harvesting systems. *Environmental Research Letters* 9 (7): 075007. https://doi.org/10.1088/1748-9326/9/7/075007.

Haque, M.M., Rahman, A., and Samali, B. (2016). Evaluation of climate change impacts on rainwater harvesting. *Journal of Cleaner Production* 137: 60–69. https://doi.org/10.1016/j.jclepro.2016.07.038.

Hengl, T., Heuvelink, G.B.M., and Rossiter, D.G. (2007). About regression-kriging: from equations to case studies. *Computers and Geosciences* 33: 1301–1315. https://doi.org/10.1016/j.cageo.2007.05.001.

Hofkes, E.H. (ed.) (1983). *Small Community Water Supplies.*. Intl. Ref. Centre of Community Water Supplies. New York: Wiley.

Khastagir, A. and Jayasuriya, N. (2010). Optimal sizing of rain water tanks for domestic water conservation. *Journal of Hydrology* 381 (3–4): 181–188. https://doi.org/10.1016/j.jhydrol.2009.11.040.

Kisakye, V., Akurut, M., and Van der Bruggen, B. (2018). Effect of climate change on reliability of rainwater harvesting systems for Kabarole District, Uganda. *Water (Switzerland)* 10 (71): 1–34. https://doi.org/10.3390/w10010071.

Klemes, V. (1979). Storage mass-curve analysis in a systems-analytic perspective. *Water Resources Research* 15 (2): 359–370.

Klemes, V. (1987). One hundred years of applied storage reservoir theory. *Water Resources Management* 1: 159–175.

Latham, B.G. (1983) *Rainwater Collection Systems: The Design of Single-Purpose Reservoirs*, MSc Thesis, Department of Civil Engineering, University of Ottawa, Canada.

Lee, K.T., Lee, C.D., Yang, M.S., and Yu, C.C. (2000). Probabilistic design of storage capacity for rainwater cistern systems. *Journal of Agricultural Engineering Research* 77 (3): 343–348. https://doi.org/10.1006/jaer.2000.0597.

Liaw, C. and Chiang, Y. (2014). Dimensionless analysis for designing domestic rainwater harvesting systems at the regional level in Northern Taiwan. *Water (Switzerland)* 6: 3913–3933. https://doi.org/10.3390/w6123913.

Liaw, C. and Tsai, Y. (2004). Optimum storage volume of rooftop rain water harvesting systems for domestic use. *Journal of the American Water Resources Association* 40 (4): 901–912. https://doi.org/10.1111/j.1752-1688.2004.tb01054.x.

Linsley, R.K. and Franzini, J.B. (1979). *Water Resources Engineering*. McGraw Hill.

Lopes, V.A.R., Marques, G.F., Dornelles, F., and Medellin-Azuara, J. (2017). Performance of rainwater harvesting systems under scenarios of non-potable water demand and roof area typologies using a stochastic approach. *Journal of Cleaner Production* 148: 304–313. https://doi.org/10.1016/j.jclepro.2017.01.132.

Loucks, D.P. and van Beek, E. (2017). *Water Resource Systems Planning and Management: An Introduction to Methods, Models, and Applications*. Springer, Deltares and UNESCO-IHE https://doi.org/10.1007/978-3-319-44234-1_1.

McMahon, T.A. and Adeloye, A.J. (2005). *Water Resources Yield*. Water Resources Publications. USA.

McMahon, T.A. and Mein, R. (1978). *Reservoir Capacity and Yield*. Elsevier.

McMahon, T.A., Pegram, G.G.S., Vogel, R.M., and Peel, M.C. (2007). Revisiting reservoir storage – yield relationships using a global streamflow database. *Advances in Water Resources* 30: 1858–1872. https://doi.org/10.1016/j.advwatres.2007.02.003.

Mitchell, V.G. (2007). How important is the selection of computational analysis method to the accuracy of rainwater tank behaviour modelling? *Hydrological Processes* 21: 2850–2861. https://doi.org/10.1002/hyp.

Moran, P.A.P. (1959). *The Theory of Storage*. Methuen & Co. Ltd., London. UK.

Ndiritu, J.G. (1992) *Comparison of Gould's Probability Matrix and the Behaviour Analysis method of Reservoir Capacity Determination*, MSc thesis, Department of Civil Engineering, University of Nairobi, Kenya.

Ndiritu, J.G. (2005). Maximising water supply system yield subject to multiple reliability constraints via simulation-optimisation. *Water SA* 31 (4): 423–434. https://doi.org/10.4314/wsa.v31i4.5133.

Ndiritu, J.G. and Sinha, P. (2009). A parsimonious trigonometric model of reservoir operating rules. In: *Proceedings, International Water, Environment, Energy and Society (WEES) Conference*, vol. 2, 1025–1033. New Delhi, India: Allied Publishers.

Ndiritu, J., Odiyo, J.O., Makungo, R. et al. (2011). Yield-reliability analysis for rural domestic water supply from combined rainwater harvesting and run-of-river abstraction. *Hydrological Sciences Journal* 56 (2): 238–248. https://doi.org/10.1080/02626667.2011.555766.

Ndiritu, J.G., McCarthy, S., and Tshirangwana, N. (2014). Probabilistic assessment of the rainwater harvesting potential of schools in South Africa. *Proceedings of the International Association of Hydrological Sciences* 364: 435–440. https://doi.org/10.5194/piahs-364-435-2014.

Ndiritu, J., Moodley, Y., and Guliwe, M. (2017a). Generalized storage-yield-reliability relationships for analysing shopping Centre rainwater harvesting systems. *Water (Switzerland)* 9 (10) https://doi.org/10.3390/w9100771.

Ndiritu, J., Odiyo, J., Makungo, R. et al. (2017b). Development of probabilistic operating rules for Hluhluwe Dam, South Africa. *Physics and Chemistry of the Earth* 100: 343–352.

Ndiritu, J., Ilemobade, A., and Kagoda, P. (2018). Guidelines for rainwater harvesting system design and assessment for the city of Johannesburg , South Africa. *Proceedings of the International Association of Hydrological Sciences* 379: 409–414. https://doi.org/10.5194/piahs-379-409-2018.

Otieno, F. and Ndiritu, J. (1997). The effect of serial correlation on reservoir capacity using the modified Gould's probability matrix method (MGPM). *Water SA* 23 (1): 63–70.

Palla, A., Gnecco, I., and Lanza, L.G. (2011). Non-dimensional design parameters and performance assessment of rainwater harvesting systems. *Journal of Hydrology* 401 (1–2): 65–76. https://doi.org/10.1016/j.jhydrol.2011.02.009.

Palla, A., Gnecco, I., Lanza, L.G., and La Barbera, P. (2012). Performance analysis of domestic rainwater harvesting systems under various European climate zones. *Resources, Conservation and Recycling* 62: 71–80. https://doi.org/10.1016/j.resconrec.2012.02.006.

Prairie, J., Nowak, K., Rajagopalan, B. et al. (2008). A stochastic nonparametric approach for streamflow generation combining observational and paleoreconstructed data. *Water Resources Research* 44: 1–11. https://doi.org/10.1029/2007WR006684.

Pretto, P.B., Chiew, F.H.S., McMahon, T.A., and Vogel, R.M. (1997). The (mis) behavior of behavior analysis storage estimates. *Water Resources Research* 33 (4): 703–709.

Raudkivi, A.J. (1979). *Hydrology. An Advanced Introduction to Hydrological Processes and Modelling*. Pergamon Press. UK.

Rippl, W. (1883). The capacity of storage-reservoirs for water-supply. In: *Minutes, Proceedings*, vol. 71, 270–278. Institute of Civil Engineers.

Sample, D.J., Liu, J., and Wang, S. (2013). Evaluating the dual benefits of rainwater harvesting systems using reliability analysis. *Journal of Hydrologic Engineering, ASCE* 18 (10): 1310–1321. https://doi.org/10.1061/(ASCE)HE.1943-5584.0000748.

Saunders, A. (2012) *Shopping Centre Water Efficiency*. Report Prepared by HFM Asset Management Perth, Australia.

Schiller, E.J. and Latham, B.G. (1992). A generalized method for designing rainwater collectors. *Canadian Water Resources Journal* 17 (2): 121–127. https://doi.org/10.4296/cwrj1702121.

Schulze, R.E. (2012) A 2011 *Perspective on Climate Change and the South African Sector*. Report No. TT518/12, Water Research Commission, South Africa.

Seo, Y., Choi, N.J., and Park, D. (2012). Effect of connecting rain barrels on the storage size reduction. *Hydrological Processes* 26 (23): 3538–3551. https://doi.org/10.1002/hyp.8430.

Shaw, E.M., Beven, K.J., Chappell, N.A., and Lamb, R. (2011). *Hydrology in Practice*, 4e. Spon Press. UK.

Shi, F., Fang, K., Xu, C., and Borgaonkar, H. (2016). Interannual to centennial variability of the South Asian summer monsoon over the past millennium. *Climate Dynamics* 49: 2803–2814. https://doi.org/10.1007/s00382-016-3493-9.

Srikanthan, R. and McMahon, T.A. (1985a). Gould's probability matrix method 1. The strating month problem. *Journal of Hydrology* 77: 125–133.

Srikanthan, R. and McMahon, T.A. (1985b). Gould's probability matrix method 2. The annual autocorrelation problem. *Journal of Hydrology* 77: 135–139.

Srikanthan, R. and McMahon, T.A. (2001). Stochastic generation of annual, monthly and daily climate data: a review. *Hydrology and Earth System Sciences* 5 (4): 653–670.

Su, M.D., Lin, C.H., Chang, L.F. et al. (2009). A probabilistic approach to rainwater harvesting systems design and evaluation. *Resources, Conservation and Recycling* 53 (7): 393–399. https://doi.org/10.1016/j.resconrec.2009.03.005.

Taffere, G.R., Beyene, A., Vuai, S.A.H. et al. (2016). Reliability analysis of roof rainwater harvesting systems in a semi-arid region of sub-Saharan Africa: case study of Mekelle, Ethiopia. *Hydrological Sciences Journal* 61 (6): 1135–1140. https://doi.org/10.1080/02626667.2015.1061195.

Tan, Q. and Xu, X. (2014). Comparative analysis of spatial interpolation methods: an experimental study. *Sensors and Transducers* 165 (2): 155–163.

Taylor, P., Feki, H., Slimani, M., and Cudennec, C. (2012). Incorporating elevation in rainfall interpolation in Tunisia using geostatistical methods. *Hydrological Sciences Journal* 57 (7): 1294–1314. https://doi.org/10.1080/02626667.2012.710334.

Thomas, R.B., Kirisits, M.J., Lye, D.J., and Kinney, K.A. (2014). Rainwater harvesting in the United States: a survey of common system practices. *Journal of Cleaner Production* 75: 166–173. https://doi.org/10.1016/j.jclepro.2014.03.073.

Tozer, C.R., Vance, T.R., Roberts, J.L. et al. (2016). An ice core derived 1013-year catchment-scale annual rainfall reconstruction in subtropical eastern Australia. *Hydrological Sciences Journal* 20: 1703–1717. https://doi.org/10.5194/hess-20-1703-2016.

Tu, M., Hsu, N., Tsai, F.T., and Yeh, W.W. (2008). Optimization of hedging rules for reservoir operations. *Journal of Water Resources Planning and Management, ASCE* 134 (1): 3–13.

Vu, T.M., Mishra, A.K., Konapala, G., and Liu, D. (2018). Evaluation of multiple stochastic rainfall generators in diverse climatic regions. *Stochastic Environmental Research and Risk Assessment* 32 (5): 1337–1353. https://doi.org/10.1007/s00477-017-1458-0.

Wang, H. and Liu, J. (2013). Reservoir operation incorporating hedging rules and operational inflow forecasts. *Water Resources Management* 27 (5): 1427–1438. https://doi.org/10.1007/s11269-012-0246-3.

Ward, S.L. (2010) *Rainwater harvesting in the UK: a strategic framework to enable transition from novel to mainstream.* PhD Thesis, University of Exeter, UK.

Ward, S., Memon, F.A., and Butler, D. (2012). Performance of a large building rainwater harvesting system. *Water Research* 46 (16): 5127–5134. https://doi.org/10.1016/j.watres.2012.06.043.

Weibull, W. (1951). A statistical distribution function of wide applicability. *Journal of Applied Mechanics* https://doi.org/citeulike-article-id:8491543.

Youn, S.G., Chung, E.S., Kang, W.G., and Sung, J.H. (2012). Probabilistic estimation of the storage capacity of a rainwater harvesting system considering climate change. *Resources, Conservation and Recycling* 65: 136–144. https://doi.org/10.1016/j.resconrec.2012.05.005.

van Zyl, H.J., Ilemobade, A.A., van Zyl, J.E., and Le Gat, Y. (2017). Statistical characterisation and estimation of non-domestic water demand. *Urban Water Journal* 14 (7): 720–726.

27

Towards Developing Generalized Equations for Calculating Potential Rainwater Savings

Monzur A. Imteaz[1], Muhammad Moniruzzaman[1], and Abdullah Yilmaz[2]

[1] Swinburne University of Technology, Melbourne, VIC, Australia
[2] La Trobe University, Bendigo, VIC, Australia

27.1 Introduction

Rainwater tanks (RWTs) have been in use for centuries, mainly in remote areas where other suitable source(s) of water is limited or scarce. Even remote communities, after some preliminary treatments, use this harvested water for potable purposes. However, in the past this alternative water supply never got the attention of urban residents/authorities, where there exists a piped water supply network or other convenient alternative source(s). As such, optimization of tank size was not a crucial factor, as users installed RWTs in a large block of land, where users had plenty of remaining spaces for other purposes. Recently, with the pressure of ever-increasing populations in urban areas and consequently water demands, stressed with depletion and uncertainty of some existing water sources, water authorities and residents in urban areas also started installing RWTs, mainly to use such water for non-potable purposes and to reduce the total demand of potable water. Some developed nations adopted gray water reuse and wastewater recycling, both of which have safety concerns and as such require higher-level treatments, which consequently become expensive and non-feasible in many cases. In comparison to these expensive and safety-concerned approaches, rainwater harvesting is easy, cheaper, and requires minimal treatment if used for non-potable purposes. Many water authorities around the world are now imposing mandatory RWT installations with every new building; e.g. in Bangladesh the authority has made it mandatory that RWTs are to be installed for every proposed residential building to be built on plots having an area above 300 m^2 (BNBC 2014). Some other authorities (e.g. BASIX in New South Wales, Australia) assign scores toward different sustainable features including RWTs to be included in new buildings, and the owners/developers are required to achieve a minimum total score to get government approval for the proposed building (BASIX 2018).

While installing RWTs in urban areas, where space is restricted and expensive; and residents do not want to dedicate big space for such RWT, size optimization has become a crucial factor. The selection of optimum RWT size is a major factor to increase the efficiency of a rainwater harvesting system, as well as to make the RWT installations economically feasible through reducing the expected payback period (Mun and Han 2012). Rahman et al. (2016) have reported that with the water price at the time of the study installing RWTs is not financially viable for Australian arid regions as the benefit–cost ratio became much smaller than 1.0, although the same study reported that with a 20 kl tank a reliability of 59–98% can be achieved for toilet and laundry use depending on the location within Australia.

Due to the lack of confidence on the economic benefit of RWTs, many users are reluctant to install this sustainable feature; e.g. a survey from the Australian Bureau of Statistics (ABS) showed that about 47% of the population of Melbourne showed reluctance in adopting a rainwater harvesting system (Rahman et al. 2012). With this emerging factor of space limitation, there have been numerous studies on RWTs, mainly focusing on water savings potential, reliability, and economic benefit. Few studies were also conducted on the quality of harvested water through RWTs (Rahman et al. 2014; Despins et al. 2009). However, as in an urban setting it is not expected that the harvested rainwater will be used for potable purposes, the quality of such water is not a crucial factor.

Handbook of Water Harvesting and Conservation: Basic Concepts and Fundamentals, First Edition. Edited by Saeid Eslamian and Faezeh Eslamian.

27.2 State of the Art

Among the wide range of studies on water savings potential, Mehrabadi et al. (2013), through a study conducted for three climatic conditions in Iran, claimed that at least 75% of the non-potable water demand could be satisfied for typical residential buildings, and this can be achieved for almost about 70% of the year by storing rainwater from larger roofs (180–240 m^2). However, for smaller roof areas (60–120 m^2), similar savings can be achieved for only 45% of the time of the year. Ghisi et al. (2007) evaluated rainwater savings potentials from residential RWTs for 195 cities located in southeastern Brazil. They have found that rainwater savings potentials widely varies (12–79%) depending on locality, demand, roof area, and tank size. They have concluded that optimum tank size has to be determined for each locality and dwelling, as it depends on other factors i.e. roof area and demand as well. Domenech and Saurí (2011), through a case study in Barcelona, Spain, concluded that despite low precipitation inputs and a high variability of precipitation, daily toilet flushing demand of a single-family house could be practically met even with a relatively small tank. Roof-collected rainwater can also meet more than 60% of the landscape irrigation demand in both single and multi-family buildings. Imteaz et al. (2015), through a case study for Adelaide, Australia, showed that for a wider range of climatic conditions (dry and wet years) in a single household with a 10 kl tank an annual water savings within a range of 80–90 kl is achievable, whereas with a 5 kl such savings would be within the range of 72–76 kl per year.

In the case of a large building (i.e. roof), and large RWT, in general the achievable efficiency is higher, as the higher stored amount can be easily used by the users, even if some users are not using it for few days, causing a lower/no loss due to overflow during continuous rainy days. Ward et al. (2012) investigated the performance of a rainwater harvesting system for a large office building in UK, through monitoring the actual performance of the system during a monitoring period of eight months. The results indicated that about 87% water-saving efficiency could be achieved for the office-based RWH system; even a similar efficiency can be achieved with a smaller tank compared to the one actually installed. Matos et al. (2015) investigated financial benefits of a large RWT connected with a large roof (36 870 m^2) of a shopping center in North Portugal. They have revealed that with a discount rate of 10% the cost of the tank can be recovered within two to six years depending on weather condition. Imteaz et al. (2011), through a case study on a university campus in Melbourne, Australia, presented that with a large tank of 110 m^3 connected with buildings having a gross roof area of 2491 m^2, payback periods in the range of 15–21 years can be achieved depending on climatic condition and future water price increase rate. These comparatively higher payback periods can be attributed to higher construction (including labor) costs and lower water price in Australia. The study of Rahman et al. (2012) supported the same hypothesis, which revealed that without government rebate household RWTs of up to 5000 m^3 are not really cost-effective for the city of Sydney, where the amount of rainfall is much higher than Melbourne.

From users' perspective, a critical factor of rainwater benefit is the reliability of intended supply, which is defined as the percentage of number days in a year when rainwater is able to supply the intended demand. In many cases (especially in urban areas), there is augmented alternate supply attached with the uses, when there is no water in the RWT; however users would prefer a higher reliability to maximize their benefits. Imteaz et al. (2012), through a case study for residential RWTs in Nigeria, presented that a 100% reliability is achievable with a 7 m^3 tank for a low demand (1.8 m^3 per month); however for a higher demand (2.45 m^3 per month) a tank size of 10 m^3 is required to achieve a 100% reliability. This great reliability is only achievable for a low rainwater use location; for developed countries/nations, household water demand is much higher, which causes a lower reliability of RWTs. Imteaz et al. (2013), through a case study for Melbourne, has presented that with a roof of 100 m^2 a reliability of 100% is not achievable even with a tank size of 10 m^3, which considered a typical non-potable water demand of 9 m^3/month. They have also presented the climatic and spatial variabilities of reliabilities for the city of Melbourne; with a roof of 100 m^2 and a tank size of 10 m^3, reliabilities vary 23–36% for a dry year and 46–74% in a wet year within different locations in Melbourne. Under the similar scenario with a roof of 200 m^2 within the same locations, the expected reliabilities vary 53–72% for a dry year and 84–100% in a wet year. Karim et al. (2015), through a case study in Dhaka, Bangladesh, have revealed that reliabilities of only 15–25% can be achieved under the wet climatic condition and roof sizes varying from 140 to 200 m^2, which were considered typical roof sizes of existing multi-story flat buildings. Such low reliabilities can be attributed to consideration of very high demand of such multi-story building, where typical populations in a building varies from 30 to 50, with an average water demand of 150 l per capita/day.

All the aforementioned and most other studies were conducted only for a particular locality/city/country. Such studies are only applicable for the specific region/location the study was conducted for, outcomes of which cannot be translated/transferred to other city, as amount and distribution of rainfalls are quite different for different cities. Some authors presented such investigations for multiple countries/cities; e.g. Souza and Ghisi (2012) for 12 cities around the world and Ghisi et al. (2007) for 195 cities located in southeast Brazil. Nonetheless, finding(s) from each city is only representative for that particular city. Even if the annual rainwater amount is same for two cities, findings cannot be transferred among the cities as the rainfall distributions are likely to be different. A generalization of such findings is at large, which requires similar types of studies to be conducted for each and every locality of world, where RWTs are used and such approach is not an efficient approach. With the aim of generalizing the potentials of rainwater savings, this study considers "seasonality index" (Eq. 27.1) as a representation of rainfall distribution pattern, in addition to the total annual rainfall, which is the salient factor of rainwater saving potential.

$$SI = \frac{1}{R}\sum_{j=1}^{12}\left|X_j - \frac{R}{12}\right| \tag{27.1}$$

where, SI is the Seasonality Index, R is the mean annual rainfall, and X_j is the mean monthly rainfall for month j.

It is to be noted that as a first step of developing a global generalized equation of potential rainwater savings, this study proposes a generalized equation considering annual rainfall and seasonality index for a particular tank size, roof size, and rainwater demand. Other studies have earlier developed generalized equations of rainwater savings considering tank size, roof size, and rainwater demand for particular cities, i.e. for Sydney (Moniruzzaman and Imteaz 2017) and for Adelaide (Imteaz et al. 2016b). For the development of an ultimate generalized equation, a robust analysis incorporating all these variables will be a theme of future study.

27.3 Methodology

To achieve generalization among different cities, six major cities were selected from Australia: Sydney, Melbourne, Adelaide, Perth, Brisbane, and Darwin. Australian cities were selected due to the availability of daily good quality data and numerous uses of such data in the recent studies (Imteaz et al. 2017a, 2016a, 2016b, 2014; Khastagir and Jayasuriya 2011). From each city, an average year having no missing data was selected for further analysis. An earlier developed daily water balance model, eTank (Imteaz et al. 2017b), was used to calculate water savings potentials under different scenarios, i.e. for different combinations of roof area, tank volume, rainwater demand, and daily rainfall amounts for different locations. It is to be noted here that eTank has the capability to calculate potential water savings simultaneously for three different climatic conditions (i.e. dry, average, and wet). However, for this study, as such climatic variability investigation was not the objective, water savings were calculated for an average year only for all the selected cities. For each of the selected cities, a typical average year was selected based on mean annual rainfall amount and having no missing data.

To assess the effect of total annual rainfall amount, daily rainfall data from each location was multiplied with proper multiplication factors to generate six sets of daily rainfall data having the same seasonality index for a particular city, but different annual rainfall amounts representing rainfall amounts of six selected cities. For example, if annual rainfall amounts (in average years) of six cities are I_A, I_B, I_C, I_D, I_E and I_F, then multiplication factors for "city A" would be, I_B/I_A, I_C/I_A, I_D/I_A, I_E/I_A, and I_F/I_A. Similarly, multiplication factors for other cities were also selected. As for Darwin, the rainfall data selected was for 1981 from Stokes Hill station, which was having annual rainfall amount of 1611 mm. However, as Darwin's mean annual rainfall is 1728 mm, the daily rainfall data from the selected station (and year) was multiplied to match the total amount to 1728 mm. As such, 36 sets of daily rainfall data were generated having six different seasonality indices and six different annual rainfalls. For the current study, only annual water savings were assessed, although eTank calculates reliability, augmented water supply used, and overflow amounts, in addition to cost analysis providing payback period of tank installation and maintenance costs. Water savings were calculated for different combinations of tank sizes, roof areas, and demands. However, being a pilot study for such generalization, for simplicity water savings were correlated with "seasonality index" for a particular scenario; tank size 5 kl, rainwater demand 200 l/day, and roof size 100 m^2.

In addition to expressing annual water savings, efficiencies of water savings were also calculated, as in a previous study (Jenkins 2007) "seasonality indices" of different Australian cities were correlated with water savings efficiencies calculated through a daily water balance model simulated for a long span of available data. Water savings efficiency is defined as the ratio of annual water savings and annual rainfall amount. In the study by Jenkins (2007), annual water savings were calculated for several years (as per the available data length), then average water savings were calculated from all the calculated annual water savings.

27.4 Study Area and Data

Figure 27.1 shows the locations of the selected cities within Australia. Daily rainfall data was collected from the Australian Bureau of Meteorology website (www.bom.gov.au/climate/data/index.shtml).

Table 27.1 shows the selected years for the daily rainfall data, annual rainfall amount for the selected year, and station ID (as provided by the Australian Bureau of Meteorology) for each selected city. Annual water savings were calculated using eTank software applying 36 sets of daily rainfall data as mentioned in Section 27.3.

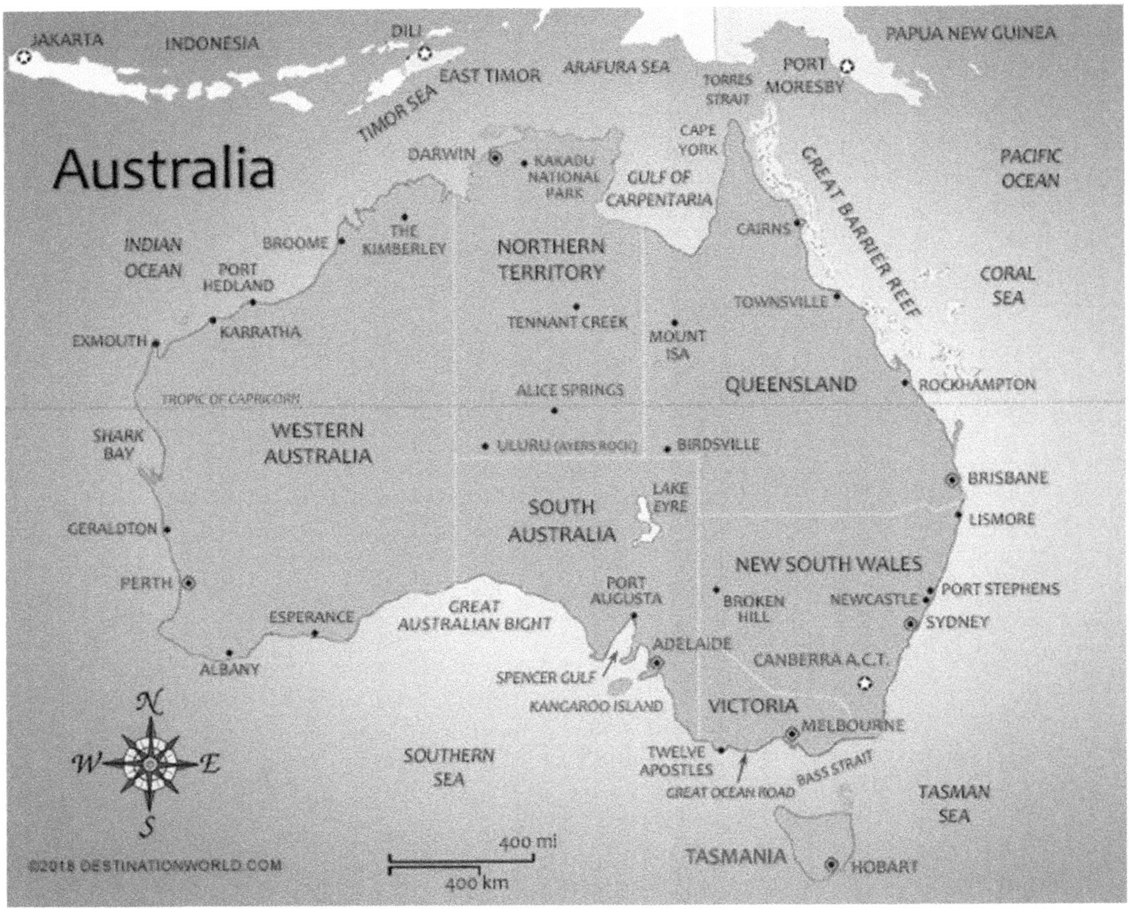

Figure 27.1 Map of Australia with the major cities.

Table 27.1 Selected rainfall stations, year, and corresponding annual rainfall amounts.

City	Station Name	Station ID	Selected year	Rainfall (mm)
Sydney	Sydney Observatory Hill	66 062	2003	1200.4
Brisbane	Brisbane regional office	40 214	1985	1140.6
Melbourne	Melbourne airport	86 282	1987	626.8
Adelaide	Adelaide airport	23 034	1987	510.4
Perth	Perth metro	9225	2002	737.8
Darwin	Stokes Hill	14 167	1981	1611

27.5 Results

Figure 27.2 shows graphs of annual water savings versus annual rainfall amounts for the six selected cities for the adopted conditions; i.e. roof area 100 m^2, tank volume 5 kl, and rainwater demand 200 l day^{-1}. From the graph it is clear that annual water savings follow a pattern with the annual rainfall amount, and patterns for all the cities are similar. With the increase of annual rainfall amount, at the lower range the water savings amount increases sharply. However, with a very high rainfall amounts, this rate of increase reduces and eventually become almost independent on annual rainfall amount. This is because of lower rainfall demand while having smaller tank size; due to smaller tank size the tank becomes full (after few consecutive rain events) with the heavy rainfall amounts and starts overflowing. As such, beyond a certain stage increase in rainfall amount does not render a higher water savings. This turning point (beyond which increase in rainfall amount does not cause increase in water savings) will shift toward higher rainfall amounts with the increase of tank size. However, in an urban residential setting, a very big tank is often difficult to achieve due to space limitations.

For this study, each selected city is represented by its "seasonality index" calculated through Eq. 27.1, using monthly rainfall data of the selected mean year. It was anticipated that the "seasonality index" will have correlations with the annual water savings, as well as rainwater saving efficiencies. Rainwater saving efficiency for each selected scenario was calculated as explained in Section 27.3. Figure 27.3 shows the graphs of calculated water savings efficiencies versus annual rainfall amounts for all the cities. It is found that with the increase of annual rainfall amounts, efficiencies decrease in all cases. This is because with the increase of rainfall amounts, tank becomes full and overflow occurs during a higher number of days, which causes efficiencies to drop.

To achieve a simple relationship between water savings efficiency and seasonality index for a particular city (i.e. particular SI) all the calculated efficiencies were averaged to get an average efficiency. Figure 27.4 shows the relationship of average rainwater saving efficiency with the seasonality index. From the figure, it is clear that the efficiency has an inverse linear relationship with the seasonality index, i.e. with the increase of seasonality index, water saving efficiency decreases. This is reasonable, as an increase in seasonality index means larger variations (from

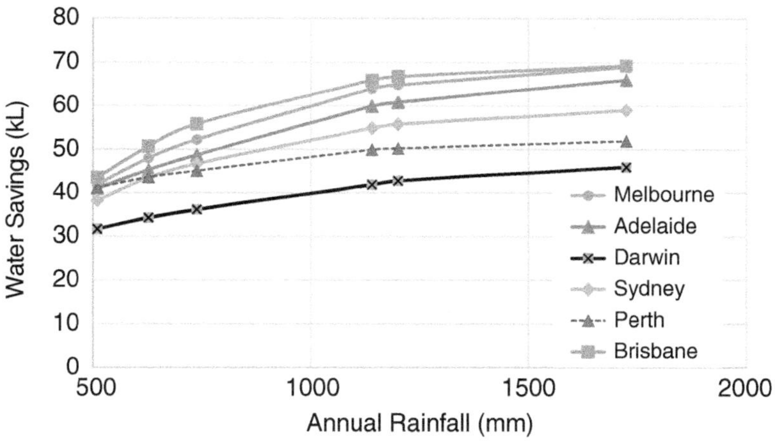

Figure 27.2 Annual water savings versus annual rainfall for six cities.

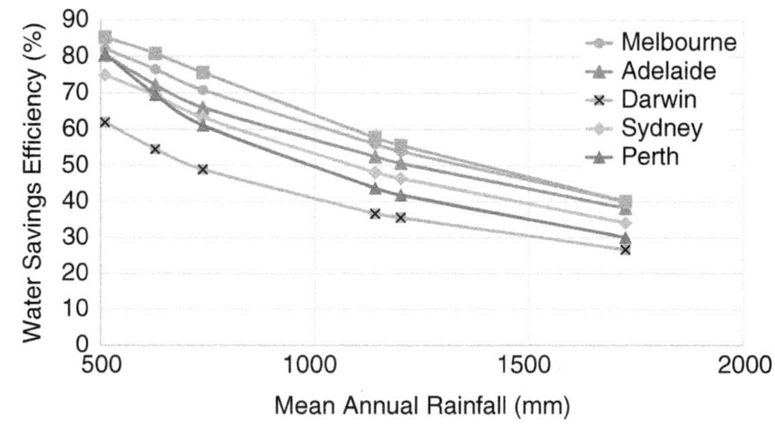

Figure 27.3 Calculated water savings efficiencies versus annual rainfall for six cities.

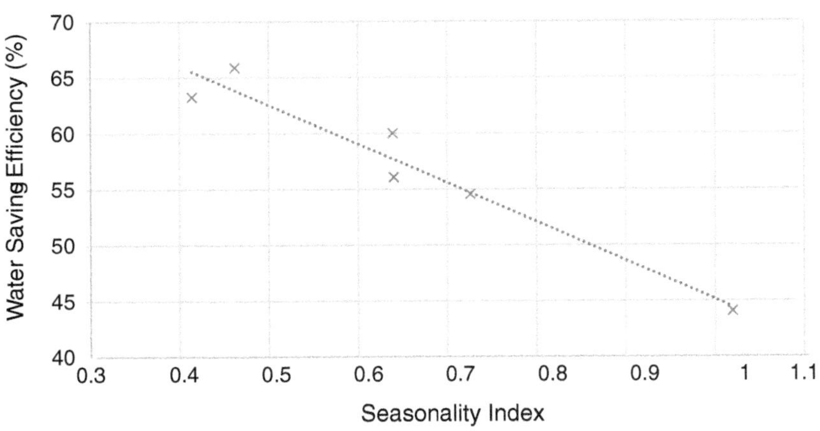

Figure 27.4 Average rainwater saving efficiency against the seasonality index.

the mean) of monthly rainfall amounts, which will cause a lower water savings and subsequently lower water-saving efficiency. The best-fit curve follows a linear pattern with a correlation coefficient (R2) of 0.94. It is to be noted here that Jenkins (2007) has presented a similar relationship with data and simulation results from 12 Australian cities, however in his study a correlation coefficient of 0.78 was achieved. The lower correlation coefficient was due to the use of 30 years of data (1961–1990), i.e. calculating water savings efficiency for each of the selected 30 years, and then taking an average of 30 years' calculated water savings. Also, the values of seasonality indices used were average values derived from the same 30 years of data. Averaging such values cause to diminish the salient feature of a particular year's rainfall time-series as well as seasonality index.

In the current study a particular year was used for a particular city, while seasonality index was calculated based on the selected year's rainfall data. Eventually, such indices were used to derive the relationship.

To assess the effect of "seasonality index" on annual water savings, calculated water savings for all the annual rainfall regimes were plotted against "seasonality indices" of all the selected cities. Figure 27.5 shows the relationships between annual water savings and seasonality index for all the annual rainfall regimes along with the best-fit line for each rainfall regime. From the figures it is clear that, similar to water savings efficiency, water savings amounts also linearly decrease with the increase of seasonality index. The coefficient of correlations for the derived best-fit equations vary from 0.80 to 0.94, while out of six equations, four equations having correlation coefficients 0.92–0.94. The reason of such inverse relationship was explained earlier for the case of water savings efficiency.

The derived best-fit equations are as follow:

For annual rainfall of 510:

$$WS = 51.01 - 17.59 * SI \tag{27.2}$$

Regression coefficient (R^2) = 0.80

For annual rainfall of 627:

$$WS = 60.53 - 25.09 * SI \tag{27.3}$$

Regression coefficient (R^2) = 0.93

For annual rainfall of 738:

$$WS = 66.80 - 29.99 * SI \tag{27.4}$$

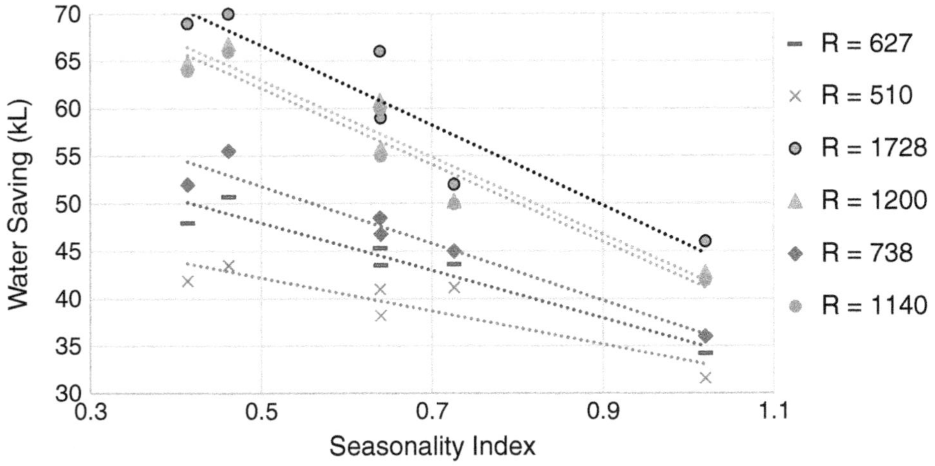

Figure 27.5 Rainwater saving efficiency against the seasonality index along with the best-fit lines ("R" in the legend represents considered annual rainfall).

Regression coefficient (R^2) = 0.94
For annual rainfall of 1140:

$$WS = 82.30 - 40.33 * SI \tag{27.5}$$

Regression coefficient (R^2) = 0.92
For annual rainfall of 1200:

$$WS = 83.19 - 40.48 * SI \tag{27.6}$$

Regression coefficient (R^2) = 0.92
For annual rainfall of 1728:

$$WS = 87.70 - 42.10 * SI \tag{27.7}$$

Regression coefficient (R^2) = 0.87
where "WS" is the annual water savings in kl and SI is the seasonality index.

Coefficients and intercepts of the above equations are expected to vary with the annual rainfall amounts. Figure 27.6 shows the relationships of coefficients and intercepts with the annual rainfall amounts. From the figure it is found that both the coefficient and intercept exhibit logarithmic relationships with the annual rainfall amount. The standard errors for the derived best-fit lines are 3.40 for the intercepts and 3.19 for the coefficients. The derived best-fit equations are:

$$\text{Coefficient} = 20.853 * \ln(R) - 109.38 \tag{27.8}$$

$$\text{Intercept} = 31.214 * \ln(R) - 140.61 \tag{27.9}$$

where R is the annual rainfall amount in mm.

Equations (27.2–27.9) can be combined to a single generalized equation as follows:

$$WS = (31.214 * \ln(R) - 140.61) - (20.853 * \ln(R) - 109.38) * SI \tag{27.10}$$

For the considered tank scenario (i.e. roof area 100 m^2, tank volume 5 kl and rainwater demand 200 l day^{-1}). Equation (27.10) can be used to calculate expected water savings for any annual rainfall amounts and any SI.

27.6 Conclusions

With the aim of developing a robust generalized equation for expected water savings applicable for any place/locality considering all the associated parameters related to rainwater savings, as a first step a generalized equation is proposed which is applicable for any annual rainfall amount and any SI for a particular tank scenario (i.e. tank volume, roof size, and rainwater demand). For the establishment of such equation, six Australian cities were selected due to data quality and easy access to daily rainfall data. To evaluate the effect of SI, six cities having different SI values were selected. Again, to evaluate the effect of annual rainfall amount, data from each city was multiplied with different multiplication factors to generate six sets of data (having different annual rainfall amounts) having same SI. In total, 36 sets of daily rainfall data were used to develop the generalized equation.

To assess the preliminary relationship, model calculated water savings for all the 36 scenarios were plotted against annual rainfall amounts. As expected, water savings increased with the increase of annual water amount; the rates of increases are higher at the lower annual rainfalls, however rates of increases decrease for higher annual rainfall amounts. During very high rainfall amounts, there are more days when tank gets full and overflow occurs, causing a reduced amount of increment in the total water saving. This phenomenon is further ascertained by calculating water savings efficiencies (ratio of water savings to total rainfall) for all the cases. It is found that for all the cases water savings efficiency decrease with the increase of annual rainfall. For each city, individual pattern of

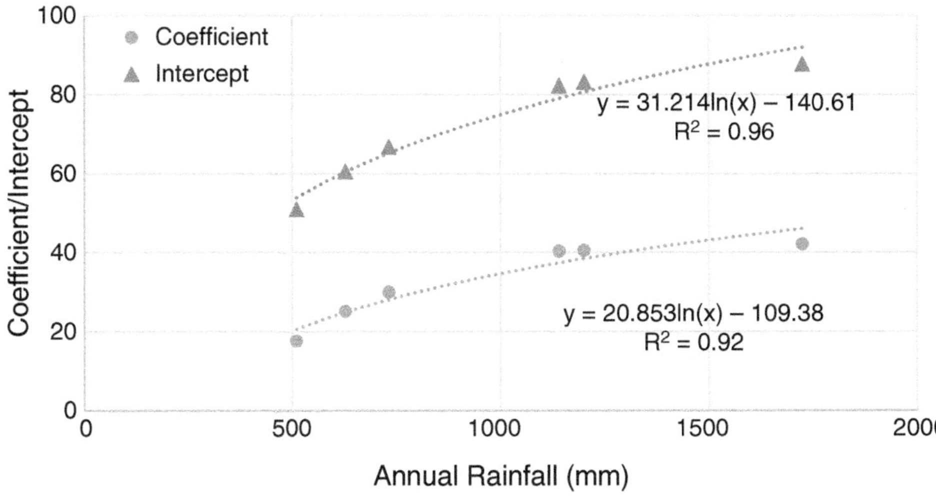

Figure 27.6 Coefficients and intercepts with the annual rainfall amounts.

such relationship is exponential. However, considering average values for each city, tanking an average of all the efficiencies for different rainfall amounts for a particular city (having a particular SI), it is found that the average efficiency linearly decreases with the increase of SI. Similarly, annual water savings amounts also linearly decrease with the increase of SI for all the studied annual rainfall amounts. This finding provided an avenue to develop a generalized equation of water savings, which can be used for any region, where the region is represented by the SI value from the rainfall pattern.

Six linear equations (one for each studied rainfall amount) were derived from the best-fit lines of the relationships between water savings and SI. Both the coefficients and intercepts of the derived equations were correlated with the annual rainfall amounts through logarithmic equations, which eventually yielded a single equation for water savings depending on annual rainfall amount and SI of the locality/region. The derived equation is valid for a particular RWT input scenario. As mentioned earlier, Moniruzzaman and Imteaz (2017) have presented generalized equation for water savings having independent variables of tank size, rainwater demand, and roof area for a particular locality. Similar technique can be applied and amalgamated with the relationship presented in this chapter, which will lead to a generalized equation valid for any locality, and any tank size, roof area, and rainwater demand. This will be a part of future study.

Acknowledgement

The authors acknowledge Tuition Fee Scholarship under Australian Government's Research Training Scheme for the second author, which enabled production of outcomes presented in this chapter.

References

Bangladesh National Building Code (2014). *Housing and Building Research Institute*. Bangladesh: Ministry of Housing and Public Works.

Building Sustainability Index (2018) http://www.basix.nsw.gov.au/iframe/getting-started/using-the-basix-assessment-tool.html, accessed on 13 March, 2018.

Despins, C., Farahbakhsh, K., and Leidl, C. (2009). Assessment of rainwater quality from rainwater harvesting systems in Ontario, Canada. *Journal of Water Supply: Research and Technology—AQUA* 58 (2): 117–134.

Domenech, L. and Saurí, D. (2011). A comparative appraisal of the use of rainwater harvesting in single and multi-family buildings of the metropolitan area of Barcelona (Spain): social experience, drinking water savings and economic costs. *Journal of Cleaner Production* 19 (6): 598–608.

Ghisi, E., Bressan, D.L., and Martini, M. (2007). Rainwater tank capacity and potential for potable water savings by using rainwater in the residential sector of southeastern Brazil. *Building and Environment* 42: 1654–1666.

Imteaz, M.A., Shanableh, A., Rahman, A., and Ahsan, A. (2011). Optimisation of rainwater tank design from large roofs: a case study in Melbourne, Australia. *Resources, Conservation and Recycling* 55 (11): 1022–1029.

Imteaz, M.A., Adeboye, O., Rayburg, S., and Shanableh, A. (2012). Rainwater harvesting potential for Southwest Nigeria using daily water balance model. *Resources, Conservation and Recycling* 62: 51–55.

Imteaz, M.A., Ahsan, A., and Shanableh, A. (2013). Reliability analysis of rainwater tanks using daily water balance model: variations within a large city. *Resources, Conservation and Recycling* 77: 37–43.

Imteaz, M.A., Matos, C., and Shanableh, A. (2014). Impacts of climatic variability on rainwater tank outcomes for an inland city, Canberra. *International Journal of Hydrology Science and Technology* 4 (3): 177–191.

Imteaz, M.A., Paudel, U., Ahsan, A., and Santos, C. (2015). Climatic and spatial variability of potential rainwater savings for a large coastal city. *Resources, Conservation and Recycling* 105: 143–147.

Imteaz, M.A., Sagar, K., Santos, C., and Ahsan, A. (2016a). Climatic and spatial variations of potential rainwater Savings for Melbourne. *International Journal of Hydrology Science and Technology* 6 (1): 45–61.

Imteaz, M.A., Paudel, U., Matos, C., and Ahsan, A. (2016b). Generalized equations for rainwater tank outcomes under different climatic conditions: a case study for Adelaide. *International Journal of Water* 10 (4): 301–314.

Imteaz, M.A., Moniruzzaman, M., and Karim, M.R. (2017a). Rainwater tank analysis tools, climatic and spatial variability: a case study for Sydney. *International Journal of Water* 11 (3): 251–265.

Imteaz, M.A., Karki, R., Shamseldin, A., and Matos, C. (2017b). eTank and contemporary online tools for rainwater tank outcomes analysis. *International Journal of Computer Aided Engineering and Technology* 9 (3): 372–384.

Jenkins, G. (2007). Use of continuous simulation for the selection of an appropriate urban rainwater tank. *Australian Journal of Water Resources* 11 (2): 231–246.

Karim, M.R., Bashar, M.Z.I., and Imteaz, M.A. (2015). Reliability and economic analysis of urban rainwater harvesting in a megacity in Bangladesh. *Resources, Conservation and Recycling* 104: 61–67.

Khastagir, A. and Jayasuriya, N. (2011). Investment evaluation of rainwater tanks. *Water Resources Management* 25 (14): 3769–3784.

Matos, C., Santos, C., Bentes, I. et al. (2015). Economic analysis of a rainwater harvesting system in a commercial building. *Water Resources Management* 29 (11): 3971–3986.

Mehrabadi, M.H.R., Saghafian, B., and Haghighi Fashi, F. (2013). Assessment of residential rainwater harvesting efficiency for meeting non-potable water demands in three climate conditions. *Resources, Conservation and Recycling* 73: 86–93.

Moniruzzaman, M. and Imteaz, M.A. (2017). Generalized equations, climatic and spatial variabilities of potential rainwater savings: a case study for Sydney. *Resources, Conservation & Recycling* 125: 139–156.

Mun, J.S. and Han, M.Y. (2012). Design and operational parameters of a rooftop rainwater harvesting system: definition, sensitivity and verification. *Journal of Environmental Management* 93 (1): 147–153.

Rahman, A., Keane, J., and Imteaz, M.A. (2012). Rainwater harvesting in greater Sydney: water savings, reliability and economic benefits. *Resources, Conservation and Recycling* 61: 16–21.

Rahman, S., Khan, M.T.R., Akib, S. et al. (2014). Sustainability of rain water harvesting system in terms of water quality: a case study. *The Scientific World Journal* 2: 10–17.

Rahman, A., Hajani, E., and Eslamian, S. (2016). Rainwater harvesting in arid regions of Australia. Chapter 26. In: *Handbook of Drought and Water Scarcity: Environmental Impacts and Analysis of Drought and Water Scarcity* (eds. S. Eslamian and F. Eslamian), 489–500. USA: CRC Press.

Souza, E.L. and Ghisi, E. (2012). Potable water savings by using rainwater for non-potable uses in houses. *Water* 4 (3): 607–628.

Ward, S., Memon, F.A., and Butler, D. (2012). Performance of a large building rainwater harvesting system. *Water Research* 46 (16): 5127–5134.

Part I

Sustainable Water Harvesting and Conservation in a Changing Climate

28

Water Harvesting, Climate Change, and Variability

Jew Das[1], *Manish Kumar Goyal*[1], *and N.V. Umamahesh*[2]

[1] *Discipline of Civil Engineering, Indian Institute of Technology, Indore, India*
[2] *Department of Civil Engineering, National Institute of Technology, Warangal, India*

28.1 Introduction

According to the United Nations World Water Development Report (WWDR) (WWAP 2015), the global water demand is significantly influenced by population growth, urbanization, and the security policies for food and energy in the developing countries. The availability of water resources varies in different areas of the globe due to the unpredictable distribution of precipitation and runoff. In addition, due to a combination of factors including climate change, inefficient water supply and distribution systems, and water pollution, an increase in the population will aggravate the distribution and availability of water resources, which in turn will affect the per capita water availability (Eslamian and Okhravi 2015). Based on the Food and Agriculture Organization (FAO) AQUASTAT dataset (2018), most of the developing countries come under the vulnerability to scarcity zone according to the per capita water availability. Due to the variability in the surface water, groundwater use has augmented and interestingly, India, China, Nepal, and Pakistan extract nearly half the world's total groundwater (WWAP 2015). In this sense, it is understood that developing countries are going to be affected by climate change and its variability. To minimize the adverse effect of water scarcity, water harvesting is one of the adaptive measures and in the present study, India, as a developing country, is considered to evaluate the possibility of water harvesting under climate change and its variability. The water scarcity not only influences the water resources of the country but it also has a significant impact on the social and cultural aspects of mankind.

The inherent vagueness and adversity of climate change bring forward the complex and extreme phenomena like floods, cyclones, high and intense precipitation, heat waves, droughts, etc., and challenges people to search for an appropriate mitigation and adaptation strategy as the climate and culture are interlinked (Pandey et al. 2003). The occurrence of extreme climate events results in population dislocation and migration to safer and productive localities (deMenocal and Peter 2001; Núñez et al. 2002). In addition, human migration as a survival strategy to climate change variability is noticed in various parts of the globe, including Africa, Australia, and South America (deMenocal and Peter 2001; Polyak and Asmerom 2001; Núñez et al. 2002; Tyson et al. 2002; Bowler et al. 2003). Verschuren et al. (2000) reported that based on the scientific evidences and oral lore there exists a strong association among cultural development, climate change, and water stress. However, Pandey et al. (2003) suggested to rehabilitate the dwelling environment through effective adaptation strategies to improve water harvesting as quoted in the Indian proverb, "Capture rain where it rains" (Pandey 2001). The progress toward sustainability and resilience requires a complete knowledge of climate variability and corresponding adaptation by human society. A comprehensive study regarding the widespread rainwater harvesting with respect to the climate variability in India from 4500 BCE to 1999 CE is detailed by Pandey et al. (2003) and the readers were advised to follow the article. In addition, Pandey et al. (2003) explored the evidence of correlation between the water harvesting structures and climate variability like aridity and drought conditions, and found it to be significant. Based on the historical evidence, it is worth mentioning that climate change and its variability affects all aspects of human life and hence we must work collectively together and stop procrastinating.

In India, rainfall is a finite and main source of freshwater that needs proper management for effective use. Realizing the urgency, India has a rich history of preserving surplus water, using water harvesting structures since 4500 BCE.

Handbook of Water Harvesting and Conservation: Basic Concepts and Fundamentals, First Edition. Edited by Saeid Eslamian and Faezeh Eslamian.

Water has been harvested from the rainfall directly, or by diverting the water from the flooded rivers, and the constructional design of the water harvesting structures depends on the variability of the precipitation and available resources. Therefore, India has its own water harvesting methodologies that imitate the geographical and cultural uniqueness. However, due to the improper management of the water harvesting structures, along with the alarming climate change variability (e.g. erratic rainfall), the problem has been manifested in many ways. In addition, due to cultural, economic, and political factors, the traditional water harvesting structures have become unused or have lost their significance in the present scenario (Pandey et al. 2003).

Water availability has significant impacts in many parts of the India, which in turn puts immense pressure on the fragile agricultural system of the country. Therefore, it is necessary to assess climate change variability to rejuvenate the existing water harvesting structures and evaluate the possibility of water harvesting to meet the demands of the communities for the present and possible future climate change scenarios. Though climate change is a global phenomenon, it has varying degrees of regional impacts (Goyal and Rao 2018). Although there are various schools of thought relating to the different climate change causes, viz. greenhouse gases emissions, aerosols, changes in albedo, and solar irradiance, climate change is real and happening right now. In the prevailing adverse consequences of climate change and its variability, it is indispensable to evaluate and analyze the change ability of available water resources for better management practices and adaptation strategies. Climate change as a potential "risk multiplier" has significant impact on developing countries, where the economical and societal development are closely associated with the agriculture and water resources (Pande et al. 2014). Within this context, the demand of water resources has increased over the years due to the increase in the population, urbanization, agricultural demand, and economic growth over India (Mall et al. 2006). Also, Mall et al. advocated that change in the land use pattern, over exploitation of water storage are modifying the hydrological cycle of many climatic regions. The changing global climate alters the hydrological cycle, which in response causes the variability in the frequency of the extreme events, availability of water, irrigation water use, and quality of freshwater resources (Simonovic 2017). According to the Fifth Assessment Report (AR5) by the Intergovernmental Panel for Climate Change (IPCC 2014), the annual mean precipitation, extreme precipitation, and monsoon precipitation are likely to intensify over India. Moreover, extreme precipitation becomes more intense with an increase in global mean surface temperature. Hence, it is essential to explore different possibilities for rainfall-runoff harvesting as the rainfall surplus and variability will have a prominent impact on the short-, medium-, and long-term future planning, operation, and management of water resources (Almazroui et al. 2017). In this sense, over the past decades, researchers have endeavored to understand and assess the climate change impact on water resources over India. These include, but are not limited to Goyal and Ojha 2012; Ghosh and Katkar 2012; Narsimlu et al. 2013; Uniyal et al. 2015; Alam et al. 2016; Das and Umamahesh 2016; Singh and Goyal 2016; Reshmidevi et al. 2018; and Sinha et al. 2018.

In addition to climate change variability, some very basic hydrological phenomena enable us to decide the scope of rainwater harvesting and are discussed here. The runoff generation from rainfall events depends upon the type of soil and land cover. Therefore, it is necessary to establish a strong correlation between runoff generation and precipitation in case of runoff harvesting. However, the correlation structure can be significantly affected due to the change in the precipitation pattern and intensity. The regions with large mean annual precipitation experience lower variability and vice versa as stated by Pisharoty (1990). Therefore, in regions with lesser mean annual precipitation, the reliability of rainwater harvesting is likely to be low (Kumar et al. 2006). The high-intensity rainfall in arid and semi-arid regions of India produces high-intensity runoff in shorter duration, limiting the effective storage capacity of the harvesting structure. In addition, past studies carried out by researchers have shown that soil infiltration, storage potential of the aquifers, and the terrain information play an important role in the deciding the scope of rainwater harvesting.

As India is an agriculture-dominated country, the availability of surface water and the possibility of groundwater recharge must be taken into account for sustainable adaptive measures. In the context of global water resources, climate change exhibits spatiotemporal variability, viz. some humid areas are likely to receive less rainfall and hence suffer from desertification, whereas some arid zones are likely to get more precipitation that leads to increase in groundwater recharge (Almazroui et al. 2017). Hence, it is necessary to conserve the surplus water for proper utilization during the water stress period and enable the possibility of groundwater recharge. Therefore, discussion toward water harvesting possibility in the context of climate change and variability is carried out. Water harvesting structures like terraces, ponds, roof collection, ditches, weirs, and small dams retain the surplus water during the wet season and hence significantly contribute to the aquifer recharge. Moreover, the IPCC (1996) reported that the variability in the magnitude and frequency of precipitation is likely to alter the recharge events as recharge occurs mainly after the storm events. In India, the irrigated agriculture is more commonly practiced than the rain-fed agriculture, which in turn imposes increasing stress on the non-renewable groundwater sources that may accelerate the water scarcity in future (Asoka et al. 2017). Hence, it is worth mentioning that water harvesting or conservation practices would be a promising option to utilize the surplus water efficiently for optimum productivity. As discussed earlier, according to the IPCC (2014), water resources of India are likely to increase by the end of twenty-first century and this in turn facilitates the improvement in water harvesting techniques, as well as restoration of agricultural activities by allocating water harvesting measures at the most suitable areas (Almazroui et al. 2017). However, climate change impact is likely to vary at a regional scale, and hence it is necessary to evaluate the water abundance or water scarcity conditions at a regional scale to assess the suitability of water harvesting structures for sustainable development.

In this sense, the present study illustrates the water harvesting possibilities under the influence of climate change and variability to assess the challenges imposed by climate change. In this regard, a case study is undertaken to evaluate the climate change impact on the precipitation under different climate scenarios and assess the future surplus precipitation with respect to the historical period. To assess the induced impact in future for better risk and resources management, global climate models (GCMs) are used as most credible tools (Goharian et al. 2016). The present and future climate is simulated by GCMs under different climate scenarios that incorporate the changes in the atmospheric forcings at horizontal (2°–4° grid resolution) and 10–20 layers in the vertical direction (Dibike and Coulibaly 2005). Water harvesting possibilities are evaluated based on the historical observations and downscaled future projection of rainfall from the GCM outputs. Prior to the case study, a general background of water harvesting techniques is briefly discussed in an Indian context.

28.2 Water Harvesting

Despite large river networks all over the country, India primarily depends on the rainfall that is the only copious and clean source of water. In addition, rainfall over India is very erratic, viz. scanty rainfall in the North West part and abundant rainfall in North East region. Due to the irregularities in the precipitation pattern, floods, drought, or both occur very frequently. Hence, India has a rich tradition of water harvesting systems to redirect or store the water and use it at the time of need. If harvested properly, even scant rainfall can meet the demand of water. According to definition, water harvesting is the collecting or storing of water to satisfy the various water demands such as drinking water, irrigation, water for livestock, etc., or to recharge the groundwater (Handia et al. 2003; Villarreal and Dixon 2005; Evans et al. 2006; Kumar et al. 2006). In general, rainwater harvesting encompasses techniques to induce, collect, conserve, and store runoff from various sources by connecting the runoff-producing area with the runoff-receiving area (Boers and Ben-Asher 1982; Rockstrom 2000; Young et al. 2002). Different water harvesting techniques involve (i) collecting the rainfall where it rains; (ii) diverting the flood water beside the riverbank; and (iii) constructing structures like dams or barrages across the river. However, the types of water harvesting structures rely on the rainfall and terrain pattern of the region. The advantages of the water harvesting techniques include: (i) control of the flood and runoff; (ii) reduction of soil erosion; (iii) decrease of silting of rivers; (iv) recharge of groundwater; (v) reduction of water losses and improvement of water utilization; (vi) increase of green cover in surrounding areas. Table 28.1 presents the types of harvesting techniques according to the eco-zone and precipitation.

Therefore, a brief discussion regarding different traditional water harvesting structures are presented, based on the eco-regions over India.

28.2.1 *Trans*-Himalayan Region

28.2.1.1 Zing

The common harvesting structures in this region is known as Zing (Figure 28.1). Zings are the small tanks. The melted water from the snow, ice, and glaciers is the only source of water in the region and hence the melted water stream is guided by the channel and is brought to the tank. Generally, the water collected by the end of the day is used for the next day.

Table 28.1 Eco-zones and water harvesting techniques.

Eco-zone type	Water harvesting techniques
Hilly and mountainous region (orographic precipitation)	Water diversion from stream to the channels
Arid and semi-arid region (seasonal precipitation)	Water diversion through channel to a tank
Flood plains (heavy precipitation)	Flood controlling structures like weir, barrage, small dams
Coastal areas (seasonal precipitation)	Primary focus in coastal plains is given to prevent the intrusion of saline water
Groundwater aquifer	Dug wells, bore wells

Figure 28.1 Zing water harvesting structure.

28.2.2 Western Himalaya

28.2.2.1 Kul

These structures (Figure 28.2a) are found mainly in the precipitous mountain areas (Himachal Pradesh and Jammu region). Kuls carry water from the glaciers to the village and if the terrain is muddy the channel is lined with stones to prevent clogging.

28.2.2.2 Naula

These structures are found in the hilly areas of Uttaranchal. Naulas (Figure 28.2b) are small ponds or wells, and are constructed across the stream by providing a stone wall. Generally, these structures are used for drinking purposes. Big shady trees are planted near the structure to reduce the evaporation. Naulas are common in Kumaon division of Uttaranchal and less developed in Garhwal due to the availability of water resources from perennial rivers.

28.2.2.3 Khatri

These are carved structures in the hard rock mountain about 6 ft deep and 10 × 12 ft in size. These structures are in Hamirpur, and Mandi district of Himachal Pradesh. Khatris (Figure 28.2c) are of two types: the rainwater collected from the roof is mainly used for washing and animals, and the seepage rainwater through the rocks is used for human consumption.

28.2.3 Eastern Himalaya

28.2.3.1 Apatani

These structures harvest both the ground and surface water and are adopted by Apatani tribe of Ziro in lower Subansiri district of Arunachal Pradesh. This system is practiced in the elevated area with gentle slope. These structures are made by terracing the valley into the plots and an earthen dam is built up to a height of 0.6 m supported by the bamboo frames. Each plot has inlet and outlet and the outlet of the upper lying plot acts as an inlet to the lower lying plot. Therefore, based on the requirements the inlets and outlets can be operated. The stream water is collected by constructing a wall of 2–4 m high and 1 m thick near forested hill slope and conveyed to the agricultural fields through channel network.

28.2.4 North Eastern Hill Ranges

28.2.4.1 Zabo

This type of water harvesting system is mainly practiced in Nagaland in Northeast India. As rainfall occurs in the hilltop of a protected forest, the water passes through various terraces along the slope as overland flow (Figure 28.3). The uses of the terraces at different slopes are different viz. the middle terrace collects water in the pond like structure followed by the terraces with cattle yard and cultivation toward the foot of the hills.

28.2.4.2 Bamboo Drip Irrigation

These systems are commonly used in Meghalaya to tap the stream and spring water with the use of bamboo pipes from the hilltop and drip irrigate the black pepper cultivation.

Figure 28.2 (a) Kul, (b) Naula, and (c) Khatri water harvesting structures.

Figure 28.3 Zabo water harvesting structures.

The flow of the water is controlled by varying the diameter of the bamboo pipe.

28.2.5 Brahmaputra Valley

28.2.5.1 Dongs

Dongs are the ponds constructed to harvest the water for irrigation by the Bodo tribes of Assam. These structures ensure provision of water for paddy cultivation. Dongs route water from the perennial water resources such as, small rivers, streams to the paddy cultivating fields. The breadth of the dong varies 7–15 ft on average and it increases gradually from the sources till the endpoint. The ponds are individually owned and there is no community involvement for digging and maintenance.

28.2.5.2 Dungs

Dungs are found in Jalpaiguri district of West Bengal. These are the links between the stream and the rice field for irrigation purposes.

28.2.6 Indo-Gangetic Plains

28.2.6.1 Ahar and Pynes

These are the floodwater harvesting structures mainly found in south Bihar. An ahar is a catchment embanked on three sides, used to provide irrigation to the winter crops after draining out the remaining excess water after summer cultivation. Pynes are the canals built in between the river and ahar to provide water from rivers.

28.2.6.2 Bengal's Inundation Channel

These inundation channels carry floodwater enriched with silt and fish to the lakes and tanks. These channels are long,

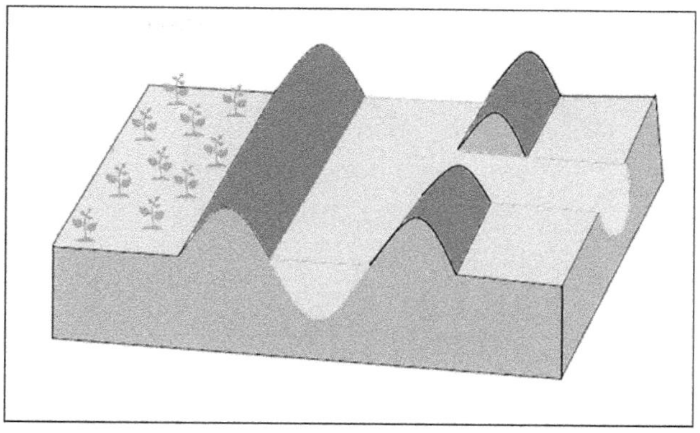

Figure 28.4 Bengal's Inundation Channel water harvesting structures.

parallel, and continuous. The irrigation to the cultivation is carried out by cutting the banks of the canals (Figure 28.4).

28.2.6.3 Dighis

Dighis are square or circular reservoirs used to store water for personal use. The water is supplied to the dighis by the canal and each dighi has its own sluice gate. These structures are made during the empire of Shahjahan (1627–58 AD).

28.2.6.4 Baolis

Baolis are the stepwells to harvest water.

28.2.7 Thar Desert

28.2.7.1 Kunds

Kunds are the circular underground wells located in western Rajasthan and Gujarat. Kunds are made to harvest rainwater for drinking purpose. The location of the kund is generally at the center and the adjacent areas gently slope toward the kund. A wire mesh is provided to restrict the entry of debris in to the well.

28.2.7.2 Kuis/Beris

Kuis (Figure 28.5) are the deep pits that are used to harvest rainfall and to collect the seepage water. Generally, the mouth of the pit is made narrow to prevent water loss due to the evaporation and as the depth increases the pit gets wider so that water can seep into a large surface area. The top of the structure is covered with planks of wood. The water harvested by the structures is used sparingly as it is the only hope for an emergency.

28.2.7.3 Baoris/Bers

Baoris are found in Rajasthan and are mainly used for drinking. The water loss from the structure by means of evaporation is negligible and hence the structure can hold the water for a long period of time. The baoris are not

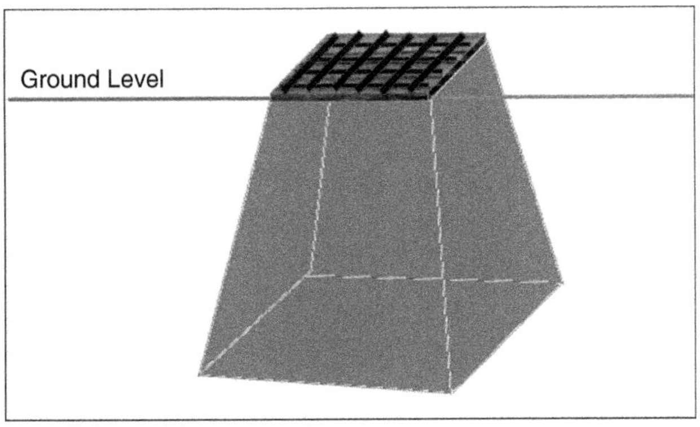

Figure 28.5 Kuis/Beris water harvesting structures.

connected to any water course and the water is stored due to the seepage from the nearby water bodies. These structures occupy minimum space and are famous in Jodhpur city. The condition of half of the baoris is good while the remaining half require maintenance.

28.2.7.4 Jhalaras

Jhalaras are mainly located in Rajasthan and Gujarat with a rectangular shape, meant for community use (but not for drinking) and for religious rites. Jhalaras are built to ensure the regular water supply to neighboring areas. The steps are built on three or even all four sides of the tank and jhalaras collect the seepage water from talab or lake located upstream.

28.2.7.5 Nadis

Nadis are the village ponds that store excess water from the adjoining catchment and the water remains in the pond for a duration of two months to one year. These structures are found in Jodhpur in Rajasthan.

28.2.7.6 Tobas

Tobas are generally constructed over a land with very low porosity with depression in a natural catchment area.

28.2.7.7 Tankas

Tankas are found in Bikaner. These are the circular tanks made in the ground to harvest rainwater for the drinking purpose. If the rainfall is less than the normal, then the tankas are filled with the water from the wells and nearby tanks.

28.2.7.8 Khadin

In this system, the farmland is used to harvest the rainwater. The khadin bund is provided to restrict the moment of water and at one side spillway allows to drain the excess water. After utilizing the water from the khadin, the saturated land is used for cultivation.

28.2.7.9 Virdas

Virdas are used to harvest rainwater during monsoons. Virdas are shallow wells dug on an undulating topography. This type of structure helps to separate fresh water from salt water and are built by nomadic Maldharis.

28.2.7.10 Paar System

It is a common water harvesting practice in western Rajasthan region to harvest rainwater. The Paar system is a combination of six to ten kuis or beris systems.

28.2.8 Central Highlands

28.2.8.1 Talab

Talabs are also known as reservoirs. They can be naturally made or man-made and are used for rainwater harvesting. They can be found in Bundelkhand and in Udaipur.

Talab is used for drinking and irrigation purposes. Talab stores water during the monsoon season and after the monsoon the bed of the talab can be used for the cultivation.

28.2.8.2 Saza Kuva

Saza kuva is a circular well pit mainly constructed for irrigation in eastern Rajasthan. The soil dug out from the pit is used to prepare the elevated platform away from the well.

28.2.8.3 Johad

Johad is an earthen check dam constructed to harvest rainwater. Also, it improves the percolation and groundwater recharge. As a result, the groundwater level rises by 6 m in Alwar district of Rajasthan.

28.2.8.4 Naada/Bandha

In this system of water harvesting, a check dam is constructed across a stream to capture the monsoon flows. The water submergence makes the land more fertile with the silt deposits and the soil also retains more water.

28.2.8.5 Pat

In this type of water harvesting system, water is diverted from the hill streams to the irrigation channel. The diversion is made by piling up the stones and the floor is made leak proof with teak leaves and mud. However, this type of structure needs regular maintenance. It takes two weeks to get the pat flowing.

28.2.8.6 Repat

The repat is a percolation tank, with a bund to impound the rainwater flowing through a watershed. The types of material are required to prepare repat depend on the height of the repat. Repat is generally used to recharge the groundwater rather than irrigating the land.

28.2.8.7 Chandela Tank

These tanks are constructed to stop the water flow between the hills by means of massive earthen embankments. These embankments are supported on the wall of coarser stones. As the tanks are made up of lime and mortar, they can survive even after a thousand years. These tanks provide drinking water to the villagers and cattle. However, silting is a major problem for this type of structure.

28.2.8.8 Bundela Tank

Though the tanks are larger in size compared to the chandela tank, these tanks are not as cost effective as chandela tanks. These tanks are also constructed to meet the water demands in the surrounding areas.

28.2.9 Eastern Highlands

28.2.9.1 Katas /Mundas/Bandhas

These harvesting structures are the primary irrigation sources in the ancient Orissa and Madhya Pradesh. An earthen embankment, curved at either end, is constructed to hold up an irregularly shaped sheet of water.

28.2.10 Deccan Plateau

28.2.10.1 Cheruvu

Cheruvu (tanks) are the reservoirs to store runoff and found all over Andhra Pradesh and Telangana. These tanks were constructed during Kakatiya dynasty. These are very popular and the major source of irrigation. Tanks are earthen bunded, constructed across the slopes of the landscape.

28.2.10.2 Kohli Tanks

These tanks were the backbone of small group of farmers (Kohlis) for the irrigation purposes, some 250–300 years ago. These tanks were of all sizes and provisions were made to bring the water to the doorsteps of villagers.

28.2.10.3 Bhanadaras

Bhanadaras are check dams made across the rivers to raise the water levels in the upstream side so that the irrigation channel can get water from the raised water level. Over a small stream, the bhanadaras can supply water for a few months after the monsoon season.

28.2.10.4 Phad

The community-managed phad irrigation system came into existence some 300–400 years ago in northwestern Maharashtra. The size of the phad varies from 10 to 200 ha. The irrigation to the phad comes from the channel diverted from bhanadaras and the villagers decide which phad to be irrigated and which to leave fallow.

28.2.10.5 Kere

Kere is the Kannada version of tanks. The kere are fed by the check dams and generally the kere are built in series. Hence, outflow from one tank is supplied to another tank, usually situated a few kilometers apart, to ensure the seepage from a higher tank will be collected in the next lower tank.

28.2.10.6 The Ramtek Model

The model is named after a town called Ramtek in Maharashtra. These tanks are extending from the foothills to the plains following a chain pattern. Once the tank near to the foothill is filled, then water flows to fill the successive tank through interconnecting channels.

28.2.11 Western Ghats

28.2.11.1 Surangam

The word surangam means tunnel. It is a horizontal wall 1.8–2.0 m thick and 0.45–0.70 m wide excavated in hard laterite rock formation. The length of surangam varies from 3 to 300 m. The surangam is used to harvest the groundwater during dry months. Also, several vertical shafts are provided to ensure the atmospheric pressure inside.

28.2.12 Western Coastal Plains

28.2.12.1 Virdas

This structure is already discussed under Section 28.2.7.9. Readers are advised to follow the same.

28.2.13 Eastern Ghats

28.2.13.1 Korambus

Korambus is a temporary dam constructed over the mouth of the channel. The korambus helps in raising the water level in the channel and diverting the water to the field canal. Also, the height of the korambus is adjusted so that the fields upstream are not submerged. The water is supplied until all the fields are irrigated.

28.2.14 Eastern Coastal Plains

28.2.14.1 Eri

Eris are tanks used to irrigate about one-third of the irrigated area of Tamil Nadu. Eris maintain the ecological harmony, and prevent soil erosion and wastage of runoff during groundwater recharge.

28.2.14.2 Ooranis

Ooranis are tanks in south Travancore and provide just enough water to cultivate the land in the absence of any large tanks.

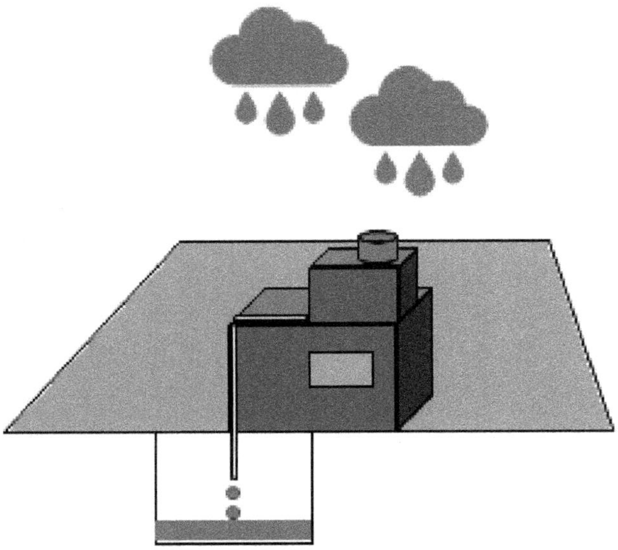

Figure 28.6 Rooftop water harvesting structures.

28.2.15 Rooftop Harvesting

Most of the above discussed water harvesting structures are found in rural areas. However, the urban areas in India are facing an ironic type of problem. In the urban cities of India, water scarcity and flooding have become two sides of the same coin. The short-duration, high-intensity precipitation causes severe flooding in the urban areas due to the impervious terrain and leaves very little water for the groundwater recharge. Hence, in the urban areas rainwater harvesting is one of the solutions to tackle both flooding and water scarcity. The commonly used water harvesting technique in the urban area is rooftop harvesting (Figure 28.6). In this method rainwater from the roof is collected using gutters and down take pipes, and finally collected in tanks for future use. The size of the storage tank depends on the rainfall availability, requirement, and available space. Moreover, the excess water can be used in groundwater recharge through shaft and dug well.

28.2.16 Perforated Pavements

These pavements are intended to reduce the stormwater and are mostly useful in urban areas. The solid concrete pavers are placed over highly permeable open aggregate. The joints are filled with open graded aggregate, making the surface almost 100% permeable. This technique facilitates more infiltration by reducing stormwater runoff even in intense precipitation.

28.2.17 Infiltration Pits

These pits (Figure 28.7) are made to collect the runoff and the collected water infiltrates to recharge the groundwater.

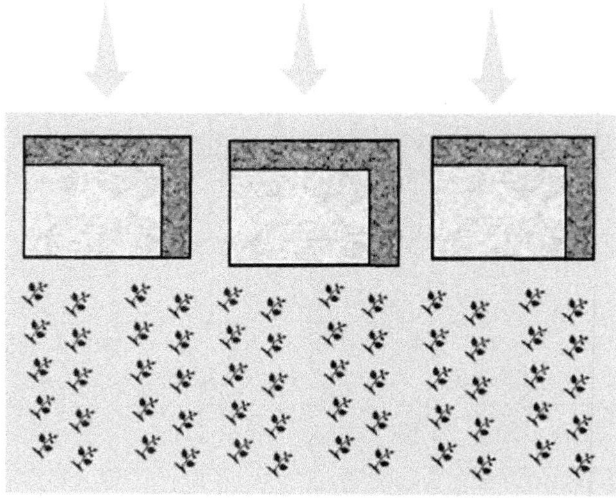

Figure 28.7 Infiltration pits water harvesting structures.

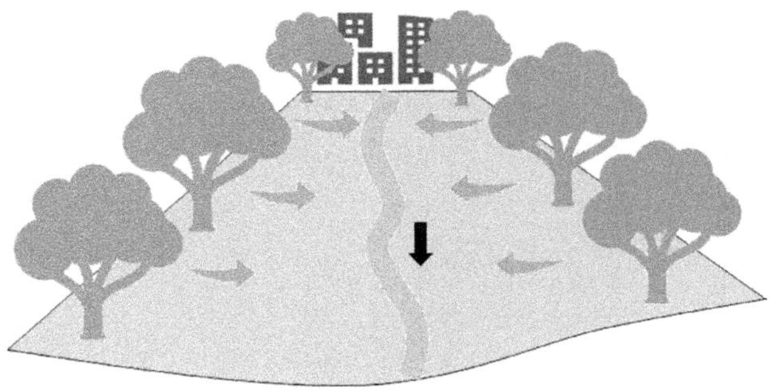

Figure 28.8 Swale water harvesting structures.

The dimension of the pits is 4 m. long, 2 m wide, and 1 m deep.

28.2.18 Swale

A swale (Figure 28.8) is a natural way made up with a gently sloping ditch or open canal with grass, which guides the water flows on the side of a road or structure. The runoff through the swale can be diverted into temporary storage to promote infiltration and groundwater recharge.

Despite of the variability in the water harvesting techniques, all these methods have three common practices as discussed by Boers and Ben-Asher (1982):

- These techniques are small-scale operations in terms of area, storage, and investment.
- They depend on the regional hydrological characteristics and do not involve storing of water in large reservoir or mining of groundwater.
- Water harvesting is applied in arid and semi-arid regions where runoff has an intermittent character and precipitation is highly erratic. Therefore, flood and drought characteristics have significant impact on agriculture.

Realizing the importance of water harvesting, it has become important to assess the possibility of water harvesting and modification of the present water harvesting structures under the impact of climate change. Hence, the historical background of water harvesting as an adaptation measure to the climate change is briefly discussed in the following section. Moreover, social and cultural effects in human life due to climate change and water harvesting are presented.

28.3 Case Study

28.3.1 Study Area

Wainganga, the largest sub-basin of the Godavari river, is considered (Figure 28.9) in the present study. Wainganga basin is located between 19°30′–22°50′ N latitude and 78°0′–81°0′ E longitude and it blankets about 49 695.40 km^2. The length of the river is about 635.40 km. Agricultural and forestlands are the predominant land use classifications over the study area. The decadal land use classifications (1985, 1995, and 2005) are studied for the Wainganga basin and it has been observed that there is no significant change in land use pattern over the years (Das and Umamahesh 2018). Therefore, the variability in the streamflow due to dynamic vegetation is assumed to remain constant in the future. The agro-climatic zone of the basin is divided into three parts: (i) Central plateau and hills region; (ii) Eastern plateau and hills region; and (iii) Western plateau and hills region. Mostly fine and medium texture soil covers the basin, and the erosivity of the soil comes in the category of moderate to severe. The elevation of the basin varies from maximum 1032 m to minimum 144 m.

28.3.2 Climate and Rainfall

The annual precipitation over the basin varies from 900 to 1600 mm with the highest rainfall during the monsoon season. The present study area falls under the tropical climate with maximum temperature varying between 39 and 47° C and minimum temperature fluctuating between 7 and 13° C during winter. The annual cycle is divided into four different seasons, viz. January to March (post-monsoon Rabi), April to June (pre-monsoon), July to September (monsoon), and October to December (post-monsoon Kharif). The spatial variability of precipitation in the four seasons during the historical period (1971–2000) is presented in Figure 28.10.

The precipitation data for the historical period is collected from the India Meteorological Department (IMD) at a resolution of 0.5° × 0.5° (~50 km × 50 km). The grid points covering the study area are superimposed over the basin and presented in Figure 28.9. It can be noted from the Figure 28.10 that the precipitation pattern varies

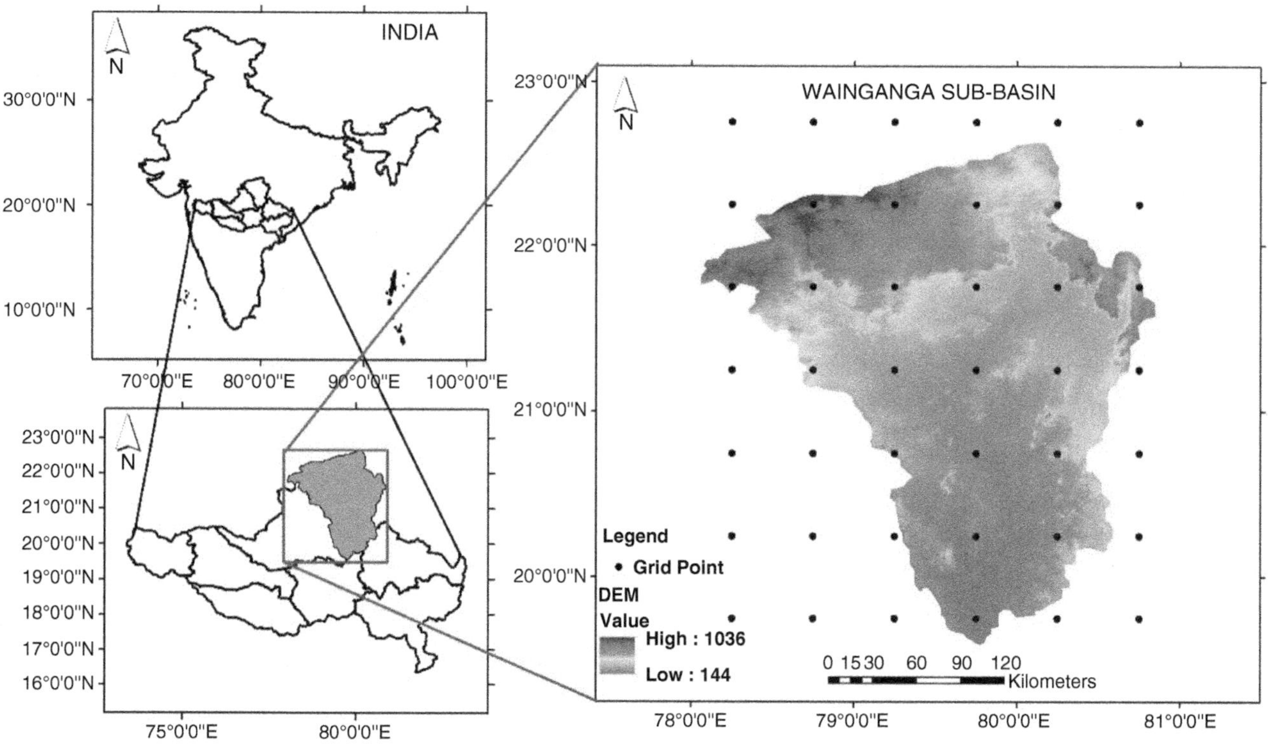

Figure 28.9 Location map of the study area.

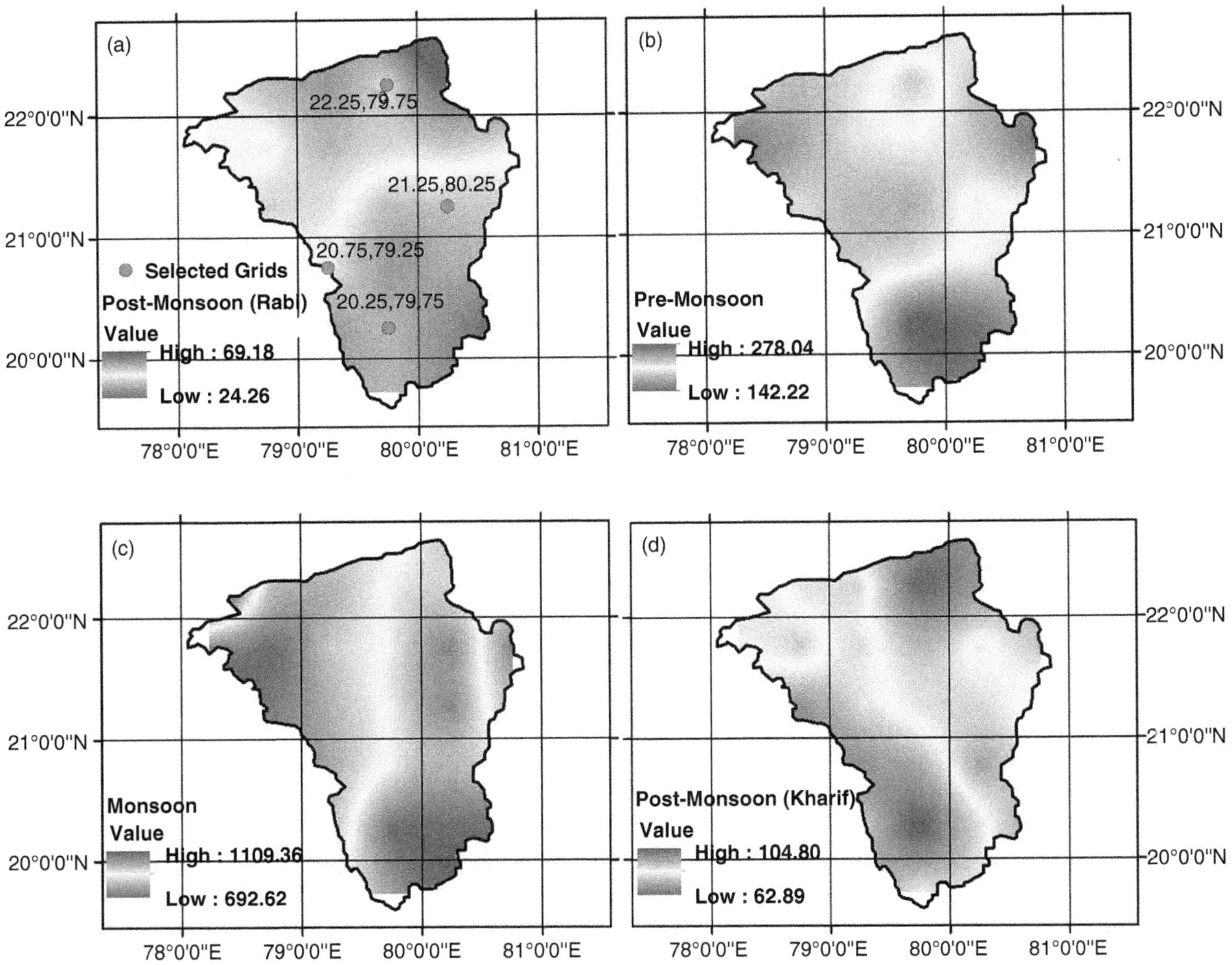

Figure 28.10 Mean seasonal variation of rainfall for the historical period over the basin.

significantly over the seasons. It is interesting to note that in different seasons different parts of the basins get high rainfall, viz. during post-monsoon Rabi northern part, during pre-monsoon southern part, during monsoon eastern and southern parts, and during post-monsoon western and southern parts receive high precipitation.

These variabilities can be attributed to the variation in the different major controls of the climate such as air circulation, altitude pressure, tropical cyclones, etc. To assess the impact of climate change and possibility of water harvesting techniques, four different points over different regions of the basin are selected and presented in Figure 28.10a.

28.3.3 GCM Projection and Scenarios

The outputs from the GCMs are used to project the future changes in the hydroclimatic variables. Hence, the GCM outputs are downloaded from the Coordinated Regional Climate Downscaling Experiment (CORDEX) for the South Asia from the Indian Institute of Tropical Meteorology, Pune (IITM) website http://cccr.tropmet.res.in/cordex/files/downloads.jsp (accessed on 12th February, 2018). For the brevity, only one GCM (ACCESS 1.0) from Commonwealth Australia Scientific and Industrial Research Organization (CSIRO) is considered under two different future scenarios, namely representative concentration pathways (RCP) 4.5 and 8.5 over the selected four different points as shown in Figure 28.10a. As the GCM output falls on the same grid points of the observed data, regridding operation is not carried out in the present study. The future data for the precipitation under the GCM are extracted for the near future (2011–2040), mid future (2041–2070), and far future (2071–2099). Based on the past literature (Rauscher et al. 2010; Argüeso et al. 2013; Teutschbein and Seibert 2013), it is advisable to perform bias correction before using GCM data in climate change studies. Therefore, quantile mapping technique is carried out to correct the bias in the future based on the correction

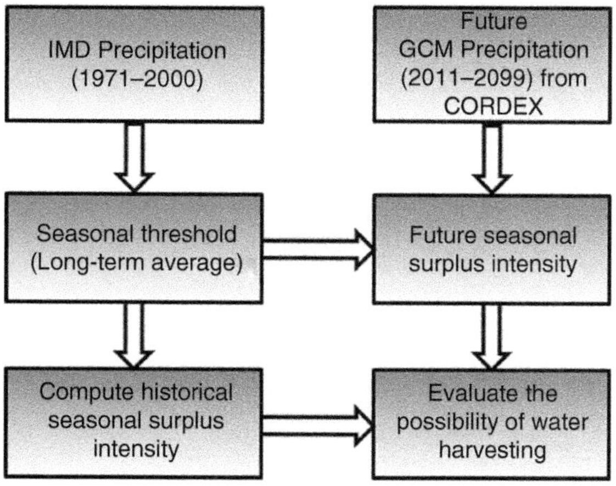

Figure 28.11 Flowchart of the present study.

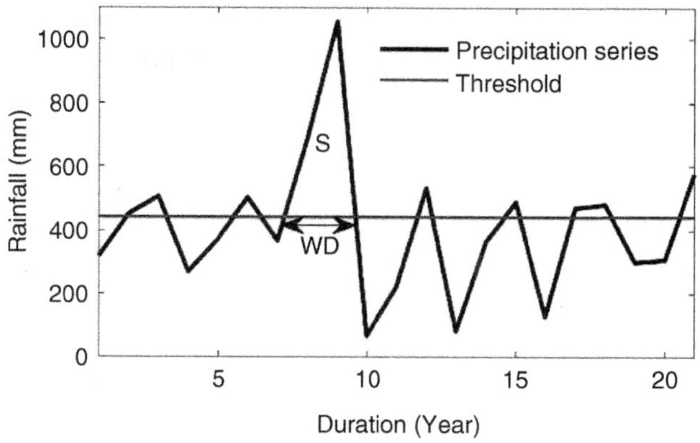

Figure 28.12 Graphical definition of S (surplus) and WD (wet duration) over a given threshold.

applied to the historical data. It is also noted that the GCM data for the historical period (1971–2000) are extracted for the selected points. As suggested by Gudmundsson et al. (2012), a non-parametric quantile mapping is used to correct the bias with respect to the historical observation. The interested readers are advised to follow the article by Gudmundsson et al. (2012).

28.3.4 Surplus Intensity

To investigate the water harvesting possibility for the future, projection of rainfall patterns for future scenarios need to be considered and hence, seasonal water harvesting possibility is presented in the study. The surplus rainfall amount for a season, above the arithmetic mean for the historical period, determines the water harvesting possibility in the area (Figure 28.11). The adopted methodology comprises the following two steps:

1. To compute the wet duration (WD) and total surplus (S) over this duration under the historical and future period. The graphical definition of wet duration and total surplus is presented in Figure 28.12.
2. To compute the surplus intensity (SI) using the equation $SI = S/WD$ (Almazroui et al. 2017). The threshold value is defined as the long-term average during the historical period.

28.4 Results and Discussion

In the present study area, the groundwater level is temporal and dynamic in nature and primarily depends on the rainfall pattern and aquifer material. In Wainganga basin, there are 263 observational wells and the distribution of the wells is quite uniform across the basin. Moreover, it is also being noticed that the groundwater fluctuation is significant during the pre- and post-monsoon period. The maximum recharge takes place during monsoon season. Hence, the variability of future precipitation amount needs to be assessed to make proper adaptation and management practices under climate change scenarios.

Figure 28.13 shows the variability of the seasonal precipitation for the selected grid points with the corresponding threshold values. The threshold value is computed based on the long-term average of seasonal rainfall during the historical period (1971–2000). The precipitation above the threshold value and corresponding duration are considered as total surplus and wet duration, respectively. It is also worth mentioning that the number of wet duration and total surplus periods are same.

As discussed earlier, for brevity only one GCM is used to evaluate the rainfall variability under RCP 4.5 and 8.5 scenarios during 2011–2099. To correct the bias in the precipitation series, non-parametric quantile mapping is carried out as the GCM simulated precipitation is not able to capture the extreme events (Das and Umamahesh 2017). This process is carried out for all selected points. After bias correction, the correction factor is applied to the future precipitation series. Figure 28.14 presents the boxplot of observed and future projected seasonal precipitation under different climate change scenarios.

It can be noted from Figure 28.14 that the monsoon precipitation for the future period is likely to decrease significantly compared to the historical observation. However, there is no significant difference noticed during the other seasons. As the monsoon precipitation contributes 70–80% of the total annual rainfall, the substantial reduction in the monsoon rainfall will affect the water resources adversely. The future projected series is considered to evaluate the future surplus index under different scenarios to assess the possibility of water harvesting. Figure 28.15

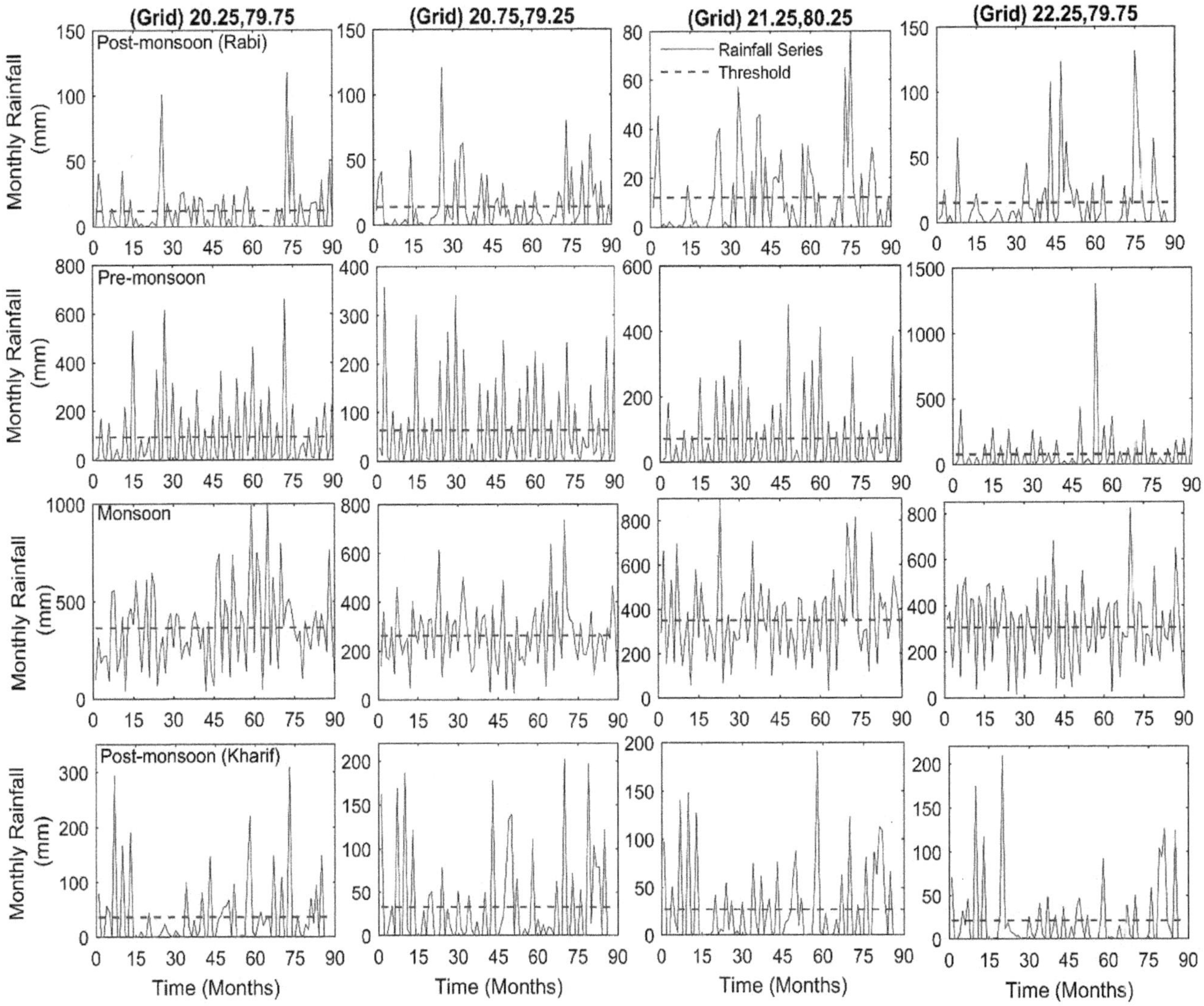

Figure 28.13 Seasonal rainfall variability with threshold value for the selected grid points and different seasons.

depicts the surplus intensity under different scenarios for the post-monsoon (Rabi) season at the selected grid points.

It can be noted from Figure 28.15 that no surplus intensity is noticed at grid point (21.25,80.25) based on the historical threshold value. However, maximum surplus is achieved during far future under RCP4.5 at (20.25,79.75); (20.75.79.25); and (22.25,79.75) grid points. Moreover, an increasing trend is observed for RCP4.5 for all the grid points except (21.25,80.25). Moreover, it is also observed that the number of surplus and wet duration periods are likely to decrease with more intense surplus index. It suggests that high-intensity precipitation is likely to occur within short period of time. Hence, this variability of precipitation under climate change will be helpful to take proper adaptation strategy for water harvesting. The surplus intensity for pre-monsoon, monsoon, and post-monsoon periods are computed and presented in Figures 28.16–28.18, respectively.

From Figures 28.16–28.18, it can be noted that the surplus intensity index for pre-monsoon, monsoon, and post-monsoon (Kharif) with respect to the historical observation is likely to reduce significantly. The single lines in the boxplots suggest only single observation for that period. Hence, the outcomes from the present study show that the declining rate of future precipitation is going to affect the groundwater of the study area near the selected points and ultimately will affect the agricultural productivity. According to the result, the long-term water harvesting project might be directed and the investigation can be used to model the groundwater fluctuation and recharge in the selected regions under climate change variability. However, the uncertainty associated with the modeling of future projection of rainfall should be considered for

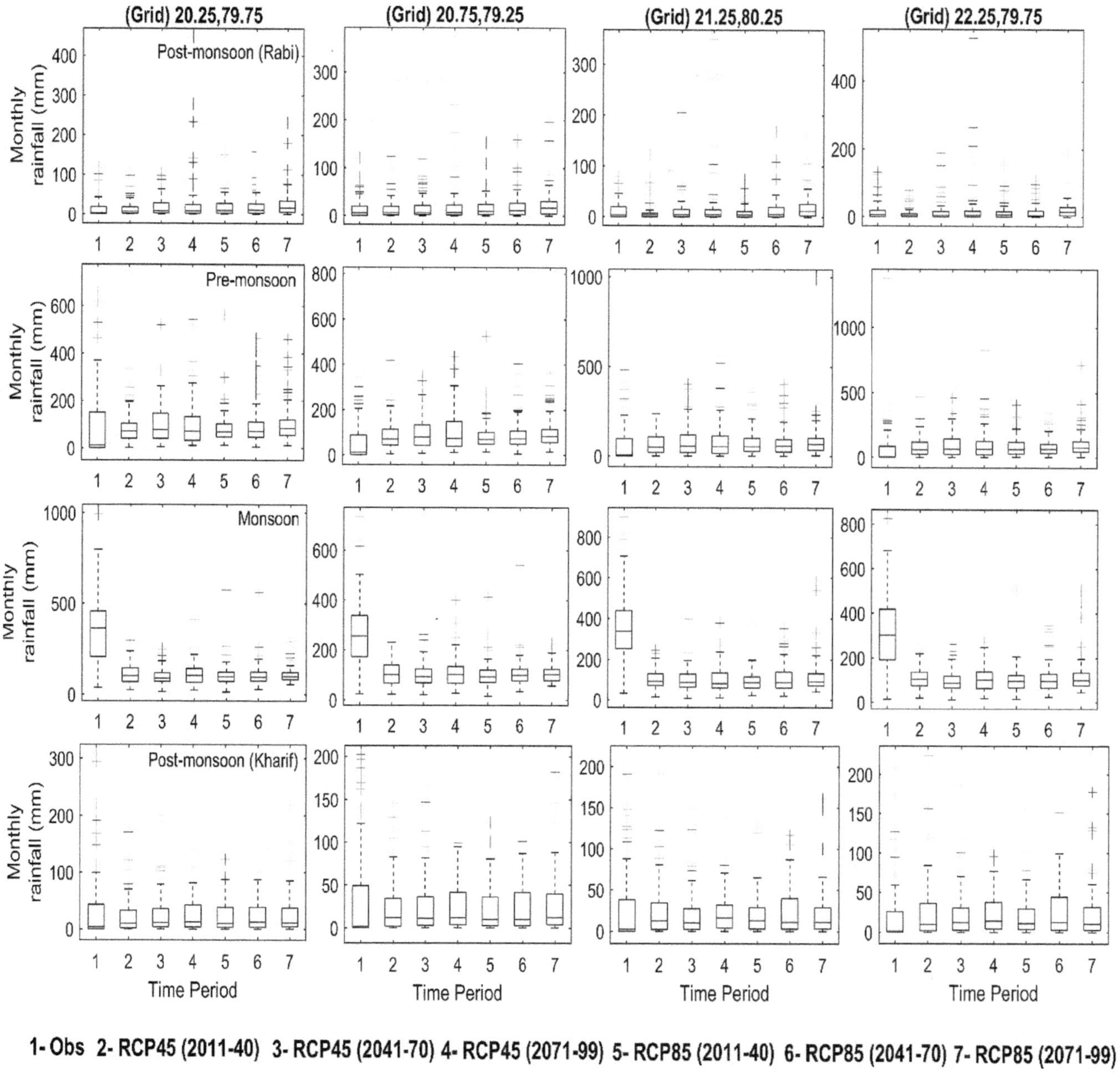

Figure 28.14 Future projection of precipitation under different scenarios for the selected grid points and different seasons.

better effective water harvesting techniques. Therefore, the uncertainties to be addressed during this type of study are discussed in the next section.

28.4.1 Understanding the Uncertainty

The projection of hydroclimatic variables are given priority over geophysical, hydrogeological, economic, and social aspects in the modeling of water harvesting techniques. Hence, it is inevitable to assess the uncertainty while projecting for the future (Eslamian et al. 2011). The uncertainties associated with the future projection are GCM and scenario uncertainties. The uncertainty associated with the GCM projections derives from the vagueness in future greenhouse emissions, and the response of the GCM to the atmospheric forcing (Tang et al. 2012). However, the scenario uncertainty stems from the unpredictability and incomplete knowledge about the future climate that arises from inadequate information and understanding about the biophysical process (New and Hulme 2000). Hence, it should be noted that scenarios are proposed to be exploratory than predictive (Brown et al. 2015). Moreover, Wilby et al. (2014) illustrated the cascade of uncertainty due to the choice of scenarios, GCM, and

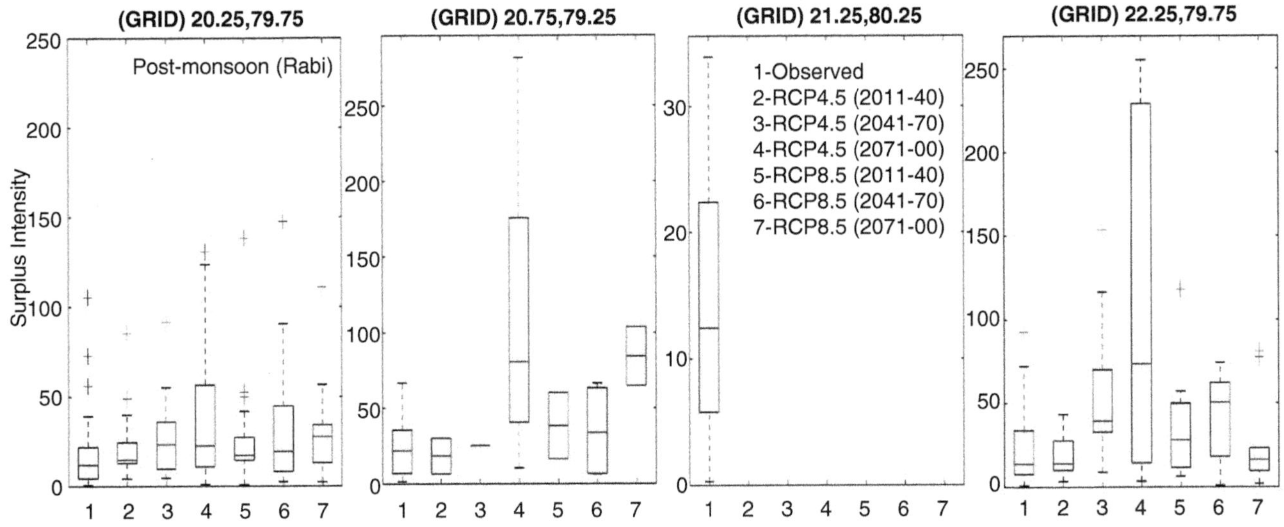

Figure 28.15 Surplus intensities during post-monsoon season under different climate scenarios.

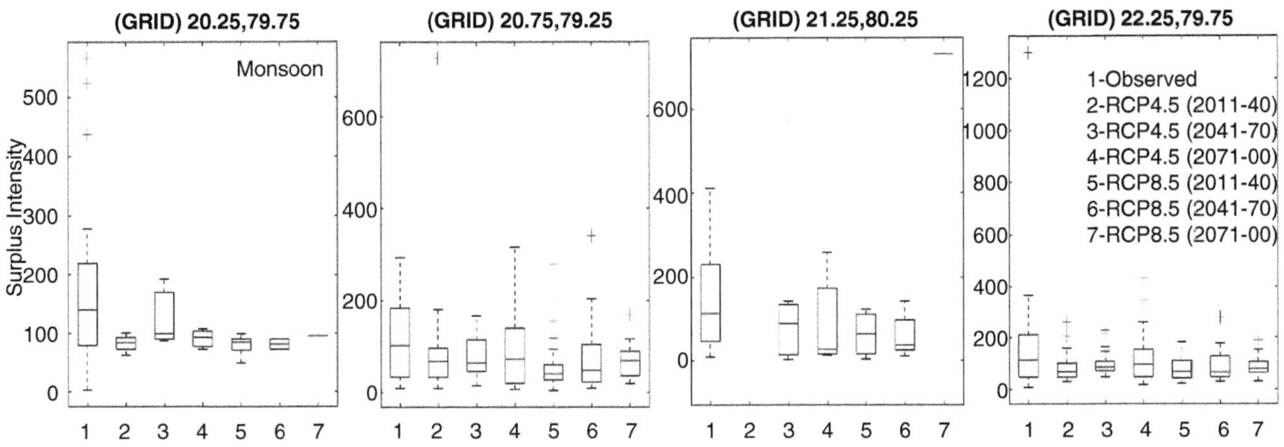

Figure 28.16 Surplus intensities during pre-monsoon season under different climate scenarios.

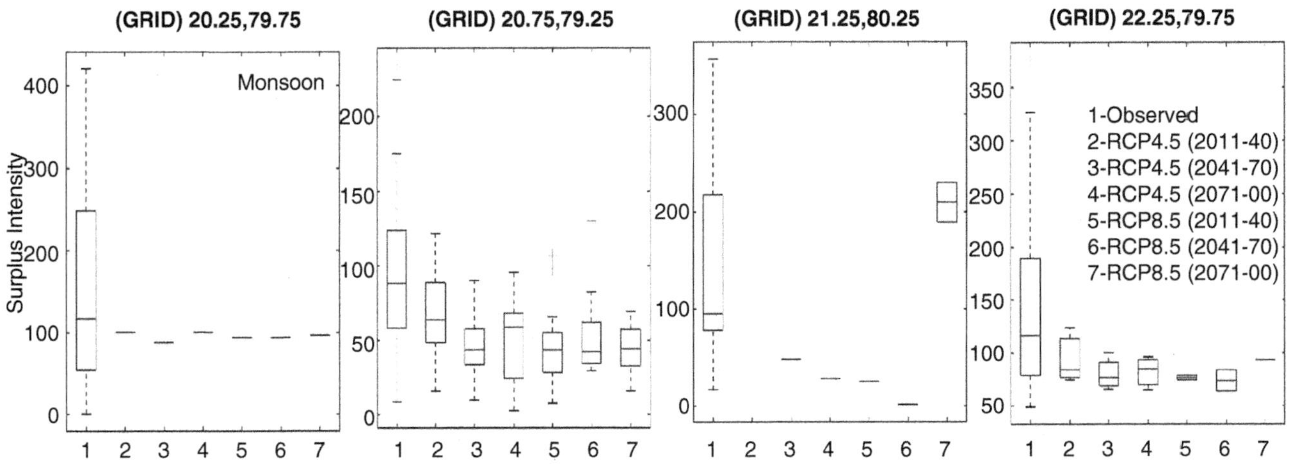

Figure 28.17 Surplus intensities during monsoon season under different climate scenarios.

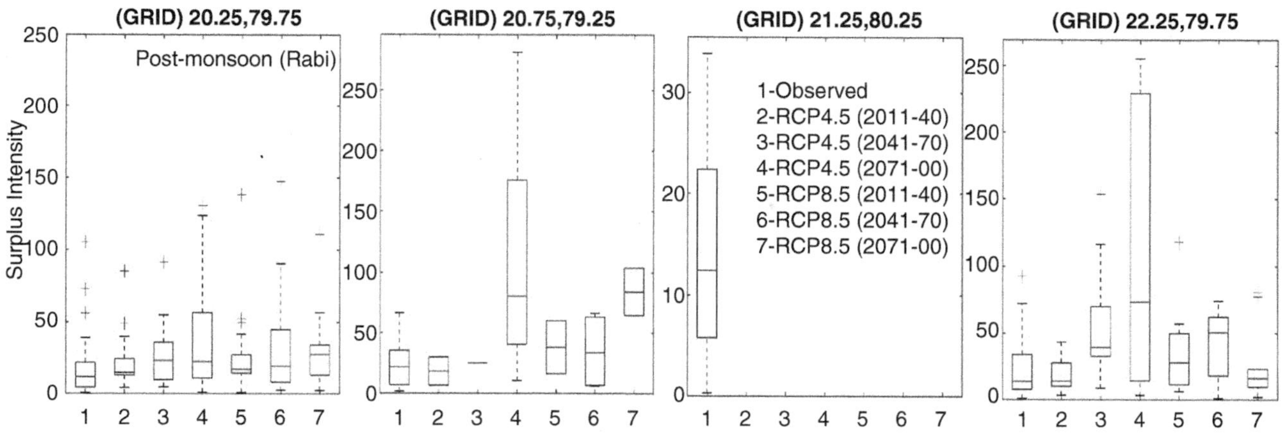

Figure 28.18 Surplus intensities during post-monsoon season under different climate scenarios.

realization of climate variability in precipitation. They stated that the uncertainty associated with scenarios will lead to even larger uncertainties in the regional climate change impact. Therefore, it is essential to incorporate the uncertainties which are stemming from the aforementioned sources for efficient and effective water harvesting. The computed surplus intensity for the future is associated with the uncertainty. Moreover, the no surplus intensity under different future projections (e.g. Figure 28.15 [grid 21.25, 80.25], and Figure 28.18 [grid 21.25, 80.25]) may be attributed to the uncertainty associated with the GCM, scenarios. Hence, in the climate change studies uncertainty should be assessed for robust future projection. However, the present study provides a direction to pursue studies in the context of water harvesting under climate change and its variability in future, incorporating different uncertainty.

28.5 Conclusion

In general, water resources are becoming more vulnerable under the influence of climate change, especially for the agricultural dominant countries in the world as they are more fragile to the climate variability. Hence, it is necessary to enhance water availability. Water harvesting is one of the alternatives to augment water availability. Moreover, water accountability is first and foremost step to begin with before any water harvesting project. In the context of climate variability, it is important to evaluate whether the basin has any surplus flow or there is a significant amount of water loss due to evaporation. This can be achieved by examining the water balance components over the basin under climate change. To capture the hydrological variabilities, water balance and water accounting studies should be accomplished for typical rainfall years. In river basins under arid conditions, water losses due to the evaporation from reservoirs, ponds, and tanks lead to water shortage. If this can be prevented, then the hydrological gain will be maximized. In addition, research should be focused toward the green and blue water. Green water represents the water in the soil that can be directly used by the vegetation in the form of transpiration and evaporation. Blue water is the water diverted from the surface and underground systems for human use. The primary focus of the water harvesting structures are to store water and use it at the time of need. At the same time, focus should be given toward the green water by evaluating the water availability as green water, crop production, and the water losses due to the evaporation. In the high rainfall regions where runoff is more in comparison to water resources, effective strategies should be made to collect the excess water through proper land use and cropping pattern to improve the green as well as blue water. Though the present analysis considers only the availability of precipitation in the future scenarios, attempts can be made to achieve the future runoff generation over a basin by using the precipitation projection as an input. Therefore, efficient and effective water harvesting under climate change and its variability investigation should be carried out in the following areas: (i) to evaluate the adverse climate change impact on the hydrology; (ii) to understand the dynamic land use changes and catchment hydrology; (iii) to increase the water utilization efficiency by minimizing the water losses from the water harvesting structure; (iv) water accountability and evaluation of water balance components for effective water harvesting; (v) to optimize the water harvesting without affecting the downstream side; (vi) to evaluate the impact of water harvesting on water resources of river basins which has undergone high degree of development; and (vii) to rejuvenate the traditional water harvesting structures, which in turn will preserve the social and cultural aspects of mankind.

References

Alam, S., Ali, M.M., and Islam, Z. (2016). Future streamflow of Brahmaputra river basin under synthetic climate change scenarios. *Journal of Hydrologic Engineering* 21 (11): 5016027. https://doi.org/10.1061/(ASCE)HE.1943-5584.0001435.

Almazroui, M., Islam, M.N., Balkhair, K.S. et al. (2017). Rainwater harvesting possibility under climate change: a basin-scale case study over western province of Saudi Arabia. *Atmospheric Research* 189: 11–23. https://doi.org/10.1016/j.atmosres.2017.01.004.

Argüeso, D., Evans, J.P., and Fita, L. (2013). Precipitation bias correction of very high resolution regional climate models. *Hydrology and Earth System Sciences* 17 (11): 4379–4388. https://doi.org/10.5194/hess-17-4379-2013.

Asoka, A., Gleeson, T., Wada, Y., and Mishra, V. (2017). Relative contribution of monsoon precipitation and pumping to changes in groundwater storage in India. *Nature Geoscience* 10 (2): 109–117. https://doi.org/10.1038/ngeo2869.

Boers, T.M. and Ben-Asher, J. (1982). A review of rainwater harvesting. *Agricultural Water Management* 5 (2): 145–158. https://doi.org/10.1016/0378-3774(82)90003-8.

Bowler, J.M., Johnston, H., Olley, J.M. et al. (2003). New ages for human occupation and climatic change at Lake Mungo, Australia. *Nature* 421 (6925): 837–840. https://doi.org/10.1038/nature01383.

Brown, C., Brown, E., Murray-Rust, D. et al. (2015). Analysing uncertainties in climate change impact assessment across sectors and scenarios. *Climatic Change* 128 (3–4): 293–306. https://doi.org/10.1007/s10584-014-1133-0.

Das, J. and Umamahesh, N.V. (2016). Downscaling monsoon rainfall over river Godavari Basin under different climate-change scenarios. *Water Resources Management* 30 (15): 5575–5587. https://doi.org/10.1007/s11269-016-1549-6.

Das, J. and Umamahesh, N.V. (2017). Uncertainty and nonstationarity in streamflow extremes under climate change scenarios over a river basin. *Journal of Hydrologic Engineering* 22 (10) https://doi.org/10.1061/(ASCE)HE.1943-5584.0001571.

Das, J. and Umamahesh, N.V. (2018). Spatio-temporal variation of water availability in a River Basin under CORDEX simulated future projections. *Water Resources Management* 32 (4): 1399–1419. https://doi.org/10.1007/s11269-017-1876-2.

deMenocal, P.B. and Peter, B. (2001). Cultural responses to climate change during the late Holocene. *Science* 292 (5517): 667–673. https://doi.org/10.1126/science.1059287.

Dibike, Y.B. and Coulibaly, P. (2005). Hydrologic impact of climate change in the Saguenay watershed: comparison of downscaling methods and hydrologic models. *Journal of Hydrology* 307: 145–163. https://doi.org/10.1016/j.jhydrol.2004.10.012.

Eslamian, S. and Okhravi, S. (2015). Quantifying glimpse to climate change and adaptation strategies. *Journal of Rainwater Catchment Systems* 3 (3): 15–26.

Eslamian, S., Gilroy, K.L., and McCuen, R.H. (2011). Climate change detection and modeling in hydrology. *Climate Change - Research and Technology for Adaptation and Mitigation*: 87–100. https://doi.org/10.5772/24550.

Evans, C.A., Coombes, P.J., and Dunstan, R.H. (2006). Wind, rain and bacteria: the effect of weather on the microbial composition of roof-harvested rainwater. *Water Research* 40 (1): 37–44. https://doi.org/10.1016/j.watres.2005.10.034.

FAO AQUASTAT. (2018) Online database. Rome, Food and Agriculture Organization of the United Nations (FAO). http://www.fao.org/nr/water/aquastat/main/index.stm.

Ghosh, S. and Katkar, S. (2012). Modeling uncertainty resulting from multiple downscaling methods in assessing hydrological impacts of climate change. *Water Resources Management* 26 (12): 3559–3579. https://doi.org/10.1007/s11269-012-0090-5.

Goharian, E., Burian, S.J., Bardsley, T., and Strong, C. (2016). Incorporating potential severity into vulnerability assessment of water supply systems under climate change conditions. *Journal of Water Resources Planning and Management* 142 (2) 04015051(1-12). doi: https://doi.org/10.1061/(ASCE)WR.1943-5452.0000579.

Goyal, M.K. and Ojha, C.S.P. (2012). Downscaling of precipitation on a Lake Basin: evaluation of rule and decision tree induction algorithms. *Hydrology Research IWA* 43 (3): 215–230.

Goyal, M.K. and Rao, Y.S. (2018). The impact of climate change on water resources in India. *ASCE- Journal of Environmental Engineering* 144 (7): 04018054.

Gudmundsson, L., Bremnes, J.B., Haugen, J.E., and Engen-Skaugen, T. (2012). Technical note: downscaling RCM precipitation to the station scale using statistical transformations – a comparison of methods. *Hydrology and Earth System Sciences* 16 (9): 3383–3390. https://doi.org/10.5194/hess-16-3383-2012.

Handia, L., Tembo, J.M., and Mwiindwa, C. (2003). Potential of rainwater harvesting in urban Zambia. *Physics and Chemistry of the Earth Parts A/B/C* 28 (20–27): 893–896. https://doi.org/10.1016/j.pce.2003.08.016.

IPCC (1996) *Climate Change 1995, The Science of Climate Change Contribution of WGI to the Second Assessment Report of the Intergovernmental Panel on Climate Change.* doi: https://doi.org/10.1017/CBO9781107415324.004.

IPCC (2014) *Climate Change 2014: Synthesis Report. Contribution of Working Groups I, II and III to the Fifth Assessment Report of the Intergovernmental Panel on Climate Change, Core Writing Team, R.K. Pachauri and L.A. Meyer*. doi: https://doi.org/10.1017/CBO9781107415324.004.

Kumar, D.M., Ghosh, S., Patel, A. et al. (2006). Rainwater harvesting in India: some critical issues for basin planning and research. *Land Use and Water Resources Research* 6: 1–17.

Mall, R.K., Gupta, A., Singh, R. et al. (2006). Water resources and climate change: an Indian perspective. *Current Science* 90 (12): 1610–1626.

Narsimlu, B., Gosain, A.K., and Chahar, B.R. (2013). Assessment of future climate change impacts on water resources of upper Sind River basin, India using SWAT model. *Water Resources Management* 27 (10): 3647–3662. https://doi.org/10.1007/s11269-013-0371-7.

New, M. and Hulme, M. (2000). Representing uncertainty in climate change scenarios: a Monte-Carlo approach. *Integrated Assessment* 1 (3): 203–213. https://doi.org/10.1023/A:1019144202120.

Núñez, L., Grosjean, M., and Cartajena, I. (2002). Human occupations and climate change in the Puna de Atacama, Chile. *Science* 298 (5594): 821–824. https://doi.org/10.1126/science.1076449.

Pande, S.K., Rayar, A.J., and Hakizimana, P. (2014). Coping with climate change through water harvesting techniques for sustainable agriculture in Rwanda. In: *Vulnerability of Agriculture, Water and Fisheries to Climate Change*, 217–239. Dordrecht: Springer Netherlands https://doi.org/10.1007/978-94-017-8962-2_14.

Pandey, D.N. (2001). A bountiful harvest of rainwater. *Science* 293 (5536): 1763–1765. https://doi.org/10.1126/science.293.5536.1763b.

Pandey, D.N., Gupta, A.K., and Anderson, D.M. (2003). Rainwater harvesting as an adaptation to climate change. *Current Science* 85 (1): 46–59.

Pisharoty, P.N. (1990). *Characteristics of Indian Rainfall*, Monograph,. Ahmedabad, India: Physical Research Laboratories.

Polyak, V.J. and Asmerom, Y. (2001). Late Holocene climate and cultural changes in the southwestern United States. *Science* 294 (5540): 148–151. https://doi.org/10.1126/science.1062771.

Rauscher, S.A., Coppola, E., Piani, C., and Giorgi, F. (2010). Resolution effects on regional climate model simulations of seasonal precipitation over Europe. *Climate Dynamics* 35 (4): 685–711. https://doi.org/10.1007/s00382-009-0607-7.

Reshmidevi, T.V., Nagesh Kumar, D., Mehrotra, R., and Sharma, A. (2018). Estimation of the climate change impact on a catchment water balance using an ensemble of GCMs. *Journal of Hydrology* 556: 1192–1204. https://doi.org/10.1016/j.jhydrol.2017.02.016.

Rockstrom, J. (2000). Water resources management in smallholder farms in Eastern and Southern Africa: an overview. *Physics and Chemistry of the Earth, Part B: Hydrology, Oceans and Atmosphere* 25 (3): 275–283. https://doi.org/10.1016/S1464-1909(00)00015-0.

Simonovic, S.P. (2017). Bringing future climatic change into water resources management practice today. *Water Resources Management* https://doi.org/10.1007/s11269-017-1704-8.

Singh, V. and Goyal, M.K. (2016). Analysis and trends of precipitation lapse rate and extreme indices over North Sikkim eastern Himalayas under CMIP5ESM-2M RCPs experiments. *Atmospheric Research* 167: 34–60. https://doi.org/10.1016/j.atmosres.2015.07.005.

Sinha, J., Sharma, A., Khan, M., and Goyal, M.K. (2018). Assessment of the impacts of climatic variability and anthropogenic stress on hydrologic resilience to warming shifts in Peninsular India. *Scientific Reports* 8 (1): 13833. https://doi.org/10.1038/s41598-018-32091-0.

Tang, C., Crosby, B.T., Wheaton, J.M., and Piechota, T.C. (2012). Assessing streamflow sensitivity to temperature increases in the Salmon River Basin, Idaho. *Global and Planetary Change*. Elsevier B.V. 88–89: 32–44. https://doi.org/10.1016/j.gloplacha.2012.03.002.

Teutschbein, C. and Seibert, J. (2013). Is bias correction of regional climate model (RCM) simulations possible for non-stationary conditions? *Hydrology and Earth System Sciences* 17 (12): 5061–5077. https://doi.org/10.5194/hess-17-5061-2013.

Tyson, P.D., Lee-Thorp, J., Holmgren, K., and Thackeray, J.F. (2002). Changing gradients of climate change in Southern Africa during the past millennium: implications for population movements. *Climatic Change* 52 (1–2): 129–135. https://doi.org/10.1023/A:1013099104598.

Uniyal, B., Jha, M.K., and Verma, A.K. (2015). Assessing climate change impact on water balance components of a River Basin using SWAT model. *Water Resources Management* 29 (13): 4767–4785. https://doi.org/10.1007/s11269-015-1089-5.

Verschuren, D., Laird, K.R., and Cumming, B.F. (2000). Rainfall and drought in equatorial East Africa during the past 1,100 years. *Nature* 403 (6768): 410–414. https://doi.org/10.1038/35000179.

Villarreal, E.L. and Dixon, A. (2005). Analysis of a rainwater collection system for domestic water supply in Ringdansen, Norrköping, Sweden. *Building and Environment* 40 (9): 1174–1184. https://doi.org/10.1016/j.buildenv.2004.10.018.

Wilby, R., Dawson, C., Murphy, C. et al. (2014). The statistical down scaling model - decision centric (SDSM-DC):

conceptual basis and applications. *Climate Research* 61 (3): 259–276. https://doi.org/10.3354/cr01254.

WWAP (United Nations World Water Assessment Programme) (2015). *The United Nations World Water Development Report 2015: Water for a Sustainable World*. Paris: France, UNESCO.

Young, M.D., Gowing, J., Wyseure, G.C., and Hatibu, N. (2002). Parched–thirst: development and validation of a process-based model of rainwater harvesting. *Agricultural Water Management* 55 (2): 121–140. https://doi.org/10.1016/S0378-3774(01)00187-1.

29

Water Harvesting and Sustainable Tourism

Neda Torabi Farsani[1], Homa Moazzen Jamshidi[2], Mohammad Mortazavi[3], and Saeid Eslamian[4]

[1] *Department of Museum and Tourism, Art University of Isfahan, Isfahan, Iran*
[2] *Department of Economics and Arts Entrepreneurship, Art University of Isfahan, Isfahan, Iran*
[3] *Department of Conservation and Restoration, Art University of Isfahan, Isfahan, Iran*
[4] *Department of Water Engineering, College of Agriculture, Isfahan University of Technology, Isfahan, Iran*

29.1 Introduction

Nowadays, global warming and irregular water resource management have affected the emergence of drought and water shortages, especially in areas located in desert biomes. Water scarcity is recognized as a global problem in many areas and water crises are developing. Some theorists in developing countries believe that if we do not manage water, we will have water wars, not water crises, in the near future (Renani 2015). In recent decades, many researchers have strived to introduce strategies and novel methods and technology for water conservation (Eslamian et al. 2015). In addition, other researches have made attempts at water saving through integrating ancient and new methods and using traditional knowledge, building skills, and local resources. Regarding this, rainwater harvesting systems can be a strategy for water conservation because rainwater harvesting can not only become a source for irrigation and reduce pressure on the main water supply, but can also decrease the risk of surface water flooding. According to Kahinda et al. (2007), rainwater harvesting describes the small-scale concentration, collection, storage, and use of rainwater runoff for productive purposes.

Obviously, promoting the tourism industry at a destination which needs a lot of freshwater sources and applying a water harvesting system for tourism purposes can be part of a solution to develop sustainable tourism at a destination. It is noteworthy that tourists' use of freshwater in developing countries' touristic islands and desert destinations was criticized.

In this context, this book chapter strives to identify water management strategies for sustainable tourism. Additionally, the authors focus on rainwater harvesting systems at museums as these attractions are important tourism destinations.

Lastly, the benefits of using rainwater harvesting system rainwater for the tourism economy are discussed.

29.2 Water Management: An Approach to Sustainable Tourism

A rapid increase in per capita income and leisure time, as well as advances in technology, have led to increased demand for recreation and holidays all over the world (Farsani et al. 2011). Inevitably, such large-scale tourism activity increases water consumption in tourism destinations. In addition, tourism activities, especially in developing countries located in arid and semi-arid climates, contribute to water losses. Therefore, implementing sustainable approaches to tourism development can also be a strategy for water management. Among sustainable water management methods, using rainwater harvesting and gray water for tourism purposes, paying particular attention to water-saving labeling for tourism products and services (such as eco-labels), and encouraging hotels, restaurants, and museums toward green principles are strategies which can contribute to water conservation and to minimize the negative impacts of tourism development on a destination's environment. In this section, the authors attempt to give an overview of water conservation schemes in the tourism sector.

One of the forms of tourism that faces challenges for water supply is nature tourism. Nature tourism, such as ecotourism, geotourism, etc., needs to have a water management plan because many tourist facilities are located in remote areas where piped water from government

Handbook of Water Harvesting and Conservation: Basic Concepts and Fundamentals, First Edition. Edited by Saeid Eslamian and Faezeh Eslamian.

supplies is not an option and therefore raises the need to drill boreholes. Hence, using a water harvesting system should provide the necessary water in these areas. Furthermore, use of water-efficient equipment and technologies (dual-system toilets, adjusting single-flush cisterns, push taps, sensors, and low-filter showerheads which use less water than conventional shower heads) are necessary in these territories (Ecotourism site 2017).

Furthermore, agritourism and rural tourism are other tourism products which use water and where there exist stress on water use for both purposes (agriculture and tourism). In order to reduce water consumption in agritourism destinations, a successful example is the case of villages in India where water harvesting systems such as bamboo drip irrigation, sand bores, etc., were applied (UNWTO 2013).

Hotels and the hospitality industry are further sectors of the tourism industry which can help global water shortage problems through sustainable water management. Since the roof area of hotels is usually big, the hospitality industry is ideal for using rainwater collection. In some hotels, rainwater is collected, filtered, and fed back into the property through a robust treatment system and the water is utilized for non-potable purposes like vehicle washing, toilet flushing, and irrigation.

Nowadays, hotel managers should be aware of this method and analyze the benefits they and the destination could gain from implementing water conservation methods. The methods which can help sustainable water management in hotels are as follows (Tuppen 2013; Plumbing 2018; Drew and Hanson 2018):

- Using drought-tolerant plants in hotel landscaping can reduce irrigation costs
- Educating hotel staff, especially laundry and kitchen staff, to reduce the overall amount of water used
- Putting rain sensors on outdoor sprinkler systems to ensure there is no unnecessary use of water
- Looking for WaterSense labeled products for hotels
- Installing meters to measure water consumption in order to assure water management in hotels
- Switching out older toilets for low-flow or high-efficiency toilets
- Keeping on top of equipment maintenance
- Using a water harvesting system for green landscape irrigation in hotels
- Using rainwater for hotel cooling towers, because they are most often the largest consumers of water
- Using rainwater or gray water for hotel laundry
- Using gray water or rainwater for toilet flushing in hotels
- Recapturing hotel swimming pool water for irrigation
- Installing push-button showers by the pool in order to reduce water use
- Covering swimming pools to prevent evaporation
- Using gray water from baths and sinks for irrigation
- Encouraging guests to shower instead of having a bath

Encouraging hotels to follow sustainability principles and joining green hotels or eco-accommodations as environmentally responsible lodging is an additional strategy for water management in tourism. According to Han et al. (2010) a green hotel is an environmentally friendly lodging property that institutes and follows ecologically sound programs/practices (e.g. water and energy savings, reduction of solid waste, and cost saving) to help protect our planet.

It is noteworthy that the awards, guidelines, labels, and formal certificates of environmental quality related to the tourism industry took on greater importance in the 1990s. These aim to distinguish companies or institutions that contribute in some way toward sustainable development in tourism (Lima and Careto 2007). Since eco-labels play an important role in the promotion of sustainable tourism marketing, using eco-labels (Figure 29.1) in tourism products and services can save water and increase water management awareness. Eco-labels on one hand help a quality product stand out in the market, and also allow customers to recognize and choose products with a low negative impact on the environment (earth-friendly products); on the other hand, they can guarantee and support entrepreneurs, small and medium-size businesses, products, and services (Farsani et al. 2011).

Using a gray water/rainwater harvesting system for tourist attractions is one more strategy for sustainable water management at destinations. Nowadays, some attractions such as fountains, especially music fountains, theme parks, etc., attract thousands of tourists, and water management in these attractions should not only maximize environmental benefits but also economic benefits.

According to Xu (2013), water planning and management in theme parks is not simple and needs environmental

Figure 29.1 Example of eco-labels on water management.

planning tools, policy, and governance support. Moreover, the benefits of recycling water in theme parks are as follows: water conservation, reducing water costs, improving the public image for the park, reducing the need for investment from the government, and improving the water quality. Xu (2013) noted that the authorities of theme parks should pay particular attention to wastewater treatment technology, rainwater tanks, and water recycling. SeaWorld, USA; Sunway Lagoon in Kuala Lumpur, Malaysia; Adventure Park in Victoria, Australia; and Wet'n'Wild Sydney, Australia, constitute good examples for implementing water conservation practices. For instance, when SeaWorld implemented a water conservation system, 30 million gallons of water was saved during a year and it received the water-saving pioneer award in 2008 (Xu 2013).

Creating a green space near hotels and restaurants and producing agricultural products for promoting agritourism, especially in arid and semi-arid climates, through rainwater harvesting can save rainwater and optimum use of runoff. Microcatchment water harvesting, promoting on-farm water harvesting such as design of contour ridges, semi-circular and trapezoidal bunds, small pits, small run-off basins, runoff strips, inter-row systems, Meskats, and contour-bench terraces are methods applied in arid and semi-arid climates for traditional runoff management (Chakoshi and Tabatabaei Yazdi 2012).

29.2.1 Water Harvesting and Museums

Today, the role and influence of museums in society and everyday life are more pronounced. Hinz, ICOM (International Council of Museums) president from 2010 to 2016, pointed out that "museums must be able to guarantee their role in safeguarding cultural heritage, given the increasing precariousness of ecosystems, situations of political instability, and the associated natural and man-made challenges that may arise" (ICOM). Therefore, the scope of activities and expectations of museums within cultural policy discourse has become ever wider (McCall and Gray 2014).

One of the missions of museums is to engage the public in environmental issues and help solve them. The water crisis is one of the most important issues today. In other words, nowadays the world, and especially the Middle East, are facing a water crisis. Museums as educators and mediators have an important role in this regard. The International Museum Day (IMD) theme for 2015 was proposed by the ICOM network as "museums for a sustainable society" which demonstrates the vital and active role of museums in the community and their mission in this field. The 2015 IMD theme puts museums at the core of sustainability. Accordingly, fostering awareness among society is one of the main approaches of museums in this field.

Croitoru (2016) believed that the Global Network of Water Museums can be a way to improve public awareness about water management and to valorize the past by educating future generations. Besides, water museums can preserve and present ancient irrigation and water supply technologies as well. Jordan Museum, which presents complex irrigation and water harvesting systems in Petra and Jawa, located in the eastern desert, and Yazd Water Museum (Iran), which showcases the region's ancient aqueduct construction, featuring tools and technology exhibits, constitute good examples. In the Yazd Water Museum, the audience can become familiar with the construction and operation of Qanats. Moreover, safeguarding and educating people in indigenous knowledge of water management in such museums can be used to foster environmental sustainability, traditional knowledge, and practices can be integrated in sustainable environment schemes and synergies sought between traditional environmental practices and high technology (Hosagrahar 2012). Here, museums help the traditional knowledge to play a decisive role in sustainable development. According to Alivizatou (2016), traditional knowledge is a valuable asset that needs to be safeguarded, but that can also serve the public benefit.

In addition, organizing tours for visiting Qanats[1] is another activity that increases public awareness about traditional water management in Iran especially in desert areas.

According to Ambrose and Paine (2006), in recent decades museums have been striving to reduce energy and water use and they are going to green building[2]; green buildings are designed to take environmental impacts and waste minimization into account (Brophy and Wylie 2013). In other words, green museums address environmental practices. Regarding sustainable water management practices, the Museum Victoria, Australia, committed to installing an underground tank to store rainwater captured from the extensive roof of the Royal Exhibition Building and from surrounding paved areas. The completed tank is 23 m × 23 m wide and 2.6 m high. It has a storage capacity of 1.35 million liters. This water provides a reliable source for the fountains, lakes, and irrigation of the garden and trees (Royal Exhibition Building 2010). Moreover, in June 2011, the Museum of London in the UK launched the project of rainwater harvesting for water management of

1 Gently sloping underground channels to transport water from an aquifer or water well to the surface for irrigation and drinking. This is an old system of water supply from a deep well with a series of vertical access shafts. Qanats still create a reliable supply of water for human settlements and irrigation in hot, arid, and semi-arid climates.

2 Sustainable or green buildings are those which use less energy, water, and natural resources, creates less waste, and are healthier for people.

the museum. The system collects rainwater from an area of over 850 m^2 of the museum's flat roof and high walkway which is filtered and stored in a 25 000-l storage tank in the basement. The rainwater harvesting is used for museum facilities such as toilets in the museum's bar and restaurant and the schools lunch space. Irrigation is also supplied by the system to the Rotunda area of the green roof and Rotunda gardens (Museum of London 2015). Furthermore, since 2008, the Mumbai Museum in India has set up a water harvesting system and until now has saved about 2400,00 l of water every year (Pandhare 2016). In addition, the Science Museum of Virginia, in the US, used a roofshed capture system for rainwater harvesting (Sample 2011). In this context, being green supports the museum's mission, saves money, and can make a positive impact on the environment while encouraging staff, board, volunteers, and visitors to do the same in their lives (Brophy and Wylie 2013).

The Science Museum can also play an important role in educating about rainwater harvesting in schools. The Santa Fe Children's Museum, in the US, organizes a vast array of water awareness projects for adults and children. It is noteworthy that the rainwater harvesting system consists of gutters and multiple downspouts feeding one tank capable of holding 10 000 gal (37 900 l) of water, located in the outdoor play area. The tank is above ground and is playfully painted to fit right into the surrounding play and learning area (Pushard 2018).

The North Bengal Science Centre in India also strives for public awareness about rainwater harvesting to address the water scarcity problem in the region due to mismanagement of water resources. In this regard, the curators of the museum organized a visit to a Model for Rain Water Harvesting in the museum for kids, children, and youths. In the water harvesting model, the rainwater from the roof of the building is collected though a polyvinylchloride (PVC) pipe line and stored in a PVC tank, in a brick masonry reservoir after draining out the first 10 minutes of rain using a first-flush device. A slow sand filter and anti-bacteria filter are connected to the 3000-l PVC reservoir to make the water potable (National Council of Science Museums 2018).

Furthermore, the architecture of some museums is eco-friendly and suitable for rainwater harvesting. The ArtScience Museum (Singapore) constitutes a good example in this regard. The ArtScience Museum is focused on the fusion of art and science and it is certainly the most environmentally sensitive building in the development. Interpreted as a lotus flower (Figure 29.2) or outreached hand, the roof deftly collects water and light for the museum's use (Michler 2018).

Figure 29.2 ArtScience Museum (Singapore) with eco-friendly architecture which is suitable for rainwater harvesting. Source: https://inhabitat.com/lotus-shaped-singapore-artmuseum-collects-rain-and-light.

These examples clearly show that museums can step forward to support communities to go through with sustainable development. Museums bring out the possibility of the use of indigenous knowledge of water management like Qanats and historic hydraulic structures for today. Moreover, rainwater harvesting and the proper use of water can become a good norm in modern society by education in museums (Figure 29.3).

Lastly, green museums not only attempt to increase public awareness through organizing water management educational programs, but also their example as attractions that incorporate sustainability in its management principles and architectural styles (Figure 29.4). Museums with

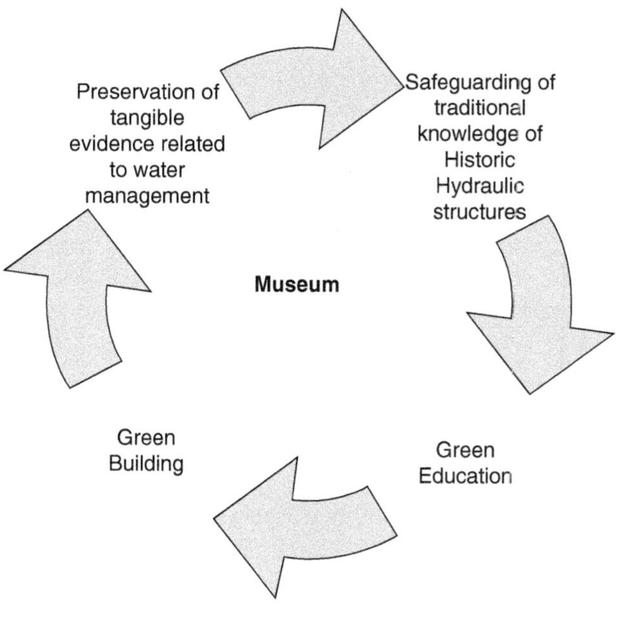

Figure 29.3 Movement of museums toward sustainability.

Figure 29.4 Water harvesting system: an approach in green museums.

green building architecture can play a leadership role in moving toward a more environmentally sustainable future.

29.3 Tourism and Water Harvesting Economy

29.3.1 The Impact of Tourism on Water Demand

There is a complex and mutual reinforcing relationship between the development of water and wastewater infrastructure in a region or country and the development of tourism. Tourism activity increases the demand for local water and sanitation infrastructure utilities considerably, putting a pressure on their sustainable use and development (Frone and Frone 2013).

Although the tourism share of global water use is not significant – in comparison to agriculture, which constitutes an estimated 70% of total water consumption, tourism is far less relevant at 1% (Gössling 2002). Still, tourism is often a major user of freshwater in areas where water is scarce or where renewal rates of aquifers are limited, and its contribution to water consumption can be nationally and regionally significant. In some tourist destinations, the permanent population of the region swells to an average of more than six or seven times with tourist arrivals during the summer or holiday times (Kotios et al. 2009; Hadjikakou 2014; Enriquez et al. 2017). Therefore, these regions encounter a significant increase in water demand, especially during the time of year with the lowest amount of rainfall. The demand pressure happens through direct and indirect uses of water in the tourism sector. Even though direct water use – including kitchens, laundry, toilets, showers, swimming pools, cooling, or the irrigation of gardens, as well as water use for various activities such as golf, diving, saunas, or spas – is considerable, indirect tourism-related water consumption is even more relevant, creating water hinterlands, i.e. regions from which "virtual" water is imported. Examples of this are the water used in construction activities, fuel production, irrigation, and food production, especially in tropical tourism in which food availability and provisions are an important part of the image of "abundance" that characterizes the tropical tourism paradise.

This reality implies the importance of a management system that can respond to the fluctuations of water demand in tourist destinations. Management responses to water scarcity can be categorized under two broad strategies: demand-side management (reducing water consumption) and supply-side management (increasing water provision) (Gössling 2016).

29.3.2 Water Harvesting as a Supply-Side Water Management Strategy

Water harvesting methods are among the supply-side solutions for dealing with the problem. Authorities in tourism destinations with water scarcity and stress can use different supply-side strategies such as constructing additional desalination plants in coastal areas, importing water from other parts of the country, or overexploitation of groundwater resources. The costs of these short-term strategies are born by the permanent inhabitants of the region and transferred back to the tourism industry in the form of high water costs and environmental damage by both greenhouse gas emissions and the burdening of the marine environment with salty discharge (Page et al. 2014). On the other hand, rainwater harvesting systems are one of the long-term supply-side strategies which have their roots in history and countries like ancient Greece, that have been using them for millennia through ancient water cisterns (Enriquez et al. 2017). Rainwater harvesting can bring about different benefits:

- The potential to increase resilience and strengthen water security by providing another source of water that could be drawn on in the dry summer months. Even when untreated, the use of rainfall runoff can alleviate some stress on water resources by providing water for secondary/non-potable uses, such as gardening and flushing (Enriquez et al. 2017).
- Substantial reductions in the need for water withdrawals and imports. Reducing water withdrawals from underground resources, which are not replaceable, can prevent their destruction and saline water intrusion, which has negative impacts on the quality of drinking

water and agricultural land leading to less production and/or increased use of fertilizers (Kotios et al. 2009).
- Reduction in floods and soil erosion. During the rainy season, rainwater is collected in large storage tanks which also helps reduce floods in some low-lying areas. Apart from this, it also helps reduce soil erosion and contamination of surface water with pesticides and fertilizers from rainwater runoff, which results in cleaner lakes and ponds.
- Reduction of water bills for households and businesses. Water usage for flushing, washing clothes, or watering the garden, which make up a considerable share of household water demand, can be supplied with a rainwater harvesting system. For businesses, this share is much higher, particularly for industries such as hotels that have a large number of regular guests in one place.
- Improving plant growth by using rainwater for irrigation because stored rainwater is free from pollutants as well as salts, minerals, and other natural and man-made contaminants.

Besides these benefits, rainwater harvesting systems have some disadvantages like unpredictable rainfall, initial high cost of the infrastructures, the need for regular maintenance, storage limits, and the possible seeping of chemicals, insects, dirt, or animal droppings in certain roof types.

29.3.3 Financial and Economic Analysis of Rainwater Harvesting Projects

Regarding advantages and disadvantages and the growing interest for rainwater harvesting projects, different studies have been performed for the financial and economic analysis of projects with different purposes such as agricultural irrigation, complementary urban and pre-urban water supply, water management for sustainable tourism, etc. (Pandey 1991; Goel and Kumar 2005; Kotios et al. 2009; Jianbing et al. 2010; Smith et al. 2010; Liang and van Dijk 2011; Victor 2011; Frone and Frone 2013; Hadjikakou 2014; Amos et al. 2016; Gössling 2016; Enriquez et al. 2017; Amos et al. 2018). Economic analysis takes into account the opportunity costs of resources employed and attempts to measure in monetary terms the private and social costs and benefits of a project to the community or economy (BusinessDictionary 2018). Financial and economic measures for such an analysis are payback period (PP), the benefit-cost ratio (BCR), and net present value (NPV). If there is a payback period, a BCR of less than one, or a positive NPV, the project is considered financially viable. The majority of researchers find rainwater harvesting systems to be financially non-viable (Amos et al. 2018; Gao et al. 2017; Ishida et al. 2011; Kumar 2004; Kumar et al. 2005; Mitchell et al. 2007; Mitchell and Rahman 2006; Rahman et al. 2007, 2012; Roebuck et al. 2011, 2012). It is noteworthy that the results of these measurements are dependent on the assumptions about model parameters and factors like the local cost of water, capital costs, consideration of subsidiary benefits, etc. A negative financial viability, thus, does not necessarily equate to a negative economic viability. The economic evaluation should take into account broader considerations such as the definition of need, and indirect benefits such as the social, environmental, and ecological benefits of the projects (Amos et al. 2018). These benefits/values that include direct and indirect use values and non-use values of natural water resources and wetlands in tourism destinations can be captured by economic valuation studies using hypothetical markets.

29.3.4 Raising Revenue for Financing Rainwater Harvesting Projects

There are some financial instruments that local authorities can use for financing water-harvesting projects, among which are the Tourism User Fee (TUF) and Tax-Exempt Green Bonds.

TUF is a tourism-based tax that could be levied on all hotel rooms or as a departure tax at the airports and seaports and then allocated to water conservation and sustainability projects. Alternatively, a portion of the revenue could be allocated to hotels and restaurants to offer special incentives to operations that invest in advanced water technologies to improve their environmental performance.

Tax-Exempt Green Bonds are fixed interest financial instruments, introduced in recent years to stimulate and coordinate public and private investments in ecologically sustainable development.

29.3.5 Rainwater Harvesting in Modern Tourism

Nowadays, there is an increasing demand for environmentally conscious and socially responsible travel that emphasizes economic benefits to host communities and the minimization of negative impacts of tourism, and the preservation of cultural and natural resources is growing in importance. A larger share of visitors are now constituted of eco-conscious millennials who will dramatically reshape consumption patterns and consumer decisions and have greater interest for modern tourism: ecotourism, geo-tourism, pro-poor tourism, cultural tourism, responsible tourism, sustainable tourism, etc. (Enriquez et al. 2017). Using environmentally friendly projects such as rainwater harvesting systems and informing tourists about water scarcity and the use of such systems in their accommodation facilities and destinations will be appealing for these kinds of visitors.

29.4 Conclusion

Nowadays, due to global warming, the emergence of drought, and the lack of water management, especially in drylands and developing countries, novel solutions and indigenous knowledge are required to resolve the problems of water resource shortages. Rainwater harvesting systems are novel strategies used to exploit the water that have received less attention. Rainwater harvesting is one of the most important techniques to deal with a water crisis in the territories which are faced with a water shortage. The basis of this method is the allocation of land for collecting rainwater and then storage for use at the time required. In fact, rainwater harvesting systems are methods for collecting and storing rainwater that can be used to provide water for domestic or commercial consumption, for small-scale farming or household purposes, or for tourism purposes. According to the definition of sustainable development, the use of existing resources should be such that future generations can benefit from the same quantity and quality. In most countries with water shortages, the application of micro and macro policies in the field of water resources management has increased. Among these strategies, using rainwater harvesting systems and water extraction from unconventional sources are highly important. Today's travel has increased due to technological advances and an increase in the number of non-working days, and therefore promoting the tourism industry in a destination has placed pressure on freshwater sources. Hence, in order to achieve sustainable tourism, destinations should manage freshwater sources.

Among the sustainable water management methods are using rainwater harvesting in some forms of nature tourism such as eco-tourism and geotourism, in which many tourist facilities are located in remote areas; using rainwater harvesting for preparing water for domestic or commercial consumption and irrigation of small-scale farming in agritourism; paying attention to water-saving labeling for tourism products and services such as (eco-labels) and encouraging hotels, restaurants, and museums toward green principles; using a gray water and rainwater harvesting system in theme parks and other attractions; and creating green space around hotels, museums, and restaurants through rainwater harvesting and gray water. These are novel strategies which can contribute to managing water shortages and can minimize the negative impacts of tourism industry on a destination environment. Lastly, it can be concluded that rainwater harvesting systems are one of the long-term planning and investment strategies for sustainable tourism purposes.

References

Alivizatou, M. (2016). *Intangible Heritage and the Museum: New Perspectives on Cultural Preservation*. London/New York, USA: Routledge.

Ambrose, T. and Paine, C. (2006). *Museums Basics*, 4–6. London/New York: Routledge Taylor/Francis Group.

Amos, C.C., Rahman, A., and Gathenya, J.M. (2018). Economic analysis of rainwater harvesting systems comparing developing and developed countries: a case study of Australia and Kenya. *Journal of Cleaner Production* 172: 196–207.

Brophy, S.S. and Wylie, E. (2013). *The Green Museum: A Primer on Environmental Practice*. New York, USA: Altamira Press.

BusinessDictionary. (2018) economic analysis. Available at: http://www.businessdictionary.com/definition/economic-analysis.html (accessed: 20 April 2018).

Chakoshi, B. and Tabatabaei Yazdi, J. (2012) Rainwater harvesting: A way to use indigenous knowledge to provide water in arid areas. The first national conference on water discharge systems, December, 2012, Mashhad, Iran.

Christian Amos, C., Rahman, A., and Mwangi Gathenya, J. (2016). Economic analysis and feasibility of rainwater harvesting systems in urban and peri-urban environments: a review of the global situation with a special focus on Australia and Kenya. *Water* 8 (4): 149.

Croitoru, V. L. (2016) The Global Network of Water Museums – another way to improve public awareness. Available at: https://iwa-connect.org/#/message/597855c101ffbc7d3ef0a8b1/the-global-network-of-water-museums-another-way-to-improve-public-awareness (accessed: 3 April 2018).

Drew, B., and Hanson, R. (2018) Rainwater Collection for Hotels and Restaurants. ECOVIE Available at: http://ecovieenvironmental.com/rainwater-collection-for-hotels-and-restaurants (accessed: 7 April 2018).

Ecotourism site. (2017) Water management and conservation for tourism facilities. Available at: https://ecotourismkenya.org/blog/water-management-and-conservation-for-tourism-facilities (accessed: 4 April 2018).

Enriquez, J., Tipping, D.C., Lee, J.J. et al. (2017). Sustainable water Management in the Tourism Economy: linking the Mediterranean's traditional rainwater cisterns to modern needs. *Water* 9 (11): 868.

Eslamian, S.S., Malekian Jabali, M. and Okhravi, S.S., (2015) Rainwater harvesting as a practical LID method for confronting climate change. International Conference on

Sustainable Development with a Focus on Agriculture, Environment and Tourism, Tabriz, Iran.

Farsani, N.T., Coelho, C., and Costa, C. (2011). *Geoparks and Geotourism: New Approaches to Sustainability for the 21st Century*. Florida, USA: Universal-Publishers.

Frone, S.M. and Frone, D.F. (2013). Sustainable tourism and water supply and sanitation development in Romania. *Journal of Tourism and Hospitality Management* 1 (3): 140–153.

Gao, H., Zhou, C., Li, F. et al. (2017). Economic and environmental analysis of five Chinese rural toilet technologies based on the economic input–output life cycle assessment. *Journal of Cleaner Production* 163: S379–S391.

Goel, A.K. and Kumar, R. (2005). Economic analysis of water harvesting in a mountainous watershed in India. *Agricultural Water Management* 71 (3): 257–266.

Gössling, S. (2002). Global environmental consequences of tourism. *Global Environmental Change* 12 (4): 283–302.

Gössling, S. (2016) Tourism and water: Interrelationships and management. Available at: Global Water Forum: www.globalwaterforum.org (accessed: 27 April, 2018).

Hadjikakou, M. (2014) Measuring the Impact of Tourism on Water Resources: alternative frameworks, Doctoral dissertation, Faculty of Engineering and Physical Sciences, University of Surrey, Guildford, England.

Han, H., Hsu, L.T.J., and Sheu, C. (2010). Application of the theory of planned behavior to green hotel choice: testing the effect of environmental friendly activities. *Tourism management* 31 (3): 325–334.

Hosagrahar, J. (2012) Culture: A Driver and an Enabler of Sustainable Development. Thematic Think Piece for UNESCO. UN System Task Team on the Post-2015 UN Development Agenda.

Ishida, C., Grantham, R., Stober, T. et al. (2011). A regional cost-benefit analysis of rainwater harvesting: sustainable economics to justify green technologies. *Proceedings of the Water Environment Federation* 5: 454–461.

Jianbing, Z., Changming, L., and Hongxing, Z. (2010). Cost–benefit analysis for urban rainwater harvesting in Beijing. *Water International* 35 (2): 195–209.

Kahinda, J.M.M., Taigbenu, A.E., and Boroto, J.R. (2007). Domestic rainwater harvesting to improve water supply in rural South Africa. *Physics and Chemistry of the Earth, Parts A/B/C* 32 (15–18): 1050–1057.

Kotios, A., Plageras, P., Galanos, G. et al. (2009). *The Impact of Tourism on Water Demand and Wetlands: Searching for a Sustainable Coexistence*. Volos, Greece: Department of Planning and Regional Development, School of Engineering, University of Thessaly, p8.

Kumar, M.D. (2004). Roof water harvesting for domestic water security: who gains and who loses? *Water International* 29 (1): 43–53.

Kumar, R., Singh, R.D., and Sharma, K.D. (2005). Water resources of India. *Current Science* 89 (5): 794–811.

Liang, X. and van Dijk, M.P. (2011). Economic and financial analysis on rainwater harvesting for agricultural irrigation in the rural areas of Beijing. *Resources, Conservation and Recycling* 55 (11): 1100–1108.

Lima, S. and Careto, H. (2007). Qualidade ambiental e certificação no turismo. In: *Turismo e Desenvolvimento Sustentável*, vol. 2 (eds. H. Careto and S. Lima), 214–260. Lisboa: GEOTA.

McCall, V. and Gray, C. (2014). Museums and the 'new museology': theory, practice and organisational change. *Museum Management and Curatorship* 29 (1): 19–35.

Michler, A. (2018) Lotus-Shaped Singapore Museum Collects Rain and Light. Available at: https://inhabitat.com/lotus-shaped-singapore-artmuseum-collects-rain-and-light (accessed: 3 April 2018).

Mitchell, C., and Rahman, A. (2006). Life cycle cost analysis of rainwater tank in a multistorey residential building in Sydney. In 30th Hydrology & Water Resources Symposium: Past, Present & Future (p. 279). Conference Design. Launceston, Australia, 4–7 December 2006.

Mitchell, V., Deletic, A., Fletcher, T.D. et al. (2007). Achieving multiple benefits from stormwater harvesting. *Water Science and Technology* 55 (4): 135–144.

Museum of London. (2015) Rainwater Harvesting at Museum of London. Available at: www.museumoflondon.org.uk/application/files/8014/5451/4697/MOL_and_Rainwater_Harvesting_-_March_2015.pdf (accessed: 3 April, 2018).

National Council of Science Museums. (2018) Rain Water Harvesting System. Available at: http://nbscsiliguri.org/attractions/rain-water-harvesting-system.html (accessed: 3 April 2018).

Page, S.J., Essex, S., and Causevic, S. (2014). Tourist attitudes towards water use in the developing world: a comparative analysis. *Tourism Management Perspectives* 10: 57–67.

Pandey, S. (1991). The economics of water harvesting and supplementary irrigation in the semi-arid tropics of India. *Agricultural Systems* 36 (2): 207–220.

Pandhare, M. (2016) Water Harvesting, Solar Energy & More at Mumbai Museum. Available at: https://www.thebetterindia.com/57560/water-harvesting-solar-energy-mumbai-museum (accessed: 3 April 2018).

Plumbing, G. (2018) 8 Water Conservation Methods for Hotels. Available at: http://www.gerberonline.com/pipeline/8-water-conservation-methods-for-hotels (accessed: 7 April 2018).

Pushard, D. (2018) Teach Your Children: Santa Fe Children's Museum. Available at: http://www.harvesth2o.com/santafekidsmuseum.shtml#.WsNkcJddLIU (accessed: 3 April 2018).

Rahman, A., Dbais, J., Mitchell, C. et al. (2007). Study of rainwater tanks as a source of alternative water supply in a multistory residential building in Sydney, Australia. In: *World Environmental and Water Resources Congress 2007: Restoring our Natural Habitat*, 1–10. American Society of Civil Engineers, Tampa, Florida. UK.

Rahman, A., Keane, J., and Imteaz, M.A. (2012). Rainwater harvesting in greater Sydney: water savings, reliability and economic benefits. *Resources, Conservation and Recycling* 61: 16–21.

Renani, M. (2015) In the next five years, we will witness a comprehensive water conflict in the country. FarsNews Available at: http://www.entekhab.ir/fa/news/203210 (accessed: 27 April, 2018).

Roebuck, R.M., Oltean-Dumbrava, C., and Tait, S. (2011). Whole life cost performance of domestic rainwater harvesting systems in the United Kingdom. *Water and Environment Journal* 25 (3): 355–365.

Roebuck, R.M., Oltean-Dumbrava, C., and Tait, S. (2012). Can simplified design methods for domestic rainwater harvesting systems produce realistic water-saving and financial predictions? *Water and Environment Journal* 26 (3): 352–360.

Royal Exhibition Building. (2010) Phase 2 - Water Harvesting. Available at: https://museumsvictoria.com.au/website/reb/history/world-heritage-world-futures/phase-2-water-harvesting/index.html (accessed: 3 April 2018).

Sample, D. (2011) Technical Aspects of Design of Rainwater Harvesting-and their Evaluation. Rutgers University Conference: Managing Stormwater Runoff from Impervious Surfaces: Green Infrastructure Solutions for New Jersey New Brunswick, NJ January 27, 2011. Available at: http://water.rutgers.edu/Conference2011/SamplePresentation.pdf.

Smith, R.B., Hildreth, L.A., and Savadago, K. (2010). Evaluating the economic impacts of water harvesting in Burkina Faso. In: *Valuation of Regulating Services of Ecosystems: Methodology and Applications* (eds. P. Kumar and M.D. Wood), 27–67. New York, USA: Routledge.

Tuppen, H. (2013) Water Management and Responsibility in Hotels. Green Hotelier Newsletter. Available at: http://www.greenhotelier.org/know-how-guides/water-management-and-responsibility-in-hotels (accessed: 7 April 2018)

UNWTO (The World Tourism Organization). (2013) Importance of Rain Water Harvesting at Agriculture Tourism Destinations. Available at: http://wtd.unwto.org/ar/node/38647 (accessed: 7 April, 2018).

Victor, O. (2011). Economic advantage of rainwater harvesting over water borehole: a sustainable development at the Extension Site Federal Polytechnic, Oko. *International Journal of Water Resources and Environmental Engineering* 3 (4): 73–76.

Xu, B. (2013) Water Planning and Management in Theme Park Tourism: A Comparison of Australian and Chinese Policy and Practice. PhD thesis, Griffith School of Environment, Griffith University, Australia.

30

Rainwater Harvesting Policy Issues in the MENA Region: Lessons Learned, Challenges, and Sustainable Recommendations

Muna Yacoub Hindiyeh[1], *Mohammed Matouq*[2], *and Saeid Eslamian*[3]

[1] *Civil and Environmental Engineering Department, School of Natural Resources Engineering and Management, German Jordanian University, Jordan*
[2] *Chemical Engineering Department, Al-Balqa Applied University, Amman, Jordan*
[3] *Department of Water Engineering, College of Agriculture, Isfahan University of Technology, Isfahan, Iran*

30.1 Introduction

The most ancient practice to meet water supply needs is rainwater harvesting (RWH). In the 1970s, RWH received renewed attention, as it is a productive water source, water savings and conservation means, and sustainable development tool.

RWH has been used for drinking water and agricultural irrigation since 11 500 BCE in North America and 4500 BCE in Asia and the Middle East (Pandey et al. 2003; Mays et al. 2013; Ghimire and Johnston 2015). "Is there anything that thou hast seen under the heavens that is better than water?" Solomon said to the Queen of Sheba (Amos et al. 2016). It is an age-old water supply technology; communities in ancient India, Egypt, Jordan, and Rome were designed with individual cisterns and paved courtyards, which captured rainwater to augment supply from the city's aqueducts.

RWH plays a major role to meet increasing water demand and cope with climate change and variability. Other potential benefits of RWH include reduced impacts on the environment and human health, reduced storm water runoff and combined sewer over flows, and economic viability (Ghimire et al. 2012; Wang and Zimmerman 2015). The implementation of RWH has remained a challenge, due to a lack of understanding of its human health and environmental impacts (criteria pollutants from material selection and energy use) as well as the lack of regulations governing RWH practices (Ghimire et al. 2017).

In modern times, with increasing concerns about water security, the RWH system with its high water-saving potential is an important area of research. There is a growing international interest, particularly in the more water-stressed countries, resulting in a significant body of research on RWH in recent years. Therefore, the aims of this chapter are to examine the policies and institutional framework's support or non-support of RWH (policy analysis). In order to systemize the assessment of the policy analysis, four criteria were discovered during the literature review: incentives; education, gender, and raising awareness; compulsion and enforcement; and rural institutions.

Reviewing the policies, strategies, and legislation approaches of the RWH systems from Europe, Australia, and the US as success stories and lessons learned will lead to identifying key issues and areas requiring further research in the Middle East and North Africa (MENA) region and other developing countries. This chapter will serve as a key reference on policies aspects of RWH to researchers, water engineers, environmentalists, town planners, and policy makers dealing with RWH issues in urban and rural environments.

Strengthen water governance among public and private actors alike: the political, social, economic, and administrative systems and processes that influence water's use and management. Essentially, who gets what water, when and how, and who has the right to water and related services, and their benefits (SIWI n.d.). As an alternative water source, RWH processes are one of the best ways to tackle the water crisis and eradicate poverty.

30.2 Definitions of RWH

There is no consensus regarding the definition. The definition of Wikipedia for RWH is "the accumulation and deposition of rainwater for reuse on-site, rather than allowing it to run off," while on Wikiversity, the definition is: "Rainwater harvesting is a technique of collection and storage of rainwater into natural reservoirs or tanks, or the infiltration of surface water into subsurface aquifers (before it is lost as

Handbook of Water Harvesting and Conservation: Basic Concepts and Fundamentals, First Edition. Edited by Saeid Eslamian and Faezeh Eslamian.

UPC Definitions – Waters for Reuse

- Harvested rainwater – stormwater that is conveyed from a building roof, stored in a cistern, and disinfected and filtered before being used for toilet flushing. It can also be used for landscape irrigation (Ecker 2007).
- Reclaimed water – water treated to domestic wastewater tertiary standards by a public agency suitable for a controlled use, including supply to water closets, urinals, and trap seal primers for floor drains and floor sinks. Reclaimed water is conveyed in purple pipes (California's purple pipe system is one of the better-known water reclamation systems).

Figure 30.1

surface runoff)." Through an internet search for "rainwater harvesting," one might see different worlds: rainwater collection, rain catchment, rainwater capture, roof water harvesting, or storm water harvesting.

Definitions are crucial when writing policies and legislations. If rainwater is defined incorrectly, rules may be invoked, which are not appropriate. Avoid classifying rainwater as stormwater, runoff, gray water, or wastewater.

For example, although a few US states and local jurisdictions have developed standards or guidelines for RWH, it is largely unaddressed by regulations and codes. Neither the Uniform Plumbing Code (UPC) nor the International Plumbing Code (IPC) directly address RWH in their potable or stormwater sections. Other reuse waters are covered by codes. The UPC's Appendix J addresses reclaimed water use for water closets and urinals and the IPC's Appendix C addresses gray water use for water closets and urinals along with subsurface irrigation (Traugott 2007). Both sections focus on treatment requirements, measures necessary to prevent cross-contamination with potable water, and appropriate signage and system labeling. However, because of a general lack of specific RWH guidance, some jurisdictions have regulated harvested rainwater as reclaimed water, resulting in more strict requirements than necessary. These issues have led to confusion as to what constitutes harvested rainwater, gray water, or reclaimed water (Ecker 2007).

30.3 Rainwater Harvesting Toward Millennium and Sustainable Development Goals

As Lehmann et al. (2010) pointed out, the adoption of RWH "makes some significant contributions to achieving the Millennium Development Goals (MDGs)." The World Bank, Organization for Economic Cooperation and Development (OECD), and several NGOs formulated eight MDGs in 2000: eradicate extreme poverty and hunger; achieve universal primary education; promote gender equality and empower women; reduce child mortality; improve maternal health; combat HIV/AIDS, malaria, and other severe diseases; ensure environmental sustainability; and develop a global partnership for development.

Lehmann et al. (2010) explained how RWH can help to reach these goals: the availability of water saves energy and time, thus labor and money, since water does not have to be carried to households from distant sources. While the harvested water leads to more reliable and greater yields, the members of the households can use their saved time to go about other work, generating more income. RWH at schools improves education, health, hygiene, and nutrition; students can spend more time learning, as they do not need to carry water to the school before preparing meals.

"Women are usually in charge of the household water supply" (Lehmann et al. 2010). RWH provides them with more time at their disposal, which can be invested in other activities and they can get paid work in the local markets. Therefore, their status as a decision-making actor in the household increases. Properly stored rainwater provides households with safe and hygienic water and therefore reduces the risk of infection and child mortality and helps combat other severe diseases. Furthermore, the quality and quantity of water at home also affects maternal health. RWH ensures environmental sustainability and can provide access to safe drinking water without threatening natural water sources.

RWH offers strong potential to contribute to the achievement of most of the targets of the 2030 Sustainable Development Goals (SDGs) ("Ensure availability and sustainable management of water and sanitation for all"), including with regard to its other targets on drinking water, sanitation, water quality, water use efficiency, and integrated water resources management (IWRM). The RWH contribution will translate into particularly striking positive impacts on other SDGs, with regard to water security for underpinning reducing multidimensional poverty (SDG 1); sustainable agriculture (SDG 2, notably Target 2.4); healthy lives (SDG 3); building resilient (water-related) infrastructure (SDG 9); sustainable urban settlements (SDGs 11); and disaster risk reduction (SDG 11 and, as related to climate change, 13).

There are strong synergies between the WASH and water-quality targets; examples of synergies include increasing access to water supply, sanitation, and hygiene (WASH) (SDG 6.1, 6.2) in homes, healthcare facilities, schools, and workplaces, complemented by wastewater treatment (SDG 6.3), as a way to reduce risk of

"… the capacity of a population to safeguard sustainable access to adequate quantities of acceptable quality water for sustaining livelihoods, human well-being, and socio-economic development, for ensuring protection against water-borne pollution and water-related disasters, and for preserving ecosystems in a climate of peace and political stability (UN Water 2013)".

Figure 30.2

water-borne disease (SDG 3.1–3.3, 3.9), malnutrition (SDG 2.2); decent work (SDG 8); and achieving gender equality through supporting access to education women's economic empowerment (SDG 5 and SDG 10). These Goal 6 targets also reduce inequalities by ensuring essential services are available to all, including marginalized groups and vulnerable people (SDG 1, SDG 10 [UN-Water 2016]). A case study in Ghana found that a 15-minute reduction in water collection time increased girls' school attendance by 8–12%. This is an example of how improving access to water supply (SDG 6.1) increases girls' participation in education (4.1, 4.2, 4.5) and addresses gender inequalities (5.1) and, consequently, supports sustainable development (UN-Water 2016).

30.4 Water Administration and Legislation

Public policy includes rules, regulations, laws, and codes, as well as tools to maintain, create, or change behavior. Backed by judicial authority to prevent injurious actions against the public or those that are contrary to the public good, the ultimate goal of many of the guidelines is to ensure public safety, to spur economic activity, or both. Water quality fit for human use and recreation is an instance of both protecting public health and valuing ecosystem integrity. Any Water Act was an attempt to solve the problem of national water degradation. The best RWH policies protect water as human right, safeguard public health, and promote rainwater use, resource conservation, and sustainability. Effective policy types include: (i) government standards and enforcement, (ii) incentive-based policies, and (iii) outcome-based policies. Featuring some combination of all three types may be useful to promote broad participation, it is uncommon to see multiple policy strategies combined (Stark and Pushard 2008).

Ineffective policy trends emerge, too. Strict policies may create barriers to execution. In one example, the code specified a particular type of "roof washer," creating financial and technical barriers. Adding expense, technical

Water funds are institutional platforms developed by cities and conservation practitioners that can address governance issues by bridging scientific, jurisdictional, financial, and implementation gaps.

Figure 30.3

knowledge and skills to meet code requirements creates obstacles to implement the policy. Another common restrictive code example is the requirement for an engineer's approval. Many financial and technical barriers may be avoided while maintaining safe rainwater use.

Water resources in most countries are considered public property, and have been managed by government agencies or municipal councils. In many countries of the MENA region, public property extends to all natural resources occurring on the surface and in the subsurface (Haddadin 2006). Government policies with respect to water resources are ordinarily drafted by the concerned government agency in charge, usually in collaboration with other government agencies or ministries directly affected by these policies. The outcome of such consultation is usually reflected in the draft, and the policy is formulated and approved by a council of ministers. The importance of submitting the draft policy to the council of ministers is that water resources touch on the economic, social, and environmental activities and have direct impact on the well-being of the people.

Legislation is the instrument for implementation of water policy. When water legislation endorsed, it reflects the water policy adopted at the time, and any changes in that are enforced through amendments of the laws in effect (Hadaddin 2006).

30.5 Policy and Regulatory Approaches to RWH Use

This section looks at the need for comprehensive water-use policies that address RWH and argues for a regulatory approach that is both effective and workable.

30.5.1 The Need for Policy

Some countries stimulate RWH through national policy, while in others regional or local governments are taking the lead. Policies range from requiring the installation of RWH systems to subsidy programs or exemptions from paying taxes. In the UK, the driving forces for RWH are the government's water strategy "Future Water," the Flood and Water Management Act, and the European Water

Framework Directive. These encourage the use of other water sources such as rainwater (Ward et al. 2012).

Countries like Japan, Singapore, the US, Germany, India, Brazil, Australia, China, and South Korea are the best examples of RWH. RWH is widely practiced in Maldives, Sri Lanka (it especially benefited tsunami-hit populations), Myanmar (more so, after the cyclone Nargis), Bhutan, Bangladesh (as an alternate drinking water source for arsenic-affected areas), and Thailand.

In developing countries, the biggest challenge to widespread RWH may fall at the center of finance, followed by the volume of water depending on erratic rainfall, calling for large water storages, the failure to link with the other urban water components, poor public perception and quality, and a lack of commitment from politicians.

Previous literature and chapters in this book have asserted that developing countries must address increasing water demand in the context of increasingly over-exploited water resources. As countries consider non-conventional water sources to supplement their supplies, RWH use is one option that is gaining interest from both consumers and policy makers. Lehmann et al. (2010) have shown that particularly water-stressed households and communities are reusing their household RWH, regardless of the legality or health risks. However, such informal RWH use often does not adequately manage the risks to health and the environment, leaving water-stressed, low-income populations facing additional health problems.

Despite the research efforts on the technical aspects of RWH use, piloting of systems suitable to the developing countries as described in earlier chapters, and the promotion of RWH use, especially among low-income rural communities by donor organizations such as the International Development Research Centre (IDRC), United Nations for Environment Program (UNEP), and the United States Agency for International Development (USAID), government policies on promoting and regulating RWH use are still weakly developed. Countries adopting centrally managed combined water supply use have been hesitant to promote RWH use, partly since it has health risks. Governments are cautious of allowing too much decentralized control of RWH because of health risks.

There was a realization that despite several pilot projects demonstrating the feasibility and potential of RWH use, there was a general lack of policy support in the MENA countries. In the region, no country has a RWH code or a regulation that explicitly permits RWH use and regulates it clearly, without placing undue burden on the householders. At first glance, this may seem odd in such a water-stressed region.

There have been some concerns voiced about the negative ecological impacts from RWH due to the possibility that "rainwater harvested upstream reduces the runoff otherwise available to others, or the environment, downstream" (Cosgrove and Rijsberman 2000). A policy on RWH should therefore seek to manage the various risks associated with its use, for e.g. the health of householders (families and children), and the needs of the environment must be addressed. Policy should also recognize the potential benefits of allowing RWH use. The risk–benefit relationship will be different in different areas, as will the cost–benefit ratio, and each of these must be interpreted within the particular social and socioeconomic context. Policies should take into account the different contexts and should develop a clear message to households, communities, and potential RWH users.

When considering policy responses to the push for more RWH use, the first question to ask is, "should RWH use be permitted at all?" Policy makers may decide that central control of drinking water supply is the best option and that the risks of household RWH use are too high. In addition, while it may be reasonable for authorities to prohibit RWH use in urban areas for drinking water purposes – for risk of contamination, particularly in areas of high population density, as well as insufficient area to use RWH in irrigation. Therefore, it may be more difficult to argue against allowing it in rural areas; particularly given the abundant evidence (cited earlier) that water-stressed communities are already using RWH. Decision makers could argue that water is a commodity that has been paid for by the householder and that the government has little claim on what the householder should do with that water. If consumers choose to water their own garden with water from their RWH tank, it is difficult to argue that this should be prohibited.

Some commentators see the need for RWH use as the result of a failure of water delivery service (either due to resource constraints, operational inefficiencies, or policy reasons like allocation). However, in the current context of the water-scarce countries in the Middle East, it is difficult to argue against facilitating and allowing RWH and household RWH use, at least in areas where there is sufficient planted area to make use of it. If central water supply networks were to become widespread and provide usable water easily and cheaply to householders, then there would be arguments for prohibiting household RWH use, to allow all the water resources to be captured and treated centrally.

No country has taken a clear view of the financial and economic benefits of RWH use in comparison with the alternatives. Even where there is an appreciation of the worth of RWH, no MENA country has developed a clear approach to its use that clearly states the responsibilities of the users and the regulatory requirements. There is a need for policy and regulatory frameworks in the MENA region to be examined in order to harmonize RWH policy with the water supply and demand policies. There is also a need to encourage authorities to send out consistent messages on RWH with clear rules and regulations to ensure that the required protection to health, water resources, and the environment is provided, while allowing communities to make use of this valuable resource.

30.5.2 Key Characteristics of Good Policy

The creation of policies – through a coordinated effort between educational institutions, government, and the public – will encourage rather than discourage RWH.

When creating public policy, we believe it is time to ask: What behavior or action do we want to encourage? With respect to rainwater policies, this could include safe use of water, conservation of both water and energy, as well as other resources, and the promotion of green building and sustainability. In formulating new policy in this growing sector, we have a unique opportunity to both enhance quality of life and support green business opportunities (Stark and Pushard 2008).

When creating new policy, ask yourself if the policy:

- Makes sense from economic, social, and environmental perspectives.
- Is outcome based (focuses on results).
- Is easy for the public to understand (written in lay terms; short as possible).
- Is simple for the public to implement (user friendly).
- Is achievable and measurable.
- Is reviewed by actual practitioners.
- Includes a way to educate the public and elected officials.
- Has a standardized and streamlined application process.

30.5.3 Framework for a Policy

This section sets out some important issues specific to the MENA region, which should be addressed in policy.

30.5.3.1 Policy Must Balance the Risks from Controlled RWH Use with the Alternatives

Water policymakers in the MENA region have to address a number of challenging issues. Priorities vary from country to country, but the goals of increasing economic growth, protecting and improving health, protecting environmental resources, addressing food security, reducing poverty, and maintaining internal and external security all have implications for water resource and supply management policy. While RWH use carries risks, there are also risks to communities with inadequate water supplies. These include economic deficits and hunger.

The primary concerns of indoor rainwater use are cross-contamination of the potable supply and human contact with pathogens that may be present in the collected rainwater. The Stockholm Framework (WHO 2001) allows the concept of relative risk, whereby the risks of both using and not using a particular intervention should be considered together. The WHO guidelines for "Greywater Management Health Consideration" (2006) allow that the tolerable burden of disease may vary from one country to another and suggests a flexible and contextually derived approach to risk management. Each country must therefore address its own risk context. In other words, a practice which carries an unacceptable risk in one country may actually reduce overall health risks in another country, or at least carry a risk which is acceptable when balanced with the benefits that the practice brings. In some areas of MENA, this may mean that allowing RWH use is the lesser evil, when compared to the results of water poverty, particularly in low-income areas. The result may be that practices which are unacceptable in other countries are promoted in some areas in the MENA region.

30.5.3.2 Policy Must Be Integrated

Ideally, RWH use should be set within an integrated part of a comprehensive water resource management framework. Policies on allocation, water supply, demand management, agricultural policy, and wastewater use should be linked and complementary. As an example, a country that has accepted the principle of water supply must consider where household RWH use fits within this and decide how this particular RWH resource can be used – at the household level or under the control of the authorities. Prohibiting use at the household level, and then not using it downstream, would go against a policy of recognizing its importance as a secondary water resource. Alternatively, a country with a highly developed water treatment-and-use policy and practice may want to restrict RWH use in areas where total rainwater and floods captured and used centrally.

30.5.3.3 Policy Should Be Simple and Incentivize RWH Use

The principle of simplicity and ease of implementation adopted by the United States Environmental Protection Agency (USEPA), Arizona, Alabama, Texas, Ohio, and Florida, follow tax incentives and refunds. If complex application requirements – such as form filling, presentation of drawings, system inspection, or water quality monitoring – are the responsibility of households, there is unlikely to be significant uptake of regulated RWH use. In countries with weak regulatory systems, and where local authorities have limited capacity, a realistic and workable regime should be adopted, whereby the requirements placed on local authorities should be minimized as much as possible. Following the USEPA example, information and guidance on risk management by householders should be well publicized and the responsibility placed with the householders to manage the system. Perhaps some pre-approved treatment systems, relevant to the MENA countries could be pre-determined, removing the burden from householders from seeking out professional advice (US EPA 2008).

Changing and unpredictable climates and droughts complicate the water stress caused by increased urbanization. For example, with Australia's "Millennium Drought" strategy and government incentive results, 34% of households have adopted RWH systems. This is the highest adoption rate in the world (Beatty and McLindin 2012). Review of this successful implementation therefore holds many lessons for the international community (Gardner et al. 2015).

30.5.3.4 Risk Management Should Be Behavior Based, Rather than Technology or Water-Quality Based

The WHO Drinking Water Quality guidelines (2011) move away from an exclusive focus on water quality for RWH. A realistic policy in a MENA country would allow means of risk mitigation such as simple treatment (filtration, bio-sand filter, aluminum sulfate, ceramic pot filter), disinfection (chlorination, SODIS, boiling), drip irrigation, protective gloves for workers, and restrictions on RWH usage and permitted crops, and may suggest requirements for food preparation and cooking of produce irrigated by RWH. If policy requires water quality to be of a particular standard, it will: (i) be too expensive to implement; (ii) be too expensive to regulate; (iii) not fully address all the important risks; (iv) likely be too conservative and restrictive. Each country should review the successful policies and guidelines and seek to address these within the country's context. Education of users should be an important target for policy makers.

Encouraging RWH and use requires enabling the practice through codes and regulations and providing incentives. State or municipal codes need to address public health concerns by stipulating water quality and cross-contamination requirements. Similar to reclaimed and gray water, specific RWH codes need to be developed. Codes should establish acceptable uses for rainwater and corresponding treatment requirements. Disinfection of rainwater for use should be considered as the standard, but recent research and policies should encourage authorities to evaluate lesser requirements for non-potable uses in water toilets and urinals. The simplification of the on-site treatment process and associated cost savings could broaden the use of RWH without increasing exposure risks.

Human conduct and level of education, reflected in the level of awareness of the relationship between water and health, hygiene and sanitation, and management and maintenance skills of the RWH systems are social factors controlling water quality of a RWH system.

30.5.3.5 Policy Development Should Include Stakeholders

In the literature review, many argue strongly that lack of participation of end users in the development of the RWH systems leads to non-optimal solutions. Policy developers should therefore take care to consult with and involve communities in developing policies, which are appropriate, understandable, and workable at the community level. The context in low-income areas is likely to be different from that which the (generally more well-to-do) policy makers have experience in. Financial analysis should also consider the various stakeholders.

30.5.3.6 Policy Must Be Clear Regarding Implementation

The Jordanian code is an example of a RWH regulatory tool developed without sufficient information as to how it should be implemented. Codes should set out clearly what potential users should do to satisfy the regulatory authorities. In the case of some (e.g. Portland, US), an application is to be made, with supporting information. In the case of Georgia, the user is automatically in compliance, provided certain basic conditions are met.

30.5.3.7 Policy Should Not Place Undue Financial Burdens on Users

No one should be penalized for responsibly using RWH. Expensive professional assistance and use of expensive materials should not be required. Application fees (if any) should take into account that the purpose of the RWH policy is to provide additional water resources to water-stressed (and possibly low-income) populations, not a revenue raiser for local government. Some of the costs cited in earlier chapters suggest that the types of RWH treatment units that are being proposed for the region are costly enough in themselves, without adding any extra expense.

Economic analysis of the RWH system plays an important role in the search for cost-effective solutions to water security, particularly in developing countries. However, economic analysis of the RWH systems has a limited presence in scientific literature. Moreover, the limited amount of studies focusing mainly on financial aspects of RWH systems has often reported conflicting results (Amos et al. 2016).

30.5.3.8 Privately Owned RWH Systems and Use Should Be Considered for Poor Communities

A characteristic of privately owned RWH means that the involvement of the owners is large. Consequently, the control of central authorities and water supply companies will be reduced. This means that a successful implementation of privately owned systems also needs policy and management innovations (Partzsch 2009) and the development of new business models (Van Der Hoek et al. 2011).

Some countries in the MENA region have adopted a centralized approach to water supply, where water is captured and treated centrally and is then distributed and sold for drinking in large networks and schemes. This approach sometimes conflicts with the idea of use RWH at household privately or even at a community level. In the context of alleviating community water scarcity, especially in low-income and rural areas, policy makers must recognize the importance of the locally managed RWH resource, and not allow a centralizing tendency to over-rule.

30.5.3.9 Policy Should Differentiate with Regard to Scale

Many jurisdictions that have drawn up legislation for RWH use have found it beneficial to differentiate between large and small users, since the implications of RWH use in each case are different and the cost and complexity of solutions are also different. In many cases described in the literature, the utilization of harvested rainwater is limited to only one or two applications, e.g. toilet flushing, landscape or garden irrigation, or laundry washing, which is logical considering the single-family building approach. Increasing size and scale to building-block or neighborhood level may render a more cost-effective system and will open doors for other applications, like cooling, street flushing, or application in energy systems. However, other applications of RWH may also be addressed in a policy.

Customers of a large hotel or high-rise building that uses the RWH from residents and staff will expect a higher degree of protection than a single household reusing its own RWH under its own control, as contamination will be on a large scale. One of the main purposes in large-usage legislation would be to provide protection to health and environment and ensure the responsible design, installation, and operation of the RWH system. Since large-usage systems may need additional treatment complexity, and therefore cost, will result.

By contrast, household systems, where RWH is used solely within the property of the household, carry less risk and should be cheap and easy to install, maintain, and regulate. Policy should clearly address these contexts separately.

30.6 Considerations When Establishing a Municipal Rainwater Harvesting Program (U.S. EPA 2008)

(1) **Establish specific codes or regulations for rainwater harvesting**
Building and plumbing codes are largely silent on RWH. Consequently, gray water requirements are often used to govern RWH systems, resulting in requirements that are more stringent than necessary. Codes should define RWH and establish its position as an acceptable stormwater management/water conservation practice.

(2) **Identify acceptable end uses and treatment standards**
Each municipality will need to consider and identify acceptable uses for harvested rainwater and the required treatment for specified uses. Rainwater is most commonly used for non-potable applications and segregated by indoor and outdoor uses.
- **Typical outdoor uses**:
  i. Irrigation; and
  ii. Vehicle washing.
- **Typical indoor uses**:
  i. Toilet flushing;
  ii. Heating and cooling; and
  iii. Equipment washing.
- Non-potable uses typically require minimal treatment.

Prescreening is only needed for outdoor uses to limit fouling of the collection system. Indoor non-potable uses do not necessarily require treatment beyond screening, although some municipalities have adopted a conservative approach and require filtration and disinfection prior to use. Harvested rainwater can be used for potable applications, although a special permitting process should be established to ensure that proper treatment (e.g. filtration and disinfection) is provided and maintained.

(3) **Detail required system components**
- Jurisdictions often delineate between rain barrels and cisterns because of the size and potential complexity of the systems. Rain barrels collect relatively small quantities of water and generally only require mosquito prevention, proper overflow, and an outlet for outdoor uses. Cisterns can be 100 to several thousand gallons in size and may be connected to various indoor plumbing and mechanical systems. Needed system requirements include:
- Pre-filtration – Filtration prior to the rain barrel or cistern should be provided to remove solids and debris.
- Storage containers – Rain barrels and cisterns should be constructed of a National Sanitation Foundation-approved storage container listed for potable water use.
- Back-flow prevention – For cisterns that require a potable water make-up for operation, back flow prevention in the form of an air gap or backflow assembly must be provided.

- Dual piping system – a separate piping system must be provided for harvested rainwater distribution. The pipe should be labeled and color coded to indicate non-potable water. Purple piping indicating reclaimed water is often used for RWH systems. Cross connections with the potable water supply system are prohibited.
- Signage – permanent signage should be provided at every outlet and point of contact indicating non-potable water not for consumption. In addition, biodegradable dyes can be injected to indicate nonpotable water.

(4)?> **Permitting**
Rain barrels should not need to be permitted provided that they are installed correctly and direct overflow to a proper location. A permit application process should be instituted for cistern systems used for non-potable uses. If harvested rainwater is used for potable water, the collection and treatment system should be inspected and approved by the public health department.

(5) **Maintenance**
Adequate design and maintenance of the cistern and piping system is the responsibility of the cistern owner.

(6) **Rates of use**
For harvesting systems to be efficient stormwater retention systems, the collected rainwater needs to be used in a timely matter to ensure maximum storage capacity for subsequent rain events. Cistern systems generally supply uses with significant demands, ensuring timely usage of the collected water. Outreach and education are critical components of rain barrel programs, however, because of the more episodic and less structured use of this collected water. Municipalities should inform homeowners of the steps needed to maximize the effectiveness of their rain barrels. Harvesting programs targeting susceptible combined sewer areas have used slow draw down of the rain barrels to delay stormwater release to the sewer system, yet ensure maximum storage capacity for subsequent rain events (U.S. EPA 2008).

30.7 Regulatory Approaches in Other Countries

Although harvested rainwater is mostly used for non-drinking purposes, in some circumstances, rainwater can be treated to be safe for human consumption. In 2009, the UNEP highlighted the growing popularity of rainwater collection techniques, and recognized its potential to reduce the number of people who do not have access to water for human consumption. This water optimization process has been widely implemented in rural areas in countries like Brazil, China, New Zealand, and Thailand.

In recent times, the RWH system, being a clean and healthy source of water, has gained much more popularity in different parts of the world where rainfalls are available in sufficient quantities, taking into consideration of potable and freshwater inadequacy. Countries like Japan, Singapore, the US, Germany, India, Brazil, Australia, China, and South Korea are the best examples of harvesting it. In this regard, according to the UNEP, some cities around the world already have the best examples of RWH and its utilization, including Singapore, Tokyo, Berlin. Developing countries can learn from these cities where RWH systems have been quite successful.

Several formal policies on RWH use have been adopted in different countries outside the MENA region, and these have been reviewed in Amos et al. (2016) (RAIN 2008). Australia and the US states of California, Texas, and Portland provide two particularly interesting approaches. The US Water Alliance (2017), Portland and San Francisco, have produced guidance on managing risk from RWH. Some countries such as Australia (Ward 2010), Canada, and the US do not have national policies or regulations for rainwater quality. In the US, Georgia and Texas, and in Canada, Alberta and Ontario, have devised guidelines for some quality parameters and encourage the harvesting and management of rainwater (Marsalek 2010).

Regulatory approaches to RWH use adopted in the US, Germany, and Australia are examined to determine how the balance is made between practical and cost-effective RWH use and risk management. The context of the MENA region is discussed, together with some key elements that locally derived policies should address.

A selection of relevant policies and guidance is discussed below.

30.7.1 Australia

Australia provides a good example of combining government code with incentive programs. This will motivate people to recycle water and utilize RWH technology, in an effort to combat the drought.

With the *Under Water for the Future* initiative, the Australian government launched the National Rainwater and Greywater Initiative, which offered generous refunds to people who installed new RWH systems or gray water systems. Up to $500 was available for households that installed rainwater collection systems, and up to $10 000 was available for surf life clubs that undertook a large water-saving project or installed a RWH system.

When the project ended, over seven million Australian dollars had been paid in household rebates, and over

$650 000 worth of rebates had been paid to surf life clubs (Australian Department of Environment 2015). The Victorian Government has also recognized that water should be "fit for purpose," noting that it is not necessary for all water to meet drinking water standards (Radcliffe 2006). Rebate schemes in Australia have been successful.

In Sydney there is the privilege of being able to freely use water from a RWH system not connected to the mains, while others are restricted by the "water wise rules" that have replaced water restrictions (Sydney Water 2016). RWH systems are listed by the real estate agents as an "eco-friendly feature," and were found to represent a premium of AU$18 000 in Perth, Australia.

30.7.2 Germany

Germany has promoted RWH practices since the 1980s, and has carried out studies to assess the practicality and viability of collecting runoff from traffic surfaces in densely populated areas (Nolde 2007).

RWH systems have been successfully introduced in the German Industrial Standards (DIN). Since 2003 the planning, installation, maintenance, and operation of such systems has been regulated via various standards, the DIN 1989 (DIN 2002) RWH Systems, the DIN 1986 Rainwater Pipes (DIN 2003), and the DIN 1988 Drinking Water Installation (DIN 1988). This has provided an effective technical and institutional framework for the application of facilities for rainwater utilization, facilitating an easy planning and installation process for the end users and service providers. The owners of such systems have to announce construction to the water supply companies, but without having to apply for a permit. This procedure facilitates direct design, planning, and construction practices by companies, and avoids bureaucratic barriers or delays for the customers and users (Schuetze 2013). The German government continues to offer financial incentives for those who disconnect stormwater from sewer discharge, and in the past has provided subsidies to promote installation of RWH systems.

A great example of the German government's support toward RWH practices is in the construction of the "Postdamer Platz" in Berlin. The permit granted by the council was subjected to meeting very strict stormwater management regulations, which ensured that the construction of this project did not overload the combined sewerage system (where rainwater and industrial water are collected in the same pipe).

The new complex was required to discharge no more than 1% rainwater to sewerage. To comply with the regulation, all 19 buildings were constructed using green roofs and the collected roof runoff is used for toilet flushing and irrigation. An artificial lake was constructed to provide a means to retain water and allow it to evaporate. Rooftop runoff is collected in three cisterns of 2550 cubic meters of capacity, which corresponds to 12% of the annual precipitation (Schmidt 2009).

30.7.3 United Kingdom

British Standard for Rainwater Harvesting BS8515:2013:

Storm savers adhere to the stipulations within BS8515 and all systems comply with its requirements in terms of sizing, manufacture, and installation. Covers the design, installation, water quality, maintenance, and risk management of RWH systems. The introduction of BS 8515:2013 means that there is clear guidance on what are the minimum acceptable standards that RWH companies and the people specifying their systems have to meet.

The best practice guidelines for RWH under the Water Supply (Water Fittings) Regulations 1999 set out the Water Regulations Advisory Scheme (WRAS) guidelines. WRAS are particularly concerned with preventing cross contamination of the mains water supply where a RWH system is connected to the property.

30.7.4 Bermuda

The Public Health Act regulates rainwater utilization systems in Bermuda and requires that catchments be whitewashed by white latex paint; the paint must be free from metals that might leach into water supplies. Roofs must be repainted every two to three years and storage tanks must be cleaned at least once every six years. It is the owner's responsibility to keep catchments, tanks, gutters, pipes, vents, and screens in good repair.

30.7.5 The Netherlands

According to the Dutch Drinking Water Decree (Drinkwaterbesluit 2011) (Geel 2003), currently application of harvested rainwater in the Netherlands is only allowed on a small scale for toilet flushing, and is used occasionally in larger office buildings (Hofman and Paalman 2014).

30.7.6 India

RWH is practiced on large scales in cities like Delhi, where RWH is a part of the state policy (Centre for Science and Environment 2010).

Tamil Nadu is the first Indian state to make RWH mandatory. On 30 May, 2014, the state government announced that it would set up 50 000 RWH structures at various parts of the capital city of Chennai.

Indian Railways has installed rooftop RWH systems at more than 2400 different locations, including station buildings across the country, in order to replenish the groundwater table and meet its growing demand for water (Hossain and Latifee 2017).

30.7.7 Indonesia

Recognizing the need to alter the drainage system, the Indonesian government introduced a regulation requiring that all buildings have an infiltration well. The regulation applies to two-thirds of the territory, including the Special Province of Yogyakarta, the Capital Special Province of Jakarta, West Java, and Central Java Province.

30.7.8 Brazil

In 2003, a program called "Programa Um Milhão de Cisternas" ("One Million Cisterns") was organized by the Articulação Semiarido Brazileiro (ASA), a community network that aims to improve conditions for the residents of Semi Arid Brazil (SAB). The project aims to provide one million homes with rooftop RWH systems where rainwater can be collected and stored until the dry season. The simple, economic, and efficient system is comprised of a gutter, a pipe, and a 16 000-l capacity collection tank. Installation of the system takes up to a week, and water can be easily extracted from the tank using a manual pump.

In 2015, the project has successfully delivered 578 336 rainwater collection systems. The impact of these tanks extends beyond eliminating the need for people to travel long distances seeking water; it has allowed residents to practice agriculture, provide water for livestock, and improve sanitary standards at facilities such as rural schools. The success of the project gained interest from the Brazilian government, who currently provides financial support to the project (Articulação Semiarido Brasileiro 2015). In other parts of Brazil, such as Porto Alegre, São Paulo, and Curitiba, buildings under construction are obliged by law to be equipped with RWH systems (Mores 2006).

A further example of the growing interest in RWH and utilization is the establishment of the Brazilian Rainwater Catchment Systems Association, which was founded in 1999.

30.7.9 China

Since the 1990s, the China government's strategy for water management in the northern region has been water harvesting, collection systems have been heavily implemented. Following a drought in 1995, the provincial government of Gansu commissioned a RWH and reuse project called "121-Project." This project's name refers to its goal: every family should have at least two rainwater collection cisterns for every one acre of agricultural land. Within a year of the project's implementation, 1.31 million people had benefited from access to water for domestic use (Guerquin 2010).

30.7.10 Capiz Province, The Philippines

In the Philippines, a RWH program was initiated in 1989 in Capiz Province with the assistance of the Canadian IDRC. About 500 rainwater storage tanks were constructed made of wire-framed ferro-cement, with capacities varying from 2 to 10 m^3. The construction of the tanks involved building a frame of steel reinforcing bars (rebar) and wire mesh on a sturdy reinforced concrete foundation. The tanks were then plastered both inside and outside, thereby reducing their susceptibility to corrosion relative to metal storage tanks.

The RWH program in Capiz Province was implemented as part of an income generation initiative. Under this arrangement, loans were provided to fund the capital cost of the tanks and related agricultural operations. Loans of US$200, repayable over a three-year period, covered not only the cost of the tank but also one or more income-generating activities such as the purchase and rearing of pigs, costing around US$25 each. Mature pigs can sell for up to US$90 each, providing an income opportunity for generating that could provide sufficient income to repay the loan. This type of innovative mechanism for financing rural water supplies can help avoid the requirement for water resources development subsidies.

30.7.11 United States

Although state regulations and statutes continue to favor RWH (Harvesth2o 2016; NCSL 2016), the USEPA reported "there are currently no federal regulations governing RWH for non-potable use, and the policies and regulations enacted at the state and local levels vary widely from one location to another" (Ghimire et al. 2017).

In the US Virgin Islands, "every new building is required by law to incorporate a cistern to store roof runoff." Tucson, Arizona, and Santa Fe County, New Mexico, both require the use of rainwater. Moreover, although not "required," in the mountains above Honolulu, hundreds of households are dependent almost exclusively on rainwater for all domestic purposes, including drinking.

Code language found in the states of Texas, Ohio, Oregon, and Washington, and in the cities of Portland and Eugene (Oregon), and Seattle, while Kentucky, Hawaii, Arizona, New Mexico, Washington, West Virginia, Texas, and the US Virgin Islands have guidelines for the practice. In San Juan County, Washington, the Department of Health and Community Services provides a checklist of what a RWH system must include. The State of Washington provides guidelines from the State Building Codes Council. Delaware provides a unique policy that not only encourages rainwater use, but also addresses water availability in times of drought and in a sustainable way: "Water utilities, both public and private, should have adequate supplies of water available, even in times of drought, to meet the present and future needs of this state on a continuing and sustainable basis."

Financial incentives are growing. Currently, Texas is leading the way, offering multiple incentives at the local and state levels. Austin, for example, provides a 30% subsidy for the cost of cisterns up to $500 and sells rain barrels below cost. The rebate application also includes assistance with tank sizing as well as information about area suppliers and contractors. Under Austin's Commercial Incentive Program, "commercial entities may be eligible for as much as a $40 000 rebate against the cost of installing new equipment and processes to save water and the state provides property and sales tax exemptions for commercial installations." Hays County, Texas, provides a $100 rebate on the application fee and a property tax exemption. In San Antonio, a 50% rebate is available for new water-saving equipment at the commercial scale. Texas supports rainwater-harvesting activities at state and higher education facilities through a task force and code. Finally, the Lone Star State promotes RWH with a code allowing performance contracting, which allows recuperation of initial investments through savings earned on utility bills. In other words, the water and energy-conserving measures are expected to pay for themselves within the contracted period (Stark and Pushard 2008).

During the 1970s, California encouraged water conservation with the California Water Conservation Tax Law. This law provided tax credits up to $3000 for implementation of rainwater, gray water, or combined storage cisterns or other water conservation devices. However, in 1982 the law was repealed.

The confusion among waters for use, and the lack of uniform national guidance, has resulted in differing use and treatment guidelines among state and local governments and presents a barrier to rainwater use. Texas promotes harvested rainwater for any use including potable uses provided appropriate treatment is installed; Portland, like many other jurisdictions, generally recommends rainwater use to the non-potable applications of irrigation, hose bibbs, water closets, and urinals.

Meanwhile, Oklahoma recently passed the "Water for 2060 Act" which initiates grants for information campaigns on capturing and using harvested rainwater. Even in Colorado where RWH was actually illegal until 2009, residential property owners are now allowed to collect rainwater as of Feb 25, 2013. Colorado residents are legally allowed up to two rain barrels for RWH systems that comply with the Colorado Water Conservation Board. Collecting more than what is allotted is still illegal as is groundwater harvesting. In 2009, two laws were passed that loosened restrictions.

A review of treatment standards among various jurisdictions shows a wide range of requirements from minimal treatment to reclaimed water standards. San Francisco allows rainwater to be used for toilet flushing without being treated to potable standards. Texas requires filtration and disinfection for non-potable indoor uses, and Portland requires filtration for residential non-potable indoor uses, but requires filtration and disinfection for multi-family and commercial applications. Treatment requirements ultimately come down to risk exposure, with the risk of bacterial exposure determining the most stringent levels of treatment. However, San Francisco's Memorandum indicates a belief in a low-exposure risk with rainwater used for toilet flushing. Likewise, testing conducted in Germany demonstrated that the risk of *Escherichia coli* contact with the human mouth from toilet flushing was virtually non-existent, resulting in the recommendation that special disinfection measures were unnecessary for rainwater dedicated to non-potable uses.

Currently, nine states have laws restricting the collection of rainwater, but the severity of those laws differ. In fact, a number of independent studies proved that letting people collect rainwater on their property actually reduces demand from water facilities and improves conservation efforts.

30.7.12 St. Thomas, US Virgin Islands

A rainwater utilization system is a mandatory requirement for a residential building permit in St. Thomas. A single-family house must have a catchment area of 112 m^2 and a storage tank with 45 m^3 capacity. There are no restrictions on the types of rooftop, water collection system, and construction materials. Many of the homes on St. Thomas are constructed so that at least part of the roof collects rainwater and transports it to storage tanks located within or below the house. Fecal coliform contamination and Hg concentration found in the samples collected from the rainwater utilization systems in St. Thomas were higher than EPA water quality standards, which limits the use of this water to non-potable applications unless adequate treatment is provided.

30.7.13 Portland

Portland's RWH "One and Two Family Dwelling" Specialty Code provides a good example of specific rainwater use stipulations. Although the code does not address multi-family residential or non-residential applications, rainwater use is permitted for these facilities, but due to the unique design of each system, commercial use systems are considered on a case-by-case basis. In addition, multi-family residential units and sleeping portions of hotels are allowed to use rainwater for irrigation only; non-residential buildings are permitted to use rainwater for irrigation, water features, water closets and urinals. In these applications, water provided for water closets and urinals must be treated with filters and UV and/or chlorination according to the City of Portland Office of Planning (2001).

Portland's code permits rainwater use for potable uses at family dwellings only through an appeals process. In addition, rainwater used only for outdoor irrigation is not covered by the code and needs no treatment prior to use. Acceptable indoor non-potable uses are hose bibbs, water closets, and urinals. The code illuminates several important issues that need to be considered when developing RWH code (U.S. EPA 2008).

- Water quality and its impact on human health is a primary concern with RWH. This issue is comprised of two components: end use of the rainwater and treatment provided. Rainwater used for residential irrigation (on the scale of rain barrel collection) does not typically require treatment. Commercial applications and non-potable indoor uses require treatment, but the type of use will determine the extent of treatment. Each jurisdiction will need to assess the level of treatment with which it is comfortable, but limiting rainwater use to water closets, urinals, and hose bibbs presents little human health risk. Each system will require some level of screening and filtration to prevent particles and debris from traveling through the plumbing system, and most jurisdictions require disinfection with UV or chlorination because of bacterial concerns. Table 30.1 provides an example of minimum water quality guidelines and suggested treatment methods for collected rainwater.
- **Cross-contamination** – Cross-contamination of the potable water system is a critical concern for any water use system. Cross-contamination measures for rainwater use systems will be similar to those for reclaimed and gray water systems. When rainwater is integrated as a significant supply source for a non-potable indoor use, a potable make-up supply line is needed for dry periods and when the collected rainwater supply is unable to meet water demands. The make-up supply to the cistern is the point of greatest risk for cross-contamination of the potable supply. Codes will require a backflow prevention assembly on the potable water supply line, an air gap, or both. In addition to backflow prevention, the use of a designated, dual piping system is also necessary. Purple pipes, indicating reused water, are most often used to convey rainwater, and are accompanied by pipe stenciling and point-of-contact signage that indicates the water is non-potable and not for consumption.
- **Maintenance and inspection** – The operation and maintenance of RWH systems are the responsibility of the property owner. Municipal inspections occur during installation and inspections of backflow prevention systems are recommended on an annual basis. For the property owner, the operation of a RWH system is similar to a private well. Especially for indoor uses, annual water testing to verify water quality is recommended as well as regular interval maintenance to replace treatment system components such as filters or UV lights. The adoption and use of RWH systems will add to the inspection responsibilities of the municipal public works department, but the type of inspection, level of effort, and documentation required will be similar to those of private potable water systems and should be readily integrated into the routine of the inspection department (US EPA 2008).

30.7.14 Singapore

Singapore has focused on creating an extensive network of drains and canals, aiming to control flooding and creating a potable water supply (Zhang et al., 2015). Today, more than half of Singapore's land area is used as catchments for rainwater collection. Catchment areas are heavily protected, and no pollution-causing activities are allowed in these areas (Khoo 2009).

30.7.15 Kenya

In 2005 the Kenyan government, in conjunction with a cross-section of the key stakeholders and in particular the Kenya-Belgium Study and Consultancy Fund, undertook a study for investigating various ways of securing water autonomy, which resulted in the Practice Manual for Water Supply Services (2005). Several simple formulas are recommended in the manual for RWH, including: (i) how to calculate rainfall yield; (ii) how to calculate monthly demand for a given number of people; and (iii) how to calculate minimum tank storage.

Table 30.1 Minimum water quality guidelines and treatment options for storm water reuse Texas Rainwater Harvesting Evaluation Committee (2006).

Use	Minimum water quality guidelines	Suggested treatment options
Potable indoor uses	• Total coliforms – 0 • Fecal coliforms – 0 • Protozoan cysts – 0 • Viruses – 0 • Turbidity <1 NTU[a)]	• Pre-filtration – first flush diverter • Cartridge filtration – 3 μm sediment filter followed by 3 μm activated carbon filter • Disinfection – chlorine residual of 0.2 ppm or UV disinfection
Non-potable indoor uses	• Total coliforms <500 cfu[b)] per 100 ml • Fecal coliforms <100 cfu per 100 ml	• Pre-filtration – first flush diverter • Cartridge filtration – 5 μm sediment filter • Disinfection – chlorination with household bleach or UV disinfection
Outdoor uses	N/A	Pre-filtration – first flush diverter

a) NTU – nephelometric turbidity units.
b) cfu – colony forming units.

Kenya has undertaken limited economic analysis of RWH systems, and this should be a priority for the government and NGOs in Kenya (Amos et al. 2016). To encourage a wider adoption of RWH systems, the legal issues embedded in the government policies in Kenya should be removed, as they are considered barriers to installing RWH systems (Amos et al. 2016).

30.7.16 Namibia

RWH is not encouraged by the Namibian National Water Supply and Sanitation Policy nor is it supported financially by the Namibian government. Woltersdorf et al. (2014) recommend that government integrate funding for RWH infrastructure and finance from private garden operation and maintenance costs into a national RWH policy, which would create the conditions to achieve the benefits of an upscale of RWH-based gardening in Namibia (Woltersdorf et al. 2014).

30.7.17 Middle East

The current regulatory framework is different in the national entities of the Middle East. According to health regulations, the Israeli Ministry of Health does not permit the use of RWH for drinking, and any storage or purification system must be under its strict supervision. This comprehensive regulation prevents any legal use of harvested water for drinking (Lange et al., 2012). In the Palestinian Authority, comparable limitations do not exist. Instead, the Palestinian Water Authority has given priority to RWH as a part of its water emergency plan (Kittani 2006).

Box 30.1

Institutional and political realities are factors that define the framework within which people can operate as stakeholders in a regional context.

Also, the Jordanian Ministry of Water and Irrigation encourages RWH, and new houses must be equipped with water storage facilities (Abdulla and Al-Shareef 2009). The green building guideline and rating system for Jordan refers to Jordan's Related Building Codes (as compulsory requirements) and international green rating systems such as LEED from the United States, BREEAM from the United Kingdom, ESTIDAMA from Abu Dhabi, Dubai green building rating system, QSAS from Qatar, and many more.

30.8 Challenges and Limitations

- Economic analysis of the RWH system plays an important role in the search for cost-effective solutions to water security, particularly in developing countries. However, economic analysis of RWH systems has a limited presence in scientific literature. Moreover, the small number of studies focusing mainly on financial aspects of RWH systems have often reported conflicting results.
- Financing the long-term operation and maintenance is a key issue. Financing with minimal external assistance is

an important consideration for non-government organizations (NGOs) hoping to provide sustainable solutions to developing countries. Lessons can be learned from a review of many projects that are still failing to address basic issues such as training communities, establishing cost recovery mechanisms, and supply of spares.
- While there are plenty of studies based on hypothetical situations, there are few studies of multiple RWH system installation economics using actual water consumption data from real-life situations.
- Unfortunately, multiple overlapping codes may discourage rainwater-harvesting activities. Some areas require permits and reviews by all listed code officials.
- Lack of existing public health-based standards and a streamlined permitting process.
- Very little research is found on the stormwater management benefits of RWH systems.

30.9 Future Recommendations for the MENA Region

- A rebate scheme for RWH systems should be a preferred option for policy makers in developing countries. There is also an indication that funding tank installations will be successful if adequate training for maintenance is also organized.
- Financial incentives may be particularly beneficial for the rural poor in water-stressed areas. Financial support such as grants, subsidies, tax exemptions, revolving funds, and income-producing activities should not be underestimated. A more sophisticated approach will combine the above suggestions and use multiple strategies in concert that change over time to meet specific contexts.
- Detailed cost-benefit analyses would be required to determine the appropriate scale of investment in both approaches, at the landscape and the household levels.
- There is a need to standardize the methods of economic analysis of RWH systems.
- Further research on RWH should focus on financial analysis covering multiple benefits, life cycle analysis incorporating energy use and greenhouse gas emission, productive water use such as boosting rural and urban agriculture, and institutional and sociopolitical support to improve acceptability of RWH.
- Education is crucial, but an area often ignored in practice. Technical support and training sessions also guide the public rather than letting them struggle.
- The success of RWH systems depends on strong collaboration between municipal utilities and public health agencies to ensure projects protect public health and meet water quality standards.
- Advance the use of onsite non-potable water systems by sharing best practices and fostering a supportive policy and regulatory environment.
- Develop a guidebook and actionable framework for regulating and managing onsite non-potable water systems based on best-in-class science by utility leaders and local public health regulators. This guidebook should have a consistent policy framework across cities, which is one of the best ways that we can integrate onsite systems in a way that protects public health and meets our water needs.
- As we develop regulatory approaches for RWH, it is important that they are guided by risk-based science.
- Link knowledge, policy, and practice to strengthen water services delivery and water resources governance.
- Place greater emphasis on the role of water in maintaining environmental health and providing ecosystem services.
- Training centers shall be initiated, reinforced, and upgraded. Cooperation with outside centers and agencies shall be promoted.
- International and regional cooperation shall be pursued in the fields of research, development, and technology transfer in RWH. Management, quality control, and economics shall be promoted. Exchange of information and experience shall be maintained with regional and international parties.
- Developing countries included in the MENA region need to address the impact of climate change on their social, economic and environmental development. Adaptation measures must ensure institutional response capacity, community education, and awareness of the risks.
- Drought management and adaptation to climate change will need to be addressed through proper policies and regulations.
- Devise monitoring systems and other tools to capture or factor in women's involvement in RWH projects.
- Place renewed emphasis on gender roles and innovation in women's roles in water management.
- Target women and girls in awareness and education campaigns, including supporting their education in science and social subjects to enable them become water sector professionals.

30.10 Conclusion

Brazilian and Chinese rural habitants have benefited from government-funded programs that allow them to capture water during the rainy season and store it for later usage.

Japan, Germany, and Australia are the leading countries regarding the production, implementation, and study of RWH systems (Ward 2010). Singapore has invested on highly efficient canals and drains, and has protected a large number of catchment in order to avoid pollution of rainwater. Australia, one of the driest continents on earth, has encouraged citizens and businesses to implement RWH technology by providing funding toward these projects.

All USA states have multiple codes such as those related to building, electric, plumbing, zoning, and stormwater. Unfortunately, multiple overlapping codes may discourage rainwater-harvesting activities. Some areas require permits and reviews by all the listed code officials (except zoning). Of course, the intent is to protect public health and safety (U.S. EPA 2008).

The simplification of the on-site treatment process and associated cost savings could widen the use of RWH without increasing exposure risks. In addition to code development, incentives for RWH should be instituted. Municipalities should review their water tariff to see if they appropriately account for the full cost of water. Pricing alternatives such as increasing block tariff, which increase the price of water with increased use, create an incentive to conserve potable water. The combined actions of establishing certain requirements for RWH systems and increasing the currently underpriced cost of water creates a complementary system that can encourage the use of alternative water sources (U.S. EPA 2008).

Donor-funded work in the MENA region has demonstrated that RWH is of interest as a means to supplement the amount of drinking water and irrigation water available to communities, particularly in low-income areas. Local regulatory agencies have been slow to address importance of RWH application through policies and regulations. With workable treatment solutions available (although perhaps more work is needed to further reduce complexity and expense) and numerous demonstration schemes having been tried out, governments must address head-on the implications and risks. Within the context of their own water resources management policies and priorities, and taking into account the needs of (particularly) rural and low-come water-stressed users, authorities must examine the relative risks and benefits of allowing RWH use and must take a view on the risk mitigation requirements that are necessary and workable in each context. On this basis, policies must be developed and communicated to the potential users, taking into account the cultural values.

The policies must then be implemented clearly and fairly. It is only after a number of years of consistent RWH use on a larger scale that the success or otherwise of many of the treatment approaches discussed in this book, together with the environmental and health-related implications, can properly be evaluated against the benefits to the householders. Ultimately, the market will tease out the costs and benefits, and communities will take their own decisions on the effectiveness of the RWH use. But governments owe their citizens a clear policy framework, against which the RWH use can be implemented, examined, investigated, and discussed.

This Chapter concludes that RWH have high potential to meet contemporary and future water resources management challenges, as reflected in the 2030 Agenda for Sustainable Development, the SDGs and their targets.

MENA Countries should be motivated to structure and implement environmentally friendly policies for RWH, and take different approaches to increase potable water availability.

Finally, in order to continue amending the legal framework, politicians can benefit from the experiences that were made in other countries. The authors suggest an interaction of policies and institutions – especially rural institutions and NGOs – that combine economic incentives with active support on a national, regional and local level as well as in the field of education and raising awareness.

References

Abdulla, F. and Al-Shareef, A. (2009). Roof rainwater harvesting systems for household water supply in Jordan. *Desalination* 243: 195–207.

Amos, C., Rahman, A., and Gathenya, J. (2016). Review economic analysis and feasibility of rainwater harvesting systems in urban and peri-urban environments: a review of the global situation with a special focus on Australia and Kenya. *Water* 8 (4): 149–170.

Articulação Semiarido Brasileiro, 2015 Ações- P1MC Programa Um Milhão de Cisternas. Available from www.asabrasil.org.br/acoes/p1mc#categoria_img [18th October 2015].

Beatty, R., and McLindin, M. (2012) Rainwater harvesting and urban design in Australia. Proc. Water Environ. Fed. 6435–6447.

Belgium Study and Consultancy Fund (2005). *Design Manual for Water Supply in Kenya*. Belgium.

Centre for Science and Environment (2010) Urban rainwater harvesting. Last accessed 4 December 2010. URL: http://www.rainwaterharvesting.org/urban/urban.htmhttp://

www.rainw http://aterharvesting.org/urban/urban.htm. Delhi.

City of Portland Office of Planning & Development Review, Rainwater Harvesting – ICC – RES/34/#1 & UPC/6/#2: One & Two Family Dwelling Specialty Code: 2000 Edition; Plumbing Specialty Code: 2000 Edition, March 13, 2001.

Cosgrove, W.J. and Rijsberman, F.R. (2000). *World Water Vision: Making Water Everybody's Business*. London: Earthscan Publications. UK,

DIN, Deutsches Institut fuer Normung e.V (1988). *DIN 1988 Drinking Water Supply Systems*. Berlin, Germany: DIN.

DIN, Deutsches Institut fuer Normung e.V (2002). *DIN 1989-1 Rainwater Harvesting Systems*. Berlin, Germany: DIN.

DIN, Deutsches Institut fuer Normung e.V (2003). *DIN 1986 Drainage Systems on Private Ground*. Berlin, Germany: DIN.

Drinkwaterbesluit 2011. Besluit van 23 mei 2011, houdende bepalingen inzake de productie en distributie van drinkwater en de organisatie van de openbare drinkwatervoorziening. Beschikbaar via: http://wetten.overheid.nl/BWBR0030111/geldigheidsdatum_03-07-2013.

Ecker S.R. (2007) Rainwater harvesting and the plumbing codes, Plumbing Engineer, March 2007, available at http://www.plumbingengineer.com/march_07/rainwater.php (accessed June 2008).

Gardner, T., Begbie, D., Sharma, A.K., and Tjandraatmadja, G. (2015). Rainwater tanks in Australia: their social/political context, a research overview, policy implications, future research needs, and application of findings to other countries. In: *Rainwater Tank Systems for Urban Water Supply: Design, Yield, Energy, Health Risks, Economics and Social Perceptions* (eds. A.K. Sharma, D. Begbie and T. Gardner), 319. London, UK: IWA Publishing.

Geel, P.L. (2003) Toepassing huishoudwater; Brief van de Staatssecretaris van volkshuisvesting, ruimtelijke ordening en milieubeheer.

Ghimire, S. and Johnston, J. (2015). Traditional knowledge of rainwater harvesting compared to five modern case studies. In: *World Environmental and Water Resources Congress 2015* (eds. K. Karvazy and V. Webster), 182–193. USA: American Society of Civil Engineers.

Ghimire, S., Watkin, D.W. Jr.,, and Li, K. (2012). Life cycle cost assessment of a rainwater harvesting system for toilet flushing. *Water Science and Technology: Water Supply* 12: 309–320.

Ghimire, S., Johnston, J., Ingwersenc, W., and Sojka, S. (2017). Life cycle assessment of a commercial rainwater harvesting system compared with a municipal water supply system. *Journal of Cleaner Production* 151: 74–86.

Guerquin, F. (2010). *World Water Actions: Making Water Flow for All*, 129. Routledge. USA,

Haddadin, M. (2006). *Water Resources in Jordan: Evolving Policies for Development, the Environment, and Conflict Resolution*, The RFF Press Water Policy Series, 1e. Washington: Resources for the Future ISBN-13: 978-1933115320.

Harvesth2o 2016. Regulations & Statutes. http://www.harvesth2o.com/statues_regulations.shtml (April 11, 2016).

Hofman J., and Paalman M. (2014) Knowledge for Climate (KFC) Rainwater harvesting, a sustainable solution for urban climate adaptation?. KfC report number KfC 142/2014. KWR Water Cycle Research Institute, The Netherland.

Hossain S., Latifee E.H. (2017) Concept of Rainwater harvesting in city areas. The Independent.

Khoo, T.C. (2009). Singapore water: yesterday, today and tomorrow. In: *Water Management in 2020 and Beyond*, 237–250. Berlin, Heidelberg, Germany: Springer.

Kittani, H. (2006) Water Harvesting in Palestine, the 2nd International Conference on Water Resources and Arid Environment, King Saud University, Riyadh, Saudi Arabia.

Lange, J., Husary, S., Gunkel, A. et al. (2012). Potentials and limits of urban rainwater harvesting in the Middle East. *Hydrology and Earth System Sciences* 16: 715–724.

Lehmann C., Sukada T., and Lourete A. (2010) Low-Cost Technologies towards Achieving the Millennium Development Goals. The Case of Rainwater Harvesting. International Policy Centre for Inclusive Growth (ed.), IPC-IG Policy Research Brief, (12).

Marsalek, J. (2010) Rainwater harvesting in Canada and the USA: Policies and Regulations. In: 3RWHM Proceedings, the 3rd International Conference on Rainwater Harvesting and Management, 2012 May 20–24. Goseong, Korea, pp. 80–82.

Mays, S.L., Antoniou, G., and Angelakis, A.N. (2013). History of water cisterns: legacies and lessons. *Water* 5 (4): 1916–1940.

Mores, F.V. (2006) Reaproveitamento da água de chuva em condomínios residenciais e comerciais na cidade de Porto Alegre.

NCSL (2016) State Rainwater Harvesting Laws and Programs. http://www.ncsl.org/research/environment-and-natural-resources/rainwater-harvesting.aspx (April 11, 2016).

Nolde, E. (2007). Possibilities of rainwater utilisation in densely populated areas including precipitation runoffs from traffic surfaces. *Desalination* 215 (1): 1–11.

Pandey, D.N., Gupta, A.K., and Anderson, D.M. (2003). Rainwater harvesting as an adaptation to climate change. *Current Science* 85 (1): 46.

Partzsch, L. (2009). Smart regulation for water innovation – the case of decentralized rainwater technology. *Journal of Cleaner Production* 17 (11): 985–991.

Radcliffe, J.C. (2006). Future directions for water recycling in Australia. *Desalination* 187 (1): 77–87.

Rainwater Harvesting Implementation Network (RAIN) (2008) RAIN Water Quality Guidelines: Guidelines and practical tools on rainwater quality, The Netherland.

Schmidt, M. (2009) Rainwater harvesting for mitigating local and global warming. In Fifth Urban Research Symposium Marseilles, France.

Schuetze, T. (2013). Rainwater harvesting and management-policy and regulations in Germany. *Water Science and Technology: Water Supply* 13 (2): 376–385.

SIWI, n.d. Strategy 2018–2021. http://www.siwi.org/publications/siwi-strategy-2018-2021.

Stark T. and Pushard D. (2008) The State of Rain Water harvesting in the United State. http://H20Harvesting.org.

Sydney Water. 2016 Water Restrictions. Available online: www.sydneywater.com.au/SW/water-theenvironment/what-we-re-doing/water-restrictions (accessed on 31 March 2016).

Texas Rainwater Harvesting Evaluation Committee, 2006 Rainwater Harvesting Potential and Guidelines for Texas, Report to the 80th Legislature, Texas Water Development Board, Austin, TX, November 2006, USA.

Traugott A. (2007) Reclaimed Water and the Codes, Consulting-Specifying Engineer, April 1, 2007, available at http://www.csemag.com/article/CA6434236.html (accessed June 2008).

U.S. EPA (2008). *Managing Wet Weather with Green Infrastructure. Municipal Handbook. Rainwater Harvesting Policies.* Low Impact Development Center Christopher Kloss. EPA-833-F-08-010.

UN Water (2013). *Water Security & the Global Water Agenda: A UN-Water Analytical Brief*. Hamilton, Ontario, Canada: United Nations University Institute for Water, Environment & Health.

UN-Water (2016). *Water and Sanitation Interlinkages across the 2030 Agenda for Sustainable Development*. Geneva: UN-Water http://www.unwater.org/app/uploads/2016/08/Water-and-Sanitation-Interlinkages.pdf.

US Water Alliance, WE & RF, and WRF (2017). *A Guidebook for Developing and Implementing Regulations for Onsite Non-potable Water Systems.* USA: The National Blue Ribbon Commission for Onsite Non Potable Water Systems.

Van Der Hoek, J.P., Hofman, J.A.M.H., and Van Someren, T.C.R. (2011). Integration and innovation of the urban water cycle: the Waternet experience. *Journal of Environmental Science & Engineering* 5: 533–544.

Wang, R. and Zimmerman, J.B. (2015). Economic and environmental assessment of fire building rainwater harvesting systems in various U.S. cities. *Environmental Science & Technology* 49: 1768–1778.

Ward, S.L. (2010). *Rainwater Harvesting in the UK: A Strategic Framework to Enable Transition from Novel to Mainstream.* UK: University of Exeter.

Ward, S., Barr, S., Butler, D., and Memon, F.A. (2012). Rainwater harvesting in the UK: socio-technical theory and practice. *Technological Forecasting and Social Change* 79: 1354–1361.

WHO (2001) An approach to risk assessment and risk management of water-related infectious diseases, developed following an expert meeting held in Stockholm by WHO in 2001, Sweden.

Woltersdorf, L., Jokisch, A., and Kluge, T. (2014). Benefits of rainwater harvesting for gardening and implications for future policy in Namibia. *Water Policy* 16: 124–143.

Zhang, D., Gersberg, R.M., Ng, W.J., and Tan, S.K. (2015). Conventional and decentralized urban stormwater management: a comparison through case studies of Singapore and Berlin, Germany. *Urban Water Journal* 14 (2): 113–124.

Index

NB Page locators containing illustrations are indicated by *italic* type.

Handbook of Water Harvesting and Conservation: Basic Concepts and Fundamentals, First Edition. Edited by Saeid Eslamian and Faezeh Eslamian.

b

d

h

j

k

l

m

n

s

x

y